Essentials of Sociology

Aaliyah Brown
ID: 2872302

amoja program

Recent Sociology Titles from W. W. Norton

The Contexts Reader, Third Edition edited by Syed Ali and Philip N. Cohen

Code of the Street by Elijah Anderson

In the Trenches: Teaching and Learning Sociology by Maxine P. Atkinson and Kathleen S. Lowney

Social Problems, Third Edition by Joel Best

The Art and Science of Social Research by Deborah Carr, Elizabeth Heger Boyle, Benjamin Cornwell, Shelley Correll, Robert Crosnoe, Jeremy Freese, and Mary C. Waters

You May Ask Yourself: An Introduction to Thinking like a Sociologist, Sixth Edition by Dalton Conley

The Family: Diversity, Inequality, and Social Change, Second Edition by Philip N. Cohen

Race in America by Matthew Desmond and Mustafa Emirbayer

Gender: Ideas, Interactions, Institutions, Second Edition by Lisa Wade and Myra Marx Ferree

The Real World: An Introduction to Sociology, Sixth Edition by Kerry Ferris and Jill Stein

Mix It Up: Popular Culture, Mass Media, and Society, Second Edition by David Grazian

Readings for Sociology, Eighth Edition edited by Garth Massey

Families as They Really Are, Second Edition edited by Barbara J. Risman and Virginia E. Rutter

Sex Matters: The Sexuality and Society Reader, Fifth Edition edited by Mindy Stombler, Dawn M. Baunach, Elisabeth O. Burgess, Wendy Simonds, and Elroi J. Windsor

Cultural Sociology: An Introductory Reader edited by Matt Wray

American Society: How It Really Works, Second Edition by Erik Olin Wright and Joel Rogers

To learn more about Norton Sociology, please visit wwnorton.com/soc

SEVENTH EDITION

Essentials of Sociology

Anthony Giddens

London School of Economics

Mitchell Duneier

Princeton University

Richard P. Appelbaum

University of California, Santa Barbara

Deborah Carr

Boston University

W. W. NORTON & COMPANY, INC.
New York · London

W. W. NORTON & COMPANY has been independent since its founding in 1923, when Wiliam Warder Norton and Mary D. Herter Norton first published lectures delivered at the People's Institute, the adult education division of New York City's Cooper Union. The firm soon expanded its program beyond the Institute, publishing books by celebrated academics from America and abroad. By midcentury, the two major pillars of Norton's publishing program—trade books and college texts—were firmly established. In the 1950s, the Norton family transferred control of the company to its employees, and today—with a staff of four hundred and a comparable number of trade, college, and professional titles published each year—W. W. Norton & Company stands as the largest and oldest publishing house owned wholly by its employees.

Editor: Sasha Levitt
Assistant Editor: Erika Nakagawa
Project Editor: Diane Cipollone
Managing Editor, College: Marian Johnson
Managing Editor, College Digital Media: Kim Yi
Production Manager: Stephen Sajdak
Media Editor: Eileen Connell
Associate Media Editor: Ariel Eaton
Media Editorial Assistant: Samuel Tang
Marketing Manager, Sociology: Julia Hall
Design Director: Hope Miller Goodell, Rubina Yeh
Photo Editor: Cat Abelman
Director of College Permissions: Megan Schindel
Permissions Assistant: Patricia Wong
Information Graphics Design: Kiss Me I'm Polish LLC, New York
Composition: Graphic World
Manufacturing: Transcontinental

Library of Congress Cataloging-in-Publication Data

Names: Giddens, Anthony, author.
Title: Essentials of sociology / Anthony Giddens [and three others].
Description: Seventh edition. | New York : W.W. Norton & Company, Inc.,
   [2019] | Includes bibliographical references and index.
Identifiers: LCCN 2018053632 | **ISBN 9780393656381 (pbk.)**
Subjects: LCSH: Sociology.
Classification: LCC HM585 .G52 2019 | DDC 301—dc23 LC record available at https://lccn.loc.gov/2018053632

W. W. Norton & Company, Inc., 500 Fifth Avenue, New York, NY 10110
wwnorton.com
W. W. Norton & Company Ltd., 15 Carlisle Street, London W1D 3BS

1 2 3 4 5 6 7 8 9 0

Contents

Chapter 2: Culture and Society 40

Chapter 3: Socialization, the Life Course, and Aging   70

Chapter 6: Deviance, Crime, and Punishment   156

Chapter 7: Stratification, Class, and Inequality

Chapter 8: Global Inequality 222

Chapter 9: Gender Inequality 248

Chapter 11: Families and Intimate Relationships 320

Chapter 12: Education and Religion   354

WHAT ARE THE FORCES BEHIND WORLD POPULATION GROWTH? — 479

HOW DO ENVIRONMENTAL CHANGES AFFECT YOUR LIFE? — 488

Chapter 16: Globalization in a Changing World — 496

HOW DOES GLOBALIZATION AFFECT SOCIAL CHANGE? — 499

WHY DOES TERRORISM SEEM TO BE ON THE RISE IN THE WORLD TODAY? — 505

WHAT ARE SOCIAL MOVEMENTS? — 506

WHAT FACTORS CONTRIBUTE TO GLOBALIZATION? — 513

Preface

We believe that sociology plays an essential role in modern intellectual culture and occupies a central place within the social sciences. We have aimed to write a book that merges classic sociological theories with up-to-the-minute social issues that interest sociologists today. We also believe that sociologists must use rigorous research methods in order to study and understand human behavior. We highlight findings from ethnographic studies to document the hows and whys of social behavior, and also present current statistical data to document important social trends. We aim to present material in a "fair and balanced" way. Although each of the authors has his or her own perspective on social theories, methods, and social policy, we have worked hard to ensure that our treatment is unbiased and non-partisan. We strive to present the most complete picture of sociology possible. Given the vast array of topics encompassed by sociology, however, we made difficult choices about what the most essential topics in sociology are today. We hope readers are engaged, intrigued, and occasionally inspired by the ideas presented in this book.

About the Essentials Edition

The Seventh Edition of *Essentials of Sociology* is based on the Eleventh Edition of our best-selling text *Introduction to Sociology*. We created the Essentials Edition for instructors and students who are looking for a briefer book that can fit into a compressed academic schedule. We have reduced the length of the book by roughly one-third, and we reduced the number of chapters from twenty to sixteen. We cut selected topics to focus the chapters on the core ideas of sociology, while still retaining the themes that have made the text a successful teaching tool.

Major Themes

The book is constructed around four basic themes that provide its character. The newest theme is applying sociology to everyday life. Sociological thinking enables self-understanding, which in turn can be focused back on an improved understanding of the social world. Studying sociology can be a liberating experience: It expands our sympathies and imagination, opens up new perspectives on the sources of our own behavior, and creates an awareness of cultural settings different from our own. Sociological ideas challenge dogma, teach appreciation of cultural variety, and allow us insight into the working of social institutions. At a more practical level, the text shows how the skills and knowledge acquired in sociology classes can be applied to far-ranging careers, from health care to law enforcement (new "Employing Your Sociological Imagination" spreads in select chapters).

Our second theme is inequalities. Throughout the text, we highlight that important resources—whether education, health, income, or social support—are not fairly or evenly distributed to all individuals. We highlight the ways that gender, race, social class, and age shape our daily lives in the United States. We also pay keen attention to global inequalities, and reveal how differences in economic and natural resources throughout the world powerfully influence even very personal experiences—including health, religion, and relationships.

A third theme of the book is that of social and historical context. Sociology was born of the transformations that wrenched the industrializing social order of the West away from the lifestyles characteristic of earlier societies. The pace of social change has continued to accelerate, and it is possible that we now stand on the threshold of transitions as significant as those that occurred in the late eighteenth and nineteenth centuries. Sociology has the prime responsibility for charting the transformations of our past and for grasping the major lines of development taking place today. Our understanding of the past also contributes to our understanding of institutions in the present and future.

The fourth fundamental theme of the book is globalization. For far too long, sociology has been dominated by the view that societies can be studied as independent entities. But even in the past, societies never really existed in isolation. Today we can see a clear acceleration in processes of global integration. This is obvious, for example, in the expansion of international trade across the world. The emphasis on globalization also connects closely with the weight given to the interdependence of the industrialized and developing worlds today. In every chapter, visually engaging full-page "Globalization by the Numbers" infographics highlight how countries across the globe compare on key metrics, such as incarceration rate, maternity leave benefits, voter turnout, and gender inequality.

Despite these interconnections, however, societies have their own distinctive attributes, traditions, and experiences. Sociology cannot be taught solely by understanding the institutions of any one particular society. While we have slanted our discussion toward the United States, we have also balanced it with a rich variety of materials drawn from other regions—especially those undergoing rapid social change, such as the Middle East, Asia, Africa, and Eastern Europe. The book also includes much more material on developing countries than has been usual in introductory texts.

All of the chapters in the book have been updated and revised to reflect the most recent available data. Each chapter opens with a contemporary news event or social trend—ranging from the most local and seemingly trivial (like an email from Yale University administrators about Halloween costumes) to the most global and profound (such as the civil war in Syria). These events are used to motivate and explain the key sociological concepts, themes, and studies that are elaborated throughout the text. Other substantive changes include:

Chapter 1 Sociology: Theory and Method The chapter opener has been updated with more current data on cyberbullying and also includes findings from the 2017 Youth Risk Behavior Study and a 2015 school climate study focused on the experiences of LGBT students. The "Developing a Global Perspective" section has been updated with the results of the 2016 American Freshman survey. The discussion of the divorce rate has been changed to reflect the newest trends. A new Employing Your Sociological Imagination feature titled "What Can You Do with a Sociology Major?" highlights the careers of Dr. Martin Luther King Jr., Michelle Obama, and Dr. Ruth Westheimer, demonstrating how the skills and knowledge acquired in sociology classes can be applied to far-ranging careers. The Digital Life box has been updated with data from a 2017 survey on online harassment and now discusses the LGBT experience of cyberbullying. The box also highlights a new location-based networking app, Islands, that is billing itself as "Slack for college students." Table 1.4 has been updated with the most current data on how countries across the globe view the United States, including data following the 2016 presidential election. A revised Globalization by the Numbers infographic, titled "Opinion of the United States," captures the considerable differences among nations regarding the proportion of the population that holds favorable attitudes toward the United States and illustrates how these attitudes have changed over time.

Chapter 2 Culture and Society The discussion of the different forms of culture has been reorganized in order to better differentiate between nonmaterial and material culture. In the discussion of norms, data on cigarette smoking have been updated. The discussion of material culture now considers how material culture is rapidly becoming globalized. The section on the nature/nurture debate has been expanded and now highlights recent research, including studies on obesity and gender differences in mathematical ability, that explore the relationship between genetics and social influences. A new Globalization by the Numbers infographic explores how different countries conceive of "national identity," highlighting the relative importance of factors such as birthplace, language, and national customs. The discussion on the emergence of industrialized societies now highlights how sociology was established in this context. The section titled "Does the Internet Promote a Global Culture?" has been substantially updated and now uses Saudi Arabia as a case study to show how the Internet can be a means of strengthening traditional cultural values. This section also explores how the Internet can function as an echo chamber as well as be used by groups such as ISIS to build a community around ideas that challenge the dominant culture.

Chapter 3 Socialization, the Life Course, and Aging The discussion of theories of child development has been expanded and now includes a section on Charles Horton Cooley and his theory of the looking-glass self. The section on families as agents of socialization now discusses Annette Lareau's study of the differing child rearing strategies employed by upper-middle-class and working-class parents. The discussion on education has been expanded and now introduces the concept of the hidden curriculum. In the section on mass media, data on media consumption have been updated, drawing on the 2017 American Time Use survey. The gender learning discussion now more fully considers the role of parents and television shows. A new Digital Life box, titled "New Apps Challenge Kids—and the Gender Binary," discusses how app developers are creating new games for kids that challenge gender norms.

Data related to the graying of U.S. society, including the size and growth of the elderly population, life expectancy, health issues, use of the Internet, and aid, have all been updated. A new Employing Your Sociological Imagination feature highlights the demand for workers with a basic knowledge of aging and the challenges facing older adults; the feature then demonstrates how an activities director at a long-term care facility would benefit from an understanding of life course sociology and sociological theories of aging. A new Globalization by the Numbers infographic, "Graying of the World," shows the current size of the elderly population in eleven different countries and the projected growth of this population as well as a breakdown of the older population in the United States by age and race.

Chapter 4 Social Interaction and Everyday Life in the Age of the Internet A new discussion of human interaction with nonhumans, including human-computer interaction, particularly interaction with virtual personal assistants such as Alexa, explores the way modern technology has changed the nature of social interaction. The section on Goffman's dramaturgical model has been expanded and the discussion of impression management now points to an incident at Harvard where students had their admissions offers rescinded for sending sexually explicit and racist messages to each other in a private Facebook group. The discussion of personal space now highlights recent research on Greyhound bus passengers as well as how people use mobile technology to "extend" personal space. The authors now consider how technology is rearranging time and space, highlighting research by danah boyd on how teens today use social media to be in touch with their local community. The Globalization by the Numbers infographic, "Who Owns a Smartphone?" compares the most current rates of smartphone ownership in different countries across the globe to provide a picture of this new digital divide. The examination of social media and harassment has been expanded and now considers how the Internet and social media have supported the growth of millennial-led social movements. A new section, titled "Interaction on the 'Digital Street,'" introduces recent ethnographic research by Jeffrey Lane on how encounters between boys and girls are being reshaped online.

Chapter 5 Groups, Networks, and Organizations The "Technology and Modern Organizations" section has been updated with the results of a Gallup survey on remote work. Data on female CEOs have been updated. The discussion of social capital has been updated with voter turnout data from the 2016 election as well as a 2017 survey of trust in government. The Globalization by the Numbers infographic on Internet connectivity throughout the world has been updated with the most recent data. The Digital Life box on crowdfunding and the strength of weak ties now looks at how Americans have recently turned to platforms such as YouCaring and GoFundMe to raise millions of dollars for victims of Hurricane Harvey and for victims of sexual assault and harassment. Drawing on interview data from the American Sociological Association, a new Employing Your Sociological Imagination feature highlights the field of organizational sociology and proposes ways in which a degree in sociology can help students climb the corporate ladder.

Chapter 6 Deviance, Crime, and Punishment This chapter now includes an entirely new "Big Question" dedicated to unpacking the great crime decline: "What were the causes and costs of the great crime decline?" Drawing on research by sociologist Patrick Sharkey, this new discussion examines three factors that have contributed to this decline—prisons, the death penalty, and policing—and also considers the costs and benefits associated with this drop in crime. Data throughout the chapter have been updated, including incarceration rates and prison populations; hate crimes; crime rates by gender, age, and race; cost of imprisonment; and numbers and rates of violent crime and property crime in the United States. The section on youth and crime features updated data on drug use as well as a discussion of the opioid epidemic and how it has been framed as a predominately white issue, which has affected the public health response to the crisis. The discussion of the use of capital punishment in the United States now presents more current statistics. The section on policing includes a new discussion of sociologist Forrest Stuart's recent ethnography *Down, Out, and Under Arrest: Policing and Everyday Life in Skid Row*, which documents the effect of intensive policing on the homeless population in downtown Los Angeles. A discussion of the benefits of the crime decline looks at the impact of lower crime rates on two areas of social life: life expectancy and school performance. The Globalization by the Numbers infographic has been updated with the most recent data on global incarceration rates. A new Employing Your Sociological Imagination feature explores the way a sociological analysis of big data can help law enforcement officers.

Chapter 7 Stratification, Class, and Inequality The story of Viviana Andazola Marquez, a young woman who was able to beat the odds and rise out of poverty and homelessness to become a student at Yale University, has been updated to reflect recent developments in her life, including the deportation of her father by ICE. The discussion of interracial marriage has been updated with more current data. Data on income distribution, mean household income by income group and race, and racial disparities in wealth have all been updated. In the section on education, data on median earnings, the growing wage premium, graduation rates, and college enrollment by race have all been updated. The table presenting the relative social prestige of various occupations in the U.S. has been refreshed with new occupations. The Globalization by the Numbers infographic on income inequality now highlights both the income share held by the top 10 percent of the population and the share held by the bottom 10 percent of the population in nine different countries. The discussion of social inequality in the United States highlights how the Black–white wealth gap has increased in recent years. The discussion of poverty in the United States has been updated with current statistics, including poverty rates by age, race, and gender and demographic data on the working poor and homeless populations, as well as the results of a 2016 public opinion poll on whether the poor are responsible for their plight. The final section highlights a recent study by Stanford economist Raj Chetty on intergenerational income mobility that found that young people entering the workforce today are considerably less likely to outearn their parents than young people born two generations before them.

Chapter 8 Global Inequality A new chapter opener uses the memorable story of eight-year-old Wang Fuman—dubbed "frost boy" by Chinese media—to highlight disparities between the United States and China, where Fuman is just one of countless "left behind" children. *Absolute poverty* and *relative poverty* are now introduced and defined. The Globalization by the Numbers infographic, titled "An Unequal World," has been completely updated with the most recent data on gross national income per capita, population, life expectancy, fertility rate, and infant mortality rate. Global Map 8.1, "Rich and Poor Countries: The World by Income in 2017," has been updated to reflect new World Bank country classifications. The discussion of daily life in rich vs. poor countries has been expanded, including the section on health; the section on hunger and malnutrition explores the link between conflict and hunger, highlighting climate change as an additional contributor to increasing violence. Data on chronic hunger, literacy rates and participation in secondary education, and child labor have been updated. The discussion of neoliberalism has been expanded and now considers how neoliberal ideas have impeded economic development in poor countries. The section on world-systems theory now highlights China as an example of a country that has taken a state-driven approach to economic development. The Apple iPhone is used as an example in the expanded discussion of global commodity chains. A new section introduces William Robinson's theory of global capitalism and also considers work by sociologist Leslie Sklair on the transnational capitalist class. The summary of dependency theory now introduces the concept of dependent development. The concluding section now considers recent developments such as the effect of slowing economic growth on the European Union, the rise of far-right nationalist parties throughout Europe, global climate change, and the Syrian refugee crisis. This section also discusses a dire 2018 report from the Center for Climate and Security on climate change and its threat to global stability.

Chapter 9 Gender Inequality   A new chapter opener considers the birth of the #MeToo movement, highlighting diverse cases—from Silicon Valley to a Chicago automobile factory—in order to emphasize how pervasive sexual harassment is in the workplace. The discussion of the difference between sex and gender has been expanded and the term *nonbinary* is now introduced. The section on gender socialization now tells the story of gender-expansive Brooklyn kindergartner Leo Davis, whose parents sued the Education Department. *Hegemonic masculinity* has been added as a key term. The discussion of "doing gender" has been expanded and now considers how nonbinary individuals strategically choose how to behave and present themselves. Danica Roem is featured as an example of rising acceptance of non-cisgender people. Estimates of the nonbinary population are now provided as well as new examples of countries and U.S. states that allow for nonbinary identification. Yemen is now discussed as an example of a country characterized by a high level of gender inequality. In the section on gender inequality in social institutions, data on women's labor force participation, the gender pay gap, and occupational segregation have all been updated. The Globalization by the Numbers infographic on gender inequality shows countries' most up-to-date ratings on the Gender Inequality Index as well as current statistics on women's labor force participation, representation in government, and participation in secondary school. The concept of the "glass escalator" is now introduced. The section on sexual harassment has been thoroughly revised and expanded and now highlights the experience of women at the Ford Motor Company as well as recent research on the effect of sexual harassment on women's long-term career prospects and financial stability. The data in the section on gender inequality in politics have been updated and a new discussion of Hillary Clinton considers the gender gap in voting and the role that sexism played in the 2016 election. Data on intimate partner violence have been updated. The section on rape now highlights recent research on the experience of young men. The discussion of rape culture now points to specific recent controversies on college campuses as well as recent research by Michael Kimmel; the term *toxic masculinity* is introduced. A new final "Big Question" considers, "How can we reduce gender-based violence?"

Chapter 10 Race, Ethnicity, and Racism   This chapter has been thoroughly revised throughout to provide a more contemporary picture of race and ethnicity in the United States, including new "Big Questions" on why racial and ethnic antagonism exist, how racism operates in American society today, and how sociologists explain the persistence of racial inequality. The opener has been updated with current data on the proportion of Americans who identify as multiracial as well as data on interracial marriages. The discussion of race has been overhauled and now traces the evolution of racial categories in the U.S. Census to highlight the concept of race as a social construction. The Globalization by the Numbers infographic has been updated to reflect the most up-to-date racial and ethnic populations in six countries. The discussion of minority groups now considers the future of the United States as a majority-minority country. A new Big Question, "Why do racial and ethnic antagonism exist?" examines both psychological theories, including prejudice and discrimination and the authoritarian personality, as well as sociological interpretations, including ethnocentrism, group closure, and resource allocation. A new Big Question "How Does Racism Operate in American Society Today?" includes an expanded discussion of institutional racism that takes an in-depth look at the Department of Justice report produced in 2015 in the aftermath of Michael Brown's death. A new section on interpersonal racism looks at both overt racism, using examples from the 2016 election, as well as Eduardo Bonilla-Silva's concept of colorblind racism. The concepts of white privilege and microaggressions are now introduced. The discussion of immigration has been updated with more current data and highlights recent trends in undocumented immigration and top sending countries. Data on the relative sizes of different racial/ethnic populations have been updated. A revamped Big Question on how race/ethnicity affect one's life chances includes updated data on racial differences in educational attainment, occupational attainment, income, residential segregation, and political power. The discussion of the Black–white gap in wealth has been expanded and includes data from a 2017 report from the Federal Reserve. The discussion of racial disparities in health has been overhauled and expanded and now considers recent research on racial discrimination in health care and maternal mortality rates for Black women. The section on residential segregation now highlights a recent report on educational segregation sixty years after the landmark passage of *Brown*. Data in the section on gender and race, including earnings by race and sex, have been updated. The chapter now concludes with a new Big Question—"How do sociologists explain racial inequality?"—that explores cultural, economic, and discrimination-based theories of inequality.

Chapter 11 Families and Intimate Relationships   The opener has been updated with the results of a 2017 Pew study of attitudes on gay marriage. The discussion of polygamy has been revised to incorporate recent trends. The section on sociological theories of families now features a dedicated discussion of symbolic interactionist approaches that highlights recent research on immigrant families. The discussion of marriage and family in the United States includes more recent data on marriage rates, age at first marriage, people living alone, and childbearing patterns. In the "Race, Ethnicity, and American Families" section, birthrates by racial group, nonmarital fertility, and Black and white family patterns have all been updated. A new figure highlights the differences between the structure of Black and white households in 2017. The section dedicated to social class now considers the effect of the opioid crisis on families. A new Employing Your Sociological Imagination feature demonstrates how a background in sociology is excellent preparation for the challenges and duties of a career as a marriage and family therapist. The discussion of child abuse has been updated with more current data. The Globalization by the Numbers infographic for this chapter, titled "Maternity Leave," has been refreshed with a handful of new countries.

Chapter 12 Education and Religion   In the opener about Malala Yousafzai, data on girls' education in Pakistan has been updated. The discussion of homeschooling has been updated with current data and more recent research on efficacy. The chapter's treatment of sociological theories of education has been broadened and now includes dedicated subsections on functionalism, symbolic interactionism, and conflict theory. Data pertaining to global literacy have been updated. A new section, titled "The Resegregation of American Schools?" examines the state of educational segregation more than sixty years after the passage of *Brown v. Board of Education*, drawing on recent reports from the Civil Rights Project and the Brookings Institution to explore the issue of continued racial segregation among schools. The section on educational reform now discusses Trump-era developments, including the expansion of school choice under new Education Secretary Betsy DeVos; this section also includes the results of a recent national survey of more than 500 K–12 teachers. The discussion of Durkheim's concept of totemism has been expanded. The section on the global rise of religious nationalism now highlights recent developments in Iran, including the 2017 reelection of Hassan Rouhani, and points to ISIS as an example of how religious nationalism can turn violent. A new Globalization by the Numbers infographic provides a global picture of religious affiliation as well as religious affiliation in the United States. The section on trends in religious affiliation now discusses the results of the 2016 American Values Atlas Religion Report, including the finding that the most youthful religious groups are all non-Christian. The discussion of Protestantism now considers the impact that Evangelicals had on the 2016 election. More current data is provided on the American Muslim population, including new projections by the Pew Research Center on the future growth of this population.

Chapter 13 Politics and Economic Life   The discussion of the welfare state now considers recent examples such as Brexit and new Trump policies. The number of democratic nations across the globe has been updated according to data from Freedom House. The section on the Internet and democratization now explores the role that the Internet and social media played in the 2016 election; data on where Americans receive their news have been updated. A new section is now dedicated to examining the rise of populist authoritarianism. A new Employing Your Sociological Imagination feature explains how sociology students can translate their degree into a successful career as a political activist or politician. The discussion of political parties has been updated to reflect more current trends in party affiliation. A discussion of voter turnout in the 2016 election, including turnout by race, sex, educational attainment, and income, has been added. A new Globalization by the Numbers infographic compares voter turnout rates in twelve different countries, highlighting recent elections in the UK, Mexico, Venezuela, France, Germany, Turkey, and Egypt; an additional section of the infographic illustrates who voted for Trump by demographic group. The voter turnout discussion now considers the impact of voter ID laws. The section on interest groups now incorporates data on campaign spending from the 2016 election. Statistics on women's participation in politics have been updated, including data on women's voting patterns in the 2016 election. The section on the power elite now introduces the term *deep state*. The discussion of military spending was updated with 2017 data. The discussion of conservatives attacking the foundation of Democracy is now highlighted to explain people's declining faith in the U.S. government. Data on labor union membership and work stoppages have been updated and a new survey of public opinion of labor unions referenced. The discussion of transnational corporations has been updated with more current data.

Chapter 14 The Sociology of the Body: Health, Illness, and Sexuality A new chapter opener uses the heart-breaking story of Dr. Shalon Irving, a Black scientist who died due to pregnancy-related complications. The opener then goes on to discuss the phenomenon of "weathering" and racial disparities in health. The discussion of eating disorders has been expanded to consider emerging research on gay and bisexual men as well as the intersection of gender and race. Data pertaining to obesity have been updated. The revamped Globalization by the Numbers now shows obesity rates in eleven different countries broken down by sex; the bottom portion now shows obesity rates in the United States by age and race. The sections on the sick role and symbolic interactionist approaches to illness have both been expanded to now consider people with disabilities. The discussion of social class–based inequalities in health now explores the recent rise in "deaths of despair." Data in the section on race-based and gender-based inequalities in health have been updated, including life expectancy and rates of hypertension. Data on malaria and HIV/AIDS have all been updated. A thoroughly revised section on the diversity of human sexuality now introduces the terms *heteronormativity* and *QUILTBAG*. Public opinion on premarital and extramarital sex has been updated. Slut-shaming is now discussed as well as the burgeoning sex-positive movement. The new Employing Your Sociological Imagination box highlights the need for health professionals to practice cultural sensitivity and be aware of health disparities. Data on the sex lives of high school students as well as men's and women's number of lifetime sexual partners have been updated. The discussion on homophobia has been broadened to explain heterosexism and transphobia and highlights examples like the new transgender military ban. A new survey of school climate and the experience of LGBTQ students has been added.

Chapter 15 Urbanization, Population, and the Environment The opener has been updated based on a new report of most polluted cities. Data throughout the chapter on urbanization have been updated based on a 2018 report from the UN Department of Economic and Social Affairs. The section on urban problems now includes a new discussion of Richard Rothstein's 2017 book, *The Color of Law*, which shows how racial segregation today is largely due to governmental housing policies. A new Employing Your Sociological Imagination spread provides an in-depth look at the occupation of demographer, illustrating how people in this job apply sophisticated statistical tools and sociological concepts to vexing social issues such as high unemployment or sex discrimination. The discussion of globalization in the Global South has been updated to reflect more up-to-date projections of urban growth. The Globalization by the Numbers infographic on urbanization has been updated to reflect the largest cities in 2018 as well as new projections for 2035. In the section on demography, data on birthrates, death rates, and infant mortality rates across the globe have been updated. The global map detailing population growth rates around the world has been updated with new data. The section on the demographic transition now includes an updated number of what the United Nations have deemed the "least developed countries" as well as more current data on hunger and malnourishment. The discussion of global warming and climate change now discusses Trump's exit from the Paris Agreement.

Chapter 16 Globalization in a Changing World The chapter opener has been updated and now more fully explains the civil war in Syria and the role that globalization played in the events of the Arab Spring. The discussion of globalization's effect on the physical environment has been expanded to consider the impact of global warming.

A new section on technology explores the important role that technology has played in social change, highlighting the "time-space compression." A revamped Big Question about why terrorism seems to be on the rise has been shifted to this chapter. The discussion of the classical theories of social movements has been streamlined. The section on new social movements now includes more recent examples, including the #MeToo and Antifa movements, and discusses the Women's March on Washington in 2017, as well as technology's role in facilitating protests. The discussion of technology and social movements now highlights how technology has facilitated the formation of cross-border networks of activists, including the International Campaign to Abolish Nuclear Weapons, which received the 2017 Nobel Peace Prize. Data on global Internet usage have been updated. The concept of a "weightless economy" is now discussed in the section on information flows, which also highlights the rise of cloud computing. The discussion of the rise of individualism now brings in a recent study of seventy-seven countries on changes in values and behaviors. *Hybridization* is now introduced as a key term. Hurricanes Harvey and Maria are now highlighted in the discussion of ecological risk. Data on global poverty and global trade have also been updated. The Globalization by the Numbers infographic paints an updated picture of wealth inequality around the world. The discussion of global inequality has been updated based on a 2017 report from Credit Suisse and also incorporates more current data from the World Bank on global poverty. Trump's recent trade deals are discussed in the section on global justice. Data on farm subsidies in the United States have been updated.

Organization

There is very little abstract discussion of basic sociological concepts at the beginning of this book. Instead, concepts are explained when they are introduced in the relevant chapters, and we have sought throughout to illustrate them by means of concrete examples. While these are usually taken from sociological research, we have also used material from other sources (such as newspaper or popular magazine articles). We have tried to keep the writing style as simple and direct as possible, while endeavoring to make the book lively and full of surprises.

The chapters follow a sequence designed to help achieve a progressive mastery of the different fields of sociology, but we have taken care to ensure that the book can be used flexibly and is easy to adapt to the needs of individual courses. Chapters can be skipped or studied in a different order without much loss. Each has been written as a fairly autonomous unit, with cross-referencing to other chapters at relevant points.

Study Aids

Every chapter in the Seventh Edition of *Essentials of Sociology* features:

- **Learning Goals** are outlined at the start of the chapter and then recur throughout the chapter in marginal notations at the beginning of the relevant sections to promote active learning.

- **"Concept Checks"** throughout each chapter help students assess their understanding of the major topics in the chapter. Each "Concept Check" has at least three questions that range from reading comprehension to more advanced critical thinking skills.

- **"Digital Life" boxes** in every chapter get students thinking critically about how the Internet and smartphones are transforming the way we date, manage our health, and even practice religion.

- **"Globalization by the Numbers" infographics** transform raw numbers into visually interesting displays that put the United States in a global context. Interactive versions in the ebook make the data dynamic and include integrated assignments that engage students with the data.

- **"Big Picture" Concept Maps** at the end of every chapter, which integrate the "Big Questions," key terms, and "Concept Checks" into a handy and visually interesting study tool, serve as both a pre-reading guide to the chapter as well as a post-reading review.

Acknowledgments

Many individuals offered us helpful comments and advice on particular chapters, and, in some cases, large parts of the text. They helped us see issues in a different light, clarified some difficult points, and allowed us to take advantage of their specialist knowledge in their respective fields. We are deeply indebted to them. Special thanks go to Jason Phillips, who worked assiduously to help us update data in all chapters and contributed significantly to editing as well; and Dmitry Khodyakov, who wrote thought-provoking Concept Check questions for each chapter.

We would like to thank the many readers of the text who have written us with comments, criticisms, and suggestions for improvements. We have adopted many of their recommendations in this new edition.

Adalberto Aguirre, University of California, Riverside

Francis O. Adeola, University of New Orleans

Patricia Ahmed, South Dakota State University

Colleen Avedikian, University of Massachusetts Dartmouth

Debbie Bishop, Lansing Community College

Sharon Bjorkman, Pikes Peak Community College

Kim Brackett, Auburn University

Joy Branch, Southern Union State Community College

Edith Brotman, Towson University

Tucker Brown, Austin Peay State University

Cecilia Casarotti, Hillsborough Community College

Susan Cody-Rydzewski, Georgia Perimeter College

Caroline Calogero, Brookdale Community College

Paul Calarco, Hudson Valley Community College

Giana Cicchelli, Fullerton College

Karen Coleman, Winona State University

Dawn Conley, Gloucester County College

Olga Custer, Oregon State University

Raymonda Dennis, Delgado Community College

Sarah DeWard, Eastern Michigan University

Jason Dixon, Walters State Community College

Jonathon Fish, Trident Technical College

Matthew Flynn, University of Texas at Austin

Clare Giesen, Delgado Community College

Ron Hammond, Utah Valley University

Nicole Hotchkiss, Washington College

Howard Housen, Broward College

Rahime-Malik Howard, El Centro College

Annie Hubbard, Northwest Vista College

Onoso Imoagene, University of Pennsylvania

Kristin Ingellis, Goodwin College

Jennifer Jordan, University of Wisconsin-Milwaukee

Ryan Kelty, Washington College

Qing Lai, Florida International University

Andrew Lash, Valencia College

Kalyna Lesyna, Palomar College

Danilo Levi, Delgado Community College

Ke Liang, Baruch College

Devin Molina, Bronx Community College

Monita H. Mungo, University of Toledo

Jayne Mooney, John Jay College of Criminal Justice

Kendra Murphy, University of Memphis

Rafael Narvaez, Winona State University

Timothy L. O'Brien, University of Wisconsin-Milwaukee

Daniel O'Leary, Old Dominion University

Takamitsu Ono, Anne Arundel Community College

Carolyn Pevey, Germanna Community College

Robert Pullen, Troy University

Kent Redding, University of Wisconsin-Milwaukee

Matt Reynolds, College of Southern Idaho

James Rice, New Mexico State University

Fernando Rivera, University of Central Florida

Dan Rose, Chattanooga State Community College

Elizabeth Scheel-Keita, St. Cloud State University

Dave Seyfert, Adjunct, Pikes Peak Community College

Luis Sfeir-Younis, University of Michigan

John M. Shandra, Stony Brook University

Rachel Stehle, Cuyahoga Community College

Larry Stern, Collin College

Daniel Steward, University of Illinois at Urbana-Champaign

Karen Stewart-Cain, Trident Technical College

Richard Sweeney, Modesto Junior College

Adrienne Trier-Bieniek, Valencia College

Jason Ulsperger, Arkansas Tech University

Thomas Waller, Tallahassee Community College

Candace Warner, Columbia State Community College

Tammy Webb, Goodwin College

Ron Westrum, Eastern Michigan University

Jeremy White, Pikes Peak Community College

Jessica Williams, Texas Woman's University

Kristi Williams, Ohio State University

Annice Yarber, Auburn University

Erica Yeager, Anne Arundel Community College

We have many others to thank as well. Holly Monteith did a marvelous job of copyediting the new edition. We are also extremely grateful to project editor Diane Cipollone, who managed the countless details involved in creating the book. Assistant editor Erika Nakagawa skillfully tracked all the moving parts that go into publishing this complicated project. Production manager Stephen Sajdak did impressive work guiding the book through production, so that it came out on time and in beautiful shape. We also thank Eileen Connell, our e-media editor, and Ariel Eaton, our associate e-media editor, for developing all of the useful supplements that accompany the book. Agnieszka Gasparska and the entire team of designers at Kiss Me I'm Polish gave the book a stunning design and also managed to digest a huge amount of data to create the Globalization by the Numbers infographics throughout *Essentials of Sociology*.

We are also grateful to our editors at Norton—Steve Dunn, Melea Seward, Karl Bakeman, and Sasha Levitt—who have made important substantive and creative contributions to the book's chapters and have ensured that we have referenced the very latest research. We also would like to register our thanks to a number of current and former graduate students—many of whom are now tenured professors at prestigious universities—whose contributions over the years have proved invaluable: Wendy Carter, Audrey Devin-Eller, Neha Gondal, Neil Gross, Black Hawk Hancock, Paul LePore, Alair MacLean, Ann Meier, Susan Munkres, Josh Rossol, Sharmila Rudrappa, Christopher Wildeman, David Yamane, and Katherina Zippel.

Anthony Giddens
Mitchell Duneier
Richard Appelbaum
Deborah Carr

SEVENTH EDITION

Essentials of Sociology

1

Sociology: Theory and Method

Opinion of the United States

p. 9

What is the "sociological imagination"?

Learn what sociology covers as a field and how everyday topics like love and romance are shaped by social and historical forces. Recognize that sociology involves developing a sociological imagination and a global perspective and understanding social change.

What theories do sociologists use?

Learn about the development of sociology as a field. Be able to name some of the leading social theorists and the concepts they contributed to sociology. Learn the different theoretical approaches modern sociologists bring to the field.

What kinds of questions can sociologists answer?

Be able to describe the different types of questions sociologists address in their research.

What are the steps of the research process?

Learn the steps of the research process and be able to complete the process yourself.

What research methods do sociologists use?

Familiarize yourself with the methods available to sociological researchers, and know the advantages and disadvantages of each. See how researchers use multiple methods in a real study.

What ethical dilemmas do sociologists face?

Recognize the ethical problems researchers may face, and identify possible solutions to these dilemmas.

How does the sociological imagination affect your life?

Understand how adopting a sociological perspective allows us to develop a richer understanding of ourselves, our significant others, and the world.

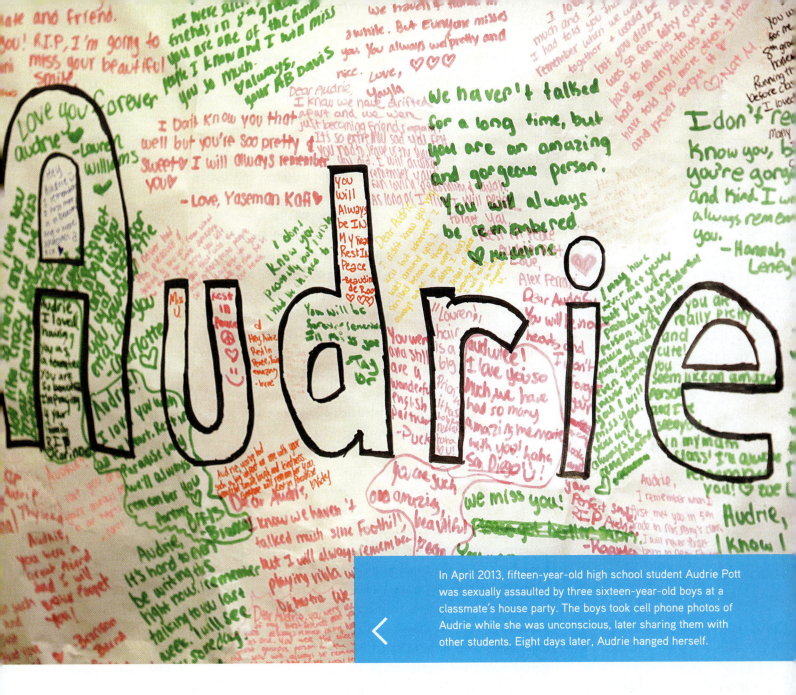

In April 2013, fifteen-year-old high school student Audrie Pott was sexually assaulted by three sixteen-year-old boys at a classmate's house party. The boys took cell phone photos of Audrie while she was unconscious, later sharing them with other students. Eight days later, Audrie hanged herself.

Sociology is the scientific study of human social life, groups, and societies. It is a dazzling and compelling enterprise, as its subject matter is our own behavior as social beings. The scope of sociological study is extremely wide, ranging from the analysis of how people establish social connections with one another to the investigation of global social processes, such as the rise of Islamic fundamentalism.

Sociology teaches us that what we regard as natural, inevitable, good, or true may not be such and that the "givens" of our life—including things we assume to be genetic or biological—are strongly influenced by historical, cultural, social, and even technological forces. Understanding the subtle yet complex and profound ways in which our individual lives reflect the contexts of our social experience is central to the sociological outlook. A brief example will provide a taste of the nature and objectives of sociology.

Anyone who has attended middle school or high school knows that bullying is a common occurrence. Through much of history, teachers, principals, and parents turned a blind eye, often believing that "boys will be boys." In recent years this cavalier attitude toward bullying has been called into question by students, teachers, and policymakers alike. A recent spate of suicides by teenagers subjected to merciless bullying has raised awareness that bullying is no longer "kid stuff" and in nearly all states is grounds for suspension, expulsion, or even more serious punishment. Over the past decade, bullying-related tragedies have been documented throughout the United States, involving teenagers of all backgrounds—male and female, Black and white, Asian and Latino, gay and straight, cisgender and transgender, rich and poor, rural and suburban.

For Devin Brown, the bullying began shortly after he started at Rothschild Middle School. Things escalated after he reported another student for carrying a knife and threatening a teacher. Rather than being regarded as a hero by his classmates, he was derided as a "snitch" and was regularly threatened and beaten up at school. In April 2013, after months of relentless harassment, Brown hanged himself at home in his closet.

That same month, fifteen-year-old high school student Audrie Pott hanged herself in her San Jose, California, home. Eight days earlier, Pott had been sexually assaulted at a classmate's house party by three sixteen-year-old boys. She woke up to find her clothes pulled off and her body covered in lewd markings. The trauma didn't end there, though. The boys took pictures of Pott while she was unconscious and shared them with other students. Just days before she took her life, a devastated Pott posted messages on Facebook that read, "My life is over" and "The whole school knows."

It's not just American teens who are killing themselves as a desperate reaction to intolerable bullying. In Japan, teen suicide rates spike each year on September 1, the nation's first day of school; experts say the specter of going back to school and facing mistreatment by their peers (along with intense academic pressure) is enough to trigger suicide among depressed teens (Wright 2015).

Brown and Pott are just two of hundreds of teenagers who have committed suicide after being bullied and humiliated by their classmates. Today, anti-bullying laws exist in all fifty states; Montana was the last state to enact such laws in 2015 (Baumann 2015). In 2011, New Jersey passed the nation's toughest anti-bullying legislation, triggered in part by the high-profile suicide of Tyler Clementi. In 2010, the eighteen-year-old Rutgers University freshman committed suicide by jumping off the George Washington Bridge, just two weeks after he started his first semester in college. The suicide came days after his roommate used a webcam to spy on Clementi during an intimate encounter with a man in his dorm room and then posted on social media about it.

Sociology helps us to understand and analyze scientifically social phenomena like bullying and suicide. American sociologist C. Wright Mills (1959) observed that social sciences enable people to "translate private troubles into public issues." What Mills meant is that individuals often believe that the problems that they (and others) face are personal, perhaps resulting from their own traits or decisions. But social scientists recognize that these seemingly "**personal**" **troubles**, if occurring in patterned ways to large numbers of individuals, reflect important "**public issues**" or consequences of social structures.

For example, Devin Brown, Audrie Pott, and Tyler Clementi all committed suicide shortly after being tormented by their peers. Some observers might think that the suicides were an isolated problem, perhaps the reaction of three teens who were depressed or emotionally unstable. However, a sociologist would look at the social context and try to understand just how common such events are and determine whether some subgroups are particularly vulnerable to such problems. Sociologists might look at historical data to track the timing of these suicides, as researchers did in Japan. This analysis showed that teen suicides weren't spread evenly across the year but rather clustered in early September (Wright 2015).

Sociologists might also consult data from national surveys, such as the 2017 Youth Risk Behavior Study, which found that nearly one in five high school students had been bullied on school property within the last year; 15 percent had been electronically bullied (Kann et al. 2018). Sociologists have also found that gay and lesbian teens are far more likely than their straight peers to be harassed at school. One survey of more than 7,500 high school students found that nearly 44 percent of gay and 40 percent of lesbian teens said they had been bullied in the previous year, compared with just 26 and 15 percent of heterosexual boys and girls, respectively (Berlan et al. 2010). A 2015 study of lesbian, gay, bisexual, and transgender (LGBT) students found that more than eight in ten LGBT students had been verbally harassed, nearly six in ten had been sexually harassed, and more than a quarter (27 percent) had been physically harassed within the past year (Kosciw et al. 2016). Studies such as these help us recognize that the anguish Clementi, Pott, and Brown experienced is hardly an isolated incident and instead reflects pervasive social problems that require far-reaching solutions. Sociology can help us understand the questions of what, why, and how public issues and personal troubles arise.

What Is the "Sociological Imagination"?

Learn what sociology covers as a field and how everyday topics like love and romance are shaped by social and historical forces. Recognize that sociology involves developing a sociological imagination and a global perspective and understanding social change.

When we learn to think sociologically, we can also better understand the most personal aspects of our own lives. For instance, have you ever been in love? Almost certainly you have. Most people who are in their teens or older know what being in love is like. Love and romance provide some of the most intense feelings we ever experience. Why do people fall in love? The answer may seem obvious: Love expresses a mutual physical and personal attachment between two individuals. These days, we might not all think that love is "forever," but falling in love, we may agree, is an experience arising from universal human emotions. It seems natural for a couple in love to want personal and sexual fulfillment in their relationship, perhaps through marriage.

Yet this pattern whereby love leads to marriage is in fact very unusual. Romantic love is not an experience all people across the world have—and where it does happen, it is rarely connected to marriage. The idea of romantic love did not become widespread until fairly recently in our society, and it has never even existed in many other cultures.

Only in modern times have love and sexuality become closely connected. In the Middle Ages and for centuries afterward, men and women married mainly to keep property in the hands of the family or to raise children to work the family farm—or, in the case of royalty, to seal political alliances. Spouses may have become close companions after marriage, but not before. People sometimes had sexual affairs outside marriage, but these inspired few of the emotions we associate with love today. Romantic love was regarded as a weakness at best and a kind of sickness at worst.

Romantic love developed in courtly circles as a characteristic of extramarital sexual adventures by members of the aristocracy. Until about two centuries ago, it was confined to

What is the origin of romantic love? Originally, romantic love was limited to affairs for medieval aristocrats like Tristan and Isolde, the subjects of a thirteenth-century court romance that inspired poems, operas, and films.

sociological imagination

The application of imaginative thought to the asking and answering of sociological questions. Someone using the sociological imagination "thinks himself away" from the familiar routines of daily life.

such circles and kept separate from marriage. Relations between husband and wife among aristocratic groups were often cool and distant. Each spouse had his or her own bedroom and servants; they may have rarely seen each other in private. Sexual compatibility was not considered relevant to marriage. Among both rich and poor, the decision of whom to marry was made by one's immediate and extended family; the individuals concerned had little or no say in the matter.

This remains true in many non-Western countries today. (Social scientists typically define "Western" countries as economically rich nations, including most in North America and Europe as well as Japan and Australia.) For example, in Afghanistan under the rule of the Taliban, men were prohibited from speaking to women to whom they were not related or married, and marriages were arranged by parents. The Taliban government saw romantic love as so offensive that it outlawed all nonreligious music and films. Like many in the non-Western world, the Taliban believed Afghanistan was being inundated by Hollywood movies and American pop music and videos, which are filled with sexual images.

Neither romantic love, then, nor its association with marriage can be understood as a natural or universal feature of human life. Rather, such love has been shaped by social and historical influences. These are the influences sociologists study.

Most of us see the world in terms of the familiar features of our own lives. Sociology demonstrates the need for a much broader view of our nature and our actions. It teaches that what we regard as "natural" in our lives is strongly influenced by historical and social forces. Understanding the subtle yet profoundly complex ways in which our individual lives reflect the contexts of our social experience is basic to the sociological outlook.

Learning to think sociologically means cultivating what sociologist C. Wright Mills (1959), in a famous phrase, called the **sociological imagination**. According to Mills, each of us lives in a very small orbit, and our worldviews are limited by the social situations we encounter on a daily basis, including the family and the small groups of which we're a part, the schools we attend, and even the dorms in which we live. All these things give rise to a certain limited perspective and point of view. Mills argued that we all need to overcome our limited perspective. What is necessary is a certain quality of mind that makes it possible to understand the larger meaning of our experiences. This quality of mind is the sociological imagination.

The sociological imagination requires us, above all, to "think ourselves away" from our daily routines in order to look at them anew. Consider the simple act of drinking a cup of coffee. What might the sociological point of view illuminate about such apparently uninteresting behavior? An enormous amount. First, coffee possesses symbolic value as part of our daily social activities. Often, the ritual associated with coffee drinking is much more important than the act itself. Two people who arrange to meet for coffee are probably more interested in getting together and chatting than in what they actually drink. Drinking and eating in all societies, in fact, promote social interaction and the enactment of rituals— rich subject matter for sociological study.

Second, coffee contains caffeine, a drug that stimulates the brain. In Western culture, coffee addicts are not regarded as drug users. Like alcohol, coffee is a socially acceptable drug, whereas cocaine and opium, for instance, are not. Yet some societies tolerate the recreational use of opium or even cocaine but frown on coffee and alcohol. Sociologists are interested in why these contrasts exist.

Third, an individual who drinks a cup of coffee is participating in a complicated set of social and economic relationships stretching across the world. The production and distribution of coffee require continuous transactions among people who may be thousands of miles away from the coffee drinker. Studying such global transactions is an important task of sociology because many aspects of our lives are now affected by worldwide social influences and communications.

Finally, the act of sipping a cup of coffee presumes a process of past social and economic development. Widespread consumption of coffee—along with other now-familiar items of Western diets, such as tea, bananas, potatoes, and white sugar—began only in the late 1800s under Western colonial expansion. Virtually all the coffee we drink today comes from areas (South America and Africa) that were colonized by Europeans; it is in no sense a "natural" part of the Western diet.

Studying Sociology

The sociological imagination allows us to see that many behaviors or feelings that we view as private and individualized actually reflect larger social issues. Try applying this sort of outlook to your own life. Consider, for instance, why you are attending college right now. You may think that you worked hard in high school, or that you have decided to go to college so that you have the academic credential required to find a good job; yet other, larger social forces may also have played a role. Many students who work hard in high school cannot attend college because their parents cannot afford to send them. Others have their schooling interrupted by large-scale events, such as wars or economic depressions. The notion that we need college to find a good job is also shaped by social context. In past eras, when most people worked in agricultural or manufacturing rather than professional jobs, college attendance was rare—rather than an expected rite of passage.

Although we are all influenced by the social contexts in which we find ourselves, none of us is simply determined in his or her behavior by those contexts. We possess and create our own individuality. It is the goal of sociology to investigate the connections between what society makes of us and what we make of ourselves. Our activities structure—give shape to—the social world around us and at the same time are structured by that social world.

Social structure is an important concept in sociology. It refers to the fact that the social contexts of our lives do not just consist of random assortments of events or actions; they are structured, or patterned, in distinct ways. There are regularities in the ways we behave and in the relationships we have with one another. But social structure is not like a physical structure, such as a building, which exists independently of human actions. Human societies are always in the process of **structuration**. They are reconstructed at every moment by the very "building blocks" that compose them—human beings like you.

Developing a Global Perspective

As we just saw in our discussion of the sociological dimensions of drinking a cup of coffee, all our local actions—the ways in which we relate to one another in face-to-face contexts—form part of larger social settings that extend around the globe. These

structuration

The two-way process by which we shape our social world through our individual actions and by which we are reshaped by society.

connections between the local and the global are quite new in human history. They have accelerated over the past forty or fifty years as a result of dramatic advances in communications, information technology, and transportation. The development of jet planes; large, speedy container ships; and other means of rapid travel has meant that people and goods can be continuously transported across the world. And our worldwide system of satellite communication, established only some fifty years ago, has made it possible for people to get in touch with one another instantaneously.

U.S. society is influenced every moment of the day by **globalization**, the growth of world interdependence—a social phenomenon that will be discussed throughout this book. Globalization should not be thought of simply as the development of worldwide networks—social and economic systems that are remote from our individual concerns. It is a local phenomenon, too. For example, in the 1950s and 1960s, most Americans had few culinary choices when they dined out at restaurants. In many U.S. towns and cities today, a single street may feature Italian, Mexican, Japanese, Thai, Ethiopian, and other types of restaurants next door to one another. In turn, the dietary decisions we make can affect food producers who may live on the other side of the world.

Do college students today have a global perspective? By at least one measure, the answer is yes. According to a survey of 137,456 first-year college students in 2016, 86 percent reported that they had discussed politics "frequently" or "occasionally" in the last year. Nearly half (46 percent) of students also reported that keeping up to date with political affairs is "very important" or "essential"—the highest proportion since 1990—while nearly three in five students (59 percent) said "improving my understanding of other countries and cultures" was very important or essential. About one-third (33 percent) of students said there was a "very good chance" that they would study abroad while in college (Eagan et al. 2017). These data reflect a pervasive awareness among college students today that globalization has a direct effect on our daily, private lives.

A global perspective not only allows us to become more aware of the ways that we are connected to people in other societies but also makes us more aware of the many problems the world now faces. The global perspective opens our eyes to the fact that our interdependence with other societies means that our actions have consequences for others and that the world's problems have consequences for us.

Understanding Social Change

The changes in human ways of life in the last 200 years, such as globalization, have been far-reaching. We have become accustomed, for example, to the fact that most of the population lives in towns and cities rather than in small agricultural communities. But this was not the case until the middle of the nineteenth century. For most of human history, the vast majority of people had to produce their own food and shelter and lived in tiny groups or in small village communities. Even at the height of the most developed traditional civilizations—such as ancient Rome or pre-industrial China—less than 10 percent of the population lived in urban areas; everyone else was engaged in food production in a rural setting. Today, in most industrialized societies,

---

globalization

The economic, political, and social interconnectedness of individuals throughout the world.

Opinion of the United States

The extent to which people hold favorable attitudes toward the United States varies considerably across nations, highlighting how macrosocial factors—migration patterns, economic factors, religion, history of military conflict—can shape individual-level attitudes. Although there are strong national and regional patterns of support for the United States, we also see considerable historical variation.

Population reporting favorable views of the United States (%), 2003–2017

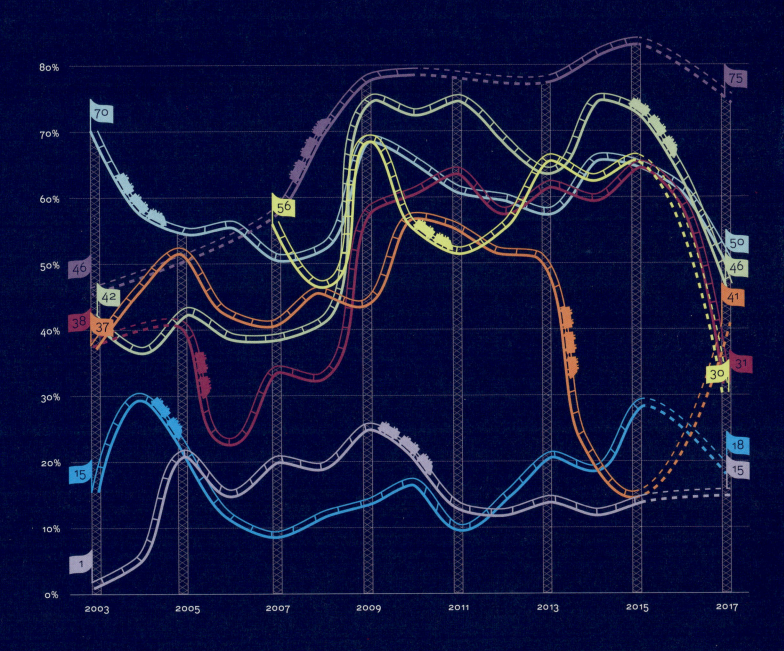

Legend:
- United Kingdom
- France
- Turkey
- Jordan
- South Korea
- Spain
- Russia
- Mexico

Source: Pew Research Center Global Indicators Database 2017.

1. How does sociology help us understand the causes of bullying?

2. Contrast public issues and personal troubles.

3. What is the sociological imagination, according to C. Wright Mills?

4. How does the concept of social structure help sociologists better understand social phenomena?

5. What is globalization? How might it affect the lives of college students today?

>

Learn about the development of sociology as a field. Be able to name some of the leading social theorists and the concepts they contributed to sociology. Learn the different theoretical approaches modern sociologists bring to the field.

these proportions have become almost completely reversed. By 2050, 68 percent of the world population is expected to live in urban areas (UN Department of Economic and Social Affairs 2018).

These sweeping social transformations have radically altered, and continue to alter, the most personal and intimate side of our daily existence. To extend a previous example, the spread of ideals of romantic love was strongly conditioned by the transition from a rural to an urban, industrialized society. As people moved into urban areas and began to work in industrial production, marriage was no longer prompted mainly by economic motives—by the need to control the inheritance of land and to work the land as a family unit. "Arranged" marriages—fixed through the negotiations of parents and relatives—became less and less common. Individuals began to initiate marriage relationships on the bases of emotional attraction and personal fulfillment. The idea of "falling in love" as a precondition for marriage was formed in this context.

Sociology was founded by thinkers who sought to understand the initial impact of transformations that accompanied industrialization in the West. Although our world today is radically different from that of former ages, the original goal of sociologists remains: to understand our world and what future it is likely to hold for us.

What Theories Do Sociologists Use?

Sociologists do more than collect facts; they also want to know why things happen. For instance, we know that industrialization has had a major influence on the emergence of modern societies. But what are the origins and preconditions of industrialization? Why is industrialization associated with changes in methods of criminal punishment or in family and marriage systems? To respond to such questions, we must construct explanatory theories.

Theories involve constructing abstract interpretations that can be used to explain a wide variety of situations. Of course, factual research and theories can never be completely separated. Sociologists aiming to document facts must begin their studies with a theory that they will evaluate. Theory helps researchers identify and frame a factual question, yet facts are needed to evaluate the strength of a theory. Conversely, once facts have been obtained, sociologists must use theory to interpret and make sense of these facts.

Theoretical thinking also must respond to general problems posed by the study of human social life, including issues that are philosophical in nature. For example, based on their theoretical and methodological orientations, sociologists hold very different beliefs about whether sociology should be modeled on the natural sciences.

Early Theorists

Humans have always been curious about why we behave as we do, but for thousands of years our attempts to understand ourselves relied on ways of thinking passed down from generation to generation, often expressed in religious rather than scientific terms. The systematic scientific study of human behavior is a relatively recent development, dating back to the late 1700s and early 1800s. The sweeping changes ushered in by the French Revolution of 1789 and the emergence of the Industrial Revolution in Europe formed the backdrop for the development of sociology. These major historical events shattered traditional ways of life and forced thinkers to develop new understandings of both the social and natural worlds.

A key development was the use of science instead of religion to understand the world. The types of questions these nineteenth-century thinkers sought to answer are the very same questions sociologists try to answer today: What is human nature? How and why do societies change?

AUGUSTE COMTE

Many scholars contributed to early sociological thinking, yet particular credit is given to the French philosopher Auguste Comte (1798–1857), if only because he invented the word *sociology*. Comte originally used the term *social physics*, but some of his intellectual rivals at the time were also making use of that term. Comte wanted to distinguish his own views from theirs, so he introduced the term *sociology* to describe the subject he wished to establish.

Comte believed that this new field could produce a knowledge of society based on scientific evidence. He regarded sociology as the last science to be developed—following physics, chemistry, and biology—but as the most significant and complex of all the sciences. Sociology, he believed, should contribute to the welfare of humanity by using science to understand, predict, and control human behavior. Late in his career, Comte drew up ambitious plans for reconstructing both French society in particular and human societies in general, based on scientific knowledge.

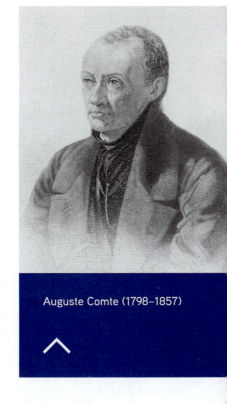

Auguste Comte (1798–1857)

HERBERT SPENCER

Herbert Spencer (1820–1903)—a British philosopher, biologist, anthropologist, and political theorist—was both highly influenced by and highly critical of Comte's writing. Spencer held that development is a natural outcome of individual achievement. In *The Study of Sociology* (1873), he argued that society can change and improve the quality of life for all people only when everyone changes his or her behavior to maximize individual potential. In other words, he believed that privileged members of society enjoyed a high quality of life because they had earned this status. He further argued that the state should not assist in improving the life chances of individuals, as doing so interferes with the natural order: The best persons succeed and the rest fall behind because of their own lack of effort or ability.

While Spencer's writings are considered an important influence on functionalist perspectives, which we will learn about later in this chapter, his ideas have fallen out of favor with many contemporary sociologists. His ideas were roundly attacked by Lester Frank Ward, the first president of the American Sociological Association (Carneiro and Perrin 2002). However, Spencer's belief in the "survival of the fittest" had a profound influence on economics and political science, especially among scholars and policymakers endorsing a laissez-faire approach.

Herbert Spencer (1820–1903)

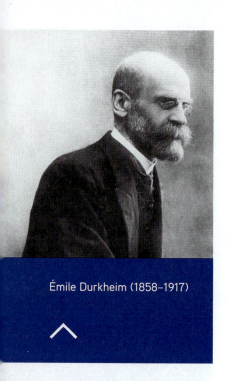

Émile Durkheim (1858–1917)

social facts

According to Émile Durkheim, the aspects of social life that shape our actions as individuals. Durkheim believed that social facts could be studied scientifically.

organic solidarity

According to Émile Durkheim, the social cohesion that results from the various parts of a society functioning as an integrated whole.

social constraint

The conditioning influence on our behavior by the groups and societies of which we are members. Social constraint was regarded by Émile Durkheim as one of the distinctive properties of social facts.

ÉMILE DURKHEIM

Another French scholar, Émile Durkheim (1858–1917), has had a much more lasting and central impact on modern sociology than either Comte or Spencer. Although he drew on aspects of Comte's work, Durkheim thought that many of his predecessors' ideas were too speculative and vague and that Comte had not successfully established a scientific basis for studying human behavior. To become a science, according to Durkheim, sociology must study **social facts**, aspects of social life that shape our actions as individuals, such as the state of the economy or the influence of religion. Durkheim believed that we must study social life with the same objectivity as scientists who study the natural world: His famous first principle of sociology was "study social facts as things!" By this he meant that social life can be analyzed as rigorously as objects or events in nature. The key task of the socio-logist, according to Durkheim, was to search for correlations among social facts to reveal laws of social structure.

Like a biologist studying the human body, Durkheim saw society as a set of indepen-dent parts, each of which could be studied separately. These ideas drew on the writings of Spencer, who also likened society to a biological organism. A body consists of specialized parts, each of which contributes to sustaining the continuing life of the organism. These parts necessarily work in harmony with one another; if they do not, the life of the organ-ism is under threat. So it is, according to Durkheim, with society. For a society to function and persist over time, its specialized institutions (such as the political system, religion, the family, and the educational system) must work in harmony with one another and function as an integrated whole. Durkheim referred to this social cohesion as **organic solidarity**. He argued that the continuation of a society thus depends on cooperation, which in turn presumes a consensus, or agreement, among its members over basic values and customs.

Another major theme Durkheim pursued, as have many others since, is that society exerts **social constraint** over the actions of its members. Durkheim argued that society is far more than the sum of individual acts; when we analyze social structures, we are studying characteristics that have "solidity" comparable to structures in the physical world. Social structure, according to Durkheim, constrains our activities in a parallel way, setting limits on what we can do as individuals. It is "external" to us, like the walls of a room.

One of Durkheim's most influential studies concerned the analysis of sui-cide (Durkheim 1897). Suicide may appear to be a purely personal act, the outcome of extreme personal unhappiness. Durkheim showed, however, that social factors exert a fundamental influence on suicidal behavior—**anomie**, a feeling of aimless-ness or despair provoked by modern social life, being one of these influences. Suicide rates show regular patterns from year to year, he argued, and these patterns must be explained sociologically. According to Durkheim, changes in the modern world are so rapid and intense that they give rise to major social difficulties, which he linked to anomie. Traditional moral controls and standards, which were supplied by religion in earlier times, are largely broken down by modern social development; this leaves indi-viduals in many societies feeling that their daily lives lack meaning. Many criticisms of Durkheim's study can be raised, but it remains a classic work that is relevant to sociology today.

KARL MARX

The ideas of the German philosopher Karl Marx (1818–1883) contrast sharply with those of Comte and Durkheim, but like these men, he sought to explain the societal changes that took place during the Industrial Revolution. When Marx was a young man, his political activities brought him into conflict with the German authorities; after a brief stay in France, he settled permanently in exile in Great Britain. Marx's viewpoint was founded on what he called the **materialist conception of history**. According to this view, it is not the ideas or values human beings hold that are the main sources of social change, as Durkheim claimed; rather, social change is prompted primarily by economic influences. Conflicts between classes—the rich versus the poor—provide the motivation for historical development. In Marx's words, "All human history thus far is the history of class struggles."

Though he wrote about many historical periods, Marx concentrated on change in modern times. For him, the most important changes were bound up with the development of **capitalism**. Capitalism is a system of production that contrasts radically with previous economic systems in history. It involves the production of goods and services sold to a wide range of consumers. Those who own capital, or factories, machines, and large sums of money, form a ruling class. The mass of the population make up the working class, or wage workers who do not own the means of their livelihood but must find employment that the owners of capital provide. Marx saw capitalism as a class system in which conflict between classes is a common occurrence because it is in the interests of the ruling class to exploit the working class and in the interests of the workers to seek to overcome that exploitation.

Marx predicted that in the future, capitalism would be supplanted by a society in which there were no classes—no divisions between rich and poor. He didn't mean that all inequalities would disappear; rather, societies would no longer be split into a small class that monopolizes economic and political power and the large mass of people who benefit little from the wealth their work creates. The economic system would come under communal ownership and a more equal society than we know at present would be established.

Marx's work had a far-reaching effect in the twentieth century. Through most of the century, until the fall of Soviet communism in the early 1990s, more than one-third of the world population lived in societies whose governments claimed to derive their inspiration from Marx's ideas. In addition, many sociologists have been influenced by Marx's ideas about class inequalities.

MAX WEBER

Like Marx, Max Weber (pronounced "VAY-ber"; 1864–1920) cannot be labeled simply a sociologist; his interests and concerns ranged across many areas. Born in Germany, where he spent most of his academic career, Weber was educated in a range of fields. Like other thinkers of his time, Weber sought to understand social change. He was influenced by Marx but was also strongly critical of some of Marx's views. He rejected the materialist conception of history and saw class conflict as less significant than Marx did. In Weber's view, economic factors are important, but ideas and values have just as much effect on social change.

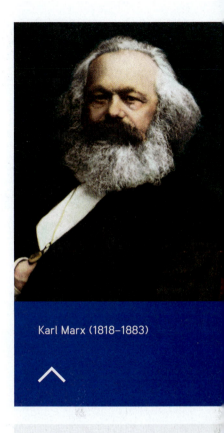

Karl Marx (1818–1883)

anomie

A concept first brought into wide usage in sociology by Émile Durkheim, referring to a situation in which social norms lose their hold over individual behavior.

materialist conception of history

The view developed by Karl Marx according to which material, or economic, factors have a prime role in determining historical change.

capitalism

An economic system based on the private ownership of wealth, which is invested and reinvested to produce profit.

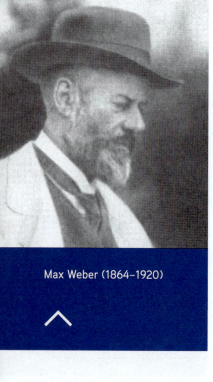

Max Weber (1864–1920)

Some of Weber's most influential writings compared the leading religious systems in China and India with those of the West. Weber concluded that certain aspects of Christian beliefs strongly influenced the rise of capitalism. He argued that the capitalist outlook of Western societies did not emerge only from economic changes, as Marx had argued. In Weber's view, cultural ideas and values help shape society and affect our individual actions.

One of the most influential aspects of Weber's work was his study of bureaucracy. A bureaucracy is a large organization that is divided into jobs based on specific functions and staffed by officials ranked according to a hierarchy. Industrial firms, government organizations, hospitals, and schools are examples of bureaucracies. Bureaucracy makes it possible for these large organizations to run efficiently, but at the same time it poses problems for effective democratic participation in modern societies. Bureaucracy involves the rule of experts, who make decisions without much consultation with those whose lives are affected by those decisions.

Weber's contributions range over many other areas, including the study of the development of cities, systems of law, types of economy, and the nature of classes. He also wrote about the overall character of sociology itself. According to Weber, humans are thinking, reasoning beings: We attach meaning and significance to most of what we do, and any discipline that deals with human behavior must acknowledge this.

Neglected Founders

Durkheim, Marx, and Weber are widely acknowledged as foundational figures in sociology, yet other important thinkers from the same period made valuable contributions to sociological thought as well. Very few women or members of racial minorities were given the opportunity to become professional sociologists during the "classical" period of the late nineteenth and early twentieth centuries. Their contributions deserve the attention of sociologists today.

HARRIET MARTINEAU

Harriet Martineau (1802–1876) was born and educated in England. She was the author of more than fifty books and numerous essays. Martineau is now credited with introducing sociology to England through her translation of Comte's founding treatise of the field, *Positive Philosophy* (Rossi 1973). She also conducted a firsthand systematic study of American society during her extensive travels throughout the United States in the 1830s, which is the subject of her book *Society in America*.

Martineau is significant to sociologists today for several reasons. First, she argued that when one studies a society, one must focus on all its aspects, including key political, religious, and social institutions. Second, she insisted that an analysis of a society must include an understanding of women's lives. Third, she was the first to turn a sociological eye on previously ignored issues, such as marriage, children, domestic and religious life, and race relations. Finally, she argued that sociologists should do more than just observe; they should also act in ways that benefit society. Martineau herself was an active proponent of women's rights and of the emancipation of slaves.

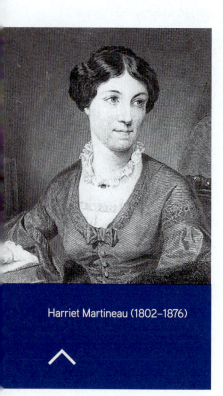

Harriet Martineau (1802–1876)

W. E. B. DU BOIS

W. E. B. Du Bois (1868–1963) was the first African American to earn a doctorate from Harvard University. Du Bois made many contributions to sociology. Perhaps most important

TABLE 1.1

Interpreting Modern Development

Durkheim	1.	The main dynamic of modern development is the **division of labor** as a basis for social cohesion and **organic solidarity**.
	2.	Durkheim believed that sociology must study **social facts** as things, just as science would analyze the natural world. His study of suicide led him to stress the important influence of social factors, qualities of a society external to the individual, on a person's actions. Durkheim argued that society exerts **social constraint** over our actions.
Marx	1.	The main dynamic of modern development is the expansion of **capitalism**. Rather than being cohesive, society is divided by class differences.
	2.	Marx believed that we must study the divisions within a society that are derived from the economic inequalities of capitalism.
Weber	1.	The main dynamic of modern development is the **rationalization** of social and economic life.
	2.	Weber focused on why Western societies developed so differently from other societies. He also emphasized the importance of cultural ideas and values on social change.

is the concept of "double consciousness," which is a way of talking about identity through the lens of the particular experiences of African Americans (Morris 2015). He argued that American society lets African Americans see themselves only through others' eyes: "It is a particular sensation, this double consciousness, this sense of always measuring one's soul by the tape of a world that looks on in amused contempt and pity. One ever feels his two-ness—an American, a Negro, two souls, two thoughts, two unreconciled strivings, two warring ideals in one dark body, whose dogged strength alone keeps it from being torn asunder" (Du Bois 1903, pp. 2–3). Du Bois made a persuasive claim that one's sense of self and one's identity are greatly influenced by historical experiences and social circumstances—in the case of African Americans, the effect of slavery and, after emancipation, segregation and prejudice.

Throughout his career, Du Bois focused on race relations in the United States. As he said in an often-repeated quote, "The problem of the twentieth century is the problem of the color line." His influence on sociology today is evidenced by continued interest in the questions that he raised, particularly his concern that sociology must explain "the contact of diverse races of men." Du Bois was also the first social researcher to trace the problems faced by African Americans to their social and economic underpinnings, a connection that most sociologists now widely accept. Finally, Du Bois became known for connecting social analysis to social reform. He was one of the founding members of the National Association for the Advancement of Colored People (NAACP) and a longtime advocate for the collective struggle of African Americans.

Later in his life, Du Bois became disenchanted by the lack of progress in American race relations. He moved to the African nation of Ghana in 1961 when he was invited by the nation's president to direct the *Encyclopedia Africana*. He died in Ghana in 1963.

W. E. B. Du Bois (1868–1963)

Although Du Bois receded from American life in his later years, his impact on American social thought and activism has been particularly profound, with many ideas of the Black Lives Matter movement being informed by his writings (Morris 2015).

Modern Theoretical Approaches

The origins of sociology were mainly European, yet the subject is now firmly established worldwide—with some of the most important developments having taken place in the United States.

SYMBOLIC INTERACTIONISM

The work of George Herbert Mead (1863–1931), a philosopher teaching at the University of Chicago, had an important influence on the development of sociological thought, in particular through a perspective called **symbolic interactionism**. Mead placed great importance on the study of language in analyzing the social world. He reasoned that language allows us to become self-conscious beings—aware of our own individuality. The key element in this process is the **symbol**, something that stands for something else. For example, the word *tree* is a symbol that represents the object tree. Once we have mastered such a concept, Mead argued, we can think of a tree even if none is visible; we have learned to think of the object symbolically. Symbolic thought frees us from being limited in our experience to what we actually see, hear, or feel.

Unlike animals, according to Mead, human beings live in a richly symbolic universe. This applies even to our very sense of self. Each of us is a self-conscious being because we learn to look at ourselves as if from the outside—we see ourselves as others see us. When a child begins to use "I" to refer to that object (herself) whom others call "you," she is exhibiting the beginnings of self-consciousness.

Virtually all interactions between individuals involve an exchange of symbols, according to symbolic interactionists. When we interact with others, we constantly look for clues to what type of behavior is appropriate in the context and how to interpret what others are doing and saying. Symbolic interactionism directs our attention to the detail of interpersonal interaction and how that detail is used to make sense of what others say and do. For instance, suppose two people are out on a date for the first time. Each is likely to spend a good part of the evening sizing the other up and assessing how the relationship is likely to develop, if at all. Both individuals are careful about their own behavior, making every effort to present themselves in a favorable light; but, knowing this, both are likely to be looking for aspects of the other's behavior that would reveal his or her true beliefs and traits. A complex and subtle process of symbolic interpretation shapes their interaction.

FUNCTIONALISM

Symbolic interactionism has been criticized for concentrating too much on things that are small in scope. Symbolic interactionists have struggled to deal with larger-scale structures and processes—the very thing that a rival tradition of thought, **functionalism**, tends to emphasize. Functionalist thinking in sociology was originally pioneered by Comte and by Spencer.

To study the function of a social activity is to analyze the contribution the activity makes to the continuation of the society as a whole. The best way to understand this idea

---

symbolic interactionism

A theoretical approach in sociology developed by George Herbert Mead that emphasizes the role of symbols and language as core elements of all human interaction.

symbol

One item used to stand for or represent another—as in the case of a flag, which symbolizes a nation.

functionalism

A theoretical perspective based on the notion that social events can best be explained in terms of the functions they perform, that is, the contributions they make to the continuity of a society.

is by analogy to the human body, a comparison Comte, Durkheim, Spencer, and other functionalist authors made. To study an organ, such as the heart, we need to show how it relates to other parts of the body. When we learn how the heart pumps blood around the body, we then understand that the heart plays a vital role in the continuation of the life of the organism. Similarly, analyzing the function of some aspect of society, such as religion, means identifying the part it plays in the continued existence and health of a society. Functionalism emphasizes the importance of moral consensus in maintaining order and stability in society. Moral consensus exists when most people in a society share the same values. Functionalists regard order and balance as the normal state of society—a social equilibrium grounded in the existence of a moral consensus among the members of society.

Functionalism became prominent in sociology in the mid-twentieth century through the writings of Talcott Parsons and Robert K. Merton, each of whom saw functionalist analysis as providing the key to the development of sociological theory and research. Merton's version of functionalism has been particularly influential. Merton distinguished between manifest and latent functions. **Manifest functions** are those known to, and intended by, the participants in a specific type of social activity. **Latent functions** are consequences of that activity of which participants are unaware. To illustrate this distinction, Merton used the example of a rain dance performed by the Hopi tribe of Arizona and New Mexico. The Hopi believe that the ceremony will bring the rain they need for their crops (manifest function). This is why they organize and participate in it. But using Durkheim's theory of religion, Merton argued that the rain dance also has the effect of promoting the cohesion of the Hopi society (latent function). A major part of sociological explanation, according to Merton, consists in uncovering the latent functions of social activities and institutions.

For much of the twentieth century, functionalist thought was considered the leading theoretical tradition in sociology, particularly in the United States. In recent years, its popularity has declined as its limitations have become apparent. Many functionalist thinkers, including Talcott Parsons, unduly stressed factors leading to social cohesion at the expense of those producing division and conflict. In addition, many critics argue that functional analysis attributes to societies qualities they do not have. Functionalists often wrote as though societies have "needs" and "purposes," even though these concepts make sense only when applied to individual human beings.

CONFLICT THEORIES

A third influential approach is conflict theory. In general, **conflict theories** underscore the role of coercion and power in producing social order. Social order is believed to be maintained by domination, with power in the hands of those with the greatest political, economic, and social resources; historically, this would include white men with ample economic and political resources. Two particular approaches typically classified under the broad heading of conflict theories are Marxism and feminist theories.

Marxism Marxists, of course, trace their views back to the writings of Karl Marx. But numerous interpretations of Marx's major ideas are possible, and today there are schools of Marxist thought that take very different theoretical positions. In all of its versions, **Marxism** differs from non-Marxist perspectives in that its adherents see it as a combination of sociological analysis and political reform. Marxism is supposed to generate a program of radical political change.

manifest functions

The functions of a particular social activity that are known to and intended by the individuals involved in the activity.

latent functions

Functional consequences that are not intended or recognized by the members of a social system in which they occur.

conflict theories

A sociological perspective that emphasizes the role of political and economic power and oppression as contributing to the existing social order.

Marxism

A body of thought deriving its main elements from Karl Marx's ideas.

power

The ability of individuals or the members of a group to achieve aims or further the interests they hold.

———

ideology

Shared ideas or beliefs that serve to justify the interests of dominant groups. Ideologies are found in all societies in which there are systematic and ingrained inequalities among groups. The concept of ideology connects closely with that of power.

———

feminism

Advocacy of the rights of women to be equal with men in all spheres of life.

———

feminist theory

A sociological perspective that emphasizes the centrality of gender in analyzing the social world and particularly the experiences of women. There are many strands of feminist theory, but they all share the intention to explain gender inequalities in society and to work to overcome them.

———

postmodernism

The belief that society is no longer governed by history or progress. Postmodern society is highly pluralistic and diverse, with no "grand narrative" guiding its development.

Moreover, Marxists place more emphasis on conflict, class divisions, power, and ideology than do many non-Marxist sociologists, especially those influenced by functionalism. The concept of power and a closely associated notion, ideology, are of great importance to Marxist sociologists and to sociology in general. **Power** refers to the ability of individuals or groups to make their own concerns or interests count, even when others resist. Power sometimes involves the direct use of force but is almost always accompanied by the development of **ideology**: ideas that are used to justify the actions of the powerful. Power, ideology, and conflict are always closely connected. Many conflicts are about power, because of the rewards it can bring. Those who hold the most power may depend mainly on the influence of ideology to retain their dominance but are usually also able to use force if necessary.

Feminism and Feminist Theory   Feminist theory is one of the most prominent areas of contemporary sociology. This is a notable development because issues of gender are nearly absent in the work of the major figures who established the discipline. The success of feminism's entry into sociology required a fundamental—and often contested—shift in the discipline's approach.

Many feminist theorists brought their experiences in the women's movement of the 1960s and 1970s to their work as sociologists. Like Marxism, **feminism** makes a link between sociological theory and political reform. Feminist sociologists often have been advocates for political and social action to remedy the inequalities between women and men in both the public and private spheres.

Feminist sociologists argue that women's lives and experiences are central to the study of society. Historically, sociology, like most academic disciplines, has presumed a male point of view. Driven by a concern with women's subordination in American society, feminist sociologists highlight gender relations and gender inequality as important determinants of social life in terms of both social interaction and social institutions, such as the family, the workplace, and the educational system. **Feminist theory** emphasizes that gender differences are not natural but socially constructed.

Today, feminist sociology often encompasses a focus on the intersection of gender, race, and class. A feminist approach to the study of inequality has influenced new academic fields, such as LGBTQ studies. Taken together, these theoretical perspectives underscore power imbalances and draw attention to the ways that social change must entail shifts in the balance of power—consistent with the overarching themes of conflict theories.

POSTMODERN THEORY

Postmodernists claim that the very foundation on which classic social thought is based has collapsed. Early thinkers were inspired by the idea that history unfolds sequentially and leads to progress. Adherents of **postmodernism** counter that there are no longer any "grand narratives" or metanarratives—overall conceptions of history or society— that make any sense (Lyotard 1985). Some go so far as to argue that there is no such thing as history.

The postmodern world is not destined, as Marx hoped, to be a socialist one. Instead, it is one dominated by the new media, which "take us out" of our past. Postmodern society is highly pluralistic and diverse. In countless films and videos, on TV shows and websites, images circulate around the world. We are exposed to many ideas and values, but these have little connection with the history of places where we live or with our own personal histories. The world is constantly in flux.

Bullying Goes Viral

Social life in the twenty-first century has gone digital—for both good and bad. For you and your college classmates, bullying often occurs online. As we saw earlier in this chapter, teens like Audrie Pott and Tyler Clementi were tormented by classmates who shared images and videos of their victims with untold numbers of people. How did this happen? How did these incidents, which happened behind closed doors, go viral for all to see? Bullying, once considered the antics of a few "bad apples," is now understood to be a more sweeping social problem—one that exemplifies the core themes of the sociological imagination.

Countless websites and apps facilitate *cyberbullying*—the use of the Internet, smartphones, or other electronic devices to embarrass or hurt another person (Sagan 2013). A 2017 survey by the Pew Research Center found that four in ten Internet users have experienced online harassment. Young Internet users are the most likely to be harassed online: Roughly two-thirds of Internet users between the ages of eighteen and twenty-nine have been the target of online harassment, with 41 percent having experienced severe harassment online, including stalking, physical threats, sexual harassment, or sustained harassment. The study also detected strong gender differences: Young women (like Audrie Pott) are much more likely than their male counterparts to experience certain forms of online harassment, including sexual harassment (21 percent vs. 9 percent) (Duggan 2017). Young adults who identify as LGBTQ are also at particularly high risk of cyberbullying: According to a survey of more than 10,500 LGBTQ students between the ages of thirteen and twenty-one, nearly half reported having been cyberbullied in the last year (Kosciw et al. 2016).

The problem is particularly widespread today because hate-spewing bullies can hide behind the anonymity of the Internet; teens who would never dream of bullying a classmate face-to-face may get lured into the cruel behavior online (Hoffman 2010). Before it was shut down in early 2017, anonymous college messaging app Yik Yak—described as a "Twitter without handles"—was criticized for enabling cyberbullying after a series of high-profile incidents at schools across the country involving racist and sexist posts. A number of universities, including the College of Idaho and Illinois College, went so far as to ban the app. More recently, a new location-based networking app, Islands, is billing itself as "Slack for college students" (McKenzie 2017). While the app does have some anonymous chat spaces, it aims to curb cyberbullying by prompting students to link to their Facebook, Snapchat, and Instagram accounts.

While the Internet and smartphones gave rise to cyberbullying, technology is also equipping today's youth with new tools for combating bullying: For example, STOPit and Stop Bullies allow users to record videos and take photos to send to campus police or school authorities, while Back Off Bully lets students book appointments with their school counselors. And the Internet can also serve as a safe space for marginalized groups: The "It Gets Better" project, created by columnist Dan Savage and his partner, has inspired more than 60,000 user-created videos that convey a message of hope to LGBTQ youth facing bullying.

Does the explosion of cyberbullying indicate that today's youth are cruel and insensitive to others' vulnerabilities? Or is there something about the current cyberculture that promotes cruelty and insensitivity? Revising Mills's notions of "personal troubles" and "public issues," how might you explain cyberbullying? Do you think anti-bullying apps can be effective, or are larger social changes needed?

Students at Lewiston Middle School attend a vigil for fellow student Anie Graham, who committed suicide in May 2017. Thirteen-year-old Graham was bullied both at school and on social media.

One of the important theorists of postmodernism is the French philosopher and sociologist Jean Baudrillard, who believes that the electronic media have destroyed our relationship to our past and created a chaotic, empty world. Baudrillard was strongly influenced by Marxism in his early years. However, he argues that the spread of electronic communication and the mass media have reversed the Marxist theorem that economic forces shape society. Rather, social life is influenced above all by signs and images.

In a media-dominated age, Baudrillard says, meaning is created by the flow of images, as in TV programs. Much of our world has become a sort of make-believe universe in which we are responding to media images rather than to real persons or places. Is "reality" television a portrayal of social "reality," or does it feature televised people who are perceived to be "real"? Do hunters in Louisiana really look and act like the Robertson family on *Duck Dynasty*, and do the tough guys in *Amish Mafia* resemble the peaceful Amish who live and work in Lancaster County, Pennsylvania? Baudrillard would say no and would describe such images as "the dissolution of life into TV."

Theoretical Thinking in Sociology

We have described four overarching theoretical approaches, which refer to broad orientations to the subject matter of sociology. Yet theoretical approaches are distinct from theories. Theories are more narrowly focused and represent attempts to explain particular social conditions or events. They are usually formed as part of the research process and in turn suggest problems to be investigated by researchers. An example would be Durkheim's theory of suicide, referred to earlier in this chapter.

Sociologists do not share a unified position on whether theories should be specific, wide ranging, or somewhere in between. Robert K. Merton (1957), for example, argues forcefully that sociologists should concentrate their attention on what he calls "middle-range theories." Middle-range theories are specific enough to be tested directly by empirical research yet are sufficiently general to cover a range of different phenomena.

Relative deprivation theory is an example of a middle-range theory. It holds that how people evaluate their circumstances depends on with whom they compare themselves. Feelings of deprivation do not necessarily correspond to the absolute level of material deprivation one experiences. A family living in a small home in a poor area where everyone is in more or less similar circumstances is likely to feel less deprived than a family living in a similar house in a neighborhood where the majority of the other homes are much larger and neighbors are wealthier.

Assessing theories, and especially theoretical approaches, in sociology is a challenging and formidable task. The fact that there is not a single theoretical approach that dominates the field of sociology might be viewed as a limitation. But this is not the case at all: The jostling of rival theoretical approaches and theories reveals the vitality of the sociological enterprise. This variety rescues us from dogma or narrow-mindedness. Human behavior is complex, and no single theoretical perspective could adequately cover all of its aspects. Diversity in theoretical thinking provides a rich source of ideas that can be drawn on in research and stimulates the imaginative capacities so essential to progress in sociological work.

microsociology

The study of human behavior in contexts of face-to-face interaction.

macrosociology

The study of large-scale groups, organizations, or social systems.

science

The disciplined marshaling of empirical data, combined with theoretical approaches and theories that illuminate or explain those data.

empirical investigation

Factual inquiry carried out in any area of sociological study.

Levels of Analysis: Microsociology and Macrosociology

One important distinction among the different theoretical perspectives we have discussed in this chapter involves the level of analysis at which each is directed. The study of everyday behavior in situations of face-to-face interaction is usually called **microsociology**. **Macrosociology**, by contrast, is the analysis of large-scale social systems, like the political system or the economy. It also includes the analysis of long-term processes of change, such as industrialization. At first glance, it may seem as though micro and macro perspectives are distinct from each other. In fact, the two are closely connected (Giddens 1984; Knorr-Cetina and Cicourel 1981).

Macro analysis is essential if we are to understand the institutional background of daily life. The ways in which people live their everyday lives are shaped by the broader institutional framework. For example, because of societal-level technological developments, we have many ways of maintaining friendships today. We may choose to call, send an email or text message, or communicate via Facebook or Skype, yet we may also choose to fly thousands of miles to spend the weekend with a friend.

Micro studies, in turn, are necessary for illuminating broad institutional patterns. Face-to-face interaction is clearly the main basis of all forms of social organization, no matter how large scale. Suppose we are interested in understanding how business corporations function. We could analyze the face-to-face interactions of directors in the boardroom, staff working in their offices, or workers on the factory floor. We would not build up a picture of the whole corporation in this way, because some of its business is transacted through printed materials, letters, the telephone, and computers; yet we would certainly gain a good understanding of how the organization works.

In later chapters, we will explore further examples of how interaction in micro contexts affects larger social processes and how macro systems in turn influence more confined settings of social life.

CONCEPT CHECKS

1. What role does theory play in sociological research?
2. According to Émile Durkheim, what makes sociology a social science? Why?
3. According to Karl Marx, what are the differences between the classes that make up a capitalist society?
4. What are the differences between symbolic interactionist and functionalist approaches to the analysis of society?
5. How are macro and micro analyses of society connected?

What Kinds of Questions Can Sociologists Answer?

Be able to describe the different types of questions sociologists address in their research.

Can we really study human social life in a scientific way? To answer this question, we must first define the word *science*.

Science is the use of systematic methods of **empirical investigation**, the analysis of data, theoretical thinking, and the logical assessment of arguments to develop a body of knowledge about a particular subject matter. Sociology is a scientific endeavor, according to this definition. It involves systematic methods of empirical investigation, analysis of data, and assessment of theories in the light of evidence and logical argument.

High-quality sociological research goes beyond surface-level descriptions of ordinary life; rather, it helps us understand our social lives in a new way. Sociologists are interested in the same questions that other people worry about and debate: Why do racism and sexism exist? How can mass starvation exist in a world that is far wealthier than it has ever been before? How does the Internet affect our lives? However, sociologists often develop answers that run counter to our commonsense beliefs—and that generate further questions. One major feature that helps distinguish science from other idea systems (such as religion) is the assumption that all scientific ideas are open to criticism and revision.

Good sociological work also tries to make the questions as precise as possible and seeks to gather factual evidence before coming to conclusions. Some of the questions that sociologists ask in their research studies are largely **factual**, or empirical, **questions**.

Factual information about one society, of course, will not always tell us whether we are dealing with an unusual case or a general set of influences. For this reason, sociologists often want to ask **comparative questions**, relating one social context within a society to another or contrasting examples drawn from different societies. A typical comparative question might be, How much do patterns of criminal behavior and law enforcement vary between the United States and Canada? Similarly, **developmental questions** ask whether patterns in a given society have shifted over time: How is the past different from the present?

Yet sociologists are interested in more than just answering factual questions, however important and interesting they may be. To obtain an understanding of human behavior, sociologists also pose broader **theoretical questions** that encompass a wide array of specific phenomena (Table 1.2). For example, a factual question may ask, To what extent do expected earnings affect one's choice of an occupation? By contrast, a theoretical question may ask, To what extent does the maximization of rewards affect human decision making?

factual questions

Questions that raise issues concerning matters of fact (rather than theoretical or moral issues).

———

comparative questions

Questions concerned with drawing comparisons among different human societies.

———

developmental questions

Questions that sociologists pose when looking at the origins and path of development of social institutions.

TABLE 1.2

A Sociologist's Line of Questioning

Factual Question	What happened?	Did the proportion of women in their forties bearing children for the first time increase, decrease, or stay the same during the 2010s?
Comparative Question	Did this happen everywhere?	Was this a global phenomenon, or did it occur just in the United States or only in a certain region of the United States?
Developmental Question	Has this happened over time?	What have been the patterns of childbearing over time?
Theoretical Question	What underlies this phenomenon?	Why are more women now waiting until their thirties and older to bear children? What factors would we look at to explain this change?

Sociologists do not strive to attain theoretical or factual knowledge simply for its own sake. Social scientists agree that personal values should not be permitted to bias conclusions but that, at the same time, research should pose questions that are relevant to real-world concerns. In this chapter, we further explore such issues by asking whether it is possible to produce objective knowledge. First, we examine the steps involved in sociological research. We then compare the most widely used research methods as we consider some actual investigations. As we shall see, there are often significant differences between the way research should ideally be carried out and real-world studies.

CONCEPT CHECKS

1. Why is sociology considered a science?

2. What are the differences between comparative and developmental questions?

What Are the Steps of the Research Process?

Learn the steps of the research process and be able to complete the process yourself.

The research process begins with the definition of a research question and ends with the dissemination of the study findings (Figure 1.1). Although researchers do not necessarily follow all seven steps in the order set forth here, these steps serve as a model for how to conduct a sociological study. Conducting research is a bit like cooking. New researchers, like novice cooks, may follow the "recipe" to a tee. Experienced cooks often don't work from recipes at all, instead relying on the skills and insights they've acquired through years of hands-on experience.

1. Define the Research Problem

All research starts from a research problem. Often, researchers strive to uncover a fact: What proportion of the U.S. population attends weekly religious services? How far does the economic position of women lag behind that of men? Do LGBTQ and straight teens differ in their levels of self-esteem?

The best sociological research begins with problems that are also puzzles. A puzzle is not just a lack of information but a gap in our understanding. The most intriguing and influential sociological research correctly identifies and solves important puzzles.

Rather than simply answering the question "What is happening?" skilled researchers contribute to our understanding by asking "Why is this phenomenon happening?" We might ask, for example, "Why are women underrepresented in science and technology jobs?" or "What are the characteristics of high schools with high levels of bullying?"

Research does not take place in a vacuum. A sociologist may discover puzzles by reading the work of other researchers in books and professional journals or by being aware of emerging trends in society.

theoretical questions

Questions posed by sociologists when seeking to explain a particular range of observed events. The asking of theoretical questions is crucial to allowing us to generalize about the nature of social life.

2. Review the Evidence

Once a research problem is identified, the next step is to review the available evidence; it's possible that other researchers have already satisfactorily clarified the problem. If not, the sociologist will need to sift through whatever related research does exist to see how

FIGURE 1.1

Steps in the Research Process

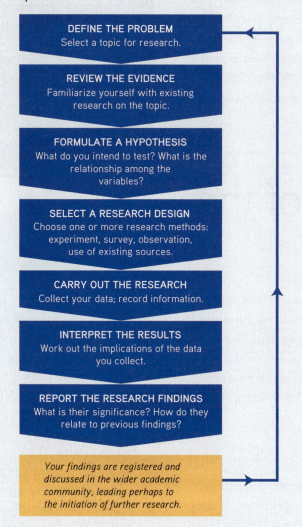

DEFINE THE PROBLEM
Select a topic for research.

REVIEW THE EVIDENCE
Familiarize yourself with existing research on the topic.

FORMULATE A HYPOTHESIS
What do you intend to test? What is the relationship among the variables?

SELECT A RESEARCH DESIGN
Choose one or more research methods: experiment, survey, observation, use of existing sources.

CARRY OUT THE RESEARCH
Collect your data; record information.

INTERPRET THE RESULTS
Work out the implications of the data you collect.

REPORT THE RESEARCH FINDINGS
What is their significance? How do they relate to previous findings?

Your findings are registered and discussed in the wider academic community, leading perhaps to the initiation of further research.

hypothesis

An idea or a guess about a given state of affairs, put forward as a basis for empirical testing.

data

Factual information used as a basis for reasoning, discussion, or calculation. Social science data often refer to individuals' responses to survey questions.

useful it is for his or her purposes. What have others found? If their findings conflict with one another, what accounts for the conflict? What aspects of the problem has their research left unanalyzed? Have they looked only at small segments of the population, such as one age group, gender, or region? Drawing on others' ideas helps the sociologist clarify the issues that may be raised and identify the appropriate research methods.

3. Make the Problem Precise

A third stage involves working out a clear formulation of the research problem. If relevant literature already exists, the researcher may have a good idea of how to approach the problem. Hunches about the nature of the problem can sometimes be turned into a definite **hypothesis**—an educated guess about what is going on—at this stage. A hypothesis must be formulated in such a way that the factual material gathered will provide evidence either supporting or disproving it.

4. Work Out a Design

The researcher must then decide how to collect the research material or **data**. Many different research methods exist, and researchers should choose the method (or methods) that are best suited to the study's overall objectives and topic. For some purposes, a survey (in which questionnaires are normally used) might be suitable. In other circumstances, interviews or an observational study may be appropriate.

5. Carry Out the Research

Researchers then proceed to carry out the plan developed in step 4. However, practical difficulties may arise, forcing the researcher to rethink his or her initial strategy. Potential subjects may not agree to answer questionnaires or participate in interviews. A business firm may not give a researcher access to its records. Yet omitting such persons or institutions from the study could bias the results, creating an inaccurate or incomplete picture of social reality. For example, it would be difficult for a researcher to answer questions about how corporations have complied with affirmative action programs if companies that have not complied do not want to be studied.

6. Interpret the Results

Once the information has been gathered, the researcher's work is not over—it is just beginning! The data must be analyzed, trends tracked, and hypotheses tested. Most important, researchers must interpret their results in such a way that the results tell a clear story and directly address the research puzzle outlined in step 1.

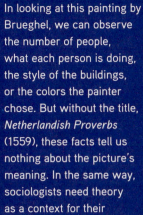

In looking at this painting by Brueghel, we can observe the number of people, what each person is doing, the style of the buildings, or the colors the painter chose. But without the title, *Netherlandish Proverbs* (1559), these facts tell us nothing about the picture's meaning. In the same way, sociologists need theory as a context for their observations.

7. Report the Findings

The research report, usually published as a book or an article in a scholarly journal, provides an account of the research question, methods, findings, and implications of the findings for social theory, public policy, or practice. This is a final stage only in terms of addressing the original research puzzle. In their written reports, most social scientists pose questions that remain unanswered and suggest new questions that might be explored in future studies. Each individual study contributes to the larger, collective process of understanding the human condition.

CONCEPT CHECKS

1. What are the seven steps of the research process?
2. What is a hypothesis?

What Research Methods Do Sociologists Use?

Sociologists have a range of methods at their disposal, both qualitative and quantitative. **Qualitative methods** can be broadly thought of as approaches that explore the deeper meaning of a particular setting. Sociologists using qualitative methods may rely on personal and collective accounts or observations of a person or situation. These observations are strictly subjective, suggesting an interpretive approach to describing actors and their social contexts. **Quantitative methods**, by contrast, use data that are objective and statistical. This type of research often focuses on documenting trends, comparing subgroups, or exploring correlations. While methods are often classified as qualitative or quantitative, scholars today are increasingly interested in mixed methods, which combine the two.

Familiarize yourself with the methods available to sociological researchers, and know the advantages and disadvantages of each. See how researchers use multiple methods in a real study.

qualitative methods

Approaches to sociological research that often rely on personal and/or collective interviews, accounts, or observations of a person or situation.

quantitative methods

Approaches to sociological research that draw on objective and statistical data and often focus on documenting trends, comparing subgroups, or exploring correlations.

ethnography

The firsthand study of people using observation, in-depth interviewing, or both. Also called "fieldwork."

participant observation

A method of research widely used in sociology and anthropology in which the researcher takes part in the activities of the group or community being studied.

survey

A method of sociological research in which questionnaires are administered to the population being studied.

Ethnography

One widely used qualitative method is **ethnography**, or firsthand studies of people using observations, interviews, or both. Here the investigator socializes, works, or lives with members of a group, organization, or community. In the case of **participant observation**, the researcher may participate directly in the activities he or she is studying. An ethnographer cannot secretly infiltrate the groups he or she studies but must explain and justify his or her presence to its members. The ethnographer must gain the cooperation of the community and sustain it over a period of time if any worthwhile results are to be achieved. Other ethnographers, by contrast, may observe at a distance and may not participate directly in the activities under observation.

For a long while, research reports based on participant observation usually omitted any account of the hazards or problems that the researcher had to overcome, but more recently, the published reminiscences and diaries of field-workers have been more honest and open. The researcher may be frustrated because the members of the group refuse to talk frankly about themselves; direct queries may be welcomed in some contexts but met with a chilly silence in others. Some types of fieldwork may be emotionally isolating or even physically dangerous, for instance, a researcher studying a street gang might be seen as a police informer or might become unwittingly embroiled in conflicts with rival gangs.

In traditional works of ethnography, accounts were presented without very much information about the observer. It was believed that ethnographers could present "objective" observations of the things they studied. More recently, ethnographers have been willing to talk and write about themselves and the nature of their connection to the people under study, even acknowledging possible sources of bias in their observations. For example, a researcher might discuss how his or her race, class, gender, or sexual orientation affected the work or how the status differences between observer and observed distorted the dialogue between them.

ADVANTAGES AND LIMITATIONS OF FIELDWORK

Where it is successful, ethnography provides rich information on the behavior of people in real-world settings. We may develop a better understanding not only of the group but of social processes that transcend the situation under study.

But fieldwork also has serious limitations. Only fairly small groups or communities can be studied, and much depends on the skill of the researcher in gaining the confidence of the individuals involved. Also, a researcher could begin to identify so closely with the group that he or she loses the perspective of an objective observer, or the researcher may reach conclusions that are more about his or her own effects on the situation than the researcher or readers ever realize. Finally, interpreting ethnographies usually involves problems of generalizability, because we cannot be sure that what we find in one context will apply in others or even that two different researchers will draw the same conclusions when studying the same group.

Surveys

Quantitative methodologists have a range of analytical tools and data resources at their disposal, but surveys are the most commonly used. When conducting a **survey**, researchers ask subjects to provide answers to structured questionnaires. The researcher may

Princeton sociologist Matthew Desmond spent more than a year doing participant observation research of tenants and their landlords in Milwaukee, Wisconsin, for his ethnography on eviction.

administer the survey in person or mail it to a study participant who will then return it by mail. Survey results—especially those based on random samples of the larger population—can often be generalized to the population at large, yet this method provides less in-depth information than the highly descriptive, nuanced slices of life obtained in fieldwork.

STANDARDIZED AND OPEN-ENDED QUESTIONS

Two types of questions are used in surveys. With "standardized," or "fixed-choice," questions, only a fixed range of responses is possible, for instance, Yes/No/Don't Know or Very Likely/Likely/Unlikely/Very Unlikely. Such questions have the advantage that responses are easy to compare and count because only a small number of categories are involved. However, the information they yield is limited because they do not allow for subtleties of opinion or verbal expression. For example, in a national survey of high school and middle school students' experiences with bullying, study participants answered Yes/No questions like "Has someone ever sent you a threatening or aggressive email, instant message, or text message?" but this question does not tell how severe the threat was or how upset a student was by this event (Lenhart 2007; Lenhart et al. 2011).

Open-ended questions, by contrast, typically provide more detailed information, because respondents may express their views in their own words. In fact, responses to open-ended survey questions are considered qualitative data, as they often convey thoughts, perceptions, and feelings. Open-ended questions allow researchers to probe more deeply into what the respondent thinks. For example, the national study of Internet bullying supplemented its survey with open-ended interviews. These data allowed researchers to understand more fully what bullying entailed. However, the lack of standardization means that answers may be difficult to compare across respondents.

In surveys, all the items must be readily understandable to interviewers and interviewees alike. Questions are usually asked in a set order. Large national surveys are conducted

"How would you like me to answer that question? As a member of my ethnic group, educational class, income group, or religious category?"

^

TABLE 1.3

Three of the Main Methods Used in Sociological Research

RESEARCH METHOD	STRENGTHS	LIMITATIONS
Ethnography	Usually generates richer and more in-depth information than other methods. Provides a broader understanding of social processes.	Can be used to study only relatively small groups or communities. Findings might apply only to groups or communities studied; they may not be easily generalizable.
Surveys	Makes possible the efficient collection of data on large numbers of individuals. Allows for precise comparisons to be made among the answers of respondents.	Material gathered may be superficial; if the questionnaire is highly standardized, important differences among respondents' viewpoints may be glossed over. Responses may be what people profess to believe rather than what they actually believe.
Experiments	Influences of specific variables can be controlled by the investigator. They are usually easier for subsequent researchers to repeat.	Many aspects of social life cannot be brought into the laboratory. Responses of those studied may be affected by the experimental situation.

pilot study

A trial run in survey research.

sampling

Studying a proportion of individuals or cases from a larger population as representative of that population as a whole.

sample

A small proportion of a larger population.

representative sample

A sample from a larger population that is statistically typical of that population.

regularly by government agencies and research organizations, with interviews carried out more or less simultaneously across the whole country. Those who conduct the interviews and those who analyze the data could not do their work effectively if they constantly had to be checking with one another about ambiguities in the questions or answers.

Survey researchers take care to ensure that respondents can easily understand both the questions and the response categories posed. For instance, a seemingly simple question like "What is your relationship status?" might baffle some people. It would be more appropriate to ask "Are you single, married, separated, divorced, or widowed?" Many survey questions are tried-and-true measures that have been used successfully in numerous prior studies. Researchers developing new survey questions often conduct a pilot study to test out new items. A **pilot study** is a trial run in which a small number of people complete a questionnaire and problematic questions are identified and revised.

Although surveys have been used primarily to obtain information on individuals, in recent years, social scientists have used surveys to learn about members of the respondent's social network and have developed techniques to link the survey reports of one individual to his or her friends, high school classmates, family members, or spouse. These complex data allow researchers to understand social networks; sociologists are increasingly interested in the ways that aspects of one's social network, such as how diverse one's friends are or how large one's networks are, shape even highly personal attributes, such as one's political attitudes or body weight (Christakis and Fowler 2009).

SAMPLING

Often sociologists are interested in the characteristics of large numbers of individuals, for example, the political attitudes of the American population as a whole. It would be impossible to study all these people directly, so researchers' solution is to use **sampling**—they concentrate on a **sample**, or small proportion, of the overall group. We can usually be confident that results from a properly chosen sample can be generalized to the total population. Studies of only 2,000–3,000 voters, for instance, can give a very accurate indication of the attitudes and voting intentions of the entire population. But to achieve such accuracy, we need a **representative sample**: The group of individuals studied must be typical, or representative, of the population as a whole.

A single best procedure for ensuring that a sample is representative is **random sampling**, in which a sample is chosen so that every member of the population has an equal probability of being included. The most sophisticated way of obtaining a random sample is to assign each member of the population a number and then use a computer to generate a random numbers list from which the sample is derived, for instance, by picking every tenth number. Random sampling is often done by researchers doing large population-based surveys aimed at capturing the behaviors or attitudes of the overall U.S. population. For qualitative researchers interested in a particular population, such as street vendors or gangsters, it simply would not make sense to try to draw a random sample.

Experiments

An **experiment** enables a researcher to test a hypothesis under highly controlled conditions established by the researcher. Experiments are often used in the natural sciences and psychology, as they are considered the best method for ascertaining "causality," or the influence of a particular factor on the study's outcome. In an experimental situation, the researcher directly controls the circumstances being studied. Because most experiments occur in laboratories, however, the scope of topics explored is quite restricted. We can bring only small groups of individuals into a laboratory setting, and in such experiments, people know they are being studied and may behave unnaturally. Experiments also neglect the macrosocial context, such as historical or political influences. Experiments are generally considered quantitative studies, because researchers often want to measure quantitatively the effect of the study's manipulation, for example, do young people commit a greater number of aggressive acts while playing a video game if they have just been exposed to a violent film clip as opposed to a peaceful film clip?

Although experiments are much more common in psychology than in sociology, several experimental studies have made important contributions to sociological knowledge. Perhaps the most infamous example is the Stanford prison experiment carried out by Philip Zimbardo (1972), who set up a make-believe prison, randomly assigning some student volunteers to the role of prison guards and others to the role of prisoners. His aim was to see how social role shaped attitudes and behavior. The results shocked the

random sampling

Sampling method in which a sample is chosen so that every member of the population has the same probability of being included.

experiment

A research method in which variables can be analyzed in a controlled and systematic way, either in an artificial situation constructed by the researcher or in naturally occurring settings.

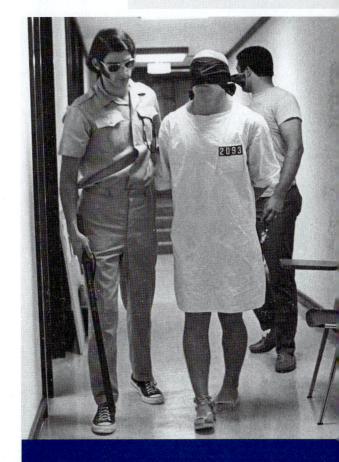

In Philip Zimbardo's make-believe prison, tension between students playing guards and students playing prisoners became dangerously real.

comparative research

Research that compares one set of findings on one society with the same types of findings on other societies.

---

oral history

Interviews with people about events they witnessed earlier in their lives.

---

triangulation

The use of multiple research methods as a way of producing more reliable empirical data than are available from any single method.

CONCEPT CHECKS

1. What are the main advantages and limitations of ethnography as a research method?

2. Contrast the two types of questions commonly used in surveys.

3. Discuss the main strengths of experiments.

4. What are the similarities and differences between comparative and historical research?

5. Why is it important to use triangulation in social research?

investigators. Students who portrayed the guards quickly assumed an authoritarian manner; they displayed genuine hostility toward the prisoners, ordering them around and verbally abusing and bullying them. The prisoners, by contrast, showed a mixture of apathy and rebelliousness—a response often noted among inmates in real prisons. These effects were so marked and the level of tension so high that the experiment had to be called off at an early stage. Zimbardo concluded that behavior in prisons is influenced more by the nature of the prison situation itself than by the individual characteristics of those involved.

Comparative and Historical Research

Comparative research is of central importance in sociology because it enables researchers to document whether social behavior varies across time and place and by one's social group memberships. Most comparative work is quantitative, in that researchers aim to document whether behaviors and attitudes change over time and place; thus, a consistent metric is required to make comparisons. For example, divorce rates rose rapidly in the United States after World War II, reaching a peak in 1979. Since then, the divorce rate has dropped by nearly one-quarter, with fewer than 17 marriages per 1,000 ending in divorce in 2016 (Hemez 2017)—a statistic that expresses profound changes taking place in the area of sexual relations and family life. Do these changes reflect specific features of American society? We can find out by comparing divorce rates in the United States with those in other countries. Although the U.S. rate is higher than the rate in most other Western societies, the overall trends are similar.

Like comparative researchers, historical analysts also care about comparing the past with the present, but they may be less concerned with documented trends and more concerned with delving deeply into particular historical periods to understand how historical context shapes individual lives. As such, historical researchers frequently focus on one narrow time period and have deep knowledge about that era; this perspective helps them make sense of the material they collect about a particular social or historical problem.

Sociologists commonly want to investigate past events by interviewing people involved in the events. Some periods of recent history can be studied in this way, such as the 1960s civil rights movement in the United States. Research in **oral history** means interviewing people about events they witnessed at some point earlier in their lives. This kind of research can stretch back in time at the most only some sixty or seventy years. To study much earlier historical periods, sociologists depend on the use of documents and written records, often held in special collections at libraries or the National Archives.

Despite the distinctive strengths of ethnography, surveys, experiments, comparative research, and historical analysis, each method has limitations. Sociologists often combine several methods in a single piece of research, using each method to supplement and check on the others. This process is known as **triangulation**. Laud Humphreys's classic *Tearoom Trade* (1970) study is an example of how researchers may use multiple methods to develop a deep understanding of social behavior. *Tearoom Trade* explored the phenomenon within the gay community involving the pursuit of impersonal sex in public restrooms. This study used surveys and observation to obtain fascinating glimpses into the secret lives of gay men. Yet, as we will see in the next section, it also revealed the important ethical challenges sociologists face.

What Ethical Dilemmas Do Sociologists Face?

In his groundbreaking *Tearoom Trade* study, Humphreys (1970) investigated "tearooms," or public restrooms where men would go to have sex with other men—often hiding their "secret" lives from their wives, children, and coworkers. Humphreys's study cast a new light on the struggles of men who were forced to keep their sexual proclivities secret. His book led to a deeper understanding of the consequences of the social stigma and legal persecution associated with gay lifestyles.

Despite the value of Humphreys's work, it is held up as a cautionary example of the ethical dilemmas that researchers face. The key ethical questions that sociologists must ask are (1) Does the research pose risks to the subjects that are greater than the risks they face in their everyday lives? (2) Do the scientific gains or "benefits" of the research balance out the "risks" to the subjects? These questions do not have easy answers, as Humphreys's work reveals.

Humphreys set out to understand what kinds of men came to the tearooms. To answer this question, he took on the role of a "lookout"—a person who loitered in the tearoom and would let the others know if an intruder, such as a police officer, was nearby. This allowed him to observe the gay men's activity. He could not easily ask questions or talk to the men in the tearoom, however, because of the norm of silence that prevailed. Humphreys also could not ask personal questions of men who wanted to remain anonymous.

Given his desire to learn more than his observations would allow, Humphreys's solution was to learn about the men in the tearooms by using survey methods. He would write down the license plate numbers of men who drove into the parking lot and who then went into the restrooms for the purpose of engaging in anonymous sex. Humphreys then gave those license plate numbers to a friend who worked at the Department of Motor Vehicles (DMV), securing the addresses of the men.

Months later, Washington University in St. Louis was conducting a door-to-door survey of sexual habits. Humphreys asked the principal investigators if he could add the names of his sample of tearoom participants. Humphreys then disguised himself as one of the investigators and went to interview these men at their homes, supposedly just to ask the survey questions but actually also to learn more about their social backgrounds and lives. He found that most of these men were married and led very conventional lives.

Humphreys later acknowledged that he was less than truthful to the men whose behavior he was studying. He didn't reveal his identity as a sociologist when observing the tearoom activities. People who came into the tearoom assumed he was there for the same reasons they were and that his presence could be accepted at face value. While he did not tell any direct lies while observing the tearoom, he also did not reveal the real reason for his presence there.

Was his behavior ethical? The study had many benefits, including moving forward scientific knowledge about gay men during a period when their behaviors were highly stigmatized. If Humphreys had been completely frank at every stage of the research, his study might not have gotten as far as it did. At the same time, the costs to the research subjects were potentially high. The observational part of his study posed only modest risk: Humphreys did not collect identifying information about the participants. What he

measures of central tendency

The ways of calculating averages.

correlation coefficient

A measure of the degree of correlation between variables.

mean

A statistical measure of central tendency, or average, based on dividing a total by the number of individual cases.

mode

The number that appears most often in a data set.

median

The number that falls halfway in a range of numbers.

standard deviation

A way of calculating the spread of a group of figures.

informed consent

The process whereby the investigator informs potential participants about the risks and benefits involved in the study.

debriefing

Following a study, informing participants about the true purpose of the study and revealing any deception that happened during the study.

knew about them was similar to what all the other people in the tearoom knew. His presence did not expose them to any more risk than they already encountered in their everyday lives.

The more problematic aspect of Humphreys's study was that he wrote down the license plate numbers of the people who came into the tearooms, obtained their home addresses from the DMV, and visited their homes under the guise of conducting a survey for Washington University. Even though Humphreys did not reveal to the men's families anything about the activities he observed in the tearooms, and even though he took great pains to keep the data confidential, the knowledge he gained could have been damaging. Because the activity he was documenting was illegal at the time, police officers might have demanded that he release information about the men's identities. A less skilled investigator might have slipped up when interviewing the subjects' families. Humphreys could have lost or misplaced his notes, which could then have ended up in the wrong hands.

Humphreys was one of the first sociologists to study the lives of gay men. His account was a humane treatment that went well beyond what little was known about gay men at that time. Although none of his subjects suffered as a result of his book, Humphreys himself later said that if he were to do the study again, he would not trace license plates or visit people's homes. Instead, after gathering his data in the public tearooms, he might try to get to know a subset of the people well enough to inform them of his goals for the study.

It is unlikely that Humphreys's tactics will be repeated in the future. Yet as research and the technologies researchers use evolve, new questions about research ethics will arise. For example, in 2013, a team of social scientists published an article exploring whether an individual's mood is affected by the content in his or her Facebook feed (Kramer, Guillory, and Hancock 2013). The researchers aimed to test a theory of "emotional contagion," or the idea that a person's own mood (as conveyed by the emotional tone of his or her Facebook posts) would be affected by the mood conveyed by the posts or news articles in the person's feed. To help the researchers test this theory, Facebook manipulated the newsfeeds of more than half a million randomly selected users, changing the number of positive and negative posts they saw in their feeds (Goel 2014). While some critics were concerned about issues of consent, others worried that the manipulated feeds could pose a risk to depressed or anxious Facebook users.

The federal government and both public and private universities maintain a number of procedures and policies to ensure that researchers conduct their research in an ethical fashion. In recent years, the federal government has become increasingly strict with universities that receive grant money for research. The National Science Foundation and the National Institutes of Health have strict requirements outlining how human subjects must be treated. In response to these requirements, American universities now have institutional review boards (IRBs) that routinely review all research involving human subjects.

The result of these review procedures has been both positive and negative. On the positive side, researchers are more aware of ethical considerations than ever before. On the negative side, many sociologists are finding it increasingly difficult to get their work done when IRBs require them to secure informed consent from research subjects before being able to establish a rapport with the subjects. **Informed consent** means that study participants are given a broad description of the study prior to agreeing to participate. After reading this summary, they are free to opt out of the research. Another safeguard used to protect subjects is **debriefing**; after the research study ends, the investigator

Research in sociology often makes use of statistical techniques in the analysis of findings. Some are highly sophisticated and complex, but those most often used are easy to understand. The most common statistics are **measures of central tendency** (ways of calculating averages) and **correlation coefficients** (measures of the degree to which one variable relates consistently to another).

There are three methods of calculating averages, each of which has certain advantages and shortcomings. Take as an example the amount of personal wealth (including all assets, such as houses, cars, bank accounts, and investments) owned by thirteen individuals. Suppose the thirteen own the following amounts:

1.	$0	8.	$80,000
2.	$5,000	9.	$100,000
3.	$10,000	10.	$150,000
4.	$20,000	11.	$200,000
5.	$40,000	12.	$400,000
6.	$40,000	13.	$10,000,000
7.	$40,000		

The **mean** corresponds to the average, arrived at by adding together the personal wealth of all thirteen people and dividing the result by thirteen. The total is $11,085,000; dividing this by thirteen, we reach a mean of $852,692.31. The mean is often a useful calculation because it is based on the whole range of data provided. However, it can be misleading where one or a small number of cases are very different from the majority. In our example, the mean is not in fact an appropriate measure of central tendency because the presence of one very large figure, $10,000,000, skews the picture. One might get the impression when using the mean to summarize these data that most of the people own far more than they actually do. In such instances, one of two other measures may be used.

The **mode** is the figure that occurs most frequently in a given set of data. In our example, it is $40,000. The problem with the mode is that it doesn't take into account the overall distribution of the data, that is, the range of figures covered. The most frequently occurring case in a set of figures is not necessarily representative of their distribution as a whole and thus may not be a useful average. In this case, $40,000 is too close to the lower end of the figures.

The third measure is the **median**, which is the middle of any set of figures; here this would be the seventh figure, again $40,000. Our example gives an odd number of figures: thirteen. If there had been an even number—for instance, twelve—the median would be calculated by taking the mean of the two middle cases, items 6 and 7. Like the mode, the median gives no idea of the actual range of the data measured.

Sometimes a researcher will use more than one measure of central tendency to avoid giving a deceptive picture of the average. More often, he or she will calculate the **standard deviation** for the data in question. This is a way of calculating the **degree of dispersal**, or the range, of a set of figures—which in this case goes from $0 to $10,000,000.

Correlation coefficients offer a useful way of expressing how closely connected two (or more) variables are. Where two variables correlate completely, we can speak of a perfect positive correlation, expressed as 1.0. Where no relation is found between two variables—they have no consistent connection at all—the coefficient is 0. A perfect negative correlation, expressed as –1.0, exists when two variables are in a completely inverse relation to each other. Perfect correlations are never found in the social sciences. Correlations of the order of 0.6 or more, whether positive or negative, are usually regarded as indicating a strong degree of connection between whatever variables are being analyzed. Positive correlations on this level might be found between, say, social class background and voting behavior.

degree of dispersal

The range or distribution of a set of figures.

You will often come across tables when reading sociological literature. They sometimes look complex, but they are easy to decipher if you follow a few basic steps, listed here; with practice, these will become automatic. (See Table 1.4 as an example.) Do not succumb to the temptation to skip over tables; they contain information in concentrated form, which can be read more quickly than would be possible if the same material were expressed in words. By becoming skilled in the interpretation of tables, you will also be able to check how justified the conclusions a writer draws actually seem.

1. **Read the title in full.** Tables frequently have long titles that represent an attempt by the researcher to state accurately the nature of the information conveyed. The title of Table 1.4 first reveals the subject of the data. Next, it indicates that the table provides material for comparison. And finally, it specifies that the data are given only for a limited number of countries.

2. **Look for explanatory comments, or notes, about the data.** A source note at the foot of Table 1.4 indicates that the data were obtained from the Pew Research Center, a large international survey organization. It also notes that data were not available for all nations for all years. Notes may say how the material was collected or why it is displayed in a particular way. If the data have not been gathered by the researcher but are based on findings originally reported elsewhere, a source will be included. The source sometimes gives you some insight into how reliable the information is likely to be, while also indicating where to find the original data. In our table, the source note makes clear that the data have been taken from one organization.

3. **Read the headings along the top and left-hand side of the table.** (Sometimes tables are arranged with "headings" at the foot rather than the top.) These tell you what type of information is contained in each row and column. In reading the table, keep each set of headings in mind as you scan the figures. In our example, the headings on the left give the countries involved, whereas those at the top refer to the proportion who hold a "favorable" opinion of the United States and the years for which data are available.

4. **Identify the units used; the figures in the body of the table may represent cases, percentages, averages, or other measures.** Sometimes it may be helpful to convert the figures to a form more useful to you: If percentages are not provided, for example, it may be worth calculating them.

5. **Consider the conclusions that might be reached from the information in the table.** Most tables are discussed by the author, and what he or she has to say should, of course, be borne in mind. But you should also ask what further issues or questions could be suggested by the data: How might you explain some of these declines? Or the sudden and precipitous drops?

CONCEPT CHECKS

1. What ethical dilemmas did Humphreys's study pose?

2. Contrast informed consent and debriefing.

discusses any concerns the subjects may have and acknowledges whether strategies such as deception were used. Despite these safeguards, there will likely never be easy solutions to vexing problems posed by research ethics, especially in an era when the Internet and high-tech firms are offering up new and innovative ways to collect and examine data.

TABLE 1.4

Opinion of the United States: Comparison of Selected Nations

Several interesting trends can be seen from the data in this table. First, the extent to which people hold favorable attitudes toward the United States varies considerably across nations. Second, although there are strong national and regional patterns of support for the United States, we also see considerable historical variation within nations. For example, in China, about one-third held favorable views in 2007, yet this proportion has steadily gone upward, reaching 50 percent in 2016. Other nations show a steep and sudden drop rather than a steady decline. While nearly two-thirds of Mexicans held a favorable review as recently as 2015, this share had plummeted to 30 percent by 2017. The opposite can be seen in Russia.

PERCENTAGE OF PERSONS WHO HOLD A "FAVORABLE" (VS. "UNFAVORABLE") OPINION OF THE UNITED STATES											
COUNTRY	2007	2008	2009	2010	2011	2012	2013	2014	2015	2016	2017
China	34	41	47	58	44	43	40	50	44	50	–
France	39	42	75	73	75	69	64	75	73	63	46
Germany	30	31	64	63	62	52	53	51	50	57	35
Indonesia	29	37	63	59	54	–	61	59	62	–	48
Japan	61	50	59	66	85	72	69	66	68	72	57
Jordan	20	19	25	21	13	12	14	12	14	–	15
Kenya	87	–	90	94	83	–	81	80	84	63	54
Mexico	56	47	69	56	52	56	66	63	66	–	30
Pakistan	15	19	16	17	12	12	11	14	22	–	–
Poland	61	68	67	74	70	69	67	73	74	74	73
Russia	41	46	44	57	56	52	51	23	15	–	41
South Korea	58	70	78	79	–	–	78	82	84	–	75
Turkey	9	12	14	17	10	15	21	19	29	–	18
United Kingdom	51	53	69	65	61	60	58	66	65	61	50
United States	80	84	88	85	79	80	81	82	83	83	85

Note: Data not available for all nations for all years.

Source: Poushter and Bialik, Pew Research Center, 2017.

How Does the Sociological Imagination Affect Your Life?

Understand how adopting a sociological perspective allows us to develop a richer understanding of ourselves, our significant others, and the world.

When we observe the world through the prism of the sociological imagination, we are affected in several important ways. First, we develop a greater awareness and understanding of cultural differences. For example, a high school guidance counselor won't easily gain the confidence of his or her students who are the targets of bullying without being sensitive to the ways that gender and sexual orientation shape students' experiences at school.

EMPLOYING YOUR SOCIOLOGICAL IMAGINATION

>

What Can You Do with a Sociology Major?

What do former First Lady Michelle Obama, the late president Ronald Reagan, sex expert Dr. Ruth Westheimer, and the late civil rights leader Martin Luther King Jr. have in common? If you guessed that they were all sociology majors, you're correct. At first blush, their careers couldn't be more different. Dr. Ruth raised eyebrows in the 1980s and 1990s, when she discussed in (often frank) detail the sex lives of Americans. Ronald Reagan was a conservative Republican president who had the backing of the religious right. Martin Luther King Jr. was a Baptist minister and leading figure in the civil rights movement. Before she became our first Black First Lady, Michelle Obama had successful careers as a lawyer, college dean, and hospital administrator.

Yet, on closer inspection, each of their careers shows evidence of essential skills obtained from an education in sociology, including critical thinking skills, a well-developed sociological imagination, and an understanding of research methods. Throughout this book, the "Employing Your Sociological Imagination" feature shows how the skills and knowledge acquired in sociology classes can be applied to far-ranging careers. According to the American Sociological Association (2015), the most common field employing sociology majors is social services and counseling, which accounts for roughly one in five recent graduates. Yet that also means that 80 percent work in other fields, revealing the remarkable breadth of opportunities that a sociology degree provides.

Consider the career of Dr. Ruth Westheimer. After earning a degree in sociology, she obtained a doctorate in education and worked at family-planning organizations like Planned Parenthood before making the leap to television and radio in the 1980s. At that time, people looking for information about their sex lives couldn't simply turn to Google.

Second, we are better able to assess the results of public policy initiatives. For instance, while all fifty states have anti-bullying laws, we will not be able to understand how effective they are unless we systematically obtain data on the levels of bullying before and after these policies were implemented.

Third, we may become more self-enlightened and may develop wise insights into our own behaviors. Have you ever bullied or picked on a classmate at school? Have you ever harassed a classmate online? If yes, why did you pick on that particular student? Were you encouraged to do so by your classmates? Did the anonymous nature of the Internet make you feel like you wouldn't get "caught"? Sociology helps us understand why we act as we do and helps us recognize that social context—such as our peers or school dynamics—also may shape our behaviors.

Fourth, developing a sociological eye toward social problems and developing rigorous research skills opens many career doors—as urban planners, social workers, and personnel

On her radio program *Sexually Speaking* and several TV shows, and in the more than forty books she wrote, Dr. Ruth offered people thoughtful, straightforward, and even humorous advice based in compelling social science and medical research. She also recognized that public policies regarding health must be evidence based. The four-foot-seven doctor and dynamo famously held her own in a tense debate on CNN's *Crossfire* in 2002, when she stood up to a passionate advocate for abstinence-only education, informing him that no data supported the efficacy of such programs (CNN 2002).

Michelle Obama has had several careers working in nonprofits dedicated to children and health care. It's not surprising, then, that one of her first missions as First Lady was to focus on the high and rising obesity rates among children in the United States. Her Let's Move (2017) program aimed to help children and their families maintain a healthy weight. The design of this program showed the clear imprint of a sociological imagination. Rather than viewing high body weight as a product of individual-level factors like lack of willpower or poor food choices, Obama recognized that child obesity is a public issue rather than a personal trouble, noting that obesity rates are especially high among Black and Latino children living in low-income neighborhoods with limited access to full-service grocery stores, affordable healthy food, and safe places to exercise. Key components of the program included increasing access to healthy foods in public schools and creating public vegetable gardens and safe playgrounds for children in urban neighborhoods. These public approaches to the obesity problem stand in stark contrast to solutions telling individual children and parents simply to "watch what you eat."

These are just two of myriad examples of how sociology can shape how we think, reason, and devise and execute solutions to social problems. In the chapters that follow, we will show even more ways that sociology can prepare you for a diverse range of challenging and satisfying careers.

As First Lady, Michelle Obama, a sociology major, focused on battling the public issue of childhood obesity.

managers, among other jobs. An understanding of society also serves those working in law, journalism, business, and medicine.

In sum, sociology is a discipline in which we often set aside our personal views and biases to explore the influences that shape our lives and those of others. Sociology emerged as an intellectual endeavor along with the development of modern societies, and the study of such societies remains its principal concern. Sociology has major practical implications for people's lives. Learning to become a sociologist is an exciting academic pursuit! The best way to make sure it is exciting is to approach the subject in an imaginative way and to relate sociological ideas and findings to your own life.

CONCEPT CHECKS

1. Describe four ways that sociology can help us in our lives.

2. What skills and perspectives do sociologists bring to their work?

CHAPTER 1

The Big Picture

Sociology:
Theory and Method

Thinking Sociologically

1. Healthy older Americans often encounter discriminatory treatment when younger people assume they are slow and thus overlook them for jobs they are fully capable of doing. How would each of the theoretical perspectives—functionalism, class conflict theory, and symbolic interactionism—explain the dynamics of prejudice against older adults?

2. Explain in some detail the advantages and disadvantages of doing comparative or historical research. What will it yield that will be better than experimentation, surveys, and ethnographic fieldwork? What are its limitations?

3. Let's suppose the dropout rate in your high school increased dramatically. The school board offers you a $500,000 grant to do a study to explain the sudden increase. Following the study procedures outlined in your text, explain how you would go about doing your research. What might be some of the hypotheses to test in your study? How would you prove or disprove them?

Learning Objectives

What Is the "Sociological Imagination"?

p. 5

Learn what sociology covers as a field and how everyday topics like love and romance are shaped by social and historical forces. Recognize that sociology involves developing a sociological imagination and a global perspective, and understanding social change.

What Theories Do Sociologists Use?

p. 10

Learn about the development of sociology as a field. Be able to name some of the leading social theorists and the concepts they contributed to sociology. Learn the different theoretical approaches modern sociologists bring to the field.

What Kinds of Questions Can Sociologists Answer?

p. 21

Be able to describe the different types of questions sociologists address in their research.

What Are the Steps of the Research Process?

p. 23

Learn the steps of the research process and be able to complete the process yourself.

What Research Methods Do Sociologists Use?

p. 25

Familiarize yourself with the methods available to sociological researchers, and know the advantages and disadvantages of each. See how researchers use multiple methods in a real study.

What Ethical Dilemmas Do Sociologists Face?

p. 31

Recognize the ethical problems researchers may face, and identify possible solutions to these dilemmas.

How Does the Sociological Imagination Affect Your Life?

p. 35

Understand how adopting a sociological perspective allows us to develop a richer understanding of ourselves, our significant others, and the world.

Terms to Know

sociology • personal troubles • public issues

sociological imagination • structuration • globalization

social facts • organic solidarity • social constraint • anomie • materialist conception of history • capitalism • symbolic interactionism • symbol • functionalism • manifest functions • latent functions • conflict theories • Marxism • power • ideology • feminism • feminist theory • postmodernism • microsociology • macrosociology

science • empirical investigation • factual questions • comparative questions • developmental questions • theoretical questions

hypothesis • data

qualitative methods • quantitative methods • ethnography • participant observation • survey • pilot study • sampling • sample • representative sample • random sampling • experiment • comparative research • oral history • triangulation

measures of central tendency • correlation coefficient • mean • mode • median • standard deviation • degree of dispersal • informed consent • debriefing

Concept Checks

1. How does sociology help us understand the causes of bullying?
2. Contrast public issues and personal troubles.
3. What is the sociological imagination, according to C. Wright Mills?
4. How does the concept of social structure help sociologists better understand social phenomena?
5. What is globalization? How might it affect the lives of college students today?

1. What role does theory play in sociological research?
2. According to Émile Durkheim, what makes sociology a social science? Why?
3. According to Karl Marx, what are the differences between the classes that make up a capitalist society?
4. What are the differences between symbolic interactionist and functionalist approaches to the analysis of society?
5. How are macro and micro analyses of society connected?

1. Why is sociology considered a science?
2. What are the differences between comparative and developmental questions?

1. What are the seven steps of the research process?
2. What is a hypothesis?

1. What are the main advantages and limitations of ethnography as a research method?
2. Contrast the two types of questions commonly used in surveys.
3. Discuss the main strengths of experiments.
4. What are the similarities and differences between comparative and historical research?
5. Why is it important to use triangulation in social research?

1. What ethical dilemmas did Humphreys's study pose?
2. Contrast informed consent and debriefing.

1. Describe four ways that sociology can help us in our lives.
2. What skills and perspectives do sociologists bring to their work?

2

Culture and Society

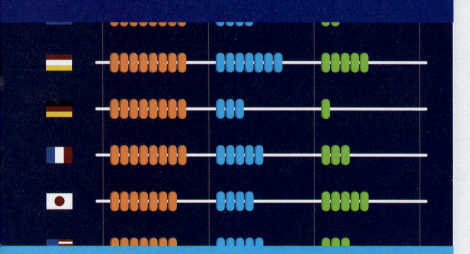

National Identity

p. 65

During fall 2015, students across the United States organized demonstrations against racism on college campuses. Controversy abounded at Yale University when an instructor publicly questioned the validity of an email that had gone out from administrators with proposed guidelines for Halloween costumes.

THE BIG QUESTIONS

What is culture?

Know what culture consists of, and recognize how it differs from society.

How does human culture develop?

Begin to understand how both biological and cultural factors influence our behavior. Learn the ideas of sociobiology and how others have tried to refute these ideas by emphasizing cultural differences.

What happened to premodern societies?

Learn how societies have changed over time.

How has industrialization shaped modern society?

Recognize the factors that transformed premodern societies, particularly how industrialization and colonialism influenced global development. Know the differences among industrialized societies, emerging economies, and developing societies and how these differences developed.

How does globalization affect contemporary culture?

Recognize the effect of globalization on your life and the lives of people around the world. Think about the effect of a growing global culture.

In October 2015, the campus of Yale University broke out in controversy over a series of emails written by administrators about Halloween. The uproar began when an initial email went out from an Intercultural Affairs Committee representing Native American, Black, Jewish, Latino, Asian American, and international students: "The end of October is quickly approaching, and along with the falling leaves and cooler nights come the Halloween celebrations on our campus and our community," the memo began. "These celebrations provide opportunities for students to socialize as well as to make positive contributions to our community. . . . However, Halloween is also unfortunately a time when the normal thoughtfulness and sensitivity of most Yale students can sometimes be forgotten and some poor decisions can be made, including wearing feathered headdresses, turbans, wearing 'war paint' or modifying skin tone or wearing blackface or redface."

While acknowledging students' right to free expression, the administrators asked students to consider how "culturally unaware or insensitive choices" might affect other groups. The memo proposed that students ask themselves a series of questions before deciding on their costumes:

1. For a funny costume, is the humor based on "making fun" of real people, human traits, or cultures?

2. For a historical costume, does it further historical inaccuracies or misinformation?

3. For a cultural costume, does it reduce cultural differences to stereotypes or jokes?

4. For a religious costume, does it mock or belittle someone's deeply held faith tradition?

A few days after the Intercultural Affairs Committee sent out this advice to students, the deputy director of one of the Yale dormitories wrote a pointed response that questioned whether it was appropriate for college administrators to police the costumes of young adults. In an email to the dorm residents, she asked, "Is there no room anymore for a child or young person to be a little bit obnoxious, a little bit inappropriate or provocative, or yes, offensive? American universities were once a safe space not only for maturation but also for a certain regressive, or even transgressive, experience; increasingly, it seems, they have become places of censure and prohibition."

This email set in motion a series of protests, with many students calling for her resignation (and that of her husband, who defended her email in his capacity as the dorm's director). Many felt that she was dismissing the power of harmful stereotypes to further degrade marginalized groups by encouraging those who take offense to a person's costume to "look away." Although the president of Yale and the Yale College dean came out in support of the dorm director and his wife keeping their jobs, she ultimately decided to resign from teaching at the college.

So why were the Yale students so upset? At the heart of this controversy over Halloween costumes is a concept that sociologists refer to as **cultural appropriation**, which occurs when members of one cultural group borrow elements of another's culture, such as when a non-Indian person dons a sari or a non-Japanese person wears a kimono. Is it always offensive to take on elements of a culture to which you don't belong? Even the most well-intentioned and seemingly benign decisions to borrow the cultural style of another group can be understood quite differently by those who come from that culture. There are no hard-and-fast rules that can resolve such conflicts. One thing we can do is be aware of what is at stake here. Sometimes cultural appropriation can reduce an entire way of life to a demeaning stereotype that exacerbates historically unequal power relations. For this reason, many schools have banned the use of Native American mascots.

Similarly, it was this sociological insight that led the Intercultural Affairs Committee to urge students to be particularly thoughtful and sensitive to others' feelings on Halloween.

As the protests at Yale demonstrate, issues related to cultural appropriation often come to a head at Halloween and at other campus parties. But culture is more than just how we dress. In this chapter, we will look at what culture is and its role in encouraging conformity to shared ways of thinking and acting. We then consider the early development of human culture, emphasizing features that distinguish human behavior from that of other species. After assessing the role of biology in shaping human behavior, we examine the aspects of culture that are essential for human society. This

cultural appropriation

When members of one cultural group borrow elements of another group's culture.

"We're a culture, not a costume" was a poster campaign launched by Ohio University that sparked a national conversation about racially insensitive Halloween costumes.

WE'RE A CULTURE, NOT A COSTUME

YOU THINK IT'S HARMLESS, BUT YOU'RE NOT THE TARGET.

leads to a discussion of cultural diversity, examining the cultural variations not only across different societies but also within a single society, such as the United States.

Cultural variations among human beings are linked to different types of society, and we will compare and contrast the main forms of society found in history. The point of doing this is to tie together closely the two aspects of human social existence: the different cultural values and products that human beings have developed and the contrasting types of society in which such cultural development has occurred. Too often, culture is discussed separately from society, as though the two were disconnected, whereas in fact, they are closely intertwined. Throughout the chapter, we concentrate on how social change has affected cultural development. One instance of this is the effect of technology and globalization on the many cultures of the world.

What Is Culture?

Know what culture consists of, and recognize how it differs from society.

The sociological study of culture began with Émile Durkheim in the nineteenth century and soon became the basis of anthropology, a social science specifically focused on the study of cultural differences and similarities among the world's peoples. Early social scientists assumed that "primitive" cultures were inferior, lagging far behind modern European "civilization." Sociologists and anthropologists now recognize that different cultures have their own distinctive characteristics. The task of social science is to understand this cultural diversity, which is best done by avoiding value judgments.

When we use the term *culture* in daily conversation, we often think of "high culture," such as fine art, literature, classical music, or ballet. From a sociological perspective, the concept includes these activities but also many more. **Culture** consists of the values held by members of a particular group, the languages they speak, the symbols they revere, the norms they follow, and the material goods they create and that become meaningful for them—bows and arrows, plows, factories and machines, computers, books, dwellings. It refers to the ways of life of the individual members or groups within a society: their apparel, marriage customs and family life, patterns of work, religious ceremonies, and leisure pursuits. We should think of culture as a "design for living" or "tool kit" of practices, knowledge, and symbols acquired—as we shall see later—through learning rather than by instinct (Kluckhohn 1949; Swidler 1986). Some elements of culture, especially the beliefs and expectations people have about one another and the world they inhabit, are a component of all social relations.

Sociologists and anthropologists distinguish between two forms of culture: "nonmaterial culture," the cultural ideas that are not themselves physical objects, and "material culture," the physical objects that a society creates.

Defining "Culture": Nonmaterial Culture

Nonmaterial culture comprises the nonphysical components of culture, including values and norms, symbols, language, and speech and writing.

VALUES AND NORMS

Values are abstract ideals. For example, monogamy—being faithful to one's sole romantic partner—is a prominent value in most Western societies. In other cultures, alternatively, a

culture

The values, norms, and material goods characteristic of a given group. The notion of culture is widely used in sociology and the other social sciences (particularly anthropology). Culture is one of the most distinctive properties of human social association.

values

Ideas held by individuals or groups about what is desirable, proper, good, and bad. What individuals value is strongly influenced by the specific culture in which they happen to live.

person may be permitted to have several wives or husbands simultaneously. Similarly, some cultures value individualism highly, whereas others place great emphasis on collectivism. A simple example makes this clear. Most pupils in the United States would be outraged to find another student cheating on an examination. In the United States, copying from someone else's paper goes against core values of individual achievement, equality of opportunity, hard work, and respect for the rules. Russian students, however, might be puzzled by this sense of outrage among their American peers. Helping one another pass an examination reflects the value Russians place on equality and on collective problem solving in the face of authority.

Within a single society or community, values may also conflict: Some groups or individuals may value traditional religious beliefs, whereas others may favor freedom of expression, individual rights, and gender-based equality. Some people may prefer material comfort and success, whereas others may favor simplicity and a quiet life. The Yale Halloween incident vividly highlights an instance when the competing values of freedom of expression and sensitivity to the feelings of minority group members came into conflict. In our changing age—filled with the global movement of people, ideas, goods, and information—it is not surprising that we encounter instances of cultural values in conflict.

Norms are widely agreed-upon principles or rules people are expected to observe; they represent the dos and don'ts of social life. Norms of behavior in marriage include, for example, how husbands and wives are supposed to behave toward their in-laws. In some societies, they are expected to develop a close relationship; in others, they keep a clear distance. Many of our everyday behaviors and habits are grounded in cultural norms. Movements, gestures, and expressions are strongly influenced by cultural factors. A clear example can be seen in the way people smile—particularly in public contexts—across different cultures. Among the Inuit (Eskimos) of Greenland, for example, one does not find the strong tradition of public smiling that exists in many areas of western Europe and North America. This does not mean that the Inuit are cold or unfriendly; it is simply not their common practice to smile at or exchange pleasantries with strangers.

Norms, like the values they reflect, vary widely both across and within cultures. Among most Americans, for example, one norm calls for direct eye contact between persons engaged in conversation; completely averting one's eyes is usually interpreted as a sign of weakness or rudeness. Yet, among the Navajo, a cultural norm calls for averting one's eyes as a sign of respect. Direct eye contact, particularly between strangers, is considered rude because it violates a norm of politeness.

Norms also change over time. For example, beginning in 1964, with the U.S. surgeon general's report "Smoking and Health," which presented definitive medical evidence linking smoking with a large number of serious health problems, the U.S. government waged a highly effective campaign to discourage people from smoking. A social norm favoring smoking—once associated with independence, sex appeal, and glamour—has given way to an equally strong antismoking social norm that depicts smoking as unhealthful, unattractive, and selfish. In 2016, the proportion of American adults who smoked was less than 16 percent, compared to 42 percent in 1964, when the surgeon general's report was issued (Centers for Disease Control and Prevention 2018).

Values and norms work together to shape how members of a culture behave within their surroundings. Even within a single culture, the norms of conduct differ by age, gender, and other important social subgroups. Gender norms are particularly powerful; women are expected to be more docile, more caring, and even more moral than men.

norms

Rules of conduct that specify appropriate behavior in a given range of social situations. A norm either prescribes a given type of behavior or forbids it. All human groups follow definite norms, which are always backed by sanctions of one kind or another—varying from informal disapproval to physical punishment.

Smoking was portrayed as sophisticated and elegant in the 1950s. Today, smokers are vilified for harming themselves and their children. What has contributed to these drastic changes in our perceptions of smokers?

LANGUAGE

Language demonstrates both the unity and the diversity of human culture because there are no cultures without language, yet there are thousands of different languages spoken in the world. Anyone who has visited a foreign country armed only with a dictionary knows how difficult it is to understand anything or to be understood. Although languages that have similar origins have words in common with one another—as do, for example, German and English—most of the world's major language groups have no words in common at all.

Language is involved in virtually all our activities. In the form of ordinary talk or speech, it is the means by which we organize most of what we do. However, language is involved not just in mundane, everyday activities but also in ceremony, religion, poetry, and many other spheres. One of the most distinctive features of human language is that it allows us to vastly extend the scope of our thought and experience. Using language, we can convey information about events remote in time or space and can discuss things we have never seen. We can develop abstract concepts, tell stories, make jokes, and express sarcasm.

Languages—indeed, all symbols—are representations of reality. The symbols we use may signify things we imagine, such as mathematical formulas or fictitious creatures, or they may represent (i.e., "re-present," or make present again in our minds) things initially experienced through our senses. Symbols even represent emotions, as the common emoticons of :) (happy) and ;) (good-natured winking) reveal. Human behavior is oriented toward the symbols we use to represent reality rather than toward the reality itself—and these symbols are determined within a particular culture. Because symbols are representations, their cultural meanings must be interpreted when they are used. When you see a four-footed furry animal, for example, you must determine which cultural symbol

language

A system of symbols that represent objects and abstract thoughts; the primary vehicle of meaning and communication in a society.

to attach to it. Do you decide to call it a dog, a wolf, or something else? If you determine it is a dog, what cultural meaning does that convey? In American culture, dogs are typically regarded as household pets and lavished with affection. Among the Akha of northern Thailand, dogs are seen as food and treated accordingly. The diversity of cultural meanings attached to the word *dog* thus requires an act of interpretation. In this way, we are freed, in a sense, from being directly tied to the physical world around us.

In the 1930s, the anthropological linguist Edward Sapir and his student Benjamin Lee Whorf advanced the **linguistic relativity hypothesis**, or the Sapir–Whorf hypothesis, which argues that the language we use influences our perceptions of the world. That is because we are much more likely to be aware of things in the world if we have words for them (Haugen 1977; Malotki 1983; Witkowski and Brown 1982). Expert skiers or snowboarders, for example, use terms such as *black ice, corn, powder,* and *packed powder* to describe different snow and ice conditions. Such terms enable them to more readily perceive potentially life-threatening situations that would escape the notice of a novice. In a sense, then, experienced winter athletes have a different perception of the world—or at least, a different perception of the alpine slopes—than novices do.

Language also helps give permanence to a culture and an identity to a people. Language outlives any particular speaker or writer, affording a sense of history and cultural continuity, a feeling of "who we are." One of the central paradoxes of our time is that despite the globalization of the English language through the Internet and other forms of global media, local attachments to language persist, often out of cultural pride. For example, the French-speaking residents of the Canadian province of Quebec are so passionate about their linguistic heritage that they often refuse to speak English, the dominant language of Canada, and periodically seek political independence from the rest of Canada.

SPEECH AND WRITING

All societies use speech as a vehicle of language. However, there are other ways of "carrying," or expressing, language—most notably, writing. The invention of writing marked a major transition in human history. Writing first began as the drawing up of lists. Marks would be made on wood, clay, or stone to keep records about significant events, objects, or people. For example, a mark, or sometimes a picture, might be drawn to represent each tract of land possessed by a particular family or set of families (Gelb 1952). Writing began as a means of storing information and as such was closely linked to the administrative needs of early civilizations. A society that possesses writing can locate itself in time and space. Documents can be accumulated that record the past, and information can be gathered about present-day events and activities.

Writing is not just the transfer of speech to paper or some other durable material. It is a phenomenon of interest in its own right. Written documents or texts have qualities in some ways quite distinct from the spoken word. The impact of speech is always by definition limited to the particular contexts in which words are uttered. Ideas and experiences can be passed down through generations in cultures without writing, but only if they are regularly repeated and passed on by word of mouth. Written texts, conversely, can endure for thousands of years, and through them, people from past ages can in a certain sense address us directly. This is, of course, why documentary research is so important to historians. By interpreting the texts that past generations leave behind, historians can reconstruct what their lives were like.

Defining "Culture": Material Culture

Material culture consists of the physical objects that individuals in society create. These objects, in turn, influence how we live. They include the food we eat; the clothes we wear; the cars we drive to the homes we live in; the tools and technologies we use to make those goods, from sewing machines to computerized factories; and the towns and cities that we build as places in which to live and work.

While the symbols expressed in speech and writing are the chief ways in which cultural meanings are formed and expressed, they are not the only ways. Material objects can also be used to generate meanings. A **signifier** is any vehicle of meaning—any set of elements used to communicate. The sounds made in speech are signifiers, as are the marks made on paper or other materials in writing. Other signifiers, however, include dress, pictures or visual signs, modes of eating, forms of building or architecture, and many other material features of culture (Hawkes 1977).

Styles of dress, for example, normally help signify differences between the sexes. Even colors can signify important aspects of culture. In contemporary society, young girls are typically dressed in pink, whereas boys are dressed in blue—but this wasn't always the case (Paoletti 2012). In the nineteenth century, both boys and girls wore frilly white clothing. A June 1918 article in *Ladies' Home Journal* stated, "The generally accepted rule is pink for the boys, and blue for the girls. The reason is that pink, being a more decided and stronger color, is more suitable for the boy, while blue, which is more delicate and dainty, is prettier for the girl" (Paoletti 2012). As we saw in the case of the Yale Halloween incident, material goods, such as costumes, can communicate very powerful social meanings. Although dressing up like a Native American might be deeply offensive on a college campus, the response might be different at a Chicago Blackhawks hockey game, where there is no expectation that students will be in a "safe space."

But material culture is not simply symbolic; it is also vital for catering to physical needs—in the tools or technology used to acquire food, make weaponry, construct dwellings, manufacture our clothing, and so forth. We have to study both the practical and the symbolic aspects of material culture to understand it completely.

Today, material culture is rapidly becoming globalized, largely through modern information technology, such as the computer and the Internet. Although the United States has been at the forefront of this technological revolution, most other industrial countries are catching up. In fact, it no longer makes sense to speak of an exclusively "U.S. technology" any more than it makes sense to speak of a U.S. car. The iPhone, for example, contains hundreds of components that are sourced from some 200 manufacturers across the planet, embodying technology developed in Japan, South Korea, Taiwan, Europe, and the United States (Menasians 2016). Another example of the globalization of material culture is the way that classrooms and department stores the world over increasingly resemble one another and the fact that McDonald's restaurants are now found on nearly every continent.

Culture and Society

"Culture" can be distinguished from "society," but these notions are closely connected. A **society** is a system of interrelationships that connects individuals together. No culture could exist without a society; equally, no society could exist without culture. Without

material culture
The physical objects that society creates that influence the ways in which people live.

signifier
Any vehicle of meaning and communication.

society
A group of people who live in a particular territory, are subject to a common system of political authority, and are aware of having a distinct identity from other groups. Some societies, like hunting-and-gathering societies, are small, numbering no more than a few dozen people. Others are large, numbering millions—modern Chinese society, for instance, has a population of more than a billion people.

In the photo on the right, members of a 1970s commune relax outdoors. On the left, a Harajuku girl poses for a photograph in Tokyo, Japan. Though their distinctive styles set them apart from mainstream society, these people are not as nonconformist as they may think they are. Both subcultures conform to the norms of their respective social groups.

>

culture, we would not be human at all, in the sense in which we usually understand that term. We would have no language in which to express ourselves, we would have no sense of self-consciousness, and our ability to think or reason would be severely limited.

Culture also serves as a society's glue because culture is an important source of conformity, providing its members with ready-made ways of thinking and acting. For example, when you say that you subscribe to a particular value, such as formal learning, you are probably voicing beliefs that conform to those of your family members, friends, teachers, or others who are significant in your life. Cultures differ, however, in how much they value conformity. Research based on surveys of more than 100,000 adults in over sixty countries shows that Japanese culture lies at one extreme in terms of valuing conformity (Hofstede 1997), while at the other extreme lies American culture, one of the least conformist, ranking among the world's highest in cherishing individualism (Hamamura 2012).

American high school and college students often see themselves as especially nonconformist. The hipsters of today, like the hippies of the 1960s and the punks of the 1980s, sport distinctive clothing styles, haircuts, and other forms of bodily adornment. Yet how individualistic are they? Are young people with piercings or tattoos really acting independently? Or are their styles perhaps as much the "uniforms" of their group as navy blue suits are among middle-aged businesspeople? There is an aspect of conformity to their behavior—conformity to their own group.

Since some degree of conformity to norms is necessary for any society to exist, one of the key challenges for all cultures is to instill in people a willingness to conform. This is accomplished in two ways (Parsons 1964). First, individuals learn the norms of their culture. While this occurs throughout a person's life, the most crucial learning occurs during childhood, and parents play a key role. When learning is successful, the norms are so thoroughly internalized that they become unquestioned ways of thinking and acting; they come to appear "normal." (Note the similarity between the words *norm* and *normal*.)

When a person fails to learn and adequately conform to a culture's norms, a second way of instilling cultural conformity comes into play: social control. Social control often involves the punishment of rule breaking. Administration of punishment includes such informal behavior as rebuking friends for minor breaches of etiquette, gossiping behind their backs, or ostracizing them from the group. Official, formal forms of discipline might range from parking tickets to imprisonment (Foucault 1979). Durkheim, one of the founders of sociology, argued that punishment serves not only to help guarantee conformity among those who would violate a culture's norms and values but also to vividly remind others what the norms and values are.

Is it possible to describe an "American" culture? Although the United States is culturally diverse, we can identify several characteristics of a uniquely American culture. First, it reflects a particular range of values shared by many, if not all, Americans, such as the belief in the merits of individual achievement or in equality of opportunity. Second, these values are connected to specific norms: For example, it is usually expected that people will work hard to achieve occupational success (Bellah et al. 1985; Parsons 1964). Third, it involves the use of material artifacts created mostly through modern industrial technology, such as cars, mass-produced food, and clothing.

CONCEPT CHECKS

1. Describe two examples of nonmaterial culture and two examples of material culture.

2. What is the linguistic relativity hypothesis?

3. What role does culture play in society?

How Does Human Culture Develop?

Human culture and human biology are closely intertwined. Understanding how culture is related to the physical evolution of the human species can help us better understand the central role that culture plays in shaping our lives.

Begin to understand how both biological and cultural factors influence our behavior. Learn the ideas of sociobiology and how others have tried to refute these ideas by emphasizing cultural differences.

Early Human Culture: Adaptation to Physical Environment

Scientists believe that the first humans evolved from apelike creatures on the African continent some 4 million years ago. Their conclusion is based on archaeological evidence and knowledge of the close similarities in blood chemistry and genetics between chimpanzees and humans. The first evidence of humanlike culture dates back only 2 million years. In these early cultures, humans fashioned stone tools, derived sustenance by hunting animals and gathering nuts and berries, harnessed the use of fire, and established a highly cooperative way of life. Because early humans planned their hunts, they must also have had some ability for abstract thought.

Culture enabled early humans to compensate for their physical limitations, such as lack of claws, sharp teeth, and running speed, relative to other animals (Deacon 1998). Culture freed humans from dependence on the instinctual and genetically determined set of responses to the environment characteristic of other species. The larger, more complex human brain permitted a greater degree of adaptive learning in dealing with major environmental changes, such as the Ice Age. For example, humans figured out how to build fires and sew clothing for warmth. Through greater flexibility, humans were able to survive unpredictable challenges in their surroundings and shape the world with their ideas and tools.

As these pictures of members of a North Pole community (left) and a desert community (right) demonstrate, culture is powerfully influenced by geographic and climate conditions.

Yet early humans were closely tied to their physical environment, because they still lacked the technological ability to modify their immediate surroundings significantly (Bennett 1976; Harris 1975, 1978, 1980). Their ability to secure food and make clothing and shelter depended largely on the physical resources that were close at hand. Cultures in different environments varied widely as a result of adaptations by which people fashioned their cultures to be suitable to specific geographic and climatic conditions. For example, the cultures developed by desert dwellers, where water and food were scarce, differed significantly from the cultures that developed in rain forests, where such natural resources abounded. Human inventiveness spawned a rich tapestry of cultures around the world. As you will see at the conclusion of this chapter, modern technology and other forces of globalization pose both challenges and opportunities for future global cultural diversity.

Nature or Nurture?

Because humans evolved as a part of the world of nature, it would seem logical to assume that human thinking and behavior are the result of biology and evolution. In fact, one of the oldest and most enduring controversies in the social sciences is the "nature/nurture" debate: Are we shaped by our biology, or are we products of learning through life's experiences, that is, of nurture? Biologists and some psychologists emphasize biological factors in explaining human thinking and behavior. Sociologists, not surprisingly, stress the role of learning and culture. They are also likely to argue that because human beings are capable of making conscious choices, neither biology nor culture wholly determines human behavior.

The "nature/nurture" debate has raged for more than a century. In the 1930s and 1940s, many social scientists focused on biological factors, with some researchers seeking (unsuccessfully), for example, to prove that a person's physique determined his or her personality. In the 1960s and 1970s, scholars in different fields emphasized culture. For example, social psychologists argued that even the most severe forms of mental illness were the result of the labels that society attaches to unusual behavior rather than the result of biochemical processes (Scheff 1966). Today, partly because of new understandings in genetics and brain neurophysiology, the pendulum is again swinging toward the side of biology.

The resurgence of biological explanations for human behavior began in the 1970s, when the evolutionary biologist Edward O. Wilson published *Sociobiology: The New*

Synthesis (1975). The term **sociobiology** refers to the application of biological principles to explain the social activities of animals, including human beings. Using studies of insects and other social creatures, Wilson argued that genes influence not only physical traits but behavior as well. In most species, for example, males are larger and more aggressive than females and tend to dominate the "weaker sex." Some suggest that genetic factors explain why, in all human societies that we know of, men tend to hold positions of greater authority than women.

One way in which sociobiologists have tried to illuminate the relations between the sexes is by means of the idea of "reproductive strategy." A reproductive strategy is a pattern of behavior, arrived at through evolutionary selection, that favors the chances of survival of offspring. The female body has a larger investment in its reproductive cells than the male body does—a fertilized egg takes nine months to develop. Thus, according to sociobiologists, women will not squander that investment and are not driven to have sexual relations with many partners; their overriding aim is the care and protection of children. Men, on the other hand, tend toward promiscuity. Their wish to have sex with many partners is sound strategy from the point of view of the species; to carry out their mission, which is to maximize the possibility of impregnation, they move from one partner to the next. In this way, it has been suggested, we can explain differences in sexual behavior and attitudes between men and women.

Sociobiologists do not argue that our genes determine 100 percent of our behavior. For example, they note that depending on the circumstances, men can choose to act in nonaggressive ways. Even though this argument would seem to open up the field of sociobiology to culture as an additional explanatory factor in describing human behavior, social scientists have roundly condemned sociobiology for claiming that a propensity for particular behaviors, such as violence, is somehow "genetically programmed" into our brains (Seville Statement on Violence 1990).

How Nature and Nurture Interact

Most sociologists today would acknowledge a role for nature in determining attitudes and behavior, but with strong qualifications. For example, babies are born with the ability to recognize faces: Babies a few minutes old turn their heads in response to patterns that resemble human faces but not in response to other patterns (Cosmides and Tooby 1997; Johnson and Morton 1991). But it is a large leap to conclude that because babies are born with basic reflexes, the behavior of adults is governed by **instincts**: inborn, biologically fixed patterns of action found in all cultures. Sociologists tend to argue strongly against **biological determinism**, or the belief that differences we observe between groups of people, such as men and women, are explained wholly by biological (rather than social) causes.

Sociologists no longer pose the question as one of nature or nurture. Instead, they ask how nature and nurture interact to produce human behavior. Recent studies exploring the relationship between genetics and social influences have generally concluded that although genetics is important, how genes might affect behavior depends largely on the social context. For example, a study of obesity among adolescents found that social and behavioral factors, such as a family's lifestyle (for example, how much time a family spends watching TV or how often a family skips meals), have a significant effect on the likelihood that children will end up overweight, even when both parents are heavy (Martin 2008). Similarly, an international study of gender differences in mathematical ability found that

---

sociobiology

An approach that attempts to explain the behavior of both animals and human beings in terms of biological principles.

instinct

A fixed pattern of behavior that has genetic origins and that appears in all normal animals within a given species.

biological determinism

The belief that differences we observe between groups of people, such as men and women, are explained wholly by biological causes.

Sociologists argue that our preferences for particular body types are not biologically ingrained but rather shaped by the cultural norms of beauty communicated through magazine ads, commercials, and movies.

such differences varied widely across countries, with differences in high mathematical performance between males and females reflecting the country's level of gender inequality rather than purely biological factors (Penner 2008). Even alcoholism is strongly affected by social context: Although a specific gene has been identified as increasing one's propensity for alcohol dependence, a strong family support system can greatly reduce that risk (Pescosolido et al. 2008).

Sociologists' main concern, therefore, is with how our different ways of thinking and acting are learned through interactions with family, friends, schools, television, and every other facet of the social environment. For example, sociologists argue that it's not an inborn biological disposition that makes American heterosexual males feel romantically attracted to a particular type of woman. Rather, it is the exposure they've had throughout their lives to tens of thousands of magazine ads, TV commercials, and film stars that emphasize specific cultural standards of female beauty.

Early child rearing is especially relevant to this kind of learning. Human babies have a large brain, requiring birth relatively early in their fetal development, before their heads have grown too large to pass through the birth canal. As a result, human babies are totally unequipped for survival on their own, compared with the young of other species, and must spend a number of years in the care of adults. This need, in turn, fosters a lengthy period of learning, during which the child is taught his or her society's culture.

Because humans think and act in so many different ways, sociologists do not believe that "biology is destiny." If biology were all-important, we would expect all cultures to be highly similar, if not identical. Yet this is hardly the case. This is not to say that human cultures have nothing in common. Surveys of thousands of different cultures have concluded that all known human cultures have such common characteristics as language, forms of emotional expression, rules that tell adults how to raise children or engage in sexual behavior, and even standards of beauty (Brown 1991). But there is enormous variety in exactly how these common characteristics play themselves out.

All cultures provide for childhood socialization, but what and how children are taught varies greatly from culture to culture. An American child learns the multiplication tables from a classroom teacher, while a child born in the forests of Borneo learns to hunt with older members of the tribe. All cultures have standards of beauty and ornamentation, but what is regarded as beautiful in one culture may be seen as ugly in another (Elias 1987;

Elias and Dunning 1987; Foucault 1988). However, some feminist scholars have argued that with global access to Western images of beauty on the Internet, cultural definitions of beauty throughout the world are growing narrower and increasingly emphasize the slender physique that is so cherished in many Western cultures (Sepúlveda and Calado 2012).

Cultural Diversity

The study of cultural differences highlights the importance of cultural learning as an influence on our behavior. Human behavior and practices—as well as beliefs—vary widely from culture to culture and often contrast radically with what people from Western societies consider normal. In the West, we eat oysters, but we do not eat kittens or puppies, both of which are regarded as delicacies in some parts of the world. Westerners regard kissing as a normal part of sexual behavior, but in other cultures the practice is either unknown or regarded as disgusting. All these different kinds of behavior are aspects of broad cultural differences that distinguish societies from one another.

SUBCULTURES

Small societies tend to be culturally uniform, but industrialized societies are themselves culturally diverse or multicultural, involving numerous **subcultures**. As you will discover in the discussion of global migration in Chapter 10, practices and social processes like slavery, colonialism, war, migration, and contemporary globalization have led to populations dispersing across borders and settling in new areas. This, in turn, has led to the emergence of societies that are cultural composites, meaning that

subcultures

Cultural groups within a wider society that hold values and norms distinct from those of the majority.

the population is made up of groups from diverse cultural and linguistic backgrounds. In modern cities, many subcultural communities live side by side. Some experts have estimated that as many as 800 different languages are regularly spoken by residents of New York City and its surrounding boroughs (Roberts 2010).

Most major European cities have become increasingly diverse in recent decades, as large numbers of persons from North Africa have arrived. As transnational migration has increased, many European societies have struggled with how to integrate persons who bring with them distinct cultural and religious backgrounds. For example, in 2011, the French government made it a punishable offense for Muslim women to wear full-face veils in public spaces (except for houses of worship and private cars). France is a country based on the values of "liberty and equality" for all, and the *niqab*, a veil that covers a woman's hair and face, leaving only the eyes clearly visible, is viewed as a cultural practice that oppresses women and deprives them of their freedoms (Erlanger 2011). This controversy

The tension between subgroup values and national values came to a head in 2011 when the French government banned Muslim women from wearing full-face veils in public. French policymakers believed that the *niqab* oppressed women and violated the nation's values of liberty and equality.

over Muslim women's veils vividly portrays the challenges when different subcultural communities live side by side.

Subculture does not refer only to people from different cultural backgrounds, or who speak different languages, within a larger society. It can also refer to any segment of the population that is distinguishable from the rest of society by its cultural patterns. Examples might include goths, computer hackers, hipsters, Rastafarians, and fans of hip-hop. Some people might identify themselves clearly with a particular subculture, whereas others may move fluidly among a number of different ones.

Culture plays an important role in perpetuating the values and norms of a society, yet it also offers important opportunities for creativity and change. Subcultures and **countercultures**—groups that largely reject the prevailing values and norms of society— can promote views that represent alternatives to the dominant culture. Social movements or groups of people sharing common lifestyles are powerful forces of change within societies. In this way, subcultures give people the freedom to express and act on their opinions, hopes, and beliefs. For example, throughout most of the twentieth century, gays and lesbians formed a distinct counterculture in opposition to dominant cultural norms. In a few cities, such as San Francisco, New York, and Chicago, gays and lesbians lived in distinct enclaves and even developed political power bases. Over time, their political claims and lifestyle became more and more acceptable to mainstream Americans, so much so that gay marriage was legalized nationwide in 2015. Today, gays and lesbians are no longer a counterculture. As the wider society has increasingly embraced their demand to be included in the institution of marriage, gays and lesbians have embraced one of the most significant institutions of mainstream society.

U.S. schoolchildren are frequently taught that the United States is a vast melting pot into which various subcultures are assimilated. **Assimilation** is the process by which different cultures are absorbed into a single mainstream culture. Although it is true that virtually all peoples living in the United States take on many common cultural characteristics, many groups strive to retain some subcultural identity. In fact, identification based on race or country of origin in the United States persists today and is particularly strong among African Americans and immigrants from Asia, Mexico, and Latin America (Totti 1987).

Given the immense cultural diversity and number of subcultures in the United States, a more appropriate metaphor than the assimilationist "melting pot" might be the culturally diverse "salad bowl," in which all the various ingredients, though mixed together, retain some of their original flavor and integrity, contributing to the richness of the salad as a whole. This viewpoint, termed **multiculturalism**, calls for respecting cultural diversity and promoting equality of different cultures. Adherents to multiculturalism acknowledge that certain central cultural values are shared by most people in a society but also that certain important differences deserve to be preserved (Anzaldua 1990).

CULTURAL IDENTITY AND ETHNOCENTRISM

Every culture displays its own unique patterns of behavior, and these may seem alien to people from other cultural backgrounds. If you have traveled abroad, you are probably familiar with the sensation that can result when you find yourself in a new culture. Everyday habits, customs, and behaviors that you take for granted in your own culture

---

countercultures

Cultural groups within a wider society that largely reject the values and norms of the majority.

assimilation

The acceptance of a minority group by a majority population, in which the new group takes on the values and norms of the dominant culture.

multiculturalism

The viewpoint according to which ethnic groups can exist separately and share equally in economic and political life.

The Secret Power of Cultural Norms and Values

Norms are widely agreed-upon principles or rules that people are expected to observe; they represent the dos and don'ts of social life. One way to illustrate the power of a social norm is to examine reactions to norm violations. Those who violate norms are often subject to the overt or subtle disapproval of others. Common reactions might include being scolded or mocked by friends for minor breaches of etiquette, being gossiped about behind our backs, or being ostracized from the social group. Yet norms are so powerful that violators often feel shame or self-criticism even in the absence of others' words or actions.

A vivid display of the power of norms is PostSecret, an ongoing community art project where people mail in their secrets anonymously on one side of a homemade postcard. The postcards are then posted online for others to view or comment on. What does it tell us about social norms when a young woman confesses, "He found my vomit in the sink. . . . I said I was fine. I LIED"? In American society, eating disorders like bulimia violate a social norm that says we shouldn't hurt ourselves. Yet it also subtly conveys another norm: Young women are expected to be thin to live up to our cultural ideals of "beauty." Using means other than the socially approved strategies of healthy diet and exercise, however, is a source of shame.

Other postcards make claims like "I've been stealing $$$ from the piggy banks of the kids I babysit to buy groceries and weed." This postcard reveals the violations of several important norms and values (and laws): that we should not steal (especially from children), that we should not do drugs, and that poverty is a stigmatized status—and one that people are often ashamed of.

A simple scroll through the postcards on the PostSecret website reveals the many social norms at play in our culture and the deep shame or fear of reprisal that comes from violating these behavioral expectations. Yet the fact that people throughout the world are willing to anonymously share their transgressions with others shows just how common norm violations are. It also shows just how powerful and even oppressive norms can be, given how many seek refuge by silently and anonymously "confessing" their wrongs on PostSecret.

A quick look at the site also reveals that the vast majority of participants are women. As we noted earlier in the chapter, some scholars have argued that norms regarding women's behavior are more rigid than those guiding men's behavior. Can you think of secrets that men may post and how they might differ from those posted by women? How might sites like PostSecret reinforce or challenge social norms, especially gender norms? Do you think a forum like PostSecret would work in a venue other than the anonymous world of the Internet?

Our secrets powerfully reveal the social norms that govern our behavior. Postcards submitted to PostSecret, an ongoing community art project, highlight how certain behaviors, such as theft, eating disorders, and infidelity, are stigmatized.

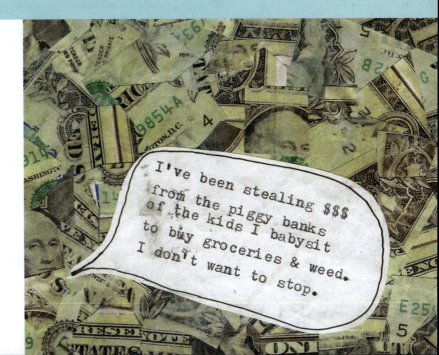

I've been stealing $$$ from the piggy banks of the kids I babysit to buy groceries & weed. I don't want to stop.

may not be part of everyday life in other parts of the world—even in countries that share the same language. The expression *culture shock* is an apt one! Often people feel disoriented when they become immersed in a new culture. This is because they have lost the familiar reference points that help them understand the world around them and have not yet learned how to navigate the new culture.

A culture must be studied in terms of its own meanings and values—a key presupposition of sociology. Sociologists endeavor as far as possible to avoid **ethnocentrism**, or judging other cultures in terms of the standards of one's own. Because human cultures vary so widely, it is not surprising that people belonging to one culture frequently find it difficult to understand the ideas or behavior of people from a different culture. In studying and practicing sociology, we must remove our own cultural blinders to see the ways of life of different peoples in an unbiased light. The practice of judging a society by its own standards is called **cultural relativism**.

Applying cultural relativism—that is, suspending your own deeply held cultural beliefs and examining a situation according to the standards of another culture—can be fraught with uncertainty and challenge. Not only can it be hard to see things from a completely different point of view but cultural relativism sometimes raises troubling issues. Consider, for example, the ritual of female genital cutting, or what opponents have called "genital mutilation." This is a painful ritual in which the clitoris and sometimes all or part of the vaginal labia of young girls are removed with a knife or a sharpened stone and the two sides of the vulva are partly sewn together as a means of controlling the young woman's sexual activity and increasing the sexual pleasure of her male partner.

In cultures where clitoridectomies have been practiced for generations, they are regarded as a normal, even expected practice. A study of 2,000 men and women in two Nigerian communities found that nine out of ten women interviewed had undergone clitoridectomies in childhood and that the large majority favored the procedure for their own daughters, primarily for cultural reasons; they would be viewed as social outcasts if they did not have the procedure. Yet a significant minority believed that the practice should be stopped (Ebomoyi 1987). Clitoridectomies are regarded with abhorrence by most people from other cultures and by a growing number of women in the cultures in which they are practiced (El Dareer 1982; Johnson-Odim 1991; Lightfoot-Klein 1989).

These differences in views can result in a clash of cultural values, especially when people from cultures where clitoridectomies are common migrate to countries where the practice is actually illegal. In France, many mothers in the North African immigrant community arrange for traditional clitoridectomies to be performed on their daughters. Some of these women have been tried and convicted under French law for mutilating their daughters. These African mothers have argued that they were only engaging in the same cultural practice that their own mothers had performed on them, that their grandmothers had performed on their mothers, and so on. They complain that the French are ethnocentric, judging traditional African rituals by French customs. Feminists from Africa and the Middle East, while themselves strongly opposed to clitoridectomies, have been critical of Europeans and Americans who sensationalize the practice by calling it backward or primitive without seeking any understanding of the cultural and economic circumstances that sustain it (Accad 1991; Johnson-Odim 1991; Mohanty 1991). In this instance, globalization has led to

ethnocentrism

The tendency to look at other cultures through the eyes of one's own culture and thereby misrepresent them.

cultural relativism

The practice of judging a society by its own standards.

a fundamental clash of cultural norms and values that has forced members of both cultures to confront some of their most deeply held beliefs. The role of the sociologist is to avoid knee-jerk responses and to examine complex questions carefully from as many different angles as possible.

Cultural Universals

Amid the diversity of human behavior, several **cultural universals** prevail. For example, there is no known culture without a grammatically complex language. All cultures possess some recognizable form of family system, in which there are values and norms associated with the care of children. The institution of **marriage** is a cultural universal, as are religious rituals and property rights. All cultures also practice some form of incest prohibition—

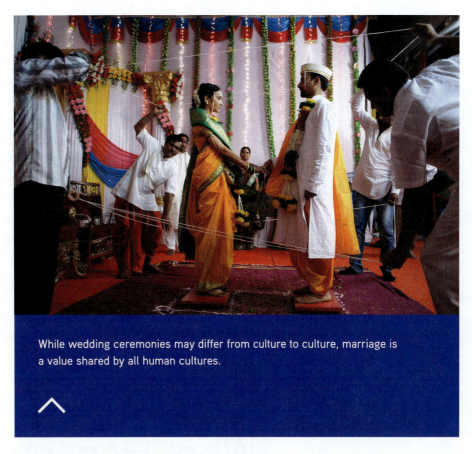

While wedding ceremonies may differ from culture to culture, marriage is a value shared by all human cultures.

the banning of sexual relations between close relatives, such as father and daughter, mother and son, and brother and sister. A variety of other cultural universals have been identified by anthropologists, including art, dancing, bodily adornment, games, gift giving, joking, and rules of hygiene. Among the cultural characteristics shared by all societies, two stand out in particular. All cultures incorporate ways of communicating and expressing meaning. All cultures also depend on material objects in daily life.

Yet there are variations within each category. Consider, for example, the prohibition against incest. Incest is typically defined as sexual relations between members of the immediate family, but in some cultures, "the family" has been expanded to include cousins and others bearing the same family name. There have also been societies in which a small proportion of the population has been permitted to engage in incestuous practices. Within the ruling class of ancient Egypt, for instance, brothers and sisters were permitted to have sex with each other.

Culture and Social Development

Cultural traits are closely related to overall patterns in the development of society. The level of material culture reached in a given society influences, although by no means completely determines, other aspects of cultural development. This is easy to see, for example, in the level of technology. Many aspects of culture characteristic of our lives today—cars, smartphones, laptops and tablets, Wi-Fi, running water, electric light—depend on technological innovations that have been made only very recently in human history.

cultural universals
Values or modes of behavior shared by all human cultures.

marriage
A socially approved sexual relationship between two individuals. Marriage normally forms the basis of a family of procreation; that is, it is expected that the married couple will produce and raise children.

CONCEPT CHECKS

1. Explain the "nature/ nurture" debate.

2. Why do sociologists disagree with the claim that biology is destiny?

3. Give examples of subcultures that are typical of American society.

4. What is the difference between ethnocentrism and cultural relativism?

5. What are two examples of cultural universals?

Learn how societies have changed over time.

The same is true at earlier phases of social development. Before the invention of the smelting of metal, for example, goods had to be made of organic or naturally occurring materials like wood or stone—a basic limitation on the artifacts that could be constructed. Variations in material culture provide the main means of distinguishing different forms of human society, but other factors are also influential. Writing is an example. As has been mentioned, not all human cultures have possessed writing—in fact, for most of human history, writing was unknown. The development of writing altered the scope of human cultural potentialities, making possible different forms of social organization than those that had previously existed. Yet writing continues to evolve even today. Think about the language you use when you send texts to your friends. If you sent your grandparents a text with acronyms like FOMO, YOLO, and SMH, would they understand what you were saying?

We now turn to analyzing the main types of society that existed in the past and that are still found in the world. In the present day, we are accustomed to societies that contain millions of people, many of them living crowded together in urban areas. But for most of human history, the Earth was much less densely populated than it is now, and it is only over the past hundred years or so that any societies have existed in which the majority of the population were city dwellers. To understand the forms of society that existed before modern industrialism, we have to call on the historical dimension of the sociological imagination.

What Happened to Premodern Societies?

Premodern societies can actually be grouped into three main categories: hunters and gatherers, larger agrarian or pastoral societies (involving agriculture or the tending of domesticated animals), and nonindustrial civilizations or traditional states. We shall look at the main characteristics of these societies in turn.

The Earliest Societies: Hunters and Gatherers

For all but a tiny part of our existence on this planet, human beings have lived in hunting-and-gathering societies, small groups or tribes often numbering no more than thirty or forty people. Hunters and gatherers gain their livelihood from hunting, fishing, and gathering edible plants growing in the wild. Hunting-and-gathering cultures continue to exist in some parts of the world, such as in a few arid parts of Africa and in the jungles of Brazil and New Guinea. Most such cultures, however, have been destroyed or absorbed by the spread of Western culture, and those that remain are unlikely to stay intact for much longer. Currently fewer than a quarter of a million people in the world support themselves through hunting and gathering—only 0.004 percent of the world's population.

Compared with larger societies—particularly modern societies, such as the United States—most hunting-and-gathering groups were egalitarian. Thus, there was little difference among members of the society in the number or kinds of material possessions; there were no divisions of rich and poor. The material goods they needed were limited to weapons for hunting, tools for digging and building, traps, and cooking utensils. Differences of position or rank tended to be limited to age and gender; men were almost always the hunters, while women gathered wild crops, cooked, and brought up the children.

Hunters and gatherers moved about a good deal, but not in a completely erratic way. They had fixed territories, around which they migrated regularly from year to year. Because they were without animal or mechanical means of transport, they could only take a few goods or possessions with them. Many hunting-and-gathering communities did not have a stable membership; people often moved among different camps, or groups split up and joined others within the same overall territory.

Hunters and gatherers had little interest in developing material wealth beyond what was needed for their basic needs. Their main concerns were with religious values and ritual activities. Members participated regularly in elaborate ceremonials and often spent a great deal of time preparing the dress, masks, paintings, or other sacred objects used in such rituals.

Hunters and gatherers are not merely primitive peoples whose ways of life no longer hold any interest for us. Studying their cultures allows us to see more clearly that some of our institutions are far from being natural features of human life. While we shouldn't idealize the circumstances in which hunters and gatherers lived, the lack of major inequalities of wealth and power and the emphasis on cooperation rather than competition are instructive reminders that the world created by modern industrial civilization is not necessarily to be equated with progress.

Pastoral and Agrarian Societies

About 15,000 years ago, some hunting-and-gathering groups turned to the raising of domesticated animals and the cultivation of fixed plots of land as their means of livelihood. **Pastoral societies** relied mainly on domesticated livestock, while **agrarian societies** grew crops (practiced agriculture). Some societies had mixed pastoral and agrarian economies.

Depending on the environment in which they lived, pastoralists reared animals, such as cattle, sheep, goats, camels, or horses. Some pastoral societies still exist in the modern world, concentrated especially in areas of Africa, the Middle East, and Central Asia. They are usually found in regions of dense grasslands or in desert or mountainous areas too poor in arable land for agriculture to be profitable.

At some point, hunting-and-gathering groups began to sow their own crops rather than simply collecting those growing in the wild. This practice first developed as what is usually called "horticulture," in which small gardens were cultivated by the use of simple hoes or digging instruments. Like pastoralism, horticulture provided for a more reliable supply of food than was possible from hunting and gathering and therefore could support larger communities. Because they were not on the move, people whose livelihood was horticulture could

pastoral societies
Societies whose subsistence derives from the rearing of domesticated animals.

agrarian societies
Societies whose means of subsistence are based on agricultural production (crop growing).

Pastoral societies still exist in certain areas of Africa, the Middle East, and Central Asia. Tibetan nomads have wandered China's Tibetan Plateau, herding yaks, for thousands of years.

develop larger stocks of material possessions than people in either hunting-and-gathering or pastoral communities. Some peoples in the world still rely primarily on horticulture for their livelihood.

Traditional Societies or Civilizations

From about 6000 BCE onward, we find evidence of societies larger than any that existed before and that contrast in distinct ways with earlier types. These societies were based on the development of cities, led to pronounced inequalities of wealth and power, and were ruled by kings or emperors. Because writing was used and science and art flourished, these societies are often called "civilizations." The earliest civilizations developed in the Middle East, usually in fertile river areas. The Chinese Empire originated in about 1800 BCE, at which time powerful states were also in existence in what are now India and Pakistan.

Most traditional (premodern) civilizations were also empires: They achieved their size through the conquest and incorporation of other peoples (Kautsky 1982). This was true, for instance, of traditional Rome and China. At its height, in the first century CE, the Roman Empire stretched from Britain in northwest Europe to beyond the Middle East. The Chinese Empire, which lasted for more than 2,000 years, up to the threshold of the twentieth century, covered most of the massive region of eastern Asia now occupied by modern China.

CONCEPT CHECKS

1. Compare the two main types of premodern societies.
2. Contrast pastoral and agrarian societies.

> Recognize the factors that transformed premodern societies, particularly how industrialization and colonialism influenced global development. Know the differences among industrialized societies, emerging economies, and developing societies and how these differences developed.

industrialization

The emergence of machine production, based on the use of inanimate power resources (such as steam or electricity).

How Has Industrialization Shaped Modern Society?

What happened to destroy the forms of society that dominated the whole of history up to two centuries ago? The answer, in a word, is **industrialization**—the emergence of machine production, based on the use of inanimate power resources (such as steam or electricity). The industrialized, or modern, societies differ in several key respects from any previous type of social order, and their development has had consequences stretching far beyond their European origins.

The Industrialized Societies

Industrialization originated in eighteenth-century Britain as a result of the Industrial Revolution, a complex set of technological changes that affected the means by which people gained their livelihood. These changes included the invention of new machines (such as the spinning jenny for weaving yarn), the harnessing of power resources (especially water and steam) for production, and the use of science to improve production methods. Because discoveries and inventions in one field lead to more in others, the pace of technological innovation in **industrialized societies** is extremely rapid compared with that of traditional social systems.

In even the most advanced of traditional civilizations, the majority of people were engaged in working on the land. The relatively low level of technological development did not permit more than a small minority to be freed from the chores of agricultural

People crowd the streets of Tokyo's entertainment district. Japan is an exemplar of an industrialized society. It is the third-largest economy, measured by GDP, lagging behind only the United States and China.

production. By contrast, a prime feature of industrialized societies today is that the large majority of the employed population work in factories, offices, or shops rather than in agriculture. And over 90 percent of people live in towns and cities, where most jobs are to be found and new job opportunities created. The largest cities are vastly greater in size than the urban settlements found in traditional civilizations. In cities, social life becomes more impersonal and anonymous than before, and many of our day-to-day encounters are with strangers. Large-scale organizations, such as business corporations or government agencies, come to influence the lives of virtually everyone.

A further feature of modern societies concerns their political systems, which are more developed and intensive than forms of government in traditional states. In traditional civilizations, the political authorities (monarchs and emperors) had little direct influence on the customs and habits of most of their subjects, who lived in fairly self-contained villages. With industrialization, transportation and communication became much more rapid, making for a more integrated "national" community.

The industrialized societies were the first **nation-states** to come into existence. Nation-states are political communities with clearly delimited borders dividing them from one another, rather than the vague frontier areas that used to separate traditional states. Nation-state governments have extensive powers over many aspects of citizens' lives, framing laws that apply to all those living within their borders. The United States is a nation-state, as are virtually all other societies in the world today.

The application of industrial technology has been by no means limited to peaceful processes of economic development. From the earliest phases of industrialization, modern production processes have been put to military use, and this has radically altered ways of waging war, creating weaponry and modes of military organization much more advanced than those of nonindustrial cultures. Together, superior economic strength, political cohesion, and military superiority account for the seemingly irresistible spread of Western ways of life across the world over the past two centuries.

industrialized societies

Highly developed nation-states in which the majority of the population work in factories or offices rather than in agriculture and in which most people live in urban areas.

nation-state

A particular type of state, characteristic of the modern world, in which a government has sovereign power within a defined territorial area and the population are citizens who know themselves to be part of a single nation.

Sociology first emerged as a discipline as industrial societies developed in Europe and North America and was strongly influenced by the changes taking place at that time. As we saw in Chapter 1, the major nineteenth-century sociological theorists (Durkheim, Marx, and Weber) all sought to explain these sweeping changes. Although they differed in their understanding and their predictions about the future, all shared a belief that industrial society was here to stay and that, as a result, the future would in many ways resemble the past.

Global Development

From the seventeenth century to the early twentieth century, the Western countries established colonies in numerous areas previously occupied by traditional societies. Although virtually all these colonies have now attained their independence, **colonialism** was central to shaping the social map of the globe as we know it today. In some regions, such as North America, Australia, and New Zealand, which were only thinly populated by hunting-and-gathering or pastoral communities, Europeans became the majority population. In other areas, including much of Asia, Africa, and South America, the local populations remained in the majority.

Societies of the first of these two types, including the United States, have become industrialized. Those in the second category are mostly at a much lower level of industrial development and are often referred to as less-developed societies, or the **developing world**. Such societies include India, most African countries (such as Nigeria, Ghana, and Algeria), and those in South America (such as Brazil, Peru, and Venezuela). Because many of these societies are situated south of the United States and Europe, they are sometimes referred to collectively as the Global South and contrasted to the wealthier, industrialized Global North.

THE GLOBAL SOUTH

The majority of countries in the Global South are in areas that underwent colonial rule. A few colonized areas gained independence early, such as Haiti, which became the first autonomous Black republic in January 1804. The Spanish colonies in South America acquired their freedom in 1810; Brazil broke away from Portuguese rule in 1822.

Some countries that were never ruled from Europe were nonetheless strongly influenced by colonial relationships. China, for example, was compelled from the seventeenth century on to enter into trading agreements with European powers, which assumed government control over certain areas, including major seaports. Hong Kong was the last of these. Most nations in the Global South have become independent states only since World War II—often following bloody anticolonial struggles. Examples include India, which, shortly after achieving self-rule, split into India and Pakistan; a range of other Asian countries (such as Myanmar, Malaysia, and Singapore); and countries in Africa (such as Kenya, Nigeria, the Democratic Republic of Congo, Tanzania, and Algeria).

Although they may include peoples living in traditional fashion, developing countries are very different from earlier forms of traditional society. Their political systems are modeled on systems first established in the societies of the West—that is to say, they are nation-states. Most of the population still live in rural areas, but many of these societies are experiencing a rapid process of city development. Although agriculture remains the main economic activity, crops are now often produced for sale in world markets rather than for local consumption. Developing countries are not merely societies that have "lagged

---

colonialism

The process whereby Western nations established their rule in parts of the world away from their home territories.

developing world

The less-developed societies, in which industrial production is either virtually nonexistent or only developed to a limited degree. The majority of the world's population lives in less developed countries.

emerging economies

Developing countries that, over the past two or three decades, have begun to develop a strong industrial base, such as Singapore and Hong Kong.

behind" the more industrialized areas. They have in large part been created by contact with Western industrialism, which has undermined the earlier, more traditional systems that were in place.

THE EMERGING ECONOMIES

Although the majority of countries in the Global South lag well behind societies of the West, some have now successfully embarked on a process of industrialization. Referred to as the **emerging economies**, they include Brazil, Mexico, Hong Kong, South Korea, Singapore, and Taiwan. Emerging economies are characterized by a great deal of industry and/or international trade. The rates of economic growth of the most successful emerging economies, such as those in East Asia, are several times those of the Western industrial economies.

The emerging economies of East Asia have shown the most sustained levels of economic prosperity. They are investing abroad as well as promoting growth at home. China is investing in mines and factories in Africa, elsewhere in East Asia, and in Latin America. South Korea's production of steel has increased by more than 30 percent in the last decade, and its shipbuilding and electronics industries are among the world's leaders (World Steel Association 2017). Singapore is becoming the major financial and commercial center of Southeast Asia. Taiwan is an important player in the manufacturing and electronics industries. All these changes have directly affected the United States, whose share of global steel production, for example, has dropped significantly since the 1970s. In fact, the "rise of the rest" (Zakaria 2008) is arguably the most important aspect of global economic change in the world today.

CONCEPT CHECKS

1. What does the concept of industrialization mean?

2. How has industrialization hurt traditional social systems?

3. Why are many African and South American societies classified as part of the Global South?

How Does Globalization Affect Contemporary Culture?

In Chapter 1 we noted that the chief focus of sociology historically has been the study of industrialized societies. As sociologists, can we thus safely ignore the Global South, leaving this as the domain of anthropology? We certainly cannot. The industrialized societies of the Global North and the developing societies of the Global South have developed in interconnection with one another and are today more closely related than ever before. Those of us living in industrialized societies depend on many raw materials and manufactured products coming from countries in the Global South to sustain our lives. Conversely, the economies of most states in the Global South depend on trading networks that bind them to industrialized countries. We can fully understand the industrialized order only against the backdrop of societies in the Global South—in which, in fact, by far the greater proportion of the world's population lives.

As the world rapidly moves toward a single, unified economy, businesses and people are moving about the globe in increasing numbers in search of new markets and economic

Recognize the effect of globalization on your life and the lives of people around the world. Think about the effect of a growing global culture.

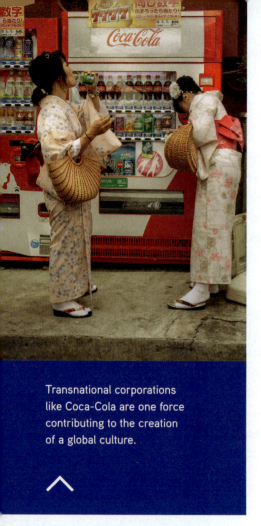

Transnational corporations like Coca-Cola are one force contributing to the creation of a global culture.

opportunities. As a result, the cultural map of the world is changing: Networks of peoples span national borders and even continents, providing cultural connections between their birthplaces and their adoptive countries (Appadurai 1986). A handful of languages come to dominate, and in some cases replace, the thousands of different languages that were once spoken on the planet.

It is increasingly impossible for cultures to exist as islands. Few, if any, places on earth are so remote as to escape radio, television, air travel (and the throngs of tourists this technology brings), or the computer. A generation ago, there were still tribes whose ways of life were completely untouched by the rest of the world. Today, these peoples use machetes and other tools made in the United States or Japan, wear T-shirts and shorts manufactured in garment factories in the Dominican Republic or Guatemala, and take medicine manufactured in Germany or Switzerland to combat diseases contracted through contact with outsiders. These people also have their stories broadcast to people around the world through satellite television and the Internet. Within a generation, or two at the most, all of the world's once-isolated cultures will be touched and transformed by global culture, despite their persistent efforts to preserve their age-old ways of life.

The forces that produce a global culture are discussed throughout this book:

■ Television, which brings U.S. culture (through networks such as MTV and shows such as *The Big Bang Theory*) into homes throughout the world daily

■ The emergence of a unified global economy, with businesses whose factories, management structures, and markets often span continents and countries

■ "Global citizens," such as managers of large corporations, who may spend as much time crisscrossing the globe as they do at home, identifying with a global, cosmopolitan culture rather than with their own nation's culture

■ A host of international organizations, including UN agencies, regional trade and mutual defense associations, multinational banks and other global financial institutions, international labor and health organizations, and global tariff and trade agreements, that are creating a global political, legal, and military framework

■ Electronic communications (via cell phone, Skype, fax, email, text message, Facebook, Twitter, and other communications on the Internet), which make instantaneous communication with almost any part of the planet an integral part of daily life in the business world

The world has become a single social system as a result of the growing interdependence, both social and economic, that now affects virtually everyone. But it would be a mistake to think of this increasing interdependence, or globalization, of the world's societies simply as the growth of world unity. The globalizing of social relations should be understood primarily as the reordering of time and distance in social life. Our lives, in other words, are increasingly and quickly influenced by events happening far away from our everyday activities.

Does the Internet Promote a Global Culture?

Many believe that the rapid worldwide growth of the Internet is hastening the spread of a global culture—one resembling the cultures of Europe and North America. Although the Internet is a truly global space, several languages prevail among Internet users. For example,

National Identity

Views on what constitutes national identity—or what it means to be an "American" or "Canadian," for example—differ widely across countries. The percentages below represent the proportion of the population in a country who believe speaking the language, sharing national customs and traditions, and having been born in the country are very important for being a true national.

	Language	Culture	Birthplace
Netherlands	84%	37%	16%
Hungary	81%	68%	52%
Germany	79%	29%	13%
France	77%	45%	25%
Japan	70%	43%	50%
U.S.	70%	45%	32%
Poland	67%	56%	42%
Sweden	66%	26%	8%
Canada	59%	54%	21%

Source: Stokes 2017.

26 percent of Internet users speak English as their main language, compared with 20 percent who speak Chinese and 8 percent who speak Spanish. In fact, ten languages alone account for more than three-quarters (77 percent) of all Internet users (Internet World Stats 2017b). Given the dominance of the English language and Western values on the Internet, belief in such values as equality between men and women, the right to speak freely, democratic participation in government, and the pursuit of pleasure through consumption may be readily diffused throughout the world over the Internet. Moreover, Internet technology itself would seem to foster such values: Global communication, seemingly unlimited (and uncensored) information, and instant gratification are all characteristics of the new technology.

Yet it may be premature to conclude that the Internet will sweep aside traditional cultures. Cyberspace is becoming increasingly global, and evidence shows that the Internet is, in many ways, compatible with traditional cultural values, perhaps even a means of strengthening them. This is especially likely to be true in countries that seek to control the Internet, censoring or blocking unwanted content and punishing those whose posts violate traditional values. One example is Saudi Arabia, a monarchy that officially enforces a highly conservative form of traditional Islam. The Saudi government not only routinely filters or blocks web content but also uses the Internet to disseminate official propaganda. Content deemed "harmful," "anti-Islamic," or "offensive" is blocked, including any criticism of the royal family, and messaging apps like Telegram and WhatsApp are restricted. To sell its iPhone in the country, Apple had to remove the device's built-in FaceTime app. Social media is heavily monitored, and cyberdissidents—for example, those who defend women's or minority rights or criticize traditional religious beliefs—are likely to receive steep fines and severe punishment. In 2016, a Saudi individual was sentenced to ten years in prison and 2,000 lashes for "spreading atheism" on Twitter (Freedom House 2016).

Yet at the same time, even under such repressive conditions, the Internet provides a space for self-expression and discussion, albeit within limits. Saudis are some of the most active social media users in the world and are the largest adopters of Twitter in the Arab world. Women, for example, who represent more than half of all Internet users in Saudi Arabia, are able to engage in Internet discussions that would be forbidden in public, such as those about women's health issues.

The Internet has sometimes been described as an echo chamber, in which people seek out like-minded others whose postings reinforce their own beliefs (Manjoo 2008; Sunstein 2012). For example, one study of a Jewish ultraorthodox religious group found that the Internet helped to strengthen the community, providing a forum for communication and for sharing ideas (Barzilai-Nahon and Barzilai 2005). In this sense, the Internet may be splitting society into what can be thought of as digitally linked tribes with their own unique cultural beliefs and values—sometimes even in conflict with the dominant culture. Examples from youth subcultures in recent years would include hippies, punks, skinheads, goths, gamers, rappers, and hipsters. Such subcultures sometimes emerge when ethnic minority youth seek to create a unique cultural identity within the dominant culture, often through music, dress, hairstyle, and bodily adornment, such as tattoos and piercings. They are often hybrids of existing cultures: For example, dancehall, like reggae, which originated as part of a youth culture in Jamaica's poor neighborhoods, today has spread to the United States, Europe, and anywhere there has been a large Caribbean migration (Niahh 2010).

Finally, of course, the Internet can be used to build a community around ideas that directly threaten the dominant culture. Al Qaeda, ISIS, and other radical Islamist jihadists rely on the Internet to spread their ideas, attract new recruits, and organize acts of violence. Suicide bombers routinely make videos celebrating their imminent deaths, videos that are posted on jihadist websites (and occasionally even on YouTube) to reach a wider audience of current and potential believers. Extremist groups from all faiths (Christian, Jewish, Muslim) have found the Internet to be a useful tool (Juergensmeyer 2009).

Globalization and Local Cultures

The influence of a growing global culture has provoked numerous reactions at the local level. Many local cultures remain strong or are experiencing rejuvenation, partly as a response to the diffusion of global culture. Such a response grows out of the concern that a global culture, dominated by North American and European cultural values, will corrupt the local culture. For example, the Taliban, an Islamic movement that controlled most of Afghanistan until 2001, historically has sought to impose traditional, tribal values throughout the country. Through its governmental Ministry for Ordering What Is Right and Forbidding What Is Wrong, the Taliban banned music, closed movie theaters, prohibited the consumption of alcohol, and required men to grow full beards. Women were ordered to cover their entire bodies with *burkas*, tentlike garments with a woven screen over the eyes; they were forbidden to work outside their homes or even to be seen in public with men who were not their spouses or relations. Violations of these rules resulted in severe punishment, sometimes death. The rise of the Taliban, and, more recently, of ISIS, can be understood at least partly as a rejection of the spread of Western culture—what Osama bin Laden referred to as "Westoxification" (Juergensmeyer 2003).

The resurgence of local cultures is sometimes seen throughout the world in the rise of **nationalism**, a sense of identification with one's country that is expressed through a common set of strongly held beliefs. Nationalism can be strongly political, involving attempts to assert the power of a nation based on a shared ethnic or racial identity over people of a different ethnicity or race. The world of the twenty-first century may well witness responses to globalization that celebrate ethnocentric nationalist beliefs, promoting intolerance and hatred rather than celebrating diversity.

New nationalisms, cultural identities, and religious practices are constantly being forged throughout the world. When you socialize with students from the same cultural background or celebrate traditional holidays with your friends and family, you are sustaining your culture. The very technology that helps foster globalization also supports local cultures: The Internet enables you to communicate with others who share your cultural identity, even when they are dispersed around the world.

Although sociologists do not yet fully understand these processes, they often conclude that despite the powerful forces of globalization operating in the world today, local cultures remain strong and, indeed, flourish. Yet local cultural and social movements can thrive and flourish only if they are allowed to do so. Given the rapid social changes in recent decades, it is still too soon to tell whether and how globalization will transform our world—whether it will result in the homogenization of the world's diverse cultures, the flourishing of many individual cultures, or both.

nationalism

A set of beliefs and symbols expressing identification with a national community.

CONCEPT CHECKS

1. What are three examples of forces that produce a global culture?

2. How can the Internet be used to perpetuate traditional cultural values?

3. What is nationalism?

CHAPTER 2

The Big Picture

Culture and Society

Thinking Sociologically

1. Mention at least two cultural traits that you would claim are universals; mention two others you would claim are culturally specific traits. Locate and use case study materials from different societies you are familiar with to show the differences between universal and specific cultural traits. Are the cultural universals you have discussed derivatives of human instincts? Explain your answer fully.

2. What does it mean to be ethnocentric? How is ethnocentrism dangerous in conducting social research? How is ethnocentrism problematic among nonresearchers in their everyday lives?

3. Think about a favorite article of clothing of yours. What values does it convey, and what messages about you or your subculture does it communicate to others?

What Is Culture?

p. 43

Know what culture consists of, and recognize how it differs from society.

How Does Human Culture Develop?

p. 49

Begin to understand how both biological and cultural factors influence our behavior. Learn the ideas of sociobiology and how others have tried to refute these ideas by emphasizing cultural differences.

What Happened to Premodern Societies?

p. 58

Learn how societies have changed over time.

How Has Industrialization Shaped Modern Society?

p. 60

Recognize the factors that transformed premodern societies, particularly how industrialization and colonialism influenced global development. Know the differences among industrialized societies, emerging economies, and developing societies, and how these differences developed.

How Does Globalization Affect Contemporary Culture?

p. 63

Recognize the effect of globalization on your life and the lives of people around the world. Think about the effect of a growing global culture.

cultural appropriation

culture • values • norms • language • linguistic relativity hypothesis • material culture • signifier • society

1. Describe two examples of nonmaterial culture and two examples of material culture.
2. What is the linguistic relativity hypothesis?
3. What role does culture play in society?

sociobiology • instinct • biological determinism • subcultures • countercultures • assimilation • multiculturalism • ethnocentrism • cultural relativism • cultural universals • marriage

1. Explain the "nature/nurture" debate.
2. Why do sociologists disagree with the claim that biology is destiny?
3. Give examples of subcultures that are typical of American society.
4. What is the difference between ethnocentrism and cultural relativism?
5. What are two examples of cultural universals?

pastoral societies • agrarian societies

1. Compare the two main types of premodern societies.
2. Contrast pastoral and agrarian societies.

industrialization • industrialized societies • nation-state • colonialism • developing world • emerging economies

1. What does the concept of industrialization mean?
2. How has industrialization hurt traditional social systems?
3. Why are many African and South American societies classified as part of the Global South?

nationalism

1. What are three examples of forces that produce a global culture?
2. How can the Internet be used to perpetuate traditional cultural values?
3. What is nationalism?

3

Socialization, the Life Course, and Aging

Transgender student Coy Mathis and her parents sued their school district to guarantee Coy the right to use the girls' bathroom. Although social institutions such as schools often encourage conformity to norms of behavior, they can change over time.

THE BIG QUESTIONS

How are children socialized?
Learn about socialization (including gender socialization), and know the most important agents of socialization.

What are the five major stages of the life course?
Learn the various stages of the life course, and see the similarities and differences among different cultures and historical periods.

How do people age?
Understand that aging is a combination of biological, psychological, and social processes. Consider key theories of aging, particularly those that focus on how society shapes the social roles of older people and that emphasize aspects of age stratification.

What are the challenges of aging in the U.S.?
Evaluate the experience of growing old in the contemporary United States. Identify the physical, emotional, and financial challenges older adults face.

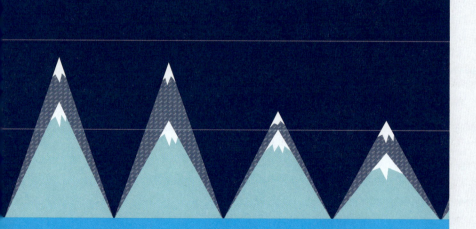

Graying of the World

p. 93

In July 2015, at the annual ESPY awards, the Arthur Ashe Courage Award was presented to Caitlyn Jenner in recognition of the former Olympic athlete's bravery in coming out as transgender. Caitlyn, now a transgender woman, was born Bruce Jenner in 1949 and attracted international glory when he won the gold medal for the grueling decathlon at the 1976 Summer Olympics in Montreal. The Arthur Ashe award—which counts Muhammed Ali, Billie Jean King, and Nelson Mandela as past winners—honors "strength in the face of adversity, courage in the face of peril and the willingness to stand up for their beliefs no matter what the cost." In her acceptance speech, Jenner vowed "to do whatever I can to reshape the landscape of how transgender people are viewed and treated" (ESPN 2015).

Nearly three years earlier, six-year-old Coy Mathis and her family were fighting their own courageous battle for transgender rights and respect. Coy, a transgender girl, was told that she could not use the girls' bathroom at her Colorado elementary school. Although she was allowed to use the gender-neutral restroom in the nurse's office and teacher's lounge, the school ruled the main girls' bathroom off-limits to Coy. She and her parents, Kathryn and Jeremy, successfully sued the school district, marking a major legal victory for transgender persons. Coy's parents said that they wanted their daughter to enjoy the same rights as any of her classmates and that relegating her to a special bathroom would make Coy vulnerable to bullying and stigmatizing (Banda and Ricciardi 2013).

Coy's battle sparked heated debates among parents, teachers, media pundits, and bloggers. Some questioned whether a young child could really have a gender identity. Others scoffed that Coy's preference for "girly" clothes and toys was just a phase. Others challenged the very existence of transgender identities. Jeff Johnston, a self-described "gender issues analyst" with the conservative organization Focus on the Family, baldly asserted that "male and female are categories of existence," denying the existence of any other gender identities (Erdely 2013).

While Kathryn and Jeremy initially thought that Coy would grow out of her predilection for all things girly, it soon became obvious to them that this was something much more than a phase. Born male—one of a triplet—Coy had rebelled against "boy" clothing and haircuts ever since she was a toddler. Refusing to wear firefighter or knight costumes, Coy instead gravitated toward princess dresses and demanded that her meals be served on pink decorative plates. Coy would cry when other children referred to her as a boy and tearfully asked her mother when they would be going to the doctor so that Coy could get her "girl parts." Coy's parents reached out to doctors, psychologists, and other parents of children who seemed uncomfortable in their own bodies. Fearful Coy would end up a statistic—a staggering 25 to 40 percent of transgender children and teens attempt suicide—Kathryn and Jeremy decided to raise Coy as a girl (Grossman and D'Augelli 2007; Haas, Rodgers, and Herman 2014). They had already seen signs of depression in young Coy; even at age three, she would become listless and sullen, refusing to put on boy's clothes and begging not to have to play outside. Coy showed sparks of happiness and joy only when allowed truly to be herself—a little girl (Erdely 2013).

The experiences of Caitlyn Jenner and Coy Mathis—and the public response (whether supportive or critical) from observers worldwide—illustrate the importance and complexities of socialization to everyday life. Sociologists are interested in the processes through which a young child such as Coy learns to become a member of society, complying with (or rejecting) society's ever-evolving expectations for how one should act, think, feel, and even dress. Social institutions—such as schools in Coy's case or sports in Jenner's case—and social actors encourage conformity to contemporary social norms through praise and discourage non-conformity through punishment and disapproval. Yet social institutions change over time, and the forces that socialize children shift accordingly. Try to imagine how Jenner's classmates and parents might have reacted if she had identified and dressed as a girl when she was a young child in the early 1950s. Contrast this with the support that Coy received from her parents and most of her classmates in the early 2010s. The study of socialization embodies a core theme of the "sociological imagination," that our lives are a product of both individual biographies and sociohistorical context (Mills 1959).

Socialization is the process whereby an infant becomes a self-aware, knowledgeable person, skilled in the ways of his or her culture. The socialization of young persons contributes to the phenomenon of **social reproduction**—the process whereby societies have structural continuity over time. During socialization, especially in the early years, children learn the ways of their parents and ancestors, thereby carrying on their values, norms, and social practices

socialization

The social processes through which we develop an awareness of social norms and values and achieve a distinct sense of self.

social reproduction

The process whereby societies have structural continuity over time. Social reproduction is an important pathway through which parents transmit or produce values, norms, and social practices among their children.

across the generations. All societies have characteristics that endure over long stretches of time, even though their members change. But at the same time, some old norms and customs die out as members of the older generation pass away, replaced with new "rules" to live by. For instance, while older generations of parents might have reprimanded their boys for being timid or playing with dolls, newer generations of parents may encourage their sons and daughters to "just be themselves."

Socialization is not limited to childhood. Throughout the life course, individuals may experience **resocialization** when their life circumstances and social roles change. Resocialization involves either learning new skills and norms appropriate to one's new roles and contexts or unlearning those skills and norms that may no longer be relevant. For example, upon retirement, one must learn to take on a new social role that is different from the role of worker, and upon release from prison, ex-convicts must relearn how to be members of mainstream society. Or in the case of sixty-six-year-old Caitlyn Jenner, she had to relearn how to walk, dress, and interact with others in her new role as a woman. Although gender is often thought of as "natural," we will soon learn (both here and in Chapter 9) that how we dress, how we speak, and even the career and family choices we make are a product of lifelong socialization.

Socialization connects the different generations to one another (Turnbull 1983). The birth of a child alters the lives of those who are responsible for his or her upbringing—who themselves undergo new learning experiences. Parenting usually ties the activities of adults to children for the remainder of their lives. Older people still remain parents when they become grandparents, thus forging another set of relationships that bond the generations. Although the process of cultural learning is much more intense in infancy and early childhood than in later life, learning and adjustment go on through the whole life course.

In the sections that follow, we continue the theme of "nature interacting with nurture" introduced in the previous chapter. We first describe the process of human development from infancy to early childhood. We compare different theoretical interpretations of how and why children develop as they do and how gender identities develop. We move on to discuss the main groups and social contexts that influence socialization throughout the life course. Finally, we focus on one distinctive stage of the life course: old age. We discuss the problems persons age sixty-five and older face, who now make up the most rapidly growing age group in the United States and the developed world.

How Are Children Socialized?

One of the most distinctive features of human beings, compared with other animals, is self-awareness—the awareness that one has an identity distinct and separate from others. During the first months of life, an infant possesses little or no understanding of differences between human beings and material objects in the environment and has no awareness of self. Children begin to use concepts such as "I," "me," and "you" at around age two or after. They gradually come to understand that others have distinct identities, consciousness, and needs separate from their own.

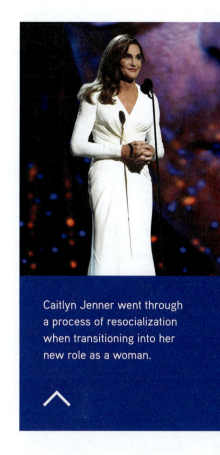

Caitlyn Jenner went through a process of resocialization when transitioning into her new role as a woman.

resocialization

The process of learning new norms, values, and behaviors when one joins a new group or takes on a new social role or when one's life circumstances change dramatically.

Learn about socialization (including gender socialization), and know the most important agents of socialization.

Theories of Child Development

The processes through which the self emerges and develops is much debated, in part because the most prominent theories about child development emphasize different aspects of socialization. The American philosopher and sociologist George Herbert Mead gave attention mainly to how children learn to use the concepts of "I" and "me." Charles Horton Cooley demonstrated the importance of other individuals for shaping a child's sense of self. Jean Piaget, the Swiss student of child behavior, focused on **cognition**—the ways in which children learn to think about themselves and their environment.

G. H. MEAD AND THE DEVELOPMENT OF SELF

Mead's ideas form the basis of a general tradition of theoretical thinking, symbolic interactionism, and have had a broad impact in sociology. Symbolic interactionism emphasizes that interaction between human beings takes place through symbols and the interpretation of meanings (see Chapter 1). Mead's work also provides an account of the main phases of child development, giving particular attention to the emergence of a sense of self.

According to Mead, infants and young children develop as social beings by imitating the actions of those around them. Play is one way in which this takes place: Young children often imitate what adults do. A toddler may make mud pies, having seen an adult cooking, or may dig in the dirt with a spoon, having observed someone gardening. Children's play evolves from simple imitation to more complicated games in which, at age four or five, they will act out an adult role. Mead called this "taking the role of the other"—learning what it is like to be in the shoes of another person. At this stage, children acquire a developed sense of self; that is, they develop an understanding of themselves as separate agents—as a "me"—by seeing themselves through the eyes of others. We achieve self-awareness, according to Mead, when we learn to distinguish the "me" from the "I." The "I" is the unsocialized infant, a bundle of spontaneous wants and desires. The "me," as Mead used the term, is the **social self**. Individuals develop **self-consciousness**, Mead argued, by coming to see themselves as others see them.

A further stage of child development, according to Mead, occurs when the child is about eight or nine years old. This is the age at which children tend to take part in organized games rather than unsystematic play. It is at this period that children begin to understand the overall values and morality that guide human behavior. To learn organized games, children must understand the rules of play and notions of fairness and equal participation. Children at this stage learn to grasp what Mead termed the **generalized other**—the general values and moral rules of the culture in which they are developing.

CHARLES HORTON COOLEY AND THE LOOKING-GLASS SELF

Charles Horton Cooley was an early-twentieth-century sociologist who studied self-concept, or the ways we view and think about ourselves. How do we come to view ourselves as humorous or cranky? Intelligent? Kind-hearted? Cooley argued that the notions we develop about ourselves reflect our interpretations of how others see us. His theory of the **looking-glass self** proposes that the reactions we elicit in social situations create a mirror in which we see ourselves. For example, if others regularly laugh at our jokes, we may perceive that they view us as funny and, in turn, view ourselves as such. Likewise, if our classmates and teachers praise us for our intelligent remarks in class, we may in turn start to view ourselves as smart.

cognition

Human thought processes involving perception, reasoning, and remembering.

social self

According to the theory of George Herbert Mead, the identity conferred upon an individual by the reactions of others. A person achieves self-consciousness by becoming aware of this social identity.

self-consciousness

Awareness of one's distinct social identity as a person separate from others. Human beings are not born with self-consciousness but acquire an awareness of self as a result of early socialization.

generalized other

A concept in the theory of George Herbert Mead, according to which the individual takes over the general values of a given group or society during the socialization process.

looking-glass self

A theory developed by Charles Horton Cooley that proposes that the reactions we elicit in social situations create a mirror in which we see ourselves.

Over time, mixed empirical evidence has led to reformulations of Cooley's classic theory. One refinement suggests that individuals take action to bring others around to their own views of themselves rather than passively accepting what others think of them. For example, a student who sees herself as very intelligent may regularly answer questions in class in an effort to ensure that her classmates also view her as very intelligent (Yeung and Martin 2003). In this way, youth are not merely passive recipients but rather active agents in shaping others' perceptions.

JEAN PIAGET AND THE STAGES OF COGNITIVE DEVELOPMENT

Piaget emphasized the child's active capability to make sense of the world. Children do not passively soak up information but instead select and interpret what they see, hear, and feel in the world around them. Piaget described several distinct stages of cognitive development during which children learn to think about themselves and their environment. Each stage involves the acquisition of new skills and depends on the successful completion of the preceding one.

Piaget called the first stage, which lasts from birth up to about age two, the **sensorimotor stage** because infants learn mainly by touching objects, manipulating them, and physically exploring their environment. Until about age four months, infants cannot differentiate themselves from their environment. Infants gradually learn to distinguish people from objects, coming to see that both have an existence independent of the infants' immediate perceptions. By the end of the sensorimotor stage, children understand that their environment has distinct and stable properties.

The next phase, called the **preoperational stage**, is the one to which Piaget devoted the bulk of his research. This stage lasts from age two to seven. Children acquire a mastery of language and an ability to use words to represent objects and images in a symbolic fashion. A four-year-old might use a sweeping hand, for example, to represent the concept "airplane." Piaget termed the stage "preoperational" because children are not yet able to use their developing mental capabilities systematically. Children in this stage are **egocentric**. As Piaget used it, this term does not refer to selfishness but to the tendency of the child to interpret the world exclusively in terms of his or her own position. For example, children at the preoperational stage cannot hold connected conversations with others. In egocentric speech, what the child says is more or less unrelated to what the other speaker said. Children talk together but not to one another in the same sense that adults do.

A third period, the **concrete operational stage**, lasts from age seven to eleven. During this phase, children can master logical but not abstract notions. They are able to handle ideas such as causality without much difficulty. They become capable of carrying out the mathematical operations of multiplication, division, and subtraction. Children by this stage are much less egocentric. In the preoperational stage, if a girl is asked "How many sisters do you have?" she may correctly answer one. But if asked "How many sisters does your sister have?" she will probably answer none because she cannot see herself from the point of view of her sister. The concrete operational child is able to answer such a question with ease.

The years from eleven to fifteen cover what Piaget called the **formal operational stage**. During adolescence, the developing child becomes able to grasp highly abstract and hypothetical ideas. When faced with a problem, children at this stage are able to review many possible ways of solving it and go through them theoretically to reach a solution.

sensorimotor stage

According to Jean Piaget, the first stage of human cognitive development, in which a child's awareness of his or her environment is dominated by perception and touch.

preoperational stage

According to Jean Piaget, the second stage of human cognitive development, in which a child has advanced sufficiently to master basic modes of logical thought.

egocentric

According to Jean Piaget, the characteristic quality of a child during the early years of life. Egocentric thinking involves understanding objects and events in the environment solely in terms of the child's own position.

concrete operational stage

The stage of human cognitive development, as formulated by Jean Piaget, in which the child's thinking is based primarily on physical perception of the world.

formal operational stage

According to Jean Piaget, the stage of human cognitive development at which the growing child becomes capable of handling abstract concepts and hypothetical situations.

According to Piaget, the first three stages of development are universal, but not all adults reach the formal operational stage. The development of formal operational thought depends in part on one's education, which may foster abstract reasoning.

Agents of Socialization

Agents of socialization are groups or social contexts in which significant processes of socialization occur. Primary socialization occurs in infancy and childhood and is the most intense period of cultural learning. It is the time when children learn language and basic behavioral patterns that form the foundation for later learning. The family is the main agent of socialization during this phase. Secondary socialization takes place later in childhood and into maturity. In this phase, schools, peer groups, social organizations (such as sports teams), the media, and eventually the workplace become socializing forces for individuals. Social interactions in these contexts help people learn the values, norms, and beliefs that make up the patterns of their culture.

FAMILIES

Because family systems vary worldwide, the range of family contacts that the infant experiences also varies widely across cultures. The mother is commonly the most important individual in the child's early life, but the nature of the relationships established between mothers and their children is influenced by the form and regularity of their contact.

In modern societies, most early socialization occurs within a small-scale or **nuclear family** context. Most American children spend their early years within a domestic unit comprising mother, father, and perhaps one or two other children, although the proportion growing up in two-parent households is lower than it was in prior decades (U.S. Bureau of the Census 2015j). In many other cultures, by contrast, aunts, uncles, and grandparents are often part of a single household and serve as caretakers even for very young infants. Even within U.S. society, family contexts vary widely. Some children are brought up in single-parent households; some are cared for by one biological and one nonbiological parent figure (for example, a divorced parent and a stepparent or parents in a same-sex relationship). The majority of mothers are now employed outside the home and return to their paid work shortly after the births of their children. Despite these variations, families typically remain the major agent of socialization from infancy to adolescence and beyond.

In most traditional societies, the family into which a person was born largely determined the individual's social position for the rest of his or her life. In modern societies, social position is not inherited at birth in this way, yet the region and social class of the family into which an individual is born affect patterns of socialization. Children pick up ways of behaving from their parents or others in their neighborhood or community. Patterns of child rearing and discipline, together with contrasting values and expectations, are found in different sectors of large-scale societies. For instance, sociologist Annette Laureau (2011) observed parents and children in their own homes and found that working-class parents emphasize "natural growth" in their children, encouraging them to play on their own; upper-middle-class parents engage in "concerted cultivation," actively fostering their kids' talents by enrolling them in a range of structured educational and extracurricular activities and closely monitoring their development. This latter approach provides children with the opportunities and skills necessary not only to succeed in school and, later, in the workforce but also to maintain their social class position.

agents of socialization

Groups or social contexts within which processes of socialization take place.

nuclear family

A family group consisting of an adult or adult couple and their dependent children.

Families are a key site of social reproduction. Children model the behavior of their parents. In this way, values and behaviors are reproduced across generations.

Of course, few, if any, children simply adopt the outlook of their parents unquestioningly. This is especially true in the modern world, in which change is so pervasive. Moreover, the wide range of socializing agents in modern societies leads to many divergences between the outlooks of children, adolescents, and the parental generation. For example, while Coy Mathis's parents supported her desire to dress, play, and identify however she wished, some of Coy's classmates' parents and media pundits held very different views that emphasized conformity to traditional gender norms rather than freedom of expression.

SCHOOLS

Another important socializing agent is the school. Schooling is a formal process: Students pursue a clearly defined curriculum of subjects. Yet schools are agents of socialization in more subtle respects. Students must be punctual, stay quiet in class, obey their teachers, and observe rules of discipline. How teachers react to their students, in turn, affects the students' views and expectations of themselves. These expectations also become linked to later job experience when students leave school. Peer groups are often formed at school, and the system of keeping children in classes according to age reinforces their impact.

Another key mechanism through which schools socialize children is the **hidden curriculum**, which refers to the subtle ways that boys and girls, middle class versus working class, and Black versus white children are exposed to different messages and curricular materials from their teachers. In Chapter 12, we delve much more fully into the ways that schools socialize children, often unwittingly perpetuating race, class, and gender inequalities.

PEER RELATIONSHIPS

Another socializing agency is the **peer group**. Peer groups consist of individuals of a similar age. The family's importance in socialization is obvious because the experience of the infant and young child is shaped more or less exclusively within it. It is less apparent, especially to those of us living in Western societies, how significant peer groups are. Children over age four or five usually spend a great deal of time in the company of friends the same age. Given the high proportion of women now in the workforce whose young children play together in day-care centers and preschool, peer relations are more important than ever before (Corsaro 1997; Harris 1998).

Peer relations are likely to have a significant effect beyond childhood and adolescence. Informal groups of people of similar ages, at work and in other situations, are usually of enduring importance in shaping individuals' attitudes and behavior. Peer groups also play an important role in changing norms, with more contemporary peer groups upholding or promoting behaviors that might not have been supported in earlier generations. While Bruce Jenner's classmates in the 1950s no doubt promoted gender conformity, Coy Mathis and her young friends may grow up to hold, and encourage in one another, much more open-minded views about gender, identity, and gender roles.

THE MASS MEDIA

Newspapers and periodicals flourished in the West from the early 1800s onward, but they were confined to a fairly small readership. It was not until a century later that such printed materials became part of the daily experience of millions of people, influencing their attitudes and opinions. The spread of mass media involving printed documents was soon

hidden curriculum

Traits of behavior or attitudes that are learned at school but not included in the formal curriculum, for example, gender differences.

peer group

A friendship group composed of individuals of similar age and social status.

YouTube and reality television personality, author, and LGBTQ rights activist Jazz Jennings uses the media to share her own experiences as a transgender teenager.

accompanied by electronic communication—radio, television, records, and videos. Today, it is a rare American who goes a day (or even an hour) without reading an article, watching a video, or listening to a podcast online. In fact, more than 90 percent of teens go online at least once a day, while nearly a quarter (24 percent) use the Internet "almost constantly" (Lenhart 2015). Americans spend a large portion of their leisure time consuming media. According to the American Time Use Survey, Americans watched an average of nearly three hours of television per day in 2017, representing more than half of their total leisure time (U.S. Bureau of Labor Statistics 2018a).

Media, in all its forms, has a powerful impact on our lives, and it is particularly influential in shaping the beliefs and behaviors of impressionable children and teens. For instance, children and adolescents often model the gender roles and practices that they see on their favorite television shows. Fashion magazines and music videos are also cited as powerful influences on girls' body image or their beliefs about "ideal" body weight and physique (Grabe, Ward, and Hyde 2008). Yet media can also teach children about topics with which their parents may be less familiar or comfortable and can provide information and even a sense of solace for children who may be lacking support in their communities. *I Am Jazz*, a reality show about the daily life of transgender teenager Jazz Jennings, has been praised for providing a role model for young children who may be conflicted about their own gender identity (*Time* 2014).

In recent years, researchers have become interested in studying the ways that video games (especially violent video games) affect children. Nearly three-quarters (72 percent) of teenagers play video games on their phones, computers, or consoles such as PlayStation, XBox, or Wii, including 84 percent of teenage boys and 59 percent of teenage girls (Lenhart 2015). Researchers are finding that violent video games may affect youth in similar ways as violent television images. For instance, rapid-action games with very violent imagery may desensitize players to violence (Engelhardt et al. 2011). Yet emerging research also shows that video games can have positive effects on children and their families. Roughly 55 percent of parents believe that playing video games helps families spend more time together (Entertainment Software Association 2011). And recent work by neuroscientists and psychologists finds that some types of fast-paced video games boost children's brain stimulation, cognitive development, spatial abilities, problem-solving skills, and even self-esteem (Granic, Lobel, and Engels 2014).

WORK

Across all cultures, work is an important setting within which socialization processes operate, although it is only in industrial societies that large numbers of people go to places of work separate from the home. In traditional communities, many people farmed the land close to where they lived or had workshops in their dwellings. "Work" in such communities was not as clearly distinct from other activities as it is for most members of the workforce in the modern West. In the industrialized countries, joining the workforce ordinarily marks a much greater transition in an individual's life than beginning work in traditional societies.

The work environment often poses unfamiliar demands, perhaps calling for major adjustments in the person's outlook or behavior. In addition to mastering the specific tasks of their job and internalizing company policies and practices, many workers also need to learn how to "feel" on the job. Sociologist Arlie Hochschild (1983) has documented the ways that workers, especially women workers, learn to feel and then display socially acceptable emotions at work. For instance, flight attendants learn to keep a calm and cool demeanor, even when dealing with a surly passenger or flying through extreme turbulence. Health care workers, police officers, firefighters, and soldiers must also learn how to manage feelings such as fear, sadness, and disgust to do their jobs. In these ways, individuals are socialized into the varied and complex skills required to be successful in the workplace.

Social Roles

Through socialization, individuals learn about **social roles**—socially defined expectations for a person in a given social position. The social role of doctor, for example, encompasses a set of behaviors that should be enacted by all individual doctors, regardless of their personal opinions or outlooks. Because all doctors share this role, it is possible to speak in general terms about the professional behavior of doctors, regardless of the specific individuals who occupy that position.

Some sociologists, particularly those associated with the functionalist school, regard social roles as fixed and relatively unchanging parts of a society's culture. According to this view, individuals learn the expectations that surround social positions in their particular culture and perform those roles largely as they have been defined. Social roles do not involve negotiation or creativity. Rather, they prescribe, contain, and direct an individual's behavior. Through socialization, individuals internalize social roles and learn how to carry them out.

This view, however, is mistaken. It suggests that individuals simply take on roles rather than creating or negotiating them. Socialization is a process in which humans can exercise agency; we are not simply passive subjects waiting to be instructed or programmed. Individuals come to understand and assume social roles through an ongoing process of social interaction.

Identity

The cultural settings in which we are born and mature to adulthood influence our behavior, but that does not mean that humans lack individuality or free will. Some sociologists do tend to write about socialization as though this were the case. But such a view is fundamentally flawed—socialization is also at the origin of our very individuality and freedom. In the course of socialization, each of us develops a sense of identity and the capacity for independent thought and action.

Identity is a multifaceted concept—it relates to the understandings people hold about who they are and what is meaningful to them. Some of the main sources of identity include gender, sexual orientation, nationality or ethnicity, and social class. Sociologists typically speak of two types of identity: social identity and self-identity (or personal identity). **Social identity** refers to the characteristics that other people attribute to an individual. These can be seen as markers that indicate who the individual is. At the same time, they place that individual in relation to other individuals who share the same attributes. Examples of social identities include student, mother, lawyer, Catholic, homeless, Asian, dyslexic,

social roles

Socially defined expectations of an individual in a given status or occupying a particular social position. In every society, individuals play a number of social roles, such as teenager, parent, worker, or political leader.

social identity

The characteristics that other people attribute to an individual.

People often exhibit multiple social identities simultaneously, sometimes seemingly conflicting ones.

∧

self-identity

The ongoing process of self-development and definition of our personal identity through which we formulate a unique sense of ourselves and our relationship to the world around us.

gender socialization

The learning of gender roles through social factors such as schooling, the media, and family.

and married. Nearly all individuals have social identities comprising more than one attribute, reflecting the many dimensions of our lives. A person could simultaneously be a mother, an engineer, a Catholic, and a city council member. Although this plurality of social identities can be a potential source of conflict for people, most individuals organize meaning and experience in their lives around a primary identity that is fairly continuous across time and place.

If social identities mark ways in which individuals are the same as others, self-identity (or personal identity) sets us apart as distinct individuals. **Self-identity** refers to the process of self-development through which we formulate a unique sense of ourselves and our relationship to the world around us. The notion of self-identity draws heavily on the work of symbolic interactionists. The individual's constant negotiation with the outside world helps create and shape his or her sense of self. Though the cultural and social environments are factors in shaping self-identity, individual agency and choice are key.

If at one time people's identities were largely informed by their membership in broad social groups, bound by class or nationality, they are now more multifaceted and less stable. Individuals have become more socially and geographically mobile due to processes such as urban growth and industrialization. This has freed people from the tightly knit, relatively homogeneous communities of the past in which patterns were passed down in a fixed way across generations. It has created the space for other sources of personal meaning, such as gender and sexual orientation, to play a greater role in people's sense of identity.

Today we have unprecedented opportunities to create our own identities. We are our own best resources in defining who we are, where we come from, and where we are going. Now that the traditional signposts of identity have become less essential, the social world confronts us with a dizzying array of choices about who to be, how to live, and what to do, without offering much guidance about which selections to make. The decisions we make in our everyday lives—about what to wear, how to behave, and how to spend our time—help make us who we are. Through our capacity as self-conscious, self-aware human beings, we constantly create and re-create our identities, patterns exemplified by the gender transitions of Caitlyn Jenner and Coy Mathis.

Gender Socialization

As we learned in the case of Coy Mathis, gender influences every aspect of daily life. How a child dresses and speaks, the toys the child plays with, the activities in which he or she engages, and how others view the child are all powerfully shaped by gender. Yet, the norms and expectations about how one "should" behave as a boy or girl must be, in part, learned. Agents of socialization play an important role in how children learn gender roles. Let's now turn to the study of **gender socialization**: the learning of gender roles through social factors such as the family and the media.

REACTIONS OF PARENTS AND ADULTS

Many studies have been carried out on the degree to which gender differences are the result of social influences. Classic studies of mother–infant interaction show differences in the treatment of boys and girls even when parents believe their reactions to both are the same. Adults asked to assess the personality of a baby give different answers according to whether they believe the child to be a girl or a boy. In one experiment, five young mothers were observed while interacting with a six-month-old named Beth. They tended to smile at her often and offer her dolls to play with. She was seen as "sweet" with a "soft cry." The reaction of a second group of mothers to a child the same age, named Adam, was noticeably different. The baby was likely to be offered a train or other "male" toys to play with. Beth and Adam were actually the same child, dressed in different clothes (Will, Self, and Datan 1976).

The case of baby Storm Stocker vividly reveals just how deeply entrenched gender and gender socialization are, even in the twenty-first century. Storm's parents, Kathy Witterick and David Stocker, decided to keep their baby's sex a secret, informing only their midwives and two older sons. They dressed Storm in gender-neutral clothing and refused to use gender-specific pronouns like *he* or *she* when describing their baby. They wanted to make sure that others did not treat their child in stereotypically gendered ways, such as those experienced by babies Beth and Adam (Will, Self, and Datan 1976). When announcing Storm's birth, Kathy and David sent out an announcement proclaiming, "We decided not to share Storm's sex for now—a tribute to freedom and choice in place of limitation, a standup to what the world could become in Storm's lifetime." This simple act was met by a firestorm of angry reactions from bloggers, media commentators, and even family and friends (Davis and James 2011).

GENDER LEARNING

Gender learning by infants is almost certainly unconscious. Before a child can accurately label itself as either a boy or a girl, it receives a range of preverbal cues. For instance, male and female adults usually handle infants differently. The cosmetics women use contain scents different from those the baby might learn to associate with males. Systematic differences in dress, hairstyle, and so on provide visual cues for the infant in the learning process. By age two, children have a partial understanding of what gender is. They know whether they are boys or girls, and they can usually categorize others accurately. Not until five or six, however, does a child know that everyone has gender and that sex differences between girls and boys are anatomically based.

Parents play a pivotal role in gender learning, often unintentionally. A child's earliest exposure to what it means to be male or female comes from his or her parents. From the time their children are newborns, parents interact with their daughters and sons differently. They may dress their sons in blue and daughters in pink or speak to girls in softer and gentler tones than they do with boys. One classic study found that parents have different expectations for their sons and daughters as early as one day after they are born, where infant girls are described as "soft" and "pretty" and boys as "energetic" and "strong" (Rubin, Provenzano, and Luria 1974). It's not surprising, then, that as children become toddlers, parents (especially fathers) engage in more rough-and-tumble play with boys and hold more give-and-take conversations with girls (Lytton and Romney 1991). Even parents who are sensitive to gender-equity issues often send subtle messages related to gender—messages that the developing child internalizes. Sex-role stereotypes and subtle messages about appropriate gender-typed behavior are so powerful that even when children are exposed to diverse attitudes and experiences, they will revert to stereotyped choices (Haslett, Geis, and Carter 1992).

New Apps Challenge Kids—and the Gender Binary

The "pink is for girls, blue is for boys" mantra is still pervasive among toy makers today. Stroll down any aisle at a major toy store, and it's clear that girls' toys are still fifty shades of fuchsia, while boys' toys typically come in more masculine colors like blue, gray, or black. But some app developers are working hard to fight this gender divide with fun activities that eschew and even challenge the gender binary.

Take, for example, the app Robot Factory. This game allows children to make their own robots by dragging different parts and limbs onto a body; imaginative users can build insects, animals, extraterrestrials, or humans. Robot Factory is extremely popular, but that wasn't always the case. The first iteration of the app was very different from the final product. The original design included only body parts for "traditional"-looking robots (similar to *Star Wars* droids). When the designers tested the app with children, they were surprised to learn that both boys and girls referred to the robots as "he." Troubled by this, the designers went back to the drawing board and gave the children myriad new options to make robots of all shapes, sizes, genders, and breeds and expanded the color palette from gray to all the hues of the rainbow.

Their reinvention was a hit with boys and girls alike. According to Raul Gutierrez, CEO of Tinybop, the company behind Robot Factory, when children played with the redesigned app, they made robots that looked like boys, girls, and children of ambiguous gender.

Another app designed to appeal to both boys and girls—and to challenge the gender binary in the process—is Toca Hair Salon 2. The app is set in a multicolored hair salon, and children have the opportunity to cut and style customers' hair however they like. They can shave it off, straighten curls, give a perm or an updo, and grow (or shave off) a beard or mustache. The hair salon clientele are a hodgepodge of men and women, boys and girls, and clients whose gender is ambiguous. As Mathilda Engman, the head of consumer products at Toca Boca, explained, hair salons are traditionally thought of as "very targeted toward girls, glamour, looks, and beauty. Ours is the opposite—it's about the creativity of cutting hair and styling hair. . . . Characters have that quirkiness so that they're inviting for everyone" (Miller 2016).

Other apps are designed to show boys and girls that they can choose whatever career they like rather than sticking with gender-typed options. For instance, the app Little Farmers shows both male and female characters using big machinery and farm equipment, while the Cool Careers Dress Up app allows users to choose outfits for women doctors, astronauts, scientists, and computer programmers rather than just fashion models (Gudmundsen 2017).

Experts believe that children's apps present a unique opportunity to challenge the gender binary in ways that other toys cannot. The "packaging" of computer-based games—the tablet or smartphone—is gender neutral, unlike the pink and blue boxes that line the shelves of brick-and-mortar toy stores, says Jess Day of Let Toys Be Toys, a British-based initiative aimed at promoting gender fluidity in children's toys and technologies. Others believe the key to having apps that appeal to boys and girls alike, and that promote gender inclusiveness, is having more apps designed by women and nonbinary persons. According to a 2016 survey of game developers, fully 72 percent identify as men and 23 percent as women; 5 percent identify as transgender, androgynous, or nonbinary or refuse to specify (Statista 2017). Initiatives like Girls Who Code aim to encourage more girls to become interested in computer programming and app design. As app designers become more diverse so, too, will the apps themselves.

Toys today are still typically packaged and marketed along strict gender lines. Some developers, however, are attempting to blur this binary and create apps that appeal to all kids regardless of gender.

Children's toys, picture books, and television programs also tend to follow stereotypical patterns. Toy stores and department stores usually categorize their products by gender. Even toys that seem neutral in terms of gender are not always so in practice. For example, toy kittens and rabbits might be thought of as appropriate for girls, whereas lions and tigers are seen as more appropriate for boys. Similarly, boys are typically expected to dress up like ninjas or superheroes for Halloween, whereas girls are expected to dress up like princesses or other highly "feminine" characters.

But these stark gender divides are fading. For example, in 2015, Target, the nation's largest retailer, stopped dividing their toy sections into "boy" and "girl" sections and also discontinued the pink- and blue-colored walls previously used to draw attention to gender-typed toys. Recognizing that nearly every product for children is gender typed, Target management also stopped dividing up other departments by gender, such as bedding. Boys and girls might be equally likely to want a *Star Wars* or a *Dora the Explorer* comforter (Luckerson 2015). The notion that gender-neutral toys are healthy for all children is rapidly spreading, with more and more manufacturers abandoning gender-typed colors and designs of their toys.

Similarly, children's books and television shows teach important, though subtle, lessons about gender. Scholarly analyses of children's books and TV shows find that girls are highly underrepresented. Sociologist Janice McCabe and colleagues (2011) examined nearly 6,000 books published from 1900 to 2000 and found that males are central characters in 57 percent of children's books published per year, whereas only 31 percent have female central characters. While adult male characters (including male animal characters) were found in every book, only one-third of children's books published in any given year featured central characters that were adult women or female animals.

In her "Pink & Blue" project, photographer JeongMee Yoon records girls' obsession with the color pink. What are the implications of the gender-typed packaging and color coding that we see in children's toys and clothing?

Although there are exceptions, analyses of children's television programs match the findings about children's books. In the most popular cartoons, most leading figures are male, and males dominate the active pursuits. Similar images appear in the commercials that air throughout the programs. For instance, researchers recently examined the ways that boys and girls are portrayed in children's programming on three networks: Disney Channel, Cartoon Network, and Nickelodeon (Hentges and Case 2013). Boy characters outnumbered girls three to two, and there was some evidence that characters were depicted behaving in stereotypical ways, where boys were more likely to be aggressive "rescuers" and girls were more likely to show affection.

RACE SOCIALIZATION

Scholars have long recognized the ways that we learn to be male or female, but how did you learn about your racial or ethnic background? Did your parents ever teach you

race socialization

The specific verbal and nonverbal messages that older generations transmit to younger generations regarding the meaning and significance of race.

Sociologists are interested in how children learn what it means to be a member of a particular racial group, especially one that is devalued.

CONCEPT CHECKS

1. What is social reproduction? What are some specific ways that the four main agents of socialization contribute to social reproduction?

2. According to Mead, how does a child develop a social self?

3. What are the four stages of cognitive development according to Piaget?

4. How do the media contribute to gender role socialization?

5. What are the main components of race socialization?

what it means to be white, Black, Asian, or Latino? Sociologists have recently explored the process of **race socialization**, which refers to the specific verbal and nonverbal messages that older generations transmit to younger generations regarding the meaning and significance of race, racial stratification, intergroup relations, and personal identity (Lesane-Brown 2006).

The research team of sociologist Tony Brown and psychologist Chase Lesane-Brown examined the messages that parents teach and the effects of this socialization on children's lives. Their work rests on the assumption that while ethnic-minority parents (especially Black parents) must socialize their children to be productive members of society, just as white parents do, they also face an additional task: raising children with the skills to survive and prosper in a society that often devalues Blackness. As part of race socialization, Black parents also prepare their children to understand their heritage, their culture, and what it means to belong to a racial group that has historically occupied a low and stigmatized status in the United States (Lesane-Brown 2006).

What exactly do Black parents teach their children about race, race stratification, and race relations? Lesane-Brown, Brown, and colleagues (2005) developed a detailed index capturing the specific messages and lessons that parents pass down to their children. Among the messages encompassed in race socialization are color blindness (e.g., "race doesn't matter"), individual pride (e.g., "I can achieve anything"), group pride (e.g., "I'm proud to be Black"), distrust of other racial or ethnic groups (e.g., "don't trust white people"), and deference to other racial or ethnic groups (e.g., "whites are better than Blacks"). The lessons that Black adolescents and college students found to be the most useful, however, were those that emphasized pride and color blindness, such as "race doesn't matter" and "with hard work, you can achieve anything regardless of race" (Lesane-Brown et al. 2005).

Understanding race socialization will become increasingly important for future cohorts of young people. In our increasingly global society, children and young people will need to develop the skills and capacities to negotiate multicultural contexts in their everyday lives (Priest et al. 2014). Parents, teachers, and other agents of socialization must also promote positive racial attitudes, counter negative attitudes, and enable effective responses to racism when it occurs. Although race socialization has historically focused on raising Black children to fit in and get ahead in a racist world, scholars today recognize that white children, too, should be socialized to recognize and fight racism when they see it unfold (Priest et al. 2014).

What Are the Five Major Stages of the Life Course?

The transitions that individuals pass through during their lives may be biologically fixed—from childhood to adulthood and eventually to death. But the stages of the human **life course** are social as well as biological. They are influenced by culture and by the material circumstances of people's lives. For example, in the modern West, death is usually thought of in relation to old age because most people enjoy a life span of seventy-five years or more. In traditional societies of the past more people died at younger ages than survived to old age.

Childhood

In modern societies, childhood is a clear and distinct stage of life between infancy and adolescence. Yet the concept of childhood, like so many other aspects of social life today, has come into being only over the past two or three centuries. In earlier societies, the young moved directly from a lengthy infancy into working roles within the community. French historian Philippe Ariès (1965) argued that "childhood," conceived of as a separate phase of development, did not exist in medieval times. In the paintings of medieval Europe, children are portrayed as little adults, with mature faces and the same style of dress as their elders.

Until the early twentieth century, in the United States and most other Western countries, children were put to work at what now seems a very young age. There are countries in the world today, in fact, where young children are engaged in full-time work, sometimes in physically demanding circumstances (for example, in coal mines). The ideas that children have distinctive rights and that child labor is morally wrong are quite recent developments.

Because of the prolonged period of childhood that we recognize today, modern societies are in some respects more child centered than traditional ones. Parents are viewed as the sole protectors of their children, and parents who behave in ways that may be considered hurtful to their children are judged harshly. For instance, not all the parents at Eagleside Elementary were supportive of Kathryn and Jeffrey Mathis's decision to allow Coy to identify as a girl.

It seems possible that as a result of changes currently occurring in modern societies, the separate character of childhood is diminishing. Some observers have suggested that children now grow up too fast. Even small children may watch the same television programs as adults, thereby becoming much more familiar early on with the adult world than preceding generations did.

The Teenager

The idea of the teenager also didn't exist until the early twentieth century, when compulsory education and child labor laws were enacted. Prior to that time, teenagers were not required to attend school, so adolescence was a time for working in fields and factories and for marrying and bearing children. Today, by contrast, adolescence is considered a time to learn, grow, and make choices about the kind of adult one wants to someday become.

The biological changes involved in puberty (the point at which a person becomes capable of adult sexual activity and reproduction) are universal. Yet in many cultures, these

Learn the various stages of the life course, and see the similarities and differences among different cultures and historical periods.

life course
The various transitions and stages people experience during their lives.

This *Madonna and Child*, painted in the thirteenth century by Duccio di Buoninsegna, depicts the infant Jesus with a mature face. Until recently, children in Western society were viewed as little adults.

physical changes do not produce the degree of emotional turmoil and uncertainty often found among teens in modern societies. In cultures that celebrate "rites of passage," or distinct ceremonies that signal a person's transition to adulthood, the process of psychosexual development generally seems easier to negotiate. Adolescents in such societies have less to "unlearn" because the pace of change is slower. There is a time when children in Western societies are required to be children no longer: to put away their toys and break with childish pursuits. In traditional cultures, where children are already working alongside adults, this process of unlearning is normally much less jarring.

In Western societies, teenagers are betwixt and between, navigating the often-complicated space between childhood and adulthood: They often try to act like adults, but they are treated by law as children. Pop culture promotes sexy clothing among teens yet frowns upon teenage sexual activity. Teens may wish to go to work and earn money as adults do, but they are required to stay in school.

Young Adulthood

Young adulthood, typically defined as roughly ages twenty to thirty, is a stage of exploration, often before one settles on a permanent job, spouse, or home. This stage of personal and sexual development is unique to modern societies (Furstenberg et al. 2004). Scholars have observed a "delayed transition to adulthood" among young people in the late twentieth and early twenty-first centuries. Particularly among more affluent groups, people in their early twenties take the time to travel and explore sexual, political, and religious affiliations; try out different careers; and date and live with several romantic partners. The importance of this postponement of the responsibilities of full adulthood is likely to increase, given the extended period of education many people now undergo.

Although it is difficult to pinpoint precisely when one makes the "transition to adulthood," one team of researchers identified five benchmarks of adulthood: leaving one's parents' home, finishing school, getting married, having a child, and achieving financial independence. In 1960, fully 65 percent of men and 77 percent of women had achieved all five benchmarks by age thirty. By contrast, only 25 percent of men and 39 percent of women had done so in 2010 (Furstenberg and Kennedy 2013; Furstenberg et al. 2004). These statistics clearly show that the transition to adulthood is being delayed today and that some benchmarks historically considered as signifiers of adulthood, such as becoming a parent, may now be less central to one's identity as an adult (Figure 3.1).

Midlife or "Middle Age"

Most young adults in the West today can look forward to a life stretching right through to old age. In premodern times, few could anticipate such a future with much confidence. Death through sickness or injury was much more frequent among all age groups than it is today, and women faced a high rate of mortality in childbirth. Given these advances in life expectancy, a "new" life course stage has been recognized in the twentieth century: midlife, or middle age (Cohen 2012).

Midlife, the stage between young adulthood and old age, is generally believed to fall between ages forty-five and sixty-five. However, midlife is distinct from other life course stages in that there is not an "official" or legal age of entry. For example, American youth become legal adults at age eighteen, whereas age sixty-five is generally believed to signify the transition to old age. One's entry to midlife, by contrast, tends to be signified by the social roles one adopts (or relinquishes). While some scholars believe that menopause, or the loss of reproductive potential, signals women's transition to midlife, others believe that for both men and women, midlife is marked by transitions such as the "empty nest" stage.

Midlife is also a psychological turning point where men and women may assess their past choices and accomplishments and make new choices that prepare them for the second half of life. Keeping a forward-looking outlook in middle age has taken on a particular importance in modern societies. Most people do not expect to be doing the same thing their whole lives, as was the case for the majority in traditional cultures. For example, women who spent their early adulthood raising a family and whose children have left home may feel free to pursue new personal goals, whereas men who stayed at financially stable jobs while supporting their young families may choose to pursue their earlier career dreams (Lachman 2001).

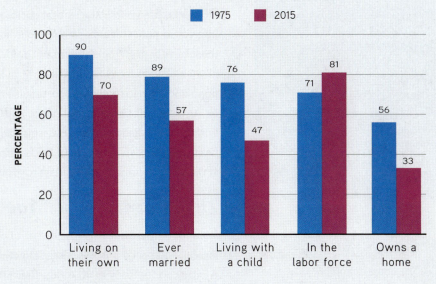

FIGURE 3.1

Thirty-Year-Olds: 1975 vs. 2015

Source: U.S. Bureau of the Census 2016b.

Later Life

Surviving until the life course stage of "elder" in a traditional culture often marked the pinnacle of an individual's status. Older people were normally accorded a great deal of respect and had a say over matters of importance to the community. Within families, the authority of both men and women typically increased with age. In industrialized societies, by contrast, older people tend to lack authority within both the family and the wider community. Having retired from the labor force, they may be poorer than ever before in their lives. At the same time, there has been a great increase in the proportion of the population over age sixty-five. In 1900, only about one in twenty-five people in the United States was sixty-five or older. Since then, the proportion of older adults has more than tripled: In 2016, about one in every seven Americans was over sixty-five.

No longer living with their children and often having retired from paid work, some older people find it difficult to make the final period of their life rewarding. It used to be thought that those who successfully cope with old age do so by turning to their inner resources, becoming less interested in material rewards. Although this may often be true, it seems likely that in a society in which many are physically healthy in old age, an outward-looking view will become more prevalent. With advances in medical technologies, older adults are living longer and healthier than ever before. These extensions in life span have been accompanied by expanded opportunities for lifelong learning, with many older adults learning new skills and pursuing new leisure activities. Those in retirement might find renewal in what has been called the "third age," in which a new phase of education begins.

CONCEPT CHECKS

1. What is meant by the term *life course*?

2. What are the five stages of the life course, and what are some defining features of each stage?

3. How is midlife different from the life course stages of childhood and later life?

How Do People Age?

Of all the life course stages that sociologists study, older adults are the group of greatest interest to policymakers. Why? Older adults, or individuals age sixty-five or older, are the most rapidly growing segment of the U.S. population (Figure 3.2); as such, they will create new challenges for American society. In 2016, older adults represented 15 percent of the U.S. population; by 2040, that proportion is expected to rise to nearly 22 percent (Administration on Aging 2018). Growing old can be a fulfilling and rewarding experience, or it can be filled with physical distress and social isolation. For most older Americans, the experience of aging lies somewhere in between. In this section, we delve into the meaning of being old and look at the ways in which people adapt to growing old, at least in the eyes of sociologists.

The Meanings of "Age"

What does it mean to age? **Aging** can be defined as the combination of biological, psychological, and social processes that affect people as they grow older (Abeles and Riley 1987; Atchley 2000; Riley et al. 1988). These three processes suggest the metaphor of three different, although interrelated, developmental "clocks": (1) a biological one, which refers to the physical body; (2) a psychological one, which refers to the mind and mental capabilities; and (3) a social one, which refers to cultural norms, values, and role expectations having to do with age. Our notions about the meaning of age are rapidly changing, both because recent research is dispelling many myths about aging and because advances in nutrition and health have enabled many people to live longer, healthier lives than ever before.

Growing Old: Trends and Competing Sociological Explanations

Social gerontologists, or social scientists who study aging, have offered a number of theories regarding the nature of aging in U.S. society. Some of the earliest theories emphasized individual adaptation to changing social roles as a person grows older. Later theories focused on how society shapes the social roles of older adults, often in inequitable ways. The most recent theories have been more multifaceted, focusing on the ways in which older persons actively create their lives within specific institutional contexts (Hendricks 1992).

THE FIRST GENERATION OF THEORIES: FUNCTIONALISM

The earliest theories of aging reflected the functionalist approach that was dominant in sociology during the 1950s and 1960s. They emphasized how individuals adjusted to changing social roles as they aged and how those roles were useful to society. The earliest theories often assumed that aging brings with it physical and psychological decline and that changing social roles have to take this decline into account (Hendricks 1992).

Talcott Parsons, one of the most influential functionalist theorists of the 1950s, argued that U.S. society needs to find roles for older persons consistent with advanced age. He expressed concern that the United States, with its emphasis on youth and its avoidance of death, had failed to provide roles that adequately drew on the potential wisdom and maturity of its older citizens. Moreover, given the graying of U.S. society that was evident even in Parsons's time, he believed that this failure could well lead to older people becoming discouraged and

---

Understand that aging is a combination of biological, psychological, and social processes. Consider key theories of aging, particularly those that focus on how society shapes the social roles of older people and that emphasize aspects of age stratification.

aging

The combination of biological, psychological, and social processes that affect people as they grow older.

social gerontologists

Social scientists who study older adults and life course influences on aging processes.

FIGURE 3.2

Growth of the Older Population in the U.S. by Age Group, 1900-2050

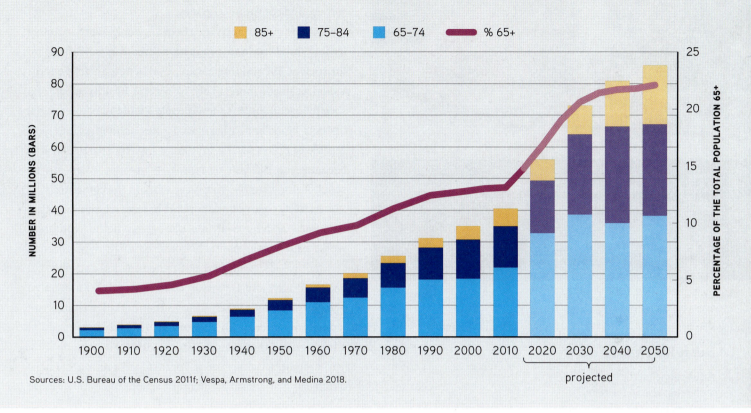

Sources: U.S. Bureau of the Census 2011f; Vespa, Armstrong, and Medina 2018.

alienated from society. To achieve a "healthy maturity," Parsons (1960) argued, older adults need to adjust psychologically to their changed circumstances, while society needs to redefine the social roles of older persons. Their former roles (such as work) have to be abandoned, while new forms of productive activity (such as volunteer service) need to be identified.

Parsons's ideas set the foundation for **disengagement theory**, the notion that it is functional for society to remove people from their traditional roles when they become older, thereby freeing up those roles for other, younger persons (Cumming and Henry 1961; Estes, Binney, and Culbertson 1992). According to this perspective, given the increasing frailty, illness, and dependency of older people, it becomes increasingly dysfunctional for them to occupy traditional social roles they are no longer capable of adequately fulfilling. Older adults, therefore, should retire from their jobs, pull back from civic life, and eventually withdraw from other activities as well. Disengagement is assumed to be functional for the larger society because it opens up roles for younger people, who presumably will carry them out with fresh energy and new skills. Disengagement is also assumed to be functional for older persons because it enables them to take on less taxing roles consistent with their advancing age and declining health.

Although there is some intuitive appeal to disengagement theory, the idea that older people should completely disengage from the larger society is based on the outdated stereotype that old age involves frailty and dependence. As a result, no sooner did the theory appear than these very assumptions were challenged, often by some of the theory's original proponents (Cumming 1963, 1975; Hendricks 1992; Henry 1965; Hochschild 1975; Maddox

disengagement theory

A functionalist theory of aging that holds that it is functional for society to remove people from their traditional roles when they become elderly, thereby freeing up those roles for others.

Rewarding activities, such as volunteer work, can enhance health and well-being in later life.

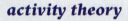

activity theory

A functionalist theory of aging that maintains that busy, engaged people are more likely to lead fulfilling and productive lives.

continuity theory

Theoretical perspective on aging that specifies that older adults fare best when they participate in activities consistent with their personalities, preferences, and activities from earlier in life.

social conflict theories of aging

Arguments that emphasize the ways in which the larger social structure helps to shape the opportunities available to older adults. Unequal opportunities are seen as creating the potential for conflict.

1965, 1970). These challenges gave rise to two distinct yet related functionalist theories of aging, which drew conclusions quite opposite to those of disengagement theory: activity and continuity theories.

According to **activity theory**, people who are busy leading fulfilling and productive lives can be functional for society. The guiding assumption is that an active individual is much more likely to remain healthy, alert, and socially useful. In this view, people should remain engaged in their work and other social roles as long as they are capable of doing so. If a time comes when a particular role becomes too difficult or taxing, then other roles can be sought—for example, volunteer work in the community.

Activity theory finds support in research showing that continued activity well into old age—whether volunteer work, paid employment, hobbies, or visits with friends and family—is associated with good mental and physical health (Birren and Bengtson 1988; Rowe and Kahn 1987; Schaie 1983). Yet critics observe that not all activities are equally valuable, giving rise to **continuity theory**. This theory specifies that older adults fare best when they participate in activities that are consistent with their personalities, preferences, and activities from earlier in life (Atchley 1989). For instance, a retired elementary school teacher may find volunteering at a local elementary school to be much more satisfying than playing bingo at a local community center.

Critics of functionalist theories of aging argue that these theories emphasize the need for older adults to adapt to existing conditions, either by disengaging from socially useful roles or by actively pursuing them, but that they do not question whether the circumstances older adults face are just. In response to this critique, another group of theorists arose—those growing out of the social conflict tradition (Hendricks 1992).

THE SECOND GENERATION OF THEORIES: SOCIAL CONFLICT

Unlike their predecessors, who emphasized the ways that older adults could be integrated into the larger society, the second generation of theorists focused on sources of **social conflict** between older persons and society (Hendricks 1992). Like other theorists who were studying social conflict in U.S. society during the 1970s and early 1980s, these theorists stressed the ways in which the larger social structure helped shape the opportunities available to older persons; unequal opportunities were seen as creating the potential for conflict.

According to this view, many of the problems of aging—such as poverty, poor health, and inadequate health care—are systematically produced by the routine operation of social institutions. A capitalist society, the reasoning goes, favors those who are most economically powerful. Although certainly some older adults have "made it" and are set for life, many have not—and these people must fight to get even a meager share of society's scarce resources. Among persons age sixty-five and older, those who fare worst tend to include women, low-income people, and ethnic minorities (Atchley 2000; Estes 1986, 1991; Hendricks 1992; Hendricks and Hendricks 1986). For example, poverty rates among older adults have plummeted over the past sixty years, with roughly 9 percent of older adults living in poverty today, compared with 35 percent in 1959 (Semega, Fontenot, and Kollar 2017). However, even today, the poverty rate among older adults is as high as 40 percent among unmarried Black and Hispanic older women (Carr 2010).

THE THIRD GENERATION OF THEORIES: LIFE COURSE PERSPECTIVES

Life course theorists reject what they regard as the one-sided emphases of both functionalist and conflict theories, where older adults are viewed either as merely adapting to the larger society (functionalism) or as victims of the stratification system (social conflict). Rather, **life course theory** maintains that older persons play an active role in determining their own physical and mental well-being, while recognizing the constraints imposed by social structural factors.

According to this theory, the aging process is shaped by historical time and place; factors such as wars, economic shifts, and the development of new technologies shape how people age. Yet this perspective also emphasizes agency, where individuals make choices that reflect both the opportunities and the constraints facing them. The most important theme of the life course perspective is that aging is a lifelong process: Relationships, events, and experiences of early life have consequences for later life.

CONCEPT CHECKS

1. What factors or processes should we keep in mind when studying aging or the meaning of being old?

2. Summarize the three theoretical frameworks used to describe the nature of aging in U.S. society.

3. What are the main criticisms of functionalism and conflict theory?

What Are the Challenges of Aging in the U.S.?

Evaluate the experience of growing old in the United States. Identify the physical, emotional, and financial challenges older adults face.

Older individuals make up a highly diverse category about whom few broad generalizations can be made. For one thing, the aged population reflects the diversity of U.S. society that we've made note of elsewhere in this textbook: They are rich, poor, and in between; they belong to all racial and ethnic groups; they live alone and in families of various sorts; they vary in their political values; and they are LGBTQ as well as heterosexual. Furthermore, like other Americans, they are diverse with respect to health: Although some suffer from mental and physical disabilities, most lead active, independent lives.

Race has a powerful influence on the lives of older persons. Whites, on average, live nearly four years longer than African Americans, largely because Blacks have much greater odds of dying in infancy, childhood, and young adulthood. Blacks also have much higher rates of poverty and, therefore, are more likely to suffer from inadequate health care compared with whites. As a result, a much higher percentage of whites have survived past age sixty-five compared with other racial groups. The combined effect of race and sex is substantial. In 2015, the life expectancy for a white woman was eighty-one, compared to seventy-two for a Black man (Centers for Disease Control and Prevention 2017a).

Currently about 14 percent of the older population in the United States are foreign born (U.S. Bureau of the Census 2018a). In California, New York, Hawaii, and other states that receive large numbers of immigrants, as many as one-fifth of the older population were born outside the United States (Federal Interagency Forum on Aging-Related Statistics 2013). Integrating older immigrants into U.S. society poses special challenges: Most either do not speak English well or do not speak it at all. Some are highly educated, but most are not. Most lack a retirement income, so they must depend on their families or public assistance.

life course theory

A perspective based on the assumptions that the aging process is shaped by historical time and place; individuals make choices that reflect both opportunities and constraints; aging is a lifelong process; and the relationships, events, and experiences of early life have consequences for later life.

young old

Sociological term for persons between the ages of sixty-five and seventy-four.

---

old old

Sociological term for persons between the ages of seventy-five and eighty-four.

---

oldest old

Sociological term for persons age eighty-five and older.

Finally, as people live to increasingly older ages, they are diverse in terms of age itself. It is useful to distinguish among different age categories of the 65+ population, such as the **young old** (ages sixty-five to seventy-four), the **old old** (ages seventy-five to eighty-four), and the **oldest old** (age eighty-five and older). The young old are most likely to be economically independent, healthy, and active; the oldest old—the fastest-growing segment of the age 65+ population—are most likely to encounter difficulties such as poor health, financial insecurity, and isolation (U.S. Bureau of the Census 2011f). Not only are these differences due to the effects of aging but they also reflect cohort differences. The young old came of age during the post–World War II period of strong economic growth and benefited as a result: They are more likely to be educated; to have acquired wealth in the form of a home, savings, or investments; and to have had many years of stable employment. These advantages are much less likely to be enjoyed by the oldest old, partly because their education and careers began at a time when economic conditions were not so favorable (Treas 1995).

What is the experience of growing old in the United States? Although older persons do face some special challenges, most older people lead relatively healthy, satisfying lives. Still, one national survey found a substantial discrepancy between what most Americans under sixty-five thought life would be like and the actual experiences of those who had. We next examine some of the common problems that older adults confront and identify the factors that put older persons at risk for these problems.

Health Problems

The prevalence of chronic disabilities among the older population has declined in recent years, and most older adults rate their health as reasonably good and free of major disabilities (Federal Interagency Forum on Aging-Related Statistics 2016). Still, older people suffer from more health problems than most younger people, and health difficulties often increase with advancing age. In 2015, nearly one-third of all noninstitutionalized persons age sixty-five and older reported suffering from arthritis, 29 percent had heart disease, and 27 percent had diabetes (Administration on Aging [AOA] 2018). The percentage of people needing help with daily activities increases with age: Only 3 percent of adults between the ages of sixty-five and seventy-four report needing help with personal care, yet this figure rises to 9 percent for people between seventy-five and eighty-four and to 22 percent for people over eighty-five (AOA 2018).

Paradoxically, there is some evidence that the fastest-growing group of the aged population, the oldest old (those eighty-five and older), tend to enjoy relative robustness, which partially accounts for their having reached their advanced age. This is possibly one of the reasons that health care costs for a person who dies at ninety are about one-third of those for a person who dies at seventy (Angier 1995). Unlike many other Americans, persons age sixty-five and older are fortunate in having access to public health insurance (Medicare) and, therefore, medical services. The United States, however, stands virtually alone among the industrialized nations in failing to provide adequately for the complete health care of its most senior citizens (Hendricks and Hatch 1993).

Nearly all (93 percent) of older Americans are covered to some extent by Medicare (AOA 2018). But because this program covers about half of the total health care expenses of individuals age sixty-five and older, 62 percent of older people supplement Medicare with another type of insurance (Barnett and Berchick 2017). The rising costs of private

Graying of the World

The world population is aging rapidly, or "graying." In 2017, nearly 9 percent of the global population was over 65; that proportion is expected to rise to 16 percent by 2050. Graying is the result of two long-term trends: people having fewer children and living longer.

Older population by country

(proportion 65+)

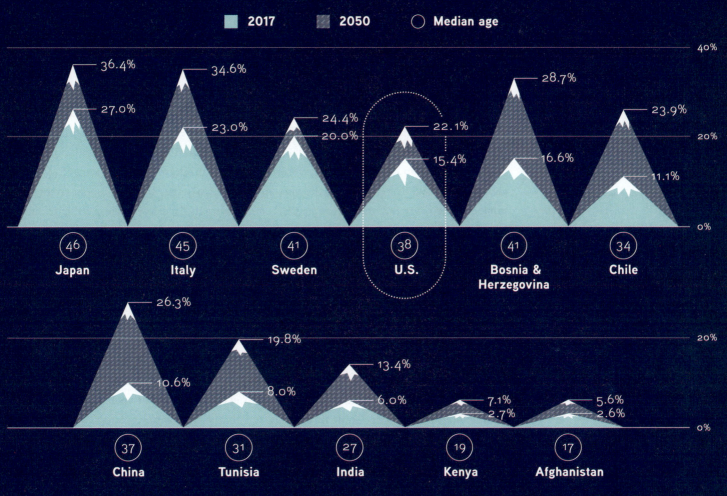

- 2017
- 2050
- ◯ Median age

Japan (46): 36.4% (2050), 27.0% (2017)
Italy (45): 34.6% (2050), 23.0% (2017)
Sweden (41): 24.4% (2050), 20.0% (2017)
U.S. (38): 22.1% (2050), 15.4% (2017)
Bosnia & Herzegovina (41): 28.7% (2050), 16.6% (2017)
Chile (34): 23.9% (2050), 11.1% (2017)

China (37): 26.3% (2050), 10.6% (2017)
Tunisia (31): 19.8% (2050), 8.0% (2017)
India (27): 13.4% (2050), 6.0% (2017)
Kenya (19): 7.1% (2050), 2.7% (2017)
Afghanistan (17): 5.6% (2050), 2.6% (2017)

Older population in the United States

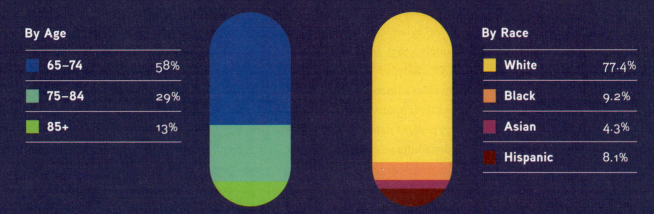

By Age

65–74	58%
75–84	29%
85+	13%

By Race

White	77.4%
Black	9.2%
Asian	4.3%
Hispanic	8.1%

Sources: UN Department of Economic and Social Affairs, Population Division 2017; U.S. Bureau of the Census 2018e.

insurance have unfortunately made this option impossible for a growing number of older adults. On average, older Americans paid nearly $6,000 out of pocket for health care in 2016—an increase of 38 percent since 2006. Despite Medicare, health care costs still compose 13 percent of older adults' total expenses (AOA 2018).

When older adults become physically unable to care for themselves, they may move into assisted-living facilities, long-term care facilities, or nursing homes. Only about 1 percent of people age sixty-five to seventy-four live in institutional settings such as nursing homes, a figure that rises to 3 percent among people seventy-five to eighty-four and to 9 percent for those over eighty-five (AOA 2018). Because the average (median) cost of a private room in a nursing home is now over $97,000 a year (Genworth 2017), the nonpoor older people who require such institutionalization may find their lifetime savings quickly depleted.

Even if the problems of social isolation, prejudice, physical abuse, and health declines affect only a relatively small proportion of all older persons, the raw numbers of people facing these challenges will increase as the large baby boom cohort enters old age. The baby boom cohort refers to the 75 million people born between 1946 and 1964 in the United States; the oldest boomers turned sixty-five in 2011. This large population will present unforeseen challenges for government-funded programs, such as Social Security and Medicare, while reinventing the very meaning of old age. The baby boom cohort is more educated than any generation that has come before it; American society will no doubt benefit by incorporating rather than isolating future cohorts of older adults and drawing on their considerable reserves of experience and talent.

Elder Abuse

Mistreatment and abuse of older adults may take many forms, including physical, sexual, emotional, or financial abuse; neglect; or abandonment. Elder mistreatment is very difficult to measure and document. Older adults who are embarrassed, ashamed, or fearful of retaliation by their abusers may be reluctant to report such experiences. As a result, official prevalence rates are low. Worldwide, it is estimated that between 4 and 6 percent of older adults experience some form of abuse at home. The National Social Life, Health, and Aging Project (NSHAP) is the first nationally representative population-based survey to ask older adults about their recent experiences of mistreatment. Laumann and colleagues (2008) found that 9 percent of older adults reported verbal mistreatment, 3.5 percent reported financial mistreatment, and less than 1 percent reported physical mistreatment by a family member. Women and persons with physical disabilities were most likely to report abuse.

It is widely believed that abuse results from the anger and resentment that adult children feel when confronted with the need to care for their infirm parents (King 1984; Steinmetz 1983). Most studies have found this to be a false stereotype, however. In the NSHAP study, most mistreatment was perpetrated by someone other than a member of the elder's immediate family. Of those who reported verbal mistreatment, 26 percent named their spouse or romantic partner as the perpetrator, 15 percent named their child, and 57 percent named someone other than a spouse, parent, or child. Similarly, 56 percent of older adults who reported financial mistreatment said that someone other than a family member was responsible; of family members, though, children were mentioned most often, while spouses were rarely named (Laumann, Leitsch, and Waite 2008).

Social Isolation

One common stereotype about older adults is that they are socially isolated. This is not true of the majority of older people: Four out of five older persons have living children, and the vast majority can rely on their children for support, if necessary (Federal Interagency Forum on Aging-Related Statistics 2013). More than nine out of ten adult children say that maintaining parental contact is important to them, including the provision of financial support if it is needed (Suitor et al. 2011). The reverse is also true: Many studies have found that older parents continue to provide support for their adult children, particularly during times of difficulty, such as divorce. Most older adults have regular contact with their children and live near them; about 85 percent of older persons live within an hour of one of their children. However, relatively few live with their children. This arrangement is exactly what they want; studies repeatedly show that older adults—even those with serious health limitations—prefer to remain independent and reside in their own homes. They want "intimacy at a distance" (Gans and Silverstein 2006).

Despite its subjective nature, loneliness is a serious problem for many older adults; it is linked to sleep problems, poor cardiovascular health, and elevated blood pressure, each of which carries long-term consequences for mortality risk (Cacioppo et al. 2002). Loneliness also may be a particularly serious social problem for older adults in future generations. Smaller families and increased rates of divorce and childlessness among future cohorts of older adults may create a context where older persons maintain objectively fewer relationships (Manning and Brown 2011). More important, however, some have argued that current cohorts of midlife adults have unrealistically high expectations for what their social relationships should provide (e.g., one's partner should be one's "soul mate"); if these lofty expectations go unfulfilled, then older adults may report higher levels of emotional loneliness as well (Carr and Moorman 2011).

In 2017, 34 percent of older women and 20 percent of men lived alone (AOA 2018). Women are more likely than men to live alone because they are more likely to outlive their spouses; they are also less likely than men to remarry following widowhood or divorce. While 70 percent of older men are married, the same can be said of only 46 percent of older women. Women are also more likely than men to be widowed. In 2017, 33 percent of older women and only 11 percent of older men were widowed (AOA 2018). Part of the reason that older women are less likely than men to remarry is the highly skewed sex ratio among older adults. In 2016, there were 126 older women for every 100 older men; for those eighty-five or older, this ratio increases to 187 women for every 100 men (AOA 2018). The fact that women outlive men means that older women are more likely to experience problems of isolation. These problems are compounded by cultural values that make growing old gracefully easier for men than for women. In U.S. culture, youth and beauty are viewed as especially desirable qualities for women. Older men, on the other hand, are more likely to be valued for their material success. As a result, older single men are much more likely to find a mate than older women because the pool of eligible mates for older men is more likely to include potential partners who are many years younger.

The mere presence of social relationships does not ward off loneliness. An estimated 29 percent of older married persons report some symptoms of loneliness; this pattern is particularly common among persons whose spouses are ill, who have a dissatisfying (or nonexistent) sexual relationship, or who have infrequent or conflicted conversations (AARP 2012; de Jong Gierveld et al. 2009). As de Jong Gierveld and Havens (2004) noted,

> ## Activities Director at a Nursing Home
>
> People are living longer than ever before in the United States and worldwide. By the year 2030, one in five Americans will be age sixty-five or older. As a result, nearly every profession—from medicine to marketing—will need at least a basic knowledge of aging. For some professions, such as geriatric medicine or nursing, it's crucial for workers to understand the biology of aging. Other professions, such as geriatric social worker, nursing home administrator, activities director at a senior center, or even a personnel officer charged with hiring older workers, require a strong grasp of the challenges facing many older adults, such as ageism and social isolation.
>
> The activities director at a nursing home or long-term care facility may find the themes and concepts of life course sociology especially relevant to his or her work. There are currently more than 30,000 long-term care facilities in the United States, which are home to more than 1 million residents (Federal Interagency Forum on Aging-Related Statistics 2016). These residents, many of whom moved in after living in their own homes for decades, may find their moves difficult. The facilities staff try to ease the transition and make life as fulfilling as possible for its older residents by providing them with a schedule of age-appropriate activities. These activities might include lectures from local college professors, outings to local museums and concerts, and book clubs for older adults with good physical and cognitive health. In dementia-care wings or for patients who are starting to experience steep physical or cognitive declines, the activities may be more basic, such as chair-based fitness classes or movie screenings. The activities director plans this full slate of activities to keep residents engaged and entertained.
>
> Think about the theories you've learned in this chapter and the research findings you've read about; these lessons are very helpful to activities directors. Sociological

loneliness depends on one's "standards as to what constitutes an optimal network of relationships." That is, it's getting less support than we want rather than the objective number of social ties that matters when it comes to loneliness.

Prejudice

ageism

Discrimination or prejudice against a person on the basis of age.

Discrimination on the basis of age, or **ageism**, is now against federal law. The Age Discrimination in Employment Act of 1967 (ADEA) protects job applicants and employees forty years of age and older from discrimination on the basis of age in hiring, firing, promotion, and pay. Nonetheless, prejudices based on false stereotypes are common. Older adults are frequently seen as perpetually lonely, sad, infirm, forgetful, dependent, senile, old-fashioned, inflexible, and embittered (Palmore 2015).

There are a number of reasons for such prejudice. The American obsession with youthfulness, reflected in popular entertainment and advertising, leads many younger people to disparage their elders, frequently dismissing them as irrelevant. The new information technology culture undoubtedly reinforces these prejudices because youthfulness and computer abilities seem to go hand in hand. In the fast-paced world of Twitter and Snapchat, young people may come to view older adults as anachronistic. These stereotypes are harmful, especially if they translate into discriminatory or ageist treatment.

studies tell us that data do not support the claims of disengagement theory, so few activities directors would insist that their oldest residents simply sit quietly all day, waiting to die. Rather, a knowledge of continuity theories helps activities directors to plan recreation and enrichment programs that match their residents' preferences. For instance, if the activities director notices that many residents are retired schoolteachers, he or she may try to arrange regular volunteer or tutoring activities at a local school. A knowledge of continuity theory also helps the activities director to advise family members as they move their loved ones into his or her new room at the facility. He or she may suggest ways to set up the furniture or decorations in the room so that it resembles the place the older adult called home prior to his or her relocation.

Yet activities directors' jobs are not all fun and games. Another key task is supervising staff. A knowledge of sociology is especially helpful when devising programs and procedures to ensure that direct care staff, such as nursing assistants and doctors, do not behave in an ageist manner toward residents. Ageism refers to discriminatory or unkind treatment on the basis of one's age. It often takes subtle forms, such as health care providers' tendency to use "elderspeak," or the infantilizing language and tone that people may use when speaking to older adults. A singsong voice, unnecessarily loud or slow explanations, overly simplistic language, and seemingly benign greetings like "how are you today, young lady?" are viewed as condescending and disingenuous by older adults and gerontologists alike (Leland 2008). Sociological studies further show that when older adults are treated like children, their mental and even physical health may decline. Sociology makes us sensitive to the fact that even seemingly harmless microaggressions can compromise the well-being of older adults. In these many ways, sociology provides an important knowledge base that will be critical in helping to meet the needs of the large and rapidly growing population of older adults in the United States and worldwide.

The U.S. population is rapidly graying, creating a demand for workers who can effectively work with and care for older adults. An activities director at a long-term care facility would benefit from an understanding of sociological theories of aging.

The actions of the aging baby boom cohort may help chip away at outdated and inaccurate notions of what old age is. Older adults are becoming an increasingly large presence online. In 2018, two-thirds of Americans sixty-five and older used the Internet, and more than one-third (37 percent) used social networking sites such as Facebook (Pew Research Center 2018b). Experts agree that baby boomers may play a critical role in further helping to dissolve stereotypes of the frail, senile older adult.

In many ways, older adults face some of the same problems experienced by Coy Mathis and Caitlyn Jenner, whom we met at the beginning of the chapter. Young boys are expected to be strong and tough—not "sissies." Girls, but not boys, are believed to like pink, wear skirts, and behave in a feminine way. Likewise, older adults are stereotyped as being old-fashioned and out of it. Even though these two examples are very different—one involving the youngest stage of the life course, the other involving the oldest—they both reveal the power of social expectations. However, social expectations can change over time, and as old "cohorts" or generations die out and are replaced with younger generations holding more contemporary beliefs, we might expect that stereotypes—whether based on age or gender—will slowly fade away.

CONCEPT CHECKS

1. Contrast the young old, old old, and oldest old.

2. Describe at least three common problems that older Americans often confront.

3. Define ageism and provide one explanation for this form of prejudice.

The Big Picture

Socialization, the Life Course, and Aging

Thinking Sociologically

1. Concisely review how an individual becomes a social person according to the two leading theorists discussed in this chapter: George Herbert Mead and Jean Piaget. Which of these two theories seems more appropriate and correct to you? Explain why.

2. Conforming to gender-typed expectations regarding clothing, hair, and other aspects of personal appearance is one of many things we do as a result of socialization. Suggest how the family, peers, schools, and mass media help establish the desire to conform with (or reject) typically "male" versus "female" expectations for appearance. Of the preceding, which force is the most persuasive? Explain.

How Are Children Socialized?

p. 73

Learn about socialization (including gender socialization), and know the most important agents of socialization.

What Are the Five Major Stages of the Life Course?

p. 85

Learn the various stages of the life course, and see the similarities and differences among different cultures and historical periods.

How Do People Age?

p. 88

Understand that aging is a combination of biological, psychological, and social processes. Consider key theories of aging, particularly those that focus on how society shapes the social roles of older people and that emphasize aspects of age stratification.

What Are the Challenges of Aging in the U.S.?

p. 91

Evaluate the experience of growing old in the contemporary United States. Identify the physical, emotional, and financial challenges older adults face.

Terms to Know

Concept Checks

socialization • social reproduction • resocialization

cognition • social self • self-consciousness • generalized other • looking-glass self • sensorimotor stage • preoperational stage • egocentric • concrete operational stage • formal operational stage • agents of socialization • nuclear family • hidden curriculum • peer group • social roles • social identity • self-identity • gender socialization • race socialization

1. What is social reproduction? What are some specific ways that the four main agents of socialization contribute to social reproduction?
2. According to Mead, how does a child develop a social self?
3. What are the four stages of cognitive development according to Piaget?
4. How do the media contribute to gender role socialization?
5. What are the main components of race socialization?

life course

1. What is meant by the term *life course*?
2. What are the five stages of the life course, and what are some defining features of each stage?
3. How is midlife different from the life course stages of childhood and later life?

aging • social gerontologists • disengagement theory • activity theory • continuity theory • social conflict theories of aging • life course theory

1. What factors or processes should we keep in mind when studying aging or the meaning of being old?
2. Summarize the three theoretical frameworks used to describe the nature of aging in U.S. society.
3. What are the main criticisms of functionalism and conflict theory?

young old • old old • oldest old • ageism

1. Contrast the young old, old old, and oldest old.
2. Describe at least three common problems that older Americans often confront.
3. Define ageism and provide one explanation for this form of prejudice.

4

Social Interaction and Everyday Life in the Age of the Internet

Who Owns a Smartphone?

p. 117

Notre Dame football star Manti Te'o was widely ridiculed when news broke that Lennay Kekua, his girlfriend of nearly a year, never existed. Kekua, constructed as part of an elaborate "catfishing" hoax by an acquaintance of Te'o's, existed only on social media.

THE BIG QUESTIONS

What is social interaction and why study it?

Understand how the subfield of microsociology contrasts with earlier sociological work. See why the study of social interaction is of major importance in sociology. Recognize the difference between focused and unfocused interaction and learn the different forms of nonverbal communication.

How do we manage impressions in daily life?

Learn about the ways you carefully choose to present yourself to others in daily interactions—both face-to-face and virtually.

What rules guide how we communicate with others?

Learn the research process of ethnomethod-ology and recognize how we use context to make sense of the world. Understand why people get upset when conventions of talk are not followed.

How do time and space affect our interactions?

Understand that interaction is situated, that it occurs in a particular place and for a particular length of time. Recognize how technology is reorganizing time and space.

How do the rules of social interaction affect your life?

See how face-to-face interactions reflect broader social factors such as social hierarchies.

For most college students, romantic relationships are just as important as their schoolwork. In the 1950s and 1960s, coeds might have met for the first time at a formal "mixer" and written long, heartfelt letters to each other during their summer months spent apart. In the 1980s and 1990s, a couple might have met while standing in line at a keg party and kept in touch over late-night phone calls during winter and summer breaks.

Today, many college students first meet their boyfriend or girlfriend in their dorm or through a mutual friend, but many also meet (and keep in touch) online, whether through meet-up websites and smartphone apps like Tinder or social networking sites like Facebook. But is it possible to maintain a meaningful romantic relationship with no face-to-face contact and only virtual exchanges with one's partner? Manti Te'o thought so.

Seemingly overnight, Manti Te'o was transformed from a national sports star to a national joke. In 2012, Te'o was a star football player for Notre Dame. An All-American and finalist for the Heisman Trophy, Te'o was a highly decorated college football player. But Te'o was more than a football star; he became a hometown hero when he led his team to victory on the same day in September 2012 that he learned of the deaths of both his grandmother and girlfriend. Despite his heartbreak, Te'o did not miss a single football game that season, telling reporters and teammates that he had promised his girlfriend, Lennay Kekua, that he would play regardless of what happened to her. Kekua, a Stanford University student, had been battling leukemia.

In January 2013, Te'o made headlines again. The sports blog Deadspin broke the shocking news that Te'o's girlfriend, Lennay Kekua, hadn't died. In fact, she had never existed at all. Kekua was entirely fictional, constructed as part of an elaborate Internet hoax by a distant acquaintance of Te'o's.

How was it possible that Te'o had maintained a nearly yearlong "relationship" with a fictional young woman? In a public statement, Te'o explained, "This is incredibly embarrassing to talk about, but over an extended period of time, I developed an emotional relationship with a woman I met online. We maintained what I thought to be an authentic relationship by communicating frequently online and on the phone, and I grew to care deeply about her" (ESPN 2013b). Te'o had previously lied to his family, teammates, and the press about meeting Kekua in person, afraid they would think he was "crazy." Was Te'o crazy, or was he simply trying to maintain a long-distance online relationship while also juggling his busy schedule as a student athlete? Is an exclusively online relationship really a form of social interaction?

Throughout most of human history, people have communicated mainly face-to-face. The U.S. Postal Service was established in the late eighteenth century, making it easier than ever to communicate through writing. Then, in the nineteenth century, the advent of the telephone revolutionized how Americans interacted with one another. In the last two decades, email, SMS, and social networking sites have once again revolutionized the way humans communicate. In this chapter, we will explore how each of these forms of communication—along with subtle, nonverbal aspects of communication—constitutes social interaction and carries important messages about how our society functions.

What Is Social Interaction and Why Study It?

Erving Goffman was a highly influential sociologist who created a new field of study focused on **social interaction**. In the 1950s and 1960s, Goffman wrote that sociologists needed to concern themselves with seemingly trivial aspects of everyday social behavior. His work on social interaction is just one example of the broader sociological subfield called **microsociology**. This term was conceived by sociologist Harold Garfinkel to describe a field of study that focused on individual interaction and communication within small groups; this subfield stood in contrast with earlier sociological work, which historically had examined large social groups and societal-level behaviors. Goffman, and eminent scholars such as George Herbert Mead and Herbert Blumer, examined seemingly small exchanges, such as conversation patterns and the ways that social actors develop a shared understanding of their social context.

social interaction

The process by which we act with and react to those around us.

microsociology

The study of human behavior in contexts of face-to-face interaction.

> Understand how the subfield of microsociology contrasts with earlier sociological work. See why the study of social interaction is of major importance in sociology. Recognize the difference between focused and unfocused interaction and learn the different forms of nonverbal communication.

The study of social interaction reveals important things about human social life. For instance, think about the last time you walked down the street and passed a stranger or shared an elevator ride with a stranger. Did you subtly try to avoid eye contact? Goffman believed that such small gestures are meaningful and rich with messages about human interaction. When passersby—either strangers or intimates—quickly glance at each other and then look away again, they demonstrate what Goffman (1967, 1971) calls **civil inattention**. Civil inattention is not the same as merely ignoring another person. Each individual indicates recognition of the other person's presence but avoids any gesture that might be taken as too intrusive. Goffman argued that the study of such apparently insignificant forms of social interaction is of major importance in sociology and, far from being uninteresting, is one of the most absorbing of all areas of sociological investigation. There are three reasons for this.

First, our ordinary routines give structure and form to what we do. We can learn a great deal about ourselves as social beings, and about social life itself, from studying them. Our lives are organized around the repetition of similar patterns of behavior from day to day, week to week, month to month, and year to year. Think of what you did yesterday, for example, and the day before that. If they were both weekdays, you probably woke up at about the same time each day (an important routine in itself). You may have gone to class fairly early in the morning, making a journey from home to school that you make virtually every weekday. You perhaps met some friends for lunch, returning to classes or private study in the afternoon. Later, you retraced your steps back home or to your dorm, possibly going out later in the evening with friends.

Of course, the routines we follow are not identical from day to day, and our patterns of activity on weekends usually contrast with those on weekdays. If we make a major change in our lives, like leaving college to take a full-time job, alterations in our daily routines are usually necessary, but then we establish a new and fairly regular set of habits again.

Second, the study of everyday life reveals to us how humans can act creatively to shape reality. Although social behavior is guided to some extent by forces such as roles, norms, and shared expectations, individuals also have **agency**, or the ability to act, think, and make choices independently (Emirbayer and Mische 1998). The ways that people perceive reality may vary widely based on their backgrounds, interests, and motivations. Because individuals are capable of creative action, they continuously shape reality through the decisions and actions they take. In other words, reality is not fixed or static—it is created through human interactions. However, as we shall see later in this chapter, even our most private or seemingly minor interactions are shaped by **structure**, or the recurrent patterned arrangements and hierarchies that influence or limit the choices and opportunities available to us.

Third, studying social interaction in everyday life sheds light on larger social structures, systems, and institutions. All large-scale social systems depend on the patterns of social interaction we engage in daily. This is easy to demonstrate. Let's reconsider the case of two strangers passing on the street. Such an event may seem to have little direct relevance to large-scale, more permanent forms of social structure. But when we take into account many such interactions, they are no longer irrelevant. In modern societies, most people live in towns and cities and constantly interact with people they do not know personally. Civil inattention is one of many mechanisms that give public life—with its bustling crowds and fleeting, impersonal contacts—its distinctive character.

As they wait to board the train, commuters engage in what Erving Goffman called civil inattention.

civil inattention

The process whereby individuals in the same physical setting demonstrate to each other that they are aware of the other's presence.

agency

The ability to think, act, and make choices independently.

structure

The recurrent patterned arrangements and hierarchies that influence or limit the choices and opportunities available to us.

unfocused interaction

Interaction occurring among people present in a particular setting but not engaged in direct face-to-face communication.

focused interaction

Interaction between individuals engaged in a common activity or in direct conversation with each other.

encounter

A meeting between two or more people in a situation of face-to-face interaction. Our daily lives can be seen in a series of different encounters strung out across the course of the day. In modern societies, many of these encounters are with strangers rather than people we know.

Focused and Unfocused Interaction

In many social situations, we engage in what Goffman calls unfocused interaction with others. **Unfocused interaction** takes place whenever individuals exhibit mutual awareness of one another's presence but do not engage in direct communication or conversation. This is usually the case anywhere large numbers of people are assembled, such as on a busy street, in a theater, or at a party. When people are in the presence of others, even if they do not directly talk to them, they continually communicate nonverbally through their posture and facial and physical gestures.

Focused interaction occurs when individuals directly attend to what others say or do. Except when someone is standing alone, say, at a party, all interaction involves both focused and unfocused exchanges. Goffman calls an instance of focused interaction an **encounter**, and much of day-to-day life consists of encounters with other people—family, friends, colleagues—frequently occurring against the background of unfocused interaction with others present. Small talk, seminar discussions, games, and routine face-to-face interactions (with ticket clerks, waiters, shop assistants, and so forth) are all examples of encounters.

Encounters always need "openings," which indicate that civil inattention is being discarded. When strangers meet and begin to talk at a party, the moment of ceasing civil inattention is always risky, because misunderstandings can easily occur about the nature of the encounter being established (Goffman 1971). Hence, the making of eye contact may first be ambiguous and tentative. A person can then act as though he or she had made no direct move if the overture is not accepted. In focused interaction, each person communicates as much by facial expression and gesture as by the words actually exchanged.

Goffman distinguishes between the expressions individuals "give" and those they "give off." The first are the words and facial expressions people use to produce certain impressions on others. The second are the clues that others may spot to check their

Placing your coffee order with a barista is an example of an encounter. Our daily lives are filled with these instances of focused interaction.

sincerity or truthfulness. For instance, a restaurant owner listens with a polite smile to the statements of customers about how much they enjoyed their meals. At the same time, he is noting how pleased they seemed to be while eating the food, whether a lot was left over, and the tone of voice they use to express their satisfaction.

Think about how Goffman's concepts of focused and unfocused interaction, developed mainly to explain face-to-face social encounters, would apply to our current age of Internet communication. Can you think of a way in which unfocused interaction occurs on Facebook and Twitter? In some small online communities, everyone can have a mutual awareness of who else is online, without being in direct contact with them. On sites like Twitter, people are constantly broadcasting status updates about what they're doing at that moment. These status updates make it possible for people in unfocused interaction to have even more control over how they are perceived than people who are merely in one another's presence. Instead of revealing their facial expressions or posture, of which they may be unconscious, people can carefully craft the messages, or tweets, they wish to broadcast.

Nonverbal Communication

Social interaction—both unfocused and focused interaction—requires many forms of **nonverbal communication**, which refers to the exchange of information and meaning through facial expressions, eye contact, gestures, and movements of the body. Nonverbal communication, sometimes referred to as "body language," often alters or expands on what is said with words. In some cases, our body language may convey a message that is discrepant with our words.

FACE, GESTURES, AND EMOTION

One major aspect of nonverbal communication is the facial expression of emotion. Paul Ekman and his colleagues developed what they call the Facial Action Coding System (FACS) for describing movements of the facial muscles that give rise to particular expressions (Ekman and Friesen 1978). By this means, they tried to inject some precision into an area notoriously open to inconsistent or contradictory interpretations—for there is little agreement about how emotions are to be identified and classified. Charles Darwin, one of the originators of evolutionary theory, claimed that basic modes of emotional expression are the same in all human beings and across all cultures. Although some have disputed the claim, Ekman's research among people from widely different cultural backgrounds seems to confirm Darwin's view. Ekman and W. V. Friesen carried out a study of an isolated community in New Guinea, whose members previously had virtually no contact with outsiders. When they were shown pictures of facial expressions conveying six emotions, the New Guineans identified the same emotions (happiness, sadness, anger, disgust, fear, surprise) we would.

According to Ekman, the results of his own and similar studies of different peoples support the view that the facial expression of emotion and its interpretation are innate in human beings. He acknowledges that his evidence does not conclusively demonstrate this and that it's possible that widely shared cultural learning experiences are involved; however, his conclusions are supported by other types of research. Irenäus Eibl-Eibesfeldt (1972) studied six children born deaf and blind to determine to what extent their facial expressions were the same as those of sighted and hearing individuals in particular emotional situations. He found that

Paul Ekman's photographs of facial expressions from a tribesman in an isolated community in New Guinea helped test the idea that basic modes of emotional expression are the same among all people. Here the instructions were to show how your face would look if you were a person in a story and (a) your friend had come and you were happy, (b) your child had died, (c) you were angry and about to fight, and (d) you saw a dead pig that had been lying there a long time.

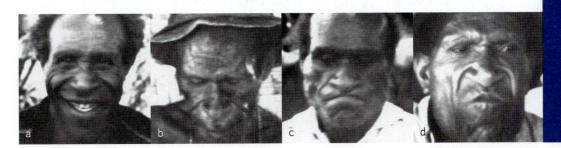

a   b   c   d

Hand gestures are one form of nonverbal communication. Unlike facial expressions, hand gestures vary widely by culture.

the children smiled when engaged in obviously pleasurable activities, raised their eyebrows in surprise when sniffing at an object with an unaccustomed smell, and frowned when repeatedly offered a disliked object. Because the children could not have seen other people behaving in these ways, it seems that these responses must be innately determined.

By contrast, there are no gestures or bodily postures that are universally known and understood in all cultures. In some societies, for instance, people nod when they mean no, the opposite of Anglo-American practice. Gestures Americans tend to use a great deal, such as pointing, seem not to exist among certain peoples (Bull 1983). Similarly, a straightened forefinger placed in the center of the cheek and rotated is used in parts of Italy as a gesture of praise but appears to be unknown elsewhere (Donadio 2013).

Like facial expressions, gestures and bodily posture are continually used to fill out utterances as well as to convey meanings when nothing is actually said. All three can be used to joke, show irony, or indicate skepticism. The nonverbal impressions that we convey may inadvertently indicate that what we say is not quite what we really mean. Blushing is perhaps the most obvious example, but innumerable other subtle indicators can be picked up by other people. Genuine facial expressions tend to evaporate after four or five seconds. A smile that lasts longer could indicate deceit. An expression of surprise that lasts too long may indicate deliberate sarcasm—to show that the individual is not in fact surprised after all.

NONVERBAL COMMUNICATION IN THE DIGITAL AGE

On the Internet, it is very difficult to capture dimensions of emotion that are present only with facial expressions. At first, the need that Internet users felt to approximate facial gestures resulted in at least two common faces:

:) or :-)

As time passed, a need for greater subtlety resulted in other widely understood variations, such as this winking smiley face:

;-)

Email may have once been devoid of facial expression, but today the average email user may insert different emotions into a message. Strongly felt sentiments might be typed in all capitals, a gesture that is considered "shouting." The strong need human beings feel to communicate with their faces has also led to other innovations, like Skype and FaceTime. But in general, people who communicate over email or the phone lack the benefit of seeing the faces of their conversational partners as they speak.

Why and how does this matter for human relationships and interactions? On the phone, whether it's a cell phone or a landline, an individual will frequently talk for a longer stretch of time than he or she would in a face-to-face conversation. Unable to see the face of a conversational partner, the speaker can't as readily adjust what he or she is saying in response to clues from the listener that he or she "gets it." Yet, the phone maintains at least some immediacy of feedback that email and text messages, to a certain extent, lack. This is why in email disputes, people who are unable to make mutual adjustments in response to verbal or facial cues will end up saying much more—communicated in the form of long messages—than they would need to say in spoken conversation.

Which is best? Would you prefer to make your point via email or text message, over the phone or Skype, or in person? Using sociological insights like these might make you prefer electronic communication at certain times and face-to-face communication at others. For example, if you are dealing with a powerful person and want to get your thoughts across, you may want to avoid a situation in which he or she can signal with facial gestures that your idea is silly and thus inhibit you from making all your points. The power to signal with facial gestures is one of the things that people do to control the flow of a conversation. On the other hand, face-to-face communication gives you an opportunity to try out an idea on someone more powerful than yourself without going too far down the road if he or she is actually unreceptive. You probably would not want to conduct an important conversation via text message, instead limiting its use to minor or immediate issues.

Nonhumans in Social Interaction

Goffman (1959) and the first social interactionists viewed nonhuman entities as props—used, on occasion, by humans to enhance their interactional performances. However, since the 1980s, a growing number of sociologists have argued that nonhumans may also be legitimate participants in social interaction (Cerulo 2009). Some scholars have argued that

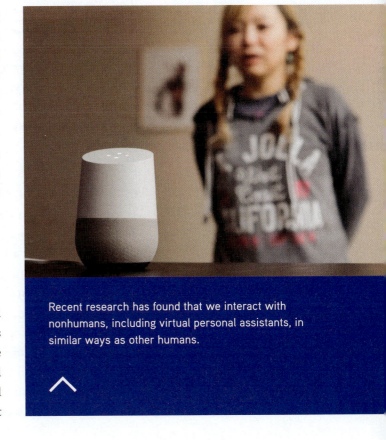

Recent research has found that we interact with nonhumans, including virtual personal assistants, in similar ways as other humans.

CONCEPT CHECKS

1. What is microsociology and how does it differ from earlier sociological work?

2. What are three reasons it is important to study social interaction?

3. What are three forms of nonverbal communication?

4. Describe several ways that individuals communicate their emotions to one another.

>

Learn about the ways you carefully choose to present yourself to others in daily interactions—both face-to-face and virtually.

status

The social honor or prestige that a particular group is accorded by other members of a society. Status groups normally display distinct styles of life—patterns of behavior that the members of a group follow. Status privilege may be positive or negative.

through "projection," humans endow nonhumans with human capacities, "[allowing] humans to legitimate nonhumans as viable others in social interaction" (Cerulo 2009: 536; see also Owens 2007). There is also compelling evidence that individuals interact with a variety of nonhumans as they would with humans. Pets, street pigeons, deities, the deceased, and technological forms like robots and video game avatars evoke human empathy and feelings of companionship and intimacy and even shape behavior (Cerulo 2009; Jerolmack 2013).

The most important development in recent years has been human–computer interaction, particularly interaction with virtual personal assistants such as Alexa, Amazon's cloud-based voice service. Many people today begin their relationships with these devices with very simple commands, such as to play music or get information about the weather or sports. But over time, many people come to interact with their personal assistants, as artificial intelligence has allowed Alexa to become more and more sophisticated in its responses. Alexa has even received marriage proposals and been the first to hear that someone was contemplating suicide. Amazon, the manufacturer of Alexa, has programmed it to respond to such messages with the names of suicide hotlines. Such interactions would most likely have been unimaginable to Goffman.

How Do We Manage Impressions in Daily Life?

Goffman and other writers on social interaction often draw on imagery from drama and theater in their analyses. Using the theatrical performance as an inspiration for looking at everyday social life, they take note of how people present themselves to those around them and how they try to control impressions during many kinds of interactions.

Social Roles

In the theater-like or dramaturgical model that Goffman employs, social life is seen as though played out by actors on a stage—or on many stages, because how we act depends on the roles we are playing at a particular time. The concept of a social role, which originated in a theatrical setting, is an important component of social interaction in the dramaturgical approach. Roles are socially defined expectations that a person in a given **status** (or **social position**) follows.

The teacher's role, for example, involves acting in specified ways toward his or her pupils. The performance depends not only on the actor but also on the audience. Thus, the teacher must depend on the students to affirm him or her in that role. The same is true of the student, who works to be affirmed by the teacher or other students in certain kinds of identities. Some students want to get high grades and therefore seek to impress the teacher, while others seek to be known as the "class clown" and need only impress fellow students. From an early age, students learn that they must sit up in their seats and appear at attention in the classroom, and most students conform to this expectation.

Front and Back Regions

Much of social life, Goffman argues, can be divided up into front regions and back regions. Front regions are social occasions or encounters in which individuals act out formal roles—these are "on-stage performances." The back regions are where they assemble the props and prepare themselves for interaction in the more formal setting. Back regions resemble the back stage of a theater or the off-camera activities of film or TV productions.

When they are safely "behind the scenes," people relax and give vent to feelings and styles of behavior that they keep in check when on the front stage. Thus, a teacher might be quite formal when standing in front of students but show sides of himself or herself that students could barely imagine in the teacher's lounge. Likewise, the students might sit upright in the classroom but engage in "sloppy sitting" during recess or in the cafeteria at lunch time. According to Goffman, back regions permit "profanity, open sexual remarks, elaborate griping, . . . rough informal dress, . . . use of dialect or substandard speech, mumbling or shouting, playful aggressivity and 'kidding'" (Goffman 1973). Teamwork is often involved in creating and preserving front region performances. Thus, a wife and husband may take care to conceal their quarrels from their children, preserving a front of harmony, but fight bitterly once the children are tucked away in bed.

Impression Management

Goffman argues that even as people go through life largely spontaneously, they are sensitive to how they are seen by others (that is, their "audience") and use many forms of **impression management** to control how others see them. This occurs through the concealment and revelation of information, including information that we might "give off" unintentionally if we are not careful. When going on a job interview, for example, a person will typically dress more formally and try to put his or her best foot forward by speaking formally; however, when going out with friends, he or she might dress down or use slang. Most interaction involves people managing impressions for audiences of people in their immediate environment as they strategically choose to conceal and reveal information at will.

Seeking approval and respect, individuals want to "save face" at every turn. In social interactions, human beings tend to collaborate with others to make sure that the encounter ends without embarrassment for anyone. Social life, like a play, involves many players, and they must collaborate to make each scene work.

Although people cooperate to help one another save face, they also endeavor individually to preserve their own dignity, autonomy, and respect. One of the ways that people do this is by arranging for "audience segregation" in their lives. In each of their roles, they act somewhat differently, and they endeavor to keep the roles both distinct and separate from one another. This means that they can have multiple selves. Frequently these selves are consistent, but sometimes they are not. People find it very stressful when boundaries break down or when they cannot reconcile their role in one part of life with their role in another. For example, some college freshmen try to distance themselves from former classmates to carve out a new "college" identity that won't be tainted by embarrassing stories from high school. Or, some people live very different lives at home and

social position

The social identity an individual has in a given group or society. Social positions may be general in nature (those associated with gender roles) or may be more specific (occupational positions).

impression management

Preparing for the presentation of one's social role.

at work. For example, due to discrimination against transgender men and women, some trans men—people designated female at birth whose gender identity is male—will not emphasize their trans identity for fear of being marginalized in the workplace (Schilt 2010). Like all people who engage in audience segregation, they show a different face to different people. Audience segregation implicitly encourages impression management.

The concepts of audience segregation and impression management help us understand some of the dilemmas of electronic communication. Consider the social situation of a copied message. You write a message to a friend asking her whether she prefers to go to the early show or the late show. You also tell your friend that you have a new boyfriend whom you hope she will like. She replies and copies the other people who are thinking of going to the movie, many of whom you never intended to tell about the new romance. Suddenly, the audience segregation you had imagined has broken down.

Some people maintain two different Facebook pages, one linked to family members or coworkers and another linked to friends and peers. Why might someone do this? Our Facebook pages are a strategy to "impression manage," allowing us to carefully and selectively portray an image of ourselves to the outside world. On your "professional" page, you might try to convey an image of a respectable student and employee by carefully curating the information and images you post. By contrast, on your personal page, you might post photos that present a more fun and carefree version of yourself.

In recent years, undergraduate students have posted pictures of themselves drinking at parties, or even naked, only to discover that future employers conducted web searches before making hiring decisions. Some students have been expelled from their colleges for posting inappropriate photos or comments on Facebook. In 2017, ten students admitted to Harvard's freshman class had their offers of admission revoked after sending sexually explicit and racist messages to one another in a private Facebook group they formed. Another example of blurring audiences is the case of sexting: A high school student may send a revealing photo of herself to her boyfriend, only to have him forward it to the entire school—whether out of cruelty or by mistake. Personal catastrophes like these occur frequently in the age of email and smartphones.

CONCEPT CHECKS

1. Identify three roles you play in social life.

2. Compare and contrast front and back regions.

3. Describe two forms of impression management.

> Learn the research process of ethnomethodology and recognize how we use context to make sense of the world. Understand why people get upset when conventions of talk are not followed.

What Rules Guide How We Communicate with Others?

Conversations are one of the main ways in which our daily lives are maintained in a stable and coherent manner. But we can make sense of what is said in conversation only if we know the social context, which does not appear in the words themselves. Take the following conversation (Heritage 1985):

> A: I have a fourteen-year-old son.
> B: Well, that's all right.
> A: I also have a dog.
> B: Oh, I'm sorry.

What do you think is happening here? What is the relation between the speakers? What if you were told that this is a conversation between a prospective tenant and a landlord? The conversation then becomes sensible: Some landlords accept children but don't permit their tenants to keep pets. Yet if we don't know the social context, the responses of individual B seem to bear no relation to the statements of A. Part of the sense is in the words, and part is in the way in which the meaning emerges from the social context.

The most inconsequential forms of daily talk presume complicated, shared knowledge brought into play by those speaking. In fact, our small talk is so complex that it has so far proved impossible to program even the most sophisticated computers to converse with human beings. The words used in ordinary talk do not always have precise meanings, and we "fix" what we want to say through the unstated assumptions that back it up. If Maria asks Tom, "What did you do yesterday?" the words in the question themselves suggest no obvious answer. A day is a long time, and it would be logical for Tom to answer, "Well, at 7:16, I woke up. At 7:18, I got out of bed, went to the bathroom, and started to brush my teeth. At 7:19, I turned on the shower . . ." We understand the type of response the question calls for by knowing Maria, what sorts of activities she and Tom consider relevant, and what Tom usually does on a particular day of the week, among other things.

Ethnomethodology

Ethnomethodology is the study of the "ethnomethods"—the folk, or lay, methods—people use to make sense of what others do and particularly of what they say. We all apply these methods, normally without having to give any conscious attention to them. This field was created by Harold Garfinkel, who, along with Goffman, was one of the most important figures in the study of microinteraction.

Garfinkel argued that to understand the way people use context to make sense of the world, sociologists need to study the "background expectancies" with which we organize ordinary conversations. He highlighted these in some experiments he undertook with student volunteers. The students were asked to engage a friend or relative in conversation and to insist that casual remarks or general comments be actively pursued to make their meaning precise. If someone said, "Have a nice day," the student was to respond, "Nice in what sense, exactly?" "Which part of the day do you mean?" and so forth. One of the exchanges that resulted ran as follows. S is the friend, E the student volunteer (Garfinkel 1963):

> S: *How are you?*
> E: *How am I in regard to what? My health, my finances, my schoolwork, my*
> *peace of mind, my . . . ?*
> S: *(red in the face and suddenly out of control) Look! I was just trying to be*
> *polite. Frankly, I don't give a damn how you are.*

Why do people get so upset when apparently minor conventions of talk are not followed? The answer is that the stability and meaningfulness of our daily social lives depend on the sharing of unstated cultural assumptions about what is said and why. If we weren't able to take these for granted, meaningful communication would be impossible. Any question or contribution to a conversation would have to be followed by a massive "search procedure" of the sort Garfinkel's subjects were told to initiate, and interaction would simply break down. What seem at first sight to be unimportant conventions

ethnomethodology

The study of how people make sense of what others say and do in the course of day-to-day social interaction. Ethnomethodology is concerned with the "ethnomethods" by which people sustain meaningful exchanges with one another.

Turning Away from Face-to-Face Interaction

How many times have you taken out a book and started reading it during a meal with friends or relatives? For most people, that would constitute an unacceptable breach of everyday etiquette. And yet, it has become completely routine to pull out a smartphone to read emails or text messages during an in-person interaction. What is the difference between reading a book and reading the messages on your smartphone? In each case, we seem to be distracted by something outside the conversation.

Until very recently, there was an expectation that people who engaged in face-to-face communication would maintain eye contact and even occasionally nod their heads while listening. While old-fashioned etiquette suggests that two or more people should be engaged in a continuous flow of focused interaction as they talk to one another, today it is increasingly acceptable for conversation to happen in a far less focused manner. Dozens of apps on the average smartphone can remove part of our attention while we are still in physical proximity to others with whom we are interacting. It is not uncommon for a person to look down and move from app to app while involved in a face-to-face interaction. A recent survey found that 89 percent of cell phone owners said that they used their phone—whether to read or send a message or take a photo—during their last social activity (Rainie and Zickuhr 2015).

What is truly new about interaction today is that people are involved in multiple conversations at one time, or they are following one story online while listening to another in person. Often, people send emails and texts with the expectation that conversations online will be as ongoing as face-to-face communication. Thus, people feel obligated to respond immediately online even if they are immersed in a face-to-face interaction.

Does using a cell phone while talking have the effect of ruining conversation? A recent study found that just the presence of a cell phone can inhibit our ability to connect with the people around us (Lin 2012). Younger users—often referred to as digital natives—often see nothing wrong with having one's attention in multiple places at the same time. They have grown up in a world in which it feels natural to interact with a cell phone in hand. They move effortlessly between face-to-face and online interactions. They also argue that having the Internet as a conversational resource raises the level and quality of interaction as they look things up that are relevant to the discussion.

Even if digital natives have a point—that they can process interaction just as well when they are multitasking—don't they have a moral imperative to look someone in the eye and make that person feel valued? Or are such expectations purely social constructions of a particular historical era, which will fall by the wayside as the older generation passes on? Is it possible that as the human mind gets accustomed to simultaneously interacting online and in person, the traditional moral claims on conversation will no longer feel reasonable?

Should that happen, will that have any enduring impact on human character? Sherry Turkle, a sociologist and clinical psychologist at MIT, has argued that our reliance on smartphones while we talk is having a detrimental impact on our capacity to put ourselves in the place of others—to experience empathy: "We suppress this capacity by putting ourselves in environments where we're not looking at each other in the eye, not sticking with the other person long enough or hard enough to follow what they're feeling" (Davis 2015b).

Today it is not uncommon to simultaneously interact with people both face-to-face and online. In what other ways does technology affect the way we interact?

of talk, therefore, turn out to be fundamental to the very fabric of social life, which is why their breach is so serious.

Interactional Vandalism

We feel most comfortable when the tacit conventions of small talk are adhered to; when they are breached, we can feel threatened, confused, and insecure. In most everyday talk, conversants are carefully attuned to the cues they get from others—such as changes in intonation, slight pauses, or gestures—to facilitate conversation smoothly. By being mutually aware, conversants "cooperate" in opening and closing interactions and in taking turns to speak. Interactions in which one party is conversationally "uncooperative," however, can create tension.

Garfinkel's students created tense situations by intentionally undermining conversational rules as part of a sociological experiment. But what about situations in the real world in which people make trouble through their conversational practices? The term **interactional vandalism** describes cases in which a subordinate person breaks the tacit rules of everyday interaction that are of value to the more powerful. For example, homeless people on the street often do conform to everyday forms of speech in their interactions with one another, local shopkeepers, the police, relatives, and acquaintances. But when they choose to, they subvert the tacit conventions for everyday talk in a way that leaves passersby disoriented. Even more than physical assaults or vulgar verbal abuse, interactional vandalism leaves victims unable to articulate what has happened.

interactional vandalism

The deliberate subversion of the tacit rules of conversation.

How might interactional vandalism play out on the Internet? Can we think of ways in which less powerful people engaged in electronic communications undermine the taken-for-granted rules of interaction that are of value to the more powerful? The very existence of the Internet creates spaces in which less powerful people can hold their superiors accountable in ways they never were before. Think of all the blogs in which workers talk anonymously about their bosses or situations in which workers forward rude messages from their boss to other employees. Because of the Internet, powerful people are less able to segregate their audiences—treating some people poorly behind the scenes and treating others nicely in public.

The concept of "trolling" might be seen as an interactional mode that shares certain, though not all, aspects of interactional vandalism. A troll is someone who disrupts the taken-for-granted purposes of an online community, such as a forum, message board, or blog. As such, he or she might post items that are deliberately provocative. Such provocations might have the effect of undermining the civility that is a foundation for the kind of communication envisioned by the site's founders. Or the controversies raised by trolls can sometimes increase traffic to the site. Some readers of a comment posted by a troll might be lured further into the interaction while others might attempt to restore normal order by dismissing the actions of the troll.

To what extent is trolling an example of interactional vandalism of the kind found in face-to-face communication on the sidewalk? Like the men on the street who act sincere as they pretend not to understand that a two-second pause is a signal to close a conversation, trolls pretend not to understand certain assumptions of the conversational world for the specific purpose of being disruptive. Trolls will write as if they are sincere members of the group who perhaps do not understand certain things, while at a deeper level they know precisely what they are doing. In interactional vandalism on the sidewalk, a less powerful

response cries

Seemingly involuntary exclamations individuals make when, for example, being taken by surprise, dropping something inadvertently, or expressing pleasure.

personal space

The physical space individuals maintain between themselves and others.

Cultural norms frequently dictate the acceptable boundaries of personal space. In the Middle East, for example, people frequently stand closer to each other than is common in the West.

person is subverting normal interaction to undermine the taken-for-granted control of someone in a subordinate position. In trolling, the parties are often anonymous, so it's not always possible to know what the actual power dynamics are.

Response Cries

Some kinds of utterances are not talk but consist of muttered exclamations, or what Goffman (1981) has called **response cries**. Consider Lucy, who exclaims, "Oops!" after knocking over a glass of water. "Oops!" seems to be merely an uninteresting reflex response to a mishap, rather like blinking your eye when a person moves a hand sharply toward your face. It is not a reflex, however, as shown by the fact that people do not usually make the exclamation when alone. "Oops!" is normally directed toward others present. The exclamation demonstrates to witnesses that the lapse is only minor and momentary, not something that should cast doubt on Lucy's command of her actions.

Expressions like "oops!" or "my bad" are used only in situations of minor failure rather than in major accidents or calamities—which also demonstrates that the exclamation is part of our controlled management of the details of social life. This may all sound contrived and exaggerated. Why bother to analyze such an inconsequential utterance in this detail? Surely we don't pay as much attention to what we say as this example suggests? Of course we don't—on a conscious level. The crucial point, however, is that we take for granted an immensely complicated, continuous control of our appearance and actions. In situations of interaction, we are never expected just to be present on the scene. Others expect, as we expect of them, that we will display what Goffman calls "controlled alertness." A fundamental part of being human is continually demonstrating to others our competence in the routines of daily life.

Personal Space

There are cultural differences in the definition of **personal space**. In Western culture, people usually maintain a distance of at least three feet when engaged in focused interaction with others; when standing side by side, they may stand closer together. In the Middle East, people often stand closer to each other than is thought acceptable in the West. Westerners visiting that part of the world might find themselves disconcerted by this unexpected physical proximity.

Edward T. Hall (1969, 1973), who has worked extensively on nonverbal communication, distinguishes four zones of personal space. Intimate distance, of up to one and a half feet, is reserved for very few social contacts. Only those involved in relationships in which regular bodily touching is permitted, such as lovers or parents and children, operate within this zone of private space. Personal distance, from one and a half to four feet, is the normal spacing for encounters with friends and close acquaintances. Some intimacy of contact is permitted, but this tends to be strictly limited. Social distance, from four to twelve feet, is the zone usually maintained in formal settings, such as interviews. The fourth zone is that of public distance, beyond twelve feet, preserved by those who are performing to an audience.

In ordinary interaction, the most fraught zones are those of intimate and personal distance. If these zones are invaded, people try to recapture their space. We may stare at the intruder as if to say, "Move away!" or elbow him or her aside. When people are forced into proximity closer than they deem desirable, they might create a kind of physical boundary: A reader at a crowded library desk might physically demarcate a private space by stacking books around its edges (Hall 1969, 1973). Similarly, Greyhound bus passengers may use facial and body language, move luggage, or wear headphones to enforce the "unspoken seat rule" that "passengers should not sit next to another person when there are more than enough open rows" (Kim 2012: 275). More generally, individuals may "extend" personal space in a social sense through the use of mobile technology (Hatuka and Toch 2016: 2203). In a recent study, Hatuka and Toch (2016) found that individuals report being interrupted less and feeling a greater sense of privacy when using a smartphone.

Eye Contact

Eye contact is yet another aspect of social interaction that illustrates important social norms and reveals (and perpetuates) power differentials. As we saw earlier in this chapter, we are guided by a powerful norm that strangers should not make eye contact. Strangers or chance acquaintances virtually never hold the gaze of another. To do so may be taken as an indication of hostile intent. It is only where two groups are strongly antagonistic to each other that strangers might indulge in such a practice—for example, when whites in the United States have been known to give a "hate stare" to Blacks walking past.

Studies show that we tend to rate a person who makes eye contact as more likable, pleasant, intelligent, credible, and dominant than a person exhibiting less or no eye contact. However, excessive eye contact may make an observer feel uncomfortable in certain situations. To look too intently might be taken as a sign of mistrust. Eye contact also reveals power relations. Looking at a colleague when speaking conveys confidence and respect. Prolonged eye contact during a debate or disagreement can signal that you're standing your ground. It also signifies your position in the hierarchy. People who are high status tend to look longer at people they're talking to, compared with others. Culture also guides how we look at each other. In many Eastern and some Caribbean cultures, meeting another's eyes is considered rude. Asians are more likely than persons from Europe or the United States to regard a person who makes eye contact as angry or unapproachable (Akechi et al. 2013).

CONCEPT CHECKS

1. What is interactional vandalism?
2. Give an example of a response cry.
3. What are the four zones of personal space?

How Do Time and Space Affect Our Interactions?

Understand that interaction is situated, that it occurs in a particular place and for a particular length of time. Recognize how technology is reorganizing time and space.

When we wrote earlier editions of this textbook, before the age of the Internet, we used to say that understanding how activities are distributed in time and space was fundamental to analyzing encounters and to understanding social life in general. It was common to say in those days that all interaction is "situated"—that it occurs in a particular place and has a specific duration in time. It made sense to claim that our actions over the course of a day tend to be "zoned" in time as well as in space. Thus, for example, most people would spend a zone—say, from 9:00 AM to 5:00 PM—of their daily time working. Their weekly time was also zoned: They would be likely to work on weekdays and spend weekends at

time-space

When and where events occur.

---

regionalization

The division of social life into different regional settings or zones.

home, altering the pattern of their activities on the weekend days. As we moved through the temporal zones of the day, we were often moving across space as well: To get to work, we might take a bus from one area of a city to another or perhaps commute in from the suburbs. When we analyzed the contexts of social interaction, therefore, it was often useful to look at people's movements across **time-space**.

The concept of **regionalization** can help us understand how social life is zoned in time-space. Take the example of a private house. A modern house is regionalized into rooms, hallways, and floors (if there is more than one story). These spaces are not just physically separate areas but are zoned in time as well. The living rooms and kitchen are used most in the daylight hours, the bedrooms at night. The interaction that occurs in these regions is bound by both spatial and temporal divisions. Some areas of the house form back regions, such as the den or the basement, where people can be themselves without worrying about what other people think. For instance, some people may leave their old furniture and children's tattered toys in the den or family room and may be slightly less vigilant about "keeping up appearances" in rooms that guests seldom visit. By contrast, the living room may display lovely furniture, well-appointed decorations, and sophisticated coffee-table books so that a family can convey to others that they are dignified and respectable. At times, the whole house can become a back region. Once again, this idea is beautifully captured by Goffman (1974):

> On a Sunday morning, a whole household can use the wall around its domestic establishment to conceal a relaxing slovenliness in dress and civil endeavor, extending to all rooms the informality that is usually restricted to the kitchen and bedrooms. So, too, in American middle-class neighborhoods, on afternoons the line between children's playground and home may be defined as backstage by mothers, who pass along it wearing jeans, loafers, and a minimum of make-up.

With the rise of mobile technologies such as the smartphone, the organization of time and space has undergone a radical reorganization. With employers expecting their workers to be constantly in touch via emails received after official working hours, we can no longer say that people spend a single zone from 9:00 AM to 5:00 PM working. With the rise of remote work, nor can we say that people are expected to be in the office on weekdays and at home on the weekends.

Indeed, the Internet makes it possible for us to interact with people we never see or meet, in any corner of the world. Such technological change rearranges space—we can interact with anyone without moving from our chair. It also alters our experience of time because communication on the Internet is almost immediate. Until about fifty years ago, most communication across space required a duration of time. If you sent a letter to someone abroad, there was a time gap while the letter was carried by ship, train, truck, or plane to the person to whom it was written.

Today, people still write letters by hand and send cards, of course, but instantaneous communication has become basic to our social world. Our lives would be almost unimaginable without it. We are so used to being able to watch our favorite TV show online or send an email to a friend in another part of the world, at any hour of the day, that it is hard for us to imagine what life would be like otherwise.

Who Owns a Smartphone?

While rates of smartphone ownership in developing countries have skyrocketed in recent years, there remains a significant digital divide, with richer countries reporting higher levels of ownership.

Percentage of adults who own smartphones

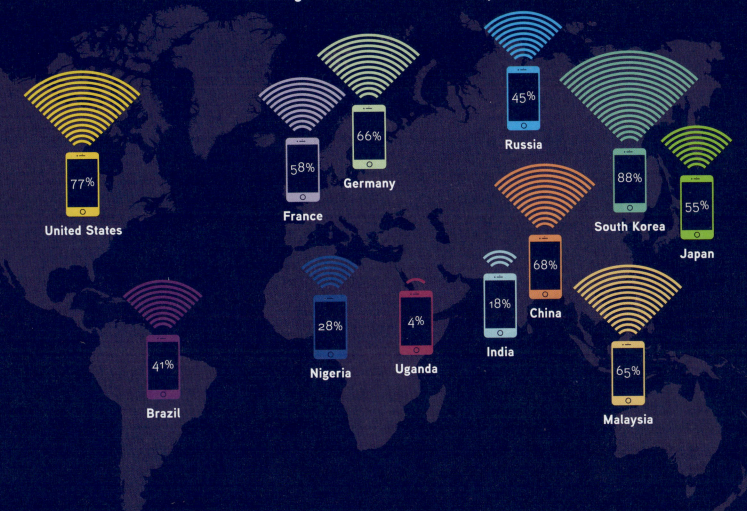

United States — 77%
France — 58%
Germany — 66%
Russia — 45%
South Korea — 88%
Japan — 55%
Brazil — 41%
Nigeria — 28%
Uganda — 4%
India — 18%
China — 68%
Malaysia — 65%

Who owns a smartphone in the United States?

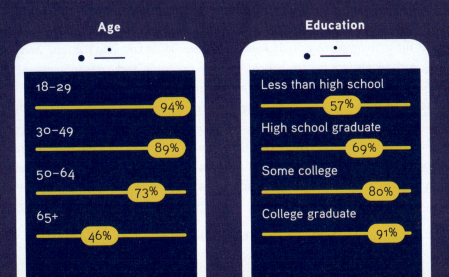

Age

18–29	94%
30–49	89%
50–64	73%
65+	46%

Education

Less than high school	57%
High school graduate	69%
Some college	80%
College graduate	91%

Income

Less than $30,000	67%
$30,000–$49,999	82%
$50,000–$74,999	83%
$75,000+	93%

Sources: Perrin 2018; Pew Research Center 2018d.

What drives people to increasingly interact over the Internet and through social media? For some, online platforms are more accessible than other public spaces. For example, sociologist danah boyd (2014) argues that when she attended high school in the 1990s, she and her friends would go on the Internet as an escape mechanism. They entered into chat rooms to connect with people who had similar interests and to escape from their local community. Today, however, boyd argues that students participate in social media to be in touch with their local community. In the 1980s and 1990s, many high school students might have gone to their local shopping mall to be around others and to know what was going on. Today, young people have less free time and geographic mobility; they have less access to increasingly regulated public spaces like malls. Platforms like Facebook and Twitter, then, enable teens to interact despite these constraints on their time and location.

Clock Time

clock time

Time as measured by the clock, in terms of hours, minutes, and seconds. Before the invention of clocks, time reckoning was based on events in the natural world, such as the rising and setting of the sun.

In modern societies, the zoning of our activities is strongly influenced by **clock time**. Without clocks and the precise timing of activities, and their resulting coordination across space, industrialized societies could not exist (Mumford 1973). Today the measuring of time by clocks is standardized across the globe, making possible the complex international transport systems and communications we now depend on. World standard time was first introduced in 1884 at a conference of nations held in Washington, D.C. The globe was then partitioned into twenty-four time zones, one hour apart, and an exact beginning of the universal day was fixed.

Fourteenth-century monasteries were the first organizations to try to schedule the activities of their inmates precisely across the day and week. Today, there is virtually no group organization that does not do so. The greater the number of people and resources involved, the more precise the scheduling must be. Eviatar Zerubavel (1979, 1982) demonstrated this in his study of the temporal structure of a large modern hospital. A hospital must operate on a twenty-four-hour basis, and coordinating the staff and resources is a highly complex matter. For instance, the nurses work for one time period in Ward A, another time period in Ward B, and so on, and are also called on to alternate between day- and night-shift work. Nurses, doctors, and other staff, plus the resources they need, must be integrated both in time and in space.

The Compulsion of Proximity

In modern societies, we are constantly interacting with others whom we may never see or meet. Rapid advances in communications technology, such as email, the Internet, and e-commerce, have only increased this tendency toward indirect interactions. According to one view, as the pace of life accelerates, people are increasingly isolating themselves; we now interact more with our televisions and computers than with our neighbors or members of the community. One study found that Internet use cuts into the time we spend socializing with people face-to-face, watching TV, and sleeping. Some researchers conclude that the substitution of email for face-to-face communication has led to a weakening of social ties and a disruption of techniques used in personal dialogue for avoiding conflict.

Furthermore, online communication seems to allow more room for misinterpretation, confusion, and abuse than more traditional forms of communication (Friedman and Currall 2003). Nationally representative survey data show that 41 percent of U.S. adults have been harassed online, while 61 percent have witnessed harassment directed toward others; nearly one in five have experienced severe forms of online harassment, including physical threats, harassment over a sustained period, sexual harassment, or stalking (Duggan 2017).

How far can electronic communication substitute for face-to-face interaction? Sociologists Deirdre Boden and Harvey Molotch (1994) argue that there is no substitute for face-to-face interaction. They argue further that humans have a true need for personal interaction, which they call the **compulsion of proximity**. People put themselves out to attend meetings, Boden and Molotch suggest, because situations of co-presence provide much richer information about how other people think and feel, and about their sincerity, than any form of electronic communication. Only by actually being in the presence of people who make decisions affecting us in important ways do we feel able to learn what is going on and confident that we can impress them with our own views and our own sincerity. And as Manti Te'o learned, electronic communication in the absence of accompanying face-to-face communication may provide a platform for highly insincere and dishonest behavior.

The G7 summit, a face-to-face meeting of the heads of governments of seven leading industrialized countries, is an example of what Molotch and Boden call the compulsion of proximity. Individuals, including world leaders, prefer face-to-face interactions because they provide richer information about how people think and feel.

However, many Internet enthusiasts defend its potential. They argue that online communication has many inherent advantages that cannot be claimed by more traditional forms of interaction, such as the telephone and face-to-face meetings. The human voice, for example, may be far superior in terms of expressing emotion and subtleties of meaning, but it can also convey information about the speaker's age, gender, ethnicity, or social position—information that could be used to the speaker's disadvantage. Electronic communication, it is noted, masks all these identifying markers and ensures that attention focuses strictly on the content of the message. This can be a great advantage for women or other traditionally disadvantaged groups whose opinions are sometimes devalued in other settings (Pascoe 2000). Electronic interaction is often presented as liberating and empowering because people can create their own online identities and speak more freely than they would elsewhere. But this is not always the case: Personal or physical characteristics are often subjects of harassment on social media—by strangers as well as acquaintances, friends, and family. In fact, 14 percent of U.S. adults say they have been harassed online because of their political views, while 9 percent have been targeted for their appearance, 9 percent for their race or ethnicity, and 8 percent for their gender (Duggan 2017: 3).

One advantage of the Internet is that people can communicate with those who don't share their geographic region; perhaps you have reconnected via Facebook with childhood friends who live hundreds of miles away. Individuals can bridge both physical and social distance on the Internet. Since the 1990s, people have forged relationships and shared information in online forums related to health issues like depression and HIV (Barak et al. 2008; Coursaris and Liu 2009). For instance, studies show that people may feel more comfortable discussing sensitive topics like sexual orientation, condom use, and HIV testing in an anonymous, online community rather than in person with members of their social circle (Taggart et al. 2015). Social media has also supported the growth of millennial-led social

compulsion of proximity

People's need to interact with others in their presence.

movements, such as the campus movement protesting sexual assault and the Black Lives Matter movement, allowing these movements to grow quickly, organizing and recruiting participants in less costly ways (Milkman 2017: 9).

Research shows that social networking may even enhance social integration and friendships (Hampton et al. 2011). A recent survey of teens and technology use found that 57 percent of teens between the ages of thirteen and seventeen had made new friends online (Lenhart 2015). Another survey found that persons who use social networking sites are more trusting, have more close relationships, receive more emotional and practical social support, and are more politically engaged than those who do not use such sites. As danah boyd's (2014) research on youth, discussed earlier, suggests, for many people, online relationships are quite meaningful.

CONCEPT CHECKS

1. How is technology rearranging space?

2. Is face-to-face inter-action, or co-presence, an important aspect of human action? Why or why not?

See how face-to-face interactions reflect broader social factors such as social hierarchies.

How Do the Rules of Social Interaction Affect Your Life?

As we saw in Chapter 1, microsociology, the study of everyday behavior in situations of face-to-face interaction, and macrosociology, the study of the broader features of society like race, class, or gender hierarchies, are closely connected. We now examine a social encounter that you may experience frequently—walking down a crowded city sidewalk—to illustrate this point. The rules of social interaction can never be understood independently of categories like race, class, and gender. All of these categories shape the social interactions in our everyday lives.

Women and Men in Public

Take, for example, a situation that may seem micro on its face: A woman walking down the street is verbally harassed by a group of men. Although the harassment of a single woman might be analyzed in microsociological terms by looking at a single interaction, it is not fruitful to view it that simply. Such harassment is typical of street talk involving men and women who are strangers (Gardner 1995; Hollaback! and IRL School 2015).

The problem of street harassment was powerfully highlighted by a recent video demonstration in which actress Shoshana Roberts was filmed walking down the streets of New York City over the course of ten hours (Rob Bliss Creative 2014). Wearing jeans and a T-shirt and instructed by the director to have her nonverbal communication indicate lack of interest, the camera captured her being hit with a barrage of constant catcalls—108 in total, not counting winks and whistles (Butler 2014). Ultimately, the video went viral and helped bring public attention to the issue of street harassment.

At the same time, some online responses to the video—including repeated threats of rape and violence directed toward Roberts (Butler 2014)—highlight the Internet as yet another public space in which sexism can play out. Additionally, many criticized the

director for the way that the video most prominently portrayed Black and Latino male harassers, as opposed to white and Asian male harassers. Addressing the issue of class and racial bias in the video's execution, Hanna Rosin (2014) wrote, "Activism is never perfectly executed. We can just conclude that they caught a small slice of catcallers, and lots of other men do it, too. But if the point of this video is to teach men about the day-to-day reality of women, then this video doesn't hit its target."

In what remains key research on gender and public harassment, Carol Brooks Gardner (1995) found that in various settings, these types of unwanted interactions occur as something women frequently experience as abusive. These kinds of interactions cannot be understood without also looking at the larger background of gender hierarchy in the United States. In this way, we can see how microanalysis and macroanalysis are connected. For example, Gardner linked the harassment of women by men to the larger system of gender inequality, represented by male privilege in public spaces, women's physical vulnerability, and the omnipresent threat of rape.

Most infamous of the settings in which Gardner (1995) studied public harassment were the edges of construction sites. Another recent controversy in Princeton, New Jersey, regarding a mall's billboard, vividly displays that street harassment is more than just a micro exchange between one man (or a group of men) and one woman walking down the street. When MarketFair Mall was undergoing renovation, a sign was erected that said, "We apologize for the whistling construction workers, but man you look good! So will we soon, please pardon our dust, dirt and other assorted inconveniences." A passerby, Elizabeth Harman, saw the billboard and took offense, posting an image of the sign on Facebook. The image went viral, spurring an online petition, a series of angry blogs, and the eventual removal of the sign.

As Harman, a philosophy professor, explained, "The issue of street harassment is really normalized in our society. . . . I didn't want my daughter to think that was a normal way to think about men yelling at women" (Karas 2012). Sociologist Gwen Sharp (2012) observed that street harassment (and the celebration of it through the MarketFair Mall sign) not only perpetuates the assumption that women are sexual objects to be admired but also that working-class men, such as construction workers, are sexist, uncouth, and unable to restrain their lustful thoughts.

INTERACTION ON THE "DIGITAL STREET"

Recently, sociologists have looked at how the proliferation of smartphones and the rise of social media are changing how boys and girls interact in public spaces. For teenagers in low-income urban areas, the street has long served as a hub of social life and dating. Street interactions between boys and girls may incorporate aspects of both courtship and the incivility that characterizes street talk between adult men and women who are strangers. Today these encounters are reshaped online, or what urban sociologist and communication scholar Jeffrey Lane refers to as the "digital street." Lane studied a cohort of teenagers in Harlem and found that boys and girls had changed the experience of public space by using social media to buffer interaction. Whereas boys were more visible and acted more dominant toward girls on the sidewalk, girls gained visibility and control online.

In his book *The Digital Street*, Lane found that whereas for some boys, social media provided *another*, alternative way to call out to girls, other boys engaged girls digitally *instead* of on the sidewalk, often using a messaging feature. Messaging girls in private rather than approaching them face-to-face on the sidewalk shielded boys from public rejection in front of their friends when advances went unmet. But private messages were not necessarily the best strategy. Christian, one of the teenagers in Lane's study, explained that "a girl hates" when a boy writes "a million messages, like, 'What's up, I'm trying to talk to you.'" A girl would rather a boy like one of her photos and then leave the girl to make the next move. The use of social media enabled ways to communicate at different distances and paces that steered teens either away from or toward meetings in person.

Lane argued that girls and boys were safer with social media because they could establish particular identities without the need to live those out in person. But traditionally gendered norms of interaction and roles also carried forward. For instance, boys and girls alike articulated the pressure on young women to objectify their bodies online. "Girls get naked for likes," said one eighteen-year-old girl in Lane's research. False identity was another issue. Unlike in physical space, appearance and person may decouple online in the case of "fake pages" or "catfish"—profiles that depict someone other than its user. The possibility of a fake page created trust issues among boys and girls that were compounded by the fact that law enforcement used profiles designed to mimic girls in the neighborhood to monitor and gather intelligence on boys of interest.

Race and the Public Sphere

In his book *Streetwise: Race, Class, and Change in an Urban Community* (1990), sociologist Elijah Anderson noted that studying everyday life sheds light on how social order is created by the individual building blocks of infinite micro-level interactions. He was particularly interested in understanding interactions when at least one party was viewed as threatening. Anderson showed that the ways many Blacks and whites interact on the streets of a northern city had a great deal to do with the structure of racial stereotypes, which is itself linked to the economic structure of society. In this way, he showed the link between micro interactions and the larger macro structures of society.

Anderson began by recalling Erving Goffman's description of how social roles and statuses come into existence in particular contexts or locations:

> *When an individual enters the presence of others, they commonly seek to acquire information about him or bring into play information already*

possessed. . . . Information about the individual helps to define the situation,
enabling others to know in advance what he will expect of them and they
may expect of him. (Anderson 1990: 166)

Following Goffman's lead, Anderson asked, What types of behavioral cues and signs make up the vocabulary of public interaction? He concluded that the people most likely to pass inspection are those who do not fall into commonly accepted stereotypes of dangerous persons: "Children readily pass inspection, while women and white men do so more slowly, black women, black men, and black male teenagers most slowly of all." In demonstrating that interactional tensions derive from outside statuses, such as race, class, and gender, Anderson shows that we cannot develop a full understanding of the situation by looking at the micro interactions themselves. This is how he makes the link between micro interactions and macro processes.

Anderson argues that people are streetwise when they develop skills, such as "the art of avoidance," to deal with the vulnerability they feel to violence and crime. According to Anderson, whites who are not streetwise cannot distinguish among different kinds of Black men (for example, middle-class youths versus gang members). They may also not know how to alter the number of paces to walk behind a suspicious person or how to bypass bad blocks at various times of day. In these ways, social science research can help you understand how a very ordinary behavior—navigating one's way through the city streets—reveals important lessons about the nature of social interaction today.

In recent years, Anderson has updated his earlier work on social interaction. Even in a society characterized by large amounts of racism and prejudice, Anderson has found reason for optimism. In *The Cosmopolitan Canopy* (2011), he argues that there are many places where people of different backgrounds actually get along. For Anderson, the racially and ethnically diverse spaces he has studied offer "a respite from the lingering tensions of urban life as well as an opportunity for diverse peoples to come together." They are what he calls "pluralistic spaces where people engage with one another in a spirit of civility, or even comity and goodwill."

On the basis of in-depth observations of public areas in his longtime home of Philadelphia, Anderson reports on various important sites in the city, including Reading Terminal Market, Rittenhouse Square, and the Galleria Mall. The first two are venues dominated by middle- and upper-middle-class norms and values, while the Galleria caters to the tastes of the Black working classes and the poor. All three sites, however, are spaces in which various kinds of people meet, agree to lay down their swords, carry on their life routines, and, in many cases, enjoy themselves.

Anderson begins with an ode to Philadelphia's indoor farmer's market—the downtown Reading Terminal. A regular at the market for decades, he paints a loving portrait of the many types of people who congregate there, including the population of Amish vendors. The patrons range from corporate executives to construction workers to senior citizens in poor health. They are all "on their best behavior" as they eat and shop for food and other items.

What is it about this space that causes people to "show a certain civility and even an openness to strangers"? To begin, the city is divided into two kinds of human beings: the open-minded people, whom he calls "cosmos" (shorthand for cosmopolitan), and the close-minded, whom he calls "ethnos" (shorthand for ethnocentric). As Anderson sees it,

Reading Terminal Market in Philadelphia is an example of what Elijah Anderson refers to as a "cosmopolitan canopy," a place where diverse groups congregate peacefully.

Reading Terminal is filled with open-minded people. Through a kind of people watching, each contributes to the creation of the cosmopolitan canopy. It is literally the sight of so many kinds of people in one another's physical presence, as well as participation in what one sees, that reinforces the idea of a "neutral space." Whites and minorities who have few opportunities for such interaction elsewhere can relax and move about with security. Blacks, however, understand that their status there is always provisional, meaning that at any moment, they are subject to dramatic situations in which whites fail to treat them with the respect they deserve.

The dynamic that Anderson highlights over and over is the self-fulfilling nature of the interaction: The interaction and the sight of it make it so. Most who come are probably repeat players, and they have long visualized different kinds of people getting along in the space. For newcomers, on the other hand, such visualization of tolerance is "infectious." In the Rittenhouse Square Park and the streets surrounding it, other social cues serve to bring about similar results. There is, for example, a fountain and a statue of a goat that attracts mothers, nannies, and children. The sight of "public mothering" is a cue that indicates that this is a civil place. A sense of safety and protection underlies good behavior and, in turn, leads to a virtuous circle of other acts of goodwill. Dog walkers are also crucial, with interaction naturally occurring between them and others (including children) as they form a critical mass in the park throughout the day.

The Galleria is a different story. Anderson describes it as the "ghetto downtown," a community of close-minded poor Blacks ("ethnos") in one mall. What makes it a canopy, albeit not a cosmopolitan one, is that various elements of the Black community—the "street" and the "decent"—can coexist here. People feel free to be themselves, "loud and boisterous and frank in their comments, released from the inhibitions they might feel among whites." The code of the street threatens to undermine the public order at any moment, but everyone is on his or her best behavior, with security guards reinforcing decorum. Nevertheless, Anderson stresses that through a negative feedback loop, this place has a self-reinforcing negative reputation among cosmopolitan whites and Blacks.

One of the great puzzles that social scientists will seek to resolve in the coming decades is whether "new" forms of social interaction, such as social networking sites, will alter gender relations and race relations. As we have seen throughout this chapter, virtual communication shares many of the same properties as face-to-face communication. For instance, we carefully impression manage in both venues. Recent research shows how social inequalities, such as those based around race, gender, and class, as Anderson documented in his *Streetwise* research, are embedded in new technologies. The websites we view online, and the knowledge we access through Google searches and other algorithmically driven software, reflect the prejudices of their creators (Daniels 2009; Noble 2018). In some cases, online communication may chip away at hierarchies based on power and status. In other cases, interactions that take place on online platforms are closely tied to and constrained by the social norms and cultural ideals of our offline communities.

Writing about the San Francisco tech community, Alice Marwick finds that social media use closely reflects its values—faith in deregulated capitalism and entrepreneurship. Marwick also describes how those who achieve status online fit a narrow model that reflects the preexisting distribution of power: In San Francisco's tech scene, successful online strategies "do not celebrate, for instance, outspoken women, discussion of race in technology, or openly gay entrepreneurs" (Marwick 2013: 18). As we have seen throughout the chapter, the expanded reach of new technologies certainly opens the possibility for wider and accelerated social change—but technology alone cannot determine the direction of that change.

CONCEPT CHECKS

1. How would sociologists explain the street harassment that women often experience?

2. How would sociologist Elijah Anderson define *streetwise*?

3. Can you identify any examples of cosmopolitan canopies in your neighborhood?

CHAPTER 4

The Big Picture

Social Interaction and Everyday Life in the Age of the Internet

Thinking Sociologically

1. Identify the elements important to the dramaturgical perspective. This chapter shows how the theory might be used to understand interactions between customers and service providers. How would you apply the theory to account for a plumber's visit to a client's home?

2. Smoking cigarettes is a pervasive habit found in many parts of the world and a habit that could be explained by both microsociological and macro-sociological forces. Give an example of each that would be relevant to explaining the proliferation of smoking. How might your suggested micro- and macro-level analyses be linked?

3. Think about the last photo you posted or the last status update you reported on Facebook. What impression were you trying to convey to others? How would Goffman characterize your impression-management goals?

Terms to Know

Concept Checks

social interaction • microsociology • civil inattention • agency • structure • unfocused interaction • focused interaction • encounter • nonverbal communication

1. What is microsociology and how does it differ from earlier sociological work?
2. What are three reasons it is important to study social interaction?
3. What are three forms of nonverbal communication?
4. Describe several ways that individuals communicate their emotions to one another.

status • social position • impression management

1. Identify three roles you play in social life.
2. Compare and contrast front and back regions.
3. Describe two forms of impression management.

ethnomethodology • interactional vandalism • response cries • personal space

1. What is interactional vandalism?
2. Give an example of a response cry.
3. What are the four zones of personal space?

time-space • regionalization • clock time • compulsion of proximity

1. How is technology rearranging space?
2. Is face-to-face interaction, or co-presence, an important aspect of human action? Why or why not?

1. How would sociologists explain the street harassment that women often experience?
2. How would sociologist Elijah Anderson define *streetwise*?
3. Can you identify any examples of cosmopolitan canopies in your neighborhood?

5

Groups, Networks, and Organizations

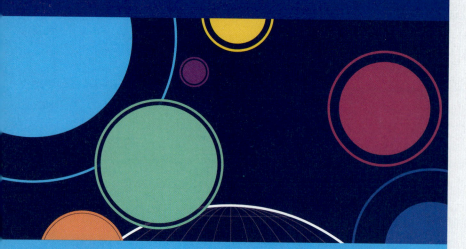

The Marching 100 was long a source of great pride at Florida A&M University. Pride turned to shock and shame in 2011 when a hazing incident led to the brutal death of drum major Robert Champion. Hazing is rampant on many college campuses and shows the tremendous power of social groups.

THE BIG QUESTIONS

What are social groups?

Learn the variety and characteristics of groups as well as the effect of groups on an individual's behavior.

How do we benefit from social networks?

Understand the importance of social networks and the advantages they confer on some people.

How do organizations function?

Know how to define an organization and understand how organizations developed over the last two centuries. Learn Max Weber's theory of organizations and view of bureaucracy.

Is bureaucracy an outdated model?

Familiarize yourself with some of the alternatives to bureaucracy that have developed in other societies or in recent times. Think about the influence of technology on how organizations operate.

How do organizations and groups affect your life?

Learn how social capital enables people to accomplish their goals and expand their influence.

Humans are social creatures. Most of us want to "belong" to some sort of group, and we cherish the friendships, the feeling of acceptance, and even the perks that go along with being a member of a group. But what are we willing to do in order to belong? Would you be willing to be beaten up? What about beating up someone else? Would you drink an entire bottle of vodka? Or stand outdoors in the January cold after being doused with cold water? Or run around in the sweltering summer heat wearing layers of heavy clothing?

Although this sounds barbaric, these types of hazing activities occur all too often on college campuses today. Hazing—or the rituals and other activities, involving harassment, abuse, or humiliation, used as a way of initiating a person into a group—occurs in an estimated 55 percent of all college fraternities and sororities today. Although some blithely view hazing as "boys being boys" or a rite of passage, others consider hazing an illegal activity that must be banished from all college campuses (Nuwer 2013).

News stories are replete with horrific incidences of hazing. In 2011, marching band members from Florida A&M University (FAMU) attacked drum major Robert Champion on a bus after a performance, paddling and beating him with musical instruments in a hazing ritual known as Crossing Bus C. The assault was so brutal that Champion died within the hour from "hemorrhagic shock caused by blunt-force trauma" (Alvarez 2013).

This senseless crime shocked the nation; the FAMU band—dubbed the Marching 100—had been one of the most celebrated in the nation. The Marching 100, like many college fraternities and sororities, had a history of hazing its members, with authorities turning a blind eye. In fact, ten years earlier, another band member had been beaten so badly that he suffered renal failure. Champion's death, though, was the breaking point, and the university acted swiftly. The university president was forced out. The band, once the shining star of FAMU, was suspended for nearly two years. Fifteen former band members were charged with manslaughter or felony hazing (Alvarez 2013). And while most would receive probation or community service, not everyone got off easy: In January 2015, the former bus leader and accused ringleader of the hazing ritual was sentenced to more than six years in prison for his role in Champion's death.

What would cause otherwise upstanding young men and women—many of whom are top students, star athletes, and accomplished musicians—to essentially torture their classmates? And why doesn't one brave soul in the group stand up to stop the abuse? Why do "pledges," those young men and women striving to join a college fraternity or sorority, subject themselves to this mistreatment?

The answers are complex but illustrate important aspects of group behavior, including conformity, or going along with the actions of the group regardless of the personal costs. In this chapter, we examine the ways in which all of us—not just fraternity brothers—are group animals. We will consider different kinds of groups, the ways group size affects our behavior in groups, and the nature of leadership. We will explore how group norms promote conformity, often to disastrous ends. We also examine the role organizations play in American society, the major theories of modern organizations, and the ways in which organizations are changing in the modern world. The increased effect of technology on organizations and the prominence of the Internet in our group life are also explored. The chapter concludes by discussing the debate over declines in social capital and social engagement in the United States today.

social group

A collection of people who regularly interact with one another on the basis of shared expectations concerning behavior and who share a sense of common identity.

social aggregate

A collection of people who happen to be together in a particular place but do not significantly interact or identify with one another.

social category

People who share a common characteristic (such as gender or occupation) but do not necessarily interact or identify with one another.

> Learn the variety and characteristics of groups as well as the effect of groups on an individual's behavior.

What Are Social Groups?

Nearly all our important interactions occur through some type of social group. You and your roommates make up a social group, as do the members of your introductory sociology class. A **social group** is a collection of people who share a common identity and regularly interact with one another on the basis of shared expectations concerning behavior. People who belong to the same social group identify with one another, expect one another to conform to certain ways of thinking and acting, and recognize the boundaries that separate them from other groups or people. Fraternities and sororities are examples of social groups, as are sports teams, musical groups, and even book groups.

Groups: Variety and Characteristics

Every day nearly all of us participate in groups and group activities. We hang out with groups of friends, study with classmates, eat dinner with family members, play team sports, and go online to meet people who share our interests (Aldrich and Marsden 1988).

But just being in one another's company does not make a collection of individuals a social group. People milling around in crowds or strolling on a beach make up a social aggregate. A **social aggregate** is a simple collection of people who happen to be together in a particular place but do not significantly interact or identify with one another. People waiting together at a bus station, for example, may be conscious of one another's presence, but they are unlikely to think of themselves as a "we"—the group waiting for the next bus to Poughkeepsie or Des Moines. By the same token, people may constitute a **social category**, sharing a common characteristic (such as gender, occupation, religion, or ethnicity) without necessarily interacting or identifying with one another. The sense of belonging to a common social group is missing.

IN-GROUPS AND OUT-GROUPS

In-groups are groups toward which one feels particular loyalty and respect—the groups that "we" belong to. **Out-groups**, on the other hand, are groups toward which one feels antagonism and contempt—"those people." The "sense of belonging" among members of the in-group is sometimes strengthened by the group's scorning the members of other groups (Sartre 1965, orig. 1948). Creating a sense of belonging in this way is especially common among racist groups, which promote their identity as "superior" by hating "inferior" groups. Jews, Catholics, African Americans, immigrants, and gay people historically—and Muslims more recently—have been the targets of such prejudice in the United States.

Most people occasionally use in-group–out-group imagery to trumpet what they believe to be their group's strengths vis-à-vis some other group's presumed weaknesses. For example, members of a fraternity or a sorority may bolster their feelings of superiority—in academics, sports, or campus image—by ridiculing the members of a different house. Similarly, a church may hold up its "truths" as the only ones, while native-born Americans may accuse immigrants—always outsiders upon arriving in a new country—of ruining the country for "real Americans."

PRIMARY AND SECONDARY GROUPS

Our lives and personalities are molded by our earliest experiences in **primary groups**, namely, our families, our peers, and our friends. Primary groups are small groups characterized by face-to-face interaction, intimacy, and a strong, enduring sense of commitment. There is also often an experience of unity, a merging of the self with the group into one personal "we." Sociologist Charles Horton Cooley (1864–1929) termed such groups *primary* because he believed that they were the basic form of association, exerting a long-lasting influence on the development of our social selves (Cooley 1964, orig. 1902).

Sports often create in-groups and out-groups by fostering loyalty for one's own team and contempt for the other team.

in-groups

Groups toward which one feels particular loyalty and respect—the groups to which "we" belong.

out-groups

Groups toward which one feels antagonism and contempt—"those people."

primary group

A group that is characterized by intense emotional ties, face-to-face interaction, intimacy, and a strong, enduring sense of commitment.

secondary group

A group characterized by its large size and by impersonal, fleeting relationships.

Secondary groups, by contrast, are large and impersonal and often involve fleeting relationships. Examples of secondary groups include business organizations, schools, work groups, athletic clubs, and governmental bodies. Secondary groups seldom involve intense emotional ties, powerful commitments to the group itself, or a feeling of unity. We seldom feel we can "be ourselves" in a secondary group; rather, we are often playing a particular role, such as employee or student. Cooley argued that while people belong to primary groups mainly because such groups are inherently fulfilling, people join secondary groups to achieve some specific goal: to earn a living, get a college degree, or compete in sports. Secondary groups may become primary groups for some of their members. For example, when coworkers begin to socialize after hours, they create bonds of friendship that constitute a primary group.

For most of human history, nearly all interactions took place within primary groups. This pattern began to change with the emergence of larger, agrarian societies, which included such secondary groups as those based on governmental roles or occupation. Some early sociologists, such as Cooley, worried about a loss of intimacy as more and more interactions revolved around large, impersonal organizations. However, what Cooley saw as the growing anonymity of modern life might also offer an increasing tolerance of individual differences. Primary groups, which often enforce strict conformity to group standards, can be stifling (Durkheim 1964, orig. 1893; Simmel 1955). Secondary groups, by contrast, are more likely than primary groups to be concerned with accomplishing a task rather than with enforcing conformity to group standards of behavior. As we will see in the Digital Life box, web-based crowdfunding efforts allow loosely knit collections of people who don't even know one another to accomplish important tasks, such as raising funds for a charity or creative project.

REFERENCE GROUPS

We often judge ourselves by how we think we appear to others, which Cooley termed the *looking-glass self*. Robert K. Merton (1938) elaborated on Cooley's concept by discussing reference groups as a standard by which we evaluate ourselves. A **reference group** is a group that provides a standard for judging one's own attitudes or behaviors (see also Hyman and Singer 1968). Family, peers, classmates, and coworkers are crucial reference groups. However, you do not have to belong to a group for it to be your reference group. For example, young people living thousands of miles away from the bright lights of Hollywood may still compare their looks and fashion choices with those of their favorite celebrity. Although most of us seldom interact socially with such reference groups as the glitterati, we may take pride in identifying with them and even imitate those people who do belong to them. This is why it is critical for children—minority children, in particular, whose groups are often represented in the media using negative stereotypes—to be exposed to reference groups that will shape their lives for the better.

reference group

A group that provides a standard for judging one's attitudes or behaviors.

The Effects of Size

Another significant way in which groups differ has to do with their size. Sociological interest in group size can be traced to the German sociologist Georg Simmel (1858–1918), who studied and theorized about the impact of small groups on people's behavior. Since Simmel's time, small-group researchers have conducted a number of laboratory experiments to

examine the effects of size on both the quality of interaction in the group and the effectiveness of the group in accomplishing certain tasks (Bales 1953, 1970; Hare, Borgatta, and Bales 1965; Homans 1950; Mills 1967).

DYADS

The simplest group, which Simmel (1955) called a **dyad**, consists of two persons. Simmel reasoned that dyads, which involve both intimacy and conflict, are likely to be simultaneously intense and unstable. To survive, they require the full attention and cooperation of both parties. If one person withdraws from the dyad, it vanishes. Dyads are typically the source of our most elementary social bonds, often constituting the group in which we are most likely to share our deepest secrets. But dyads can be very fragile. That is why, Simmel believed, a variety of cultural and legal supports for marriage—an example of a dyad—are found in societies where marriage is regarded as an important source of social stability.

TRIADS

According to Simmel, **triads**, or three-person groups, are more stable than dyads because the third person relieves some of the pressure on the other two to always get along and energize the relationship. In a triad, one person can temporarily withdraw attention from the relationship without necessarily threatening it. In addition, if two of the members have a disagreement, the third can play the role of mediator, as when you try to patch up a falling-out between two of your friends. Yet triads are not without potential problems. Alliances (sometimes termed *coalitions*) may form between two members of a triad, enabling them to gang up on the third and thereby destabilize the group.

dyad
A group consisting of two persons.

triad
A group consisting of three persons.

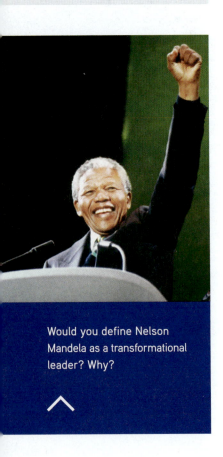

Would you define Nelson Mandela as a transformational leader? Why?

LARGER GROUPS

Simmel identified an important aspect of groups: As group size increases, their intensity decreases, while their stability and exclusivity increase. Larger groups have less intense interactions, simply because a larger number of potential smaller-group relationships exist as outlets for individuals who are not getting along with other members of the group. In a dyad, only a single relationship between two people is possible; in a triad, three different two-person relationships can occur. In a ten-person group, the number of possible two-person relationships explodes to forty-five. When one relationship doesn't work out to your liking, you can easily move on to another, as you probably often do at large parties.

Large groups also tend to be more stable than smaller ones because the withdrawal of some members does not usually threaten the group's survival. A marriage or romantic relationship falls apart if one person leaves, whereas a sports team or drama club routinely survives—though it may sometimes temporarily suffer from—the loss of its graduating seniors. Larger groups also tend to be more exclusive, because it is easier for their members to limit their social relationships to the group itself and avoid relationships with nonmembers.

Beyond a certain size, perhaps a dozen people, groups tend to develop a formal structure. Formal leadership roles may arise, such as president or secretary, and official rules may be developed to govern what the group does. We discuss formal organizations later in this chapter.

Types of Leadership

A **leader** is a person who is able to influence the behavior of other members of a group. All groups have leaders, even if the leader is not formally recognized as such. Some leaders are especially effective in motivating group members, inspiring them to achievements that might not ordinarily be accomplished. Such **transformational leaders** go beyond the merely routine, instilling in the members of their group a sense of mission or higher purpose and thereby changing the nature of the group itself (Burns 1978; Kanter 1983). They can also be a vital inspiration for social change in the world. For example, Nelson Mandela, the late South African leader who spent twenty-seven years in prison, successfully helped dismantle South Africa's system of apartheid, or racial segregation. He led his African National Congress party to victory and was elected president—leader—of the entire country. His leadership transcended national boundaries.

Most leaders are not as visionary as Mandela. Leaders who simply get the job done are termed **transactional leaders**. They are concerned with accomplishing the group's tasks, getting group members to do their jobs, and ensuring that the group achieves its goals. Transactional leadership is routine leadership. For example, the teacher who simply gets through the lesson plan each day—rather than making the classroom a place where students explore new ways of thinking and behaving—is exercising transactional leadership.

Conformity

Pressures to conform to the latest styles are especially strong among teenagers and young adults, for whom the need for group acceptance is often acute. Although sporting tattoos or the latest fashion trend—or rigidly conforming to corporate workplace policies—may seem relatively harmless, conformity to group pressure can also lead to extremely destructive behavior, such as drug abuse or even murder. For this reason, sociologists and social psychologists have long sought to understand why most people tend to go along with others and under what circumstances they do not.

GOING ALONG WITH THE GROUP: ASCH'S RESEARCH

More than sixty years ago, psychologist Solomon Asch (1952) conducted some of the most influential studies of conformity to group pressures. In one of his classic experiments, Asch asked individual subjects to decide which of three lines of different length most closely matched the length of a fourth line (Figure 5.1). The differences were obvious; subjects had no difficulty in making the correct match.

FIGURE 5.1

The Asch Task

In the Asch task, participants were shown a standard line (left) and then three comparison lines. Their task was simply to say which of the three lines matched the standard. When confederates gave false answers first, one-third of participants conformed by also giving the wrong answer.

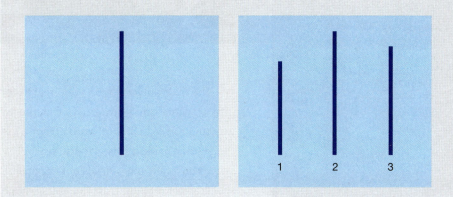

Asch then arranged a version of the experiment in which the subjects were asked to make the matches in a group setting, with each person calling out the answer in turn. In this setting, all but one of the subjects were actually Asch's secret accomplices, and these accomplices all practiced a deception on that one true subject. Each accomplice picked a line as a match that was clearly unequal to the fourth line. The unwitting subject, one of the last to call out an answer, felt enormous group pressure to make the same error. Amazingly, at least half the time the experiment was conducted, one-third of these subjects gave the same wrong answer as the others in the group. They sometimes stammered and fidgeted when doing so, but they nonetheless yielded to the unspoken pressure to conform to the group's decision. Asch's experiments clearly showed that many people are willing to go along with the group consensus, even if they believe it is incorrect.

Although the Asch study was conducted decades ago, its findings provide some insight into why hazing happens. Just like the participants in Asch's study who bowed to group pressure, members of the FAMU marching band might have found it difficult to stop the abuse of their drum major when they perceived (perhaps incorrectly) that the majority was in strong agreement about the rites of initiation.

OBEDIENCE TO AUTHORITY: MILGRAM'S RESEARCH

Another classic study of conformity was conducted by Stanley Milgram (1963). Milgram wanted to see how far a person would go when ordered by a scientist to give another person increasingly powerful electric shocks. He did so by setting up an experiment that he told the subjects was about memorizing pairs of words. In reality, it was about obedience to authority.

a b c

The male subjects who volunteered for the study were supposedly randomly divided into "teachers" and "learners." In fact, the learners were Milgram's confederates. The teacher was told to read pairs of words from a list that the learner was to memorize. Whenever the learner made a mistake, the teacher was to give him an electric shock by flipping a switch on a fake but official-looking machine. The control board on the machine indicated shock levels ranging from "15 volts—slight shock" to "450 volts—danger, severe shock." For each mistake, the voltage of the shock was to be increased, until it eventually reached the highest level. As the experiment progressed, the learner began to scream out in pain for the teacher to stop delivering shocks. Milgram's assistant, who was administering the experiment, exercised his authority as a scientist and ordered the teacher to continue administering shocks if the teacher tried to quit. (In reality, the learner, who was usually carefully concealed from the teacher by a screen, never received any electric shocks, and his "screams" had actually been prerecorded on a tape.)

The teacher was confronted with a major moral decision: Should he obey the scientist and go along with the experiment, even if it meant injuring another human being? Much to Milgram's surprise, more than half the subjects in the study kept on administering electric shocks. They continued even until the maximum voltage was reached and the learner's screams had subsided into an eerie silence as he had presumably died of a heart attack. How could ordinary people so easily conform to orders that would turn them into possible accomplices to murder?

The answer, Milgram found, was deceptively simple: Ordinary citizens will conform to orders given by someone in a position of power or authority—even if those orders have horrible consequences. From this, we can learn something about Nazi atrocities during World War II, which were Milgram's original concern. Many ordinary Germans who participated in the mass execution of Jews in Nazi concentration camps did so on the grounds that they were "just following orders." Milgram's research has sobering implications for anyone who thinks that only "others" will bend to authority, but "not me" (Zimbardo, Ebbeson, and Maslach 1977).

GROUPTHINK AND GROUP PRESSURES TO CONFORM: JANIS'S RESEARCH

The pressure to conform to group opinions may occasionally lead to bad decisions rather than creative new solutions to problems. Irving L. Janis (1972, 1989; Janis and Mann 1977) called this phenomenon **groupthink**, a process by which the members of a group ignore those ideas, suggestions, and plans of action that go against the group consensus.

groupthink

A process by which the members of a group ignore ways of thinking and plans of action that go against the group consensus.

Groupthink may embarrass potential dissenters into conforming and may also produce a shift in perceptions so that alternative possibilities are ruled out without being seriously considered. Groupthink may facilitate reaching a quick consensus, but the consensus may also be ill chosen. It may even be downright dangerous. Social scientists attribute a range of bad (and, in some cases, catastrophic) decisions to groupthink, including the space shuttle *Challenger* disaster in 1986 and the Bay of Pigs invasion in 1961 (Haig 2011; Janis and Mann 1977).

Janis engaged in historical research to see if groupthink had characterized U.S. foreign policy decisions. He examined several critical decisions, including that behind the infamous Bay of Pigs invasion of Cuba in 1961. John F. Kennedy, then the newly elected president, inherited a plan from the previous administration to help Cuban exiles liberate Cuba from the Communist government of Fidel Castro. The plan called for U.S. supplies and air cover to assist an invasion by an ill-prepared army of exiles at Cuba's Bay of Pigs. As history now shows, the invasion was a disaster. The army of exiles, after parachuting into a swamp nowhere near its intended drop zone, was immediately defeated, and Kennedy suffered a great deal of public embarrassment.

Kennedy's advisers were smart, strong willed, and well educated. How could they have failed to voice their concerns about the proposed invasion? Janis identified a number of possible reasons. First, the advisers were hesitant to disagree with the president lest they lose his favor. Second, they did not want to diminish group harmony in a crisis situation where teamwork was all-important. Third, they faced intense time pressure and had little opportunity to consult outside experts who might have offered radically different perspectives. All these circumstances contributed to a single-minded pursuit of the president's initial ideas rather than an effort to generate effective alternatives. To avoid groupthink, the group must ensure the full and open expression of all opinions, even strong dissent.

CONCEPT CHECKS

1. What is the difference between social aggregates and social groups? Provide examples of each.

2. Describe the main characteristics of primary and secondary groups.

3. When groups become large, why does their intensity decrease but their stability increase?

4. What is groupthink? How can it be used to explain why some decisions made by a group can lead to negative consequences?

How Do We Benefit from Social Networks?

Understand the importance of social networks and the advantages they confer on some people.

"*Whom* you know is often as important as *what* you know." This adage expresses the value of having good connections. Sociologists refer to such connections as **networks**—all the direct and indirect connections that link a person or a group with other people or groups. Your personal networks thus include people you know directly (such as your friends) as well as people you know indirectly (such as your friends' friends). The groups and organizations you belong to also may be networked. For example, all the chapters of Gamma Phi Beta or Hillel are linked, as are alumni from your college or university, thus connecting members to like-minded individuals throughout the United States and the world.

Networks serve us in many ways. You are likely to rely on your networks for a broad range of contacts, from obtaining access to your congressperson or senator to scoring a summer internship. Sociologist Mark Granovetter (1973) demonstrated that there can be enormous strength in weak ties, particularly among higher socioeconomic groups.

network

A set of informal and formal social ties that links people to one another.

Upper-level professional and managerial employees are likely to hear about new jobs through connections such as distant relatives or remote acquaintances. Such weak ties can be of great benefit because relatives or acquaintances tend to have very different sets of connections from one's closer friends, whose social contacts are likely to be similar to one's own. Among lower socioeconomic groups, Granovetter argued, weak ties are not necessarily bridges to other networks and so do not really widen one's opportunities (see also Knoke 1990; Marsden and Lin 1982; Wellman, Carrington, and Hall 1988).

Most people rely on their personal networks to gain advantages, but not everyone has equal access to powerful networks. In general, whites and men have more advantageous social networks than do ethnic minorities and women. Some sociologists argue, for example, that women's business, professional, and political networks are fewer and weaker than men's, so that women's power in these spheres is reduced (Brass 1985). Yet as more and more women move up into higher-level occupational and political positions, the resulting networks can foster further advancement. One study found that women are more likely to be hired or promoted into job levels that already have a high proportion of women (Cohen, Broschak, and Haveman 1998).

The Internet as Social Network

Our opportunities to belong to and access social networks have skyrocketed in recent years due to the Internet. Until the early 1990s, when the World Wide Web was developed, there were few Internet users outside of university and scientific communities. By the end of 2017, however, an estimated 287 million Americans were using the Internet (Internet World Stats 2017a), and while 52 percent of American adults used the Internet in 2000, 89 percent were online in 2018 (Pew Research Center 2018a). With such rapid communication and global reach, it is now possible to radically extend one's personal networks. Fully 57 percent of American teens have made new friends online (Lenhart 2015). It also enables people who might otherwise lack contact with others to become part of global networks. For example, people too ill to leave their homes can join online social networks or consult message boards, people in small rural communities can now take online college courses (Lewin 2012), and long-lost high school friends can reconnect via Facebook.

The Internet fosters the creation of new relationships, often without the emotional and social baggage or constraints that go along with face-to-face encounters. In the absence of the usual physical and social cues, such as skin color or residential address, people can get together electronically on the basis of shared interests, such as gaming, rather than similar social characteristics. Factors such as social position, wealth, race, ethnicity, gender, and physical disability are less likely to cloud the social interaction (Coate 1994; Jones 1995; Kollock and Smith 1996). In fact, technologies like Twitter allow people from all walks of life to catch glimpses into the lives of celebrities (as well as noncelebs).

One limitation of Internet-based social networks is that not everyone has equal access to the Internet. Lower-income persons and ethnic minorities are less likely than wealthier persons and whites to have Internet access. But while a digital divide remains, the gaps have narrowed considerably in recent years. For example, in 2000, 81 percent of American adults in households earning $75,000 or more a year used the Internet, compared to just

Internet Connectivity

While cyberspace is becoming increasingly global, there remains a digital divide between individuals with access to the Internet and those without. While 95 percent of the population of North America is using the Internet, only 35 percent of the population of Africa is online.

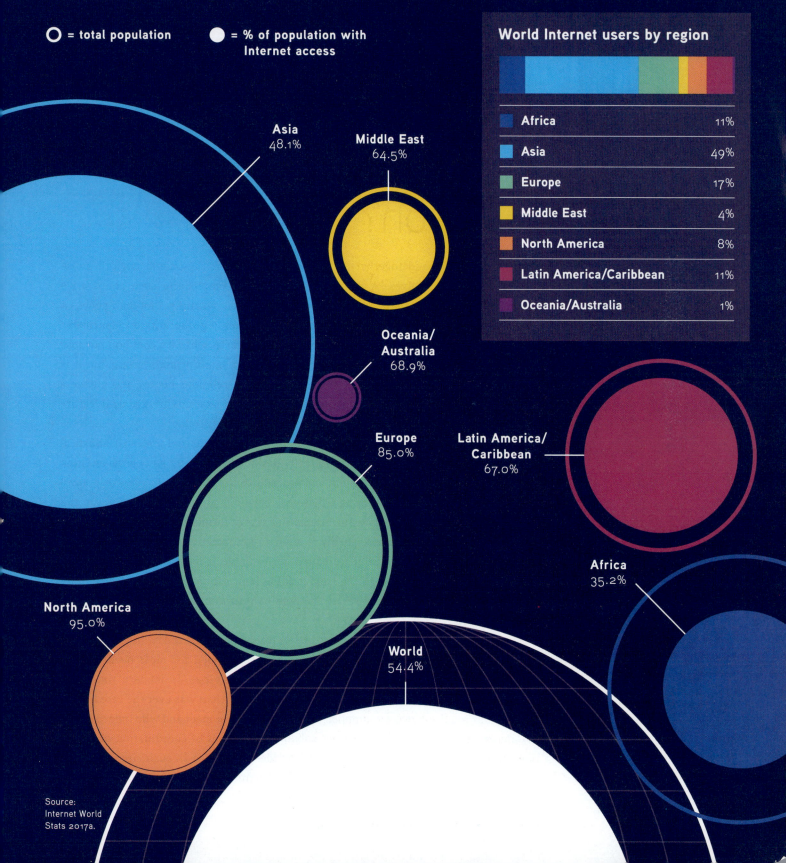

○ = total population

● = % of population with Internet access

Asia
48.1%

Middle East
64.5%

Oceania/
Australia
68.9%

Europe
85.0%

Latin America/
Caribbean
67.0%

North America
95.0%

Africa
35.2%

World
54.4%

World Internet users by region

Region	%
Africa	11%
Asia	49%
Europe	17%
Middle East	4%
North America	8%
Latin America/Caribbean	11%
Oceania/Australia	1%

Source:
Internet World
Stats 2017a.

CONCEPT CHECKS

1. According to Granovetter, what are the benefits of weak ties? Why?

2. How do men's and women's weak ties differ?

Know how to define an organization and understand how organizations developed over the last two centuries. Learn Max Weber's theory of organizations and view of bureaucracy.

organization

A large group of individuals with a definite set of authority relations. Many types of organizations exist in industrialized societies, influencing most aspects of our lives. While not all organizations are bureaucratic, there are close links between the development of organizations and bureaucratic tendencies.

34 percent of those who made less than $30,000. By 2018, however, this nearly 50 percent gap had narrowed to 17 percent, with 81 percent of those who make less than $30,000 per year now using the Internet (Pew Research Center 2018a). There remains a larger gap in usage by level of education: While 97 percent of adults with a college degree are Internet users, that proportion drops to 65 percent for those with less than a high school education. A similar gap in Internet use exists between young adults (ages eighteen to twenty-nine) and older adults (ages sixty-five and older): While 98 percent of young adults are using the Internet, the same can be said of only 66 percent of older adults.

This pattern is not limited to the United States; rates of Internet use are creeping up across the globe, enabling individuals to connect with anyone in the world who shares their interests.

How Do Organizations Function?

People frequently band together to pursue activities that they could not otherwise accomplish by themselves. A principal means for accomplishing such cooperative actions—whether it's raising money for cancer research, winning a football game, or becoming a profitable corporation—is the **organization**, a group with an identifiable membership that engages in concerted collective actions to achieve a common purpose (Aldrich and Marsden 1988). An organization can be a small primary group, but it is more likely to be a larger, secondary one: Universities, religious bodies, and business corporations are all examples of organizations. Such organizations are a central feature of all societies, and their study is a core concern of sociology today.

Organizations tend to be highly formal in modern industrial and postindustrial societies. A **formal organization** is rationally designed to achieve its objectives, often by means of explicit rules, regulations, and procedures. As Max Weber (1979, orig. 1921) first recognized almost a century ago, there has been a long-term trend in Europe and North America toward formal organizations. This rise of formality in organizations is in part the result of the fact that formality is often a requirement for legal standing. For a college or university to be legally accredited, for example, it must satisfy explicit written standards governing everything from grading policy to faculty performance to fire safety. Today, formal organizations are the dominant form of organization throughout the entire world.

It is easy to see why organizations are so important to us today. In the premodern world, families, close relatives, and neighbors provided for most needs—food, the instruction of children, work, and leisure-time activities. In modern times, the majority of the population is much more interdependent than was ever the case before. Many of our requirements are supplied by people we never meet and who indeed might live many thousands of miles away. A substantial amount of coordination of activities and resources—which organizations provide—is needed in such circumstances. A downside, however, is that organizations take things out of our own hands and put them under the control of officials or experts over whom we have little influence. For instance, we are all required to do certain things the government tells us to do—pay taxes, abide by laws, go off to fight wars—or face punishment.

Theories of Organizations

Max Weber developed the first systematic interpretation of the rise of modern organizations. Organizations, he argued, are ways of coordinating the activities of human beings, or the goods they produce, in a stable manner across space and time. Weber emphasized that the development of organizations depends on the control of information, and he stressed the central importance of writing in this process: An organization needs written rules to function and files in which its "memory" is stored. Weber saw organizations as strongly hierarchical, with power tending to be concentrated at the top.

Was Weber right? If he was, it matters a great deal to us all. For Weber detected a clash as well as a connection between modern organizations and democracy that he believed had far-reaching consequences for social life.

© Wiley Ink, inc./Distributed by Universal Uclick via Cartoonstock

Bureaucracies are often perceived as inefficient and fraught with red tape.

BUREAUCRACY

All large-scale organizations, according to Weber, tend to be bureaucratic in nature. The word *bureaucracy* was coined by Monsieur de Gournay in 1745, who combined the word *bureau*, meaning both an office and a writing table, with the suffix *cracy*, derived from the Greek verb meaning "to rule." **Bureaucracy** is thus the rule of officials. The term was first applied only to government officials, but it was gradually extended to refer to large organizations in general. Perceptions of "bureaucracy" range from highly negative—fraught with red tape, inefficiency, and wastefulness—to quite positive—a model of carefulness, precision, and effective administration.

Weber's account of bureaucracy steers between these two extremes. He argued that the expansion of bureaucracy is inevitable in modern societies; bureaucratic authority is the only way of coping with the administrative requirements of large-scale social systems. Yet he also conceded that bureaucracy exhibits a number of major failings that have important implications for the nature of modern social life.

To study the origins and nature of the expansion of bureaucratic organizations, Weber constructed an ideal type of bureaucracy. *Ideal* here refers not to what is most desirable but to a pure form of bureaucratic organization. An **ideal type** is an abstract description constructed by accentuating certain features of real cases to pinpoint their most essential characteristics. Weber (1921) listed several characteristics of the ideal type of bureaucracy:

1. **A clear-cut hierarchy of authority, such that tasks in the organization are distributed as "official duties."** Each higher office controls and supervises the one below it in the hierarchy, thus making coordinated decision making possible.

2. **Written rules govern the conduct of officials at all levels of the organization.** The higher the office, the more the rules tend to encompass a wide variety of cases and demand flexibility in their interpretation.

formal organization

Means by which a group is rationally designed to achieve its objectives, often using explicit rules, regulations, and procedures.

bureaucracy

A type of organization marked by a clear hierarchy of authority and the existence of written rules of procedure and staffed by full-time, salaried officials.

ideal type

A "pure type," constructed by emphasizing certain traits of a social item that do not necessarily exist in reality. An example is Max Weber's ideal type of bureaucratic organization.

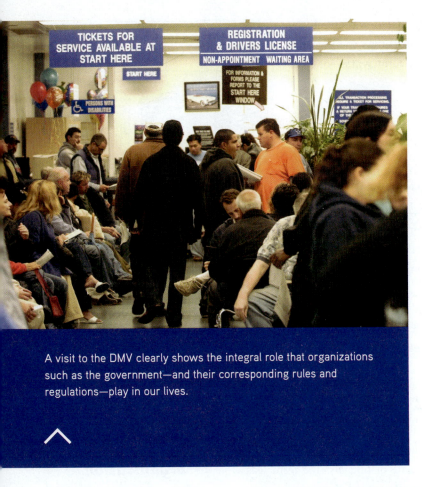

A visit to the DMV clearly shows the integral role that organizations such as the government—and their corresponding rules and regulations—play in our lives.

3. **Officials are full time and salaried.** Each job in the hierarchy has a definite and fixed salary attached to it. Promotion is possible on the basis of capability, seniority, or a mixture of the two.

4. **There is a separation between the tasks of an official within the organization and his or her life outside.**

5. **No members of the organization own the material resources with which they operate.** The development of bureaucracy, according to Weber, separates workers from the control of their means of production; officials do not own the offices they work in, the desks they sit at, or the office machinery they use.

Weber believed that the more an organization approaches the ideal type of bureaucracy, the more effective it will be in pursuing the objectives for which it was established. Yet he recognized that bureaucracy could be inefficient and accepted that many bureaucratic jobs are dull, offering little opportunity for the exercise of creative capabilities. While Weber feared that the bureaucratization of society could have negative consequences, he concluded that bureaucratic routine and the authority of officialdom over our lives are prices we pay for the technical effectiveness of bureaucratic organizations.

Since Weber's time, the bureaucratization of society has become more widespread. Critics of this development who share Weber's initial concerns have questioned whether the efficiency of rational organizations comes at a price greater than Weber could have imagined. The most prominent of these critiques refers to the "McDonaldization of society," discussed later in this chapter.

FORMAL AND INFORMAL RELATIONS WITHIN BUREAUCRACIES

Weber's analysis of bureaucracy gave prime place to **formal relations** within organizations, or the relations between people as stated in the rules of the organization. Weber had little to say about the informal connections and small-group relations that may exist in all organizations. But in bureaucracies, informal ways of doing things often allow for a flexibility that couldn't otherwise be achieved.

Informal networks tend to develop at all levels of organizations. At the very top, personal ties and connections may be more important than the formal situations in which decisions are supposed to be made. For example, meetings of boards of directors and shareholders supposedly determine the policies of business corporations. In practice, a few members of the board often really run the corporation, making their decisions informally and expecting the board to approve them. Informal networks of this sort can also stretch across different corporations. Business leaders from different firms frequently consult one another in an informal way and may belong to the same clubs and social circles.

formal relations

Relations that exist in groups and organizations, laid down by the norms, or rules, of the official system of authority.

informal networks

Relations that exist in groups and organizations developed on the basis of personal connections; ways of doing things that depart from formally recognized modes of procedure.

John Meyer and Brian Rowan (1977) argue that formal rules and procedures in organizations are usually quite distant from the practices actually adopted by the organizations' members. Formal rules, in their view, are often "myths" that people profess to follow but that have little substance in reality. They serve to legitimize—to justify—ways in which tasks are carried out, even while these ways may diverge greatly from how things are supposed to be done, according to the rules.

Deciding how far informal procedures generally help or hinder the effectiveness of organizations is not a simple matter. Systems that resemble Weber's ideal type tend to give rise to a forest of unofficial ways of doing things. This is partly because the flexibility that is lacking ends up being achieved by unofficial tinkering with formal rules. For those in dull jobs, informal procedures often also help create a more satisfying work environment. Informal connections between officials in higher positions may be effective in ways that aid the organization as a whole. On the other hand, these officials may be more concerned with advancing or protecting their own interests than with furthering those of the overall organization.

BUREAUCRACY AND DEMOCRACY

The diminishing of democracy with the advance of modern forms of organization was another problem that worried Weber a great deal (see also Chapter 13). What especially disturbed him was the prospect of rule by faceless bureaucrats. How can democracy be anything other than a meaningless slogan in the face of the increasing power bureaucratic organizations are wielding over us? After all, Weber reasoned, bureaucracies are necessarily specialized and hierarchical. Those near the bottom of the organization inevitably find themselves reduced to carrying out mundane tasks and have no power over what they do; power passes to those at the top. Weber's student Robert Michels (1967, orig. 1911) invented a phrase to refer to this loss of power that has since become famous: In large-scale organizations, and more generally in a society dominated by organizations, he argued, there is an **iron law of oligarchy**. **Oligarchy** means "rule by the few." According to Michels, the flow of power toward the top is simply an inevitable part of an increasingly bureaucratized world—hence the term *iron law*.

Was Michels right? It surely is correct to say that large-scale organizations involve the centralizing of power. Yet there is good reason to suppose that the iron law of oligarchy is not quite as hard-and-fast as Michels claimed. Unequal power is not just a function of size. In modest-sized groups, there can be marked differences of power. In a small business, for instance, where the activities of employees are directly visible to the directors, much tighter control might be exerted than in offices in larger organizations. Furthermore, in many modern organizations, power is also quite often openly delegated downward from superiors to subordinates. In many large companies, corporate heads are so busy coordinating different departments, coping with crises, and analyzing budget and forecast figures that they have little time for original thinking.

GENDER AND ORGANIZATIONS

Until some two decades ago, organizational studies did not devote very much attention to the question of gender. The rise of feminist scholarship in the 1970s, however, led to examinations of gender relations in all the main institutions in society, including organizations and bureaucracy. Feminist sociologists, most notably

iron law of oligarchy

A term coined by Weber's student Robert Michels meaning that large organizations tend toward centralization of power, making democracy difficult.

oligarchy

Rule by a small minority within an organization or society.

Mary T. Barra has served as CEO of General Motors since 2014. In 2018, women represented just 5 percent of the CEOs of the largest 500 public companies.

FIGURE 5.2

Female CEOs at S&P 500 Companies

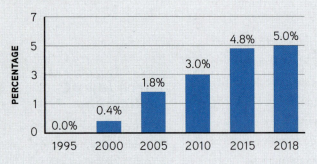

Source: Catalyst 2018a.

Rosabeth Moss Kanter in her classic book *Men and Women of the Corporation* (1977), focused on the imbalance of gender roles within organizations and the ways in which modern organizations themselves had developed in a specifically gendered way.

Feminists have argued that the emergence of the modern organization and the bureaucratic career depended on a particular gender configuration. They pointed to two main ways in which gender is embedded in the very structure of modern organizations. First, bureaucracies are characterized by occupational gender segregation. As women began to enter the labor market in greater numbers, they tended to be segregated into categories of occupations that were low paying and involved routine work. These positions were subordinate to those occupied by men and did not provide opportunities for women to be promoted. Women were used as a source of cheap, reliable labor but were not granted the same opportunities as men to build careers.

Second, the idea of a bureaucratic career was in fact a male career in which women played a crucial supporting role. In the workplace, women performed the routine tasks—as clerks, secretaries, and office managers—thereby freeing up men to advance their careers. Men could concentrate on obtaining promotions or landing big accounts because the female support staff handled much of the busywork. In the domestic sphere, women also supported the bureaucratic career by caring for the home, the children, and the man's day-to-day well-being. Women allowed male bureaucrats to work long hours, travel, and focus solely on their work without concern about personal or domestic issues.

As a result of these two tendencies, early feminist writers argued, modern organizations have developed as male-dominated preserves in which women are excluded from power, denied opportunities to advance their careers, and victimized on the basis of their gender through sexual harassment and discrimination. The #MeToo movement has made painfully clear that modern organizations—from Hollywood movie studios to automobile factories—remain troubled spaces for women. We explore this issue more fully in Chapter 9 ("Gender Inequality").

Women have made tremendous strides in politics, work, education, and most other domains since the 1970s. However, concerns about unequal pay, discrimination, and the male hold on power persist today. Furthermore, women in corporate power may not necessarily implement policies that help other women up the corporate ladder. For example, in February 2013, Marissa Mayer, newly appointed president and CEO of Yahoo, banned telecommuting at the tech giant. This highly controversial policy change was criticized for making life harder for working mothers. Just a few months later, however, Yahoo changed its parental leave policy, increasing paid leave for both moms and dads and providing new parents with cash bonuses. This new parental leave policy may help ensure that working women (and especially mothers) can remain in and ultimately ascend the ranks at their corporate employers.

CONCEPT CHECKS

1. What role do organizations play in contemporary society?

2. What does the term *bureaucracy* mean?

3. Describe five characteristics of an ideal type of bureaucracy.

4. Explain how modern organizations have developed in a gendered way.

Is Bureaucracy an Outdated Model?

For quite a long while in the development of Western societies, Weber's model held well. In government, hospital administration, universities, and business organizations, bureaucracy seemed to be dominant. Although informal social groups always develop in bureaucratic settings and tend to function effectively in the workplace, it seemed as though the future might be just what Weber had anticipated: constantly increasing bureaucratization.

Bureaucracies still exist aplenty in the West, but Weber's idea that a clear hierarchy of authority, with power and knowledge concentrated at the top, is the only way to run a large organization is starting to look archaic. Numerous organizations are overhauling themselves to become less, rather than more, hierarchical. Traditional bureaucratic structures are now believed to stifle innovation and creativity in cutting-edge industries. Departing from rigid vertical command structures, many organizations are turning to "horizontal," collaborative models to become more flexible and responsive to fluctuating markets. In this section, we examine some of the main forces behind these shifts, including globalization and the growth of information technology, and consider some of the ways in which modern organizations are reinventing themselves in light of changing circumstances.

The Transformation of Management

Traditional Western forms of management are hierarchical and authoritarian, whereas corporations in Japan, for example, typically focus on management–worker relations and try to ensure that employees at all levels feel a personal attachment to the company. The Japanese emphasis on teamwork, consensus-building approaches, and broad-based employee participation has been demonstrated to yield more productive and competitive workers. As a result, in the 1980s, many Western organizations began to introduce new management techniques to rival the productivity and competitiveness of their Japanese counterparts.

Two popular branches of management theory—human resource management and the corporate culture approach—have since been adopted by Western organizations. **Human resource management** is a style of management that regards a company's workforce as vital to its economic competitiveness: If the employees are not completely dedicated to the firm and its product, the firm will never be a leader in its field. To generate employee enthusiasm and commitment, the entire organizational culture must be retooled so that workers feel they have an investment in the workplace and in the work process.

The second management trend—creating a distinctive **corporate culture**—is closely related to human resource management. To promote loyalty to the company and pride in its work, the company's management works with employees to build an organizational culture involving rituals, events, or traditions unique to that company. These cultural activities are designed to draw all members of the firm—from the most senior managers to the newest employee—together to strengthen group solidarity. Company picnics, casual Fridays, and company-sponsored community service projects are examples of techniques for building a corporate culture.

Google, for example, has a distinctive corporate culture that is designed to help foster creativity and collaboration. The company encourages employees to design their own

Familiarize yourself with some of the alternatives to bureaucracy that have developed in other societies or in recent times. Think about the influence of technology on how organizations operate.

human resource management

A style of management that regards a company's workforce as vital to its economic competitiveness.

corporate culture

An organizational culture involving rituals, events, or traditions that are unique to a specific company.

desks to fit their personal work styles. Scooters are kept on hand in the office so employees can zoom quickly to the other side of the building for a quick conversation with coworkers. Employees can take breaks at Lego stations or grab a bite at gourmet cafeterias that serve free breakfast, lunch, and dinner. These quirky perks help Google to define and uphold its own unique culture. "The philosophy is very simple," said Google engineering director Craig Neville-Manning. "Google's success depends on innovation and collaboration. Everything we did was geared toward making it easy to talk" (Stewart 2013).

Technology and Modern Organizations

The development of **information technology**—computers and electronic communication media, such as the Internet—is another factor currently influencing organizational structures (Attaran 2004; Bresnahan, Brynjolfsson, and Hitt 2002; Castells 2000, 2001; Kanter 1991; Kobrin 1997; Zuboff 1988). Because data can be processed instantaneously in any part of the world linked to a computer-based communications system, there is no need for physical proximity between those involved. As a result, the introduction of new technology has allowed many companies to reengineer their organizational structures.

Telecommuting is an example of how large organizations have become more decentralized as the more routine tasks disappear, reinforcing the tendency toward smaller, more flexible types of enterprises (Burris 1998). A good deal of office work, for instance, can be carried out by remote employees who use the Internet and smartphones to do their work at home or somewhere other than their employer's primary office. According to a Gallup (2017) survey, 43 percent of workers in 2016 spent some time working remotely, up from 39 percent in 2012. Nearly a third of these employees work remotely 80 percent or more of the time; a fifth work remotely 100 percent of the time. Remote workers are much more likely to have a college or advanced degree and to work in managerial or professional positions (Noonan and Glass 2012).

Telecommuting poses both advantages and disadvantages for workers and their employers. One explanation for why telecommuting increases productivity is that it eliminates time spent by workers commuting to and from the office, permitting greater concentration of energy on work-related tasks (Hartig, Johansson, and Kylin 2003). However, these flexible new work arrangements have repercussions. First, the employees lose the human side of work; computer terminals are not an attractive substitute for face-to-face interaction with colleagues and friends at work. The flexibility of telework creates new types of stress stemming from isolation, distraction, and conflicting demands of work and home responsibilities (Ammons and Markham 2004; Raghuram and Wiesenfeld 2004). Remote workers are more likely to work overtime, working between 5 and 7 more hours per week than nontelecommuters (Noonan and Glass 2012).

On the other hand, management cannot easily monitor the activities of employees not under direct supervision (Dimitrova 2003; Kling 1996). While this may create problems for employers, it allows employees greater flexibility in managing their nonwork roles, thus contributing to increased worker satisfaction (Davis and Polonko 2001). Telecommuting also creates new possibilities for older workers and those with physical limitations to remain independent, productive, and socially connected (Bouma et al. 2004; Bricout 2004).

The growth of telecommuting is sparking profound changes in many social realms. It is restructuring business management practices and authority hierarchies within businesses (Illegems and Verbeke 2004; Spinks and Wood 1996) as well as contributing

For many, the word *corporation* conjures up images of a boring gray office. Yet, firms like Google are creating their own corporate cultures, offering perks like yoga classes and free food.

information technology

Forms of technology based on information processing and requiring microelectronic circuitry.

Crowdfunding and the Strength of Weak Ties

Imagine that a major symphony orchestra or prominent art museum is holding a fund-raiser. You are probably imagining wealthy men and women of a certain age, dressed in tuxedos and gowns, sipping champagne and making small talk about investments. They might vacation in the Hamptons together, or their children were friends at prep school. These images are perhaps an over-the-top stereotype, but one component of this description is probably true: Many of the attendees likely know one another.

But how might middle- and working-class people raise money for their favorite causes—whether it's to provide relief for hurricane victims, help the homeless, or even launch a company? Over the past five years, the Internet has exploded with crowdfunding sites. The term *crowdfunding* refers to the collective effort of many people who pool their money to support another person or group's cause. Crowdfunding relies on many smaller donations rather than a few large ones, like we might see at a museum fund-raiser attended by very wealthy people. For example, on Watsi—a global crowdfunding platform that allows people to directly fund low-cost medical care for people in developing nations—donors can give as little as $5 to help a patient in need (LaPorte 2013).

One of the most fascinating aspects of crowdfunding is that it allows people to raise money by relying primarily on their "weak ties," or friends of friends of friends. By spreading the word about a venture, sharing information about the web link and the project, crowdfunding involves many participants from all walks of life. Often, these donors know one another through loose social ties; perhaps they learn about a particular charity because they belong to the same Facebook group. Other times, though, donors to a particular project may belong to quite different social networks. They come together through crowdfunding platforms like Kickstarter, GoFundMe, YouCaring, and Indiegogo, which provide a point of entry for making donations to specific charities and projects.

These different platforms cater to different types of projects. For example, in the aftermath of Hurricane Harvey, thousands of Americans turned to YouCaring for help; the more than 6,000 campaigns brought in about $44 million for victims of the storm (Chan 2017b). With the help of a few A-list celebrities, the Time's Up GoFundMe campaign raised nearly $17 million in its first month to cover legal fees for victims of sexual harassment and assault (Langone 2018). Other times, crowdfunding projects are purely recreational. One of the most successful crowdfunded projects to date is *Star Citizen*, an online video game that has raised nearly $200 million on Kickstarter since 2012 (Gault 2018).

The "crowds" that support these projects would hardly constitute a primary group, as most of the members are not close significant others. In most cases, they do not even consist of secondary groups, because the crowd members may not even know one another. Do you think that the "crowd" that supports such projects constitutes a group? How might this type of group differ from the collection of people who attend fund-raising galas like the one described earlier? What makes crowdfunding successful? Have you ever used a crowdfunding site to raise money? What kind of people do you imagine might contribute to your project?

More than 20,000 people have contributed to the GoFundMe campaign for Time's Up, which has raised over $20 million to help victims of sexual assault and harassment advance their careers.

to new trends in housing and residential development that prioritize spatial and techno-logical requirements for telework in homes, which are built at increasing distances from city centers (Hartig, Johansson, and Kylin 2003).

The experiences of telecommuters clearly show how organizational adaptations to new technologies can have both positive and negative consequences for workers. While computerization has resulted in a reduction in hierarchy, it has created a two-tiered occupational structure composed of technical "experts" and less-skilled production or clerical workers. In these restructured organizations, jobs are redefined based more on technical skill than on rank or position. For expert professionals, traditional bureaucratic constraints are relaxed to allow for creativity and flexibility, but other workers have limited autonomy (Burris 1993). Although professionals benefit more from this expanded autonomy, computerization makes production and service workers more visible and vulnerable to supervision (Wellman et al. 1996; Zuboff 1988).

The "McDonaldization" of Society

Not everyone agrees that our society and its organizations are moving away from the Weberian view of rigid, orderly bureaucracies. The idea that we are witnessing a process of debureaucratization, they argue, is overstated.

In a contribution to the debate over debureaucratization, George Ritzer (1993) has developed a vivid metaphor to express his view of the transformations taking place in industrialized societies. He argues that although some tendencies toward debureaucratiza-tion have indeed emerged, on the whole, what we are witnessing is the "McDonaldization" of society. According to Ritzer, McDonaldization is "the process by which the principles of the fast-food restaurants are coming to dominate more and more sectors of American society as well as the rest of the world." Ritzer uses the four guiding principles for McDonald's restaurants—efficiency, calculability, uniformity, and control through automation—to show that our society is becoming ever more rationalized with time.

If you have ever visited McDonald's restaurants in two different locations, you will have noticed that there are very few differences between them. The interior decoration may vary slightly and the language spoken will most likely differ from country to country, but the layout, the menu, the procedure for ordering, the staff uniforms, the tables, the packaging, and the "service with a smile" are virtually identical. The McDonald's system is deliberately constructed to maximize efficiency and minimize human responsibility and involvement in the process. Except for certain key tasks, such as taking orders and pushing the start and stop buttons on cooking equipment, the restaurants' functions are highly automated and largely run themselves.

Ritzer argues that society as a whole is moving toward this highly standardized and regulated model for getting things done. Many aspects of our daily lives, for example, now involve automated systems and computers instead of human beings. E-commerce is threatening to overtake trips to the store, ATMs outnumber bank tellers, and prepackaged meals provide a quicker option than cooking. Ritzer, like Weber before him, is fearful of the harmful effects of bureaucratization on the human spirit and creativity. He argues that McDonaldization is making social life more homogeneous, more rigid, and less personal.

CONCEPT CHECKS

1. How has the Japanese model influenced the Western approach to management?

2. Explain how the development of information technology has changed the ways people live and work.

3. According to George Ritzer, what are the four guiding principles used in McDonald's restaurants? What does he mean by the "McDonaldization" of society?

What is McDonaldization? What are the consequences of highly standardized experiences?

>

How Do Organizations and Groups Affect Your Life?

Social Capital: The Ties That Bind

Most people join organizations to gain connections and increase their influence. The time and energy invested in an organization can yield valuable rewards. Parents who belong to the PTA, for example, are more likely to be able to influence school policy than those who do not belong. The members know whom to call, what to say, and how to exert pressure on school officials.

Sociologists call these benefits of organizational membership **social capital**, the social knowledge and connections that enable people to accomplish their goals and extend their influence (Coleman 1988, 1990; Loury 1987; Putnam 1993, 1995, 2000). Social capital is a broad concept and encompasses useful social networks, a sense of mutual obligation and trustworthiness, an understanding of the norms that govern effective behavior, and other social resources that enable people to act effectively. College students often become active in student government or the campus newspaper partly because they

Learn how social capital enables people to accomplish their goals and expand their influence.

social capital

The social knowledge and connections that enable people to accomplish their goals and extend their influence.

From Organizational Consultant to CEO

The field of organizational sociology has been central to the discipline ever since Max Weber formulated his theory of bureaucratic organizations. But organizational sociology has special relevance today, as organizations are being reshaped by the twin forces of information technology and globalization.

Businesses from small high-tech startups to Walmart have gone global, using information technology to create worldwide supply chains. Nonprofits from Human Rights Watch to Greenpeace rely on their global networks to more effectively press for a just and sustainable world. Organizations today are more likely to be both multiethnic and multinational—two features that are central to sociological understanding. As one observer has noted, "In a diversified and globalized corporate world, in which one might work with people of various races, sexualities, nationalities, and cultures, training as a sociologist can develop the cultural perspective and critical thinking skills necessary to succeed today" (Crossman 2017).

Sociologists are well positioned to understand the transformative forces that are reshaping organizations. The sociological study of organizations today focuses on the many ways in which organizations are shaped by—and in turn shape—the larger environments in which they are found (Scott 2004). This broad-based approach, which combines the study of organizations with a "big picture" understanding of the larger societal forces in which they are embedded, is a hallmark of sociology—and it begins at the undergraduate level.

The American Sociological Association, which interviewed nearly 800 sociology majors a year and a half after they had graduated, found that among those who held full-time jobs, 30 percent "provided administrative support and management skills in a wide variety of organizations" from business to government (Spalter-Roth and Van

hope to learn social skills and make connections that will pay off when they graduate. They may, for example, get to interact with professors, administrators, or even successful alumni, who will then, they hope, go to bat for them when they are looking for a job or applying to graduate school. Differences in social capital mirror larger social inequalities. In general, men have more capital than women, whites more than nonwhites, the wealthy more than the poor.

Robert Putnam (2000), a political scientist and author of the famous book *Bowling Alone*, distinguishes two types of social capital: *bridging*, which is outward looking and inclusive, and *bonding*, which is inward looking and exclusive. Bridging social capital unifies people across social cleavages, as exemplified by interfaith religious organizations or the civil rights movement, which brought Blacks and whites together in the struggle for racial equality. Bonding social capital reinforces exclusive identities and homogeneous groups; it can be found in ethnic fraternal organizations, church-based women's reading groups, and elite country clubs.

Vooren 2008a). The study concluded that by the time they graduate, sociology students have learned to develop evidence-based arguments, think abstractly, write effectively, formulate empirically testable questions, understand and perform statistical analysis, comprehend group dynamics and processes, and develop analytical skills, particularly the ability to locate issues within a larger "macro" perspective (Matchett 2009; see also Spalter-Roth and Van Vooren 2008b).

Jenny Chan, who majored in sociology at the University of Hong Kong, went on to help found SACOM, a Hong Kong nonprofit that played a key role in exposing labor abuses in China, including those in a giant factory where workers, despondent about their jobs, were committing suicide in large numbers. Thanks to Chan's exposé, which went viral around the world, Apple—whose iPhones were assembled in the factory—was forced to join the Fair Labor Association and address the problems that caused the suicides (Heffernan 2013; Chan 2017a). Helen Singer, to take another example, serves on the global executive team of Ashoka, an organization that supports a global network of social entrepreneurs—"individuals with innovative solutions to society's most pressing social, cultural, and environmental challenges." Singer (2018) comments that "as a sociologist, I'm always dedicated to the themes of democracy and social innovation."

On the business side, Stephen Schwartzman, co-founder of the Blackstone Group (the world's largest alternative investment firm), was an interdisciplinary major at Yale, which he credits with combining "psychology, sociology, anthropology and biology, which is really sort of the study of the human being." Christopher Connor, former CEO of the building materials giant Sherwin-Williams, majored in sociology at Ohio State (Cutrone and Nisen 2012).

From entry level to the corner office, by enabling you to better understand the relationship between organizations and society, a sociology degree provides an excellent starting point for a career in a wide variety of organizational settings.

Amazon has recently come under fire for its workplace culture. Sociology students are well positioned to help solve some of the issues plaguing modern corporations, which are being transformed by technology and globalization.

People who actively belong to organizations are more likely to feel connected; they feel engaged, able to somehow make a difference. Democracy flourishes when social capital is strong. Historically, declines in organizational membership, neighborliness, and trust in organizations and corporations have been paralleled by a decline in democratic participation and trust in the government.

Although scholars have bemoaned the fact that political involvement, club membership, and other forms of social and civic engagement that bind Americans to one another eroded significantly in the late twentieth and early twenty-first centuries, the high levels of voter turnout in the most recent presidential elections provide a glimmer of optimism. In the 2012 and 2016 elections, 59 and 60 percent of eligible voters, respectively, went to the polls. Both figures represent high levels of participation relative to presidential elections conducted over the last forty-four years. Even more telling, however, is the stark increase in the turnout of youthful voters. Fully 51 percent of voters between the ages of eighteen and twenty-nine voted in the 2008 election, and 46 percent showed up to vote in 2016. Contrast those proportions with turnout rates of youthful voters of less than 40 percent in 1996 and 40 percent in 2000 (File 2017).

Other indicators of social participation do not tell such an encouraging story. Attendance at public meetings concerning education or civic affairs has dropped sharply since the 1970s. Even more pessimistic are recent surveys asking Americans "How much of the time do you think you can trust the government in Washington to do what is right?" In 2017, trust in government was at near-historic lows, with just 3 percent of Americans saying "just about always" and 15 percent saying "most of the time" (Pew Research Center 2017c). However, trust in the U.S. government ebbs and flows as the world around us changes. For instance, following the terrorist attacks of September 11, 2001, researchers witnessed a resurgence of trust, with those saying that they trust the government always or most of the time doubling to nearly 60 percent (Pew Research Center 2017c). In times of crisis, Americans tend to pull together, and social cohesion increases—even if just temporarily.

Some sociologists believe that another indicator of the weakening of social ties in the United States is membership in clubs and social organizations. Research on declining membership in organizations such as the Sierra Club and the National Organization for Women is even more discouraging. The vast majority of these organizations' members simply pay their annual dues and receive a newsletter. Very few members actively participate, failing to develop the social capital Putnam regards as an important underpinning of democracy. Many of the most popular organizations today, such as twelve-step programs or weight loss groups, emphasize personal growth and health rather than collective goals to benefit society as a whole.

There are undoubtedly many reasons for these declines. For one, women, who were traditionally active in volunteer organizations, are more likely to hold a job than ever before. Furthermore, the commuting that results from flight to the suburbs uses up time and energy that might have been available for civic activities. But the principal source of declining civic participation, according to Putnam, is simple: television. The many hours Americans spend at home alone watching TV have replaced social engagement in the community.

CONCEPT CHECKS

1. What is social capital?
2. Describe the difference between bridging social capital and bonding social capital.

Conclusion

The primary groups of your earliest years were crucial in shaping your sense of self—a sense that changed very slowly thereafter. Throughout life, groups also instill in their members norms and values that enable and enrich social life. You may have found that close-knit, democratic groups with fair-minded leaders are better equipped to achieve their goals than less close-knit groups or those with dictatorial or narrow-minded leaders.

Although groups remain central in our lives, group affiliation in the United States is rapidly changing. As you have seen in this chapter, conventional groups appear to be losing ground in our daily lives. For example, today's college students are less likely to join civic groups and organizations than were their parents, a decline that may well signal a weaker commitment to their communities. Some sociologists worry that this signals a weakening of society itself, which could bring about social instability. Yet others argue that group life has been redefined, as young people belong to virtual groups and communities via social networking websites like Facebook and LinkedIn.

The global economy and information technology are also redefining group life in many diverse ways. For instance, your parents are likely to spend much of their careers in a handful of long-lasting, bureaucratic organizations; you are much more likely to be part of a larger number of networked, "flexible" ones. Many of your group affiliations are probably created through the Internet; in the future, your social ties may be created through other forms of communication that today can barely be envisioned. It will become increasingly easy to connect with like-minded people anywhere, creating geographically dispersed groups that span the planet—and whose members may never meet one another face-to-face.

How will these trends affect the quality of your social relationships? For nearly all of human history, most people interacted exclusively with others who were close at hand. The Industrial Revolution, which facilitated the rise of large, impersonal bureaucracies where people knew one another only casually, if at all, changed social interaction. Today, the information revolution is once again changing human interaction. Tomorrow's groups and organizations could provide a renewed sense of communication and social intimacy—or they could spell further isolation and social distance.

CHAPTER 5

The Big Picture

Groups, Networks, and Organizations

Thinking Sociologically

1. According to Georg Simmel, what are the primary differences between dyads and triads? Explain, according to his theory, how the addition of a child would alter the relationship between spouses. Does the theory fit this situation?

2. The advent of computers and the computerization of the workplace has changed our organizations and relationships with coworkers. Explain how you see modern organizations changing with the adaptation of newer information technologies.

Terms to Know

social group • social aggregate • social category • in-groups • out-groups • primary group • secondary group • reference group • dyad • triad • leader • transformational leaders • transactional leaders • groupthink

network

organization • formal organization • bureaucracy • ideal type • formal relations • informal networks • iron law of oligarchy • oligarchy

human resource management • corporate culture • information technology

social capital

Concept Checks

1. What is the difference between social aggregates and social groups? Provide examples of each.
2. Describe the main characteristics of primary and secondary groups.
3. When groups become large, why does their intensity decrease but their stability increase?
4. What is groupthink? How can it be used to explain why some decisions made by a group lead to negative consequences?

1. According to Granovetter, what are the benefits of weak ties? Why?
2. How do men's and women's weak ties differ?

1. What role do organizations play in contemporary society?
2. What does the term *bureaucracy* mean?
3. Describe five characteristics of an ideal type of bureaucracy.
4. Explain how modern organizations have developed in a gendered way.

1. How has the Japanese model influenced the Western approach to management?
2. Explain how the development of information technology has changed the ways people live and work.
3. According to George Ritzer, what are the four guiding principles used in McDonald's restaurants? What does he mean by the "McDonaldization" of society?

1. What is social capital?
2. Describe the difference between bridging social capital and bonding social capital.

6

Deviance, Crime, and Punishment

Incarceration Rates

p. 179

Protesters and police gather on the one-year anniversary of the fatal shooting of eighteen-year-old Michael Brown. Ferguson, Missouri, is just one of many cities that has been rocked by a police shooting of an unarmed Black man.

THE BIG QUESTIONS

What is deviance?

Learn how sociologists define deviance and how it is closely related to social power and social class. See the ways in which conformity is encouraged.

Why do people commit deviant acts?

Know the leading sociological, psychological, and biological theories of deviance and how each is useful in understanding crime.

How do we document crime?

Recognize the usefulness and limitations of crime statistics. Learn some important differences between men and women related to crime. Familiarize yourself with some of the varieties of crime.

Who are the perpetrators?

Understand why members of some social groups are more likely to commit or be the victims of crime.

What were the causes and costs of the great crime decline?

Consider some of the factors that have caused a decades-long decline in crime rates. Understand the social costs associated with the great crime decline.

How do crime and deviance affect your life?

Understand the costs and functions of crime and deviance.

At 12:01 p.m. on August 9, 2014, Officer Darren Wilson was driving down Canfield Drive in Ferguson, a suburb of St. Louis, Missouri, when he came across two Black men walking in the street. By 12:04 p.m., when other officers arrived on the scene, one of the two men—eighteen-year-old Michael Brown—was dead. What exactly transpired between Wilson and Brown during those three minutes is anything but clear, but ultimately, in November of the same year, a St. Louis grand jury decided not to indict the twenty-eight-year-old officer for Brown's death.

One of the most controversial phenomena in recent American life has been a series of violent encounters between police and citizens of African American communities. Brown's death, along with other high-profile incidents across the country, sparked nationwide protests—and inspired a movement. Born online, Black Lives Matter—which draws its name from a Twitter hashtag posted in the wake of George Zimmerman's acquittal in the 2012 shooting death of Trayvon Martin—gained momentum in Ferguson, Missouri, where it held its first in-person protest after Brown was shot and killed. With dozens of chapters across the country, the movement has found its way onto college campuses, on the campaign trail, even into the White House (Altman 2015).

This string of police shootings of unarmed Black men raises countless questions for sociologists, including questions about crime, deviance, and violence. Has the incidence of such violence increased in recent years, or has the rise of cell phones with video capability made it possible to document behaviors that have long been prevalent but difficult to prove? Why does the criminal justice system fail to live up to the ideal of treating all equally? How do cultural factors contribute to the perpetuation and acceptance of violence? Can we ever say that murder is "justifiable"?

Crime and punishment are tightly woven into the fabric of life in the United States. The U.S. rate of incarceration is five to eight times higher than those of Canada and the countries of Western Europe. The United States is home to less than 5 percent of the world's population but almost a quarter of the world's prisoners (Walmsley 2016). The high rate of imprisonment in the United States in recent decades is due in part to "three strikes" laws, which became popular in the 1990s. These laws require state courts to hand down mandatory and often lengthy prison sentences to persons who have been convicted of a serious criminal offense three or more times. Yet, as we will see later in this chapter, not all people who commit a criminal offense are treated the same by agents of the criminal justice and legal systems.

Rates of imprisonment, in turn, have a profound impact on U.S. society. When individuals are in prison, they are not part of the labor force and thus are not counted in the rates of unemployment reported by the government. As a result, estimates of unemployment among some subgroups, such as African American men, may be understated. At the same time, incarceration increases the long-term chances of unemployment even after someone is released from prison (Western and Beckett 1999).

But the story of crime and justice in America is not simply one of a glass half empty. Beginning in the early 1990s, the murder rate in U.S. cities began a significant decline, and by the time that Michael Brown was murdered in 2014, the country was safer from crime than during any year in the half-century since reliable crime data have been kept (Sharkey 2018). The "great crime decline," as it came to be known, occurred alongside the string of police shootings of unarmed Black men, raising another important question for sociologists: What are the causes of the crime decline, and what social costs have been incurred by the measures that have brought down crime?

In this chapter, we will be taking up these and other questions relating to crime, which is just one category of a much larger field of study called "deviance" or "deviant behavior." Deviants are those individuals who do not live by the rules that the majority of us follow. Some do so by choice; others are incapable of following the rules because they lack the resources to do so. Sometimes they're violent criminals, drug addicts, or down-and-outs who don't fit in with what most people would define as normal standards of acceptability. These are the cases that seem easy to identify. Yet things are not quite as they appear—a lesson sociology often teaches us, for it encourages us to look beyond the obvious. The concept of the deviant, as we shall see, is actually not easy to define.

We have learned in previous chapters that social life is governed by rules or norms. **Norms**, which we discussed in Chapter 2, are clearly defined and established principles or rules people are expected to observe; they represent the dos and don'ts of society. However, some norms are more powerful and important than others. Early twentieth-century sociologist William Graham Sumner identified two types of norms. **Mores** (pronounced "morays") are norms that

norms

Rules of conduct that specify appropriate behavior in a given range of social situations. A norm either prescribes a given type of behavior or forbids it. All human groups follow norms, which are always backed by sanctions of one kind or another—varying from informal disapproval to physical punishment.

mores

Norms that are widely adhered to and have great moral or social significance. Violations are generally sanctioned strongly.

are widely adhered to and have great social and moral significance. **Folkways**, by contrast, are the norms that guide our everyday actions. For example, cutting in front of someone in line at a coffee shop would be a violation of a folkway, whereas harassing the hardworking barista would be a violation of a more.

Norms affect every aspect of our lives. Orderly behavior on the highway, for example, would be impossible if drivers didn't observe the rule of driving on the right. No deviants here, you might think, except perhaps for the drunken or reckless driver. If you did think this, you would be incorrect. When we drive, most of us are not merely deviants but criminals. Most of us regularly drive at well above the legal speed limit, and some of us even text while driving—assuming there isn't a police car in sight. In such cases, breaking the law is normal behavior!

We are all rule breakers as well as conformists. We are all also rule creators. Most American drivers may break the law on the freeway, but in fact they've evolved informal rules that are superimposed on the legal rules. When the legal speed limit on the highway is 65 miles per hour, most drivers don't go above 75 or so, and they drive slower when passing through urban areas. When we begin the study of deviant behavior, we must consider which rules people are observing and which they are breaking. Nobody breaks all rules, just as no one conforms to all rules. As we shall see throughout this chapter, understanding who is or is not classified as deviant, and why, is a fascinating question at the core of sociology.

What Is Deviance?

The study of deviant behavior is one of the most intriguing yet complex areas of sociology. It teaches us that none of us is quite as normal as we might like to think. It also helps us see that people whose behavior might appear incomprehensible or odd can be viewed as rational beings when we understand why they act as they do. The study of deviance, like other fields of sociology, directs our attention to social power, which encompasses gender, race, and social class. When we look at deviance from or conformity to social rules or norms, we always have to bear in mind the question, "Whose rules?" As we shall see, social norms are strongly influenced by divisions of power and class.

Deviance may be defined as nonconformity to a given set of norms that are accepted by a significant number of people in a community or society. These people have the power to enforce their definitions of what counts as normal. No society can be divided up in a simple way between those who deviate from norms and those who conform to them. Most of us on some occasions violate generally accepted rules of behavior. Although a large share of all deviant behavior (such as committing assault or murder) is also criminal and violates the law, many deviant behaviors—ranging from bizarre fashion choices to joining a religious cult—are not criminal. By the same token, many behaviors that are technically "crimes," such as underage drinking or exceeding the speed limit, are not considered deviant because they are quite normative (see Figure 6.1). Sociologists tend to focus much of their research on behaviors that are both criminal and deviant, as such behaviors have importance for the safety and well-being of our nation.

Although most of us associate the word *deviant* with behaviors that we view as dangerous or unsavory, the same deviant act can also be the basis of membership in conventional society. Kevin Mitnick has been described as the "world's most celebrated

folkways

Norms that guide casual or everyday interactions. Violations are sanctioned subtly or not at all.

Learn how sociologists define deviance and how it is closely related to social power and social class. See the ways in which conformity is encouraged.

deviance

Modes of action that do not conform to the norms or values held by members of a group or society who can enforce their definitions. What is regarded as deviant is as variable as the norms and values that distinguish different cultures and subcultures from one another.

FIGURE 6.1

Intersection of Deviance and Crime

DEVIANCE   CRIME

Nudity and bizarre clothing

Murder and sexual assault

Exceeding the speed limit and underage drinking

computer hacker." To computer hackers everywhere, Mitnick is a pathbreaking genius whose five-year imprisonment in a U.S. penitentiary was unjust and unwarranted—proof of how misunderstood computer hacking has become with the spread of information technology. To U.S. authorities and high-tech corporations, Mitnick is one of the world's most dangerous men. Mitnick was captured by the FBI in 1995 and later convicted of downloading source code and stealing software allegedly worth millions of dollars from companies such as Motorola and Sun Microsystems. As a condition of his release from prison in January 2000, Mitnick was barred from using any communications technology other than a landline telephone. He successfully fought this legal decision and gained access to the Internet. Today, the very same skills that made Mitnick a reviled deviant thought to threaten the very stability of the information age have led to a burgeoning new career in what he calls "ethical hacking." As the founder of a consulting company, he tries to help governments and corporations test their security.

Deviance does not refer only to individual behavior; it concerns the activities of groups as well. Heaven's Gate was a religious group whose beliefs and practices were different from those of the majority of Americans. The cult was established in the early 1970s when Marshall Herff Applewhite made his way around the West and Midwest of the United States preaching his beliefs, ultimately advertising on the Internet his belief that civilization was doomed and that the only way people could be saved was to kill themselves so their souls could be rescued by a UFO. On March 26, 1997, thirty-nine members of the cult followed his advice in a mass suicide at a wealthy estate in Rancho Santa Fe, California.

The Heaven's Gate cult represents an example of a **deviant subculture**. Its members were able to survive fairly easily within the wider society, supporting themselves by running a website business and recruiting new members by emailing people they thought might be interested in their beliefs. They had plenty of money and lived together in an expensive home in a wealthy Southern California suburb.

Norms and Sanctions

We most often follow social norms because, as a result of socialization, we are accustomed to doing so. Individuals become committed to social norms through interactions with people who obey the law and mainstream values. Through these interactions, we learn self-control. The more numerous and frequent our interactions, the fewer opportunities we have to deviate from conventional norms. And, over time, the longer we interact in ways that are conventional, the more we stand to lose by not conforming (Gottfredson and Hirschi 1990).

All social norms are accompanied by sanctions that promote conformity and protect against nonconformity. A **sanction** is any reaction from others to the behavior of an individual or group that is meant to ensure that the person or group complies with a given norm. Sanctions may be positive (the offering of rewards for conformity) or negative

deviant subculture

A subculture whose members hold values that differ substantially from those of the majority.

sanction

A mode of reward or punishment that reinforces socially expected forms of behavior.

(punishment for behavior that does not conform). They can also be formal or informal. Formal sanctions are applied by a specific body of people or an agency to ensure that a particular set of norms is followed, such as a speeding ticket or expulsion from school for cheating. Informal sanctions are less organized and more spontaneous reactions to nonconformity, such as when a student is teasingly accused by friends of being a nerd for deciding to stay home and study rather than go to a party.

The main types of formal sanctions in modern societies are those represented by the courts and prisons. The police, of course, are the agency charged with bringing offenders to trial and possible imprisonment. **Laws** are norms defined by governments as principles that their citizens must follow; sanctions are used against people who do not conform to them. Where there are laws, there are also **crimes**, since crime can most simply be defined as any type of behavior that breaks a law.

Why Do People Commit Deviant Acts?

One of the most vexing puzzles asked by social scientists and laypeople alike is "Why are people deviant?" Part of the morbid allure of TV shows like *Snapped* is that we are truly puzzled and seek answers when we learn about the vicious beatings of innocent victims. Answers to the question vary widely, however, depending on one's academic discipline and even, within sociology, one's theoretical approach. We will briefly review biological and psychological explanations for deviance and then turn to the three sociological approaches that have been developed to interpret and analyze deviance: functionalist theories, interactionist theories, and conflict theories.

The Biological View of Deviance

Some of the first attempts to explain crime emphasized biological factors. The Italian criminologist Cesare Lombroso, working in the 1870s, believed that criminal types could be identified by the shape of the skull. He accepted that social learning could influence the development of criminal behavior, but he regarded most criminals as biologically degenerate or defective. Lombroso's ideas were later thoroughly discredited, but similar views have repeatedly been suggested. One such theory distinguished three main types of human physique and claimed that one type was directly associated with delinquency. Muscular, active types (mesomorphs) were considered more likely to become delinquent than those of thin physique (ectomorphs) or more fleshy people (endomorphs) (Glueck and Glueck 1956; Sheldon 1949).

Most biological theories have been widely criticized on methodological grounds. Even if there were a correlation between body type and delinquency, this would not necessarily reveal that one's body type "causes" criminal behavior. For instance, people who engage in criminal activities may need to develop more muscular physiques to protect themselves on the streets. Moreover, nearly all studies in this field have been restricted to delinquents in

Know the leading biological, psychological, and sociological theories of deviance and how each is useful in understanding crime.

CONCEPT CHECKS

1. How do sociologists define deviance?

2. Is all crime deviant? Is all deviance criminal? Why?

3. Contrast positive and negative sanctions.

laws

Rules of behavior established by a political authority and backed by state power.

crimes

Any actions that contravene the laws established by a political authority.

reform schools, and it may be that the tougher, athletic-looking delinquents are more likely to be sent to such schools than fragile-looking, skinny ones.

More recent, methodologically rigorous research has sought to rekindle the argument that deviance has a biological or genetic basis. In a study of New Zealand children, researchers investigated whether a child's propensity for aggressive behavior was linked to biological factors present at birth (Moffitt 1996). Rather than viewing biology as deterministic, this new breed of research emphasizes that biological factors, when combined with certain social factors, such as home environment, could lead to social situations involving crime. This perspective, which emphasizes gene–environment interaction, reasons that one's genes may "select" or draw a person into a particular behavior, such as aggression. At the same time, the social environment may strengthen or weaken the link between genetics and deviant behavior. For instance, even if a baby is born with a genetic predisposition for alcoholism, that baby will not likely become a problem drinker if his or her social environment provides few opportunities to drink.

The Psychological View of Deviance

Like biological interpretations, psychological theories of crime associate criminality with particular personality types. Some have suggested that in a minority of individuals, an amoral, or psychopathic, personality develops. **Psychopaths** are withdrawn, emotionless characters who delight in violence for its own sake.

Individuals with psychopathic traits do sometimes commit violent crimes, but there are major problems with the concept of the psychopath. It isn't at all clear that psychopathic traits are inevitably criminal. Nearly all studies of people said to possess these characteristics have been of convicted prisoners, and their personalities inevitably tend to be presented negatively. If we describe the same traits positively, the personality type sounds quite different, and there seems to be no reason that people of this sort should be inherently criminal. Such people might be explorers, spies, gamblers, or just bored with the routines of day-to-day life. They might be prepared to contemplate criminal adventures but could be just as likely to look for challenges in socially respectable ways.

Psychological theories of criminality can at best explain only some aspects of crime. While some criminals may possess personality characteristics distinct from characteristics of the remainder of the population, it is highly improbable that the majority of criminals do. There are all kinds of crimes, and it is implausible that those who commit them share some specific psychological characteristics. Some crimes are carried out by lone individuals, whereas others are the work of organized groups. It is not likely that the psychological makeup of people who are loners will have much in common with that of the members of a close-knit gang. Observational studies also can't discount the possibility that becoming involved with criminal groups influences people's outlooks, rather than that the outlooks actually produce criminal behavior in the first place.

Both biological and psychological approaches to criminality presume that deviance is a sign of something "wrong" with the individual rather than with society. They see crime and deviance as caused by factors outside an individual's control, embedded either in the body or in the mind. Often, scholars working in this tradition consider deviance to be caused by biological factors that require treatment, such as mental illness or a genetic tendency toward violence. These early approaches to criminology came under great criticism from later generations of scholars, who argued that any satisfactory account of the

psychopath

A specific personality type; such individuals lack the moral sense and concern for others held by most normal people.

nature of crime must be sociological, for what crime is depends on the social institutions of a society.

Sociological Perspectives on Deviance

Contemporary sociological thinking about crime emphasizes that definitions of conformity and deviance vary based on one's social context. Modern societies contain many different subcultures, and behavior that conforms to the norms of one particular subculture may be regarded as deviant outside it; for instance, there may be strong pressure on a gang member to prove himself or herself by stealing a car. Moreover, there are wide divergences of wealth and power in society that greatly influence opportunities open to different groups. Theft and burglary, not surprisingly, are carried out mainly by people from the poorer segments of the population; embezzling and tax evasion are by definition limited to persons in positions of some affluence.

FUNCTIONALIST THEORIES

Functionalist theories see crime and deviance as resulting from structural tensions and a lack of moral regulation within society. If the aspirations held by individuals and groups in society do not coincide with available rewards, this disparity between desires and fulfillment will lead to deviant behavior.

Crime and Anomie: Durkheim and Merton   As we saw in Chapter 1, the notion of **anomie** was first introduced by Émile Durkheim, who suggested that in modern societies, social norms may lose their hold over individual behavior. Anomie exists when there are no clear standards to guide behavior in a given area of social life. Under such circumstances, Durkheim believed, people feel disoriented and anxious; anomie is therefore one of the social factors influencing dispositions to suicide.

Durkheim saw crime and deviance as social facts; he believed both to be inevitable and necessary elements in modern societies. According to Durkheim, people in the modern age are less constrained by social expectations than they were in traditional societies. Because there is more room for individual choice in the modern world, nonconformity is inevitable. Durkheim recognized that modern society would never be in complete consensus about the norms and values that govern it.

Deviance is also necessary for society, according to Durkheim; it fulfills two important functions. First, deviance has an adaptive function. By introducing new ideas and challenges into society, deviance is an innovative force. It brings about change. Second, deviance promotes boundary maintenance between "good" and "bad" behaviors in society. A criminal act can ultimately enhance group solidarity and clarify social norms.

Deviance can also contribute to the stability of society. In his classic essay on prostitution, functionalist theorist Kingsley Davis (1937) wrote that even though prostitution is illegal, it is functional for society because it allows married men to fulfill their sexual urges with new partners without threatening their marriages. By contrast, a married man who forms an emotional attachment with a woman with whom he is having a "legal" though clandestine relationship can threaten both his and her marriages. Prostitution, Davis argued, indirectly contributes to the stability of the family.

Early functionalist perspectives on crime and deviance helped shift attention from explanations focused on the problems of individuals to explanations focused on social

anomie

A concept first brought into wide usage in sociology by Durkheim, referring to a situation in which social norms lose their hold over individual behavior.

FIGURE 6.2

Merton's Deviance Typology

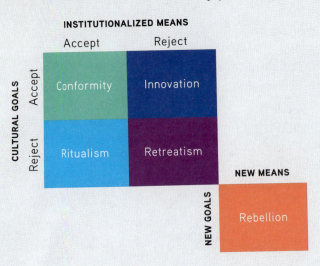

forces. Durkheim's notion of anomie was drawn on by American sociologist Robert K. Merton (1957), who constructed a highly influential theory of deviance that located the source of crime within the very structure of American society. In what became known as strain theory, Merton modified the concept of anomie to refer to the strain put on individuals' behavior when accepted norms conflict with social reality. In American society—and to some degree in other industrial societies—generally held values emphasize material success, and the means of achieving success are supposed to be self-discipline and hard work. Accordingly, it is believed that people who work hard can succeed regardless of their starting point in life. This idea is not in fact valid, because most of the disadvantaged have very few conventional opportunities for advancement, such as a high-quality education. Yet those who do not "succeed" find themselves condemned for their apparent inability to make material progress. In this situation, there is great pressure to try to get ahead by any means, legitimate or illegitimate. According to Merton, then, deviance is a by-product of economic inequalities.

Merton identifies five possible reactions to the tensions between socially endorsed values and the limited means of achieving them (see Figure 6.2). Conformists accept both societal values and the conventional means of realizing them, regardless of whether they meet with success. The majority of the population falls into this category. Innovators accept socially approved values but use illegitimate or illegal means to follow them. Criminals who acquire wealth through illegal activities exemplify this type.

Ritualists conform to socially accepted standards, although they have lost sight of the values behind these standards. They follow rules for their own sake without a broader end in view, in a compulsive way. A ritualist might remain in a boring job even though it has no career prospects and provides few rewards. Retreatists have abandoned the competitive outlook altogether, thus rejecting both the dominant values and the approved means of achieving them. An example would be the members of a self-supporting commune. Finally, rebels reject both the existing values and the means of pursuing them but wish actively to substitute new values and reconstruct the social system. The members of radical political and religious groups, such as the Heaven's Gate cult, fall into this category.

Merton's writings addressed one of the main puzzles in the study of criminology: At a time when society as a whole is becoming more affluent, why do crime rates continue to rise? By emphasizing the contrast between rising aspirations and persistent inequalities, Merton points to a sense of **relative deprivation**, or the recognition that one has less than his or her peers, as an important element in deviant behavior.

relative deprivation

The recognition that one has less than his or her peers.

Subcultural Explanations Later researchers located deviance in terms of subcultural groups that adopt norms that encourage or reward criminal behavior. Like Merton, Albert Cohen saw the contradictions within American society as the main cause of crime. However, Cohen saw the responses occurring collectively, through subcultures, while Merton emphasized individual responses. In *Delinquent Boys* (1955), Cohen argued that boys in the lower working class who are frustrated with their positions in life often join together in delinquent subcultures, such as gangs. These subcultures reject middle-class values and replace them with norms that celebrate defiance, such as delinquency and other acts of nonconformity.

Richard A. Cloward and Lloyd E. Ohlin (1960) argued further that such gangs arise in subcultural communities where the chances of achieving success legitimately are slim, such as among deprived ethnic minorities. Cloward and Ohlin's work emphasizes connections between conformity and deviance: Individuals follow rules when they have the opportunity to do so and break rules when they do not. As a result, they develop subcultures with deviant values in response to the lack of legitimate opportunities for success as defined by the wider society. This lack of opportunity is the differentiating factor between those who engage in criminal behavior and those who do not.

Functionalist theories rightly emphasize connections between conformity and deviance in different social contexts. We should be cautious, however, about the idea that people in poorer communities aspire to the same level of success

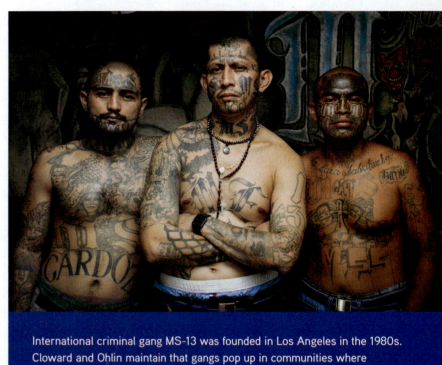

International criminal gang MS-13 was founded in Los Angeles in the 1980s. Cloward and Ohlin maintain that gangs pop up in communities where opportunities to succeed through legitimate means are limited.

as more affluent people. Most tend to adjust their aspirations to what they see as the reality of their situation. Merton, Cohen, and Cloward and Ohlin can all be criticized for presuming that middle-class values have been accepted throughout society. It would also be wrong to suppose that a mismatch of aspirations and opportunities is confined to the less privileged. There are pressures toward criminal activity among other groups, too, as indicated by the so-called white-collar crimes of embezzlement, fraud, and tax evasion.

INTERACTIONIST THEORIES

Sociologists studying crime and deviance in the interactionist tradition focus on deviance as a socially constructed phenomenon. They reject the idea that there are types of conduct that are inherently "deviant." Rather, interactionists ask how behaviors initially come to be defined as deviant and why certain groups and not others are labeled as deviant.

Learned Deviance: Differential Association One of the earliest writers to suggest that deviance is learned through interaction with others was Edwin H. Sutherland. In 1949, Sutherland advanced a notion that influenced much of the later interactionist work: He linked crime to what he called **differential association**. Differential association theory argues that we learn deviant behavior in precisely the same way we learn about conventional behavior: from our contacts with primary groups such as peers, family members, and coworkers. The term *differential* refers to the ratio of deviant to conventional social contacts. We become deviant when exposed to a higher level of deviant persons and influences, compared with conventional influences. In a society that contains a variety of subcultures, some individuals have greater exposure to social environments that encourage illegal activities.

differential association

An interpretation of the development of criminal behavior proposed by Edwin H. Sutherland, according to whom criminal behavior is learned through association with others who regularly engage in crime.

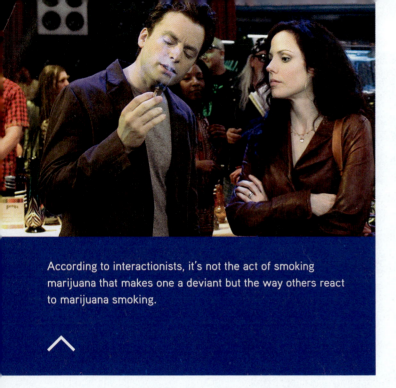

According to interactionists, it's not the act of smoking marijuana that makes one a deviant but the way others react to marijuana smoking.

Labeling Theory

Labeling Theory One of the most important interactionist approaches to understanding criminality is **labeling theory**. An early work based on labeling theory is Howard S. Becker's (1963) study of marijuana smokers. In the early 1960s, marijuana use was a marginal activity carried on by subcultures rather than the lifestyle choice—that is, an activity accepted by many in the mainstream of society—it is today (Hathaway 1997). Becker found that becoming a marijuana smoker depended on one's acceptance into the subculture, close association with experienced users, and one's attitudes toward nonusers. Labeling theorists like Becker interpret deviance not as a set of characteristics of individuals or groups but as a process of interaction between deviants and nondeviants. In other words, it is not the act of marijuana smoking that makes one a deviant but the way others react to marijuana smoking.

While other sociological perspectives are focused on why people are deviant, labeling theorists seek to understand why some people become tagged with a deviant label. In short, persons with the greatest social and economic power tend to place labels on those with less social power. Furthermore, the labels that create categories of deviance thus express the power structure of society; such rules are framed by the wealthy for the poor, by men for women, by older people for younger people, and by ethnic majorities for minority groups. For example, many children wander into other people's gardens, steal fruit, or play truant. In an affluent neighborhood, these might be regarded by parents, teachers, and police alike as relatively innocent pastimes of childhood. In poor areas, they might be seen as evidence of tendencies toward juvenile delinquency.

Once a child is labeled a delinquent, he or she is stigmatized as a deviant and is likely to be considered untrustworthy by teachers and prospective employers. The child then relapses into further criminal behavior, widening the gulf with orthodox social conventions. Edwin Lemert (1972) called the initial act of rule breaking **primary deviance**. **Secondary deviance** occurs when the individual comes to accept the label and sees himself or herself as deviant. The "self-fulfilling prophecy" may occur, where the labeled person begins to behave in such a way that perpetuates the deviant behavior. Research has shown that how we think of ourselves and how we believe others perceive us influence our propensity for committing crime. One study of a random national sample of young men showed that such negative self-appraisals are strongly tied to levels of criminality; in other words, the perception that one is deviant may in fact motivate deviant behavior (Matsueda 1992).

Labeling theory is important because it begins from the assumption that no act is intrinsically deviant. Rather, to be "deviant," one must be labeled as such. In the case of criminal activity, definitions of criminality are established by the powerful through the formulation of laws and their interpretation by police, courts, and correctional institutions. Critics of labeling theory have sometimes argued that certain acts, such as murder, rape, and robbery, are consistently prohibited across virtually all cultures. This view is surely incorrect. Even within our own culture, killing is not always regarded as murder. In a highly controversial case, George Zimmerman, a neighborhood watch volunteer who shot and killed an unarmed seventeen-year-old high school student named Trayvon Martin, was ultimately found not

---

labeling theory

An approach to the study of deviance that suggests that people become "deviant" because certain labels are attached to their behavior by political authorities and others.

primary deviance

According to Edwin Lemert, the actions that cause others to label one as a deviant.

secondary deviance

According to Edwin Lemert, following the act of primary deviance, secondary deviation occurs when an individual accepts the label of deviant and acts accordingly.

guilty of murder by a Florida jury. The jury ruled that Zimmerman acted in self-defense and was therefore protected under the state's stand-your-ground law, which permits an individual to use deadly force when faced with the risk of great bodily harm. To provide another example, in times of war, killing of the enemy is positively approved.

We can more convincingly criticize labeling theory on other grounds. First, in emphasizing the active process of labeling, labeling theorists neglect the processes that lead to acts defined as deviant. Labeling certain activities as deviant is not completely arbitrary; differences in socialization, attitudes, and opportunities influence how far people engage in behavior likely to be labeled deviant. For instance, children from deprived backgrounds are on average more likely to steal from shops than richer children are. It is not the labeling that leads them to steal in the first place so much as their background. Second, it is not clear whether labeling actually does have the effect of increasing deviant conduct. Delinquent behavior tends to increase following a conviction, but is this the result of the labeling itself? Other factors, including increased interaction with other delinquents or learning about new criminal opportunities, may be involved.

Control Theory   Control theory posits that crime occurs as a result of an imbalance between impulses toward criminal activity and the social or physical controls that deter it. Core assumptions are that people act rationally and that, given the opportunity, everyone would engage in deviant acts. Many types of crime, it is argued, are a result of "situational decisions"—a person sees an opportunity and is motivated to act.

One of the best-known control theorists, Travis Hirschi, has argued that humans are fundamentally rational beings who make calculated decisions about whether to engage in criminal activity by weighing the potential benefits and risks of doing so. In *Causes of Delinquency* (1969), Hirschi claims that there are four types of bonds that link people to society and law-abiding behavior: attachment, commitment, involvement, and belief. *Attachment* refers to emotional and social ties to persons who accept conventional norms, such as a peer group of students who value good grades and hard work. *Commitment* refers to the rewards obtained by participating in conventional activities and pursuits. For example, a high school dropout has little to lose by being arrested, whereas a dedicated student may lose his or her chance of going to college. *Involvement* refers to one's participation in conventional activities such as paid employment, school, or community activities. Finally, *beliefs* involve upholding morals and values that are consistent with conventional tenets of society.

When sufficiently strong, these four elements help maintain social control and conformity by rendering people unfree to break rules. If these bonds with society are weak, however, delinquency and deviance may result. Hirschi's approach suggests that delinquents are often individuals whose low levels of self-control are a result of inadequate socialization at home or at school (Gottfredson and Hirschi 1990).

CONFLICT THEORY

Adherents of conflict theory seek to identify why people commit crime. Conflict theorists draw on elements of Marxist thought to argue that deviance is deliberately chosen and often political in nature. Conflict theorists reject the idea that deviance is "determined" by factors such as biology, personality, anomie, social disorganization, or labels. Rather, individuals purposively engage in deviant behavior in response to the inequalities of the capitalist system. For example, many of the protesters who were arrested at Occupy Wall Street rallies were engaging in political acts that challenged the social order.

CONCEPT CHECKS

1. What are the main similarities and differences between biological and psychological views of deviance?

2. How do Merton's and Durkheim's definitions of anomie differ?

3. According to subcultural explanations, how does criminal behavior get transmitted from one group to another?

4. What is the core idea behind differential association theory?

5. What are two criticisms of labeling theory?

6. What are the root causes of crime according to conflict theorists?

Conflict theorists frame their analysis of crime and deviance in terms of the structure of society and the preservation of power among the ruling class. For example, they argue that laws are tools used by the powerful to maintain their own privileged positions. They reject the idea that laws are neutral and are applied evenly across the population. Instead, they claim that as inequalities increase between the ruling class and the working class, law becomes an ever more important instrument for the powerful to maintain order. This dynamic can be seen in the workings of the criminal justice system, which has become increasingly oppressive toward working-class "offenders," or in tax legislation that disproportionately favors the wealthy.

This power imbalance is not restricted to the creation of laws, however. The powerful also break laws, but they are rarely caught. These crimes on the whole are much more significant than the everyday crime and delinquency that attract the most attention. But fearful of the implications of pursuing white-collar criminals, law enforcement instead focuses its efforts on less powerful members of society, such as prostitutes, drug users, and petty thieves (Chambliss 1988; Pearce 1976). Studies by Chambliss, Pearce, and others have played an important role in widening the debate about crime and deviance to include questions of social justice, power, and politics. They emphasize that crime occurs at all levels of society and must be understood in the context of inequalities and competing interests among social groups.

Theoretical Conclusions

Whether someone engages in a criminal act or comes to be regarded as a criminal is influenced fundamentally by social learning and social surroundings. The way in which crime is understood directly affects the policies developed to combat it. For example, if crime is seen as the product of deprivation or social disorganization, policies might be aimed at reducing poverty and strengthening social services. If criminality is seen as freely chosen by individuals, attempts to counter it will take a different form. Now let's look directly at the nature of the criminal activities occurring in modern societies, paying particular attention to crime in the United States.

How Do We Document Crime?

Crime statistics are a constant focus of attention in the media. Most TV and newspaper reporting is based on official statistics on crime, collected by the police and published by the government. Most of these reports are based on two sources: Uniform Crime Reports (UCR) and victimization studies. Each has its own limitations and offers only a partial portrait of crime in American life.

Uniform Crime Reports (UCR) contain official data on crime that is reported to law enforcement agencies across the country that then provide the data to the FBI (see Figure 6.3). UCR focus on "index crimes," which include serious crimes, such as murder and nonnegligent manslaughter, robbery, forcible rape, aggravated assault, burglary, larceny/

FIGURE 6.3

Crime Rates in the United States, 1997–2016

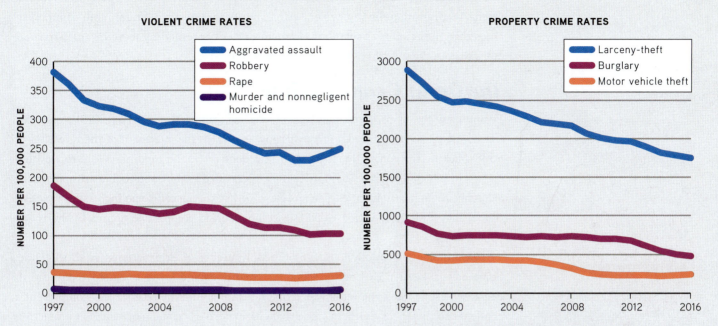

VIOLENT CRIME RATES

Legend:
- Aggravated assault
- Robbery
- Rape
- Murder and nonnegligent homicide

Y-axis: NUMBER PER 100,000 PEOPLE (0, 50, 100, 150, 200, 250, 300, 350, 400)
X-axis: 1997, 2000, 2004, 2008, 2012, 2016

PROPERTY CRIME RATES

Legend:
- Larceny-theft
- Burglary
- Motor vehicle theft

Y-axis: NUMBER PER 100,000 PEOPLE (0, 500, 1000, 1500, 2000, 2500, 3000)
X-axis: 1997, 2000, 2004, 2008, 2012, 2016

Note: The murder and nonnegligent homicides that occurred as a result of the events of September 11, 2001, are not included in this table.

Source: Federal Bureau of Investigation 2017b.

theft, motor vehicle theft, and arson. Critics of UCR note that the reports do not accurately reflect crime rates because they include only those crimes reported to law enforcement agencies; they don't, for example, include crimes reported to other agencies, such as the IRS. Furthermore, the index crimes do not include less serious crimes. Some argue that by excluding crimes that are traditionally committed by middle-class persons, such as fraud and embezzlement, UCR reify the belief that crime is an activity of ethnic minorities and the poor.

Because the UCR program focuses narrowly on crimes reported to the police, criminologists also rely on self-reports, or reports provided by the crime victims themselves. This second source of data is essential, as some criminologists think that about half of all serious crimes, such as robbery with violence, go unreported. The proportion of less serious crimes, especially small thefts, that don't come to the attention of the police is even higher. Since 1973, the Census Bureau has been interviewing households across the country to find out how many members were the victims of particular crimes over the previous six months. This procedure, which is called the National Crime Victimization Survey, has confirmed that the overall rate of crime is higher than the reported crime index. Crimes where victims may feel stigmatized are most likely to go unreported. For instance, in 2016, just 23 percent of rape or sexual assault victimizations were reported, compared with 54 percent of robberies (Morgan and Kena 2017).

Public concern in the United States tends to focus on crimes of violence—murder, assault, and rape—even though less than 14 percent of all crimes are violent (Federal Bureau of Investigation 2017b). To put this in perspective, in 2016, roughly 3.6 million people age twelve or older—or 1.3 percent of the population—were victims of at least one

Uniform Crime Reports (UCR)

Documents that contain official data on crime that is reported to law enforcement agencies that then provide the data to the FBI.

violent crime. In comparison, 11.7 million people, or nearly 9 percent of the population, were victims of a property crime, including household burglary and motor vehicle theft (Morgan and Kena 2017). In general, whether indexed by police statistics or by the National Crime Victimization Survey, violent crime, burglary, and car theft are more common in cities than in the suburbs surrounding them, and they are slightly more common in rural areas than in the suburbs (Morgan and Kena 2017).

The Great Crime Decline

Beginning in the 1990s, crime rates began to decline nationwide. In 2011, the FBI announced that crime rates had reached a forty-year low—even in the aftermath of the recession of 2008, bucking the conventional wisdom that poor economic conditions lead to elevated crime rates. Incredibly, rates of murder, rape, aggravated assault, and robbery dropped considerably, though certain cities (such as New York) experienced some increases in violent crime. That said, these increases pale in comparison to comparable figures in the 1990s (Oppel 2011).

Criminologists nationwide have offered a number of explanations for this decline. Some have suggested that better economic conditions and lower unemployment lead to decreased crime rates, though lower crime rates following the recent recession present a challenge to this theory. Others argue that citizens have become more adept at protecting themselves against crime through the use of sophisticated home security systems, while policing has become more targeted and disciplined, with police now using "hot-spot policing" to station officers around areas in which they know crime rates are relatively high. Still more scholars have suggested that the drop in crime may be related to decreasing cocaine and illegal drug usage or lower levels of lead in Americans' blood, which has been linked to higher levels of aggression in children (Wilson 2011).

The United States has relatively high rates of violent crime compared to other industrialized nations. One reason often given to explain the relatively high violent crime rate is the widespread availability of handguns and other firearms. The belief that one has a personal right to "bear arms" is widespread in American culture. But gun control laws alone would not be sufficient to tackle violent crime in the United States. Switzerland has very low rates of violent crime, yet firearms are easily accessible. All Swiss males are members of the citizen army and keep weapons in their homes, including rifles, revolvers, and sometimes other automatic weapons, plus ammunition; and gun licenses in Switzerland are easy to obtain (Bachmann 2012).

The most likely explanation for the high level of violent crime in the United States is a combination of the availability of firearms, the general influence of the "frontier tradition," and the subcultures of violence in large cities. Violence by frontiersmen and vigilantes is an honored part of American history. Also, some of the first established immigrant areas in cities developed their own informal modes of neighborhood control, backed by violence or the threat of violence. Similarly, some young people in African American and Hispanic communities today have developed subcultures of manliness and honor associated with rituals of violence, and some belong to gangs whose everyday life is one of drug dealing, territory protection, and violence (Venkatesh 2008). Many states uphold a stand-your-ground law, which allows a person to use deadly force in self-defense without first attempting to retreat. This controversial law is believed by many social scientists to normalize or justify the use of guns and violence (Vedantam and Schultz 2013).

Criminal Victimization

Even as the rate of crime declines nationwide, it is important to understand that research and crime statistics show that crime and victimization are not random occurrences across the population. Men, young persons, and African Americans are more likely than women, older persons, and whites to be both crime victims and perpetrators. Young African American men face a triple disadvantage in the United States: Being young, Black, and male are all associated with an elevated death rate due to murder.

Emerging evidence suggests that sexual minorities, including gays, lesbians, and transgender persons, may also have a higher-than-average risk of victimization. A **hate crime** is a criminal act motivated by some bias, such as racism, sexism, or homophobia. In 2016, 6,121 hate crimes were reported in the United States. Over half (58 percent) were motivated by a race, ethnicity, or ancestry bias; 21 percent resulted from a religious bias; 18 percent by a sexual orientation bias; and the remaining criminal incidents were prompted by a gender, gender identity, or disability bias (Federal Bureau of Investigation 2017a).

The likelihood of someone becoming a victim of crime is not linked only to their personal characteristics. Rather, victimization rates also vary based on where a person lives. Individuals living in poor inner-city neighborhoods run a much greater risk of becoming victims of crime than residents of more affluent suburban areas. That ethnic minorities are concentrated disproportionately in inner-city regions appears to be a significant factor in their higher rates of victimization.

CONCEPT CHECKS

1. What are the main sources of crime data in the United States?

2. Contrast Uniform Crime Reports and the National Crime Victimization Survey.

3. Describe crime trends in the 1970s through today.

4. How do sociologists explain the high rate of violent crime in the United States?

Who Are the Perpetrators?

Are some individuals and groups more likely to commit crimes? Social science research and crime statistics show that criminal acts are not randomly distributed across the population. While it is well known that the poorest groups in a society will usually experience the highest crime rates, what is less well known is that crime is also connected to—and committed by—a range of other groups, which we explore here. Thus, as we aim to show in this section, the commonsense connections between crime and certain groups are sometimes inaccurate and rarely as straightforward as we think. Here we explore various perpetrators of crime, such as youth and people in professional jobs.

Understand why members of some social groups are more likely to commit or be the victims of crime.

Gender and Crime

Sociological studies of crime and deviance have traditionally ignored half the population. Feminists have been correct in criticizing criminology for being a male-dominated discipline in which women are largely invisible in both theoretical considerations and empirical studies. Since the 1970s, many important feminist works have drawn attention to the way in which criminal transgressions by women occur in different contexts from those by men and to how women's experiences with the criminal justice system are influenced by certain gendered assumptions about appropriate male and female roles. Feminists have also highlighted the prevalence of violence against women, both at home and in public.

hate crime

A criminal act by an offender who is motivated by some bias, such as racism, sexism, or homophobia.

The statistics on gender and crime are startling. Of all crimes reported in 2016, an overwhelming 73 percent of arrestees were men (Federal Bureau of Investigation 2017b). Men drastically outnumber women in prison, not only in the United States but in all industrialized countries. In 2018, women made up just 7 percent of the U.S. prison population (Carson 2018). Men and women also vary in the types of crimes they commit; women rarely engage in violent crime and instead tend to commit less serious offenses. Petty thefts like shoplifting and public order offenses such as public drunkenness and prostitution are typical female crimes. Recent research by feminist scholars, however, reveals that violence is not exclusively a characteristic of male criminality. By studying girl gangs, female terrorists, and women prisoners, scholars have demonstrated that women do in fact participate in violent crime (albeit less often than men). Moreover, men and women are often quite similar in their motivations for turning to criminal behavior.

One thing is clear, though: Female rates of criminality are consistently lower than those of men. A controversial argument set forth in the 1950s proposed that the gender gap in crime may be less vast than statistics suggest. Otto Pollak (1950) argued that women's crimes may go undetected or unreported, or they may be treated more leniently by (male) police officers, who adopt a "chivalrous" attitude toward them. There is some evidence that female lawbreakers often avoid coming before the courts because they are able to persuade the police or other authorities to see their actions in a particular light. They invoke what has been called the "gender contract"—the implicit contract between men and women whereby to be a woman is to be erratic and impulsive, on the one hand, and in need of protection, on the other (Worrall 1990). Recent research shows that young female offenders, particularly Black girls, have increasingly been charged with assault since the 1980s (Stevens, Morash, and Chesney-Lind 2011). Nevertheless, as seen in probation officers' reliance on gendered scripts—reading girls as needy victims, not criminally dangerous—the "gender contract" may persist in criminal processing (Mallicoat 2007).

Yet differential treatment alone could hardly account for the vast difference between male and female rates of crime. The reasons are almost certainly the same as those that explain gender differences in other spheres, although the gender crime gap is impacted by additional socioeconomic factors. For example, in Chicago, researchers find that as neighborhood disadvantage—poverty, unemployment, and racial segregation—increases, the gender gap in violent crime decreases (Zimmerman and Messner 2010). This narrowing of the gap occurs because increased exposure to peer violence has a greater impact on young women's violent behavior than on young men (Zimmerman and Messner 2010).

Control theory may offer insights into the gender gap in crime. Because women are usually the primary caregivers to their children and other relatives, they

Although women still comprise a small proportion of the prison population—7 percent in 2016—the number of incarcerated women has risen significantly over the past quarter century.

may have attachments and commitments that deter them from engaging in deviant acts. Imprisonment would have very high and undesirable costs both to women and to their kin. Many "male crimes" remain "male" because of gendered differences in socialization and in opportunity, wherein men's activities and involvements are still more nondomestic and connected to greater resources than those of most women (for example, see Steffensmeier, Schwartz, and Roche 2013). As the boundaries between men's and women's social roles increasingly blur, criminologists have predicted that gender equality will reduce or eliminate the differences in criminality between men and women. As of yet, however, crime remains a gendered phenomenon.

Youth and Crime

Popular fear about crime centers on offenses such as theft, burglary, assault, and rape—street crimes that are largely seen as the domain of young working-class males. Media coverage often focuses on moral breakdown among young people and highlights such issues as vandalism, school truancy, and drug use to illustrate the increasing permissiveness in society. This equation of youth with criminal activity is not a new one, according to some sociologists. Young people are often taken as an indicator of the health and welfare of society itself.

Official crime statistics do reveal high rates of offense among young people: Over one-fifth (21.7 percent) of all offenders arrested for criminal offenses in 2016 were age twenty-one or younger (Federal Bureau of Investigation 2017b). The share of arrests appears to peak around age eighteen and to remain fairly constant through the twenties, declining thereafter. Control theory has been used to explain this pattern, called the "age–crime curve." As young people gradually transition into adulthood, they acquire those social attachments and commitments that make "conventional" behavior rewarding. As they marry, have children, find jobs, and set up their own homes, the "cost" of deviance rises; rational actors would not want to risk losing their families and homes and thus avoid deviant acts.

Although criminologists have demonstrated persuasively that most youthful deviants go on to lead perfectly happy, healthy, law-abiding lives, widespread panic about youth criminality persists. Importantly, this panic may not accurately reflect social reality. An isolated event involving young people and crime can be transformed symbolically into a full-blown crisis of childhood, demanding tough law-and-order responses. The high-profile mass murders at Columbine High School in 1999, Virginia Tech University in 2007, Sandy Hook Elementary School in 2012, and Marjory Stoneman Douglas High School in 2018 are examples of how moral outrage can deflect attention from larger societal issues. Columbine was a watershed event in media portrayals of youth crime, and some have speculated that it led to "copycat" school killings in high schools across the United States. Even though the number of shooting incidents involving students has declined over the past quarter century, attention to these high-profile mass murders has led many to think that every school is vulnerable to violent threats (Nicodemo and Petronio 2018).

Similar caution can be expressed about the popular view of drug use by teenagers. Every year, the U.S. Department of Health and Human Services conducts the National Survey of Drug Use and Health about drug-use habits. In 2016, it surveyed more than 67,000 people over the age of twelve and found that 8 percent of respondents between the ages of twelve and seventeen were current users of illicit drugs—the most common of which was marijuana—compared to 23 percent of respondents ages eighteen to twenty-five (Ahrnsbrak et al. 2017). The most common drug used by youth between the ages of twelve and seventeen is marijuana, which many states have legalized or decriminalized in recent years. According to the survey, 7 percent of young people between the ages of twelve and seventeen and 21 percent of young adults ages eighteen to twenty-five had used marijuana in the past month (Ahrnsbrak et al. 2017). Trends in drug use have shifted away from hard drugs, such as heroin and cocaine, and toward combinations of substances such as amphetamines, prescription pain relievers like Oxycontin, alcohol, and Ritalin and other stimulants. The war on drugs, some have argued, criminalizes large segments of the youth population who are generally law abiding (Muncie 1999).

Taking illegal drugs, like other forms of socially deviant behavior, is often defined in racial, class, and cultural terms; different drugs come to be associated with different groups and behaviors. When crack cocaine appeared in the 1980s, it was quickly defined by the media as the drug of choice for Black inner-city kids who listened to hip-hop. Perhaps as a result, jail sentences for crack possession were set at higher levels than sentences for possession of cocaine, which was associated more with white and suburban users. Ecstasy has, until recently, had similar white and middle- or upper-class associations. The current opioid epidemic has been characterized also as a predominately white issue, likely due to the systematic undertreatment of Black relative to white pain with pain relief prescriptions (Singhal, Tien, and Hsia 2016; Hoffman et al. 2016). The corresponding public health response has centered more sympathetically on treatment and rehabilitation than on criminalization and punishment (HBO 2015; Lopez 2017; Netherland and Hansen 2017).

Crimes of the Powerful

Although there are connections between crime and poverty, it would be a mistake to assume that crime is concentrated among the poor. Crimes carried out by people in positions of power and wealth can have farther-reaching consequences than the often-petty crimes of the poor. One of the most devastating events of the early twenty-first century was the discovery that then-trusted investment adviser Bernie Madoff had defrauded his clients—many of them senior citizens and charitable organizations—robbing them of more than $18 billion. Madoff had run an elaborate Ponzi scheme, which left many of his investors destitute and nearly bankrupted charitable organizations such as the Elie Wiesel Foundation and Stony Brook University Foundation (Creswell and Thomas 2009). This case revealed just how devastating the effects of white-collar crime can be.

The term **white-collar crime**, first introduced by Edwin Sutherland (1949), refers to crime carried out by people in professional jobs. This category of criminal activity includes tax fraud, antitrust violations, illegal sales practices, securities and land fraud, embezzlement, the manufacture or sale of dangerous products, and illegal environmental pollution, as well as straightforward theft. The distribution of white-collar crime is even harder to measure than that of other types of crime; most do not appear in the official statistics at all.

Efforts to detect white-collar crime are ordinarily limited, and it is only on rare occasions that those who are caught go to jail. Although the authorities regard white-collar

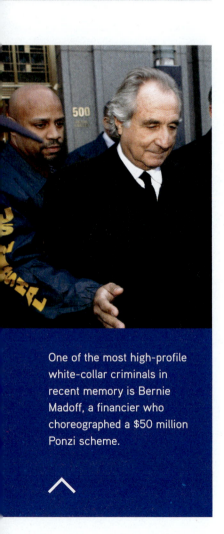

white-collar crime

Criminal activities carried out by those in white-collar, or professional, jobs.

One of the most high-profile white-collar criminals in recent memory is Bernie Madoff, a financier who choreographed a $50 million Ponzi scheme.

crime in a more tolerant light than crimes of the less privileged, it has been calculated that the amount of money involved in white-collar crime in the United States is forty times greater than the amount involved in crimes against property, such as robberies, burglaries, larceny, forgeries, and car thefts (President's Commission on Organized Crime 1986). Some forms of white-collar crime, moreover, affect more people than lower-class criminality. An embezzler might rob thousands—or today, via computer fraud, millions—of people. In recent years, white-collar criminals have victimized an estimated 36 percent of businesses and 25 percent of households, while rates of traditional property and violent crime are much lower—around 8 and 1 percent, respectively (Cliff and Wall-Parker 2017).

CORPORATE CRIME

Corporate crime refers to criminal offenses committed by large corporations across the globe. Pollution, product mislabeling, and violations of health and safety regulations affect much larger numbers of people than petty criminality does. Both quantitative and qualitative studies of corporate crime have concluded that a large number of corporations do not adhere to the legal regulations that apply to them (Slapper and Tombs 1999). Corporate crime is not confined to a few bad apples but is instead pervasive and widespread. Studies have revealed six types of violations linked to large corporations: administrative (paperwork or noncompliance), environmental (pollution, permit violations), financial (tax violations, illegal payments), labor (working conditions, hiring practices), manufacturing (product safety, labeling), and unfair trade practices (anticompetition, false advertising).

Sometimes there are obvious victims, as in environmental disasters like the 1984 spill at the Bhopal chemical plant in India and the health dangers posed to women by silicone breast implants. One of the most devastating examples in recent years was the collapse of an eight-story commercial building, Rana Plaza, in Bangladesh in April 2013.

corporate crime

Offenses committed by large corporations in society, including pollution, false advertising, and violations of health and safety regulations.

The death toll topped 1,100 and thousands of others were injured in what is considered the deadliest disaster in the garment industry. Rana Plaza housed several garment manufacturers. Although building inspectors had found cracks in the building days earlier and recommended that the building be evacuated and shut down, many of the garment workers were forced to return to work. Managers at some of the companies even threatened to withhold a month's pay from workers who refused to come to work (Manik and Yardley 2013).

As the Rana Plaza tragedy demonstrates, the hazards of corporate crime are all too real. But very often, victims of corporate crime do not see themselves as such. This is because in "traditional" crimes, the proximity between victim and offender is much closer; it is difficult not to realize that you have been mugged! In the case of corporate crime, greater distances in time and space mean that victims may not realize they have been victimized or may not know how to seek redress for the crime.

The effects of corporate crime are often experienced unevenly within society. Those who are disadvantaged by other types of socioeconomic inequalities tend to suffer disproportionately. For example, safety and health risks in the workplace tend to be concentrated most heavily in low-paying occupations. Many of the risks from health care products and pharmaceuticals have had a greater impact on women than on men, as is the case with contraceptives or fertility treatments with harmful side effects (Slapper and Tombs 1999).

Organized Crime

Organized crime refers to forms of activity that have some of the characteristics of orthodox business but that are illegal. Organized crime embraces illegal gambling, drug dealing, prostitution, large-scale theft, and protection rackets, among other activities. In *End of Millennium* (1998), Manuel Castells argues that the activities of organized crime groups are becoming increasingly international in scope. The coordination of criminal activities across borders—with the help of new information technologies—is becoming a central feature of the new global economy. Involved in activities ranging from the narcotics trade to counterfeiting to smuggling immigrants and human organs, organized crime groups are now operating in flexible international networks rather than within their own territorial realms. According to Castells, criminal groups set up strategic alliances with one another. The international narcotics trade, weapons trafficking, the sale of nuclear material, and money laundering have all become linked across borders and crime groups. The flexible nature of this networked crime makes it relatively easy for crime groups to evade the reach of law-enforcement initiatives.

Despite numerous campaigns by governments and police, the United Nations Office on Drugs and Crime (2017) estimates that the global trade in illegal drugs makes up between one-fifth and one-third of total income from transnational organized crime—higher than annual global trade in coffee, grains, or meat (2013). However, as organized crime groups grow increasingly agile in structure and diverse in the markets they rely on, the relative importance of drugs is declining (United Nations Office on Drugs and Crime 2017).

organized crime

Criminal activities carried out by organizations established as businesses.

CONCEPT CHECKS

1. Contrast the following two explanations for the gender gap in crime: differences in opportunity and biases in reporting.

2. What is the age–crime curve, and what factors have contributed to this pattern?

3. What are some of the consequences of white-collar crime?

4. Give one example of an activity classified as organized crime.

What Were the Causes and Costs of the Great Crime Decline?

Consider some of the factors that have caused a decades-long decline in crime rates. Understand the social costs associated with the great crime decline.

We began this chapter with a statistic that might be surprising to some: The rate of violent crime is now lower than at any point since reliable data have been kept. With very few exceptions, violence has gone down in almost every city; at the time that the Black Lives Matter movement started in 2014, the crime rate was the lowest of any year in modern history. Explaining this phenomenon, known as the great crime decline, has been of significant interest to sociologists. Many factors must be understood as important in contributing to this decline, and in this section, we concentrate on three: prisons, the death penalty, and policing. At the same time, we ask a less obvious question: What are the social costs and benefits associated with the great crime decline?

Prisons

The decline in crime can, in part, be attributed to one of the biggest transformations in U.S. social policy over the past few decades: the rise of mass incarceration. This is mainly due to the impact this policy has on removing criminals and potential criminals from society (Sharkey 2018). The United States locks up more people per capita than any other country and has by far the most punitive justice system in the world. The so-called prison boom began in the 1970s, with the number of inmates nearly quintupling—from roughly 100 inmates per 100,000 residents for most of the twentieth century to 486 inmates per 100,000 residents by 2004 (Pager 2007). At the end of 2016, more than 6.6 million people were under the supervision of the U.S. correctional system, including more than 1.5 million people in state and federal prisons, 740,700 people in local jails, and more than 4.5 million on probation or parole (Kaeble and Cowhig 2018). The U.S. incarceration rate has actually declined since 2009 and is currently at its lowest rate since 1996 (Kaeble and Cowhig 2018).

Although sociologists believe that the rise of incarceration has indeed contributed to lowering crime rates, there is no agreement on how much it has contributed to the great crime decline. As a policy intervention, this is particularly problematic because both the price and social cost of imprisoning an individual are enormous: In 2015, it cost an average of $31,978 to keep a prisoner in the federal prison system for a year (Prisons Bureau 2017) and more—$33,274—in the average state prison (Vera 2016). It has also become partially privatized, with private companies building and administering prisons to accommodate the growing inmate population. Some argue that this is not money well spent: Only about one-fifth of all serious crimes result in arrest—and this is of crimes known to the police, an underestimate of the true rate of crime. And no more than half of the arrests for serious crimes result in a conviction. Even so, America's prisons are so overcrowded that the average convict serves only one-third of his or her sentence.

While we might suppose that imprisoning large numbers of people or stiffening sentences would deter individuals from committing crimes, there is little evidence to

FIGURE 6.4

State and Federal Prison Population, 1978–2016

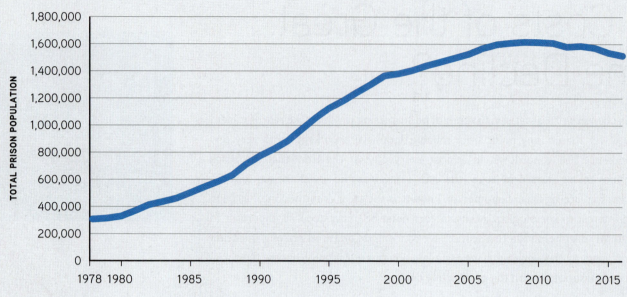

Source: Carson 2018.

support this supposition. In fact, sociological studies have demonstrated that prisons can easily become schools for crime. The rate of relapse into crime, otherwise known as recidivism, is almost 50 percent within eight years of release from federal prison but close to one-third for those released directly to a probationary sentence (Hunt and Dumville 2016). These rates suggest that instead of preventing people from committing crimes, prisons often actually make them more hardened criminals. This pattern is consistent with the key theme of differential association theory, discussed earlier. Deviance is learned from deviant peers. The more harsh and oppressive prison conditions are, the more likely inmates are to be brutalized by the experience. Yet if prisons were made into attractive and pleasant places to live, would they have a deterrent effect?

Mass incarceration has had a particularly deleterious effect on Black communities. African Americans make up around 33 percent of the current prison population, though they represent only 13 percent of the U.S. population (Carson 2018; Rastogi et al. 2011). In *The New Jim Crow*, legal scholar Michelle Alexander (2012) argues that mass incarceration creates a kind of caste system in the United States. According to Alexander, understanding mass incarceration means understanding not only the criminal justice system but also the entire structure of policies and practices that stigmatize and marginalize those who are considered criminals.

THE MARK OF A CRIMINAL RECORD

To illustrate the high cost that society pays for mass incarceration, Devah Pager (2003) conducted an experiment that showed the long-term consequences of prison on the lives of felons. Pager had pairs of young Black and white men apply for real entry-level job openings throughout the city of Milwaukee. The applicant pairs were matched by appearance, by interpersonal style, and, most important, by all job-related characteristics, such as education

Incarceration Rates

More than 10.3 million people are currently being held in penal institutions across the globe. Although the United States is home to less than 5 percent of the global population, it accounts for more than 20 percent of the world's prisoners.

of prisoners per 100,000 residents

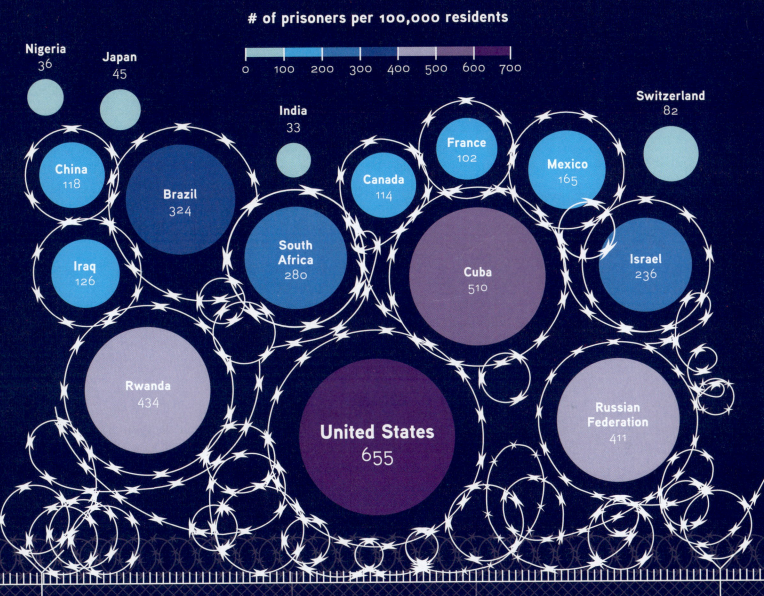

0   100   200   300   400   500   600   700

Nigeria
36

Japan
45

India
33

Switzerland
82

China
118

Brazil
324

Canada
114

France
102

Mexico
165

Iraq
126

South
Africa
280

Cuba
510

Israel
236

Rwanda
434

United States
655

Russian
Federation
411

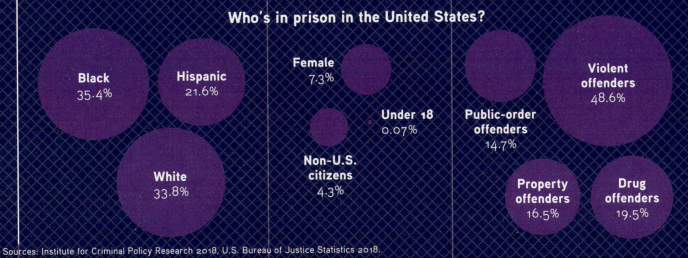

Who's in prison in the United States?

Black
35.4%

Hispanic
21.6%

Female
7.3%

Violent
offenders
48.6%

White
33.8%

Under 18
0.07%

Public-order
offenders
14.7%

Non-U.S.
citizens
4.3%

Property
offenders
16.5%

Drug
offenders
19.5%

Sources: Institute for Criminal Policy Research 2018, U.S. Bureau of Justice Statistics 2018.

level and prior work experience. In addition to varying the race of the applicant pairs, Pager also had applicants alternate presenting themselves to employers as having criminal records. One member of each of the applicant pairs would check the box "yes" on the applicant form in answer to the question "Have you ever been convicted of a crime?" The pair alternated each week which young man would play the role of the ex-offender. The experimental design allowed Pager to make the applicant pairs identical on all job-relevant characteristics so that she could know for sure that any differences she saw were the result of discrimination against felons rather than other qualifications or weaknesses of the applicant.

Pager's study revealed some striking findings. First, whites were much preferred over Blacks, and nonoffenders were much preferred over ex-offenders. Whites with a felony conviction were half as likely to be considered by employers as equally qualified nonoffenders. For Blacks, the effects were even larger: Black ex-offenders were only one-third as likely to receive a call back compared with nonoffenders. Even more surprising was the comparison of these two effects: Blacks with no criminal history fared no better than whites with a felony conviction. These results suggest that the experience of being a Black male in America today is comparable with the experience of being a convicted white criminal, at least in the eyes of Milwaukee employers. For those who believe that race no longer represents a major barrier to opportunity, these results represent a powerful challenge. Being a Black felon is a particularly tough obstacle to overcome.

Although prisons do keep some dangerous men (and a tiny minority of dangerous women) off the streets, evidence suggests that we need to find other means to deter crime. A sociological interpretation of crime makes clear that there are no quick fixes. The causes of crime, especially violent crimes, are bound up with structural conditions of American society, including widespread poverty, the condition of the inner cities, and the deteriorating life circumstances of many young men.

The Death Penalty

Many people wonder about the role of the death penalty in the great crime decline. Interestingly, there is no evidence whatsoever that the death penalty has contributed to lower crime rates in states that use it. In fact, states that use the death penalty consistently have higher murder rates than states that do not. Yet, like its mass incarceration rates, the use of capital punishment (the death penalty) makes the United States an unusual case compared to other liberal democratic nations. It remains one of the last Western countries to legally permit the practice. At the end of 2016, there were 2,814 inmates on death row in the United States, with just three states—California, Florida, and Texas—accounting for 49 percent of inmates sentenced to death (Davis and Snell 2018). While death sentences are occasionally handed down, they are rarely carried out, due to lengthy appeals. Five states executed 20 prisoners in 2016, the fewest number of executions since 1991 (Davis and Snell 2018; Gramlich 2016).

Support for capital punishment remains high in the United States, though it has been declining in the last two decades: In 2016, approximately 49 percent of adults surveyed said they believed in capital punishment, while 42 percent opposed it. This represents a significant shift from a peak in the mid-1990s, when in 1994, 80 percent of those surveyed supported the death penalty and just 16 percent were opposed (Oliphant 2016). Although American support for capital punishment is thought to be unusual in an international political climate that has increasingly condemned the use of the death penalty, the differences may not be as great as some would think. Even in countries that have eliminated the death penalty, the majority tend to support it. The difference is not really popular opinion but government structure: European constitutions give leaders the right to impose their views on the majority, while in the United States, these policies are made on a state-by-state basis. The idea of states' rights is so popular in the United States that even the Supreme Court has deferred to the states on the issue of capital punishment (Garland 2010).

Policing

After considering the uncertain impact of mass incarceration and the lack of evidence that the death penalty contributes to crime reduction, we are left with what is perhaps the biggest contributing factor in the decline of crime: policing. During the 1990s, police forces began to grow at significant rates, and many scholars believe that this accounts for 10 to 20 percent of the overall crime decline. However, there were other factors that complemented the police, including a doubling of private security guards and a massive increase in surveillance cameras (Sharkey 2018). These techniques were not only useful for deterring crime and catching criminals but were also reassuring for the public. Such activities are consistent with the perception that the police are actively engaged in controlling crime, investigating offenses, and supporting the criminal justice system.

At the same time, like the rise of mass incarceration, the consequences of increased police presence are also socially costly. In recent years, several high-profile court cases have attracted mainstream media attention around some of the most ruthless and arbitrary police practices, such as "stop-and-frisk" policies in major American cities. These policies work to put a visible police presence in communities with high crime rates by temporarily detaining and questioning individuals at an officer's discretion. In a decision declaring New York's stop-and-frisk policy to be unconstitutional, U.S. district judge Shira A. Scheindlin declared that the frequent stops made by New York police were in violation of the Fourth Amendment, which protects citizens from unreasonable searches and seizures. Judge Scheindlin's decision asks us to consider the effects of such policies in communities where individuals of color—overwhelmingly young men of color—are targeted by police officials on a near-daily basis.

In *Punished: Policing the Lives of Black and Latino Boys* (2011), sociologist Victor Rios describes the lives of young Black and Hispanic men in Oakland, California. Rios documents the strain that temporary detainment policies like stop and frisk place on heavily policed communities and how young men respond to the pervasive presence of police and other authority figures in their schools and neighborhoods. For the men in Rios's study, negative interactions with police officers are a regular occurrence. These policies affect not only their daily lives but also the way in which they perceive themselves and their long-term life trajectories. Punitive policing created a culture of mistrust and resistance to authority, and even those who seldom broke the rules were perceived negatively by others in their community. In this setting, teachers and potential employers often interpreted innocuous behavior as acts of deviance or criminal activity and denied the young men access to the resources that could have helped them grow in positive ways.

In *Down, Out, and Under Arrest: Policing and Everyday Life in Skid Row* (2016), sociologist Forrest Stuart details the relationship between police and the urban poor of Skid Row, Los Angeles. Like Rios, Stuart shows how policing shapes culture. On Skid Row, heavily surveilled residents develop a shared cognitive framework, "cop wisdom," in which

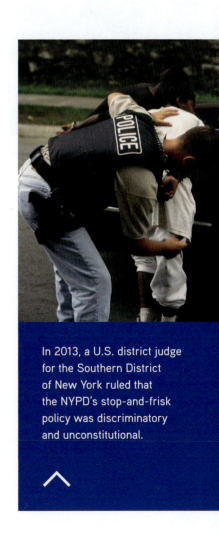

In 2013, a U.S. district judge for the Southern District of New York ruled that the NYPD's stop-and-frisk policy was discriminatory and unconstitutional.

On July 17, 2014, Staten Island resident Eric Garner was placed in a chokehold and dragged to the ground by police officers attempting to arrest him. The fatal encounter was recorded on a cell phone, and Garner's last words, "I can't breathe," became a rallying cry for the Black Lives Matter movement.

Using Cameras to Police the Police

In recent years, video technology on cell phones has led to an increased awareness of police misconduct. Although the government generates many statistics to monitor the police, including records of stops and frisks, these forms of official data ultimately rely on the police to process complaints. But if you're a resident of a heavily policed community, you want the police to be accountable independent of internal processes for reporting and resolving complaints.

In light of a string of high-profile cases of police brutality, there has been an explosion of efforts to record police–civilian interactions using cell phones and cameras. Watchdog groups known as cop-watches have sprung up in communities across the United States with the goal of documenting—and exposing—instances of police misconduct. We Copwatch is a Berkeley-based group that trains people on best practices for monitoring the police. After the shooting death of Michael Brown, the group jumped onto the scene, donating cameras to residents of the apartment complex in Ferguson, Missouri, near the spot where Brown was killed, and hosting informational sessions (Rentz and Donovan 2015; Sanburn 2014). Groups like Cop Block, Communities United Against Police Brutality, and Peaceful Streets Project are similarly dedicated to holding police officers accountable through the use of video technology and educating civilians about their civil rights when it comes to interacting with law enforcement.

Numerous apps have been developed to make it easier for people to document police abuse. The American Civil Liberties Union's Mobile Justice app allows users to discreetly record both video and audio and even send a copy of the recording directly to the ACLU. "You have the right to film the police, and everyone should feel empowered to exercise that right," said Udi Ofer, executive director of ACLU-NJ (American Civil Liberties Union 2015). With the Cop Watch Video Recorder app, created by the Network for the Elimination of Police Violence, users can set the app to automatically upload recordings to YouTube. And while most apps are focused on recording police encounters gone wrong, Five-O aims to capture the good with the bad. The app, which was developed by three teenagers from Atlanta, prompts users to rate officers with whom they've interacted and provide details of their encounters in order to collect data that can then be used to promote positive change (The Economist 2015a).

These apps have both fans and critics. Some opponents maintain that civilian videographers can interfere with police work and scare off witnesses and that video often omits important context (Calvert and Bauerlein 2015). Other opponents cite privacy concerns. Whereas advocates say that these new technologies can help prevent police brutality, even those within the movement are sometimes skeptical. They point to the many cases in which video was used to bring cops to trial, only to result in jury verdicts that exonerated the police. Yet advocates argue that it's not merely about jury verdicts; these technologies put police on notice and change the balance of power in significant ways. Some cops have been known to retaliate, grabbing cameras or even finding ways to arrest the people taking video. Courts, however, have generally upheld a person's constitutional right to film police activity.

Have you ever been afraid in the presence of a police officer? Have you ever used a cop-watch app? Would you? Do you think cop-watches and recording apps are effective ways to hold police accountable for their actions? Some police departments, including the NYPD, are implementing body camera programs to increase transparency and accountability. Do you think all police officers should have to wear body cameras?

knowledge of policing practices circumscribes their understanding of themselves, their communities, and their available options—regardless of an immediate police presence. Practicing what Stuart calls "therapeutic policing," police officers deploy "a paternalistic brand of spatial, behavioral, and moral discipline designed to 'cure' those at the bottom of the social hierarchy" (Stuart 2016, p. 6). Stuart shows that for Skid Row residents threatened with arrest for lifting weights outdoors as a group or for sitting on the sidewalk, intensive law enforcement does something besides potentially lower crime rates. Instead, such regulation might further restrict access to shelter, work, and meaningful social relations.

Today, governments eager to appear decisive on crime favor work to increase the number and resources of the police. But while a greater police presence translates into lower crime rates, the social costs might not be worth the benefits. And even when police are involved, there is more to lowering crime rates than simply having police involved. Preventing crime and reducing fear of crime are both important paths to rebuilding strong communities. One of the most significant innovations in criminology in recent years has been the discovery that the decay of day-to-day civility relates directly to criminality. Although sociologists and criminologists in earlier decades focused almost exclusively on serious crime—robbery, assault, and other violent crime—they have since discovered that minor crimes and public disorder have a powerful effect on neighborhoods. When asked to describe their problems, residents of troubled neighborhoods mention seemingly minor concerns, such as abandoned cars, graffiti, youth gangs, and similar phenomena.

People act on their anxieties about these issues: They move out of these neighborhoods (if they can afford to), buy heavy locks for their doors and bars for their windows, abandon public places like parks, and even avoid healthy behaviors like jogging and walking, out of fear. As they withdraw physically, they also withdraw from roles of mutual support with fellow citizens, thereby relinquishing the social controls that formerly helped to maintain civility within the community.

BROKEN WINDOWS THEORY

The recognition that even seemingly small acts of crime and disorder can threaten a neighborhood is based on the **broken windows theory** (Wilson and Kelling 1982). This sociological theory evolved from an innovative study conducted by the social psychologist Philip Zimbardo. He abandoned cars without license plates and with their hoods up in two entirely different social settings: the wealthy community of Palo Alto, California, and a poor neighborhood in the Bronx, New York. In both places, the cars were vandalized once passersby, regardless of class or race, sensed that the cars were abandoned and that "no one cared" (Zimbardo 1969). Any sign of social disorder in a community, even one unrepaired broken window, is a sign that no one cares. Breaking more windows—that is, committing more serious crimes—is a rational response by criminals to this situation of social disorder. Minor acts of deviance can lead to a spiral of crime and social decay.

In the late 1980s and 1990s, the broken windows theory served as the basis for new policing strategies that aggressively focused on minor crimes, such as traffic violations and drinking or using drugs in public. Studies have shown that proactive policing directed at maintaining public order can have a positive effect on reducing more serious crimes, such as robbery (Sampson and Cohen 1988). However, one flaw of the broken windows theory is that the police are left to identify "social disorder" however they wish. Without

broken windows theory

A theory proposing that even small acts of crime, disorder, and vandalism can threaten a neighborhood and render it unsafe.

a systematic definition of disorder, the police are authorized to see almost anything as a sign of disorder and anyone as a threat. In fact, as crime rates fell throughout the 1990s, the number of complaints of police abuse and harassment went up, particularly by young, urban, Black men who fit the "profile" of a potential criminal. In response to these limitations, criminologists and policy makers have developed alternative strategies for crime prevention, including community policing.

COMMUNITY POLICING

One idea that has grown in popularity in recent years is that the police should work closely with citizens to improve local community standards and civil behavior, using education, persuasion, and counseling instead of incarceration (Sharkey 2018). **Community policing** implies not only drawing in citizens themselves but changing the characteristic outlook of police forces. A renewed emphasis on crime prevention rather than law enforcement can go hand in hand with the reintegration of policing within the community and reduce the siege mentality that develops when police have little regular contact with ordinary citizens.

In order to work, partnerships among government agencies, the criminal justice system, local associations, and community organizations have to be inclusive; all economic and ethnic groups must be involved (Kelling and Coles 1997). Government and business can act together to help repair urban decay. One model is the creation of urban enterprise zones, which provide tax breaks for corporations that participate in strategic planning and invest in designated areas. To be successful, such schemes demand a long-term commitment to social objectives.

Emphasizing these strategies does not mean denying the links among unemployment, poverty, and crime. Rather, the struggle against these social problems should be coordinated with community-based approaches to crime prevention. These approaches can in fact contribute directly and indirectly to furthering social justice. Where social order has decayed along with public services, other opportunities, such as new jobs, decline as well. Improving the quality of life in a neighborhood can revive them. The decline in crime and violence in America has likely been a function of the way that local communities—police and community residents—have together taken on violence (Sharkey 2018: 181).

The Benefits of the Crime Decline

What you have read so far reflects some ambivalence among sociologists about the great crime decline. We have emphasized the causes of the downward trend in crime and we have seen that some of what contributed to the decline was the rise of mass incarceration. This is a phenomenon of modern society that has had a disproportionately negative effect on poor communities, and particularly communities of color. Yet one question that gets asked too infrequently is how poor communities have benefited from the great crime decline. The sociologist Patrick Sharkey (2018) addressed that question by looking at two areas of social life that were most likely influenced by lower crime rates: life expectancy and school performance.

Sharkey finds that in communities where crime has gone down, the amount of violence has also decreased. He argues that during the era of high crime rates, violence was one of the most prevalent public health problems faced by poor communities, on par with heart disease and cancer. Despite the fact that so many people die from violence in inner cities, violence is usually not taken seriously as a "public health" problem. Sharkey argues

---

community policing

A renewed emphasis on crime prevention rather than law enforcement to reintegrate policing within the community.

that unlike well-known diseases with massive expenditures for research, violence is a cause of death that largely goes unnoticed. It disproportionately affects the youngest parts of poor communities, particularly Black males between fifteen and thirty years of age. He found that an average Black boy who was born in 2012 could expect to live three-quarters of a year longer than a Black boy who was born when crime was at its height, in 1991. While that might not seem like a lot, he points out that it is in fact equivalent to what would happen if we were to eliminate the obesity epidemic altogether. Thus, "the decline in violence…means that thousands of young people…no longer have their lives cut short by violence. It is about human life that is preserved" (Sharkey 2018: 66).

Sharkey also found that the murder rate is only part of the story of how poor communities have benefited as a result of declining violence. This is because violence does not only affect those who are themselves on the receiving end of physical harm but also the masses of children who live in fear every time they see or hear about the violence in their communities. Living in fear causes stress for such youth, making it much harder for them to concentrate on a test or control their impulses. Sharkey says that what he found most surprising was how much students suffer academically after a violent shooting: "It was as if children who were [tested] right after a local homicide had missed the previous two years of schooling and regressed back to their level of cognitive performance from years earlier" (2018: 86). It is not only that certain kids from poor communities are exposed to more violence, but also that the schools they attend have fewer support services to meet their needs in these situations.

Thus, the discussion of deviance, crime, and punishment requires that we be prepared to confront complexity. The United States is safer than it has ever been and many of the people who have benefited from those changes are living in the very communities that have paid the biggest prices for those changes. It is perhaps most crucial, then, that we listen carefully to those most affected as we think about the future of U.S. policy in this area.

CONCEPT CHECKS

1. How can differential association theory help explain high rates of recidivism?

2. What are some of the social costs of policing?

3. What are the main criticisms of broken windows theory?

4. Name at least two specific ways that community members can combat local crime.

How Do Crime and Deviance Affect Your Life?

The Costs of Crime

Understand the costs and functions of crime and deviance.

Crime can take a toll on the financial and emotional well-being of even those people whose only contact with the criminal justice system is watching reruns of *Law and Order* or *CSI*. As we learned earlier in the chapter, corporate crime can affect everything from the quality of the food we eat to the safety of the cars we drive and the cleanliness of the air we breathe. Even those of us who live in safe and quiet neighborhoods may find our lives touched by the criminal acts of corporations, in the form of air pollution or tainted foods or medicines.

Our lives are also affected by the high fiscal costs of street crime. Maintaining local, state, and national criminal justice systems is costly—and growing costlier by the minute. State government is having a difficult time finding enough money to house, feed, and provide medical care for all these inmates. Spending on corrections has risen dramatically over the past three decades. For example, spending has varied between 2.5 and 2.9 percent

Law Enforcement: Police Officer and Civilian Employee

In this chapter, we learned that the largest factor driving the decline of crime may be policing, with the growth of police forces contributing between 10 and 20 percent of the total crime decline (Sharkey 2018). Policing has been aided by improvements in surveillance technology, including now-pervasive security cameras, and the proliferation of digital records, as governments and businesses gather information on millions of people. Increasingly, in police departments from Los Angeles to Salt Lake City, decision making by police officers systematically depends on the analysis of this so-called big data (Brayne 2017; Palantir 2018).

For example, using historical data, law enforcement agencies can identity where and when future crimes are most likely to occur and deploy officers accordingly. Civilian employees and police officers can also use models of people's social networks to identify individuals, places, and belongings related to a person of interest—for instance, siblings, cohabitators, and coworkers; addresses of previous and current residence; and vehicles. In the past, data sets and surveillance systems were analyzed in isolation. Big data analysis involves synthesizing this previously separate data, such as combining Automatic License Plate Reader data collected by law enforcement and private agencies like repossession companies (Brayne 2017).

In order to keep up with these developments, sociology is rethinking the statistical training that it offers students. In fact, some sociology departments now integrate computer science into their statistics training. Ultimately, sociology majors with such training are not only able to write algorithms to make predictions out of big data but to develop predictions informed by sociological theories of crime and deviance.

of state outlays over the last decade, with states spending $48.5 billion in 2010 as compared with $15 billion in 1982 (U.S. Bureau of Justice Statistics 2012c).

Lawmakers have few options for footing this large bill. Tax hikes are one option, but that would mean higher income taxes, property taxes, and sales taxes for everyone. In the absence of tax hikes, lawmakers may find themselves forced to cut back on other important social programs, including transportation, education, and health care (Pew Center on the States 2008). Corrections accounted for 3 percent of total state expenditures in 2017, and if this proportion increases, it could touch the lives of Americans using the many other state programs that compete for valuable tax dollars (National Association of State Budget Officers 2018).

The Functions of Deviance

The deviant acts of others also affect our personal behaviors in powerful ways. As noted earlier, deviants help us understand what is considered "right" and "wrong" among our peers, friends, and community members. Most of us try very hard to avoid the sanctions

Sociologists are also particularly well positioned to question the social implications of big data analysis in law enforcement. For several years, sociologist Sarah Brayne studied the Los Angeles Police Department, interviewing and observing seventy-five police officers and civilian employees in order to understand how the department's pioneering adoption of big data analytics amplified and transformed their police surveillance (American Sociological Association 2017). Brayne found that big data analysis "is not always as objective as it suggests. In terms of individual policing, big data aims to supplement officers' discretion with algorithm-based, quantified criminal risk assessments" (American Sociological Association 2017).

Still, bias can alter such assessments. For example, one detective told Brayne that he tends to view a person as suspicious if he enters their name in a database and sees that others have previously searched for their name as well. "Just because you haven't been arrested doesn't mean you haven't been caught," the detective told her (Brayne, 2017, p. 992). Brayne reports:

> When you start to codify or bake in police practices as objective crime data, you sort of get into this feedback loop or self-fulfilling prophecy. It puts individuals who are already under suspicion under new and deeper and quantified forms of surveillance, masked by objectivity or as one officer described it, "just math." (American Sociological Association 2017)

Recall what we learned about labeling theory in this chapter. Whose information is being captured and integrated? How do police officers view those individuals captured in their newly expanded databases—such as the roommate identified through network modeling? Sociology teaches us both the methods of big data analysis and the theories that can guide us to deploy them for the benefit of the communities served by law enforcement officers.

The NYPD relies on big data generated from the more than 4,000 cameras and license plate readers mounted throughout Lower Manhattan. A sociological imagination can help us think critically about how new forms of surveillance might affect different groups of people.

that result from doing "wrong," and we make our daily choices accordingly. For example, most of us don't want to be socially ostracized, so we may choose clothes, hobbies, romantic partners, and even our future career paths to fit in with peers. To be considered "deviant" often means being treated as a social outcast.

At a more serious level, though, most of us know what the punishments are for even minor violations, such as speeding or running a red light. By learning about the fees and punishments levied on those who break the rules, most of us will behave in accordance with the law to avoid having our driver's licenses suspended or spending a night in jail. Public punishments—whether locking horse thieves into "stocks" in town squares and making an adulteress wear a scarlet letter *A* around her neck in the colonial United States, or publicizing the names and addresses of registered sex offenders and televising "perp walks" in the contemporary United States—are designed not only to punish the "guilty" but also to prevent others from behaving in a similar way. These public humiliations affect us because they make us rethink whether it's really worthwhile to try to get away with a crime.

CHAPTER 6

The Big Picture

Deviance, Crime, and Punishment

Thinking Sociologically

1. Briefly summarize several leading theories explaining crime and deviance presented in this chapter: differential association, anomie, labeling, conflict, and control theories. Which theory appeals to you? Explain why.

2. Explain how differences in power and social influence can play a significant role in defining and sanctioning deviant behavior.

3. What do you consider the most harmful consequences of violent crime? Of white-collar crime? How do you think the "average" American views each of these different types of crime?

norms • mores • folkways

deviance • deviant subculture • sanction •
laws • crimes

1. How do sociologists define deviance?
2. Is all crime deviant? Is all deviance criminal? Why?
3. Contrast positive and negative sanctions.

psychopath • anomie • relative deprivation •
differential association • labeling theory •
primary deviance • secondary deviance

1. What are the main similarities and differences between biological and
   psychological views of deviance?
2. How do Merton's and Durkheim's definitions of anomie differ?
3. According to subcultural explanations, how does criminal behavior get
   transmitted from one group to another?
4. What is the core idea behind differential association theory?
5. What are two criticisms of labeling theory?
6. What are the root causes of crime according to conflict theorists?

Uniform Crime Reports (UCR) • hate crime

1. What are the main sources of crime data in the United States?
2. Contrast Uniform Crime Reports and the National Crime Victimization Survey.
3. Describe crime trends in the 1970s through today.
4. How do sociologists explain the high rate of violent crime in the
   United States?

white-collar crime • corporate crime •
organized crime

1. Contrast the following two explanations for the gender gap in crime:
   differences in opportunity and biases in reporting.
2. What is the age-crime curve, and what factors have contributed to this pattern?
3. What are some of the consequences of white-collar crime?
4. Give one example of an activity classified as organized crime.

broken windows theory • community policing

1. How can differential association theory help explain high rates of recidivism?
2. What are some of the social costs of policing?
3. What are the main criticisms of broken windows theory?
4. Name at least two specific ways that community members can combat
   local crime.

7

Stratification, Class, and Inequality

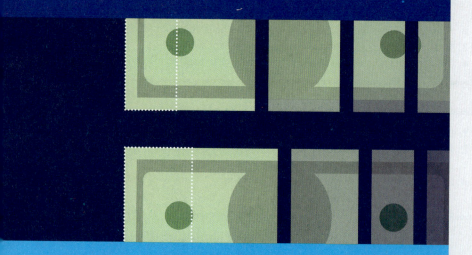

Thanks to her own tenacity in the face of social and economic barriers, Viviana Andazola Marquez managed to overcome a difficult childhood, leaving behind motels for Yale University. What does her story say about mobility in American society?

THE BIG QUESTIONS

What is social stratification?

Learn about social stratification and how social background affects one's life chances. Become acquainted with the most influential theories of stratification.

How is social class defined in the United States?

Understand the social causes and consequences of social class in U.S. society as well as the complexities and challenges of defining class.

What are the causes and consequences of social inequality in the United States?

Recognize why and how the gap between rich and poor has increased in recent decades. Understand social mobility, and think about your own mobility.

How does poverty affect individuals?

Learn about poverty in the United States today, explanations for why it exists, and means for combating it. Learn how people become marginalized in a society and the forms that this marginalization takes.

How does social inequality affect your life?

Learn how changes in the American economy have led to growing inequality since the 1970s.

When she was a middle school student in Thornton, Colorado, Viviana Andazola Marquez struggled to find a place to do her homework, traveling around town in search of open Wi-Fi networks. Her single mother, an undocumented Mexican immigrant, struggled to make ends meet, and Viviana and her three siblings often found themselves homeless. Because of her mother's legal status, not even homeless shelters welcomed Viviana and her family. As a result, Viviana spent her early teenage years negotiating for late payment with motel managers and sleeping on the floors of kindhearted strangers.

Viviana's life reached a low point when she was thirteen. Her mother was arrested for disturbing the peace, jailed for two weeks by Immigration and Customs Enforcement (ICE), then placed under house arrest for six months. Viviana was forced to grow up fast. As she put it in a recent interview, "I saw so many examples of what my life could end up being. . . . I thought one way or another, I have to leave this place" (Bronner 2015). Viviana started researching which colleges and universities offered the best financial aid packages for students with no financial resources whatsoever. Her college essay—which was reprinted in the *New York Times*—tells a powerful story of endless struggle:

> *After a frustrating drive through the neighborhood and careful identification of a network, success is stated simply: Connected. It is a brief moment of victory, but short-lived as I race against the clock to complete my stack of assignments. Sure, it would be ideal to have my own Wi-Fi, but I'd be satisfied if my family obtained a home first. Every day there is a new challenge; it is a game of adaptation: I beat each situation before it beats me. Just as in any game, I endure losses and gains. . . . I moved myself around the game board. I carried my family on my back.*

Clearly her story struck a chord: Viviana was offered full scholarships to Harvard, Princeton, and Yale. Ultimately, she decided on Yale, making her not only the first in her family to attend college but the first from her entire school district ever to attend an Ivy League university. She graduated in 2018 with a degree in ethnicity, race, and migration. That same year, she was back in the headlines, writing an emotional op-ed for the *New York Times* after her father was detained by ICE (Andazola Marquez 2018). He was ultimately deported and banned from reentering the United States for twenty years.

For most Americans, one's socioeconomic background powerfully shapes one's future. Yet in Viviana's case, early-life poverty didn't stop her from excelling in high school and eventually going on to an Ivy League university. How did she transcend her roots? Viviana was smart, tenacious, and determined to escape her circumstances in order to live a better life. She also had strong support from her middle and high schools, including advisers and teachers who not only mentored and encouraged her but sometimes even picked her up and drove her to school from wherever she had spent the night (Bronner 2015).

While such a combination of factors may sometimes enable a person to overcome great obstacles and rise out of poverty and even homelessness, as we shall see in this chapter this is far more likely to be the exception than the rule. Understanding the complex interplay of personal effort and social circumstance—of biography and history—remains one of the key challenges of sociological understanding.

Sociologists speak of **social stratification** to describe inequalities among individuals and groups within human societies. Often we think of stratification in terms of wealth or property, but it can also occur on the basis of other attributes, such as gender, age, race, or religious affiliation. An important area of research in social stratification is social mobility, or one's movement up or down social class strata; an extreme example of this is Viviana. The three key aspects of social stratification are class, status, and power (Weber 1947). Although they frequently overlap, this is not always the case. The "rich and famous" often enjoy high status; their wealth often provides political influence. Yet there are exceptions. While often wealthy and powerful, drug lords, for example, usually enjoy low status.

In this chapter, we focus on stratification in terms of inequalities based on wealth and income, status, and power. In later chapters, we will look at the ways in which gender (Chapter 9) and ethnicity and race (Chapter 10) all play a role in stratification.

social stratification

The existence of structured inequalities among groups in society in terms of their access to material or symbolic rewards.

What Is Social Stratification?

All socially stratified systems share three characteristics:

1. **The rankings apply to social categories of people who share a common characteristic, such as gender or ethnicity.** Women may be ranked differently from men, wealthy people differently from the poor. This does not mean that individuals from a particular category cannot change their rank; however, it does mean that the category continues to exist even if individuals move out of it and into another category.

2. **People's life experiences and opportunities depend heavily on how their social category is ranked.** Being male or female, Black or white, upper class or working class, makes a big difference in terms of your life chances—often as big a difference as personal effort or good fortune.

3. **The ranks of different social categories tend to change very slowly over time.** In U.S. society, for example, only during the last half-century have women begun to achieve economic equality with men (see Chapter 9). Similarly, only since the 1970s have significant numbers of African Americans begun to obtain economic and political equality with whites—even though slavery was abolished more than a century and a half ago and discrimination was declared illegal in the 1950s and 1960s (see Chapter 10).

As you saw in Chapter 2, stratified societies have changed throughout human history. The earliest human societies, which were based on hunting and gathering, had very little social stratification—mainly because there were few resources to be divided up. The development of agriculture produced considerably more wealth and, as a result, a great increase in stratification. Social stratification in agricultural societies came to resemble a pyramid, with a large number of people at the bottom and a successively smaller number of people as one moved toward the top. Today, advanced industrial societies are extremely complex; their stratification is more likely to resemble a teardrop, with a large number of people in the middle and lower-middle ranks (the so-called middle class), a slightly smaller number of people at the bottom, and very few people as one moves toward the top.

But before turning to stratification in modern societies, let's first review the three basic systems of stratification: slavery, caste, and class.

Slavery

Slavery is an extreme form of inequality in which certain people are owned as property by others. Sometimes slaves have been deprived of almost all rights by law, as was the case on Southern plantations in the United States. In other societies, their position was more akin to that of servants. For example, in the ancient Greek city-state of Athens, some slaves occupied positions of great responsibility.

Systems of slave labor have tended to be unstable, because slaves have historically fought back against their subjection. Slavery also is not economically efficient, as it requires constant supervision and often involves severe punishment, which impedes worker productivity. Moreover, from about the eighteenth century on, many people in Europe and America came to see slavery as morally wrong. Today, slavery is illegal in every country

Learn about social stratification and how social background affects one's life chances. Become acquainted with the most influential theories of stratification.

slavery

A form of social stratification in which some people are owned by others as their property.

of the world, but it still exists in some places. Recent research has documented that people are still taken by force and held against their will—from enslaved brickmakers in Pakistan to sex slaves in Thailand and domestic slaves in France. The United States is not immune to such injustice. News reports of teenage girls coerced into prostitution, maids locked up and forced to work by wealthy clients, and a recent case of immigrants forced to work at convenience stores underscore that marginalized persons who lack social power can still be exploited at the hands of cruel individuals (CNN 2013).

Caste Systems

A **caste system** is a social system in which one's social status is given for life. This means that all individuals must remain at the social level of their birth throughout their lifetime. In this system, social status is based on personal characteristics—such as perceived race or ethnicity (often based on such physical characteristics as skin color), parental religion, or parental caste—that are accidents of birth and are therefore believed to be unchangeable. Caste societies can be seen as a special type of class society—in which class position is ascribed at birth rather than achieved through personal accomplishment. In caste systems, intimate contact with members of other castes is strongly discouraged. Such "purity" of a caste is often maintained by rules of **endogamy**, marriage within one's social group as required by custom or law.

Before modern times, caste systems were found throughout the world. In modern times, caste systems have typically been found in agricultural societies that have not yet developed industrial capitalist economies, such as rural India or South Africa prior to the end of white rule in 1992. The Indian caste system, for example, reflects Hindu religious beliefs and is more than 2,000 years old. According to Hindu beliefs, there are four major castes, each roughly associated with broad occupational groupings. Below the four castes are those known as the "untouchables" or Dalits (oppressed people), who—as their name suggests—are to be avoided at all costs. Untouchables are limited to the worst jobs in society, such as removing human waste, and they often resort to begging. India made it illegal to discriminate on the basis of caste in 1949, but aspects of the caste system remain in full force today, particularly in rural areas.

The few remaining caste systems in the world are being challenged further by globalization. For example, as India's modern capitalist economy brings people of different castes together, whether in the same workplace, airplane, or restaurant, it is increasingly difficult to maintain the rigid barriers required to sustain the caste system.

Class

The concept of class is most important for analyzing stratification in industrialized societies like the United States. Everyone has heard of class, but most people in everyday talk use the word in a vague way. As employed in sociology, it has some precision.

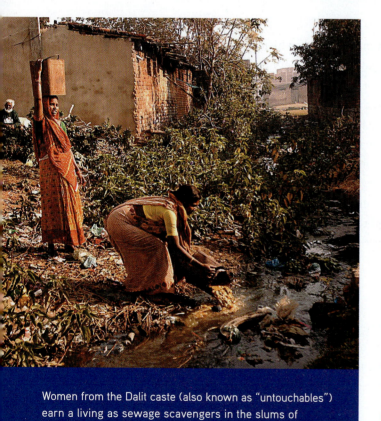

Women from the Dalit caste (also known as "untouchables") earn a living as sewage scavengers in the slums of Ranchi, India.

A social **class** is a large group of people who occupy a similar economic position in the wider society. The concept of life chances, introduced by Max Weber, is the best way to understand what class means. Your **life chances** are the opportunities you have for achieving economic prosperity. A person from a humble background, for example, has less chance of ending up wealthy than someone from a more prosperous one. And the best chance an individual has of being wealthy is to start off as wealthy in the first place.

The United States, it has been said, is the land of opportunity. For some, this is so. There are many examples of people who have risen from humble means to positions of great wealth and power. And yet there are many more cases of people who have not, including a disproportionate share of women and minorities. The idea of life chances is important because it emphasizes that although class is a powerful influence on what happens in our lives, it is not completely determining. Class divisions affect which neighborhoods we live in, what lifestyles we follow, and even which romantic partners we choose (Mare 1991; Massey 1996). Yet they don't fix people for life in specific social positions, as the older systems of stratification did. A person born into a caste position has no opportunity of escaping from it; the same isn't true of class.

Class systems differ from slavery and castes in four main respects:

1. **Class systems are fluid.** Unlike the other types of strata, classes are not established by legal or religious provisions. The boundaries between classes are never clear-cut. There are no formal restrictions on intermarriage between people from different classes.

2. **Class positions are in some part achieved.** An individual's class is not simply assigned at birth, as is the case in the other types of stratification systems. Social mobility—movement upward and downward in the class structure—is relatively common.

3. **Class is economically based.** Classes depend on inequalities in the possession of material resources. In the other types of stratification systems, noneconomic factors (such as race in the former South African caste system) are generally most important.

4. **Class systems are large scale and impersonal.** In the other types of stratification systems, inequalities are expressed primarily in personal relationships of duty or obligation—between slave and master or lower- and higher-caste individuals. Class systems, by contrast, operate mainly through large-scale, impersonal associations, such as pay or working conditions.

ARE CLASS BOUNDARIES WEAKENING?

Viviana's story might lead us to believe that social class boundaries are permeable, where the daughter of an undocumented immigrant, who bounced from motel to motel, could ultimately end up at an Ivy League university. Although Viviana's case is certainly inspiring, is it common? Is it unique to the United States and societies with fluid class strata? How much does social class mold our lives, and has its impact changed over time? Stratification scholars currently grapple with two important debates about the declining importance of social class. First, they ask whether caste systems will give way to

class

Although it is one of the most frequently used concepts in sociology, there is no clear agreement about how the term should be defined. Most sociologists use the term to refer to socioeconomic variations among groups of individuals that create variations in their material prosperity and power.

———

life chances

A term introduced by Max Weber to signify a person's opportunities for achieving economic prosperity.

class systems against the backdrop of globalization. Second, scholars question whether inequality is declining in class-based societies due in part to educational expansion and other social policies.

To address the first question, there is some evidence that globalization will hasten the end of legally sanctioned caste systems throughout the world. Most official caste systems have already given way to class-based ones in industrial capitalist societies. Modern industrial production requires that people move about freely, work at whatever jobs they are suited or able to do, and change jobs frequently according to economic conditions. The rigid restrictions found in caste systems interfere with this necessary freedom. Nonetheless, elements of caste systems persist even in advanced industrial societies. For example, some Indian immigrants to the United States seek to arrange traditional marriages for their children along caste lines, while the relatively small proportion of U.S. marriages that are interracial suggests the strength of racial barriers. In 2015, 10 percent of all marriages were between members of different races. This proportion, however, jumps to 17 percent for newlyweds (Livingston and Brown 2017).

To address the second question, some evidence suggests that at least until recently, mature capitalist societies have been increasingly open to movement between classes, thereby reducing the level of inequality. For example, studies of European countries, the United States, and Canada suggested that inequality peaked in these places before World War II, declined through the 1950s, and remained roughly the same through the 1970s (Berger 1986; Nielsen 1994). Lowered postwar inequality was due in part to economic expansion in industrial societies, which created opportunities for people at the bottom to move up, and because of government health insurance, welfare, and other programs aimed at reducing inequality. As you will see later in this chapter and in Chapter 13 (where we discuss the changing nature of the American economy), however, this trend has reversed in recent years: Inequality has actually been increasing in the United States since the 1970s.

Theories of Stratification in Modern Societies

In this section, we look at some broad theories regarding stratification. Karl Marx and Max Weber developed the most influential theoretical approaches to studying stratification. Most subsequent theories of stratification are heavily indebted to their ideas.

MARX: MEANS OF PRODUCTION AND THE ANALYSIS OF CLASS

For Marx, the term *class* refers to people who stand in a common relationship to the **means of production**—the means by which they gain a livelihood. In modern societies, the two main classes are the bourgeoisie and proletariat. The **bourgeoisie**, or capitalists, own the means of production. Members of the **proletariat**, or proletarians, by contrast, earn their living by selling their labor to the capitalists. The relationship between classes, according to Marx, is an exploitative one. In the course of the working day, Marx reasoned, workers produce more than is actually needed by employers to repay the cost of hiring them. This **surplus value** is the source of profit, which capitalists are able to put to their own use. A group of workers in a clothing factory, say, might be able to produce a hundred suits a day. Selling half the suits provides enough income for the manufacturer

means of production

The means whereby the production of material goods is carried on in a society, including not just technology but the social relations among producers.

bourgeoisie

People who own companies, land, or stocks (shares) and use these to generate economic returns, according to Marx.

proletariat

People who sell their labor for wages, according to Marx.

surplus value

In Marxist theory, the value of a worker's labor power left over when an employer has repaid the cost of hiring the worker.

Does the Digital Divide Still Matter?

Only a few years ago, the digital divide was a chasm—a large gap in Internet use that reflected differences in socioeconomic status. At the turn of the millennium, only 34 percent of those with an annual income under $30,000 used the Internet, compared with 81 percent of households earning more than $79,000. An even larger gap resulted from differences in education, with barely one out of every five high school dropouts using the Internet, compared with nearly four out of every five college graduates (Rainie 2015).

By 2015, these gaps had narrowed considerably. While nearly all households earning more than $75,000 today use the Internet, three-quarters of those earning under $30,000 do so as well. The educational gap has fallen by a comparable amount, and even 60 percent of those over sixty-five are now online. The Internet has become as commonplace as the telephone (Rainie 2015).

Or has it? Not all forms of Internet use are equal. While almost two-thirds of American adults now own smartphones, there remains a sizable age gap in ownership as well as smaller (but significant) gaps based on education and income. For example, 85 percent of young adults (those between the ages of eighteen and twenty-nine) own smartphones, compared with only 27 percent of adults over sixty-five. Gaps remain between college grads (78 percent of whom own smartphones) and those with a high school education or less (52 percent) and between those with household incomes of over $75,000 (84 percent) and those earning under $30,000 (50 percent); 44 percent of the latter, in fact, report having lost their service at one time because of financial constraints (Smith 2015a).

Such disparities matter because Americans are increasingly relying on smartphones for more than texting their friends or posting photos on Instagram. For example, more than half (53 percent) of young adults report having used a smartphone in a job search, and those who are college educated or have higher incomes are the most likely to do so. While it may not come as a surprise that most job-seeking smartphone use is for such basic activities as browsing job listings or contacting employers, nearly half have also used their smartphones to actually fill out a job application. Interestingly, those who lack a college education are far more likely to use their smartphones to fill out a job application (61 percent) than those who have a college degree (37 percent). Similarly, one-third of those with a high school education (or less) rely on their smartphone to create a résumé or cover letter, compared with only one-tenth of those with college degrees. The reason for these differences is simple: Job-seekers who never went to college are far less likely to have broadband Internet at home (Smith 2015b).

The proliferation of smartphones may indeed level the playing field, enabling poorer, less-educated adults to compensate for lack of broadband access. On the other hand, because those who are lower on the socioeconomic ladder are also less likely to own a functioning smartphone, they can be disadvantaged when seeking a job. Moreover, there are clear challenges to using a smartphone to fill out a job application, assemble a résumé, or write an effective job letter. The digital divide may be narrowing, but inequalities remain.

Although more people now have access to the Internet thanks to smartphones, how might the use of these technologies differ according to someone's income and education?

›

to pay the workers' wages. Income from the sale of the remainder of the garments is taken as profit.

Marx believed that the maturing of industrial capitalism would bring about an increasing gap between the wealth of the capitalist minority and the poverty of the large proletarian population. In his view, the wages of the working class could never rise far above subsistence level, while wealth would pile up in the hands of those owning capital. In addition, laborers would face work that is physically wearing and mentally tedious, as is the situation in many factories. At the lowest levels of society, particularly among those frequently or permanently unemployed, there would develop an "accumulation of misery, agony of labor, slavery, ignorance, brutality, moral degradation" (Marx 1977, orig. 1864).

Marx was right about the persistence of poverty in industrialized countries and in anticipating that large inequalities of wealth and income would endure. He was wrong in supposing that the income of most of the population would remain extremely low. Most people in Western countries today are much better off materially than comparable groups were in Marx's day.

WEBER: CLASS, STATUS, AND POWER

There are three main differences between Weber's theory and Marx's. First, according to Weber, class divisions derive not only from control or lack of control of the means of production but also from economic differences that have nothing directly to do with property. Such resources include especially people's skills and credentials, or qualifications. Those in managerial or professional occupations earn more and enjoy more favorable conditions at work, for example, than people in blue-collar jobs. The qualifications they possess—such as degrees, diplomas, and the skills they have acquired—make them more "marketable" than others without such qualifications.

Second, Weber distinguished another aspect of stratification besides class, which he called "status." According to Weber, **status** refers to differences among groups in the social honor, or prestige, they are accorded by others. Status distinctions can vary independent of class divisions. Social honor may be either positive or negative. For instance, doctors and lawyers have high prestige in American society. **Pariah groups**, on the other hand, are negatively privileged status groups subject to discrimination that prevents them from taking advantage of opportunities open to others. For example, members of the "untouchables" caste in India would be treated as pariahs; they are relegated to low-paying work and historically were barred from entering the homes of higher-caste persons. Possession of wealth normally tends to confer high status, but there are exceptions to this principle, such as Hollywood starlets who earn high salaries but lack the education or refinement typically associated with "status." Importantly, status depends on people's subjective evaluations of social differences, whereas class is an objective measure.

Third, Weber recognized that social classes also differ with respect to their **power**, or ability to enact change, command resources, or make decisions. Power is distinct from status and class, but these three dimensions often overlap. For example, on most college campuses, the president or provost has much greater power to change campus policies than a cafeteria worker does. Weber's writings on stratification are important because they show that other dimensions of stratification besides class strongly influence people's lives.

status

The social honor or prestige that a particular group is accorded by other members of a society. Status groups normally display distinct styles of life—patterns of behavior that the members of a group follow. Status privilege may be positive or negative.

pariah groups

Groups that suffer from negative status discrimination—they are looked down on by most other members of society.

power

The ability of individuals or the members of a group to achieve aims or further the interests they hold. Power is a pervasive element in all human relationships. Many conflicts in society are struggles over power, because how much power an individual or group is able to obtain governs how far they are able to put their wishes into practice.

Most sociologists hold that Weber's scheme offers a more flexible and sophisticated basis for analyzing stratification than that provided by Marx.

DAVIS AND MOORE: THE FUNCTIONS OF STRATIFICATION

Kingsley Davis and Wilbert E. Moore (1945) provided a functionalist explanation of stratification, arguing that it has beneficial consequences for society. They claimed that certain positions or roles in society, such as brain surgeons, are functionally more important than others, and these positions require special skills for their performance. However, only a limited number of individuals in any society have the talents or experience appropriate to these positions. To attract the most qualified people, rewards need to be offered, such as money, power, and prestige. Davis and Moore determined that because the benefits of different positions in any society must be unequal, then all societies must be stratified. They concluded that social stratification and social inequality are functional for society because they ensure that the most qualified people, attracted by lucrative rewards, fill the roles that are most important to a smoothly functioning society.

Davis and Moore's theory suggests that a person's social position is based solely on his or her innate talents and efforts. Not surprisingly, their theory has been criticized by other sociologists. The United States is not entirely a meritocratic society. Those at the top tend to have unequal access to economic and cultural resources, such as the highest-quality education, which help the upper classes transmit their privileged status from one generation to the next. For those without access to these resources, even those with superior talents, social inequality is a barrier to reaching their full potential.

CONCEPT CHECKS

1. What are the three shared characteristics of socially stratified systems?

2. How is the concept of class different from that of caste?

3. According to Karl Marx, what are the two main classes, and how do they relate to each other?

4. What are the three main differences between Max Weber's and Karl Marx's theories of social stratification?

5. How does social stratification contribute to the functioning of society? What is wrong with this argument?

How Is Social Class Defined in the United States?

Understand the social causes and consequences of social class in U.S. society as well as the complexities and challenges of defining class.

Social class in the United States is typically defined by some combination of one's income, wealth, educational attainment, and occupational status. In this section, we describe each of these attributes and describe how they are distributed throughout the U.S. population. We also compare and contrast the five major social class groups in the United States. One's income, wealth, education, and occupational status also vary widely based on personal characteristics like ethnicity and race as well as gender. We will delve further into inequality based on gender and race in Chapters 9 and 10, respectively. We will also examine differences in wealth and power among countries across the globe in Chapter 8.

Income

Income refers to wages and salaries earned from paid occupations, plus unearned money (or interest) from investments. One of the most significant changes occurring in Western countries over the past century has been the rising real income of the majority of the working population. (Real income is income excluding increases owing to inflation, which provides a fixed standard of comparison from year to year.) One of the most important reasons for this is increasing productivity—output per worker—through technological

income

Payment, usually derived from wages, salaries, or investments.

FIGURE 7.1

Mean Household Income by Income Group, 1967–2016

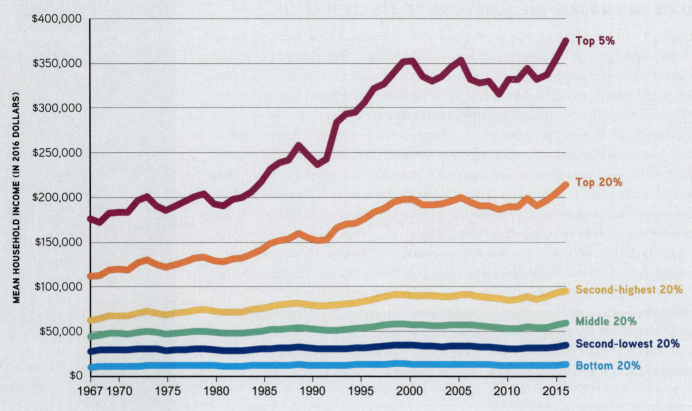

Source: Semega, Fontenot, and Kollar 2017.

development in industry. Another reason is almost everything we consume is now made in countries where wages are extremely low, keeping costs (and therefore prices) down.

Even though real income has risen in the past century, these earnings have not been distributed evenly across all groups. In 2016, the top 5 percent of all U.S. households received 22.5 percent of total income, the top 20 percent obtained 51.5 percent, and the bottom 20 percent received only 3.1 percent. This gap between the top and bottom tiers of the U.S. class structure has grown dramatically since the 1970s. Average household earnings (calculated in 2016 dollars), meaning the combined incomes of all persons living in a single household, of the bottom 20 percent of people in the United States was essentially unchanged from $12,036 in 1977 to $12,943 in 2016 (see Figure 7.1). During the same period, the richest 20 percent saw their incomes grow by 67 percent, while for the richest 5 percent of the population, income rose by more than 91 percent (Semega, Fontenot, and Kollar 2017).

Wealth

Wealth is usually measured in terms of net worth: all the assets one owns (for example, cash; savings and checking accounts; investments in stocks, bonds, and real estate properties) minus one's debts (for example, home mortgages, credit card balances, loans that need to be repaid). While most people earn their income from their work, the wealthy often derive the bulk of theirs from interest on their investments, some of them inherited. Some

wealth

Money and material possessions held by an individual or group.

scholars argue that wealth—not income—is the real indicator of social class. While income can vary from year to year based on the number of hours one worked or whether one took leave or was temporarily laid off, wealth tends to be a more enduring measure that is less susceptible to annual fluctuations.

Today, the average net worth of the bottom 90 percent of American families is only $84,000, while the average net worth of the top 1 percent has grown to $13.8 million, and that of the top 0.1 percent, $72.8 million. Stated somewhat differently, the wealthiest 0.1 percent of Americans (160,000 families) have as much wealth as the bottom 90 percent (145 million families) (Saez and Zucman 2014). There are significant differences in wealth by race. The median net worth of white households was $171,000 in 2016, compared to $20,700 for Hispanic households and $17,600 for Black households (Dettling et al. 2017).

What are some of the reasons for the racial disparity in wealth? Do Blacks simply have less money with which to purchase assets? To some degree, the answer is yes. The old adage "it takes money to make money" is a fact of life for those who start with little or no wealth. Because whites, on average, have enjoyed higher incomes and levels of wealth than Blacks, many whites are able to accrue even more wealth, which they then are able to pass on to their children (Conley 1999).

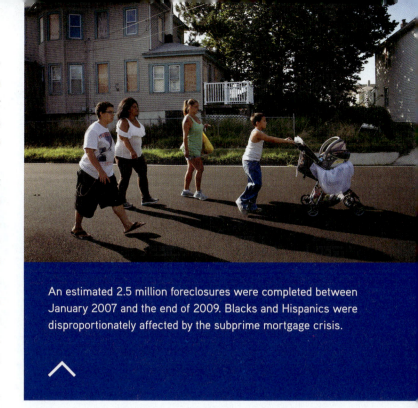

An estimated 2.5 million foreclosures were completed between January 2007 and the end of 2009. Blacks and Hispanics were disproportionately affected by the subprime mortgage crisis.

But family advantages are not the only factors. Oliver and Shapiro (1995) argued that it is easier for whites to obtain assets even when they have fewer resources than Blacks because discrimination plays a major role in the racial gap in homeownership. Blacks are rejected for mortgages 60 percent more often than whites, even when they have the same qualifications and creditworthiness. When Blacks do receive mortgages, they are more likely to take "subprime" mortgage loans, which charge much higher interest rates. Subprime lenders focus on minority communities, whereas the prime lenders are unable or unwilling to lend in those communities (Avery and Canner 2005). In 2006, of those who took out home loans, 30 percent of Blacks took out subprime loans, compared with 24 percent of Hispanics and 18 percent of whites. Blacks and Hispanics were therefore especially hard hit by the recent recession; many were forced to default on their mortgage payments and in many cases lost their homes. These issues are particularly important because homeownership constitutes American families' primary means for accumulating wealth.

Education

Sociologists also believe that education, or the number of years of schooling a person has completed, is an important dimension of social stratification. As we will see later in this chapter, how much education one receives is often influenced by the social class of one's parents. Although Viviana, the daughter of a homeless single mother, went on to attend Yale, her case is the exception to the rule.

Education is one of the strongest predictors of one's occupation, income, and wealth later in life. As shown in Figure 7.2, the median earnings of Millennials between the

FIGURE 7.2

Median Earnings of Young Adults,* 2016

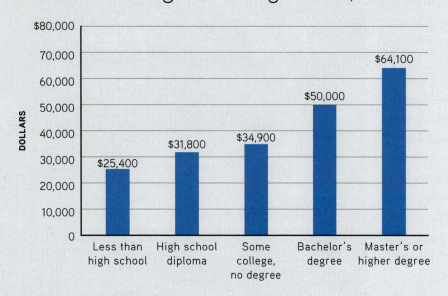

Represents median annual earnings of full-time, year-round workers ages 25–34.

Source: McFarland et al. 2018.

ages of twenty-five and thirty-four with bachelor's degrees was $50,000 in 2016, 57 percent higher than the median earnings of those with just a high school diploma ($31,800) (McFarland et al. 2018). Yet, even college graduates are highly stratified with respect to their earnings potential: Persons whose undergraduate degrees required quantitative skills, such as engineering and computer science, tend to have the highest lifetime earnings, while those with degrees that train students to work with children or provide counseling tend to have the lowest earnings (Hershbein and Kearney 2014).

The economic benefits of a college education have increased considerably over time: In 1977, for example, the gap between the hourly wages of college graduates and high school graduates was only 28 percent; by 2017, the gap had widened to nearly 50 percent (Economic Policy Institute 2018). The typical college graduate will earn more than twice as much as a typical high school graduate over his or her working life—nearly $1.2 million for a college graduate compared to $580,000 for a high school graduate (Hershbein and Kearney 2014). Although this growing "wage premium" has encouraged more Americans to go to college—34 percent of American adults had bachelor's degrees in 2017, compared to 16 percent in 1979 (U.S. Bureau of the Census 2017a)—it has also helped widen the gap between the wealthiest and the poorest workers.

Racial differences in levels of education persist, which partly explain why racial differences in income and wealth also persist. In 2016, the high school graduation rate was 88 percent for white students but just 76 percent for Black students and 79 percent for Hispanic students. And in 2017, of all people ages twenty-five and older, 94 percent of whites, 91 percent of Asian Americans, and 87 percent of Blacks had completed high school, compared to only 71 percent of Latinos (U.S. Bureau of the Census 2017a). A higher percentage of Asian and white young adults then go on to attend college: In 2015, 63 percent of Asians between the ages of eighteen and twenty-four were enrolled in college, compared to 42 percent of whites, 37 percent of Hispanics, and 35 percent of Blacks (Musu-Gillette 2017).

Occupation

In the United States and other industrialized societies, occupation is an important indicator of one's social standing. Occupational status depends heavily on one's level of educational attainment. In fact, in studies where persons are asked to rate jobs in terms of how "prestigious" they are, those requiring the most education are often—but not always—ranked most highly (see Table 7.1). The top-ranked occupations appear to share one of two characteristics: They require either a fair amount of education or a fair amount of public

TABLE 7.1

Relative Social Prestige of Select U.S. Occupations

OCCUPATION	PRESTIGE SCORE	OCCUPATION	PRESTIGE SCORE	OCCUPATION	PRESTIGE SCORE
Physician	7.6	Actor	5.7	Auto body repairperson	4.3
Architect	6.7	Firefighter	5.7	Bank teller	4.2
Dentist	6.7	Musician in an orchestra	5.6	Salesperson in a store	3.9
Registered nurse	6.5	Journalist	5.5	Day-care aide	3.6
Lawyer	6.4	Dental hygienist	5.3	Waiter/waitress	3.6
Veterinarian	6.4	Electrician	5.2	Farm laborer	3.5
High school teacher	6.1	Real estate agent	4.9	Taxi driver	3.2
Police officer	5.9	Carpenter	4.6	Janitor	3.0
Member of the clergy	5.8	Machinist	4.4	Door-to-door salesperson	2.9

Note: Respondents were asked to rank the occupations' prestige on a scale of 1–9, with 1 as the least prestigious and 9 as the most prestigious.

Source: Smith and Son 2014.

service. These rankings have been fairly consistent for nearly four decades (Griswold 2014). There are some interesting differences by age, however. Millennials seem more inclined than older Americans to value fame: Professional athletes, actors, and entertainers move up in the rankings (Harris 2014).

A Picture of the U.S. Class Structure

As we have seen so far, social class is a multifaceted concept, comprising how far we've gone in school, how much we earn, what we do for a living, and how many assets we possess. It is partly for this reason that it can be difficult to define precisely what social classes like upper, middle, and lower class mean in the United States. There can be wide differences in the lifestyles and personal characteristics of people even within a single social class group. Some scholars have gone so far as to argue that social class is a problematic concept because members of even a single social class "do not share distinct similar, life-defining experiences" (Kingston 2001). Despite this important critique, we can highlight some general characteristics that distinguish the major social strata. The purpose of the following discussion is to describe broad class differences in the United States. Bear in mind that there are no sharply defined boundaries between the classes, and there is no real agreement among sociologists about where the boundaries should fall.

THE UPPER CLASS

The **upper class** consists of the richest Americans—those households earning more than $200,000, or approximately 5 percent of all American households. Most Americans in the upper class are wealthy but not superrich. They are likely to own a large suburban home as

upper class

A social class broadly composed of the more affluent members of society, especially those who have inherited wealth, own businesses, or hold large numbers of stocks (shares).

well as a vacation home, drive expensive cars, vacation abroad, and educate their children in private schools and colleges. At the lower levels of this group, a large part of income may come from salaried earnings. This group would include many professionals, from some doctors and lawyers to university administrators and possibly even a few highly compensated professors. There are roughly 7 million households that fall into this category (U.S. Bureau of the Census 2015c).

At the very top of the upper class are the superrich—people who have accumulated vast fortunes that permit them to enjoy a lifestyle unimaginable to most Americans. If one uses a cutoff of the richest 0.1 percent in terms of income, these are people whose income tops $2 million (Lenzer 2011; Pomerleau 2013). Their wealth stems in large part from their substantial investments, including stocks and bonds and real estate, and the interest income derived from those investments. They include people who acquired their wealth in a variety of ways: celebrities, professional athletes, heads of major corporations, people who have made large amounts of money through investments or real estate, those fortunate enough to have inherited great wealth from their parents.

The superrich are highly self-conscious of their unique and privileged social class position; some give generously to such worthy causes as the fine arts, hospitals, and charities. Their common class identity is strengthened by such things as having attended the same exclusive private secondary schools (to which they also send their children). They sit on the same corporate boards of directors and belong to the same private clubs. They contribute large sums of money to their favorite politicians and may be on a first-name basis with members of Congress and perhaps even with the president. Because they are able to give large donations to political campaigns, they often have a significant influence on American politics (Domhoff 2013).

The turn of the twenty-first century saw extraordinary opportunities for the accumulation of such wealth. Globalization is one reason. Those entrepreneurs who are able to invest globally often prosper, both by selling products to foreign consumers and by making profits cheaply by using low-wage labor in developing countries. Young entrepreneurs with startup high-tech companies, such as Facebook founder Mark Zuckerberg and Yahoo cofounder Jerry Yang, made legendary fortunes. In 2018, Zuckerberg was the fifth wealthiest person in the world, with an estimated net worth of $71 billion (Kroll and Dolan 2018).

As a consequence of globalization and the information revolution, the number of superrich Americans has exploded in recent years. At the end of World War II, there were only 13,000 people worth $1 million or more in the United States. In 2017, there were more than 11 million millionaire households in the United States (Spectrem Group 2018) and 585 billionaires (Kroll and Dolan 2018). Unlike "old-money" families such as the Rockefellers or the Vanderbilts, who accumulated their wealth in earlier generations and thus are viewed as a sort of American aristocracy, much of this "new wealth" is held by entrepreneurs, including such recent arrivals as Evan Spiegel, cofounder and CEO of Snapchat, whose net worth is estimated at more than $4 billion.

THE MIDDLE CLASS

The **middle class** is a catchall for a diverse group of occupations, lifestyles, and people who earn stable and sometimes substantial incomes at primarily white-collar jobs and highly skilled blue-collar jobs. It is generally considered to include households with income between $40,000 and $200,000. The American middle class grew throughout much

middle class

A social class composed broadly of those working in white-collar and highly skilled blue-collar jobs.

Income Inequality

Income inequality refers to the unequal distribution of income across the population, or the gap between a country's richest and poorest citizens. One measure of income inequality is the income share held by the top 10 percent of the population. The greater the share, the higher the level of income inequality.

Income shares held by the top 10% and bottom 10% of population

☐ Top 10%   ☐ Bottom 10%

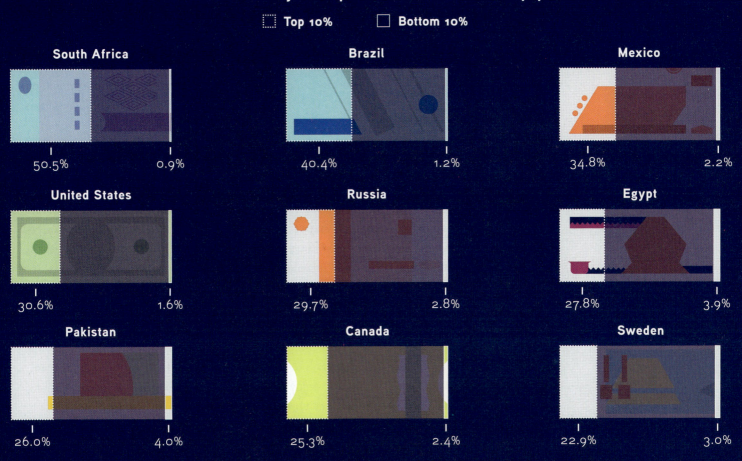

Country	Top 10%	Bottom 10%
South Africa	50.5%	0.9%
Brazil	40.4%	1.2%
Mexico	34.8%	2.2%
United States	30.6%	1.6%
Russia	29.7%	2.8%
Egypt	27.8%	3.9%
Pakistan	26.0%	4.0%
Canada	25.3%	2.4%
Sweden	22.9%	3.0%

Distribution of income in the United States

Population	1967	2016
☐ Top 5%	17.2%	22.6%
☐ Top 20%	43.6%	51.5%
☐ Second-highest 20%	24.2%	22.9%
☐ Middle 20%	17.3%	14.2%
☐ Second-lowest 20%	10.8%	8.3%
☐ Bottom 20%	4.0%	3.1%

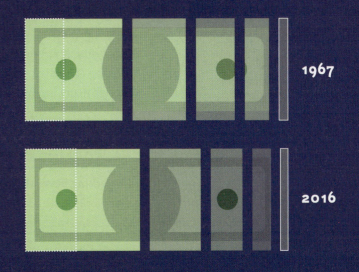

1967

2016

Sources: World Bank 2018b; Semega, Fontenot, and Kollar 2017.

of the first three-quarters of the twentieth century then shrank during most of the past four decades. While the middle class was once largely white, today it is increasingly diverse, both racially and culturally, including African Americans, Asian Americans, and Latinos.

For many years, when Americans were asked to identify their social class, the majority claimed to be middle class. The reason was partly the American cultural belief that the United States is relatively free of class distinctions; few people want to be identified as being too rich or too poor. Most Americans seem to think that others are not very different from their immediate family, friends, and coworkers (Kelley and Evans 1995; Simpson, Stark, and Jackson 1988; Vanneman and Cannon 1987). Because people rarely interact with those outside of their social class, they tend to see themselves as like "most other people," whom they then regard as being "middle class" (Kelley and Evans 1995). While the proportion of Americans who identify as middle class declined in the years after the 2008 recession, by 2017, it was back up to pre-recession levels, with 62 percent of Americans identifying as upper middle or middle class (Newport 2017).

The American middle class can be subdivided into two groups: the upper middle class and the lower middle class.

The Upper Middle Class The upper middle class consists of highly educated professionals (for example, doctors, lawyers, engineers, and professors), mid-level corporate managers, people who own or manage small businesses and retail shops, and some large-farm owners. Household incomes range quite widely, from about $100,000 to perhaps $200,000. The lower end of the income category would include college professors, for example, while the higher end would include corporate managers and small business owners. The upper middle class includes approximately 20 percent of all American households (Elwell 2014). Its members are likely to be college educated (as are their children) with advanced degrees. They own comfortable homes, drive expensive late-model cars, have some savings and investments, and are often active in local politics and civic organizations. However, they tend not to enjoy the same high-end luxuries, social connections, or extravagancies as members of the upper class.

The Lower Middle Class The lower middle class consists of trained office workers (for example, secretaries and bookkeepers), elementary and high school teachers, nurses, salespeople, police officers, firefighters, and others who provide skilled services. This group, which includes about 40 percent of American households, is the most varied of the social class strata and may include college-educated persons with relatively modest earnings, such as public elementary school teachers, as well as quite highly paid persons with high school diplomas only, such as skilled craftsmen (e.g., plumbers) and civil servants with many years of seniority. Household incomes in this group range from about $40,000 to $100,000 (Elwell 2014). Members of the lower middle class may own a modest house, although many live in rental units. Almost all have a high school education, and some have college degrees. They are rarely politically active beyond exercising their right to vote.

THE WORKING CLASS

The **working class**, about 20 percent of all American households, includes primarily **blue-collar** workers, such as factory workers and mechanics, and **pink-collar** laborers, such as clerical aides and sales clerks, and others who earn a modest weekly paycheck at a job that offers

working class

A social class broadly composed of people working in blue-collar, or manual, occupations.

———

blue- and pink-collar jobs

Jobs that typically pay low wages and often involve manual or low-skill labor. Blue-collar jobs typically are held by men (e.g., factory worker), whereas pink-collar jobs are typically held by women (e.g., clerical assistant).

little control over the size of one's income or working conditions. Household incomes range from about $20,000 to $40,000 (Elwell 2014), and at least two household members work to make ends meet. Family income is just enough to pay the rent or the mortgage, put food on the table, and perhaps save for a summer vacation. As you will see later in this chapter, many blue-collar jobs in the United States are threatened by globalization, so members of the working class today are likely to feel insecure about their own and their family's future.

The working class is racially and ethnically diverse. While older members of the working class may own a home that was bought several years ago, younger members are likely to rent. The home or apartment is likely to be in a lower-income suburb or a city neighborhood. The household car, a lower-priced model, is unlikely to be new. Children who graduate from high school are unlikely to go to college and will attempt to get a job immediately instead. Most members of the working class are not likely to be politically active even in their own community, although they may vote in some elections.

THE LOWER CLASS

The **lower class**, roughly 15 percent of American households, includes those who work part-time or not at all; household income is typically lower than $20,000 (Elwell 2014). Most lower-class individuals are found in cities, although some live in rural areas and earn a little money as farmers or part-time workers. Some manage to find employment in semiskilled or unskilled manufacturing or service jobs, ranging from making clothing in sweatshops to cleaning houses. Their jobs, when they can find them, are dead-end ones, meaning that years of work are unlikely to lead to promotion or substantially higher income. Their work is probably part-time and highly unstable, without benefits such as medical insurance, disability, or Social Security. Even if they are fortunate enough to find a full-time job, there are no guarantees that it will be around next month or even next week. Many people in the lower class live in poverty. Very few own their own homes. Most of the lower class rent, and some are homeless. If they own a car at all, it is likely to be a used car. A higher percentage of the lower class is nonwhite than is true of other social classes. Its members do not participate in politics, and they seldom vote.

THE "UNDERCLASS"

Within the lower class, some sociologists have identified a group that Swedish sociologist Gunnar Myrdal (1963) originally referred to as the **underclass** because they are "beneath" the class system, in that they lack access to the world of work and mainstream patterns of behavior. Located in the highest-poverty neighborhoods of the inner city, the underclass is sometimes called the "new urban poor."

The underclass includes many African Americans who have been trapped for more than one generation in a cycle of poverty from which there is little possibility of escape (Wacquant 1993, 1996; Wacquant and Wilson 1993; Wilson 1996). These are the poorest of the poor. Their numbers have grown rapidly over the past quarter-century and today include unskilled and unemployed men, young single mothers and their children on welfare, teenagers from welfare-dependent families, and many homeless people. They live in poor neighborhoods troubled by drugs, gangs, and high levels of violence. They are the truly disadvantaged, people with extremely difficult lives who have little realistic hope of ever making it out of poverty.

lower class

A social class composed of those who work part-time or not at all and whose household income is typically low.

underclass

A class of individuals situated at the bottom of the class system, often composed of people from ethnic minority backgrounds.

CONCEPT CHECKS

1. Name at least three components of social class. How do Blacks and whites differ along these components?

2. How do we explain the enduring racial disparity in wealth?

3. What are the major social class groups in the United States? Describe at least two ways (other than income) that these groups differ from one another.

What Are the Causes and Consequences of Social Inequality in the United States?

Recognize why and how the gap between rich and poor has increased in recent decades. Understand social mobility, and think about your own mobility.

The United States prides itself on being a nation of equals. But as we touched on earlier in the chapter, during the past thirty years, the gap between the rich and the poor in the United States has started to grow. The rich have gotten much, much richer, while middle-class incomes have stagnated and the poor have grown in number. The current gap between the rich and the poor in the United States is the largest since the Census Bureau started measuring it in 1947 (U.S. Bureau of the Census 2015k). One statistical analysis found that the United States had the most unequal distribution of household income among all twenty-one industrial countries studied (Sweden had the most equal; Organization for Economic Co-operation and Development 2010).

Ethnic Minorities Versus White Americans

There are substantial differences in income based on race and ethnicity, since minorities in the United States are more likely to hold the lowest-paying jobs. The income gaps between white households and Black and Hispanic households have persisted for more than four decades (Figure 7.3). In that time, the Black-white income gap has narrowed only modestly, while the income of Hispanics has fallen relative to whites (Kochhar and Cilluffo 2018). In 2016, the median income of white households was $65,041, compared to $39,490 for Black households and $47,675 for Hispanic households (Semega, Fontenot, and Kollar 2017). For Blacks, this is a slight improvement over previous years, as a growing number have gone to college and moved into middle-class occupations. For Latinos, however, the situation has worsened, as recent immigrants from rural areas in Mexico and Central America find themselves working at low-wage jobs (U.S. Bureau of the Census 2001).

Oliver and Shapiro (1995) found that the "wealth gap" between Blacks and whites is even greater than the income gap. Recent data show that even though family wealth rose for both Blacks and whites from 2013 to 2016, the Black–white gap increased. In 2016, white families had a

The Occupy Wall Street protests of 2011 drew attention to rising income inequality in the United States.

FIGURE 7.3

Black and Latino Income as Percentage of White Income

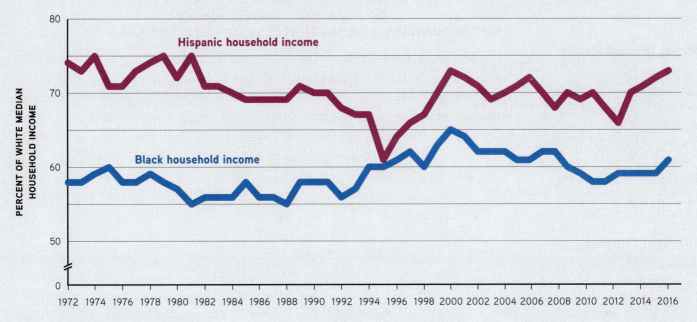

Source: U.S. Bureau of the Census 2017h.

median net worth of $171,000—nearly ten times the median net worth of Black families ($17,600) (Dettling et al. 2017). Oliver and Shapiro also found that when Blacks attained educational or occupational levels comparable with that of whites, the wealth gap still did not disappear.

Oliver and Shapiro (1995) argued that Blacks in the United States have encountered many barriers to acquiring wealth throughout history. After the Civil War ended slavery in 1865, legal discrimination (such as mandatory segregation in the South and separate schools) tied the vast majority of Blacks to the lowest rungs of the economic ladder. Racial discrimination was made illegal by the Civil Rights Act of 1964; nonetheless, discrimination has remained, and although some Blacks have moved into middle-class occupations, many have remained poor or in low-wage jobs where the opportunities for accumulating wealth are nonexistent. Less wealth means less social and cultural capital: fewer dollars to invest in schooling for one's children, a business, or the stock market—investments that in the long run would create greater wealth for future investments. We will further explore issues of racial inequality in Chapter 10.

Social Mobility

Social mobility refers to the upward or downward movement of individuals and groups among different class positions as a result of changes in occupation, wealth, or income. There are two ways of studying social mobility. First, we can look at people's own careers—how far they move up or down the socioeconomic scale in the course of their working lives. This is called **intragenerational mobility**. Alternatively, we can analyze where children are on the scale compared with their parents or grandparents. Mobility across the generations is called **intergenerational mobility**. Sociologists have long studied

social mobility

Upward or downward movement of individuals or groups among different social positions.

___

intragenerational mobility

Movement up or down a social stratification hierarchy within the course of a personal career.

___

intergenerational mobility

Movement up or down a social stratification hierarchy from one generation to another.

both types of mobility, with increasingly sophisticated methods. Unfortunately, with the exception of some recent studies, much of this research has been limited to male mobility, particularly of white males. We look at some of the research in this section.

OPPORTUNITIES FOR MOBILITY: WHO GETS AHEAD?

Is it possible for a young person from a working-class background to transcend class roots and become an upper-class professional? Is the case of Viviana Andazola Marquez a fairy tale or reality? If a reality, what factors contributed to her ascent up the social ladder? Sociologists have sought to answer this question by trying to understand which social factors are most influential in determining an individual's status or position in society. Most research shows that while the forces of social reproduction are very powerful, it is possible for people to transcend their roots; **social reproduction** refers to the processes whereby parents pass down to their children a range of resources, including both financial and cultural capital.

In a classic study of intergenerational mobility in the United States, sociologists Peter Blau and Otis Dudley Duncan (1967) found that long-range intergenerational mobility—that is, from working class to upper middle class—was rare. Why? Blau and Duncan concluded that the key factor behind occupational status was educational attainment. A child's education is influenced by family social status; this, in turn, affects the child's social position later in life. Sociologists William Sewell and Robert Hauser (1980) later confirmed Blau and Duncan's conclusions. They added to the argument by claiming that the connection between family background and educational attainment occurs because parents, teachers, and friends influence the educational and career aspirations of the child and that these aspirations then become an important influence on the schooling and careers obtained throughout the child's life. In other words, aspirations are reproduced from generation to generation because parents and children share the social location and social ties that may shape aspirations.

French sociologist Pierre Bourdieu (1984, 1988) has also been a major figure in examining the importance of family background to social status, but his emphasis is on the cultural advantages that parents can provide to their children. Bourdieu argued that among the factors responsible for social status, the most important is the transmission of **cultural capital**, or the cultural advantages that coming from a "good home" confers. Wealthier families are able to afford to send their children to better schools, an economic advantage that benefits the children's social status as adults. Parents from the upper and middle classes are mostly highly educated themselves and tend to be more involved in their children's education—reading to them, helping with homework, purchasing books and learning materials, and encouraging their progress. Bourdieu noted that working-class parents are concerned about their children's education, but they lack the economic and cultural capital to make a difference.

Although Bourdieu focused on social status in France, the socioeconomic order in the United States is similar. Those who already hold positions of wealth and power can ensure their children have the best available education, often leading them into the best jobs. Studies consistently show that the large majority of people who have "made money" did so on the basis of inheriting or being given at least a modest amount initially—which they then used to make more. In U.S. society, it's better to start at the top than at the bottom (Duncan et al. 1998; Jaher 1973; Rubinstein 1986); through social reproduction processes, those who start at the top are able to pass their economic and cultural resources down to their children.

Race and education play a major part in determining upward mobility. Sixty-three percent of Black children born into the bottom fourth of the U.S. income distribution remained

social reproduction

The process whereby societies have structural continuity over time. Social reproduction is an important pathway through which parents transmit or produce values, norms, and social practices among their children.

cultural capital

Noneconomic or cultural resources that parents pass down to their children, such as language or knowledge. These resources contribute to the process of social reproduction, according to Bourdieu.

downward mobility

Social mobility in which individuals' wealth, income, or status is lower than what they or their parents once had.

short-range downward mobility

Social mobility that occurs when an individual moves from one position in the class structure to another of nearly equal status.

absolute poverty

The minimal requirements necessary to sustain a healthy existence.

in the bottom fourth, while only 4 percent made it into the top fourth. Among white children, 32 percent of those born into the bottom fourth remained there, while 14 percent made it into the top fourth. In other words, while upward mobility is not high for anyone, it is far lower for Blacks than it is for whites. Differences in education account for at least part of the racial differences: because schools remain highly segregated by race in many parts of the country, poor Black children often do not have the same educational opportunities as whites (Hertz 2006).

DOWNWARD MOBILITY

Downward mobility occurs when one's own wealth, income, or occupational status is lower than what one's parents had. Downward mobility is less common than upward mobility; nevertheless, an estimated one-third of all Americans raised in the middle class—defined as households between the thirtieth and seventieth percentiles of the income distribution—fall out of the middle class when they become adults (Acs 2011). A person with **short-range downward mobility** moves from one job to another that is similar in pay and prestige—for example, from a routine office job to semiskilled blue-collar work. Downward intragenerational mobility is often associated with psychological problems and anxieties. Some people are simply unable to sustain the lifestyle into which they were born. But another source of downward mobility among individuals arises through no fault of their own. During the late 1980s and early 1990s, and again in the late 2000s, corporate America was flooded with instances in which middle-aged men lost their jobs because of company mergers, takeovers, or bankruptcies. These executives either had difficulty finding new jobs or could only find jobs that paid less than their previous jobs.

CONCEPT CHECKS

1. Contrast intragenerational and intergenerational mobility.

2. According to classic studies of mobility in the United States, how does family background affect one's social class in adulthood?

3. According to Pierre Bourdieu, how does the family contribute to the transmission of social class from generation to generation?

4. Describe at least two reasons for downward mobility.

How Does Poverty Affect Individuals?

At the bottom of the class system in the United States are the millions of people who live in poverty. Many do not maintain a proper diet and live in miserable conditions; their average life expectancy is lower than that of the majority of the population. In defining poverty, a distinction is usually made between absolute and relative poverty. **Absolute poverty** means that a person or family simply can't get enough to eat. People living in absolute poverty are undernourished and, in situations of famine, may actually starve to death. Absolute poverty is common in the poorer developing countries.

In the industrial countries, **relative poverty** is essentially a measure of inequality. It means being poor as compared with the standards of living of the majority. It is reasonable to call a person poor in the United States if he or she lacks the basic resources needed to maintain a decent standard of housing and healthy living conditions.

Measuring Poverty

When President Lyndon B. Johnson began the War on Poverty in 1964, around 36 million Americans lived in poverty—about 19 percent of the population at that time. In 2016, this number sat at more than 40 million people, or roughly 13 percent of the population

Learn about poverty in the United States today, explanations for why it exists, and means for combating it. Learn how people become marginalized in a society and the forms that this marginalization takes.

relative poverty

Poverty defined according to the living standards of the majority in any given society.

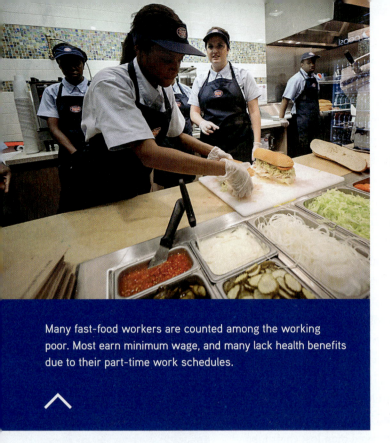

Many fast-food workers are counted among the working poor. Most earn minimum wage, and many lack health benefits due to their part-time work schedules.

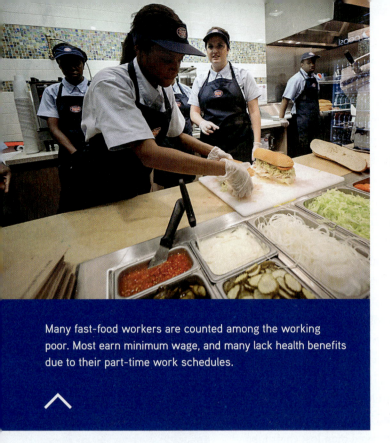

poverty line

An official government measure to define those living in poverty in the United States.

(Semega, Fontenot, and Kollar 2017). The rate of child poverty is even worse: 18 percent of children live in a household with income levels beneath the poverty line. A recent UNICEF study reported that among the thirty-five wealthiest nations in the world, the United States has the second-highest child poverty rate, falling just behind Romania (UNICEF 2012). The largest concentrations of poverty in the United States are found in the South and the Southwest, in inner cities, and in rural areas. Among the poor, 18.5 million Americans (or nearly 6 percent of the U.S. population) live in deep poverty: Their incomes are only half of the official poverty level, meaning that they live at near-starvation levels (Semega, Fontenot, and Kollar 2017).

What does it mean to be poor in the world's richest nation? The U.S. government currently calculates a **poverty line** based on cost estimates for families of different sizes. This results in a strict, no-frills budget, which for a family of four in 2016 works out to an annual cash income of about $24,400, or around $2,000 a month to cover all expenses (Semega, Fontenot, and Kollar 2017).

But how realistic is this formula? Some critics believe it overestimates the amount of poverty. They point out that the current standard fails to take into account noncash forms of income available to the poor, such as food stamps, Medicare, Medicaid, and public housing subsidies, as well as under-the-table pay obtained from work at odd jobs that is concealed from the government. Others counter that the government's formula greatly underestimates the amount of poverty because it overemphasizes the proportion of a family budget spent on food and severely underestimates the share spent on housing. According to some estimates, fully 70 percent of U.S. families whose income is less than $15,000 a year (about what would be earned under the federal minimum wage) are spending more than half of their income on housing (Joint Center for Housing Studies of Harvard University 2017). Still others observe that this formula dramatically underestimates the proportion of older adults (ages sixty-five and older) who live in poverty because they spend a relatively small proportion of their income on food yet are faced with high health care costs (Carr 2010).

Who Are the Poor?

Most Americans think of the poor as people who are unemployed or on welfare. Data on who the poor actually are show that Blacks and Latinos are more likely than whites to live in poverty, but poverty strikes members of all ethnic and racial backgrounds. Surveys show that Americans are split on whether the poor are responsible for their plight. A Pew Research Center poll in 2016 found that 53 percent believe that poverty is the result of circumstances beyond people's control, while 34 percent believe it is due to a lack of effort (Smith 2017). This represents a significant shift in opinion from twenty years earlier, when fewer than a third of respondents believed poverty to be caused by conditions beyond one's control and more than half believed that the poor were responsible for their plight. This shift in opinion may well reflect the lingering aftermath of the recent recession, which affected many working- and middle-class Americans, revealing to them that circumstances beyond one's control can adversely affect one's livelihood.

THE WORKING POOR

Many Americans fall into the category of the **working poor**—that is, people who work at least twenty-seven weeks a year but whose earnings are not high enough to lift them above the poverty line. The federal minimum wage, the legal floor for wages in the United States, was first set in 1938 at $0.25 an hour. Set on July 24, 2009, the federal minimum wage is currently $7.25 per hour, although individual states can set higher minimum wages than the federal standard. As of July 2018, the District of Columbia had the highest minimum wage, at $13.25 per hour; Washington had a minimum wage of $11.50, while California and Massachusetts had a minimum wage of $11 per hour. Although the federal minimum wage has increased over the years, since 1965, it has failed to keep up with inflation.

In 2015, there were an estimated 8.6 million individuals among the working poor, or about 6 percent of the labor force. The working poor are disproportionately nonwhite and immigrant: Blacks and Hispanics are more than twice as likely as whites and Asians to fall into this category. Women and young workers are also more likely to be classified as working poor than men and older workers (U.S. Bureau of Labor Statistics 2017f). The more education an individual has, the less likely he or she is to be among the working poor: While 16 percent of workers with less than a high school diploma are working poor, less than 2 percent of college graduates fall into this category.

Most poor people, contrary to popular belief, do not receive welfare payments because they earn too much to qualify. Only 5 percent of all low-income families with a full-time, full-year worker receive welfare benefits, and more than half rely on public health insurance rather than employer-sponsored insurance. Research on low-wage fast-food workers further reveals that many working poor lack adequate education, do not have health insurance, and are trying to support families on poverty-level wages (Newman 2000).

POVERTY, RACE, AND ETHNICITY

Poverty rates in the United States are much higher among most minority groups than among whites, even though fully 43 percent of the poor are white. As Figure 7.4 shows, Blacks and Latinos experience more than double the poverty rate of whites. This is because they often work at the lowest-paying jobs and because of racial discrimination. Asian Americans have the highest income of any group, but their poverty rate is slightly higher than that of whites, reflecting the recent influx of relatively poor Asian immigrant groups.

Latinos have somewhat higher incomes than Blacks, although their poverty rate is comparable. Nonetheless, the number of Blacks living in poverty has declined considerably in recent

> ### *working poor*
> People who work but whose earnings are not enough to lift them above the poverty line.

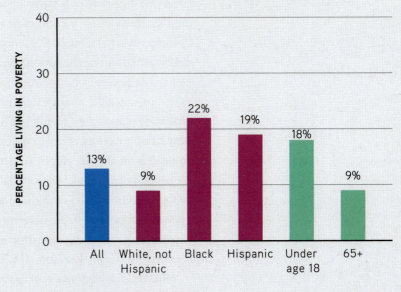

FIGURE 7.4

Americans Living in Poverty, 2016

Source: Semega, Fontenot, and Kollar 2017.

years. In 1959, 55 percent of Blacks were living in poverty; by 2016, that figure had dropped to 22 percent. A similar pattern holds for Latinos: Poverty grew steadily between 1972 and 1994, peaking at almost 31 percent of the Latino population. By 2016, however, the poverty rate for Latinos had fallen to 19 percent (Semega, Fontenot, and Kollar 2017).

THE FEMINIZATION OF POVERTY

Much of the growth in poverty is associated with the **feminization of poverty**, an increase in the proportion of the poor who are female. Growing rates of divorce, separation, and single-parent families have placed women at a particular disadvantage; it is extremely difficult for unskilled or semiskilled, low-income, poorly educated women to raise children by themselves while also holding down a job that pays enough to raise them out of poverty. As a result, in 2016, 36 percent of all single-parent families with children headed by women were poor, compared with only 7 percent of married couples with children (U.S. Bureau of the Census 2018d).

The feminization of poverty is particularly acute among families headed by Hispanic women. Although the rate has declined significantly since its peak in the mid-1980s, 41 percent of all female-headed Hispanic families with children lived in poverty in 2016. A similar proportion—39 percent—of female-headed African American families with children also live in poverty, both considerably higher than either white (30 percent) or Asian (30 percent) female-headed households (U.S. Bureau of the Census 2018d).

A single woman attempting to raise children alone is caught in a vicious circle (Edin and Kefalas 2005). If she has a job, she must find someone to take care of her children because she cannot afford to hire a babysitter or pay for day care. From her standpoint, she will take in more money if she accepts welfare payments from the government and tries to find illegal part-time jobs that pay cash not reported to the government rather than find a regular full-time job paying minimum wage. Even though welfare will not get her out of poverty, if she finds a regular job, she will lose her welfare altogether, and she and her family may end up worse off economically.

CHILDREN IN POVERTY

Given the high rates of poverty among families headed by single women, it follows that children are the principal victims of poverty in the United States. In 2016, 18 percent of children lived in poverty (Semega, Fontenot, and Kollar 2017). As noted earlier, the United States ranks second among the nation's wealthiest countries with respect to its child poverty rates (defined as poverty among people under eighteen). Nonetheless, the child poverty rate has varied considerably over the last forty years, declining when the economy expands or the government increases spending on antipoverty programs and rising when the economy slows and government antipoverty spending falls. The child poverty rate declined from 27 percent of all children in 1959 to 14 percent in 1973—a period associated with both economic growth and Johnson's War on Poverty (1963–1969). During the late 1970s and 1980s, as economic growth slowed and cutbacks were made in government antipoverty programs, child poverty grew, exceeding 20 percent during much of the period. The economic expansion of the 1990s saw a drop in child poverty rates, and by 2000, the rate had fallen to 16 percent, a twenty-year low (U.S. Bureau of the Census 2003b).

The child poverty rate rose again as a result of the 2008 recession, swelling to 22 percent in 2010. A study by the Annie E. Casey Foundation (2017) found that in 2015, 29 percent

of children lived in families where no parent had full-time, year-round employment. The economic well-being of racial minority children and children of single mothers is even more dire. In 2016, 11 percent of white children were poor compared with 31 percent of Black children and 27 percent of Hispanic children; fully 42 percent of children in single-parent families headed by a woman are in poverty (U.S. Bureau of the Census 2017f).

THE ELDERLY IN POVERTY

Although relatively few persons ages sixty-five and older live in poverty (9 percent in 2016), this aggregate statistic conceals vast gender, race, and marital status differences in the economic well-being of older adults. Elderly poverty rates in 2016 ranged from just 3 percent among white married men to an astounding 28 percent for Black women who live alone and 36 percent for Hispanic women living alone (U.S. Bureau of the Census 2018c). As we noted earlier, these figures may underestimate how widespread elderly poverty is because poverty rates fail to consider the high (and rising) costs of medical care, which disproportionately strike older adults (Carr 2010).

Because older people have for the most part retired from paid work, their income is based primarily on **Social Security** and private retirement programs. Social Security and **Medicare** have been especially important in lifting many elderly people out of poverty. Yet people who depend solely on these two programs for income and health care coverage are likely to live modestly at best. In December 2016, some 41 million retired workers were receiving Social Security benefits; their average monthly payment was about $1,360 (or just over $16,000 a year) (Social Security Administration 2018). Social Security accounts for only about 33 percent of the income of the typical retiree; most of the remainder comes from earnings, investments, and private pension funds.

Explaining Poverty: The Sociological Debate

Explanations of poverty can be grouped under two main headings: theories that see poor individuals as responsible for their status and theories that view poverty as produced and reproduced by structural forces in society. These competing approaches are sometimes

Social Security

A government program that provides economic assistance to persons faced with unemployment, disability, or old age.

Medicare

A program under the U.S. Social Security Administration that reimburses hospitals and physicians for medical care provided to qualifying people over sixty-five years old.

described as "blame the victim" and "blame the system" theories, respectively. We briefly examine each in turn.

There is a long history of attitudes that hold the poor responsible for their own disadvantaged positions. Early efforts to address the effects of poverty, such as the poorhouses of the nineteenth century, were grounded in a belief that poverty was the result of an inadequacy or pathology of individuals. The poor were seen as those who were unable—due to lack of skills, moral or physical weakness, absence of motivation, or below-average ability—to succeed in society. Social standing was taken as a reflection of a person's talent and effort; those who deserved to succeed did so, while others less capable were doomed to fail. The existence of winners and losers was regarded as a fact of life.

Such outlooks enjoyed a renaissance, beginning in the 1970s and 1980s, as the political emphasis on individual ambition rewarded those who "succeeded" in society and held those who did not succeed responsible for the circumstances in which they found themselves. Often, explanations for poverty were sought in the lifestyles of poor people, along with the attitudes and outlooks they supposedly espoused. Oscar Lewis (1968) set forth one of the most influential of such theories, arguing that a **culture of poverty** exists among many poor people. According to Lewis, poverty is not a result of individual inadequacies but is a result of a larger social and cultural milieu into which poor children are socialized. The culture of poverty is transmitted across generations because young people from an early age see little point in aspiring to something more. Instead, they resign themselves fatalistically to a life of impoverishment.

The culture-of-poverty thesis has been taken further by American political scientist Charles Murray. According to Murray (1984), individuals who are poor through "no fault of their own"—such as widows or widowers, orphans, or the disabled—fall into a different category from those who are part of the **dependency culture**. By this term, Murray refers to poor people who rely on government welfare provision rather than entering the labor market. He argues that the growth of the welfare state has created a subculture that undermines personal ambition and the capacity for self-help. Rather than orienting themselves toward the future and striving to achieve a better life, those dependent on welfare are content to accept handouts. Welfare, he argues, has eroded people's incentive to work.

An opposite approach to explaining poverty emphasizes larger social processes that produce conditions of poverty that are difficult for individuals to overcome. According to such a view, structural forces within society—factors like class, gender, ethnicity, occupational position, and education attainment—shape the way in which resources are distributed (Wilson 1996). Advocates of structural explanations for poverty argue that the perceived lack of ambition among the poor is in fact a consequence of their constrained situations, not a cause of it. Reducing poverty is not a matter of changing individual outlooks, they claim, but instead requires policy measures aimed at distributing income and resources more equally throughout society. Childcare subsidies, a minimum hourly wage, and guaranteed income levels for families are examples of policy measures that have sought to redress persistent social inequalities.

Both theories have enjoyed broad support, and social scientists consistently encourage variations of each view in public debates about poverty. Critics of the culture-of-poverty view accuse its advocates of "individualizing" poverty and blaming the poor for circumstances largely beyond their control. They see the poor as victims, not as freeloaders who are abusing the system. Most sociologists emphasize the systemic or structural causes of

culture of poverty

The thesis, popularized by Oscar Lewis, that poverty is not a result of individual inadequacies but is instead the outcome of a larger social and cultural atmosphere into which successive generations of children are socialized. The culture of poverty refers to the values, beliefs, lifestyles, habits, and traditions that are common among people living under conditions of material deprivation.

dependency culture

A term popularized by Charles Murray to describe individuals who rely on state welfare provision rather than entering the labor market. The dependency culture is seen as the outcome of the "paternalistic" welfare state that undermines individual ambition and people's capacity for self-help.

poverty. While individual initiative obviously plays a part, as we have seen in this chapter, there are major advantages conferred by being born higher up on the income and wealth ladder—and major disadvantages from being born at the bottom.

Social Exclusion

What are the social processes that lead to large numbers of people being marginalized in a society? The idea of **social exclusion** refers to new sources of inequality—ways in which individuals may become cut off from involvement in the wider society. It is a broader concept than that of the underclass and has the advantage that it emphasizes processes— mechanisms of exclusion. Social exclusion can take a number of forms. It may occur in isolated rural communities cut off from many services and opportunities or in inner-city neighborhoods marked by high crime rates and substandard housing. Exclusion and inclusion may be seen in economic terms, political terms, and social terms.

The concept of social exclusion raises the question of agency. **Agency** refers to our ability to act independently and to use free will. When dealing with social exclusion, however, the word *exclusion* implies that someone or something is being shut out by another and is beyond an individual's control. Certainly in some instances individuals are excluded through decisions that lie outside their own control. Insurance companies might reject an application for a policy on the basis of an applicant's personal history and background. An employee laid off later in life may be refused further jobs on the basis of his or her age.

But social exclusion can also result from people excluding themselves from aspects of mainstream society. Individuals can choose to drop out of school, to turn down a job opportunity and become economically inactive, or to abstain from voting in political elections. In considering the phenomenon of social exclusion, we must once again be conscious of the interaction between human agency and responsibility, on the one hand, and the role of social forces in shaping people's circumstances, on the other hand.

social exclusion

The outcome of multiple deprivations that prevent individuals or groups from participating fully in the economic, social, and political life of the society in which they live.

agency

The ability to think, act, and make choices independently.

HOMELESS PERSONS

No discussion of social exclusion is complete without reference to the people who are traditionally seen as at the very bottom of the social hierarchy: homeless persons. The growing problem of homelessness is one of the most distressing signs of changes in the American stratification system. Homeless people are a common sight in nearly every U.S. city and town and are increasingly found in rural areas as well. On a single night in January 2017, more than half a million (553,742) people were homeless (U.S. Department of Housing and Urban Development 2017). Two generations ago, homeless populations were mainly elderly, alcoholic men who were found on the skid rows of the largest metropolitan areas. Today they are primarily young single men, often of working age.

The fastest-growing group of homeless people, however, consists of people in families with children, who

More than half a million people are homeless on any given night in the United States, 60 percent of whom are men.

make up a third (33 percent) of those currently homeless. In 2017, men comprised 61 percent of the homeless population. Despite accounting for only 13 percent of the U.S. population, Blacks make up 41 percent of the homeless population. Less than half (47 percent) of homeless people are white, 22 percent are Hispanic, and 7 percent are multiracial (U.S. Department of Housing and Urban Development 2017). Only a small proportion of the homeless population are Latino or Asian American immigrants, possibly because these groups enjoy close-knit family and community ties that provide a measure of security against homelessness (Waxman and Hinderliter 1996).

There are many reasons that people become homeless. A survey of twenty-five cities by the United States Conference of Mayors (2008) identified a lack of affordable housing, poverty, and unemployment as the leading causes of homelessness among families. For single individuals, substance abuse, lack of affordable housing, and mental illness were identified as leading causes of homelessness. One reason for the widespread incidence of such problems among homeless people is that many public mental hospitals have closed their doors. The number of beds in state mental hospitals has declined by as many as half a million since the early 1960s, leaving many mentally ill people with no institutional alternative to a life on the streets or in homeless shelters. Such problems are compounded by the fact that many homeless people lack family, relatives, or other social networks to provide support.

The rising cost of housing is another factor, particularly in light of the increased poverty noted elsewhere in this chapter. Declining incomes at the bottom, along with rising rents, create an affordability gap between the cost of housing and what poor people can pay in rents (Dreier and Appelbaum 1992). Nearly half of all renters today (48 percent) are cost-burned, meaning they spend more than 30 percent of their income on rent (Joint Center for Housing Studies of Harvard University 2017). The burden of paying rent is extremely difficult for low-income families whose heads work for minimum wage or slightly higher. Paying so much for rent leaves them barely a paycheck away from a missed rental payment and possible eviction (National Low Income Housing Coalition 2000).

CONCEPT CHECKS

1. What is the poverty line, and how does the U.S. government calculate this statistic?

2. Describe the demographic characteristics of the poor in the United States.

3. Why are women and children at a high risk of becoming impoverished in the United States today?

4. Contrast the culture-of-poverty argument and structural explanations for poverty.

5. Describe the demographic characteristics of the homeless population in the United States today. What are the main reasons people become homeless?

>

Learn how changes in the American economy have led to growing inequality since the 1970s.

How Does Social Inequality Affect Your Life?

Throughout this chapter, we have shown how changes in the American economy affect social stratification, emphasizing the importance of both globalization and changes in information technology. We have pointed out that the global spread of an industrial capitalist economy, driven in part by the information revolution, has helped break down closed caste systems around the world and replace them with more open class systems. The degree to which this process will result in greater equality in countries undergoing capitalist development will be explored in the next chapter.

What do these changes hold in store for you? On the one hand, new jobs are opening up, particularly in high-technology fields that require special training and skills and pay high wages. A flood of new products is flowing into the United States, many made with cheap labor that has lowered their costs. This has enabled consumers such as yourself to buy everything from laptops to cars to athletic shoes at costs lower than you otherwise would have paid, thereby contributing to a rising standard of living.

But these benefits come with potentially significant costs. Given high levels of job loss in the recent recession and Americans' demands for low-cost products, you may find yourself competing for jobs with workers in other countries who work for lower wages. This has already been the case for the manufacturing jobs that once provided the economic foundation for the working class and segments of the middle class. Companies that once produced in the United States now use factories around the world, taking advantage of labor costs that are a fraction of those in the United States. Will the same hold true for other, more highly skilled jobs—jobs in the information economy itself? Many jobs that require the use of computers—from graphic design to software engineering—can be done by anyone with a high-speed computer connection, anywhere in the world. The global spread of tech companies will open up vastly expanded job opportunities for those with the necessary skills and training—but it will also open up equally expanded global competition for those jobs.

Partly as a result of these forces, inequality has increased in the United States since the early 1970s, resulting in a growing gap between the rich and the poor. The global economy has permitted the accumulation of vast fortunes at the same time that it has contributed to declining wages, economic hardship, and poverty in the United States. Although the working class is especially vulnerable to these changes, the middle class is not exempt: A growing number of middle-class households experienced downward mobility from the late 1970s through the mid-1990s, until a decade of economic growth benefited all segments of American society. The 2008 recession contributed to the downward mobility of middle-class Americans, at the time leaving many college graduates with high levels of debt and few prospects for rewarding employment (Demos 2010). The slow economic recovery since that time has improved prospects somewhat, although not equally for all Americans.

Research findings on social stratification and mobility may have important implications for your economic future. A new study of intergenerational income mobility by Stanford researchers found that young people entering the workforce today are considerably less likely to outearn their parents than young people born two generations before them. While 90 percent of young people born in the 1940s went on to earn more than their parents, the same is true of just 50 percent of young people born in the 1980s (Wong 2016, Chetty et al. 2017). Although the rags-to-riches story of Viviana Andazola Marquez is uplifting and inspiring, fewer and fewer young adults will be able to transcend childhoods of disadvantage unless the economy provides opportunities for education, gainful employment, and safe and secure housing.

CONCEPT CHECKS

1. How has globalization affected the life chances of young adults in the United States today?

2. How has the recent recession affected the life chances of young adults in the United States today?

CHAPTER 7

The Big Picture

Stratification, Class, and Inequality

Thinking Sociologically

1. If you were doing your own study of status differences in your community, how would you measure people's social class? Base your answer on the textbook's discussion of these matters to explain why you would take the particular measurement approach you've chosen. What would be its value(s) and shortcoming(s) compared with those of alternative measurement procedures?

2. Using occupation and occupational change as your mobility criteria, view the social mobility within your family for three generations. As you discuss the differences in jobs between your paternal grandfather, your father, and yourself, apply all these terms correctly: *vertical* and *horizontal mobility, upward* and *downward mobility, intragenerational* and *intergenerational mobility.* Explain fully why you think people in your family have moved up, moved down, or remained at the same status level.

What Is Social Stratification?

p. 193

Learn about social stratification and how social background affects one's life chances. Become acquainted with the most influential theories of stratification.

How Is Social Class Defined in the United States?

p. 199

Understand the social causes and consequences of social class in U.S. society as well as the complexities and challenges of defining class.

What Are the Causes and Consequences of Social Inequality in the United States?

p. 208

Recognize why and how the gap between rich and poor has increased in recent decades. Understand social mobility, and think about your own mobility.

How Does Poverty Affect Individuals?

p. 211

Learn about poverty in the United States today, explanations for why it exists, and means for combating it. Learn how people become marginalized in a society and the forms that this marginalization takes.

How Does Social Inequality Affect Your Life?

p. 218

Learn how changes in the American economy have led to growing inequality since the 1970s.

Terms to Know	Concept Checks
social stratification	**1.** What are the three shared characteristics of socially stratified systems? **2.** How is the concept of class different from that of caste? **3.** According to Karl Marx, what are the two main classes, and how do they relate to each other? **4.** What are the three main differences between Max Weber's and Karl Marx's theories of social stratification? **5.** How does social stratification contribute to the functioning of society? What is wrong with this argument?
slavery • caste system • endogamy • class • life chances • means of production • bourgeoisie • proletariat • surplus value • status • pariah groups • power	
income • wealth • upper class • middle class • working class • blue- and pink-collar jobs • lower class • underclass	**1.** Name at least three components of social class. How do Blacks and whites differ along these components? **2.** How do we explain the enduring racial disparity in wealth? **3.** What are the major social class groups in the United States? Describe at least two ways (other than income) that these groups differ from one another.
social mobility • intragenerational mobility • intergenerational mobility • social reproduction cultural capital • downward mobility • short-range downward mobility	**1.** Contrast intragenerational and intergenerational mobility. **2.** According to classic studies of mobility in the United States, how does family background affect one's social class in adulthood? **3.** According to Pierre Bourdieu, how does the family contribute to the transmission of social class from generation to generation? **4.** Describe at least two reasons for downward mobility.
absolute poverty • relative poverty • poverty line • working poor • feminization of poverty • Social Security • Medicare • culture of poverty • dependency culture • social exclusion • agency	**1.** What is the poverty line, and how does the U.S. government calculate this statistic? **2.** Describe the demographic characteristics of the poor in the United States. **3.** Why are women and children at a high risk of becoming impoverished in the United States today? **4.** Contrast the culture-of-poverty argument and structural explanations for poverty. **5.** Describe the demographic characteristics of the homeless population in the United States today. What are the main reasons people become homeless?
	1. How has globalization affected the life chances of young adults in the United States today? **2.** How has the recent recession affected the life chances of young adults in the United States today?

8

Global Inequality

Dubbed "frost boy" by the Chinese media after an image of him with icicles in his hair went viral, eight-year-old Wang Fuman is one of China's many "left behind" children.

THE BIG QUESTIONS

What is global inequality?
Understand the systematic differences in wealth and power among countries.

What is daily life like in rich vs poor countries?
Recognize the impact of different economic standards of living on people throughout the world.

Can poor countries become rich?
Analyze the success of the emerging economies.

How do sociological theories explain global inequality?
Learn several sociological theories explaining why some societies are wealthier than others as well as how global inequality can be overcome.

What does rapid globalization mean for the future of global inequality?
Learn about a few possible scenarios for the future of global inequality. Understand how the digital divide and climate change may increase the gap between rich and poor.

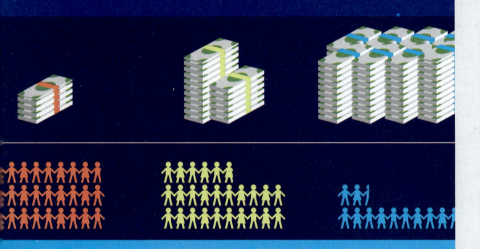

As December 2017 came to an end, powerful winter storms pummeled the United States, resulting in one of the coldest Januaries in memory. Temperatures fell to −36 degrees Fahrenheit in parts of Minnesota, hovering in the teens and twenties along the East Coast. Snow blanketed the East Coast from Maine to Florida, accompanied by strong winds and icy flooding. Massachusetts, New York, New Jersey, and North Carolina declared states of emergency; more than 300,000 people lost power. Hundreds of flights were cancelled, with many more delayed; icy roadways were closed. The so-called bomb cyclone was followed by a series of winter storms that ravaged the East Coast—including a powerful March rainstorm with gale-force winds that knocked out power for 2 million people, flooded neighborhoods, and resulted in suspended train service and 3,300 flight cancellations. In many states, schools were closed, and businesses were forced to shut down.

The winter of 2017–2018 was a cold, wet, and miserable experience for many. For some, the storms brought more than discomfort and inconvenience: As many as two dozen deaths were reported as a result of the storms. Yet, at the same time that more than 120 million Northeasterners were struggling with an arduous winter, half a world away, the residents of Xinjie County in China's southwest Yunan Province were facing their own weather challenges. In this remote, agricultural area, temperatures had dropped to below freezing, covering roads and farmlands with frost and ice. Forty years ago, four out of every five people in China lived in rural areas. Today, that figure has fallen to two out of five, but there are still nearly a half-billion Chinese in rural areas (World Bank 2018h). Many of these areas are impoverished, far from the economic expansion that has produced glittering skylines and raised hundreds of millions of Chinese into the middle class.

Although the average yearly wage in China now approaches $11,000, nearly 500 million people (40 percent of China's population) survive on less than $5.50 a day—$2,000 a year (Trading Economics 2018; Hernández 2017). If freezing temperatures were largely an inconvenience for the majority of Americans in New England and the Mid-Atlantic states, they posed a brutal hardship to millions of Chinese farmers living in ramshackle houses with few modern amenities.

The disparity between China and the United States—indeed, between the old China and the new China—was brought home by eight-year-old Wang Fuman. Dubbed "frost boy" by Chinese media, Fuman (his first name) arrived one day in January 2018 at Zhuanshanbao primary school in Yunan's Xinje County with icicles in his hair—a white halo of snow framing red, chapped cheeks. Fuman had trudged through icy mountains and streams for nearly two hours, hatless and gloveless, before arriving at his school three miles away. The compelling images of frost boy—which included photos of his swollen and blistered hands alongside a paper quiz with a near-perfect score—went viral on the Chinese Internet and were picked up by news media around the world, including the *New York Times* and *BBC News*.

While Fuman was hailed as an adorable hero in China—a symbol of the "great strength and effort of the Chinese nation," as one newspaper put it—his pluck and plight tell another story as well: that tens of millions of rural children have been left behind to live with their grandparents, their parents having moved to distant cities in the hope of earning a living (Li and Li 2018; Hernández 2018). Fuman's father works as a construction worker in a town 250 miles from home, and Fuman's mother left the family, so Fuman and his sister live with their grandmother (Hernández 2018).

Fuman is not alone in his wintry plight. China's "left behind" children suffer from malnutrition, live in run-down homes, and lack access to transportation. Many rural schools have closed, forcing students to walk long distances. While Fuman may be a heroic example of the personal drive that has helped to elevate millions of Chinese out of poverty, many rural children, confronted by so many challenges, drop out of school (Hernández 2018). According to a number of studies, some 3 million rural Chinese teenagers—one out of every three— leave school every year. One study of 50,000 students found that by grade 12, two-thirds of rural students had dropped out (Caixin Media 2016). In the United States, by way of comparison, only 19 percent of the population live in rural areas (U.S. Bureau of the Census 2016d), and among rural students, 80 percent complete high school (National Center for Education Statistics 2013).

China's president Xi Jinping has declared a "war on poverty," an ambitious effort to move 70 million people above China's poverty line ($1.17 a day, or $427 a year) by the end of 2020. Impoverished farmers are being moved from rural mud-and-brick huts to newly built housing in newly built villages; government officials are held responsible for the newly created jobs the farmers hope to find (Schmitz 2017). Whether this effort to create thousands of new jobs will be successful remains to be seen, but China's past efforts to raise people out of poverty have been impressive. According to the World Bank, since China began market-oriented economic reforms in 1978, more than 800 million people have been lifted out of poverty (World Bank 2018a).

The experience of "frost boy" in rural China, in comparison with the mostly minor hardships Americans experienced during the winter storms, is illustrative of gross inequities in health and well-being throughout the globe. **Globalization**—the increased economic, political, and social interconnectedness of the world—has produced opportunities for unthinkable wealth and technological development, as can be found in China's coastal cities. Yet at the same time, globalization has left many, like Fuman and his family, behind. Apart from millions of other children in rural China, there is widespread poverty and suffering in many countries in much of Asia, Africa, Latin America, and the Caribbean.

In the previous chapter, we examined the American class structure, noting vast differences among individuals' income, wealth, jobs, and quality of life. The same is true in the world as a whole: Just as we can speak of rich or poor individuals within a country, we can also talk about rich or poor countries in the world system. A country's position in the global economy affects how its people live, work, and even die. In this chapter, we look closely at differences in wealth and power among countries in the late twentieth and early twenty-first centuries. We examine how differences in economic standards of living affect people throughout the world. We then turn to the emerging economies of the world to understand which countries are improving their fortunes, and why. This will lead us to a discussion of different theories that attempt to explain why global inequality exists and what can be done about it. We conclude by speculating on the future of economic inequality in a global world.

What Is Global Inequality?

Global inequality refers to the systematic differences in wealth and power that have resulted from globalization. Sociology's challenge is not merely to identify all such differences but to explain why they occur—and how they might be overcome. While we see the most pronounced differences between rich countries and poor countries, globalization has also resulted in growing numbers of poor people in many rich countries, while at the same time producing some superrich people in poorer countries.

In defining poverty, a distinction is usually made between absolute and relative poverty. **Absolute poverty** means that a person or family simply can't get enough to eat. People living in absolute poverty are undernourished and, in situations of famine, may actually starve to death. Absolute poverty is common in the poorer countries of the Global South, including African countries such as Nigeria, the Democratic Republic of Congo, and the Central African Republic. In contrast, **relative poverty** is essentially a measure of inequality. It means being poor as compared with the standards of living of the majority. It is reasonable to call a person poor in the United States if he or she lacks the basic resources needed to maintain a decent standard of housing and healthy living conditions. Although the United States and other industrialized countries have their share of poor people, poverty in these countries looks very different than poverty in the Global South.

globalization

The development of social and economic relationships stretching worldwide. In current times, we are all influenced by organizations and social networks located thousands of miles away. A key part of the study of globalization is the emergence of a world system—for some purposes, we need to regard the world as forming a single social order.

global inequality

The systematic differences in wealth and power among countries.

<

Understand the systematic differences in wealth and power among countries.

absolute poverty

The minimal requirements necessary to sustain a healthy existence.

relative poverty

Poverty defined according to the living standards of the majority in any given society.

There are many ways to classify countries in terms of inequality. One simple way is to compare the wealth produced by each country per average citizen. This approach measures the value of a country's yearly output of goods and services produced by its total population and then divides that total by the number of people in the country. The resulting measure is termed the *per-person gross national income* (GNI), a measure of the country's yearly output of goods and services per person. The World Bank, an international lending organization that provides loans for development projects in poorer countries, uses this measure to classify countries as high income (an annual per-person GNI in 2017 of $12,056 or more), upper middle income ($3,896–$12,055), lower middle income ($996–$3,895), or low income ($995 or less) (World Bank 2018e).

The infographic on the next page shows how the World Bank (2018b) divides 217 countries and economies, containing more than 7.5 billion people, into four economic classes. The infographic shows that while nearly half of the world's population lives in low-income and lower-middle-income countries, slightly more than half lives in upper-middle-income or high-income countries. Bear in mind that this classification is based on average income for each country; therefore, it masks income inequality within each country. Such differences can be significant, although we do not focus on them in this chapter. For example, the World Bank classifies India as a lower-middle-income country because its per-person GNI in 2017 was only $1,820. Yet, despite widespread poverty, India also boasts a large and growing middle class. China, on the other hand, was classified as upper middle income because its GNI per capita in 2017 was $8,690; it nonetheless has hundreds of millions of people living in poverty (World Bank 2018j).

Comparing countries on the basis of economic output alone can also be misleading because GNI includes only goods and services that are produced for cash sale. Many people in low-income countries are farmers or herders who produce for their own families or for barter, involving noncash transactions. The value of their crops and animals is not taken into account in the statistics. Furthermore, economic output is not the sole indicator of a country's worth: Poor countries are no less rich in history, culture, and traditions than their wealthier neighbors, but the lives of their people are much harsher.

High-Income Countries

High-income countries are generally those that were the first to industrialize, a process that began in England some 250 years ago and then spread to Europe, the United States, and Canada. In the 1970s, Japan joined the ranks of high-income, industrialized nations, while Singapore, Hong Kong, and Taiwan moved into this category only within the last decade or so. The reasons for the success of these Asian latecomers to industrialization are much debated by sociologists and economists; we will look at them later in the chapter.

High-income countries are home to 17 percent of the world's population, or about 1.2 billion people, yet they lay claim to 64 percent of the world's annual output. High-income countries had an average per-person GNI of $40,136 in 2017 (World Bank 2018j). High-income countries offer adequate housing and food, drinkable water, and other comforts unknown in many parts of the world. Although these countries often have large numbers of poor people, most of their inhabitants enjoy a standard of living unimaginable by the majority of the world's people.

Global Inequality

Global inequality refers to the systematic differences in wealth and power that exist among countries. The World Bank uses per person gross national income (GNI) to classify countries into four economic classes: low income, lower middle income, upper middle income, and high income.

	Low income	Lower middle income	Upper middle income	High income
Gross national income per capita (current U.S. $)	$744	$2,118	$8,192	$40,136
Total population (in millions)	732	2,973	2,576	1,249
Annual population growth	2.7%	1.6%	0.8%	0.7%
Life expectancy at birth (in years)	63	68	75	80
Fertility rate (average # of births per woman)	4.6	2.8	1.8	1.7
Infant mortality rate (# of infant deaths per 1,000 births)	51	38	12	5

Source: World Bank 2018b.

Middle-Income Countries

The middle-income countries (including lower middle and upper middle) are located primarily in East and Southeast Asia and include the oil-rich countries of the Middle East and North Africa, a few countries in the Americas (Mexico, Central America, Cuba and other countries in the Caribbean, and South America), and the once-communist republics that formerly made up the Soviet Union and its Eastern European allies (Global Map 8.1). Most of these countries began to industrialize relatively late in the twentieth century and therefore are not yet as industrially developed (or as wealthy) as the high-income countries. Russia and the other countries that once composed the Soviet Union, on the other hand, are highly industrialized. While Russia's living standards initially eroded as a result of the collapse of the Soviet Union (and, with it, their communist economies) in 1991, within a decade, they began to grow again. Between 2000 and 2013, Russia's per-person GNI increased more than eightfold, from $1,710 to $14,840, although in the past three years, it has declined by one-third, to $9,720, in 2016. The fortunes of the Russian economy reflect the risk of depending excessively on a single product—oil. When oil prices were high, the Russian economy benefited; in recent years, with oil prices declining, Russia has suffered.

In 2017, middle-income countries were home to 74 percent of the world's population (more than 5.5 billion people) but accounted for only 35 percent of the output produced in that year; their average per-person GNI was $4,940, just 12 percent that of high-income countries (World Bank 2018b). Although many people in these countries are substantially better off than their neighbors in low-income countries, most do not enjoy anything resembling the standard of living common in high-income countries. They often live in crowded urban neighborhoods, including large slums that lack reliable water and sewage services; their cities may suffer from high levels of air and water pollution; and some still live in rural areas on small farms that provide only a basic living standard.

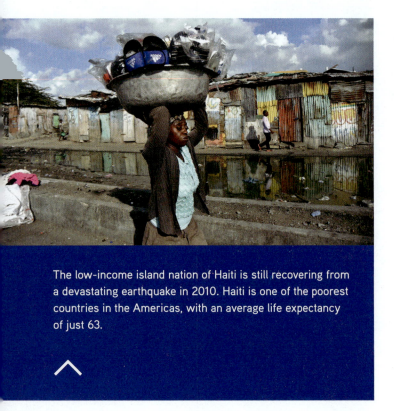

The low-income island nation of Haiti is still recovering from a devastating earthquake in 2010. Haiti is one of the poorest countries in the Americas, with an average life expectancy of just 63.

Low-Income Countries

Finally, the low-income countries include much of eastern, western, and sub-Saharan Africa; North Korea in East Asia; Nepal in South Asia; and Haiti in the Caribbean. These countries have mostly agricultural economies and have only recently begun to industrialize. Scholars debate the reasons for their late industrialization and widespread poverty, as we will see later in this chapter.

In 2017, the low-income countries accounted for less than 10 percent of the world's population (732 million people) and produced only 0.7 percent of the world's yearly output of wealth; their average per-person GNI was only $744 (World Bank 2018b). Fertility rates are much higher in low-income countries than elsewhere, with large families providing additional farm labor or otherwise contributing to family income. (In wealthy industrial societies, where children are more likely to be in school than on the farm, the economic benefit of large families declines, so people tend to have fewer children.) In fact, there is an inverse relationship between income

Rich and Poor Countries: The World by Income in 2017

Like individuals in a country, the countries of the world as a whole can be seen as economically stratified. In general, those countries that experienced industrialization the earliest are the richest, while those that remain agricultural are the poorest. An enormous—and growing—gulf separates the two groups.

■ High: $12,056 or more   ■ Upper middle: $3,896–$12,055   ■ Lower middle: $996–$3,895   ■ Low: $995 or less

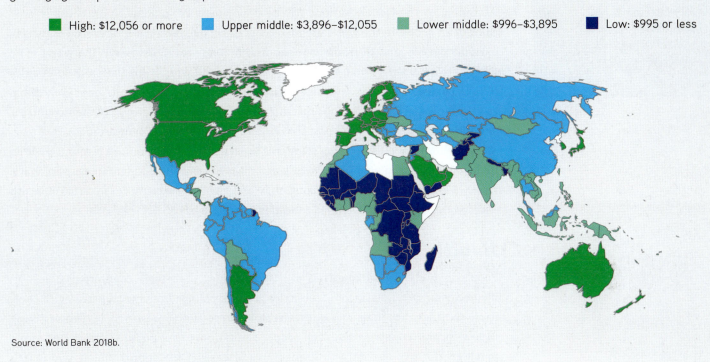

Source: World Bank 2018b.

level and population growth: In general, the poorer the country, the faster the growth in population. Between 2000 and 2017, the population of high-income countries increased 12 percent; upper-middle-income countries, 14 percent; lower-middle-income countries, 30 percent; and low-income countries, 57 percent (World Bank 2018b).

In many of these low-income countries, people struggle with poverty, malnutrition, and even starvation. Most people live in rural areas, although this is rapidly changing: Hundreds of millions of people are moving to huge, densely populated cities, where they live either in dilapidated housing or on the open streets (see Chapter 15). Yet during the last thirty years, the overall standard of living in the world has risen slowly. The average global citizen is better off today than ever before. Illiteracy rates are down, infant mortality and malnutrition are less common, people are living longer, average income is higher, and poverty is down. Because these figures are overall averages, however, they hide the substantial differences among countries: The gap between rich and poor countries has grown. Between 1986 and 2017, for example, average per-person GNI increased by 253 percent in high-income countries but increased by only 149 percent in low-income countries, widening the global gap between rich and poor. In 2017, the average person in a typical high-income country earned fifty-four times as much as his or her counterpart in a low-income country (World Bank 2018b).

CONCEPT CHECKS

1. Explain how the World Bank measures global inequality, and discuss some of the problems associated with measuring global inequality.

2. Compare and contrast high-income, middle-income, and low-income countries.

What Is Daily Life Like in Rich vs. Poor Countries?

Recognize the impact of different economic standards of living on people throughout the world.

According to the United Nations (UN) Development Programme, approximately one-third of all people in the world are undernourished, one person out of every ten lives in extreme poverty, and the same number are illiterate (Wulfhorst 2017). The world is no longer predominantly rural: In 2018, 55 percent of the world's population lived in cities (UN Department of Economic and Social Affairs 2018). Many of the poor come from racial, ethnic, or religious groups that differ from the dominant groups of their countries, and their poverty is at least in part the result of discrimination (Narayan 1999).

An enormous gulf in living standards separates most people in rich countries from their counterparts in poor ones. Wealth and poverty make life different in a host of ways. Here we look at the differences among high- and low-income countries in terms of health and disease, hunger and nutrition, education, and child labor.

Health

People in high-income countries are far healthier than their counterparts in low-income countries. Low-income countries generally suffer from inadequate health facilities, and their hospitals and clinics seldom serve the poorest people. People living in low-income countries also lack proper sanitation, drink polluted water, and run a much greater risk of contracting infectious diseases. Only approximately one-fourth of people living in low-income countries have access to improved sanitation facilities, compared with virtually everyone (99 percent) in high-income countries (World Bank 2018d). In 2016, 36 percent of all children under five in low-income countries suffered from stunted growth, in comparison with only 2.5 percent in high-income countries (World Bank 2018k); newborn children in low-income countries are more than ten times as likely to die than children in high-income countries (World Bank 2018g). Children often die of illnesses that are readily treated in wealthier countries, such as measles or diarrhea.

All these factors contribute to physical weakness and poor health, making people in low-income countries susceptible to illness and disease. One important indicator of overall health is life expectancy at birth. In 2016, the average person in high-income countries could expect to live 80 years, compared with only 63 years in low-income countries (World Bank 2018g). There is growing evidence that the high rates of HIV/AIDS infection found in many African countries and Haiti are due in part to the weakened health of impoverished people, especially children (Mody et al. 2014).

One chilling example of the relationship between global poverty and disease is the Ebola epidemic that broke out in West Africa in 2014. Ebola is a deadly disease that is spread through contact with the blood or bodily fluids of infected persons who are showing symptoms. The illness spread rapidly partly because it was new to this region of Africa and so went unrecognized: Caregivers, from family members to professional health workers, initially believed it to be malaria or some other disease that is transmitted by mosquitoes rather than human contact. Ebola also spread rapidly in West Africa because

Can Apps Heal Global Inequalities?

Think about how you use your smartphone. Maybe you check the latest news headlines or sports scores. Or maybe you like to post photos on Facebook. But can you think of ways that you might use your smartphone to improve the well-being of the millions of people living in poor nations? A number of app developers are doing just that. Despite an enduring digital divide, smartphone ownership in developing countries has jumped from just 21 percent in 2013 to 37 percent in 2015, making it possible for these developers to reach even more people across the globe (Poushter 2016).

In 2015, developing regions accounted for approximately 99 percent of all global maternal deaths (WHO 2015a). This is partly due to the fact that nearly half of all women in low-income countries give birth without the help of a skilled health care worker. The Safe Delivery app aims to reduce maternal mortality in developing countries by teaching birth attendants how to deal with emergency childbirth situations. Video guides and step-by-step instructions educate users on how to prevent infection as well as deal with problems such as prolonged labor and hypertension (Kweifio-Okai 2015).

Smartphones have also emerged as a particularly effective way to diagnose eye disorders. According to the World Health Organization, more than 280 million people around the world have vision problems or are blind; an estimated 90 percent of these sight-impaired people live in poor nations. However, many lack access to vision care. Developed in Kenya by a British ophthalmologist, the Peek (Portable Eye Examination Kit) app enables health care workers to perform detailed eye exams in the field with just a smartphone (Kuo 2016).

The easy-to-dispense Peek eye test displays the letter "E" in varying orientations. Patients, who don't need to be able to read English, just indicate the direction the letter is facing. The test giver then swipes the screen in that direction or shakes the phone if the patient can't tell. The test takes about a minute, and results are available instantaneously (Sohn 2015). In addition to visual acuity tests, the app can detect cataracts, glaucoma, and signs of nerve disease (CNN 2016). A clip-on camera adapter, which can be made with a 3D-printer, allows users to take high-resolution images of a person's retina that can then be sent to doctors to diagnose remotely (Kuo 2016).

Other entrepreneurs are developing apps to aid agricultural production in Africa. A competition called Apps4Africa encouraged young people to develop apps that address the impact of climate change on various communities. The Grainy Bunch, which was developed in Tanzania, features a national grain supply chain management system that monitors the purchase, storage, distribution, and consumption of grain across the entire nation. Similarly, Agro Universe, developed by a team from Uganda, creates a regional marketplace, helping communities prepare for pest- and drought-induced food shortages by linking communities to farmers with available produce (Fenner 2012).

Do you believe that apps can play an effective role in solving some of the problems of global inequality? Why or why not? Can you think of an app that would help solve some of the problems of global inequality that you read about in this chapter?

Peek (Portable Eye Examination Kit), which uses a phone's camera to scan people's eyes for cataracts and other problems, has been used successfully in Kenya, Botswana, and India. Smartphone apps like Peek are bringing vision care to poor, remote communities in the developing world.

there were no health care facilities capable of dealing with the large and growing number of Ebola patients, who need to be completely isolated and treated by trained medical personnel wearing special (and costly) protective suits.

Many victims were in remote rural areas that lacked usable roads or other infrastructure, making it difficult to identify, isolate, and treat them; when they flooded into crowded cities in search of treatment, the disease quickly spread. The borders between Liberia, Sierra Leone, and Guinea are open in many places, making containment of the disease difficult. And years of war and corrupt governments in this region meant that a concerted state-led response was unlikely (Fox 2014). As a result, during the first months of the outbreak, as many as three-quarters of infected victims died. When Ebola is diagnosed early and treated adequately, the mortality rate can be as low as 30 percent (NPR 2014).

Still, conditions are improving somewhat. Between 1990 and 2016, for example, the infant mortality rate dropped from 112 per 1,000 live births to 51 in low-income countries. Life expectancy has also improved, from just 51 years in low-income countries in 1990 to 63 years in 2016 (World Bank 2018g).

Hunger and Malnutrition

People in low-income countries are more likely to suffer malnutrition, starvation, and famine, which are also major global sources of poor health. Problems of inadequate food are nothing new. What seems to be new is their pervasiveness—the fact that so many people in the world today appear to be on the brink of starvation.

The Food and Agriculture Organization (2017) estimates that 815 million people, just over one in nine people in the world, suffer from chronic hunger, up from 777 million a year earlier—the vast majority of whom live in developing countries. More than half of the world's hungry are found in Asia and the Pacific; approximately one-quarter are in sub-Saharan Africa. In Southern Asia alone, nearly 272 million people are undernourished, representing 14 percent of the population; 23 percent of the population of sub-Saharan Africa is undernourished. According to UNICEF (2015), nearly half of all deaths of children under the age of five are attributable to undernutrition. As of 2016, the UN's Food and Agriculture Organization had listed thirty-nine countries in Africa, the Middle East, Central and South Asia, the Pacific, and the Caribbean as in need of external food assistance.

Most hunger today is the result of a combination of natural and social forces. The majority of people suffering from chronic hunger live in countries struggling with violence and conflict. These countries are concentrated in the Near East and North Africa, northern sub-Saharan Africa, Central America, and Eastern Europe. In many of these countries, climate change and natural disasters have both contributed to the conflicts and worsened problems of chronic hunger. In Syria, for example, a long and violent civil war has left 85 percent of the population in poverty; nearly 7 million people are "acutely food insecure" and in need of urgent assistance (Food and Agriculture Organization 2017). Moreover, conditions are predicted to worsen considerably as a result of global climate change, which will significantly impact agricultural production, especially in the world's poorest countries. Absent a concerted effort by all countries to limit greenhouse gas production, the worst-case scenario estimates that an additional 165 million people would be plunged into extreme poverty by 2030 (Food and Agriculture Organization 2016).

The countries affected by famine and starvation are for the most part too poor to pay for new technologies that would increase their food production. Nor can they afford to purchase sufficient food imports from elsewhere in the world. As world hunger grows, the largest increases in food production occur where hunger is not a pervasive problem, while the poorest countries—those most in need of food—suffer from the lowest food production gains (Bjerga 2017). In much of Africa, for example, food production per person has declined. Surplus food produced in high-income countries like the United States is seldom affordable to the countries that need it most.

Education and Literacy

Education and literacy are important routes to economic development. Lower-income countries are disadvantaged because they lack high-quality public education systems. Thus, children in high-income countries get more schooling, and adults in those countries are more likely to be literate, or able to read and write. While virtually all high school–age males and females attend secondary school in high-income countries, only two-thirds in middle-income countries, and just over one-third in low-income countries, do so. In low-income countries, only 61 percent of men and women over the age of fifteen are able to read and write, with significant differences by sex (53 percent of women and 69 percent of men are literate). By way of comparison, virtually everyone in high-income countries is literate (UNESCO 2018a). One reason for these differences is that high-income countries spend a much larger percentage of their national income on education than low-income countries do.

Education is important for several reasons. First, it contributes to economic growth, because people with advanced schooling provide the skilled workforce necessary for higher-wage industries. Second, education may offer hope for escaping the cycle of harsh working conditions and poverty, because poorly educated people are condemned to low-wage, unskilled jobs. But if a country is not creating jobs—if its economy is stagnant, if it is suffering from drought or other extreme climate events, or if it is embroiled in conflict—then education will have the perverse effect of producing large numbers of overqualified (and unemployed) people. Finally, at least in higher-income countries, educated people have tended to have fewer children, thus slowing the global population explosion that has contributed to global poverty (see Chapter 15).

Child Labor

In low-income countries, children are often forced to work because of a combination of family poverty, lack of education, and traditional indifference among some people in many countries to the plight of those who are poor or who are ethnic minorities. According to the International Labour Organization (2017), more than 114 million

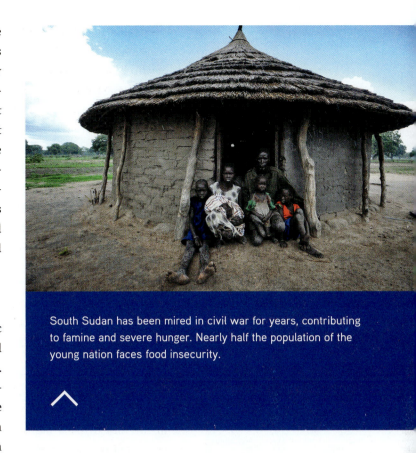

South Sudan has been mired in civil war for years, contributing to famine and severe hunger. Nearly half the population of the young nation faces food insecurity.

> Analyze the success of the emerging economies.

emerging economies

Developing countries that over the past two or three decades have begun to develop a strong industrial base, such as Singapore and Hong Kong.

children between the ages of five and fourteen are engaged in child labor worldwide, more than 35 million of whom are engaged in hazardous work. The incidence of child labor is highest in low-income countries: 19 percent of all children in low-income countries are engaged in child labor, compared to 9 percent for lower-middle-income countries and 7 percent for upper-middle-income countries (ILO 2017). Most working children labor in agriculture, with the rest in manufacturing, wholesale and retail trade, restaurants and hotels, and a variety of services, including working as servants in wealthy households. At best, these children work for long hours with little pay and are therefore unable to go to school and develop the skills that might eventually enable them to escape their lives of poverty. Many, however, work at hazardous and exploitative jobs under slavelike conditions, suffering a variety of illnesses and injuries.

Can Poor Countries Become Rich?

In the mid-1970s, a number of low-income countries in East Asia were undergoing a process of industrialization that appeared to threaten the global economic dominance of the United States and Europe (Amsden 1989). This process began with Japan in the 1950s, followed by Hong Kong in the 1960s and Taiwan, South Korea, and Singapore in the 1970s and 1980s. Other Asian countries began to follow in the 1980s and early 1990s, most notably China, but also Malaysia, Thailand, and Indonesia. Today, most of these **emerging economies** are middle income, and some—such as Hong Kong, South Korea, Taiwan, and Singapore—have moved up to the high-income category.

China, the world's most populous country, has one of the most rapidly growing economies on the planet. At an average annual growth rate of 10 percent between 1980 and 2010, the Chinese economy more than doubled and today is the world's second-largest economy behind the United States, having surpassed Japan in 2010. Since 2010, however, China's annual growth rate has declined, falling to slightly below 7 percent in 2016 and 2017 (World Bank 2017a). Still, in comparison with high-income countries (whose GDP growth averaged under 2 percent in 2017), China's performance remains impressive (World Bank 2018j). The once low-income (now middle-income) economies of East Asia as a whole averaged 7.7 percent growth per year during much of the 1980s and 1990s—a rate that has slowed slightly in recent years but still remains extraordinary by world standards (World Bank 2018c). Today, the per-person GNI in Singapore is virtually the same as that in the United States.

Economic growth in East Asia was often accompanied by important social problems. These have included the sometimes violent repression of labor and civil rights, terrible factory conditions, the exploitation of an increasingly female workforce and immigrant workers from impoverished neighboring countries, and widespread environmental degradation. Many of these atrocities continue today; Foxconn and Pegatron, two of Apple's biggest suppliers, have been found to badly mistreat workers in their Chinese factories

Hong Kong (shown here), along with Taiwan, South Korea, and Singapore, has become one of the most rapidly growing economies on earth.

(Neate 2013). Nonetheless, owing to the sacrifices of past generations of workers, large numbers of people in these countries are prospering.

The economic success of the East Asian emerging economies can be attributed to a combination of factors. Some of these factors are historical, including those stemming from world political and economic shifts. Some are cultural. Still others have to do with the ways these countries pursued economic growth. Sociologists cite five main reasons for the recent economic advances of the East Asian emerging economies. First, most were part of colonial situations that, while imposing many hardships, also helped to pave the way for economic growth. For example, Hong Kong and Singapore were former British colonies; Britain encouraged industrial development, constructed roads and other transportation systems, built relatively efficient governmental bureaucracies, and actively developed both Hong Kong and Singapore as trading centers (Cumings 1987; Gold 1986).

Second, the East Asian region benefited from a long period of world economic growth. Between the 1950s and the mid-1970s, the growing economies of Europe and the United States provided a big market for the clothing, footwear, and electronics that were increasingly being made in East Asia, creating a window of opportunity for economic development (Henderson and Appelbaum 1992). Third, economic growth in this region took off at the high point of the Cold War, when the United States and its allies, in erecting a defense against Communist China, provided generous economic aid that fueled investment in new technologies, such as transistors, semiconductors, and other electronics, contributing to the development of local industries (Amsden 1989; Castells 1992; Cumings 1987, 1997; Deyo 1987; Evans 1987; Haggard 1990; Henderson 1989; Mirza 1986). Fourth, many of the East Asian governments enacted strong policies that favored economic growth by keeping labor costs low, encouraging economic development through tax breaks and other economic policies, and offering free public education.

Finally, some have argued that cultural traditions, including a shared Confucian philosophy, contributed to these economic advances. Scholars such as Weber (1977, orig. 1904), who viewed the rise of capitalism in Western Europe as a function of the Protestant belief

CONCEPT CHECKS

1. What are the five factors that have facilitated the economic success of the East Asian emerging economies?

2. What are potential obstacles to the continued economic success of the emerging economies?

in thrift, frugality, and hard work, have observed a similar process in Asian economic history. Confucianism, it is argued, inculcates respect for one's elders and superiors, education, hard work, and proven accomplishments as keys to advancement, as well as a willingness to sacrifice today to earn a greater reward tomorrow. As a result of these values, the Weberian argument goes, Asian workers and managers are highly loyal to their companies, submissive to authority, hardworking, and success oriented. Workers and capitalists alike are said to be frugal. Instead of living lavishly, they are likely to reinvest their wealth in further economic growth (Berger 1986; Berger and Hsiao 1988; Helm 1992; Redding 1990; Wong 1986).

Recent social and cultural changes may undermine the influence of traditional values on Asian economic development. For example, thrift, a central Confucian cultural value, appears to be on the decline in Japan and the East Asian emerging economies, as young people—raised in the booming prosperity of recent years—increasingly value conspicuous consumption over austerity and investment (Helm 1992).

How Do Sociological Theories Explain Global Inequality?

> Learn several sociological theories explaining why some societies are wealthier than others as well as how global inequality can be overcome.

What causes global inequality? How can it be overcome? In this section, we examine four theories that have been advanced over the years to explain global inequality: market-oriented theories, dependency theories, world-systems theory, and global capitalism. Each of these theories has strengths and weaknesses. One shortcoming of all of them is that they frequently give short shrift to the role of women in economic development; they also emphasize economic factors at the expense of cultural or religious ones.

By drawing on all four theories together, however, we should be able to answer a key question facing the 84 percent of the world's population living outside high-income countries: How can they move up in the world economy?

Market-Oriented Theories

market-oriented theories

Theories about economic development that assume that the best possible economic consequences will result if individuals are free to make their own economic decisions, uninhibited by governmental constraint.

Fifty years ago, the most influential theories of global inequality advanced by American economists and sociologists were **market-oriented theories**, and their basic underlying assumptions remain prevalent among economists and corporations today. These theories assume that the best possible economic consequences will result if individuals are free, uninhibited by any form of governmental constraint, to make their own economic decisions. Unrestricted capitalism, if allowed to develop fully, is said to be the avenue to economic growth. Government bureaucracy should not dictate which goods to produce, what prices to charge, or how much workers should be paid. According to market-oriented theorists, governmental direction of the economies of low-income countries results only in blockages to economic development, thereby stifling economic growth and contributing to inequality both within and between nations (Berger 1986; Ranis and Mahmood 1992; Ranis 1996; Rostow 1961; Warren 1980). While market-oriented theories remain generally accepted among economists, sociologists are unlikely to accept the notion that the "invisible hand" of the marketplace, by itself, will lead to desirable outcomes for the world's poor (Slater and Tonkiss 2013).

Market-oriented theories inspired U.S. government foreign-aid programs that attempted to spur economic development in low-income countries by providing money, expert advisers, and technology, paving the way for U.S. corporations to make investments in these countries. One of the most influential early proponents of such theories was W. W. Rostow, an economic adviser to former U.S. president John F. Kennedy, whose ideas helped shape U.S. foreign policy toward Latin America during the 1960s. Rostow's explanation is one version of a market-oriented approach, termed **modernization theory**, which argued that low-income societies can develop economically only if they give up their traditional ways and adopt modern economic institutions, technologies, and cultural values that emphasize savings and productive investment.

According to Rostow (1961), the traditional cultural values and social institutions of low-income countries impede their economic effectiveness. For example, many people in low-income countries, in Rostow's view, would rather consume today than invest for the future. But to modernization theorists, the problems in low-income countries run even deeper. The cultures of such countries, according to the theory, tend to support "fatalism"—a value system that views hardship and suffering as the unavoidable plight of life. Acceptance of one's lot in life discourages people from working hard to overcome their fate. In this view, then, a country's poverty is due largely to the cultural failings of the people themselves. Such failings are reinforced by government policies that set wages and control prices and generally interfere in the operation of the economy.

Rostow's belief that free-market forces should guide economic policies in the developing world remains influential. **Neoliberalism**, the prevailing view among economists today, takes market-oriented theories one step further and argues that virtually all government involvement in the market is detrimental to economic growth. According to neoliberalism, government regulation of business should be eliminated, and government expenditures on social services, education, and health care should be dramatically curtailed. Neoliberal economists maintain that global free trade will enable all countries of the world to prosper, and thus they call for an end to restrictions on trade. They often challenge minimum wage and other labor laws as well as environmental restrictions on business. The World Trade Organization (WTO), currently comprising 164 member countries, is based on neoliberal principles: It manages a global system of free trade, ensures that free trade rules are followed by member countries, helps negotiate free trade agreements, and settles trade disputes when a member claims that free trade rules are being violated (WTO 2018).

Neoliberal ideas have proven to be detrimental to economic development in poor countries. During the latter decades of the twentieth century, international lending institutions, such as the International Monetary Fund (IMF), made loans to developing nations that were suffering from high rates of inflation, stagnant economic growth, and governments that were broke. The loans were made conditional on the adoption of free-market policies: reducing government to balance the budget and privatizing state-owned industries. Significantly reduced spending on education, health, and other social programs, along with higher taxes, were intended to bring government spending under control and reduce government benefits. Such structural adjustment policies imposed severe austerity on the

modernization theory

A version of market-oriented development theory that argues that low-income societies develop economically only if they give up their traditional ways and adopt modern economic institutions, technologies, and cultural values that emphasize savings and productive investment.

---

neoliberalism

The economic belief that free-market forces, achieved by minimizing or, ideally, eliminating government restrictions on business, provide the only route to economic growth.

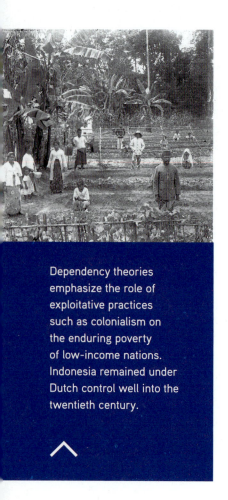

Dependency theories emphasize the role of exploitative practices such as colonialism on the enduring poverty of low-income nations. Indonesia remained under Dutch control well into the twentieth century.

poor, for whom these policies were understandably unpopular. They were also unsuccessful in promoting economic development or alleviating poverty (Afshar and Dennis 1992; Donkor 1997; Sahn, Dorosh, and Younger 1999; Easterly 2006; Stiglitz 2017).

Dependency Theories

During the 1960s, a number of theorists questioned market-oriented explanations of global inequality. Many of these critics were sociologists and economists from the low-income countries of Latin America and Africa who rejected the idea that their countries' economic underdevelopment was due to their own cultural or institutional failings. Instead, they built on the theories of Karl Marx, who argued that world capitalism would create a class of countries manipulated by more powerful countries, just as capitalism within countries leads to the exploitation of workers. **Dependency theories**, as they are called, argue that the poverty of low-income countries stems from their exploitation by wealthy countries and the multinational corporations that are based in wealthy countries. In this view, global capitalism locked many countries into a downward spiral of exploitation and poverty.

Dependency theorists argue that this exploitation began with **colonialism**, a political-economic system under which powerful countries establish, for their own profit, rule over weaker peoples or countries. Powerful nations have colonized other countries usually to procure the raw materials needed for their factories and to control markets for the products manufactured in those factories. Although colonialism typically involved European countries establishing colonies in North and South America, Africa, and Asia, some Asian countries (such as Japan) had colonies as well.

Colonialism ended throughout most of the world after World War II, but the exploitation did not: Transnational corporations continued to reap enormous profits from their branches in low-income countries. According to dependency theory, these global companies, often with the support of the powerful banks and governments of rich countries, established factories in poor countries, using cheap labor and raw materials to maximize production costs without governmental interference. In turn, the low prices set for labor and raw materials prevented poor countries from accumulating the profit necessary to industrialize themselves. Local businesses that might compete with foreign corporations were prevented from doing so. In this view, poor countries are forced to borrow from rich countries, increasing their economic dependency.

Low-income countries are thus seen not as underdeveloped but rather as misdeveloped (Amin 1974; Emmanuel 1972; Frank 1966, 1969a, 1969b, 1979; Prebisch 1967, 1971). Because dependency theorists believe that exploitation has kept their countries from achieving economic growth, they typically call for revolutionary changes that would push foreign corporations out of their countries altogether (Frank 1966, 1969a, 1969b). Dependency theorists regard the exercise of political and military power as central to enforcing unequal economic relationships. According to this theory, whenever local leaders question such unequal arrangements, their voices are quickly suppressed. When people elect a government opposing these policies, that government is likely to be overthrown by the country's military, often backed by the armed forces of the industrialized countries themselves. Dependency theorists point to many examples, including the role of the CIA in overthrowing the Marxist government of Guatemala in 1954 and the socialist government in Chile in 1973 and in undermining support for the leftist government in Nicaragua in the 1980s. In the view of dependency theory, global economic inequality is thus backed up by force.

World-Systems Theory

During the last quarter-century, sociologists have increasingly seen the world as a single (although often conflict-ridden) economic system. While dependency theories hold that individual countries are economically tied to one another, **world-systems theory** argues that the world capitalist economic system is not merely a collection of independent countries engaged in diplomatic and economic relations with one another but rather must be understood as a single unit. The world-systems approach is most closely identified with the work of Immanuel Wallerstein and his colleagues. Wallerstein showed that capitalism has long existed as a global economic system, beginning with the extension of markets and trade in Europe in the fifteenth and sixteenth centuries (Hopkins and Wallerstein 1996; Wallerstein 1974a, 1974b, 1979, 1990, 1996a, 1996b, 2004). The world system is seen as comprising four overlapping elements (Chase-Dunn 1989):

1. A world market for goods and labor

2. The division of the population into different economic classes, particularly capitalists and workers

3. An international system of formal and informal political relations among the most powerful countries, whose competition with one another helps shape the world economy

4. The carving up of the world into three unequal economic zones, with the wealthier zones exploiting the poorer ones

World-systems theorists term these three economic zones *core*, *periphery*, and *semiperiphery*. All countries in the world system are said to fall into one of the three categories. **Core countries** are the most advanced industrial countries, taking the lion's share of profits in the world economic system. These include Japan, the United States, and the countries of Western Europe. The **peripheral countries** comprise low-income, largely agricultural countries that are often manipulated by core countries for their own economic advantage. Examples of peripheral countries are found throughout Africa and, to a lesser extent, in Latin America and Asia. Natural resources, such as agricultural products, minerals, and other raw materials, flow from periphery to core—as do the profits. The core, in turn, sells finished goods to the periphery, also at a profit. World-systems theorists argue that core countries have made themselves wealthy with this unequal trade, while at the same time limiting the economic development of peripheral countries.

Finally, the **semiperipheral countries** occupy an intermediate position: These are semi-industrialized, middle-income countries that extract profits from the more peripheral countries and in turn yield profits to the core countries. Examples of semiperipheral countries include Mexico in North America; Brazil, Argentina, and Chile in South America; and the emerging economies of East Asia. The semiperiphery, though to some degree controlled by the core, is thus also able to exploit the periphery. Moreover, the greater economic success of the semiperiphery holds out to the periphery the promise of similar development. Although the world system tends to change very slowly, once-powerful countries eventually lose their economic power, and others take their place.

For world-systems theorists, the economic activities of corporations (which are increasingly global in nature) often conflict with political efforts (which remain national)

world-systems theory

Pioneered by Immanuel Wallerstein, a theory that emphasizes the interconnections among countries based on the expansion of a capitalist world economy. This economy is made up of core countries, semiperipheral countries, and peripheral countries.

core countries

According to world-systems theory, the most advanced industrial countries, which take the lion's share of profits in the world economic system.

peripheral countries

Countries that have a marginal role in the world economy and are thus dependent on the core producing societies for their trading relationships.

semiperipheral countries

Countries that supply sources of labor and raw materials to the core industrial countries and the world economy but are not themselves fully industrialized societies.

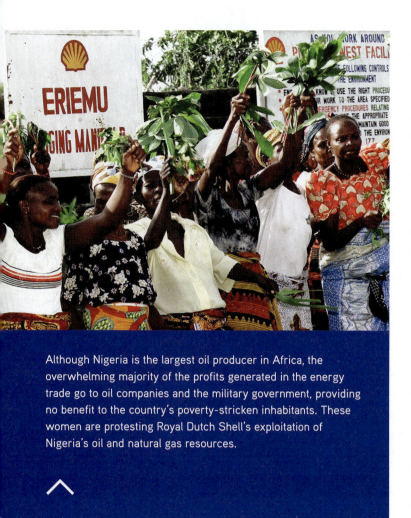

Although Nigeria is the largest oil producer in Africa, the overwhelming majority of the profits generated in the energy trade go to oil companies and the military government, providing no benefit to the country's poverty-stricken inhabitants. These women are protesting Royal Dutch Shell's exploitation of Nigeria's oil and natural gas resources.

global commodity chains

Worldwide networks of labor and production processes yielding a finished product.

to ensure that corporate behavior is socially responsible—for example, in terms of workers' rights or environmental impact. On the other hand, governments can enact policies intended to make their leading corporations competitive and successful in the global economy. In some cases, this may promote economic growth in the home country. While pro-business state policies are most likely to be effective in core countries, they have also proven to be successful in the past in promoting economic development in the semiperiphery. For example, during the 1980s and 1990s, government policies in East Asia played a central role in encouraging investment by foreign corporations, which in turn resulted in economic growth throughout the region. Apart from providing tax breaks or low-cost loans to foreign firms and subsidizing cheap worker housing (low rents meant wages could also be kept low), the governments of Taiwan, South Korea, and Singapore also outlawed trade unions, banned strikes, and jailed labor leaders in an effort to attract foreign firms looking for a supply of cheap labor (Amsden 1994; Cumings 1997; Evans 1995; Henderson and Appelbaum 1992).

Today, the Chinese government has played the central role in the economic rise of China, which regularly sets forth its goals in ambitious five- and fifteen-year plans. Trillions of dollars have been invested in high-speed trains and modern highways, seaports, and airports; research and education in science and technology; science parks and innovation hubs; and sparkling urban centers. In world-systems terms, whether these state-led investments will enable China to become a core world economy—perhaps outcompeting the United States, Europe, and Japan—remains an open question. But China's state-driven approach to development has proven quite effective in raising China from the periphery to the high semiperiphery over the past several decades, while raising hundreds of millions of Chinese into middle-class status (Appelbaum et al. 2018). In China's case, a strong state has been able to confer advantage on its leading companies through government subsidies, public investment in new technologies, and restrictions on foreign competition. Whether other countries in the periphery are able to follow China's example remains to be seen: China, with 1.3 billion people, a strong government, and a large and growing market, has been able to attract foreign corporations on highly favorable terms—for example, requiring Chinese partners with foreign firms, which assures that profits (and possibly new technologies) remain in China. Many countries in the periphery do not have this advantage and therefore are more likely, world-systems theory would argue, to be exploited.

An important offshoot of the world-systems approach is a related theory that emphasizes the global nature of economic activities. **Global commodity chains** (sometimes called "global production networks") are worldwide networks of labor and production processes yielding a finished product. These networks consist of all pivotal production activities that form a tightly interlocked "chain" extending from the raw materials needed to create the product to its final consumer (Appelbaum and Christerson 1997; Gereffi 1995, 1996; Hopkins and Wallerstein 1996).

The commodity-chain approach sees manufacturing as becoming increasingly globalized. Merchandise exports increased from $10.6 trillion in 2005 to $16.1 trillion in 2016 (World Bank 2018i). While most of the merchandise exports originate in high-income countries, their share has declined as manufacturing has shifted to middle-income countries, such as China. In 2005, high-income countries accounted for 74 percent of all merchandise exports; by 2016, that share had dropped to 67 percent. Among middle-income countries, the reverse has been the case: In only eleven years, merchandise exports increased by nearly one-quarter, from 26 percent in 2005 to 32 percent in 2016 (World Bank 2018i). The "rise of the rest" (Zakaria 2008) has been fueled by export-oriented industrialization, in which a growing number of countries now manufacture goods for world consumption. China, which moved from the ranks of low income to upper middle income largely because of its exports of manufactured goods, partly accounts for this trend. Yet the most profitable activities in the commodity chain—engineering, design, and advertising—usually occur in core countries, whereas the least profitable activities, such as factory production, occur in peripheral countries.

In the view of commodity-chain theorists, for a country to advance economically, it has to move up the value chain—from low-wage manufacturing to higher-skilled activities and eventually to those activities where the greatest economic returns can be found: global brand recognition through innovative product design and marketing. This is the path that the East Asian economies followed and that China is now pursuing: Instead of manufacturing products for U.S., European, and Japanese firms, the challenge is for Chinese firms to design and market the products themselves. To take one example, the iPhone has helped make Apple one of the most recognized and profitable companies in the world; yet the iPhone is assembled in China and other lower-wage countries, which realize only a small fraction of Apple's total revenues. China hopes to move up the commodity chain, using the technological know-how it has developed assembling products for Apple, Samsung, and other electronics companies to design and market its own technologically superior smartphone (Appelbaum et al. 2018).

The commodity-chain approach emphasizes the global nature of economic activities. Manufacturing has shifted to countries like China.

Global Capitalism Theory

More recently, sociologist William Robinson (2004, 2007, 2014, 2017) has proposed an important update of Wallerstein's world-systems theory. Robinson's theory of global capitalism claims that the capitalist system, in its current phase, is increasingly characterized by a globalized production and financial system, a transnational capitalist class, a transnational state, and a rising global police state. Contrary to world-systems theory, which views the state as a national actor serving to advance the interests of its national business class, the global capitalism approach sees stateless corporations as the key actors in the global economy: According to global capitalism theory, transnational corporations are increasingly powerful, able to bend national politicians to their goals. Individual state actors are thus replaced by global actors—a transnational capitalist class—with limited or no state loyalties.

Sociologist Leslie Sklair (2000a, 2000b) argues that the transnational capitalist class has four increasingly global factions: those who own and run transnational corporations, the bureaucrats and politicians who serve their interests, professionals (such as engineers, managers, administrators, accountants, financial advisers, and researchers) who provide their services for transnational corporations, and the merchants and media that promote global consumerism. The transnational capitalist class is said to be globalized not only because its members share global (rather than national) economic interests but also because they share similar lifestyles: They are wealthy. They travel the world, often own many homes in different countries, belong to the same exclusive clubs, and send their children to the same exclusive private schools. In effect, they constitute a global power elite that is increasingly more powerful than national power elites.

The emergence of a transnational capitalist class is accompanied by the emergence of transnational state apparatuses through which they exert power. These include informal networks (such as the World Economic Forum, which meets every year in Davos, Switzerland); formal business and financial organizations, such as the WTO, IMF, and World Bank; and regional governing bodies, such as the European Union. It is through these networks and organizations that the transnational capitalist class seeks to exercise global political authority (Robinson 2004, 2007, 2014, 2017).

Evaluating Global Theories of Inequality

Each of the four theories of global inequality has strengths and weaknesses. Together, they enable us to better understand the causes of and cures for global inequality:

1. **Market-oriented theories** recommend the adoption of modern capitalist institutions to promote economic development. They further argue that countries can develop economically only if they open their borders to trade, and they cite evidence to support this argument. But market-oriented theories overlook economic ties between poor countries and wealthy ones—ties that can impede economic growth under some conditions and enhance it under others. They blame low-income countries for their poverty rather than acknowledging outside factors, such as the business operations of more powerful nations. Market-oriented theories also ignore the ways government can work with the private sector to spur economic development. Finally, they fail to explain why some countries take off economically while others remain grounded in poverty and underdevelopment.

2. **Dependency theories** emphasize how wealthy nations have exploited poor ones. Although these theories account for the lack of consistent economic growth in Latin America and Africa, they cannot explain the occasional success stories, such as Brazil, Argentina, and Mexico, or the rapidly expanding economies of China and East Asia that have risen economically despite the presence of multinational corporations. To take these developments into account, an offshoot of dependency theory—called "dependent development"—argues that under certain circumstances (e.g., when there is an emergent working-class movement, a middle class oriented toward domestic development, the availability of abundant natural resources), some development is possible, although it is always constrained by more powerful core nations (Cardoso and Faletto 1979).

3. **World-systems theory** analyzes the world economy as a whole, looking at the complex global web of political and economic relationships that influence development and inequality in poor and rich nations alike. It is thus well suited to understanding the global economy at a time when businesses are increasingly free to set up operations anywhere, acquiring an economic importance rivaling that of many countries. One challenge world-systems theory faces lies in the difficulty

of modeling a complex and interdependent world economy. It has also been criticized for emphasizing economic and political forces at the expense of cultural ones, such as the combination of nationalism and religious belief that is currently reshaping the Middle East. Finally, world-systems theory has been said to place too much emphasis on the role of nation-states in a world economy increasingly shaped by transnational corporations that operate independently of national borders (Robinson 2004; Sklair 2002b).

Global commodity chains theory—an offshoot of world-systems theory—focuses on global businesses and their activities rather than on relationships among countries. While this approach provides important insights into how different countries and regions are affected (positively or negatively) by the ways in which they connect with global commodity chains, it also tends to emphasize the importance of business decisions over other factors, such as the role of workers and governments in shaping a country's economy (Bair 2009).

4. **Global capitalism theory**, like dependency theories and world-systems theory, argues that capitalism is fundamentally exploitative. But unlike the other two theories, global capitalism theory highlights the emergence of a transnational capitalist class and the accompanying transnational state apparatuses, institutions, and organizations through which it exerts power. Importantly, this approach alerts us to the rising power of transnational corporations, particularly in relation to the power of national governments. However, global capitalism theory has been criticized for overstating the role of transnational capital and understating the power of nation-states (Webber 2009; Harris 2008; Cioffi 2007). Core countries like the United States still exert considerable economic, political, and military power in the national interest; and China, a rapidly emerging global power, owes its success to state-driven economic planning. Global capitalism theory fails to account for countries in which state power continues to play an important role.

CONCEPT CHECKS

1. Describe the main assumptions of market-oriented theories of global inequality.

2. Why are dependency theories of global inequalities often criticized?

3. Compare and contrast core, peripheral, and semiperipheral nations.

4. How have strong governments of some East Asian nations contributed to the economic development of that region?

What Does Rapid Globalization Mean for the Future of Global Inequality?

Today the social and economic forces leading to a single global capitalist economy appear to be irreversible. What does rapid globalization mean for the future of global inequality? No sociologist knows for certain, but many possible scenarios exist. In one, our world might be dominated by large global corporations, with workers everywhere competing with one another at a global wage. Such a scenario might predict falling wages for large numbers of people in today's high-income countries and rising wages for a few in low-income countries—a development that could adversely affect your own job prospects. There might be a general leveling out of average income around the world, although at a level much lower

Learn about a few possible scenarios for the future of global inequality. Understand how the digital divide and climate change may increase the gap between rich and poor.

than that currently enjoyed in the United States and other industrialized nations. In this scenario, the polarization between the haves and the have-nots would grow, as the whole world would be increasingly divided into those who benefit from the global economy and those who do not. While hopefully your career would enable you to fall into the category of those who benefit, can you say with certainty that this would be the case? Moreover, such polarization could fuel conflict among ethnic groups and even among nations, as those suffering from economic globalization would blame others for their plight (Hirst and Thompson 1992; Wagar 1992). Many European countries and the United States are now experiencing such polarization, as seen in the rise of anti-immigrant and nationalist movements.

In another, more hopeful scenario, a global economy could mean greater opportunity for everyone, as the benefits of modern technology stimulate worldwide economic growth. In the most optimistic view, the republics of the former Soviet Union as well as the formerly socialist countries of Eastern Europe will eventually advance into the ranks of the high-income countries. Economic growth will spread to Latin America, Africa, and the rest of the world. Because capitalism requires that workers be mobile, the remaining caste societies around the world will be replaced by class-based societies. These societies will experience enhanced opportunities for upward mobility. According to this more optimistic scenario, the more economically successful East Asian emerging economies, such as Hong Kong, Taiwan, South Korea, Singapore, and China, are only a sign of things to come.

One challenge to this more hopeful scenario is the technology gap that divides rich and poor countries, making it even more difficult for poor countries to catch up. As many as 4 billion people in the world remain unconnected to the Internet at a time when economic gains increasingly depend on such access (Shenglin et al. 2018). This "digital divide" is a result of the disparity in wealth between nations, but it also reinforces those disparities. Poor countries cannot easily afford modern technology—yet, in the absence of modern technology, they face major barriers to overcoming poverty. They are caught in a vicious downward spiral from which it is difficult to escape. As access to the Internet and the use of cell phones become increasingly widespread, the technology gap between rich and poor nations may well be reduced, to the benefit of the latter, contributing to a rising economic tide for a growing number of people around the world.

What, then, is the future of global inequality? It is difficult to be entirely optimistic. Slowing economic growth coupled with rising inequality has affected the European Union, once thought to be a pillar of the global economy; concerns about inequality in Europe have threatened its future. Britain, in 2016, voted to leave the EU (the so-called Brexit) when a majority of voters concluded that EU economic and immigration policies were not in their best interest. The emergence of far-right nationalist parties throughout Europe, many of which have achieved electoral success in recent years, suggests that Britain may not be the only country to consider such an action. In March 2018, the party that received the largest number of votes in Italy's national elections was the Five Star Party, a youthful party that is strongly environmentalist, populist, critical of what it sees as the destructive aspects of globalization, and skeptical of the EU. Strongly nationalist parties also earned a large percentage of the votes, leading observers to conclude that the elections may result in Italy also rethinking its relationship to the EU (Monti 2018).

Another concern is that environmental problems, resulting from global climate change, will result in violent conflicts in many poor countries, displacing millions and creating immigration challenges in Europe and the United States. Yemen and African countries such as Nigeria, Somalia, and South Sudan face their worst drought in seventy years, threatening mass starvation. A recent study by the UN High Commission on Refugees (2017) reported that "20 million people live in areas where harvests have failed and malnutrition rates are increasing, particularly among young children. One million people are now on the brink of famine." The brutal civil war in Syria, which began in 2011, has many causes, but important among them is climate change: The combination of a growing population, drought, and water shortages has led to agricultural failures and growing food insecurity, raising tensions and contributing to the conflict (Glieck 2014). As of early 2018, the war had claimed an estimated 400,000 lives and created more than 11 million refugees (CNN 2018; World Vision 2018; UN High Commission on Refugees 2018). Environmental changes have contributed to other conflicts in the Middle East and sub-Saharan Africa. One detailed study concluded that low-income areas will be hardest hit by climate change—and least able to mitigate the effects of rising temperatures, drought, and water shortages (Calvin et al. 2016). Many of those affected by drought and warfare have sought refuge in Europe; this in turn has spurred an anti-immigrant backlash and the rise of nationalist political parties in many countries.

In 2018, the Center for Climate and Security—a nonpartisan group of U.S.-based military, national security, homeland security, intelligence, and foreign policy experts—issued a report titled *A Responsibility to Prepare: Strengthening National and Homeland Security in the Face of a Changing Climate*. The provocative title is a response to U.S. Defense Secretary James Mattis's concern that "changes in the climate pose direct threats, such as sea level rise and increased storm surges, which could inundate coastal military and civilian infrastructure. Dramatic changes in food, water and energy availability also increase the likelihood of instability and state failure across the globe" (Center for Climate and Security 2018). The poor are least able to cope with the effects of climate change, increasing the gap between rich and poor, particularly in the poorest countries. Poverty and growing inequality are a prescription for civil strife, which is why Mattis sees global instability abroad, as much as coastal flooding at home, as threatening national security.

In China, global warming may prove of some benefit to "left behind" children; warmer winters might mean fewer "frost boys" to catch the world's attention. But climate change may also affect crops, such as wheat, buckwheat, and potatoes, that are adapted to current climates, resulting in increased rural poverty and worsening inequality.

The future of global inequality remains an open question—one whose answer will depend, in large part, on whether global economic expansion can be sustained in the face of ecological constraints and a global economy that has proven to be surprisingly fragile. It remains to be seen whether the countries of the world will learn from one another and work together to create better lives for their peoples. Technological advances, including widespread use of the Internet and frequent media reports of tragedy in poorer parts of the world, including the story of Wang Fuman, may raise awareness of the startling economic and health inequalities in the world today. A well-informed awareness of global inequality may be an essential step toward trying to eradicate the vast gap between the haves and have-nots and developing social programs to eradicate the problems of hunger and disease that plague poorer societies.

CONCEPT CHECKS

1. What is the role of technology in deepening existing global inequalities?

2. How might climate change exacerbate global inequality?

CHAPTER 8

The Big Picture

Global Inequality

Thinking Sociologically

1. Concisely review the four theories offered in this chapter that explain why there are gaps between nations' economic developments and resulting global inequality: market-oriented theories, dependency theories, world-systems, and global capitalism theory. Briefly discuss the distinctive characteristics of each theory and how each differs from the others. Which theory do you feel offers the most explanatory power to addressing economic developmental gaps?

2. This chapter states that global economic inequality has personal relevance and importance to people in advanced, affluent economies. Briefly review this argument. Explain carefully whether you were persuaded by it.

Terms to Know

Concept Checks

globalization

global inequality • absolute poverty • relative poverty

1. Explain how the World Bank measures global inequality, and discuss some of the problems associated with measuring global inequality.
2. Compare and contrast high-income, middle-income, and low-income countries.

1. Why do people who live in high-income countries have better health than those who live in low-income countries?
2. What is one global cause of poor health?
3. What are two causes of global hunger today?

emerging economies

1. What are the five factors that have facilitated the economic success of the East Asian emerging economies?
2. What are potential obstacles to the continued economic success of the emerging economies?

market-oriented theories • modernization theory • neoliberalism • dependency theories • colonialism • world-systems theory • core countries • peripheral countries • semiperipheral countries • global commodity chains

1. Describe the main assumptions of market-oriented theories of global inequality.
2. Why are dependency theories of global inequalities often criticized?
3. Compare and contrast core, peripheral, and semiperipheral nations.
4. How have strong governments of some East Asian nations contributed to the economic development of that region?

1. What is the role of technology in deepening existing global inequalities?
2. How might climate change exacerbate global inequality?

9

Gender Inequality

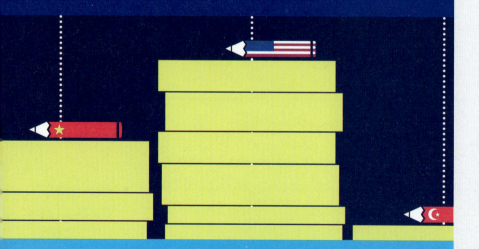

Gender Inequality

p. 267

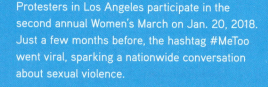

Protesters in Los Angeles participate in the second annual Women's March on Jan. 20, 2018. Just a few months before, the hashtag #MeToo went viral, sparking a nationwide conversation about sexual violence.

THE BIG QUESTIONS

Are gender differences due to nature, nurture, or both?

Evaluate the extent to which differences between women and men are the result of biological factors or social and cultural influences. Understand the concept of the gender binary, and learn what it means to identify as nonbinary.

How do gender inequalities play out in social institutions?

Recognize that gender differences are a part of our social structure and create inequalities between women and men. Learn the forms these inequalities take in social institutions such as the workplace, the family, the educational system, and the political system in the United States and globally.

Why are women the target of violence?

Learn about the specific ways that women are the target of physical and sexual violence in the United States and globally.

How does social theory explain gender inequality?

Think about various explanations for gender inequality. Learn some feminist theories about how to achieve gender equality.

How can we reduce gender-based aggression?

Learn how women and men are challenging sexism and sexual violence in the workplace and on college campuses.

The entertainment industry was rocked in 2017 when dozens of women, including high-powered actresses Salma Hayek, Angelina Jolie, Lupita Nyong'o, and Gwyneth Paltrow, came forward with horrifying tales of sexual harassment and abuse at the hands of star-making Hollywood producer Harvey Weinstein. More than eighty women have since come forward with allegations that the Miramax entertainment company cofounder behaved inappropriately, with some reporting rapes or attempted rapes; others found their careers derailed when they rebuffed his unwanted advances (Saad 2017). The public confessions of these actresses emboldened countless other women (and men) to report their experiences of mistreatment at the hands of powerful actors like Kevin Spacey and media stars like *Today Show* host Matt Lauer.

But sexual harassment in the workplace isn't limited to the glamorous world of entertainment. Nearly every industry, ranging from the hospitality business to Silicon Valley to Wall Street, from pathbreaking start-ups like Uber to the assembly line at Ford Motor Company, is the site of rampant sexual harassment and discrimination. For instance, in August 2017, the Equal Employment Opportunity Commission (EEOC) reached a $10 million settlement with Ford Motor Company for sexual and racial harassment at two Chicago plants (Chira and Einhorn 2017). Women assembly-line workers, most of whom are Black or Latino, were berated as "fresh meat," had their breasts and buttocks grabbed by male coworkers, and were given the devastating choice of having sex with their supervisor or else finding themselves fired. As line worker Miyoshi Morris told the *New York Times*, "I slept with [my supervisor] because I needed my job" to support her children.

That same year, Sarah Fowler, a twenty-six-year-old engineer at Uber, blew the whistle on her CEO and colleagues, who created a culture in which the women engineers (who make up just 6 percent of their workforce) were regularly propositioned or harassed at work. Reports to human resources staff fell on deaf ears, and the women were simply told they could quit (Dowd 2017). Fowler sought justice by publishing a tell-all blog on her experience, which ultimately led to the firing of twenty Uber corporate employees and the resignation of the company's CEO. In the wake of these events, the #MeToo hashtag spread like wildfire on social media, revealing just how widespread sexual harassment and abuse are, especially in the workplace.

As we will see later in the chapter, the demeaning, demoralizing, and even dangerous treatment of women in almost every industry may partly explain why women account for less than 30 percent of all senior officers and only 6 percent of CEOs in finance (Catalyst 2017), less than 10 percent of all film directors (Lauzen 2017), and just 17 percent of managers in the automotive industry (U.S. Equal Employment Opportunity Commission 2017). A casual observer who sees statistics like these might conclude that women are not cut out for high-powered positions or jobs in cutthroat industries, or that they simply prefer to work in other fields. Yet sociology helps us look beyond individual skills and preferences and enables us to understand the systematic ways that factors like gender shape one's access to power, resources, and opportunities. Explaining the differences and inequalities between women and men in a society is now one of the most central topics in sociology.

In this chapter, we take a sociological approach to the exploration of gender differences and gender inequality. Gender is a way for society to divide people into two categories: "men" and "women." Not all persons, however, fit neatly into one of these two categories, as we will see later in this chapter (Davis 2015a; Heine 2013). According to this socially created division, men and women have different identities and social roles. Men and women are expected to think and act in certain ways across most life domains. Gender also serves as a social status; in almost all societies, men's roles are valued more than women's roles (Bem 1993). Men and women are not only different but also unequal in terms of power, prestige, and wealth.

Despite the advances that many women have made in the United States and other Western societies, this remains true today. Sociologists are interested in explaining how society differentiates between women and men and how these differences serve as the basis for social inequalities (Chafetz 1990). Yet sociologists recognize that gender alone does not shape our life experiences. Rather, there are pronounced differences in women's and men's lives on the basis of race, social class, age, birth cohort, religion, nation of origin, sexual orientation, and even one's marital or parental status (Choo and Ferree 2010). The challenges women face in wealthy Western nations also vary markedly from those experienced by women in the Global South, underscoring the importance of **intersectionality**, or the ways that women's (and men's) multiple identities and social locations shape their experiences (Mohanty 2013).

In this chapter, we examine the origins of gender differences, assessing the debate over the role of biological factors versus social influences in the formation of gender roles. We also

intersectionality

A sociological perspective that holds that our multiple group memberships affect our lives in ways that are distinct from single group membership. For example, the experience of a Black female may be distinct from that of a white female or a Black male.

look to other cultures for evidence on this debate. Then we review the various forms of gender inequality that exist in U.S. society and throughout the globe. In this section, we focus on major social institutions, including the educational system, the workplace, the family, and the government. Next, we examine how and why women are more likely than men to be the targets of sexual violence, whether in the family, at work, or at the hands of strangers. We review the various forms of feminism and assess prospects for future change toward a gender-equal society. We conclude the chapter by examining some theories of gender inequality and applying them to the lives of women like Sarah Fowler and Miyoshi Morris.

Are Gender Differences Due to Nature, Nurture, or Both?

Are differences between boys and girls, and between men and women, due to nature, nurture, or some combination of the two? As we first noted in Chapter 2, scholars have long debated the degree to which innate biological characteristics have an enduring impact on our gender identities as "feminine" or "masculine" and the social roles based on those identities. In recent decades, scholars have further challenged the notion that gender identities fit neatly into the masculine–feminine dichotomy. Rather, gender scholars are striving to better understand persons who do not conform with the **gender binary**, instead blending two gender identities, alternating fluidly between them, or eschewing a single gender identity altogether (Olson-Kennedy et al. 2016). No one would argue that our behaviors are purely instinctive or hardwired. Yet scholars disagree as to the extent to which they believe gender differences are the product of learning and socialization.

Before we review the relative influences of nature and nurture, we first need to make an important distinction between sex and gender. **Sex** refers to physical differences of the body, whereas **gender** concerns the psychological, social, and cultural differences between males and females. This distinction is fundamental because many differences between males and females are not biological in origin. While sex is something we are born with, gender is something that we both learn and do. Sex and gender historically have been viewed as a binary, where the two categories of male and female, or masculine and feminine, were viewed as distinctive and nonoverlapping (Lorber 1996). However, sex and gender can be fuzzy and overlapping categories, and the boundaries demarcating "male" and "female" behaviors, traits, and even bodies are fluid and evolving. Small yet rising numbers of people now identify as **nonbinary** and possess gender identities that are not exclusively masculine or feminine.

The Role of Biology

How much are differences in the behavior of women and men the result of biological differences? Some researchers hold that innate differences in behavior appear in some form in all cultures and that the findings of sociobiology point strongly in this direction. Such

gender binary

The classification of sex and gender into two discrete, opposite, and nonoverlapping forms of masculine and feminine.

———

sex

The biological and anatomical differences distinguishing females from males.

Evaluate the extent to which differences between women and men are the result of biological factors or social and cultural differences. Understand the concept of the gender binary, and learn what it means to identify as nonbinary.

gender

Social expectations about behavior regarded as appropriate for the members of each sex. Gender refers not to the physical attributes distinguishing men and women but to socially formed traits of masculinity and femininity.

———

nonbinary

A gender identity that does not fit squarely into the male–female gender binary classification.

researchers are likely to draw attention to the fact, for example, that in almost all cultures, men rather than women take part in hunting and warfare. Surely, they argue, this indicates that men possess biologically based tendencies toward aggression that women lack. In looking at the kinds of jobs that women and men typically hold, they might point out that women are better suited than men for jobs like store cashier or clerical assistant. Ringing up purchases and assisting with office tasks are more passive occupations than being a stock handler or store security guard, positions that require more physical strength and aggressiveness.

Most social scientists are unconvinced by these arguments and even view them as potentially dangerous. In her classic book *The Lenses of Gender*, Sandra Lipsitz Bem (1993) notes that this kind of **biological essentialism** rationalizes and legitimizes gender differences as the natural and inevitable consequences of the intrinsic biological natures of women and men. As such, social influences are neglected or minimized. Social scientists who renounce biological essentialism would argue that the level of aggressiveness of men varies widely across cultures and that women are expected to be more passive or gentle in some cultures than in others (Elshtain 1981).

Furthermore, some argue that women are just as aggressive as men; however, women use strategies that are consistent with gender role socialization. For instance, women will use "interpersonal aggression," such as malicious gossip or "bad mouthing," rather than engaging in physical fights (Bjorkqvist, Lagerspetz, and Osterman 2006). Other data suggest that as gender roles change over time, girls may become more physically aggressive. A national study conducted by the Substance Abuse and Mental Health Services Administration (2010) found that 19 percent of adolescent females got into a serious fight at school or work in the past year, 14 percent participated in a group-against-group fight, and nearly 6 percent attacked another person with the intent to seriously hurt that person. Theories of "natural difference" are often grounded in data on animal behavior, critics point out, rather than in anthropological or historical evidence about human behavior, which reveals variation over time and place.

Despite critiques of the "nature" perspective, the hypothesis that biological factors shape some behavior patterns cannot be wholly dismissed. Studies document persuasively that biological factors—including genetics, hormones, and brain physiology—differ by sex and that these biological differences are associated with some social behaviors and aptitudes, including language skills, interpersonal interactions, and physical strength. However, nearly all social scientists agree that theories based solely on an innate predisposition neglect the vital role of social interaction and social contexts in shaping human behavior.

What does the research show? Some studies show differences in hormonal makeup between the sexes, with the male sex hormone, testosterone, associated with a propensity to violence (Rutter and Giller 1984). Yet the production of testosterone is shaped by social context; for instance, animal studies show that providing monkeys with opportunities to dominate others actually increases testosterone levels. This means that aggressive behavior may affect the production of the hormone rather than the hormone causing increased aggression—thus underscoring the importance of social context. In other words, there might be slight biological differences between men and women, but these small differences may be exacerbated and amplified by social contexts that promote behaviors that are consistent with gendered stereotypes and expectations.

biological essentialism

The view that differences between men and women are natural and inevitable consequences of the intrinsic biological natures of men and women.

Another source of information comes from the experience of identical twins. Identical twins derive from a single egg and have exactly the same genetic makeup. In one particularly infamous case in the 1960s, Bruce Reimer, one of a pair of identical male twins, was seriously injured while being circumcised, while his brother Brian was not. Psychologist John Money convinced the twins' parents that it would be best to reconstruct Bruce's genitals as a female. He was thereafter raised as a girl, Brenda. The twins at six years old demonstrated typical male and female traits as found in Western culture, and for a brief time, this case was treated as a conclusive demonstration of the overriding influence of social learning on gender differences. However, when Brenda was a teenager, she was interviewed during a television program, and the interview showed that she felt profound unease about the gender identity she had been assigned after the botched surgery. Brenda eventually made the decision to go back to living as a male (taking the name David), marrying a woman and adopting children. But after struggling with marital woes, financial instability, and the grief of losing his brother Brian to suicide, David ultimately killed himself (Colapinto 2001).

Technological advances in the last two decades have provided a new source of data for understanding biological influences on gender: brain imaging research has identified several key differences between men's and women's brains (Brizendine 2006, 2010). For example, Burman and colleagues (2007) found that the brains of school-age girls were more highly "activated," or worked harder, than the brains of school-age boys when presented with spelling and writing tasks. This greater level of activation has been associated with greater accuracy in performing such tasks. The authors do not conclude that girls' language skills are superior to those of boys, yet they do argue that their data show that girls and boys learn language in different ways. A mounting body of research concludes that gender differences in brain functioning may contribute, in part, to a wide range of social outcomes, including communication style, empathy, depression, anxiety, and fear. However, most scholars conducting this research are careful to point out that biological differences are almost always exacerbated or fostered by social contexts and norms (McCarthy 2015).

Gender Socialization

Another explanation for gender differences is **gender socialization**, or ways that individuals learn gender roles from socializing agents such as the family, peers, schools, and the media (as discussed in Chapter 4). Through contact with various agents of socialization, children gradually internalize the social norms and expectations that are seen to correspond with their sex. In other words, gender differences are not biologically determined; they are culturally produced. The concept of gender socialization teaches us that gender inequalities result because men and women are socialized into different roles.

People create gender through social interactions with others, such as family members, friends, and colleagues. This process begins at birth when doctors, nurses, and family members—the first to see an infant—assign the person to a gender category on the basis of physical characteristics. Babies are immediately dressed in a way that marks the sex category; for instance, a girl may wear a little pink bow while a boy may wear a sailor suit. "Parents don't want to be constantly asked if their child is a boy or a girl" (Lorber 1994). Once the child is marked as male or female, everyone who interacts with the child will treat it in accordance with its gender. They do so on the basis of society's assumptions, which lead people to treat women and men differently, even as opposites, reifying the gender binary (Zosuls et al. 2009).

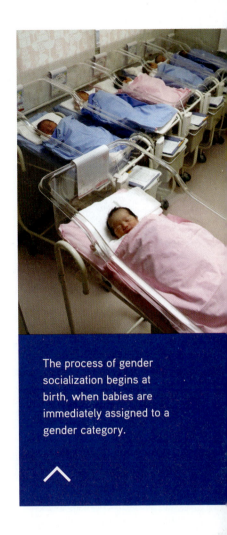

The process of gender socialization begins at birth, when babies are immediately assigned to a gender category.

gender socialization

The learning of gender roles through social factors such as schooling, peers, the media, and family.

Gender socialization is very powerful, and challenges to it can upset one's sense of order. Think about the controversy surrounding Leo Davis. This Brooklyn, New York, kindergartner was born male but has always been drawn to Barbie dolls, pastel colors, and girls' clothing. His parents describe him as "gender expansive" and encourage him as he chooses to wear dresses, a beloved pink track suit, or sparkly clothes to school (Harris 2017). Yet, once he got to school, he found that his classmates and teachers created what his parents considered a hostile environment, questioning Leo's gender, his choice of which bathroom to use, and the clothes he wore. His parents ultimately sued the Education Department to help ensure that their child would feel safe and protected at school. As sociological studies repeatedly show, once a gender is "assigned," society expects individuals to act like "females" and "males." These expectations are fulfilled and reproduced in the practices of everyday life, and challenges to these expectations are not yet widely accepted (Bourdieu 1990; Lorber 1994). Violations of these expectations may lead to confusion, if not outright hostility.

The Social Construction of Gender

In recent years, socialization and gender role theories have been criticized by a growing number of sociologists. Rather than seeing sex as biologically determined and gender as culturally learned, they argue that we should view both sex and gender as socially constructed products. Theorists who believe in the **social construction of gender** reject biological bases for gender differences. Gender identities emerge, they argue, in relation to perceived sex differences in societies and cultures, which in turn shape and even perpetuate those differences. For example, a society in which cultural ideas of masculinity are characterized by physical strength, stoicism, and tough attitudes will encourage men to cultivate a specific body image and set of mannerisms (Butler 1989; Connell 1987; Scott and Morgan 1993). Men who fail to comply with what scholars call **hegemonic masculinity**, or the social norms dictating that men should be strong, self-reliant, competitive, and unemotional, may be subtly sanctioned for not enacting gender roles in ways that are consistent with prevailing cultural norms (Messerschmidt and Messner 2018).

DOING GENDER

According to social constructionist perspectives, gender is not something that we are but something that we "do" (West and Zimmerman 1987) or a role that we perform. That means that we learn how to present ourselves as "male," "female," or nonbinary through our choice of behaviors, clothing, hairstyle, stance, body language, and tone of voice (Westbrook and Schilt 2014). For example, a number of scholars have uncovered the discouraging finding that some young heterosexual women "play dumb" both because they believe it is consistent with gendered expectations for how girls should act and because they believe that doing so may help bolster feelings of masculinity among the boys they are hoping to attract as romantic partners (Gove, Hughes, and Geerken 1980). At the same time, young men feel great pressure to be strong, confident, and funny to attract a partner. As sociologist Maria do Mar Pereira discovered in her 2014 qualitative study of fourteen-year-old boys and girls, "one of the pressures is that young men must be more dominant—cleverer, stronger, taller, funnier—than young women, and that being in a relationship with a woman who is more intelligent will undermine their masculinity" (University of Warwick 2014).

social construction of gender

A perspective holding that gender differences are a product of social and cultural norms and expectations rather than biology.

---

hegemonic masculinity

Social norms dictating that men should be strong, self-reliant, and unemotional.

Persons who identify as nonbinary also are strategic in choosing how to behave and present themselves. For instance, studies of "selfies" show that depending on the social context, some may choose to "pass" as male or female or emphasize gender ambiguity in their appearance. For instance, some called attention to their hairstyle, which often was short-cropped or buzzed in the back or on the sides, with longer or colorful hair on top or in the front. As one person wrote in the caption to their selfie, "So I tried short 'guy' hair and hated it. I tried longer 'girl' hair and liked it better but not the best. Then I tried a compromise <3" (Darwin 2017, p. 327).

How we "do gender" varies further by race, social class, and social context, underscoring the importance of intersectionality. We selectively choose to enact different aspects of gender expectations based on what we think will work best in a particular setting. For example, sociologist Nikki Jones (2009) found that young inner-city African American women would adjust their voices, stances, walks, and styles of speaking in different situations, thus giving off the impression of being "aggressive," "good," or "pretty" when they thought a particular type of femininity would "pay off." Jones described the ways that twenty-two-year-old Kiara would "do gender." Kiara was hoping to collect signatures for a petition to stop a new development in a poor neighborhood adjacent to her own. As Jones (2009: 90) describes,

> [Kiara] confidently, assertively, even aggressively approaches men on the street to sign her petition and then draws on normative expectations of manhood and femininity to encourage them to add their names to the list: Babies and women are in danger, she tells them, letting the implication that real men would sign up to protect babies and women hang in the air. She switches from aggressive to demure just long enough to flirt with a man passing by on the street and then to defiant when she passes the police station on the corner. "They don't give a fuck!" she declares loudly.

Kiara's behavior shows the complexities of gendered expectations in contemporary social life.

Cross-Cultural and Historical Findings

If gender differences were mostly the result of biology, we could expect gender roles not to vary much from culture to culture or across historical time periods. However, one set of findings that helps show how gender roles are in fact socially constructed comes from anthropologists who have studied gender in other eras and cultures.

NEW GUINEA

In her classic New Guinea study *Sex and Temperament in Three Primitive Societies*, Margaret Mead (1935/1963) observed wide variability among gender role prescriptions—and such marked differences from those in the United States—that any claims to the universality of gender roles had to be rejected. Mead studied three separate tribes in New Guinea. In Arapesh society, both males and females generally exhibited characteristics and behaviors that would typically be associated with the Western female role. Both sexes among the Arapesh were passive, gentle, unaggressive, and emotionally responsive to the needs of others. In contrast,

Mehran Rafaat (left) dresses and acts like a boy as a way to circumvent the barriers faced by girls in Afghan culture. Girls who live their lives disguised as boys are known as *bacha posh*.

Mead found that in another New Guinea group, the Mundugumor, both males and females were characteristically aggressive, suspicious, and, from a Western observer's perspective, excessively cruel, especially toward children. In both cultures, however, men and women were expected to behave very similarly. In a third group, the Tchambuli tribe of New Guinea, gender roles were almost exactly reversed from the roles traditionally assigned to males and females in Western society. Women "managed the business affairs of life," while "the men . . . painted, gossiped and had temper tantrums" (Mead 1972).

THE !KUNG

Among the !Kung of the Kalahari Desert, who refer to themselves as *zhun/twasi* or "the real people," it is very common for both men and women to engage in childcare (Shostak 1981). Owing to the nonconfrontational parenting practices of the !Kung, who oppose violent conflict and physical punishment, children learn that aggressive behavior will not be tolerated by either men or women. Although the !Kung abide by the seemingly traditional arrangement where "men hunt and women gather," the vast majority of their food actually comes from the gathering activities of women (see Draper, as cited in Renzetti and Curran 2003). Women return from their gathering expeditions armed not only with food for the community but also with valuable information for hunters.

THE BACHA POSH IN AFGHANISTAN

In contemporary Afghanistan, boys are so highly prized that families with only daughters often experience shame and pity; as a result, some transform one young daughter into a son. The parents cut the girl's hair short, dress her in boys' clothes, change her name to a boy's name, and encourage her to participate in "boys' activities" like bicycling and playing cricket. These children are called *bacha posh*, which translates into "dressed up as a boy."

Parents of bacha posh believe that boys are afforded so many advantages in Afghan culture that it is helpful, rather than cruel, to transform their girls into boys. A bacha posh can more easily receive an education, work outside the home, and escort her sisters in public, allowing freedoms that are unheard of for girls in a society that strictly segregates men and women. In most cases, a return to womanhood takes place when the child enters puberty, a decision almost always made by her parents (Nordberg 2010). Although historical research shows that this practice may have originated as early as 100 years ago, with women disguising themselves as men to fight during periods of wartime, the practice is believed to be much more common today. Families that are struggling financially often need to send a child to work; because girls are not allowed to work, the family may opt to turn one of their daughters into a bacha posh.

BLURRING THE BOUNDARIES BETWEEN THE GENDERS

The gender binary, or the belief that only two genders (i.e., male and female) exist, is not universal. The Spaniards who came to both North and South America in the seventeenth century noticed men in the native tribes who had taken on the mannerisms of women and women who occupied male roles. Several Native American cultures hold a special honor for persons of "integrated genders." For example, the Navajo term *nádleehí* literally means "one who constantly transforms" and refers to a male-bodied person with a feminine nature, a special gift according to Navajo culture. The Navajo believe that to maintain harmony, there must be a balanced interrelationship between the feminine and the masculine within a single individual. Native activists working to renew their cultural heritage adopted the English term *two-spirit* as useful shorthand to describe the entire spectrum of gender and sexual expression (Nibley 2011).

In the contemporary United States, growing numbers of young adults—even those as young as five-year-old Leo Davis—are challenging the male–female dichotomy and embracing both genders or switching between the two. Eschewing labels such as "male" or "female," a small but growing number of people are instead choosing to identify as nonbinary, adopting diverse terms like "androgyne," "genderqueer," "genderfluid," "bigender," "agender," or "non-cis" (Schulman 2013). "Non-cis" is shorthand for "non-cisgender." The term *cis* is Latin for "on the same side as"; thus, cisgender refers to a person whose gender identity matches his or her biological sex, for example, a person born male who identifies as a man. Some ultimately choose to identify as a gender different than the one assigned at birth; these people, usually described as **transgender**, may encompass those who live as a person of the opposite gender or who use medical assistance to physically transition from one category to the other. Celebrities like Olympic athlete Caitlyn (formerly Bruce) Jenner and Jazz Jennings, the young transgender girl who is the focus of the reality show *I Am Jazz*, as well as recently elected state legislator Danica Roem, have drawn national and international attention to the experiences of transgender persons.

Although sociologists do not know for certain precisely how many individuals define their gender in ways beyond the male–female dichotomy, researchers are making important strides. The Centers for Disease Control and Prevention plans to measure diverse nonbinary gender identities for the first time in its Youth Risk Behavior Survey (YRBS) of U.S. high school students (Center for American Progress 2016). Studies based on smaller or regional samples estimate that 1.5 percent of people under eighteen and just 0.6 percent of adults ages eighteen to sixty-four identify as nonbinary, although surveys have also found that as many as 5 percent answered "don't know" when asked about their gender identity (Almeida et al. 2009; Meerwijk and Sevelius 2017).

In November 2017, Danica Roem won a historic election to Virginia's House of Delegates, becoming the first openly transgender state legislator.

transgender

A person who identifies as or expresses a gender identity that differs from their sex at birth. Transgender persons differ from nonbinary persons, who may have a fluid identity that shifts between male and female or who may identify as neither male nor female.

CONCEPT CHECKS

1. What is the difference between sex and gender?

2. How do both biology and gender socialization contribute to differences between men and women?

3. What does it mean to say that gender is something we "do"? Give an example of a way that you have done gender in your daily life.

4. How can studies of gender in other cultures contribute to the argument that gender is socially constructed?

5. What is a nonbinary gender identity? How does it challenge the male–female sex dichotomy?

Other nations are ahead of the United States in this regard and have begun to collect official statistics on persons who identify as "third gender" or "third sex," terms that encompass diverse experiences, including identifying as transgender or nonbinary or being **intersex**. Intersex persons encompass those possessing both male and female genitalia or those with ambiguous genitalia. In 2011, Nepal became the first country to include a third gender category in its national census; India soon followed (Bochenek and Knight 2012). In 2015, New Zealand added the third gender category of "gender diverse" to its national statistics system (Price 2015).

In Germany, parents now have the option of not specifying a child's sex in birth registries. The intention is to allow babies born with biological characteristics of both sexes to make a choice about who they are once they get older. Under this new law, "individuals can . . . opt to remain outside the gender binary altogether" (Heine 2013). Some Canadian territories allow a nonbinary identification on birth certificates, while the nation allows residents to indicate their sex as such on their passports (Busby 2017). In 2017, both the District of Columbia and the state of Oregon began allowing residents to indicate their gender as nonbinary, using an "X" rather than the standard "M/F" on their driver's licenses. That same year, California became the first U.S. state to offer a nonbinary category on birth certificates (Caron 2017).

Taken together, anthropological and sociological studies of gender reveal that culture, not biology, underlies gender differences. Sociologists have noted that while society teaches "masculine" and "feminine" gender roles, such an approach does not explain where these roles come from or how they can be changed. For this, we need to delve into classic and contemporary theoretical perspectives that shed light on how gender roles and gendered inequalities are built into social institutions (Lorber 1994).

How Do Gender Inequalities Play Out in Social Institutions?

Recognize that gender differences are a part of our social structure and create inequalities between women and men. Learn the forms these inequalities take in social institutions such as the workplace, the family, the educational system, and the political system in the United States and globally.

Anthropologists and historians have found that most groups, collectives, and societies throughout history differentiate between women's and men's roles and afford more status, respect, and prestige to the latter. Of course, there are specific instances across most societies in which women have more social, economic, or political power than men. A key theme of intersectionality is that multiple overlapping identities like race, class, and gender shape opportunities and experiences. For instance, in 2015, white women ages twenty-five to thirty-five were more than twice as likely as Black men to have a college degree (Reeves and Guyot 2017). However, on the whole, men tend to have more power and privilege than women across most domains and across most societies.

Furthermore, the gender divide almost universally holds women responsible for child rearing and the maintenance of the home, while political and military activities tend to be resoundingly male. Nowhere in the world do men have primary responsibility for the rearing of children. Although men certainly care for children, stay-at-home dads make up a small statistical minority, barely reaching 4 percent in the United States today (Kramer, Kelly, and McCulloch 2015). Just because women and men perform different tasks or have different responsibilities in societies does not necessarily mean that women are unequal to men. However, if the work and activities of women and men are valued differently, then the division of labor between them can become the basis for unequal gender relations.

Male dominance in a society is usually referred to as **patriarchy**. Although men are favored in almost all of the world's societies, the degree of patriarchy varies widely across societies. In the United States, women have made tremendous progress in several realms, especially education and work. Yet, throughout the world, many cultures exist where women suffer tremendous disadvantages relative to men. For instance, Yemen has been ranked last in terms of women's equality by the World Economic Forum. The nation has no female members of parliament, and less than 10 percent of ministerial positions are held by women, while the gender gap in literacy and school attendance rates is among the widest in the world. In sharp contrast, Iceland has been ranked first, with its cultural and political emphasis on gender equality, excellent women's health, and near–gender parity on key economic and educational indicators (World Economic Forum 2017).

Sociologists define **gender inequality** as the differences in the status, power, and prestige women and men have in groups, collectives, and societies. In thinking about gender inequality between men and women, we can ask the following questions: Do women and men have equal access to valued societal resources, for example, food, money, power, and time? Second, do women and men have similar life options? Third, are women's and men's roles and activities valued similarly? We examine the various forms of gender inequality in educational systems, the workplace, the home, and politics in the following sections. As you read, keep these questions in mind.

Education

If you look around your college campus, you might notice roughly equal numbers of men and women and may think that gender no longer affects whether and how one receives an education. There is some truth to this. College campuses today are roughly fifty-fifty when it comes to the number of men and women filling undergraduate classrooms; in fact, women slightly outnumber men on college campuses today. This gender gap is much larger among Blacks and Latinos than among whites (Musu-Gillette et al. 2017). Yet aggregate numbers are only part of the story. As we will see next, subtle dynamics, starting in primary school, teach boys and girls different skills and direct young men and women into divergent career paths.

UNEQUAL TREATMENT IN THE CLASSROOM

Sociologists have found that schools help foster gender differences in outlook and behavior. Studies document that teachers interact differently—and often inequitably—with their male and female students. These interactions differ in at least two ways: the frequency of teacher–student interactions and the content of those interactions. Both of the patterns are based on—and perpetuate—traditional assumptions about male and female behavior and traits.

intersex

An individual possessing both male and female genitalia. Although statistically rare, this subpopulation is of great interest to gender scholars.

patriarchy

The dominance of and privilege afforded to men over women. All known societies are patriarchal, although there are variations in the degree and nature of the power men exercise and are bestowed, relative to women.

gender inequality

The inequality between men and women in terms of wealth, income, and status.

Male students get more attention from their teachers than female students. By treating their male and female students differently, teachers—and schools—reinforce traditional gender roles.

One study showed that regardless of the sex of the teacher, male students interacted more with their teachers than female students did. Boys received more teacher attention and instructional time than girls did. This was due in part to the fact that boys were more demanding than girls (American Association of University Women 1992). Another study reported that boys were eight times more likely to call out answers in class, thus grabbing their teachers' attention. This research also showed that even when boys did not voluntarily participate in class, teachers were more likely to solicit information from them than from girls. When girls tried to bring attention to themselves by calling out in class without raising their hands, they were reprimanded (Sadker and Sadker 1994). Boys were also disadvantaged in important ways. Because of their rowdy behavior, they were more often scolded and punished than the female students. These patterns can have long-term effects; punishment, especially the most severe forms like school suspension, are linked with poorer grades, lower graduation rates, and ultimately poorer prospects for gainful employment (e.g., Shollenberger 2014).

This differential treatment of boys and girls perpetuates stereotypic gender role behavior. Girls are trained to be quiet and well behaved and to turn to others for answers, whereas boys are encouraged to be inquisitive, outspoken, active problem solvers.

THE GENDERING OF COLLEGE MAJORS

College is a time of exploration, when students take both general education classes and specialized classes within their chosen majors that prepare them for a career after graduation. Men and women differ dramatically in the majors they choose, opting for fields that are consistent with gender-typed socialization; even in the twenty-first century, women are more likely to focus on fields associated with caring and nurturing, whereas men tend to pursue fields that emphasize logic and analysis. Yet the majors women tend to choose are precisely those fields that garner the lowest earnings after graduation, whereas men are channeled into majors with high economic returns.

Researchers at Georgetown University documented gender differences in college majors using national data collected by the Census Bureau (Carnevale, Strohl, and Melton 2011) (Table 9.1). They found that the college majors with the highest proportion of female students were those in the education and health fields. For instance, 97 percent of all persons majoring in early childhood education were women, and more than 90 percent of all majors in nursing, elementary education, library science, and school counseling were female. By contrast, more than 90 percent of engineering majors were men.

Feminist scholars note that this stark gender segregation among college majors is one important reason for the persistent gender gap in pay. When the Georgetown researchers ranked college majors based on the earnings of graduates, they found that the two lowest-earning majors were counseling psychology (with median annual earnings of $29,000) and early childhood education (with median annual earnings of $36,000). Both of these fields are dominated by women. By contrast, of the ten highest-earning majors, eight were engineering fields heavily dominated by men, including petroleum engineering (with median

TABLE 9.1

The Gendering of College Majors

MAJORS WITH HIGHEST CONCENTRATION OF WOMEN			MAJORS WITH HIGHEST CONCENTRATION OF MEN		
	Median Earnings	Percentage Women		Median Earnings	Percentage Men
Early Childhood Education	$36,000	97	Naval Architecture and Marine Engineering	$82,000	97
Medical Assisting Services	$56,000	96	Mechanical Engineering Related Technologies	$80,000	94
Communication Disorders Sciences and Services	$40,000	94	Construction Services	$70,000	92
Family and Consumer Sciences	$40,000	93	Electrical and Mechanic Repairs and Technologies	$57,000	91
Nursing	$60,000	92	Industrial Production Technologies	$65,000	91
Elementary Education	$40,000	91	Mechanical Engineering	$80,000	90
Nutrition Sciences	$46,000	89	Mining and Mineral Engineering	$80,000	90
Special Needs Education	$42,000	88	Electrical Engineering Technology	$68,000	90

Source: Carnevale, Strohl, and Melton 2011.

annual earnings of $120,000) and aerospace engineering (with median annual earnings of $87,000). Research on how young people choose their majors consistently shows that subtle forces, including input from parents, friends, and guidance counselors; a lack of same-sex role models; active encouragement (or discouragement) from teachers; and limited exposure to particular fields of study tend to channel women into female-typed majors and men into male-typed majors (Morgan, Gelbgiser, and Weeden 2013; Porter and Umbach 2006).

Women and the Workplace

Rates of employment of women outside the home, for all social classes, were quite low until well into the twentieth century in the United States. Even as late as 1910 in the United States, more than one-third of gainfully employed women were maids or house servants. The female labor force consisted mainly of young single women and children. When women or girls worked in factories or offices, employers often sent their wages straight home to their parents. When they married, they withdrew from the labor force.

Since the turn of the twentieth century, women's participation in the paid labor force has risen more or less continuously, especially in the past fifty years (see Figure 9.1). In 2017, 57 percent of women ages sixteen and older were in the labor force, compared to 38 percent in 1960. An even greater change in the rate of labor force participation has occurred among married mothers of young children. In 1975, only 39 percent of married

FIGURE 9.1

Women's Participation in the Labor Force in the United States

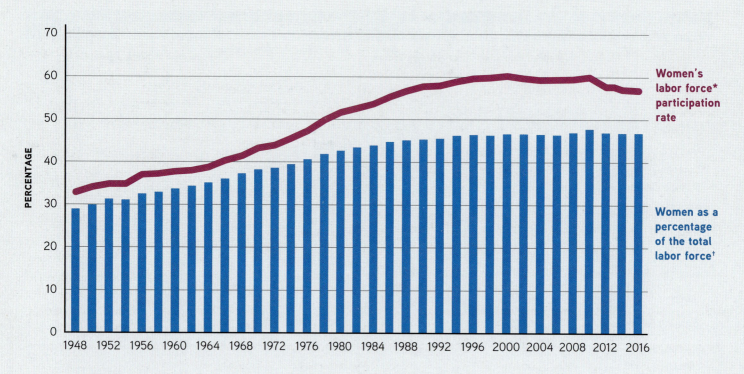

*Women in the labor force as a percent of all civilian women ages sixteen and over.

†Women in the labor force as a percent of the total labor force (both men and women) ages sixteen and over.

Source: U.S. Bureau of Labor Statistics 2018e.

women with preschool-age children (under six years old) were in the labor force; this figure had increased to 63 percent by 2016 (U.S. Bureau of Labor Statistics 2017e).

How can we explain this increase? One force behind women's increased entry into the labor force was the increase in demand, since 1940, for clerical and service workers, as the U.S. economy expanded and changed (Oppenheimer 1970). From 1940 until the mid- to late 1960s, labor force activity increased among women who were past their prime child-rearing years. During the 1970s and 1980s, as the marriage age rose, birthrates declined, and women's educational attainment increased, the growth in labor force participation spread to younger women. Many women now postpone family formation to complete their education and establish themselves in the labor force. Despite family obligations, today a majority of women of all educational levels now work outside the home during their child-rearing years (Damaske and Frech 2016).

INEQUALITIES AT WORK

Until recently, women workers were overwhelmingly concentrated in routine, poorly paid occupations. The fate of the occupation of clerk (office worker) provides a good illustration. In 1850 in the United States, clerks held responsible positions, requiring accountancy skills and carrying managerial responsibilities; less than 1 percent were women. The twentieth century saw a general mechanization of office work (starting with the introduction of the typewriter in the late nineteenth century), accompanied by a marked downgrading of the status of clerk—together with a related occupation, secretary—into a routine, low-paid occupation. Women filled these occupations as the pay and prestige of such jobs declined. Today, most secretaries and clerks are women. Once an occupation has become **gender typed**—once it is seen as mainly a "man's job" or a "woman's job"—inertia sets in.

Women have recently made some inroads into occupations once defined as "men's jobs." By the 1990s, women constituted a majority of workers in previously male-dominated professions such as accounting, journalism, psychology, public service, and bartending. In fields such as law, medicine, and engineering, their proportion has risen substantially since 1970. While women's employment in professional and managerial occupations has steadily increased to be the largest occupational category for women workers (31.4 million in 2017), significant numbers of women workers are still employed in sales and office administrative support occupations (20.6 million in 2017) (U.S. Bureau of Labor Statistics 2018c).

Another important economic trend since the 1970s has been the narrowing of the gender gap in earnings (see Figure 9.2). Between 1979 and 2017, the ratio of women's to men's median weekly earnings among full-time, year-round workers increased from 62 percent to 82 percent. Moreover, this ratio increased among all races and ethnic groups (U.S. Bureau of Labor Statistics 2018b). Some researchers have noted that the narrowing of the gender gap is less a reflection of improvement in women's economic standing than a decline in men's economic standing. The recent recession was dubbed the "he-cession" or "man-cession" because the types of jobs and industries hardest hit were those in which men were overrepresented, such as construction and finance (Rampell 2009). As men's earnings erode, the female–male earnings ratio starts to inch upward.

Sociologists have identified many reasons why a gender pay gap persists. Although direct discrimination is certainly one explanation, there are other, more subtle reasons why women typically earn less than their male counterparts. Many sociologists point to **sex segregation**—or the concentration of men and women in different occupations—as an important cause of the gender gap in earnings. While the Equal Pay Act of 1963 holds that men and women must earn equal pay for performing equal work, women tend to hold different jobs, typically jobs that are dominated by women. These jobs, on average, pay less than occupations dominated by men. For instance, in 2017, occupations with the highest proportion of women included preschool and elementary school teacher (98 percent), dental hygienist (95 percent), secretary and administrative assistant (95 percent), childcare worker (94 percent), and hairdresser

gender typing

Designation of occupations as male or female, with "women's" occupations, such as secretarial and retail positions, having lower status and pay and "men's" occupations, such as managerial and professional positions, having higher status and pay.

—

sex segregation

The concentration of men and women in different jobs. These differences are believed to contribute to the gender pay gap.

FIGURE 9.2

The Gender Pay Gap, 1979–2017

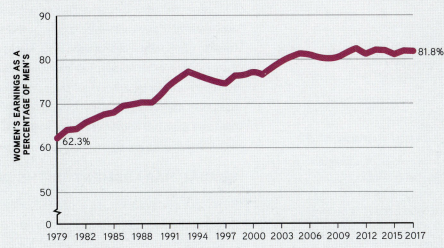

Note: Data relate to annual averages of median usual weekly earnings for full-time workers.

Source: U.S. Bureau of Labor Statistics 2018b.

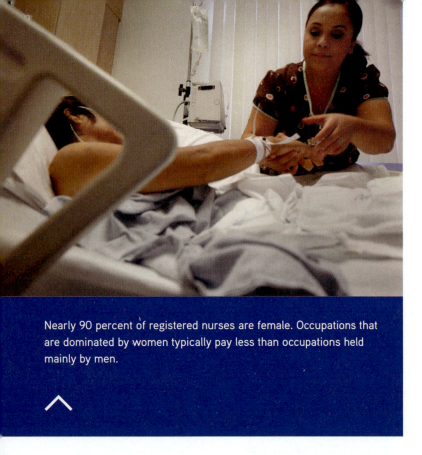

Nearly 90 percent of registered nurses are female. Occupations that are dominated by women typically pay less than occupations held mainly by men.

(93 percent) (U.S. Bureau of Labor Statistics 2018c). This is not surprising, given what we learned earlier about the concentration of women in health- and education-related college majors.

THE GLASS CEILING

Although women are increasingly moving into "traditionally male" jobs, their entry into such jobs is not necessarily accompanied by increases in pay—and increases in occupational mobility—because of the "glass ceiling." The **glass ceiling** is a promotion barrier that prevents a woman's upward mobility within an organization. The glass ceiling is particularly problematic for women who work in male-dominated occupations and professions. Women's progress is blocked not by virtue of innate inability or lack of basic qualifications but by not having the sponsorship of well-placed, powerful senior colleagues to articulate their value to the organization or profession (Baker and Cangemi 2016). As a result, women tend to progress to mid-level management positions, but they do not, in proportionate numbers, move beyond mid-management ranks. The obstacles women face as they strive to rise through the career ranks are exemplified by recent high-visibility lawsuits on Wall Street where women have successfully sued their employers for discrimination. For instance, Judy Calibuso, a financial adviser at Merrill Lynch, and her colleagues charged that their employer gave higher-profile and more profitable clients to male colleagues, leading to bigger annual bonuses and more rapid ascents up the corporate ladder. Despite their requests for more challenging and lucrative opportunities, Calibuso and her coworkers hit a glass ceiling and were forced to watch their male colleagues climb past them in the corporate ranks (McGeehan 2013).

Even in female-dominated fields, women are often left behind in less prestigious and lower-paying jobs while their male colleagues ascend the corporate ladder. Sociologist Christine Williams (1992, 1993) has written about the "glass escalator," whereby men—especially heterosexual white men—in traditionally female occupations are placed on a fast track to promotion. For instance, male teachers may be more likely to become principals, and male cashiers or restaurant workers are tracked into assistant manager or manager positions at higher and faster rates than their female colleagues. Men who are fast-tracked through the ranks are not necessarily looking for or asking for promotions; rather, the promotions often are based on employers' stereotypical beliefs regarding men's superior leadership capabilities or dedication to their careers over their family responsibilities (Goudreau 2012).

SEXUAL HARASSMENT IN THE WORKPLACE

Economic disadvantage, lack of mentorship, barriers to promotion, and daunting challenges in balancing work and family are not the only challenges women workers face. Another pervasive obstacle is sexual harassment, a topic brought into sharp focus by the #MeToo movement. **Sexual harassment** is unwanted or repeated sexual advances, remarks, or behaviors that are offensive to the recipient and cause discomfort or interfere with job

glass ceiling

A promotion barrier that prevents a woman's upward mobility within an organization.

sexual harassment

The making of unwanted sexual advances by one individual toward another, in which the first person persists even though it is clear that the other party is resistant.

performance. Power imbalances facilitate harassment; even though women can and do sexually harass subordinates, because men usually hold positions of authority, it is more common for men to harass women (Uggen and Blackstone 2004). Those power imbalances were stunningly apparent in the Harvey Weinstein sexual harassment case. A Hollywood magnate with the power to launch or destroy careers, Weinstein allegedly "blacklisted" actresses, including Oscar winner Mira Sorvino after she rejected his advances, telling other directors not to hire her.

Harassment clearly takes a toll on its victims' emotional well-being, yet it also can do long-lasting harm to their careers. A recent mixed-method study tracked participants in the multiwave Youth Development Study and carried out open-ended interviews. The researchers discovered that many women left their jobs to escape the harassment and the menacing work environment; these workplace exits, in turn, increased the women's financial stress and hurt their career prospects in the longer term (McLaughlin, Uggen, and Blackstone 2017). The researchers spoke to a former advertising agency project manager named Lisa, who told them that after she was harassed repeatedly by a coworker: "I quit, and I didn't have a job. That's it, I'm outta here. I'll eat rice and live in the dark if I have to" (p. 345). Other women quit rewarding jobs to take less desirable positions that felt safer, or even jobs where they would have minimal contact with other people to reduce their risk of further mistreatment. As Pam, a former accountant, told the interviewers, she switched to a job in tech even though "I had no interest in computer hardware whatsoever. . . . I went to a position where I am pretty much solitary. I work by myself. Which is the way that I want it" (p. 347). Yet these strategies, seen as self-protective in the short term, ultimately took a toll on Pam's and Lisa's career prospects and financial stability.

The U.S. courts have identified two types of sexual harassment. One is the quid pro quo, in which a supervisor demands sexual acts from a worker as a job condition or promises work-related benefits in exchange for sexual acts. That was the experience of Ford Motor Company worker Miyoshi Morris. The young mother's shift started at 6:00 AM, and she hadn't been able to find a day care center open early enough to care for her children at that hour. Her manager in the paint department told Miyoshi that he would rearrange her work schedule—in exchange for sex. As Miyoshi told the *New York Times*, "I was so lost, afraid, and realizing I had children to care for." She recalled thinking at that time, "Where else are you going to go and make this kind of money?" (Chira and Einhorn 2017).

The other type of sexual harassment is the "hostile work environment," in which a pattern of sexual language, lewd posters, or sexual advances makes a worker so uncomfortable that it is difficult for the worker to do his or her job (Padavic and Reskin 2002). Many of the women plaintiffs in the Ford Motor suit experienced intolerable hostility in the workplace: repeatedly being groped or bitten by coworkers, being called lewd nicknames, having male coworkers rub up against them, or witnessing male coworkers grab their crotches when their female coworkers walked by. Making matters worse, women who reported these indignities to their supervisors were ignored or even punished, losing basic privileges like bathroom breaks or being threatened with demotion or job loss (Chira and Einhortn 2017).

Historically, women have been reluctant to report these incidents. Sociologists have observed that "the great majority of women who are abused by behavior that fits legal definitions of sexual harassment—and who are traumatized by the experience—do not label what has happened to them as sexual harassment" (Paludi and Barickman 1991). Women's reluctance to report harassment is due to the following factors: (1) Many still do not recognize that sexual harassment is an actionable offense; (2) victims may be reluctant to come forward with complaints, fearing that they will not be believed, that their charges will not be taken seriously, or that they will be subject to reprisals; and (3) it may be difficult to differentiate between harassment and joking on the job (Giuffre and Williams 1994). The #MeToo movement may be a watershed moment, ushering in a time when women feel supported and empowered enough to report experiences of workplace harassment.

ECONOMIC INEQUALITY IN GLOBAL PERSPECTIVE

The United States is not alone in having a history of gender inequality in the workplace. Across the globe, men outpace women in most workplace and economic indicators. Most nations, however, like the United States, have witnessed tremendous strides in women's economic progress in recent decades. Some scholars and activists argue that women's economic empowerment has contributed in large part to China's meteoric rise as an economic power. An estimated 80 percent of the factory workers in China's Guangdong province are female, and 38 percent of the richest self-made women in the world are Chinese (Kristof and WuDunn 2009; Forbes 2017).

The International Labour Organization (2016) found that the gap between the labor force participation rates of men and women decreased only slightly between 1995 and 2015, mainly because both women's and men's participation rates have fallen. In 2015, 76 percent of men but just 50 percent of women participated in the global workforce. Women have high labor force participation rates in high-income countries, such as most European nations, where more than two-thirds of the female adult population participates in the labor market and the male–female gap in labor force participation rates is less than 15 percent on average. This is especially true in nations with extensive social benefits (such as paid maternity leave) and where part-time work is possible. In North Africa and Southern Asia, by contrast, the gender gap in labor force participation is nearly 50 percent.

The feminization of the global workforce has brought with it the increased exploitation of young, uneducated, largely rural women around the world. These women labor under conditions that are often unsafe and unhealthy, at low pay and with nonexistent job security. For instance, the collapse of the Rana Plaza building in Bangladesh in 2013 was a tragic depiction of the unsafe work conditions facing garment workers, most of whom are women, in parts of the developing world. More than 1,000 workers died in the building collapse, many of whom had been toiling for low wages in a building that had been deemed an unsafe structure.

At the other end of the occupational spectrum, a recent study by the International Labour Organization concludes that women throughout the world still encounter a glass ceiling that restricts their movement into top positions. Globally, 24 percent of senior roles are held by women (Grant Thornton 2018). In recognition of these low rates, several national governments have recently passed legislation to increase women's participation in the highest echelons of business. In Japan, for example, women have been particularly likely to face barriers to career advancement, especially in professional and managerial positions (Cunningham 2013; Simms 2013). However, policymakers in Japan have recently recognized that this is a tremendous loss of worker potential, especially when low birthrates mean that the nation may soon face a dearth of young workers. The Japanese government has formally stated goals to increase women's representation in management. Their target is to have women hold 30 percent of all upper-management positions in major corporations, a near-tripling of the current rate of 11 percent (Cunningham 2013).

Gender Inequality

The Gender Inequality Index (GII), which is used to compare gender inequality across countries, looks at women's educational attainment, labor force participation, and representation in governmental bodies, among other metrics. In the graphic below, the countries' GII rankings are displayed in the circles.

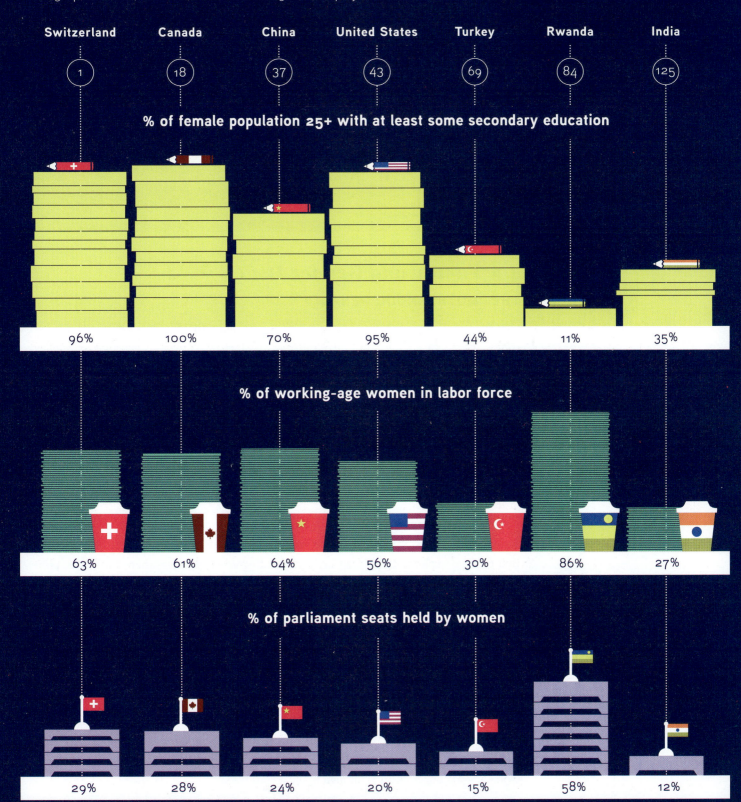

Switzerland	Canada	China	United States	Turkey	Rwanda	India
1	18	37	43	69	84	125

% of female population 25+ with at least some secondary education

| 96% | 100% | 70% | 95% | 44% | 11% | 35% |

% of working-age women in labor force

| 63% | 61% | 64% | 56% | 30% | 86% | 27% |

% of parliament seats held by women

| 29% | 28% | 24% | 20% | 15% | 58% | 12% |

Source: United Nations Development Programme 2017.

Female workers watch the launch of a ship into China's Yangtze River. Thanks to the feminization of the global workforce, many women have achieved a degree of economic independence—but often at the price of unsafe labor conditions, low pay, and nonexistent job security.

The Family and Gender Issues

THE "MOTHERHOOD PENALTY"

One key factor affecting women's careers is the perception that for female employees, work comes second to having children. Research by Stanford University sociologist Shelley Correll and colleagues (2007) found that mothers are 44 percent less likely to be hired than childless women with the same work experience and qualifications. Not only that, but mothers are offered significantly lower starting pay for the same job than equally qualified women without children (an average of $11,000 lower in this study). This earnings gap between the two groups—referred to as the "motherhood penalty"—reflects a belief by employers in traditional gender roles and notions of parenthood. Because women are still considered the primary caregiver, they are perceived by employers as less reliable and less productive workers.

Public policies may be effective in counteracting the obstacles imposed by employers' stereotypical views of mothers in the workplace. A team of sociologists led by Michelle Budig (2012) explored the ways that attitudes toward working mothers and public policies such as parental leave or public childcare affected the earnings of women workers in twenty-two countries. They found that the "motherhood penalty" varied widely across countries; gaps were smallest in nations where cultural attitudes supported maternal employment and the government provided job-protected parental leave and publicly funded childcare.

HOUSEWORK AND THE SECOND SHIFT

The struggles facing women workers do not end once they set foot in their homes after work. Women throughout the world also perform housework and childcare at the end of the paid work day, dubbed the **"second shift"** by sociologist Arlie Hochschild. As a result, women work longer hours than men in most countries.

second shift

The excessive work hours borne by women relative to men; these hours are typically spent on domestic chores following the end of a day of work outside the home.

Although there have been revolutionary changes in women's status in recent decades in the United States, including the entry of women into male-dominated professions, one area of work has lagged far behind: **housework**. Because of the increase of married women in the workforce and the resulting change in status, it was presumed that men would contribute more to housework. And while men now do more housework than they did three decades ago, a large gender gap persists. As Figure 9.3 shows, women still put in significantly more time than their male counterparts (Pew Research Center 2013a).

Further investigation shows that it is the intersection of gender, marital status, and parental status that most powerfully shapes housework. A recent study showed that whereas women save their husbands an hour of housework a week, husbands create an additional seven hours of housework for their wives every week. Childless women do an average of ten hours of housework a week before marriage and seventeen hours after marriage. Childless men, by contrast, do eight hours before marriage and seven hours afterward. Married women with more than three kids are the most overworked, reporting an average of about twenty-eight hours per week, while married men with more than three kids logged only ten hours (University of Michigan Institute for Social Research 2008).

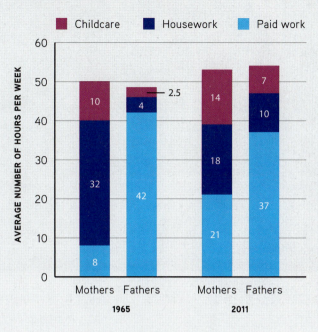

FIGURE 9.3

The Second Shift

Source: Pew Research Center 2013a.

Some sociologists have suggested that this phenomenon is best explained as a result of economic forces: Household work is exchanged for economic support. Because women typically earn less than men, they are more likely to remain economically dependent on their husbands and thus perform the bulk of the housework. Until the earnings gap is narrowed, women in heterosexual unions will likely remain in their dependent position. Recent evidence suggests same-sex couples maintain a much more egalitarian division of labor in the home, regarding both housework and childcare, perhaps reflecting the fact that partners in same-sex relations tend to have less of a pronounced earnings gap than do partners in heterosexual partnerships (Goldberg, Smith, and Parry-Jenkins 2012). For example, one recent study of 103 same-sex and 122 straight couples found that same-sex couples tend to share more duties and assign chores based on each partner's personal preference, whereas straight couples tend to revert to traditional gender roles, with women, lower earners, and those with fewer work hours bearing the brunt of stereotypical female chores (Matos 2015).

In the coming decades, experts anticipate that we may see greater gender equity when it comes to housework and childcare. A recent study of young adults between the ages of eighteen and thirty-two in the United States found that the majority of respondents—regardless of gender or education level—aspired to a romantic relationship in which they would share earning and household/caregiving responsibilities equally with their partners (Pedulla and Thébaud 2015).

housework

Unpaid work carried on in the home, usually by women; domestic chores such as cooking, cleaning, and shopping. Also called "domestic labor."

Gender Inequality in Politics

Women are playing an increasingly important role in U.S. politics, although they are still far from achieving full equality. Before 1993, there were only two women in the U.S. Senate (two of 100 Senate members) and twenty-nine in the U.S. House of

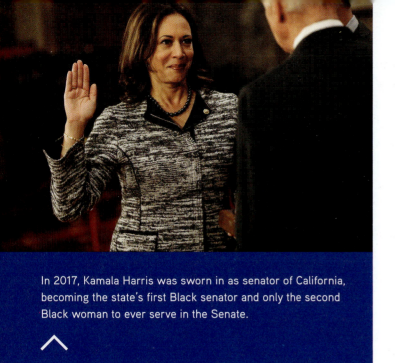

In 2017, Kamala Harris was sworn in as senator of California, becoming the state's first Black senator and only the second Black woman to ever serve in the Senate.

Representatives (out of 435). Less than a decade later—in 2001—there were thirteen women in the Senate and fifty-nine in the House. In 2018, 107 women held seats in the U.S. Congress, comprising 20 percent of the 535 members; twenty-three women (23 percent) serve in the Senate, and eighty-four women (19 percent) serve in the House. Only six state governors are women (Center for American Women and Politics 2018b). The U.S. Supreme Court had its first woman justice appointed in 1981 and its second twelve years later. Three women currently occupy seats on the Supreme Court—Ruth Bader Ginsburg, Elena Kagan, and Sonia Sotomayor—marking an all-time high.

Hopes for electing the first female president of the United States were dashed in November 2016 when Hillary Clinton, the Democratic candidate for president, lost to Republican candidate Donald Trump. The reasons behind this outcome are many and complex, yet compelling data point to sexism as an important contributor. The gender gap in voting was much wider than in past presidential elections, with fully 54 percent of women but just 41 percent of men supporting Clinton (Zillman 2016). Political scientists have also examined other characteristics of those who voted for Clinton versus Trump and found that attitudes on a "hostile sexism" scale correlated strongly with support for Trump. For instance, people more likely to endorse beliefs like "Many women are actually seeking special favors, such as hiring policies that favor women over men, under the guise of asking for equality" were more likely to support Trump over Clinton (Schaffner, MacWilliams, and Nteta 2017).

Like Hillary Clinton, the vast majority of women politicians are affiliated with the Democratic Party. In the U.S. Congress, 73 percent of women are Democrats, and 61 percent of women in state legislatures are Democrats (Center for American Women and Politics 2018a). However, the Republican Party prominently features women leaders, including Alaska senator Lisa Murkowski and Maine senator Susan Collins.

Typically, the more local the political office, the more likely it is to be occupied by a woman. Men outnumber women in politics at all levels, but the gender gap is smaller among mayors and elected members of city and county governing boards. In most states, women are more likely to serve as representatives at the local level than the state level and even less likely to serve as senators or members of the House of Representatives. The farther from home the political office, the more likely it is to be regarded as "man's work," providing a living wage, full-time employment, and a lifetime career. The costs of running for local office are typically far lower than a campaign for a higher office, which can require a war chest beyond the reach of many women (Conway 2004).

The number of women running for public office increased dramatically starting in 2017. Many of these women were newcomers to politics who were disappointed by Hillary Clinton's loss, infuriated by boastful claims by President Trump that he had grabbed women's genitals against their will, and inspired by women's marches to work toward gender-equitable policies. According to the Center for American Women and Politics, the number of women exploring runs for governor on the heels of the Trump election eclipsed the all-time high set in 1994. They also estimated that the number of Democratic women intending to challenge incumbents in the U.S. House of Representatives spiked during the same period (Alter 2018).

GENDER AND POLITICS: GLOBAL PERSPECTIVE

Women play an increasing role in politics throughout the world. Yet of the 192 countries that belong to the United Nations, only seventeen were headed by women in January 2017 (UN Women 2017). Since World War II, thirty-eight countries have been headed by women. As of the beginning of 2018, women made up only 23 percent of the combined membership of the national legislatures throughout the world (Inter-Parliamentary Union 2018). In Rwanda, women represent 56 percent of the national legislature—the highest proportion of any country. Regionally, female representation in national legislatures is highest in the Nordic countries (41 percent); in the Arab states, the figure is just 18 percent (Inter-Parliamentary Union 2018).

The United Nations ranks countries according to a measure of gender inequality, called the Gender Inequality Index (GII), which covers three dimensions of inequality: reproductive health, including maternal mortality rate; empowerment, including shares of seats in parliament held by women; and participation in the labor force (see the "Globalization by the Numbers" infographic on p. 267). By this measure, in 2015, the United States ranked 43rd out of 188 countries—behind the Scandinavian and other northern and western European countries as well as Canada, the United Kingdom, New Zealand, Japan, and China.

Why Are Women the Target of Violence?

Violence directed against women is found in many societies, including the United States. One out of three women worldwide has been beaten, coerced into sex, or abused in some other way—most often by someone she knows, including her husband or a male relative (World Health Organization [WHO] 2017). And as we read earlier in this chapter, sexual violence in the workplace is pervasive yet has only recently captured national attention.

Although sexual assault at the hands of strangers triggers panic and public concern, intimate partner violence (IPV), or violence committed against women by their romantic partners, is the most common form of violence against women; we discuss this topic more fully in Chapter 11. The prevalence of intimate partner violence is highest in Southeast Asia at 38 percent, compared to 23 percent in high-income countries (WHO 2017). In 2012 alone, an estimated 8,233 women were killed in dowry-related deaths in India (*Times of India* 2013); these are deaths of young women who are murdered or driven to suicide by continuous harassment and torture by husbands and in-laws in an effort to extort more resources and property from a bride when she marries. According to the UN Population Fund, an estimated 5,000 women are the victims of "honor killings" per year worldwide

CONCEPT CHECKS

1. What are two explanations for the stark gender segregation in college majors? How does this segregation contribute to the gender pay gap?

2. Describe at least three examples of how gender inequalities emerge in the workplace. How would a sociologist explain these inequities?

3. How do inequalities in the home, especially with regard to housework and childcare, reflect larger gender inequities in society?

4. What are some important differences between men's and women's political participation in the United States?

Learn about the specific ways that women are the target of physical and sexual violence in the United States and globally.

(United Nations 2011). Globally, it is estimated that as many as 38 percent of murdered women are killed by their intimate partners (WHO 2017).

Ritualized violence is also a common experience among women throughout the world. For example, more than 200 million girls and women alive today have been subjected to genital cutting, a practice done to control women's sexuality (UNICEF 2016). In addition, millions of women and girls throughout the world are "missing," partly as the result of female **infanticide** in cultures where boys are more highly valued than girls (World Health Organization 2000).

The trafficking of women for forced prostitution, which has been characterized as the largest slave trade in recent history, appears to be a growing problem (Basu 2014). The United Nations Global Initiative to Fight Human Trafficking (2008) estimates that at any given time, 2.5 million people worldwide are subject to forced labor as a result of human trafficking. Forty-three percent are exploited for forced sexual exploitation, and 98 percent of those persons are women or girls.

In the United States, many scholars argue that the increased depiction of violence in movies, on television, and elsewhere in American popular culture contributes to a climate that normalizes male aggression against women. The most common manifestation of violence against women is sexual assault, although stalking, cyberstalking, and sexual harassment are increasingly seen as forms of psychological (if not physical) violence as well.

Rape

Rape can be sociologically defined as the forcing of nonconsensual vaginal, oral, or anal intercourse. As one researcher observed, between consensual sex and rape lies "a continuum of pressure, threat, coercion, and force" (Kelly 1987). Common to all forms of rape is the lack of consent: At least in principle, "no" means "no" when it comes to sexual relations in most courts of law in the United States. The vast majority of sexual assaults are committed by men against women, although mounting evidence suggests that men also are victims, and gay and transgender men are particularly vulnerable. Early research documented cases where women take sexual advantage of young men who may be insecure, intoxicated, or of a lower-status position (Anderson and Struckman-Johnson 1998).

It is difficult to know with accuracy how many rapes actually occur, since rapes so often go unreported. According to the National Intimate Partner and Sexual Violence Survey, nearly one in five women (19.3 percent) and 1.7 percent of men have been raped in their lifetimes. When a broader definition of victimization is used to encompass all forms of sexual violence, including unwanted sexual contact, these statistics jump to 44 percent of women and 23 percent of men (Smith et al. 2017). Although all sexual assault is taboo, young men are particularly reluctant to come forth and report their experiences. Many feel stifled because they have internalized the belief that men should be strong, stoic, and able to defend themselves, while those assaulted by other men fear the stigma associated with sex between two men (Kassie 2015).

Rape is an act of violence, committed to wield power and control over its victim. Sexual assault is often carefully planned rather than performed on the spur of the moment to satisfy some uncontrollable sexual desire. Many rapes involve beatings, knifings, and even murder. In some instances, sexual assault is facilitated by alcohol or women having their drinks spiked with the sedative Rohypnol (i.e., "roofies") or drugs referred to as

infanticide

The intentional killing of a newborn. Female babies are more likely than male babies to be murdered in cultures that devalue women.

rape

The forcing of nonconsensual vaginal, oral, or anal intercourse.

date-rape drugs (Michigan Department of Community Health 2010). Even when rape leaves no physical wounds, it is a highly traumatic violation that leaves long-lasting psychological scars.

Most rapes are committed by relatives (fathers or stepfathers, brothers, uncles), partners, or acquaintances. Among college students, most rapes are likely to be committed by boyfriends, former boyfriends, or classmates. The National Institute of Justice Campus Sexual Assault study presents a chilling picture of violence against women on campuses across the country (Krebs et al. 2007). The study asked college women about their experience with rape, coerced sex, and unwanted sexual contact. Over 3 percent (3.4 percent) of female students surveyed reported having been raped since entering college. A previous study conducted by the same organization found that for both completed and attempted rapes, nine out of ten offenders were known to the victim. The incidence of other forms of victimization reported in the study was substantially higher than that of rape. Overall, 29 percent of female students reported having experienced attempted or completed sexual assault (Krebs et al. 2007).

Sexual Violence against Women: Evidence of "Rape Culture"?

Some feminist scholars claim that men are socialized to regard women as sex objects, to feel a sense of sexual entitlement, and to instill fear in women by dominating them (Brownmiller 1986). This socialization context, described as a **rape culture** by Susan Brownmiller (1986), may make men insensitive to the difference between consensual and nonconsensual sex and thus contribute to the high levels of sexual violence against women. Rape culture is closely tied to the concept of **toxic masculinity**, or a cluster of potentially destructive values or behaviors that have historically been a part of boys' socialization, including "domination, the devaluation of women, homophobia and wanton violence" (Kupers 2005, p. 714).

The fact that "acquaintance rapes" occur suggests that at least some men are likely to feel entitled to sexual access if they already know the woman. A survey of nearly 270,000 first-year college students reported that 55 percent of male students agreed with the statement "If two people really like each other, it's all right for them to have sex even if they've known each other only for a very short time." Only 31 percent of female students were in agreement, suggesting a rather large gender gap concerning notions of sexual entitlement (American Council on Education 2001). When a man goes out on a date with sexual conquest on his mind,

rape culture

Social context in which attitudes and norms perpetuate the treatment of women as sexual objects and instill in men a sense of sexual entitlement.

—

toxic masculinity

A cluster of potentially destructive values or behaviors that historically have been part of boys' socialization, such as the devaluation of and aggression toward women.

Emma Sulkowicz, a student at Columbia University, carried a fifty-pound mattress for her entire senior year, drawing attention to the issue of rape on college campuses.

CONCEPT CHECKS

1. Name three different kinds of violence against women.

2. How common is violence against women?

3. Why are women more likely than men to be the targets of sexual violence?

4. What is rape culture, and why is it so pervasive in contemporary society?

>

Think about various explanations for gender inequality. Learn some feminist theories about how to achieve gender equality.

he may force his attentions on an unwilling partner, overcoming her resistance through the use of alcohol, persistence, or both. While such an act may not be legally defined as rape, many women would experience it as such.

In the past decade, college campuses have witnessed myriad acts that reveal the pervasiveness of rape culture. For instance, in September 2014, a picture from a frat party at a large southern public university went viral. The photo featured a large banner with the following words in bright red letters: "no means yes, yes means anal" (Kingkade 2014). When outrage followed on campus and administrators reached out to the frat brothers, their half-hearted apology letter read, "What was written on the prop for the party was an inappropriate attempt at humor by 18- to 21-year-old men. In no way do I believe that the sign reflected our views towards our female guests or rape culture in general" (Sharp et al. 2017). Yet this is not an isolated incident or just a boyish prank; rather, it exemplifies some men's socialization into the norms of toxic masculinity. In his book *Guyland* (2008), gender scholar Michael Kimmel explains why rape culture is so rampant, even among students at competitive colleges today. Young men are insecure with their own masculinity and even threatened by the rising numbers of highly competent women on campus. These men, says Kimmel, "are endlessly trying to prove their masculinity to the other guys." Taking sexual advantage of women enables insecure men to demonstrate their masculinity while maintaining power over women (Kimmel 2010).

How Does Social Theory Explain Gender Inequality?

Investigating and accounting for gender inequality has become a central concern of sociologists. Many theoretical perspectives have been advanced to explain men's enduring dominance over women—in the realm of economics, politics, the family, and elsewhere. In this section, we will review the main theoretical approaches to explaining the nature of gender inequality at the level of society.

Functionalist Approaches

As we saw in Chapter 1, the functionalist approach sees society as a system of interlinked parts that, when in balance, operate smoothly to produce social solidarity. Thus, functionalist and functionalist-inspired perspectives on gender seek to show that gender differences contribute to social stability and integration. Though such views once commanded great support, they have been heavily criticized for neglecting social tensions at the expense of consensus and for promoting a conservative view of the social world.

Talcott Parsons, a leading functionalist thinker, concerned himself with the role of the family in industrial societies (Parsons and Bales 1955). He was particularly interested in the socialization of children and believed that stable, supportive families are the key to successful socialization. In Parsons's view, the family operates most efficiently with a clear-cut sexual division of labor in which females act in expressive roles, providing care and security to children and offering them emotional support, and men perform instrumental roles, namely, being the breadwinner in the family. This complementary division of labor, springing from a biological distinction between the sexes, would ensure the solidarity of the family, according to Parsons.

Feminists have sharply criticized claims of a biological basis to the sexual division of labor, arguing that there is nothing natural or inevitable about the allocation of tasks in society. Women are not prevented from pursuing occupations on the basis of any biological features; rather, humans are socialized into roles that are culturally expected of them. Parsons's notions of the "expressive" female have been attacked by feminists and other sociologists who see his views as condoning the subordination of women in the home. There is no basis to the belief that the "expressive" female is necessary for the smooth operation of the family—rather, it is a role that is promoted largely for the convenience of men.

In addition, cross-cultural and historical studies show that even though most societies distinguish between men's and women's roles, the degree to which they differentiate tasks as exclusively male or female and assign different tasks and responsibilities to women and men can vary greatly across time and place (Coltrane 1992). Thus, gender inequalities do not seem to be fixed or static.

Feminist Approaches

The feminist movement has given rise to theoretical approaches that attempt to explain gender inequalities and set forth agendas for overcoming those inequalities. As we learned in Chapter 1, **feminist theories** related to gender inequality contrast markedly with one another. Feminist writers are all concerned with women's unequal position in society, but their explanations for it vary substantially. Competing schools of feminism have sought to explain gender inequalities through a variety of deeply embedded social processes, such as sexism, patriarchy, capitalism, and racism. In the following sections, we look at the arguments behind four main feminist perspectives: liberal, radical, Black, and transnational feminism.

LIBERAL FEMINISM

Liberal feminism looks for explanations of gender inequalities in social and cultural attitudes. Unlike radical feminists, liberal feminists do not see women's subordination as part of a larger system or structure. Instead, they draw attention to many separate factors that contribute to inequalities between men and women. For example, liberal feminists are concerned with sexism and discrimination against women in the workplace, educational institutions, and the media. They tend to focus their energies on establishing and protecting equal opportunities for women through legislation and other democratic means. Legal advances such as the Equal Pay Act of 1963 and the Sex Discrimination Act of 1984 were actively supported by liberal feminists, who argued that enshrining equality in law is key to eliminating discrimination against women. Liberal feminists seek to work through the existing system to bring about reforms in a gradual way. In this respect, they are more moderate in their aims and methods than radical feminists, who call for an overthrow of the existing system.

While liberal feminists have contributed greatly to the advancement of women over the past century, critics charge that they are unsuccessful in dealing with the root cause of gender inequality and do not acknowledge the systemic nature of women's oppression in society. They say that by focusing on the independent deprivations that women suffer— sexism, discrimination, the glass ceiling and glass escalator, unequal pay—liberal feminists

feminist theories

A sociological perspective that emphasizes the centrality of gender in analyzing the social world and particularly the experience of women. There are many strands of feminist theory, but they all seek to explain gender inequalities in society and to work to overcome them.

liberal feminism

Form of feminist theory that believes that gender inequality is produced by unequal access to civil rights and certain social resources, such as education and employment, based on sex. Liberal feminists tend to seek solutions through changes in legislation that ensure that the rights of individuals are protected.

"His" and "Hers" Apps?

Our face-to-face interactions at school, at work, in the family, and in our everyday lives are powerfully shaped by gender. But how does gender shape our digital lives? Do men and women use the same apps? Researchers are only beginning to document the digital lives of men and women, yet most of the evidence shows that men and women aren't all that different from one another.

A recent study by Pew (Purcell 2011) asked American adults what kinds of apps they had downloaded. Overall, the most popular were apps that provided regular updates on news, weather, sports, or finances (74 percent); that helped people communicate with family and friends (67 percent); and that helped them learn about something they were interested in (64 percent). Not surprisingly, apps with more specific functions were less popular: Only 48 percent of people had downloaded apps that helped them with work-related tasks, 46 percent used apps that helped them shop, 43 percent watched movies or TV on their smartphones, and only 29 percent used apps that helped them manage their health.

The study found gender differences on just two dimensions. Men are more likely than women to use apps that help them with work-related tasks (56 percent vs. 39 percent) and that advise them in making a purchase (51 percent vs. 42 percent). The results of the Pew study suggest that men and women are more similar than different when it comes to their digital lives. Research firm Flurry Analytics similarly found that when it comes to the most popular apps, the breakdown of users is roughly fifty-fifty; women and men are equally likely to use Facebook and to share their thoughts on Twitter (Bonnington 2012; *Huffington Post* 2012).

If we delve more deeply, however, we see that the specific ways that people use social networking sites and play games online differ by gender in ways that are consistent with gender role socialization. As a 2012 *Time* magazine article proclaimed, "Men are from Google+, women are from Pinterest" (Wagstaff 2012). Men outnumber women on high-tech sites like Google+ and outnumber women four to one on "male-themed" video games that feature things like gangs, mobsters, and war. By contrast, puzzles and word games like Words with Friends or "family-themed" games are downloaded more frequently by women.

Men and women also differ with respect to interest-based social networking apps. Men slightly outpace women on music-streaming and -sharing sites like Spotify and far outnumber women on Reddit, a site where people can generate their own news stories. By contrast, women are much more likely to use apps associated with fashion and home design. For example, 44 percent of online women use Pinterest, compared with 16 percent of men (Greenwood, Perrin, and Duggan 2016).

What are your favorite apps? Do you think that your identity as male or female has shaped your preferences for particular apps? Why or why not?

Men are more likely than women to visit high-tech websites and play violent video games, whereas women favor puzzles, word games, and fashion-oriented sites and apps.

draw only a partial picture of gender inequality. Radical feminists accuse liberal feminists of encouraging women to accept an unequal society and its competitive character.

RADICAL FEMINISM

At the heart of **radical feminism** is the belief that men are responsible for and benefit from the exploitation of women. The analysis of patriarchy—the systematic domination of females by males—is of central concern to this branch of feminism. Patriarchy is viewed as a universal phenomenon that has existed across time and cultures. Radical feminists often concentrate on the family as one of the primary sources of women's oppression in society. They argue that men exploit women by relying on the free domestic labor that women provide in the home and that, as a group, men deny women access to positions of power and influence in society.

Radical feminists differ in their interpretations of the basis of patriarchy, but most agree that it involves the appropriation of women's bodies and sexuality in some form. Because women are biologically able to give birth to children, they become dependent on men for protection and livelihood. As such, the nuclear family is viewed as the site that generates "biological inequality" between women and men. Other radical feminists point to male violence against women as central to male supremacy. According to such a view, domestic violence, rape, and sexual harassment are all part of the systematic oppression of women rather than isolated cases with their own psychological or criminal roots.

Radical feminists believe that gender equality can be attained only by overthrowing the patriarchal order, because patriarchy is a systemic phenomenon. The use of patriarchy as a concept for explaining gender inequality has been popular with many feminist theorists. In asserting that "the personal is political," radical feminists have drawn widespread attention to the many linked dimensions of women's oppression.

Many objections can be raised to radical feminist views. The main one, perhaps, is that the concept of patriarchy as it has been used is inadequate as a general explanation for women's oppression. Radical feminists have tended to claim that patriarchy has existed throughout history and across cultures—that it is a universal phenomenon. Critics argue, however, that such a conception of patriarchy does not leave room for historical or cultural variations. It also ignores the important influence that race, class, or ethnicity may have on the nature of women's subordination. In other words, it is not possible to see patriarchy as a universal phenomenon; doing so risks biological reductionism—attributing all the complexities of gender inequality to a simple distinction between men and women.

BLACK FEMINISM AND TRANSNATIONAL FEMINISM

Do the versions of feminism outlined above apply equally to the experiences of both white and nonwhite women? Many Black feminists and feminists from the Global South claim they do not. They argue that ethnic and cross-national differences among women are not considered by the main feminist schools of thought, which are oriented to the dilemmas of white, predominantly middle-class women living in industrialized societies. It is not valid, they claim, to generalize theories about women's subordination as a whole from the experience of a specific group of women. These views exemplify the themes of intersectionality: Women's and men's experiences are inextricably linked to their race, ethnicity, region, and socioeconomic location.

This dissatisfaction has led to the emergence of a **Black feminism** focused on the particular problems facing Black women. The writings of African American feminists emphasize the influence of the powerful legacy of slavery, segregation, and the civil rights

radical feminism

Form of feminist theory that believes that gender inequality is the result of male domination in all aspects of social and economic life.

Black feminism

A strand of feminist theory that highlights the multiple disadvantages of gender, class, and race that shape the experiences of nonwhite women. Black feminists reject the idea of a single, unified gender oppression that is experienced evenly by all women.

CONCEPT CHECKS

1. Contrast functionalist and feminist approaches to understanding gender inequality.

2. What are the key ideas of liberal feminism? What are the critiques of this perspective?

3. What are the key ideas of radical feminism? What are the critiques of this perspective?

4. What are the key ideas of Black feminism? What are the critiques of this perspective?

>

Learn how women and men are challenging sexism and sexual violence in the workplace and on college campuses.

suffragettes

Members of early women's movements who pressed for equal voting rights for women and men.

movement on gender inequalities in the Black community. They point out that early Black **suffragettes** supported the campaign for women's rights but realized that the question of race could not be ignored. Black feminists contend, therefore, that any theory of gender equality that does not take racism into account cannot be expected to explain Black women's oppression adequately. Some also argue that Black women are multiply disadvantaged on the basis of their color, their sex, and their class position. When these three factors interact, they reinforce and intensify one another (Brewer 1993).

Transnational feminism, by contrast, focuses primarily on intersections among nationhood, race, gender, sexuality, and economic exploitation against the contemporary backdrop of global capitalism. This perspective recognizes that global processes, including colonialism, racism, and imperialism, shape gender relations and hierarchies in powerful ways (Mohanty 2003). Pioneers of transnational feminism recognize that the key themes of liberal feminism, such as concerns about equal pay for equal work or the division of household labor, are not relevant for many women in the Global South. Scholars working in this tradition often have a strong human rights orientation and see research as integral to social change. For instance, by understanding the processes through which female agricultural workers in Brazil are subordinated, transnational feminists can work to increase these women's bargaining power (Thayer 2010).

How Can We Reduce Gender-Based Aggression?

As we have seen throughout this chapter, women tend to fare worse than men with respect to education, earnings, power, risk of sexual violence, workplace discrimination and harassment, and political representation. Although the severity of the gender gap varies widely across nations, and even across subgroups within a single nation, the evidence and theories we have reviewed clearly reveal that gender inequalities are widespread. The processes through which inequalities are perpetuated may be subtle—so subtle, in fact, that we may not easily detect them in our everyday lives. For instance, an off-color joke at a frat party may at first blush seem an "inappropriate attempt at humor" yet, upon further analysis, may perpetuate a culture that normalizes sexual aggression against women.

The experiences of Ford Motor Company workers like Miyoshi Morris and Hollywood actresses like Mira Sorvino also underscore the ways that power differentials are at the root of gender-based discrimination and mistreatment. Morris's boss in the Ford paint department had the power to adjust her work schedule so that she could meet her childcare obligations, yet he exploited this power by demanding sex. Likewise, Hollywood producer Harvey Weinstein exerted power over his victims; his casting of young actresses in plum movie roles hinged on whether they would have sexual relations with him. Those who begged off lost the role not only in Weinstein's films but in the films of other directors over whom Weinstein had control.

Yet while vast power differentials exist on the basis of gender in the United States and worldwide, the intensity and breadth of these differentials depend on factors beyond gender, such as one's social class, race, or parental status. In this way, the concept of intersectionality helps us understand the diversity of men's and women's experiences. Wealthy and well-connected actresses might have a financial cushion that allows them to quit an acting role should the professional climate become uncomfortable or unsafe. Women with fewer economic resources find that their choices are much more constrained; women like Miyoshi Morris, who worried about feeding her children, simply could not afford to lose their jobs, rendering them particularly vulnerable to the threats of a predatory boss. Creating an environment in which those who have been victimized or exploited feel empowered to speak up, know that their words will be heeded, and receive support from their employers and industries is a critical step toward eradicating gender violence. Initiatives like the Time's Up movement, which set up a legal defense fund to support those who have been victimized by workplace sexual harassment, are an indication that cultural change is possible.

Standing up to gender-based violence and exploitation isn't limited to the assembly line or Hollywood. Students and administrators on college campuses also are saying "time's up" and are learning about safe and effective ways to call out and stop gender violence (Zimmerman 2016). About a decade ago, the University of Kentucky implemented its Green Dot Bystander Intervention program. The program shows students how to intervene and defuse the situation if they see an uncomfortable or potentially violent interaction on campus. Rather than assuming that someone else will step in, students learn how to get away from (or help a friend escape) a potential sexual aggressor or to get help from a campus security guard if they feel a situation spiraling out of control. Researchers have found that these kinds of programs work: A study at the University of Kentucky found that gender-based violence on campus—including sexual assault, harassment, stalking, and dating violence—was cut in half for those who participated in the training program (Coker et al. 2011).

Yet some critics have pointed out that it's not enough to teach women how to avoid sexual assault. The more important goal is to change male students' beliefs about sexual violence and entitlement. One approach that some schools are taking is to target student athletes, recognizing that some sports normalize and thrive on aggression and other behaviors that exemplify hegemonic masculinity. The NCAA (National Collegiate Athletic Association) has ramped up its efforts to reform a culture of gendered violence in college sports. In August 2017, the association implemented a new policy requiring coaches and student-athletes to undergo annual training in sexual violence prevention. All coaches and athletes at the more than 1,100 NCAA member schools also must have information about their schools' policies regarding sexual violence prevention, and the athletic department must demonstrate knowledge of and compliance with those policies. Experts like Jessica Luther, author of *Unsportsmanlike Conduct: College Football and the Politics of Rape*, believe that more needs to be done but that the NCAA's move "helps to legitimize [sexual violence] as something the public should care about" (Gibbs 2017).

The sociological study of gender has helped us understand how power imbalances on the basis of gender emerge and at the same time provides a road map for how these imbalances may be resolved. By recognizing the sources of gendered power imbalances, the distinctive pressures placed on men and women to comply with often-outdated gender expectations, and the way these inequities affect our lives, we are using our sociological imaginations to understand and ultimately fight gender inequalities.

transnational feminism

A branch of feminist theory that highlights the way that global processes—including colonialism, racism, and imperialism—shape gender relations and hierarchies.

CONCEPT CHECKS

1. How did gender inequality in the workplace affect the women workers in the Ford Motor Company plant versus actresses in Hollywood?

2. Describe at least one campus-based program designed to fight sexual violence at colleges and universities.

CHAPTER 9

The Big Picture

Gender Inequality

Thinking Sociologically

1. What does cross-cultural evidence from tribal societies suggest about the differences in gender roles? Explain.

2. Why are minority women likely to think very differently about gender inequality from how white women think about it? Explain.

3. Do you think Judy Calibuso had a compelling case in the Merrill Lynch lawsuit? Why or why not? What kind of evidence would be needed to make a reasonable judgment in that case?

Terms to Know

Intersectionality

gender binary • sex • gender • nonbinary • biological essentialism • gender socialization • social construction of gender • hegemonic masculinity • transgender • intersex

patriarchy • gender inequality • gender typing • sex segregation • glass ceiling • sexual harassment • second shift • housework

infanticide • rape • rape culture • toxic masculinity

feminist theories • liberal feminism • radical feminism • Black feminism • suffragettes • transnational feminism

Concept Checks

1. What is the difference between sex and gender?
2. How do both biology and gender socialization contribute to differences between men and women?
3. What does it mean to say that gender is something we "do"? Give an example of a way that you have done gender in your daily life.
4. How can studies of gender in other cultures contribute to the argument that gender is socially constructed?
5. What is a nonbinary gender identity? How does it challenge the male-female sex dichotomy?

1. What are two explanations for the stark gender segregation in college majors? How does this segregation contribute to the gender pay gap?
2. Describe at least three examples of how gender inequalities emerge in the workplace. How would a sociologist explain these inequities?
3. How do inequalities in the home, especially with regard to housework and childcare, reflect larger gender inequities in society?
4. What are some important differences between men's and women's political participation in the United States?

1. Name three different kinds of violence against women.
2. How common is violence against women?
3. Why are women more likely than men to be the targets of sexual violence?
4. What is rape culture, and why is it so pervasive in contemporary society?

1. Contrast functionalist and feminist approaches to understanding gender inequality.
2. What are the key ideas of liberal feminism? What are the critiques of this perspective?
3. What are the key ideas of radical feminism? What are the critiques of this perspective?
4. What are the key ideas of Black feminism? What are the critiques of this perspective?

1. How did gender inequality in the workplace affect the women workers in the Ford Motor Company plant versus actresses in Hollywood?
2. Describe at least one campus-based program designed to fight sexual violence at colleges and universities.

10

Race, Ethnicity, and Racism

Racial & Ethnic Populations

p. 287

THE BIG QUESTIONS

What are race and ethnicity?

Understand that race is a social and political construction and how it differs from ethnicity. Learn what constitutes a minority group according to the sociological perspective.

Why do racial and ethnic antagonism exist?

Learn the leading psychological theories and sociological interpretations of prejudice and discrimination. Recognize the importance of the historical roots, particularly in the expansion of Western colonialism, of ethnic conflict. Understand the different models for a multiethnic society.

How does racism operate in American society today?

Understand how racism is not only enacted by individuals but embedded in our institutions. Learn how racial inequality is maintained by both overt acts of racial hatred and color blindness. Understand the concepts of white privilege and microaggressions.

What are the origins and nature of ethnic diversity in the United States?

Familiarize yourself with the history and social dimensions of ethnic relations in America.

How do race and ethnicity affect the life chances of different groups?

Learn how racial and ethnic inequality is reflected in terms of educational and occupational attainment, income, health, residential segregation, and political power.

How do sociologists explain racial inequality?

Learn the leading theories—cultural, economic, and discrimination—sociologists use to understand the sources of ethnic and racial inequality.

The multiracial family of Bill de Blasio, his wife, Chirlane McCray, and his children, Chiara and Dante, became a topic of interest during his campaign for mayor of New York City and was highlighted even further by the public backlash to a Cheerios commercial featuring the daughter of parents of different races. What do the advertisement, its viewer response, and de Blasio's election say about race relations in America today?

When Cheerios unveiled a new television commercial in 2013, ad execs did not predict the firestorm that would follow. In the commercial, an adorable young girl, about five years old, asks her mother if Cheerios "are good for your heart." Her mother reads the cereal box and notes that Cheerios' whole-grain oats are "heart healthy." The young girl runs over to her father, who is asleep on the sofa, and sprinkles Cheerios on his heart.

At first blush, the ad shows a little girl who wants her father to stay healthy by eating a low-fat cereal. But some people saw an entirely different scenario. The little girl, with her caramel complexion and bouncy ringlets, was apparently the daughter of a white mother and a Black father (Elliott 2013). This simple fact triggered an onslaught of racist comments—so much so that YouTube promptly shut down the "comments" section on the video's web page (Goyette 2013). Angry bloggers noted that the ad was "disgusting" and made them "want to vomit." Others, still, made snide comments about the rarity of an African American man sticking around to raise his children (Goyette 2013).

But the reaction to these mean-spirited and narrow-minded comments was even more powerful. Many wrote poignant responses that this Cheerios ad was the first time they had ever seen a multiracial family—much like their own—on national television. Chirlane McCray, wife of New York City mayor Bill de Blasio, explained in an email to supporters, "Nineteen years of marriage and two children later, this is the first TV commercial I've seen with a family that looks a little bit like ours."

General Mills Foods, the maker of Cheerios, stood by its ad, acknowledging that the family represented millions of American families and noting that the number of multiracial families and children in the United States would continue to rise in the future. They even issued a follow-up ad in which the father breaks the news to his daughter that she will soon have a baby sibling, while his beaming pregnant wife looks on. Despite the outcry following the earlier ad, General Mills remained committed to depicting the new American family. "Cheerios is recognizing the changing face of America," McCray noted, "and celebrating that our differences make us stronger." TV networks followed General Mills's lead: Later that year, ABC premiered *Black-ish* and, soon after, *Fresh Off the Boat* in an effort to increase the diversity of families presented on television.

Scientific studies similarly document that for growing numbers of Americans, "race" is not a simple or monolithic identity. In 2017, nearly 9 million Americans selected two or more racial categories when questioned by the U.S. Census, which first gave people the option to choose more than one racial category in 2000. Between 2000 and 2010, the proportion of Americans who identify as multiracial increased by nearly one-third (Jones and Bullock 2012). Experts predict that the number of Americans who identify as multiracial will continue to increase steeply in coming decades. About one in six (17 percent) new marriages in 2015 were between spouses of different races (Livingston and Brown 2017). In addition, one in seven babies born in the United States in 2015 were multiracial—up from 1 percent in 1970 (Livingston 2017). Young adults today are far more accepting of interracial marriage than their parents were. For example, in 2013, 87 percent of Americans thought it was fine for Blacks and whites to marry, up from 4 percent in 1958. This statistic varied widely by age, however. Nearly all (96 percent) eighteen- to twenty-nine-year-olds support interracial marriage, whereas 70 percent of those aged sixty-five and older felt the same (Newport 2013).

While the number of Americans of multiracial identity is at an all-time high and rising, these individuals continue to negotiate their identities with observers who cling to the view that "race" is a monolithic construct. For example, Michelle López-Mullins, former president of the University of Maryland's Multiracial and Biracial Student Association (MBSA), gets tired of hearing the question "What are you?"

→ **NOTE: Please answer BOTH Question 8 about Hispanic origin and Question 9 about race. For this census, Hispanic origins are not races.**

8. Is Person 1 of Hispanic, Latino, or Spanish origin?

☐ **No,** not of Hispanic, Latino, or Spanish origin
☐ Yes, Mexican, Mexican Am., Chicano
☐ Yes, Puerto Rican
☐ Yes, Cuban
☐ Yes, another Hispanic, Latino, or Spanish origin — *Print origin, for example, Argentinean, Colombian, Dominican, Nicaraguan, Salvadoran, Spaniard, and so on.* ↗

9. What is Person 1's race? *Mark* ☒ *one or more boxes.*

☐ White
☐ Black, African Am., or Negro
☐ American Indian or Alaska Native — *Print name of enrolled or principal tribe.* ↗

☐ Asian Indian   ☐ Japanese   ☐ Native Hawaiian
☐ Chinese   ☐ Korean   ☐ Guamanian or Chamorro
☐ Filipino   ☐ Vietnamese   ☐ Samoan
☐ Other Asian — *Print race, for example, Hmong, Laotian, Thai, Pakistani, Cambodian, and so on.* ↗   ☐ Other Pacific Islander — *Print race, for example, Fijian, Tongan, and so on.* ↗

☐ Some other race — *Print race.* ↗

In 2000, for the first time, the U.S. Census gave people the option to select more than one racial category.

⌃

(Saulny 2011). Her father is Chinese and Peruvian, and her mother is white and Native American. López-Mullins recalls that when she was growing up, "I was always having to explain where my parents are from. . . . Saying 'I'm an American' wasn't enough." Although she found this frustrating when she was a child, she now embraces her mixed heritage. "Now when people ask what I am, I say, 'How much time do you have?' Race will not automatically tell you my story."

While in past generations, mixed-race persons often tolerated negative labels like "mulatto" (that is, a person with one Black parent and one white parent), López-Mullins and her friends from MBSA are proud of their backgrounds and embrace all aspects of their ethnicities. As Laura Wood, former vice president of MBSA, says, "It's really important to acknowledge who you are. . . . If someone tries to call me Black, I say, 'Yes—and white.'"

The experiences of Laura Wood and Michelle López-Mullins—along with a mounting body of sociological research—illustrate just how difficult it is to pinpoint the conditions of racial and ethnic-group membership for some individuals of multiracial heritage. In recent decades, a number of sociologists have turned their attention to this topic of multiracial identity and racial classification schemes. They have argued that a "static measure of race" is not useful for individuals of multiracial heritage who may assert different identities in varied social contexts (Cheng and Lee 2009; Harris 2003; Harris and Sim 2000). As López-Mullins told a *New York Times* reporter, "I'm pretty much checking everything. . . . Hispanic, white, Asian American, Native American" when filling out surveys like the U.S. Census. As we will see throughout this chapter, race and ethnicity are complex identities with powerful implications for our everyday lives.

What Are Race and Ethnicity?

In your daily life, you have no doubt used the terms *race* and *ethnicity* many times, but do you know what they mean? In fact, defining these terms is very difficult, not least because of the contradiction between their everyday usage and their scientific basis (or absence thereof).

Race

When the Census Bureau asks people to classify themselves, many respondents mistakenly think that humans can be readily separated into biologically different races. Yet, in biological terms, there are no clear-cut races. We therefore define **race** as a socially constructed category based on the belief in fundamental human differences associated with phenotype and ancestry (Monk 2016).

Differences in physical type among groups of human beings arise from population inbreeding, which varies according to the degree of contact among different cultural or social groups. Human population groups are a continuum. The genetic diversity within populations that share visible physical traits is as great as the diversity among them. Thus, the racial differences recognized by the Census Bureau (Black, white, Asian) should be understood as physical variations singled out by the members of a community or society as socially

Understand that race is a social and political construction and how it differs from ethnicity. Learn what constitutes a minority group according to the sociological perspective.

race

Differences in human physical characteristics used to categorize large numbers of individuals.

Four boys represent the "racial scale" in South Africa—Black, Indian, "half-caste" (mixed ethnicity), and white.

theory of racial formation

The process by which social, economic, and political forces determine the content and importance of racial categories.

significant and meaningful. Differences in skin color are treated as significant, for example, whereas differences in eye color and height are not. Racial categories are always nationally and historically specific (Fredrickson 2002) and can vary significantly from place to place.

Ever since 1790, the Census Bureau has classified the population by race, and the way it has done so powerfully illustrates that race is not a biological reality but rather a social and political construction. For the first century and a half of the census, these classifications were done by census workers who determined the race of a person by sight. Such classifications, no doubt, were quite arbitrary and often inaccurate. In 1960, the census moved to a system in which the people being counted self-reported their own race by choosing among predetermined categories.

Over the history of the census, the specific racial categories used have undergone major changes, illustrating that categories we take to be natural in one era are actually socially constructed. People from South Asia were long classified as white in the census, for example, but by the 1980s were reclassified as Asian. Mexicans were classified as white in the nineteenth century, as nonwhite in the 1930s, again as white in the 1940s, and then as Hispanic in the 1970s, all depending on demands for labor and the amount of prejudice in the country at a given moment. Today, there are ongoing debates about whether people from parts of the Middle East should continue to be classified as white, because many are not seen by themselves or others in that way.

Racial categories themselves also go in and out of fashion. In the 1890 census, "quadroon" was the racial category for people who were one-quarter Black, and "octoroon" was the racial category for people who were one-eighth Black. Ten years later, those categories were eliminated and a new, all-encompassing "Negro" category was added to the census. In 1970, "Negro or Black" was added to the census, and by 2000, it was "Black, African American, or Negro." In 2013, the Census Bureau stopped using the category "Negro" because many respondents found it offensive; an older generation that identified with that term had largely passed on.

One major sociological perspective on race is known as the **theory of racial formation** (Omi and Winant 1994); this is "the process by which social, economic and political forces determine the content and importance of racial categories, and by which they are in turn shaped by racial meanings." The main point of this theory is that ideas about race get created and re-created by governments and large-scale institutions but also by individual human beings in their everyday lives. On the one hand, political actors have a big influence on creating racial categories and ideologies, which become the basis for people's everyday understandings and social interactions related to race. They do so by treating race as a key aspect of everyday life, as when the Census uses racial categories to count people every decade or when the president of the United States refers to immigrants in a demeaning way. Either of these acts will shape understandings of racial identity in powerful ways, reinforcing society's understandings. At the same time, the theory of racial formation is attentive to how human beings create understandings in their individual families and group life, and to the relationship between these two levels of analysis.

Some social scientists argue that race is nothing more than an ideological construct whose use in academic circles perpetuates the commonly held belief that it has a basis in reality (Miles 1993). For this reason, they argue, it should be abandoned. Others disagree, claiming that race still has meaning for many people and cannot be ignored. In historical terms, race has been an extremely important concept that powerful social groups have

Racial & Ethnic Populations

The racial and ethnic categories that are relevant in a particular nation change over time and vary widely among countries.

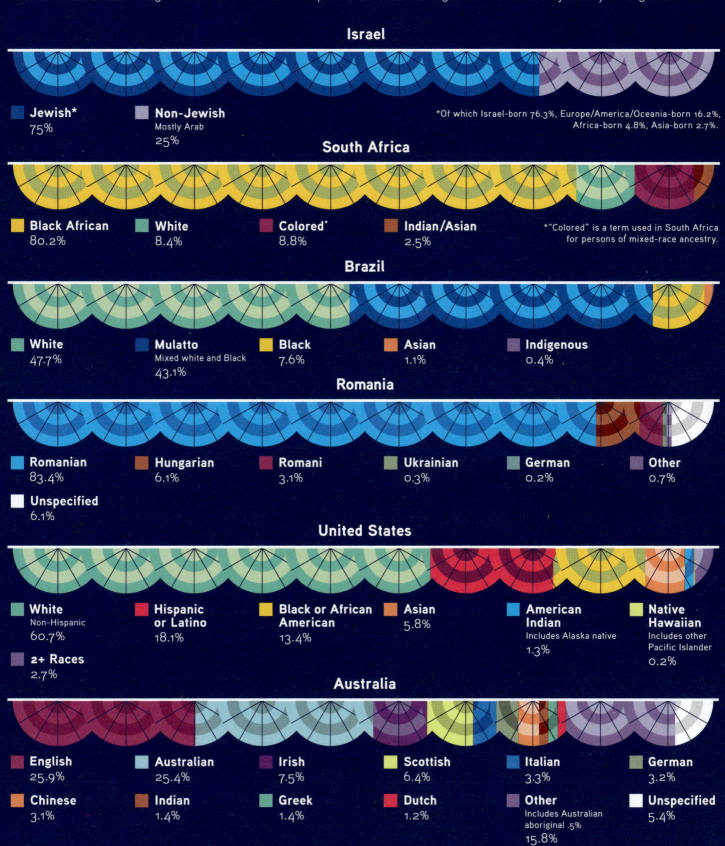

Israel

Jewish* 75%

Non-Jewish Mostly Arab 25%

*Of which Israel-born 76.3%, Europe/America/Oceania-born 16.2%, Africa-born 4.8%, Asia-born 2.7%.

South Africa

Black African 80.2%

White 8.4%

Colored* 8.8%

Indian/Asian 2.5%

*"Colored" is a term used in South Africa for persons of mixed-race ancestry.

Brazil

White 47.7%

Mulatto Mixed white and Black 43.1%

Black 7.6%

Asian 1.1%

Indigenous 0.4%

Romania

Romanian 83.4%

Hungarian 6.1%

Romani 3.1%

Ukrainian 0.3%

German 0.2%

Other 0.7%

Unspecified 6.1%

United States

White Non-Hispanic 60.7%

Hispanic or Latino 18.1%

Black or African American 13.4%

Asian 5.8%

American Indian Includes Alaska native 1.3%

Native Hawaiian Includes other Pacific Islander 0.2%

2+ Races 2.7%

Australia

English 25.9%

Australian 25.4%

Irish 7.5%

Scottish 6.4%

Italian 3.3%

German 3.2%

Chinese 3.1%

Indian 1.4%

Greek 1.4%

Dutch 1.2%

Other Includes Australian aboriginal .5% 15.8%

Unspecified 5.4%

Sources: Central Intelligence Agency 2017, U.S. Bureau of the Census 2018e.

Taiwanese Americans perform during the National Asian Heritage Festival in Washington, D.C. (left). A member of the Tlaxcala tribe of central Mexico dances at Carnival in New York (right). Styles of dress are one way that ethnic groups distinguish themselves from one another.

>

used as part of strategies of domination (Spencer 2014). For example, the contemporary situation of African Americans in the United States cannot be understood without reference to the slave trade, racial segregation, and persistent racial ideologies (Wacquant 2010). Racial distinctions are more than ways of describing differences; they are also important factors in the reproduction of patterns of power and inequality.

Ethnicity

Whereas the idea of race implies something fixed and biological, ethnicity is a source of identity based on society and culture. **Ethnicity** refers to a type of social identity related to ancestry (perceived or real) and cultural differences, which become effective or active in certain contexts. Members of ethnic groups see themselves as culturally distinct from other groups in a society and are seen by those other groups to be so in return. Different characteristics may serve to distinguish ethnic groups from one another, but the most common are language, history or ancestry (real or imagined), religion, and styles of dress or adornment.

In the United States, some of the first sociological research took place on ethnic groups, such as Italian Americans, Irish Americans, Polish Americans, and German Americans, though the Irish and the Italians were sometimes thought of as a race as well. As the United States has become more diverse, many races have come to see themselves as comprising distinct ethnicities. East Asians encompass Chinese, Koreans, and Japanese, among many other ethnic groups, and South Asians include Pakistanis, Nepalis, and Sri Lankans, while Blacks include West Indians and West Africans, among many others.

Ethnic differences are mainly learned, a point that seems self-evident until we remember how often some groups have been regarded as "born to rule" or "lazy," "unintelligent," and so forth. Indeed, when people use the term *ethnicity*, very often they do so (as with *race*) when referring to inherent characteristics, such as skin color or blood ties. Yet there is nothing innate about ethnicity; it is a social phenomenon that is produced and reproduced over time.

For many people, ethnicity is central to their individual and group identities, but for others, it is irrelevant and, for still others, seems significant only during holidays. Ethnicity can provide an important thread of continuity with the past and is often kept alive through the practices of cultural traditions. For instance, third-generation Americans of Irish descent may proudly identify themselves as Irish despite having lived their entire lives in the United States.

ethnicity

Cultural values and norms that distinguish the members of a given group from others. An ethnic group is one whose members share a distinct awareness of a common cultural identity, separating them from other groups.

Minority Groups

The term **minority group** as used in everyday life can be quite confusing. This is because the term refers to political power and is not simply a numerical distinction. There are many minorities in a statistical sense, such as people with red hair, but these are not minorities according to the sociological concept. In sociology, members of a minority group are disadvantaged as compared with members of the **dominant group** (a group possessing more wealth, power, and prestige) and have some sense of group solidarity, of belonging together. The experience of being subject to prejudice and discrimination is usually grounds for being considered disadvantaged. Being part of a minority group usually heightens feelings of common loyalty and interests.

Members of minority groups tend to see themselves as a people separated or distinct from the majority. Minority groups are sometimes, but not always, physically and socially isolated from the larger community. Although they tend to be concentrated in certain neighborhoods, cities, or regions of a country, their children may often intermarry with members of the dominant group. People who belong to minority groups (for example, Jews) sometimes actively promote endogamy (marriage within the group) to keep their cultural distinctiveness alive, although this practice has declined among less religious Jews.

The idea of a minority group is more confusing today than ever before. Some groups that were once clearly identified as minorities, such as Asians and Jews, now have more resources, intermarry at greater rates, and experience less discrimination than they did when they were originally conceived of as minority groups. Moreover, according to the U.S. Census, the United States will soon be a "majority-minority" nation. By this, sociologists mean a country in which non-Hispanic whites will no longer be in the majority, a nation in which nonwhites will be in the majority. That said, sociologists still expect whites to dominate the power structure of American society. These examples highlight the fact that the concept of a minority group is really about disadvantage rather than being a numerical distinction.

minority group
A group of people who, because of their distinct physical or cultural characteristics, find themselves in situations of inequality within that society.

dominant group
The group that possesses more wealth, power, and prestige in a society.

CONCEPT CHECKS

1. What are race and ethnicity? How are these two concepts alike, and how do they differ?
2. How do political actors and institutions participate in racial formation?
3. What differentiates a minority group from a statistical minority?

Why Do Racial and Ethnic Antagonism Exist?

To explain why racial differences become the focus of inequalities, and of conflicts, we need to make use of both psychological and sociological concepts. Psychological theories help explain why ethnic differences become so emotionally charged and clarify aspects of the nature of prejudiced attitudes. Sociological interpretation is necessary to show *how* and *why* ethnic divisions become institutionalized as forms of discrimination "built into" a given society.

Psychological Theories

Two types of psychological approaches to understanding ethnic hostilities are important. One puts forward a number of general psychological mechanisms relevant to analyzing prejudice in general. The other concentrates on the idea that there is a particular type of person who is most prone to hold prejudiced attitudes against minority groups.

Learn the leading psychological theories and sociological interpretations of prejudice and discrimination. Recognize the importance of the historical roots, particularly in the expansion of Western colonialism, of ethnic conflict. Understand the different models for a multiethnic society.

PREJUDICE AND DISCRIMINATION

prejudice

The holding of preconceived ideas about an individual or group, ideas that are resistant to change even in the face of new information. Prejudice may be either positive or negative.

―――

racism

The attribution of characteristics of superiority or inferiority to a population sharing certain physically inherited characteristics.

―――

stereotype

A fixed and inflexible category.

―――

displacement

The transferring of ideas or emotions from their true source to another object.

―――

scapegoats

Individuals or groups blamed for wrongs that were not of their doing.

―――

discrimination

Behavior that denies to the members of a particular group resources or rewards that can be obtained by others. Discrimination must be distinguished from prejudice: Individuals who are prejudiced against others may not engage in discriminatory practices; conversely, people may act in a discriminatory fashion toward a group even though they are not prejudiced against that group.

Prejudice refers to opinions or attitudes—positive or negative—held by members of one group toward another. These preconceived views are often based on hearsay and are resistant to change even in the face of direct evidence or new information. People may harbor favorable prejudices toward groups with which they identify and negative prejudices against others. One widespread form of prejudice is racism, which we will tackle in the next section. **Racism** is prejudice based on socially significant physical distinctions.

Prejudice operates mainly through **stereotyping**, which means thinking in terms of fixed and inflexible categories. All thought involves categories, by means of which we classify our experience. Sometimes, however, these categories are both ill informed and rigid. A person may have a view of Blacks or Jews, for example, that is based upon a few firmly held ideas through which information about them, or encounters with them, are interpreted. Stereotypical thinking may be harmless if it is "neutral" in terms of emotional content and distant from the interests of the individuals concerned. Americans might have stereotypical views of what the French are like, for example—such as that they all talk with clipped accents or wear striped shirts—but these may be of little consequence for most people of either nationality. Where stereotypes are associated with anxiety or fear, the situation is likely to be quite different. Stereotypes in such circumstances are commonly infused with attitudes of hostility or hatred toward the group in question.

Stereotyping is often closely linked to the psychological mechanism of **displacement**, in which feelings of hostility or anger become directed against objects that are not the real origin of those anxieties. Stereotyping leads people to blame **scapegoats** for problems that are not their fault. Scapegoating is common in circumstances in which two deprived ethnic groups come into competition with one another for economic rewards. People who direct racial attacks against poor Mexicans or African Americans, for example, are often in a similar economic position to them. They blame Blacks or Mexicans for grievances whose real causes lie elsewhere. Scapegoating is normally directed against groups that are clearly distinctive and relatively powerless, because they form a fairly easy target. Minority groups show both these characteristics. Protestants, Catholics, Jews, Italians, racial minorities, and very many others have played the unwilling role of scapegoats at various times throughout Western history.

In contrast to prejudice, **discrimination** refers to actual behavior that denies to members of a particular group resources or rewards that others can obtain. College admissions officers have been known to discriminate against members of ethnic or racial groups when they do not conform with stereotypes of those groups. One investigation of the Princeton University admissions office, for example, uncovered a rejected Hispanic applicant with this comment written on her file: "Tough to see putting her ahead of others. No cultural flavor in app." Such attitudes are racist when they function to keep Hispanics out of college or ensure that only certain "kinds" of Hispanics will gain access to institutions of higher learning.

THE AUTHORITARIAN PERSONALITY

It is possible that some types of people, as a result of early socialization, are particularly prone to stereotypical thinking and projection on the basis of repressed anxieties. A famous piece of research carried out by Theodor Adorno and his associates in the 1940s diagnosed a character type they term the "authoritarian personality" (Adorno et al. 1950).

The researchers developed several measurement scales, each for a particular area of social attitudes, for assessing levels of prejudice. On one scale, for instance, people were asked to agree or disagree with a series of statements expressing rigid, particularly anti-Semitic views. Those who were diagnosed as prejudiced on one scale also tended to be so on the others; prejudice against Jews went along with the expression of negative attitudes toward other minorities. People with an authoritarian personality, the investigators concluded, tend to be rigidly conformist, submissive to those seen as their superiors, and dismissive toward inferiors. Such people are also highly intolerant in their religious and sexual attitudes.

Adorno's research and the conclusions drawn from it have been subjected to a barrage of criticism. Some have doubted the value of the measurement scales used. Others have argued that authoritarianism is not a characteristic of personality but reflects the values and norms of particular subcultures within the wider society. The investigation may be more valuable as a contribution to understanding authoritarian patterns of thought in general rather than for distinguishing a particular personality type (Wellman 1977).

Sociological Interpretations

The psychological mechanisms mentioned earlier—stereotypical thinking, displacement, and projection—are universal in nature. They are found among members of all societies and are relevant to explaining why ethnic and racial antagonism is such a common element in cultures of many different types. However, they explain little about the social processes involved in discrimination. To study such processes, we must bring into play three sociological ideas.

ETHNOCENTRISM, GROUP CLOSURE, AND RESOURCE ALLOCATION

Sociological concepts relevant to ethnic conflicts on a general level are those of ethnocentrism, ethnic-group closure, and resource allocation. Ethnocentrism—a suspicion of outsiders combined with a tendency to evaluate the cultures of others in terms of one's own culture—is a concept we have encountered previously (see Chapter 2). Virtually all cultures have been ethnocentric to a greater or lesser degree, and it is easy to see how ethnocentrism combines with stereotypical thought. "Outsiders" are thought of as aliens, as barbarians, or as morally and mentally inferior. This was how most civilizations viewed the members of smaller cultures, for example, and it has helped to fuel innumerable ethnic clashes throughout history.

Ethnocentrism and group closure frequently go together. The term *closure* means the process whereby groups maintain boundaries separating themselves from others. Such devices include limiting or prohibiting intermarriage between the groups, restrictions on social contact or economic relationships like trading, and the physical separation of groups from one another (as in the case of racial ghettos). American Blacks have experienced all three exclusion devices at one time or another: Racial intermarriage has been illegal in some states, economic and social segregation was enforced by law in the South, and segregated Black ghettos still exist in most major U.S. cities.

Sometimes groups of equal power may mutually enforce principles of closure: Their members keep separate from each other, but neither group dominates the other. More commonly, however, the members of one group are in a position of power over another. In these circumstances, ethnic-group closure coincides with the allocation of resources, in other

The segregation of whites and Blacks was the rule of law until the passage of the Civil Rights Act in 1964. While lunch counters such as this one in Virginia and other public places are no longer segregated by law, *de facto* segregation, or segregation "in fact," persists to this day.

∧

immigration

The movement of people into one country from another for the purpose of settlement.

emigration

The movement of people out of one country to settle in another.

words, with inequalities in the distribution of wealth and material goods. There are many contexts in which this may happen, for example, through the military conquest of one group by another or the emergence of an ethnic group as economically dominant over others. Ethnic-group closure provides a means of defending the economic position of the dominant group.

Some of the fiercest conflicts between ethnic or racial groups center on the lines of closure between them, precisely because these lines usually signal inequalities in the distribution of wealth, power, or social standing. The concept of ethnic-group closure helps us understand both the dramatic and the more insidious differences that separate communities or categories of people from one another—not just why the members of some groups get shot, lynched, beaten up, or harassed but also why they don't get good jobs, a good education, or a desirable place to live. Wealth, power, and social status are scarce resources—some groups have more of them than others. To hold on to their distinctive positions or possessions, privileged groups are sometimes prepared to undertake extreme acts of violence against others. The members of underprivileged groups may also turn to violence as a means of trying to improve their own situation. The combination of group closure with ethnic prejudice and racism is frequently explosive.

Ethnic Antagonism: A Historical Perspective

In an age of globalization and rapid social change, the rich benefits and complex challenges of ethnic diversity are confronting a growing number of states. **Immigration**, the movement of people into a country to settle, and **emigration**, the process by which people leave a country to settle in another, combine to produce global migration patterns linking countries of origin and countries of destination. Migratory movements add to ethnic and cultural diversity in many societies and help shape demographic, economic, and social dynamics. International migration is accelerating with the further integration of the global economy. Meanwhile, ethnic tensions and conflicts continue to flare in societies around the world, threatening to lead to the disintegration of some multiethnic states and hinting at protracted violence in others. How can ethnic diversity be accommodated and the outbreak of ethnic conflict averted? Within multiethnic societies, what should be the relation between ethnic-minority groups and the majority population?

To fully analyze ethnic relations in current times, we must first take a historical and comparative perspective. It is impossible to understand ethnic divisions today without giving prime place to the impact of the expansion of Western colonialism on the rest of the world. Global migratory movements resulting from colonialism helped create ethnic divisions by placing different peoples in close proximity. From the fifteenth century onward, Europeans began to venture into previously uncharted seas and unexplored landmasses, not only pursuing the aims of exploration and trade but also conquering and subduing native peoples. In the shape of the slave trade, they also occasioned a large-scale movement of people from Africa to the Americas.

These population flows formed the basis of the current ethnic composition of the United States, Canada, the countries of Central and South America, South Africa, Australia, and New Zealand. In all these societies, the indigenous populations were decimated by disease, war, and genocide and subjected to European rule. They are now impoverished ethnic minorities. Since the Europeans were from diverse national and ethnic origins, they transplanted various ethnic hierarchies and divisions to their new homelands. At the height of the colonial era, in the nineteenth and early twentieth centuries, Europeans also ruled over native populations in many other regions: South Asia, East Asia, the South Pacific, and the Middle East.

For most of the period of European expansion, ethnocentric attitudes were rife among the colonists, many of whom were convinced that, as Christians, they were on a civilizing mission to the rest of the world. Europeans of all political persuasions believed themselves to be superior to the peoples they colonized and conquered. The early period of colonization coincided with the rise of **scientific racism**, or the misuse of science to support racist assumptions. During the sixteenth century, Europeans began to classify animals, people, and the material culture that they collected as they explored the world. In 1735, Swedish botanist Carolus Linnaeus published what is recognized as the first version of a modern classification scheme of human populations. He grouped human beings into four basic categories: Europaeus, Americanus, Asiaticus, and Africanus. Linnaeus assumed that each subgroup had qualities of behavior or temperament that were innate and could not be altered. He acquired much of his data from the writings, descriptions, commentaries, and beliefs of plantation owners, missionaries, slave traders, explorers, and travelers. Thus, his scientific data were shaped by the prejudices of Europeans (Smedley 1993).

During the sixteenth century, and ever since then, the legacy of European colonization has generated ethnic divisions that have occupied a central place in regional and global conflicts. In particular, racist views distinguishing the descendants of Europeans from those of Africans became central to European racist attitudes.

FORMS OF ETHNIC CONFLICT

The most extreme and devastating form of group relations in human history involves **genocide**, the systematic, planned destruction of a particular group on the grounds of group members' ethnicity, religion, culture, or political views. The most horrific recent instance of brutal destructiveness against such a group was the massacre of 6 million Jews in the German concentration camps during World War II. The Holocaust is not the only example of mass genocide in the twentieth century. Between 1915 and 1923, more than a million Armenians were killed by the Ottoman Turkish government. In the late 1970s, 2 million Cambodians died in the Khmer Rouge's killing fields. During the 1990s, in the African country of Rwanda, hundreds of thousands of the minority Tutsis were massacred by the dominant Hutu group. And in the former Yugoslavia, Bosnian and Kosovar Muslims were summarily executed by the Serb majority.

Exploitation of minority groups has been an ugly part of many countries' histories. The separation of the minority from the majority has been institutionalized in the form of **segregation**, a practice whereby racial and ethnic groups are kept physically separate by law, thereby maintaining the superior position of the dominant group. For instance, in apartheid-era South Africa, laws forced Blacks to live separately from whites and forbade sexual relations between races. In the United States, African Americans have also experienced

scientific racism

The use of scientific research or data to justify or reify beliefs about the superiority or inferiority of particular racial groups. Much of the "data" used to justify such claims are flawed or biased.

genocide

The systematic, planned destruction of a racial, ethnic, religious, political, or cultural group.

segregation

The practice of keeping racial and ethnic groups physically separate.

1. What is the difference between prejudice and discrimination?

2. Provide two examples of ways ethnic groups maintain closure.

3. How did Western colonialism contribute to the creation of ethnic divisions today?

4. What are two forms of ethnic conflict?

5. What are the four primary models of ethnic integration? Describe each one.

>

Understand how racism is not only enacted by individuals but embedded in our institutions. Learn how racial inequality is maintained by both overt acts of racial hatred and color blindness. Understand the concepts of white privilege and microaggressions.

assimilation

The acceptance of a minority group by a majority population in which the new group takes on the values and norms of the dominant culture.

legal forms of segregation. It was only in 1967 that the Supreme Court ruled in the case of *Loving v. Virginia* that the prohibition of interracial marriage violated the right to privacy.

Models of Ethnic Integration

Four primary models of ethnic integration have been adopted by multiethnic societies: assimilation, the "melting pot," pluralism, and multiculturalism. For many years, the two most common positive models of political ethnic harmony in the United States were assimilation and the melting pot. **Assimilation** meant that new immigrant groups would assume the attitudes and language of the dominant white community. The idea of the **melting pot** was different—it meant merging different cultures and outlooks by stirring them all together.

A newer model of ethnic relations is **pluralism**, in which ethnic cultures maintain their unique practices and communities yet also participate in the larger society's economic and political life. A recent outgrowth of pluralism is **multiculturalism**, in which ethnic groups exist separately and equally. It does seem at least possible to create a society in which ethnic groups are separate but equal, as is demonstrated by Switzerland, where French, German, and Italian groups coexist in the same society. But this situation is unusual, and it seems unlikely that the United States could come close to mirroring this achievement in the near future.

How Does Racism Operate in American Society Today?

Some see racism as operating in the individual consciousness. An individual may profess racist beliefs or may join in with a group, such as a white supremacist organization, that promotes a racist agenda. Yet many have argued that racism is more than simply the ideas held by a small number of bigoted individuals; rather, they argue that racism is a system of domination operating in official social institutions, such as admissions offices or police departments. We will describe both forms of racism in turn.

Institutional Racism

The idea of **institutional racism** was developed in the United States in the late 1960s by Black Power activists Stokely Carmichael and Charles Hamilton and was taken up by civil rights campaigners. It is defined as the idea that racism occurs through the respected and established institutions of society rather than through the hateful actions of some bad people—that racism pervades all of society's structures in a systematic way. Those who focus on institutional racism study how social institutions, such as schools, hospitals, police departments, and businesses, have practices supporting white supremacy built into the very fabric of their operations. These institutions structure social relations in ways that are less obvious than overt discrimination.

The concept of institutional racism is well illustrated by a document the U.S. Department of Justice (DOJ) produced in 2015 in response to a grand jury's exoneration of a white police officer in Ferguson, Missouri, after he fatally shot an unarmed Black teenager named Michael Brown. Using its full investigative powers, the DOJ conducted a study of the charges of discrimination against the Ferguson Police Department. The DOJ gained access to all the department's administrative records and conducted interviews with department staff, government officials, and citizens within the Black and white populations of Ferguson. Ferguson is one of numerous small municipalities in the St. Louis area. It is a city with a population of 21,000 that was once known for being a racist white suburb in which Blacks were not welcome. Today, 67 percent of the Ferguson population is Black. Some blocks are integrated, while many poor Blacks are segregated in the Oakmont, Northwinds, and Canfield Green complexes, where Michael Brown lived.

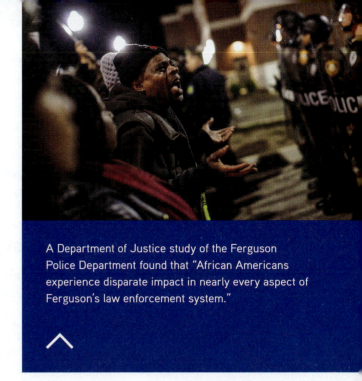

A Department of Justice study of the Ferguson Police Department found that "African Americans experience disparate impact in nearly every aspect of Ferguson's law enforcement system."

The report sets the scene of institutional discrimination quite well with statistics like these: Black drivers are more than twice as likely as white drivers to be searched during a vehicle stop (even after controlling for variables such as the reason the car was stopped), but they are found in possession of contraband 26 percent less often than white drivers. Blacks are more likely to receive multiple citations during a single incident. Notably, when speeding charges are made, Blacks fare much worse when citations are issued on the basis of the officer's visual assessment rather than radar or laser. The data demonstrate quite clearly that the disproportionate burden on African Americans cannot be explained by any difference in the rate at which people of different races violate the law. Yet, without the data collected by the DOJ, it is very hard to see an overt racial issue in any individual act of arrest. What we can see in Ferguson is that discrimination occurs through the routine practices of respected institutions like the police. It is the institution of courts and police as a whole that perpetuates a system that places such a burden on Blacks. Mass misdemeanors, citations, and summonses structure social relations in ways that are far less obvious than overt acts of discrimination.

Another way to understand how institutional racism operates in the United States is to recall that, until the 1970s, most Blacks lived either in urban ghettos or in southern states that were still marked by the remnants of Jim Crow segregation. In the urban areas marked by ghettos, it was very difficult for Blacks to find housing outside these neighborhoods, because of either racial discrimination or violence against those Blacks who tried to move into white neighborhoods. In the South, long after the Civil War, there was still a sharp "color line" that separated the races in schools, housing, and public facilities. In both the North and the South, Blacks of all socioeconomic classes led lives that were separate from whites. Middle- and upper-class Blacks were relegated to the same neighborhoods as lower-class Blacks. In other words, if you were Black, then you were Black, and that largely defined your life chances.

When we refer to the ghetto of the United States, or to southern Jim Crow, we are recognizing that racism is embedded in the structures of our political, economic, and social institutions. Even though there are many individual acts of discrimination, the racial system is not first and foremost kept alive by these individual acts of discrimination. It is kept alive and perpetuated by a larger system of segregation that was established long before those affected were even born and that exists independently of individual acts of hatred or prejudice. In fact, once people live in a highly segregated

melting pot

The idea that ethnic differences can be combined to create new patterns of behavior drawing on diverse cultural sources.

pluralism

A model for ethnic relations in which all ethnic groups in a society retain their separate identities yet share equally in the rights and powers of citizenship.

multiculturalism

The viewpoint according to which ethnic groups can exist separately and share equally in economic and political life.

institutional racism

Patterns of discrimination based on ethnicity that have become structured into social institutions.

society, it is possible for them to suffer great disadvantage based on their race without ever personally experiencing discrimination on a one-to-one basis.

Interpersonal Racism

Racial domination does not only occur at the levels of institutions; it is also enacted by individuals. Although interpersonal racism can take the form of blatant bigotry, it also has much more subtle, less obvious forms, including color-blind racism and microaggressions (Desmond and Emirbayer 2016).

OVERT RACISM: RACISM WITH RACISTS

Although many people are good at keeping their racist thoughts to themselves, it would be a mistake to conclude that overt racist acts have disappeared or that large numbers of people are not victimimized by them. In everyday life, racism can be expressed in the ideas held by bigoted individuals. It is expressed overtly through individual attitudes, perceptions, and beliefs and is sustained by the ideologically racist statements of political leaders.

The 2016 U.S. presidential election gave voice to the kind of overt racism that has been highly unusual in American politics during the lifetimes of most students reading this book. The following list gives just a few examples of overt racism, beginning with the campaign of Donald Trump and then emanating from his presidency:

1. Trump began his presidential campaign on June 16, 2015, by attacking Mexican migrants to the United States as "rapists" and "criminals," saying, "When Mexico sends its people, they're not sending their best. . . . They're sending people that have lots of problems, and they're bringing those problems with us. They're bringing drugs. They're bringing crime. They're rapists. And some, I assume, are good people." Trump's rhetoric was aimed at painting Mexicans with one broad brush and suggesting that most of the migrants were dangerous. These claims were the basis of his arguments for building a giant border wall between the United States and Mexico.

2. After the Democratic National Convention, Donald Trump attacked Khizr and Ghazela Khan, the Pakistani American parents of a Muslim U.S. Army officer who died in the Iraq War. The parents spoke at the convention, accusing Trump of not understanding the U.S. Constitution: "Have you ever been to Arlington Cemetery? Go look at the graves of brave patriots who died defending the United States of America," said the soldier's father. "You will see all faiths, genders, and ethnicities. You have sacrificed nothing—and no one." Later Trump lashed out at the couple, suggesting that Mrs. Khan had been silent during the speech because Muslim women are held in an inferior position.

3. Trump refused to disavow white supremacists when they made complimentary comments about him and his campaign. David Duke, the former leader of the KKK, stated on his radio show that not voting for Trump is "really treason to your heritage."

4. Trump gave support to white supremacist protesters who objected to the removal of a statue of Confederate general Robert E. Lee in Charlottesville, Virginia.

A young girl joins members of the Ku Klux Klan at a demonstration against the Martin Luther King Day holiday in Pulaski, Tennessee.

Even after one white supremacist drove a car into a crowd of counterprotesters—killing one and wounding at least thirty-four others—the president said the rally had some "very fine" protesters and "blame on both sides." He added, "You had a group on one side that was bad and you had a group on the other side that was also very violent." (Merica 2017; *Al Jazeera* 2017; Astor, Caron, and Victor 2017).

5. While the Trump administration was criticized by local leaders for slow and limited assistance during and after Hurricane Maria struck Puerto Rico, the president's self-defense was based on demeaning stereotypes of Puerto Ricans as lazy: He tweeted, "Such poor leadership ability by the mayor of San Juan, and others in Puerto Rico, who are not able to get their workers to help. They want everything to be done for them when it should be a community effort" (Segarra 2017).

6. On January 11, in a bipartisan meeting with lawmakers to discuss policy on immigration from Haiti, El Salvador, and African countries, Trump allegedly asked, "Why are we having all these people from shithole countries come here?" According to the *Washington Post*, "Trump then suggested that the United States should instead bring more people from countries like Norway" (Dawsey 2018).

In August 2017, white nationalists gathered in Charlottesville, Virginia, for a Unite the Right rally to protest the removal of a Confederate statue.

Although President Trump claimed to be "the least racist person that you have ever met," he has continued to make one bigoted statement after another. These statements tend to paint one or another racial group with a broad brush. While many sociologists believed we had entered an era in which racism would only be expressed with great subtlety—that people in positions of power understood the significance of speaking as though they did not see race at all—Trump's presidency upended that sort of understanding, forcing sociologists to confront a new era of overt racism.

COLOR-BLIND RACISM: RACISM WITHOUT RACISTS

Over the past several decades, some sociologists have argued that racial inequality is maintained less by overt acts of racial hatred than by color blindness itself. Sociologist Eduardo Bonilla-Silva (2006) defines color blindness as a means of maintaining racial inequality without appearing racist. First, many whites believe that they are above racism and incapable of perpetuating discrimination. They are thus unaware of the ways in which their insensitivity is psychologically damaging to racial minorities. Second, by attempting to act as if race does not exist, they perpetuate inequalities that can only be addressed by explicitly attention to racial differences. Third, many whites who do make subtle or even explicitly racial distinctions have become quite adept at maintaining an appearance of neutrality. In all these ways, much of racial inequality is maintained through the appearance of color-blind processes.

One significant aspect of color blindness is how much whites can take for granted. Many of those who profess to be "color-blind," for example, don't recognize the many ways in which they benefit from their whiteness. Just as many Blacks must take it for granted that racism pervades all actions in a systemic way, so it is that many whites can take for

white privilege

The unacknowledged and unearned assets that benefit whites in their everyday lives.

___

racial microaggressions

Small slights, indignities, or acts of disrespect that are hurtful to people of color even though they are often perpetuated by well-meaning whites.

CONCEPT CHECKS

1. Give two examples of how institutional racism operates today.

2. Relate the theory of racial formation to any example of Donald Trump's overt racism.

3. How does "color blindness" perpetuate racial inequalities?

4. What is white privilege?

5. How would you explain the harmful effects of racial microaggressions to someone who thinks you are making "a mountain out of a molehill"?

granted that they will benefit from white privilege. **White privilege** refers to the unacknowledged and unearned assets that benefit whites in their everyday lives. It manifests itself in the most taken-for-granted conditions of everyday life.

In a powerful metaphor, women's studies scholar Peggy McIntosh (1988) likens white privilege to "an invisible weightless knapsack of special provisions, assurances, tools, maps, guides, codebooks, passports, visas, clothes, compass, emergency gear, and blank checks" (1988). McIntosh then goes on to unpack this invisible knapsack by detailing more than forty "special circumstances and provisions" she experiences as a white person that her African American counterparts cannot similarly expect in their day-to-day lives:

1. Make arrangements to hang out with people of one's own race most of the time.

2. Rent or purchase housing in an area one can afford and where one wants to live.

3. Assume that the people living next door will treat one with respect.

4. Go shopping alone without being followed around the store or harassed.

5. Turn on the TV and expect to see other people of one's race most of the time (1988).

In everyday life, racial minorities often experience brief interactions that send demeaning messages and appear to the victims to be based on their race. Unlike acts of overt racism, these interactions are perpetuated by people who are often well meaning and well intentioned. They even wish to be color-blind. Whereas the racial minority member experiences the interaction as an insult, the white perpetrator can be shocked to discover there has been any incident at all. At times, the perpetrator will claim that the minority has misunderstood an "innocent" comment or is making a "mountain out of a molehill." Often these interactions are experienced silently, with the victim never expressing the outrage he or she silently feels.

The idea of microaggression was originally proposed in the 1970s by Chester M. Pierce, an African American psychiatrist at Harvard (Pierce and Dimsdale 1986). In recent years, it has caught on due to the further work of Derald Wing Sue, an Asian American psychologist at Columbia. **Racial microaggressions** are small slights, indignities, or acts of disrespect that are hurtful to people of color even though they are often perpetuated by well-meaning whites. Among the kinds of incidents Sue cited as examples of microaggressions are people asking Asian Americans where they were born or telling them that they "speak good English." Such comments suggest that they are immigrants, even when they and their families have been in this country for generations (Sue 2010).

While perpetrators might question the existence of racial microaggressions, their impact on people of color is severe and deleterious. Scholars from psychology to sociology have documented how experiencing racial microaggressions creates cumulative psychological and physiological stress, negatively impacting mental health (e.g., Feagin and McKinney 2003; Nadal et al. 2014; Ong et al. 2013; Smith, Allen, and Danley 2007; Sue 2010; Embrick, Domínguez, and Karsak 2017). Research from schools and universities emphasizes how structures and institutions can facilitate racial microaggressions through, for example, campus policing practices (Embrick, Domínguez, and Karsak 2017).

What Are the Origins and Nature of Ethnic Diversity in the United States?

Familiarize yourself with the history and social dimensions of ethnic relations in America.

More than most other societies in the world, the United States is peopled almost entirely by immigrants. Only a tiny minority, less than 1 percent, of the population today are Native Americans—those whom Christopher Columbus, erroneously supposing he had arrived in India, called "Indians." Before the American Revolution, British, French, and Dutch settlers established colonies in what is now the United States. Some descendants of the French colonists are still to be found in parts of Louisiana. Millions of slaves were brought over from Africa to North America. Huge waves of European, Asian, and Latin American immigrants have washed across the country at different periods since then.

The United States is one of the most ethnically diverse countries on the face of the globe. In this section we will pay particular attention to the divisions that have separated whites and nonwhite minority groups, such as African Americans and Asians. The emphasis is on struggle. Members of these groups have made repeated efforts to defend the integrity of their cultures and advance their social position in the face of persistent prejudice and discrimination from the wider social environment.

Early Colonization

The first European colonists in what was to become the United States were actually of a quite homogeneous background. At the time of the Declaration of Independence, the majority of the colonial population was of British descent, and almost everyone was Protestant. Settlers from outside the British Isles were at first admitted only with reluctance, but the desire for economic expansion meant having to attract immigrants from other areas. Most came from countries in northwest Europe, such as Holland, Germany, and Sweden; such migration into North America dates initially from around 1820. In the century following, about 33 million immigrants entered the United States. No migrant movement on such a scale had ever been documented, nor has such a migration occurred since.

The early waves of immigrants came mostly from the same countries of origin as the groups already established in the United States. They left Europe to escape economic hardship and religious and political oppression and for the opportunity to acquire land as the drive westward gained momentum. As a result of successive potato famines that had produced widespread starvation, 1.5 million people migrated from Ireland. The Irish were accustomed to a life of hardship and despair. In contrast with other immigrants from rural backgrounds, most Irish settled in urban industrial areas, where they sought work.

A major new influx of immigrants arrived in the 1880s and 1890s, this time mainly from southern and eastern Europe—the Austro-Hungarian Empire, Russia, and Italy. Each successive group of immigrants was subject to considerable discrimination on the part of people previously established in the country. Negative views of the Irish, for example, emphasized their supposedly low level of intelligence and drunken behavior. But as

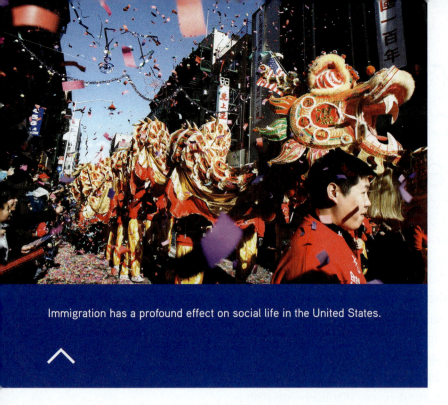

Immigration has a profound effect on social life in the United States.

they were concentrated within the cities, the Irish Americans were able to organize to protect their interests and gained a strong influence over political life. The Italians and Polish, when they reached America, were in turn discriminated against by the Irish.

Asian immigrants first arrived in the United States in large numbers in the late nineteenth century, encouraged by employers who needed cheap labor in the developing industries of the West. Some 200,000 Chinese immigrated to the United States during this period. Most were men, who came with the idea of saving money to send back to their families in China, anticipating that they would also later return there. Bitter conflicts broke out between white workers and the Chinese when employment opportunities diminished. The Chinese Exclusion Act, passed in 1882, cut down further immigration to a trickle until after World War II.

Japanese immigrants began to arrive not long after the Chinese Exclusion Act was passed. They were also subject to great hostility from whites. Opposition to Japanese immigration intensified in the early part of the twentieth century, leading to strict limits, or quotas, being placed on the numbers allowed to enter the United States. Most immigrant groups in the early twentieth century settled in urban areas and engaged in the developing industrial economy. They also tended to cluster in ethnic neighborhoods of their own. Chinatowns, Little Italys, and other clearly defined areas became features of most large cities. The very size of the influx provoked backlash from the Anglo-Saxon segment of the population. One result was the new immigration quotas of the 1920s, which restricted immigration from southern and eastern Europe. Many immigrants found the conditions of life in their new land little better and sometimes worse than in their homelands.

Immigrant America in the Twentieth and Twenty-First Centuries

If globalization is understood as the emergence of new patterns of interconnection among the world's peoples and cultures, then surely one of the most significant aspects of globalization is the changing racial and ethnic composition of societies worldwide. In the United States, shifting patterns of immigration since the end of World War II have altered the demographic structure of many regions, affecting social and cultural life in ways that can hardly be overstated. Although the United States has always been a nation of immigrants (with the obvious exception of Native Americans), most of those who arrived here prior to the early 1960s were European.

As we just discussed, throughout the nineteenth and early twentieth centuries, vast numbers of people from Ireland, Italy, Germany, Russia, and other European countries flocked to America in search of a new life, giving a distinctive European bent to American culture. (Of course, until 1808, another significant group of immigrants—Africans—came

FIGURE 10.1

Racial and Ethnic Composition of the United States, 1900–2050

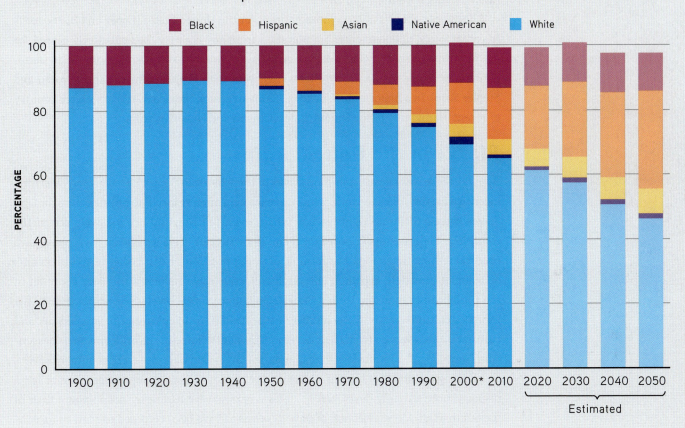

Legend: Black · Hispanic · Asian · Native American · White

Total exceeds 100 percent because starting in 2000, respondents were allowed to identify themselves as belonging to more than one racial category.

Sources: U.S. Bureau of the Census 2004, 2010f.

not because America was a land of opportunity but because they had been enslaved.) In part because of changes in immigration policy, however, more than three-quarters of the nearly 59 million immigrants admitted since 1965 have been Asian or Hispanic. This surge in immigration from Latin America and Asia has significantly altered the racial and ethnic composition of the United States (see Figure 10.1): The Hispanic share of the U.S. population jumped from 3.5 percent in 1960 to nearly 18 percent in 2015, while the Asian share rose from less than 1 percent in 1960 to 5.3 percent in 2015 (Flores 2017).

In 2016, there were nearly 44 million foreign-born individuals in the United States, composing over 13 percent of the total population. This represents a fourfold increase since 1960, when immigrants represented just over 5 percent of the U.S. population (Zong, Batalova, and Hallock 2018). In contrast to the major wave of immigration in the 1880s and 1890s, just 11 percent of the immigrant population today is of European origin. In fact, 51 percent have come from Latin America, including 26 percent from Mexico, and another 31 percent have come from Asia (Migration Policy Institute 2018). This change is attributed to two government acts: the 1965 Immigration and Nationality Act Amendments, which abolished preference for northern and western European immigrants and gave preference to "family reunification"—rather than occupational skills—as a reason for accepting immigrants, and the 1986 Immigration Reform and Control Act, which provided amnesty for many illegal immigrants.

Each year, about a million immigrants enter the United States (López and Bialik 2017). In recent years, undocumented immigration from any country "has effectively ceased for the first time in six decades" (Massey 2012, p. 13). Notably, since 2010, more Asian immigrants have come to the United States each year than Hispanic immigrants (Brown and Stepler 2016). In a historic shift, in 2013, China overtook Mexico as the top source country of recent immigrants to the United States (Chishti and Hipsman 2015).

Blacks in the United States

By 1780, there were nearly 4 million slaves in the American South. Because there was little incentive for the slaves to work, white slave owners often resorted to physical punishment. Slaves had virtually no rights in law whatsoever. But they did not passively accept the conditions their masters imposed on them. The struggles of slaves against their oppressive conditions sometimes took the form of direct opposition or disobedience to orders and, occasionally, outright rebellion (although collective slave revolts were more common in the Caribbean than in the United States). On a more subtle level, their response took the form of a cultural creativity—a mixing of aspects of African cultures, Christian ideals, and cultural threads woven from their new environments. Some of the art forms their descendants developed, as in music—for example, the invention of jazz—were genuinely new.

Feelings of hostility toward Blacks on the part of the white population were in some respects more strongly developed in states where slavery had never been known than in the South itself. Moral rejection of slavery seems to have been confined to a few more educated groups. The formal abolition of slavery changed the real conditions of life for African Americans in the South relatively little. The "Black Codes"—laws limiting the rights of Blacks—placed restrictions on the behavior of the former slaves and punished their transgressions in much the same way as under slavery. Acts were also passed legalizing segregation of Blacks from whites in public places. One kind of slavery was thus replaced by another, based on social, political, and economic discrimination.

INTERNAL MIGRATION FROM SOUTH TO NORTH

Industrial development in the North, combined with the mechanization of agriculture in the South, produced a progressive movement of African Americans northward from the turn of the century on. In 1900, more than 90 percent of African Americans lived in the South, mostly in rural areas. Today, less than half of the Black population remains in the South; three-quarters now live in northern urban areas. African Americans used to be farm laborers and domestic servants, but over a period of little more than two generations, they have become mainly urban, industrial, and service-economy workers. But African Americans have not become assimilated into the wider society in the same way the successive groups of white immigrants were. They have for the most part been unable to break free from the conditions of neighborhood segregation and poverty that other immigrants faced on arrival. Together with those of Anglo-Saxon origin, African Americans have lived in the United States far longer than most other immigrant groups. What was a transitional experience for most of the later white immigrants has become a seemingly permanent experience for Blacks.

In the majority of cities, in both the South and the North, Blacks and whites live in separate neighborhoods and are educated in different schools. Demographers have developed a statistic called the index of dissimilarity, which tells us the proportion of people who would need to move out of their neighborhood into a new neighborhood for the distribution of people in neighborhoods to approximate the overall racial breakdown of the United States. According to 2010 Census data, roughly 63 percent of either Blacks or whites would have to move to desegregate housing fully in the average American city (Nasser 2010).

Hispanics and Latinos in the United States

The wars of conquest that created the boundaries of the contemporary United States were directed not only against the Native American population but also against Mexico. The territory that later became California, Nevada, Arizona, New Mexico, and Utah—along with a quarter of a million Mexicans—was taken by the United States in 1848 as a result of the U.S. war with Mexico. The terms *Mexican American* and *Chicano* include the descendants of these people, together with subsequent immigrants from Mexico. The term *Latino* refers to people descended from Latin America, whereas *Hispanic* tends to refer to anyone living in the United States descended from Spanish-speaking regions.

In 2016, the more than 57 million Hispanics in the United States represented 18 percent of the population. The four largest groups of Latinos in the United States today are Mexican Americans (35.8 million), Puerto Ricans (5.4 million), Salvadorans (2.2 million), and Cubans (2.1 million) (Flores 2017). After the passage of the Immigration and Nationality Act of 1965, the Hispanic population grew at an extraordinary rate—by 53 percent between 1980 and 1990, 58 percent between 1990 and 2000, and 44 percent between 2000 and 2010—mainly as a result of the large-scale flow of new immigrants across the Mexican border (Flores 2017; U.S. Bureau of the Census 2011g). In fact, the growth of the Hispanic population accounts for half of total national population growth since 2000 (Flores 2017).

In recent years, Hispanic population growth has slowed as immigration from Latin America, especially Mexico, has decreased (Flores 2017). While the number of temporary legal worker entries has surged, legal immigration appears stable, and for the first time, undocumented migration has effectively ceased. Studies show that the halt in undocumented migration has resulted not from enhanced border enforcement but mainly from "larger shifts in the North American political economy" (Massey 2012, p. 13). New temporary, legal means for Mexicans to enter the United States, declining U.S. labor demand, deaccelerating Mexican population growth, and the relative stability of the Mexican economy better explain the sharp decline in rates of Latin American immigration (Massey 2012).

MEXICAN AMERICANS

Mexican Americans reside mainly in the West and Southwest, with more than half living in California or Texas (Zong, Batalova, and Hallock 2018). The majority work at low-paying jobs. In the post–World War II period up to the early 1960s, Mexican workers were admitted without much restriction. This was followed by a phase of quotas on legal immigrants and deportations of undocumented immigrants. Undocumented immigrants can be employed more cheaply than other workers, and they perform jobs that most of the rest of the population would not accept. Legislation passed by Congress in 1986 enabled undocumented immigrants living in the United States for at least five years to claim legal residence. In the past decade, overall immigration from Mexico has dropped significantly; more

Mexicans have left the United States than entered the country since the Great Recession ended (Gonzalez-Barrera 2015; Chishti and Hipsman 2015).

Mexicans in the United States typically have levels of economic well-being and educational attainment far below those of native-born Americans. In 2015, 23 percent lived below the poverty line (Flores 2017). More than two-thirds (69 percent) of Mexicans in the United States are proficient in English; however, only around 11 percent hold bachelor's degrees. Social scientists anticipate that Mexican immigrants and their children may become increasingly assimilated into life in the United States in coming decades, due in part to policies that help them obtain an affordable college education. As of 2018, seventeen states have passed laws permitting certain undocumented students who have attended and graduated from their primary and secondary schools to pay in-state tuition at state colleges. Given that about half of undocumented immigrants in the United States hail from Mexico, these policies—the source of ongoing legislative contention in several states—will have a major impact on the lives of young Mexican immigrants.

PUERTO RICANS

Puerto Rico was acquired by the United States through war, and Puerto Ricans have been American citizens since 1917. The island is poor, and many of its inhabitants have migrated to the mainland United States to improve their conditions of life. Puerto Ricans originally settled in New York City, but since the 1960s, they have moved elsewhere. A reverse migration of Puerto Ricans back to the island began in the 1970s. In recent years, however, record numbers of Puerto Ricans have been migrating to the United States to escape the island's decade-long recession. Owing to migration to the mainland as well as declining fertility rates among Puerto Rican women, there are more Puerto Ricans living in the United States than on the island (Flores 2017).

One of the most important issues facing Puerto Rican activists is the political destiny of their homeland. Puerto Rico is at present a commonwealth of the United States. As such, Puerto Ricans residing on the island are U.S. citizens, yet they do not pay federal income tax, nor can they vote for president of the United States. For years, Puerto Ricans have been divided about whether the island should retain its present status, opt for independence, or attempt to become the fifty-first state of the Union. In June 2017, 97 percent of Puerto Ricans voted in favor of statehood in a nonbinding referendum; however, less than one-quarter of registered voters actually cast ballots due to boycotts. The vote came just a few weeks after the country declared a form of bankruptcy (Robles 2017).

CUBANS

A third Latino group in the United States, Cubans, differs from the two others in key respects. Half a million Cubans fled communism following the rise of Fidel Castro in 1959, and the majority settled in Florida. Unlike other Latino immigrants, they were mainly educated people from white-collar and professional backgrounds. They have managed to thrive within the United States, many finding positions comparable with those they abandoned in Cuba.

A further wave of Cuban immigrants, from less affluent origins, arrived in 1980. Lacking the qualifications held by the first wave, these people tend to live in circumstances closer to those of the rest of the Latino communities in the United States. Both sets of

Cuban immigrants are mainly political **refugees** rather than economic migrants. The later immigrants to a large extent have become the "working class" for the earlier immigrants. They are paid low wages, but Cuban employers tend to take them on in preference to members of other ethnic groups.

The Asian Connection

About 6 percent of the population of the United States is of Asian origin. Chinese, Indian-origin Asians, and Filipinos (immigrants from the Philippines) form the largest groups, but there are also significant numbers of Vietnamese, Koreans, and Japanese living in the United States.

Most of the early Chinese immigrants settled in California, where they were employed mainly in heavy industries, such as mining and railroad construction. The retreat of the Chinese into distinct Chinatowns was not primarily their choice but was made necessary by the hostility they faced. The early Japanese immigrants also settled in California and the other Pacific states. During World War II, following the attack on Pearl Harbor by Japan, all Japanese Americans in the United States were made to report to "relocation centers," which were effectively concentration camps. Despite the fact that most of these people were American citizens, they were compelled to live in the hastily established camps for the duration of the war. Paradoxically, this situation eventually led to their greater integration within the wider society, because, following the war, Japanese Americans did not return to the separate neighborhoods in which they had previously lived. They have become extremely successful in reaching high levels of education and income, marginally outstripping whites.

Following the passage of a new immigration act in 1965, large-scale immigration of Asians into the United States again took place. Between 2000 and 2015, the U.S. Asian population grew 72 percent, faster than any other major racial or ethnic group, including Hispanics (López, Ruiz, and Patten 2017). Foreign-born Chinese Americans today outnumber those brought up in the United States. The newly arrived Chinese have avoided the Chinatowns in which the long-established Chinese have tended to remain, mostly moving into other neighborhoods.

refugees
People who have fled their homes due to a political, economic, or natural crisis.

CONCEPT CHECKS

1. Describe patterns of immigration to the United States before and after the 1960s.

2. How has the racial and ethnic composition of the United States changed since the 1960s?

3. Briefly contrast the immigration experiences of whites, African Americans, Latinos, and Asians to the United States.

4. Using the examples of African Americans and Asians, explain how neighborhood segregation may result from and perpetuate racial inequality.

How Do Race and Ethnicity Affect the Life Chances of Different Groups?

Learn how racial and ethnic inequality is reflected in terms of educational and occupational attainment, income, health, residential segregation, and political power.

Since the civil rights movement of the 1960s, has real progress been made in eliminating racial inequality? One of the driving questions of sociology is whether racial and ethnic inequality is primarily the result of factors associated with race and ethnicity or whether it is mainly a product of social and economic class position. For example, are U.S. Blacks disproportionately poor because of factors associated with race? Or do socioeconomic factors better explain why some of the Black population lives in poverty?

In this section, we will first examine the facts: how racial and ethnic inequality is expressed in terms of educational and occupational attainment, income and wealth, health, residential segregation, and political power. We will then look at the range of outcomes found within the largest racial and ethnic groups. But first, some warnings: Any comparisons of the kind undertaken here can be misleading. Racial groups, such as Blacks, Asians, and whites, are characterized by significant variation, and no single statistic can accurately represent the whole or tell you about any individual you encounter. Likewise, Hispanics are not merely diverse in their outcomes but also consist of both whites and nonwhites and therefore cannot easily be compared with groups that are more clearly racial.

Furthermore, the very names we use for ethnic and racial groups can sometimes mask complications and political choices. In this section, racial and ethnic categories are taken from the U.S. Census, a primary data source for measures of inequality. "Asian" encompasses people with origins in East and Southeast Asia and the Indian subcontinent. Sometimes referred to as Caucasians, "white" refers to people with origins in Europe, the Middle East, and North Africa. "Hispanic" refers to people descending from Spanish-speaking countries, though those from Latin America may also be called "Latinos." Hispanics and Latinos may be of any race, thus complicating comparisons among racial groups. "Black" encompasses African Americans and others of the African diaspora. Although we refer to whites in lowercase, we capitalize the term *Black* in recognition of the respect it confers on a minority group with a self-identity. As W. E. B. Du Bois argued in 1925, U.S. Blacks have "just as much right to a capital letter as 'Jew' or 'Irish' or 'Aryan' or 'Caucasian' or a dozen other similar words" (W. E. B. Du Bois 1925).

Educational Attainment

Differences between Blacks and whites in levels of educational attainment have decreased, but these seem more the result of long-established trends rather than the direct outcome of the struggles of the 1960s. After steadily improving their levels of educational attainment for the last fifty years, African Americans are for the first time close to whites in terms of finishing high school (see Figure 10.2). The proportion of Black adults with at least a high school education has increased from about 20 percent in 1960 to 87 percent in 2017. Similarly, 91 percent of Asian adults and 94 percent of non-Hispanic white adults have completed high school. However, disparities in higher educational attainment persist: 55 percent of Asians twenty-five and over have at least a bachelor's degree. This is significantly higher than the rates for both non-Hispanic white adults (38 percent) and Black adults (24 percent) (U.S. Bureau of the Census 2017b).

The situation for Hispanic adults is also striking. Just 71 percent of Hispanic adults of any race have a high school education. They have

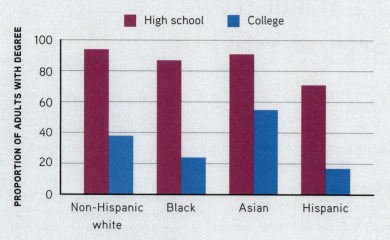

FIGURE 10.2

Educational Attainment by Race and Ethnicity in 2017

Source: U.S. Bureau of the Census 2017b.

the highest high school dropout rate of any group in the United States: 9.2 percent compared to 4.6 percent for whites and 6.5 percent for Blacks (McFarland et al. 2017). While rates of college attendance and success in graduation have gradually improved for other groups, the rate for Hispanics has held relatively steady since the mid-1980s. In 2017, just 17 percent of Hispanic adults held bachelor's or more advanced degrees (U.S. Bureau of the Census 2017b). It is possible that these poor results can be attributed to the large number of poorly educated immigrants from Latin America who have come to the United States in the last two decades. Many of these immigrants have poor English language skills, and their children encounter difficulties in schools.

Employment and Income

How does this educational attainment translate into the workplace? Let us now consider racial inequality in employment and income. Since 1970, Black men and women's unemployment rate has remained around two times that of white men and women (U.S. Bureau of the Census 1999; Fairlie and Sundstrom 1999). This remains true today: In 2017, 8.1 percent of Black men were unemployed compared to just 3.8 percent of white men (U.S. Bureau of Labor Statistics 2018e). However, this gap is considerably smaller among more educated persons aged twenty-five years and older. In 2017, the unemployment rates for adults with a bachelor's degree or more were 2.1 percent, 3.6 percent, and 2.9 percent for whites, Blacks, and Hispanics, respectively (U.S. Bureau of Labor Statistics 2018f).

Especially in the wake of the recent economic recession, there has been some debate about whether employment opportunities for minorities have improved or worsened. Statistics on unemployment don't adequately measure economic opportunity, because they count only those known to be looking for work. A higher proportion of Blacks and Hispanics have opted out of the occupational system, neither working nor looking for work. They have become disillusioned by the frustration of searching for employment that is not there (U.S. Bureau of Labor Statistics 2017d). Unemployment figures also do not reflect the increasing numbers of young men from minority groups who have been incarcerated (see also Chapter 6). Finally, of the 7.2 million jobs lost during the Great Recession, 5.6 million were jobs for workers with a high school diploma or less; of the 11.6 million jobs added since the bottom of the recession, 99 percent of them have gone to workers with at least some college education (Carnevale, Jayasundera, and Gulish 2016). As we just saw, Blacks and Hispanics are particularly underrepresented among college graduates.

The Black–white gap in household income remains relatively unchanged. In 2016, median household income for Blacks was $39,490, just over 60 percent of median household income for whites ($65,041) (Semega, Fontenot, and Kollar 2017). Increasingly larger than both the wage or income gap is the gap in household wealth: Both the median and average household wealth of Black families in 2016 were less than 15 percent those of white families—Black median household wealth was $17,600 compared to $171,000 for whites (Dettling et al. 2017). Differences in homeownership drive part of the household wealth gap: 73 percent of white households own homes, compared to just 45 percent of Black households. Rates of family inheritance, vehicle ownership, retirement accounts, business ownership, equity ownership, and debt also diverge by race (Dettling et al. 2017).

Having been systematically targeted for risky subprime loans, minority households were therefore disproportionately impacted by the recent foreclosure crisis (Rugh

What Race Am I?

Pakistani	Asian
Chinese	Puerto Rican
South American	South Asian

Apps like Don't Guess My Race highlight the fluidity and complexity of racial identity.

What *Are* You, Anyway?

The number of Americans who identify as multiracial has increased dramatically over the past decade and will continue to increase in the future as more and more young adults marry and have children with partners of a race different from their own. As we learned earlier in this chapter, multiracial individuals and families challenge us to reexamine the ways we think about race. What does a "Black" person look like? What does a "Latino" person look like? Is it even possible to determine one's racial identity based on his or her physical features?

Developed by Interactive Diversity Solutions, the Don't Guess My Race web app was designed to help users better understand and question the complexities of race. Don't Guess My Race is a game where a user is presented with photos of people's faces. For each photo, users are asked to click the one answer (of six possible options) that they believe best describes how the person in the picture racially self-identifies. The app then reveals the "correct" answer, which is accompanied by a quote from the person explaining why the person identifies as he or she does. For example, one user explained, "I consider myself African American, . . . but most people think I'm Asian because of the shape of my eyes." Readers are then directed to more information about the history of race identity in the United States and throughout the world, as well as demographic information based on the 2010 Census. According to the app's website, the program was developed to challenge users' assumptions about individuals and about race (Interactive Diversity Solutions 2018). The app's creators, cultural anthropologist Michael Baran and game producer Michael Handelman, believe that the app will demonstrate how "something that is considered natural and biological is actually a result of complex historical and cultural constructions" (Interactive Diversity Solutions 2018).

Baran also believes it is important for app users to interrogate the concept of whiteness. Although many people who are light complexioned may believe they are simply "white" and that the issue of race does not apply to them, Don't Guess My Race gives users six different options for people who appear to be white—often capturing different ways that whiteness is discussed, such as "elite" or "redneck" (Latour 2011).

Both consumer reviews of the game and assessments by race scholars have been largely positive, noting that the game forces users to think about the complexities of race, to challenge their own assumptions about what different races "look like," and to learn about the history of racial stratification in the United States. One user criticized the app on the grounds that it conflated race with ethnicity; for example, one of the options included along with a photo of a white person was the adjective *Polish*, which is technically an ethnicity rather than a race (Matthews 2011). However, most believe that the app is a clever way to reveal how race is socially constructed. What do you think? What race or races do you identify with, and why? Do you think that app users looking at your photograph would correctly identify your self-perceived racial identity? Why or why not?

and Massey 2010). During the collapse of the housing market bubble and subsequent Great Recession, Black and Hispanic household wealth were impacted severely—falling 53 percent and 66 percent, respectively—while white households experienced just a 16 percent decrease in wealth (Kochhar, Fry, and Taylor 2011). As such, wealth gaps widened between Black and white households and Hispanic and white households (Kochhar, Fry, and Taylor 2011).

In 2016, median Hispanic household income reached $47,675, above that of Black households ($39,490) but below white and Asian households ($65,041 and $81,431, respectively)—divergences that have remained constant since the 1960s (Semega, Fontenot, and Kollar 2017). This household-level difference between Hispanics and Blacks is likely due to larger Hispanic households containing a greater number of income providers (Flores, López, and Radford 2017).

Health

Disparities in health and health care among officially recognized racial and ethnic categories of the U.S. population are well documented (U.S. Department of Health and Human Services 2016). These range from infant mortality and preterm birthrates (both highest among Black babies and lowest among Asians or Pacific Islanders) to hypertension (most prevalent among Black men and women, at a rate of 42 percent, and lowest among Hispanic men and Asian women, at 28 percent and 25 percent, respectively) (U.S. Department of Health and Human Services 2016). Progress has been made (Kochanek, Arias, and Anderson 2015), but alarming racial disparities do remain—explained not by poverty alone but also by the persistence of racism.

For example, analyzing a nationally representative sample of native-born Black households, Monk (2015) finds that skin tone significantly predicts multiple types of perceived discrimination, which, in turn, significantly predicts important health outcomes. The evidence is clear that discrimination compromises quality of health and health care: Controlling for socioeconomic and access-related variables, such as insurance and income, researchers find that "racial and ethnic minorities tend to receive a lower quality of health care than non-minorities" (Committee on Understanding and Eliminating Racial and Ethnic Disparities in Health Care 2003, p. 8).

Before the turn of the century, depending on where they resided, Black women were two to six times more likely to die from complications of pregnancy than white women (American Medical Association 1999). In some parts of the country, this ratio has only grown. In addition to higher rates of known health risks and poverty among Black women, the cumulative effects of institutional and everyday racial discrimination—chiefly, stress and lower-quality care—drive the racial disparity in maternal mortality.

For example, a study conducted by the New York City Department of Health and Mental Hygiene (2016) found that between 2006 and 2010, Black women were twelve times more likely to die from pregnancy-related causes than white women; this is up from seven times more likely between 2001 and 2005 (Fields 2017). Even after controlling for risk factors like education, neighborhood poverty level, and prepregnancy obesity, from 2006 to 2010, New York City–based Black women were still three times more susceptible than white women to potentially life-threatening complications of pregnancy (New York City Department of Health and Mental Hygiene 2016). Similarly, a national-level study of Black and white women with five common pregnancy complications between 1988 and 1999 found no racial difference in their prevalence, but Black women with those conditions were two to three times more likely than white women to die from them (Tucker et al. 2007).

Residential Segregation

Neighborhood segregation has declined little over the past quarter-century, remaining especially high in metropolitan areas (Turner et al. 2013). A nationwide study of metropolitan areas shows that Blacks are considerably more segregated from whites than are Asians and Hispanics (Turner et al. 2013, p. xxii). A number of studies examine mechanisms underpinning that segregation, showing that discriminatory practices toward Black and white clients in the housing market continue (Desmond 2016; Pager and Shepard 2008). Still, while explicit "door slamming" discrimination may have fallen, Hispanic, Asian, and Black renters and homebuyers are all more likely than whites to experience discrimination when soliciting information on available units, viewing units, and renting or purchasing a home, extending the time and cost of housing searches (Turner et al. 2013).

In *American Apartheid* (1993), Douglas Massey and Nancy A. Denton argue that the history of racial segregation and its specific urban form, the Black ghetto, are responsible for the perpetuation of Black poverty and the continued polarization of Black and white. The persistence of segregation, they say, is not a result of impersonal market forces. Even many middle-class Blacks still find themselves segregated from white society. For them, as for poor Blacks, this becomes a self-perpetuating cycle. Affluent Blacks who can afford to live in comfortable, predominantly white neighborhoods may deliberately choose not to because of the struggle for acceptance they know they would face. The Black ghetto, the authors conclude, was constructed through a series of well-defined institutional practices of racial discrimination—private behavior and public policies by which whites sought to contain growing urban Black populations. Until policymakers, social scientists, and private citizens recognize the crucial role of such institutional discrimination in perpetuating urban poverty and racial injustice, the United States will remain a deeply divided and troubled society.

Residential segregation is connected to educational segregation. Over the past half-century, the South—currently the least segregated region for Black students—has lost all gains in desegregation made since the 1954 *Brown v. Board of Education* decision (Orfield and Frankenberg 2014). Nationally, segregation typically means that Black and Hispanic students tend to study in majority-poor schools, while Asian and white students generally study in middle-class schools (Orfield and Frankenberg 2014, p. 2). Black and white children now attend the same schools in most rural areas of the South and in many small and medium-sized cities throughout the country. In contrast, educational segregation is most acute in the central cities of large metropolitan areas, followed by their suburbs (Orfield and Frankenberg 2014, p. 2).

Together, a number of institutionalized factors—including lack of oversight or withdrawal of desegregation plans, school choice funding incentives, housing segregation and the drawing of school districts, and incentives for educators to move from segregated-minority schools to majority-white or integrated schools—worsen education segregation. What are its consequences? Unfortunately, students from all backgrounds miss out on "substantial benefits for educational and later life outcomes" from desegregated schooling (Orfield and Frankenberg 2014, p. 2).

Political Power

Barack Obama made history when he was elected the first Black president of the United States in 2008 and was reelected in 2012. While symbolizing progress and hope for many, Obama's presidency was also marked by "reticence [to discuss or] to carry out any race-based initiative" (Bonilla-Silva 2015, p. 1367). Toward the end of Obama's second term, Black and Latino socioeconomic status had further declined relative to that of whites (Bonilla-Silva 2015, p. 1367). Significantly, while Obama's election was heralded as demonstrating a decline in racial prejudice among the U.S. electorate, the 2016 election of Donald Trump serves to counter this notion.

Nevertheless, Obama's presidency exists within a larger trend of Blacks making tremendous gains in holding elected offices since 1970 (Joint Center for Political and Economics Studies 2000, in PBS 2018). Blacks have been voted into every major political office, including in districts where white voters predominate. The number of Black public officials surpassed 10,500 in 2010, an increase from 9,101 in 2000 and from 40 in 1960 (Bositis 2001; Joint Center for Political and Economic Studies 2011). Hispanic and Asian public officials number around 6,000 and 1,000, respectively, according to the most recent available data (Brown-Dean et al. 2015).

Despite these gains in representation, Black, Hispanic, and Asian demographics are still underrepresented among elected officials at every level (Brown-Dean et al. 2015). This discrepancy is least pronounced in Congress, in part following the 1992 reshaping of congressional districts to give minority candidates more opportunity. However, in all its history, the United States has only elected ten Black senators. The divergence remains most evident at the local level; in city councils, the local positions with most consistent data, Blacks represent 5.7 percent, Hispanics 3.3 percent, and Asians 0.4 percent of public officials, far below their shares of the national population (Brown-Dean et al. 2015). Moreover, government responsiveness to electorate needs likely varies by race: Evidence suggests that local public offices may racially discriminate in providing access to services, responding less frequently and less cordially to Black inquiries than white inquiries, regardless of socioeconomic status (Giulietti, Tonin, and Vlassopoulos 2017).

Gender and Race

The status of minority women in the United States is especially plagued by inequalities. Gender discrimination and race discrimination combined make it particularly difficult for these women to escape conditions of poverty. They share the legacy of past discrimination against members of minority groups and women in general. Until about twenty-five years ago, most minority women worked in low-paying occupations, such as household work, farm work, or low-wage manufacturing jobs. Changes in the law and gains in education have allowed for more minority women to enter white-collar professions, and their economic and occupational status has improved.

Between 1979 and 2017, the median usual weekly earnings of full-time and salaried white women grew by 37 percent, while Black women's earnings grew by 23 percent and Hispanic women's earnings grew by 22 percent (U.S. Bureau of Labor Statistics

Barack Obama became the first African American president of the United States in the historic election of 2008.

FIGURE 10.4

Earnings by Race and Sex in 2017*

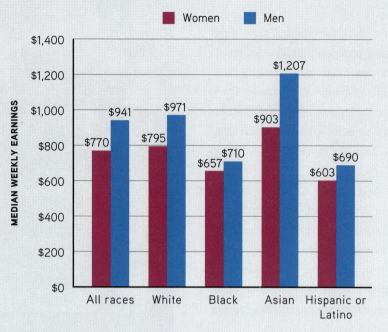

*Includes full-time and salary workers.

Source: U.S. Bureau of Labor Statistics 2018b.

2018b). Although women have made strides in earnings in the past three decades, stark race and gender earnings disparities persist (see Figure 10.4). Black women earn just 83 percent that of white women, and Hispanic women earn just 76 percent that of white women, while Asian women earn 114 percent that of white women (U.S. Bureau of Labor Statistics 2018g). In 2017, among full-time workers, white women earned about 82 percent as much as white men; Black, Hispanic, and Asian women earned 68 percent, 62 percent, and 93 percent that of white men, respectively, while Black, Hispanic, and Asian men earned 73 percent, 71 percent, and 124 percent that of white men, respectively (U.S. Bureau of Labor Statistics 2018g).

However unequal their status and pay, minority women play a critical role in their communities. They are often the major or sole wage earners in their families. Yet their incomes are not always sufficient to maintain a family. Poverty rates are higher among female-headed families than male-headed or married-couple households—in 2016, more than one-third of female-headed families lived in poverty (National Women's Law Center 2016). These rates are highest for families headed by Black and Latina women, at 39 and 41 percent, respectively (National Women's Law Center 2016).

Divergent Fortunes

When we survey the development and current position of the major ethnic groups in America, one conclusion that emerges is that they have achieved varying levels of success. Despite initially facing prejudice and discrimination upon immigrating to the United States, European immigrants managed to assimilate into the wider society. This, however, has not been the case for other minority groups. These latter groups include two minorities who have lived in North America for centuries, Native Americans and African Americans, as well as Mexicans and Puerto Ricans.

THE ECONOMIC DIVIDE WITHIN THE AFRICAN AMERICAN COMMUNITY

The situation of Blacks is the most conspicuous case of divergent fortunes. A division has opened up between the minority of Blacks who have obtained white-collar, managerial, or professional jobs—who form a small Black middle class—and the majority, whose living conditions have not improved. In 1960, most of the non-manual-labor jobs open to Blacks were those serving the Black community—a small proportion of Blacks could work as teachers, social workers, or, less often, lawyers or doctors. No more than about 13 percent of Blacks held white-collar jobs, compared with 44 percent of whites. Although there has been significant progress in the fortunes of Blacks over the past five decades, pronounced racial differences persist. For example, African Americans are still underrepresented in white-collar jobs. Although Blacks account for roughly 13 percent of the U.S. population, they hold just 9 percent of all managerial, professional, or related occupations, and only 3 percent of chief executives are Black; one-quarter of all Blacks are employed in service occupations (U.S. Bureau of Labor Statistics 2017d).

THE ASIAN SUCCESS STORY

Unlike African Americans, other minority groups have outlasted the open prejudice and discrimination they once faced. The changing fate of Asian Americans is especially remarkable. Until about half a century ago, the level of prejudice and discrimination experienced by the Chinese and Japanese in North America was greater than for any other group of non-Black immigrants. Since that time, Asian Americans have achieved a steadily increasing prosperity and no longer face the same levels of antagonism. The median income of Asian Americans is now higher than that of whites.

The category "Asian" encompasses broadly varied demographics. Notably, the majority (64 percent) of Asian American and Pacific Islanders are foreign born (U.S. Department of Labor 2016). This is due to a recent rise in immigration; between 2000 and 2015, the U.S. Asian population grew from 11.9 million to 20.4 million, a 72 percent increase (López, Ruiz, and Patten 2017). Over 60 percent of Asian American and Pacific Islanders live in just five states: one-third in California, followed by New York, Texas, Hawaii, and New Jersey (U.S. Department of Labor 2016).

Some have referred to the Asian American "success story" as a prime example of what minorities can achieve in the United States. This myth of the "model minority," however, masks big discrepancies between and within different Asian groups; many Asian Americans, including those whose families have resided in the United States for generations, still live in poverty. For example, although 12 percent of all Asian Americans lived in poverty in 2015, rates vary widely among subgroups, for example, from 35 percent among Burmese Americans to 17 percent among Native Hawaiians and Pacific Islanders to 8 percent among Filipino Americans (U.S. Department of Labor 2016; López, Ruiz, and Patten 2017). And whereas 76 percent of Indian Americans and 60 percent of Korean Americans have a bachelor's degree or more, the same can be said of only 26 percent of Native Hawaiians and other Pacific Islanders (U.S. Department of Labor 2016).

In addition to concealing important differences among Asian American subgroups, holding up Asian Americans as a model minority is problematic for other reasons. Other minorities, such as Blacks, are often compared to Asians even though they have had very different histories in this country. The misguided assumption is made that if another minority group does not succeed in the United States, it must be because they have not worked as hard as Asians.

Sociologists Jennifer Lee and Min Zhou (Lee 2012; Lee and Zhou 2015) have argued that some Asian Americans benefit from these positive stereotypes about their racial group. Lee contends that some Asian Americans experience what she calls "stereotype promise," a phenomenon whereby being viewed through the lens of a positive stereotype—in this case, as smart, hardworking, and disciplined—can actually lead

Hundreds of thousands of people march in Los Angeles to demand basic rights for immigrants.

a person to act in a way that affirms the stereotype (Lee 2012). In other words, when teachers assume that their Asian American students possess certain positive traits, these preconceived ideas can actually boost students' performance, thereby reinforcing ideas about Asian Americans being intrinsically smarter and more hardworking. There are unintended consequences to these ideas about "Asian American exceptionalism": In addition to putting enormous pressure on Asian American students to succeed in school, these ideas also disadvantage Asian American students during the college admissions process (Espenshade and Radford 2009) and may marginalize those who do not reach such cultural ideals (Lee and Zhou 2014).

LATINOS: THE DEBATE ABOUT MEXICAN AMERICANS

The largest Latino group in the United States today is also the largest immigrant group: Mexican Americans. Some sociologists believe that they are a much better test case than Asians Americans for understanding the American Dream because, on the whole, they are not as well educated as many Asian Americans. Yet there is no agreement among sociologists about how to view the Mexican American experience of assimilation and mobility. Some argue that this population is stagnating. Others think that the stagnation is only to be found in certain locations and that Mexican Americans vary significantly in their long-term intergenerational outcomes.

The pessimists who believe that Mexican Americans are stagnating tend to focus on what they call **racialization**. They believe that Mexicans are becoming a racialized minority that can be compared with Blacks. This is because, like Blacks who live in a ghetto, Mexicans in cities like Los Angeles and San Antonio end up living in segregated, high-poverty communities in which institutions, such as schools, have inferior resources and foster low expectations of their students. It is not so much that these Mexican Americans experience discrimination once they enter employment. Instead, as a result of the poor education they receive, they end up being qualified for only the lowest rungs of the labor market (Telles and Ortiz 2009).

If some sociologists are more optimistic about Mexican American mobility, it is because not all members of this immigrant group remain in cities like Los Angeles and San Antonio, where their grandparents originally settled. Many of those who move out end up marrying whites and therefore have higher rates of assimilation and upward mobility. In addition, those new Mexican migrants who move to small towns and cities usually live among a smaller number of Mexicans and a larger population of whites. They therefore end up learning English and intermarrying at much higher rates (Jiménez 2010).

What turns out to be significant is the importance of physical place in these outcomes, and no one of these experiences tells the whole story.

CONCEPT CHECKS

1. What is one major driver of the Black–white gap in household wealth?

2. Explain two different ways that racism affects health.

3. Provide three explanations for the persistence of residential segregation.

4. Why is it problematic to refer to Asian Americans as model minorities?

5. What does the case of Mexican Americans teach us about space and assimilation?

How Do Sociologists Explain Racial Inequality?

One of the enduring questions in American sociology has been why there is inequality between whites and other racial and ethnic groups. As we saw in the previous section, disparities exist among major racial and ethnic groups across a range of outcomes, such as education, housing, health, and income and wealth. Much sociological theorizing has focused on the most extreme inequalities: those between Blacks and whites. While there are many possible explanations, no one explanation is sufficient. We begin with IQ-based explanations, which carry the least weight with sociologists. We then turn to cultural, economic, and discrimination theories of ethnic and racial inequality.

IQ-Based Explanations

In the past, scholars looked to intelligence to explain and to justify inequality among racial groups. One of the most controversial explanations for this inequality has been racial differences in IQ, for as a group, Blacks have long had lower IQ scores than whites (Herrnstein and Murray 1994). One common theory to explain this gap is that these IQ differences are based on the genetic makeup of the two races, for example, that Blacks have smaller brains than whites or that IQs are inherited. However, brain size does not explain differences in intellect. For instance, while men and women have significantly greater brain-size differences than do Blacks and whites, average IQ scores do not vary by gender (Nisbet 2010).

For a long time, the hereditability of IQ (as shown in twin studies) was thought to preclude explanations that are based on the social context in which people live. However, this turned out not to be the case: Environmental influences also affect individuals' IQs. For example, the average child in an upper-middle-class family will hear substantially more words per day than someone in a poor family. That early exposure is linked to vocabulary, a significant determinant of how people perform on IQ tests. And over the past half-century, the average IQ in the United States has risen by fifteen to twenty points. That is just the average for Americans as a whole. It's impossible that the nation's genetics have shifted that much during the past fifty years (Nisbet 2010).

Thus, we understand that something about the environment in the United States affects African Americans categorically. During the same fifty-year period, the average difference between Black and white IQs decreased significantly—from fifteen points in 1945 to nine points today. These changes in IQ correspond to improvements in the Black population's standard of living, such as nutrition and prenatal care, relative to that of the white population. It is also interesting to note that the average Black person today has a higher IQ score than the average white person in 1950 (Nisbet 2010).

It turns out that IQ differences between Blacks and whites have very little to do with the genetic makeup of the races. These differences are much more influenced by social factors.

Learn the leading theories—cultural, economic, and discrimination—sociologists use to understand the sources of ethnic and racial inequality.

racialization

The process by which understandings of race are used to classify individuals or groups of people.

Cultural Explanations

Whereas IQ-based explanations are both controversial and carry little weight with sociologists, cultural explanations for racial inequality are controversial but have more adherents. These theories claim that the inequality between Blacks and whites is more determined by the cultures of different groups. One common argument is that poor Blacks have the "wrong" beliefs, values, and habits, which have the effect of holding them back—even in the face of opportunity. Just as those who believe in IQ-based explanations argue that poverty is passed on through the wrong genes, those who believe in cultural explanations argue that the wrong values are passed on from generation to generation. They also argue that poor Blacks are socially isolated from the values of mainstream society. These kinds of explanations are controversial because they tend to blame the poor themselves for their conditions.

One of the most common cultural explanations for poverty has been the nature of family life among poor Blacks. Perhaps the most controversial of these explanations came in the Moynihan Report, which was written for President Lyndon Johnson in 1965. The author, Daniel Patrick Moynihan, argued that the Black family was crumbling because a large number of illegitimate births led to a situation in which fathers were only marginally involved in the upbringing of children, and mothers were ill equipped to raise children (particularly boys) on their own. Moynihan argued that family life was so central to overcoming poverty that until the Black family could be reinvigorated, trying to solve the problem would be throwing money down a well. The firestorm caused by the Moynihan Report included a response from Martin Luther King Jr., who argued that poverty was the root of the problem.

Economic Explanations

When Martin Luther King Jr. fired back at the Moynihan Report, he was taking up an explanation for Black–white inequality that had great credibility among sociologists. The cause of poverty is not bad values or broken families but that subsequent generations of poor Blacks experience the same lack of economic opportunity as the ones who came before. Thus, young children start out idealistic and optimistic. When they see failures in their parents and grandparents, they vow that they will be different. But once they come to share the same experiences as their parents, they find themselves in similar situations. Thus, if their outcomes look the same, it is not due to culture or intelligence but to opportunity (Liebow 1967).

By the 1980s, another major sociological theory argued that the economic conditions of Blacks had become central to their life chances (Wilson 1978). The decline of manufacturing jobs was occurring at precisely the moment when the civil rights movement had expanded legal opportunity for Blacks. America was transitioning to a service economy, and the kinds of skills necessary to fill these jobs were quite different from those that most poor Blacks had. Had American jobs not migrated abroad, young Black men would have been qualified to take them. William Julius Wilson has argued that for the United States to bridge the Black–white gap in inequality, the federal government would need to create massive numbers of jobs.

Other economic explanations point to differences in wealth between Blacks and whites. Even Blacks who have achieved a measure of upward mobility into the professional middle class often cannot depend on inheriting a home or savings. Nor can they, as many more whites do, expect to receive a loan from parents or grandparents to make a down payment on a home. Consequently, even Blacks and whites with similar levels of education and income will often end up with unequal qualities of life (Oliver and Shapiro 1995; Conley 1999).

Racial Discrimination–Based Explanations

Many sociologists do not accept the idea that economic explanations are the major cause of inequality between Blacks and whites. These scholars point to anti-Black racial discrimination as a continuing cause of inequality that should not be underestimated. They refer to historical discrimination going back to slavery as well as present-day discrimination. In labor markets after the Civil War, Blacks have famously been the last hired and the first fired. Rather than hiring ex-slaves in the period from 1865 to World War I, companies advocated for immigration policies that opened the borders to large numbers of white Europeans, including Irish, Italians, and Jews. In the 1940s, sociologists demonstrated that employers would only hire Blacks when labor markets were so tight that they had no choice (Drake and Cayton 1945). By the 1980s, employers were actively seeking Latinos over Blacks in many labor markets, a trend that has continued into the present day. Experimental studies of labor market discrimination have demonstrated over and over again that less qualified whites are strongly preferred over more qualified Blacks (Pager 2003).

But perhaps the single most significant race-based explanation of Black–white inequality is the restriction of Blacks in physical space known as the ghetto. This residential segregation was created by racially explicit policies across levels of government (Rothstein 2017) as well as the discriminatory decisions of property owners and other individuals in interaction. Beginning in the 1920s in U.S. cities, white property owners entered into private agreements to stipulate that none of the homes in an entire neighborhood would be rented to Blacks. In later years, when these agreements were ruled unconstitutional, the federal government built massive housing projects in many of the areas that had previously been occupied by poor Blacks. After the civil rights movement, with the emergence of a Black middle class that could move out of the ghettos, these communities increasingly became home to the poorest Blacks who were left behind.

Because these poor Black neighborhoods were created through restrictive covenants and then reinforced by the federal government, they became unlike other poor communities, such as immigrant enclaves and slums. Inhabitants of these later neighborhoods tended to move up and out within a generation, but for many Blacks, the ghetto became a permanent place of restriction. These differences led to both an inferior quality of life and far more external control from the wider society. Within the realms of education, work, family life, violence, and local politics, Black ghettos became vicious cycles where space plays a crucial role. Conditions in each of these realms came to symbolize the Black way of life, and the conditions became the rationalization for further discrimination and segregation (Duneier 2016).

CONCEPT CHECKS

1. Provide one example of an environmental influence on a person's IQ.

2. Why might a cultural explanation of racial or ethnic inequality not explain its root cause?

3. Identify and describe two economic explanations for inequality between Blacks and whites.

4. How has racial segregation been maintained over the past century?

CHAPTER 10

The Big Picture

Race, Ethnicity, and Racism

Thinking Sociologically

1. Review the discussion of the assimilation of different American minorities, and then compare the different experiences of Asians and Latinos. Identify the criteria for assimilation and discuss which group has assimilated most readily. Then explain the sociological reasons for the difference in assimilation.

2. Does affirmative action still have a future in the United States? On the one hand, increasing numbers of African Americans have joined the middle class by earning college degrees, gaining professional jobs, and buying new homes. Yet Blacks are still far more likely than whites to live in poverty, attend poor schools, and lack economic opportunity. Given these differences, do we still need affirmative action?

3. How would you explain the recent rise in intermarriage in the U.S.? What do you think the implications are of the growing mixed-race population? Will it lead to more or less racial stratification and prejudice?

Learning Objectives

What Are Race and Ethnicity?

p. 285

Understand that race is a social and political construction and how it differs from ethnicity. Learn what constitutes a minority group according to the sociological perspective.

Why Do Racial and Ethnic Antagonism Exist?

p. 289

Learn the leading psychological theories and sociological interpretations of prejudice and discrimination. Recognize the importance of the historical roots, particularly in the expansion of Western colonialism, of ethnic conflict. Understand the different models for a multiethnic society.

How Does Racism Operate in American Society Today?

p. 294

Understand how racism is not only enacted by individuals but embedded in our institutions. Learn how racial inequality is maintained by both overt acts of racial hatred and color blindness. Understand the concepts of white privilege and microaggressions.

What Are the Origins and Nature of Ethnic Diversity in the United States?

p. 299

Familiarize yourself with the history and social dimensions of ethnic relations in America.

How Do Race and Ethnicity Affect the Life Chances of Different Groups?

p. 305

Learn how racial and ethnic inequality is reflected in terms of educational and occupational attainment, income, health, residential segregation, and political power.

How Do Sociologists Explain Racial Inequality?

p. 315

Learn the leading theories—cultural, economic, and discrimination—sociologists use to understand the sources of ethnic and racial inequality.

Terms to Know	Concept Checks

race • theory of racial formation • ethnicity • minority group • dominant group

1. What are race and ethnicity? How are these two concepts alike, and how do they differ?
2. How do political actors and institutions participate in racial formation?
3. What differentiates a minority group from a statistical minority?

prejudice • racism • stereotype • displacement • scapegoats • discrimination • immigration • emigration • scientific racism • genocide • segregation • assimilation • melting pot • pluralism • multiculturalism

1. What is the difference between prejudice and discrimination?
2. Provide two examples of ways ethnic groups maintain closure.
3. How did Western colonialism contribute to the creation of ethnic divisions today?
4. What are two forms of ethnic conflict?
5. What are the four primary models of ethnic integration? Describe each one.

institutional racism • white privilege • racial microaggressions

1. Give two examples of how institutional racism operates today.
2. Relate the theory of racial formation to any example of Donald Trump's overt racism.
3. How does "color blindness" perpetuate racial inequalities?
4. What is white privilege?
5. How would you explain the harmful effects of racial microaggressions to someone who thinks you are making "a mountain out of a molehill"?

refugees

1. Describe patterns of immigration to the United States before and after the 1960s.
2. How has the racial and ethnic composition of the United States changed since the 1960s?
3. Briefly contrast the immigration experiences of whites, African Americans, Latinos, and Asians to the United States.
4. Using the examples of African Americans and Asians, explain how neighborhood segregation may result from and perpetuate racial inequality.

racialization

1. What is one major driver of the Black–white gap in household wealth?
2. Explain two different ways that racism affects health.
3. Provide three explanations for the persistence of residential segregation.
4. Why is it problematic to refer to Asian Americans as model minorities?
5. What does the case of Mexican Americans teach us about space and assimilation?

1. Provide one example of an environmental influence on a person's IQ.
2. Why might a cultural explanation of racial or ethnic inequality not explain its root cause?
3. Identify and describe two economic explanations of inequality between Blacks and whites.
4. How has racial segregation been maintained over the past century?

11

Families and Intimate Relationships

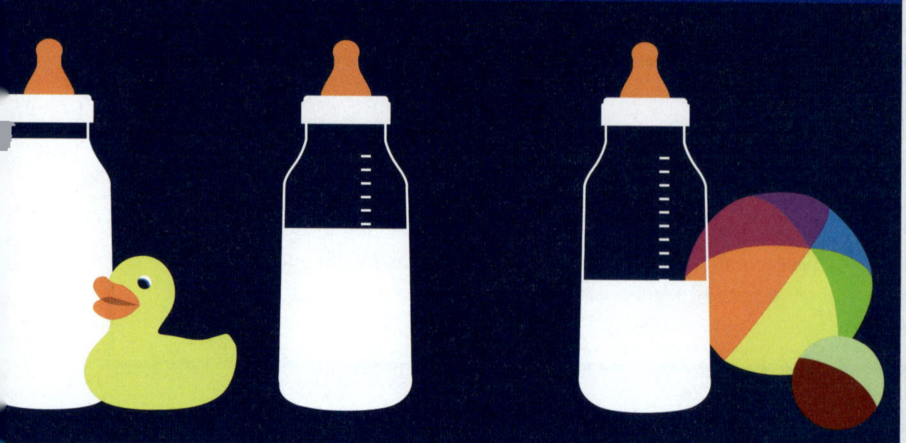

Maternity Leave

p. 347

Same-sex marriage supporters celebrate outside the Supreme Court building in Washington, D.C. On June 26, 2015, the Court ruled in *Obergefell v. Hodges* that same-sex couples have the right to marry nationwide.

THE BIG QUESTIONS

How do sociological theories characterize families?

Review the development of sociological thinking about families and family life.

How have families changed over time?

Understand how families have changed over the last 300 years. See that although a diversity of family forms exists in different societies today, widespread changes are occurring that relate to the spread of globalization.

What do marriage and family in the United States look like today?

Learn about patterns of marriage, childbearing, divorce, remarriage, and child-free families. Analyze how different these patterns are today compared with other time periods.

Why does family violence happen?

Learn about sexual abuse and violence within families.

How do new family forms affect your life?

Learn about some alternatives to traditional marriage and family patterns that are becoming more widespread.

In the early 1960s, Edith Windsor was a young woman living and working in New York. With her master's degree in math from New York University, she was the rare woman working at IBM as a computer programmer. Edith was also a woman in love. She had met Thea Spyer, a clinical psychologist, at Portofino, a restaurant in New York's Greenwich Village that was a popular hangout for gay women. After several years of dating, Thea proposed to Edith in 1967, offering her a brooch, rather than an engagement ring, to symbolize their commitment. Even though Thea and Edith couldn't legally marry at that time, they went on to live together as a loving couple for more than four decades.

In 2007, Thea's health declined, and doctors told the couple that Thea had only a short time to live. The couple wanted to formalize their union before Thea died, so they promptly flew to Toronto, Canada—one of the few places where a same-sex couple could marry at the time—and tied the knot. Just two years later, Thea died (Gabbatt 2013).

Edith Windsor, plaintiff in the *United States v. Windsor* case, celebrates the Supreme Court's decision to overturn the Defense of Marriage Act.

family

A group of individuals related to one another by blood ties, marriage, or adoption, who form an economic unit, the adult members of which are often responsible for the upbringing of children.

kinship

A relation that links individuals through blood ties, marriage, or adoption.

marriage

A socially and legally approved sexual relationship between two individuals.

Edith, then eighty years old and widowed with health problems of her own, soon learned that the IRS was ordering her to pay $363,053 in federal estate taxes on her inheritance from Thea. Edith—knowing full well that heterosexual couples who were legally married did not have to pay a comparable tax—was angry at the injustice. Married couples, according to the federal tax code, are allowed to transfer money or property from spouse to spouse upon death without having to pay estate taxes—a rule referred to as "unlimited marital deduction" (Coplan 2011). Although the state of New York, where Edith lived, recognized her marriage to Thea, the federal government refused to treat them the same way as other married couples because of a federal law called the Defense of Marriage Act (DOMA), which defines marriage as "a legal union between one man and one woman." This injustice impelled Edith to challenge the constitutionality of DOMA and seek a refund of the estate tax she had been forced to pay. In 2010, Edith sued the federal government.

In June 2013, after more than two years of appeals and legal red tape, the U.S. Supreme Court declared DOMA to be unconstitutional in the landmark *United States v. Windsor*. The ruling meant that married same-sex couples must receive the same federal benefits, rights, and privileges afforded to all other Americans. LGBTQ Americans celebrated another big victory that day when the Supreme Court cleared the way for same-sex marriage in California.

An even more important and pathbreaking Supreme Court decision was handed down on June 26, 2015, when the Court ruled by a 5–4 vote that the U.S. Constitution guarantees individuals the right to same-sex marriage. Technically, the ruling says that states cannot prohibit the issuing of marriage licenses to same-sex couples or deny recognition of lawfully performed out-of-state marriage licenses to same-sex couples. This ruling invalidated same-sex marriage bans and effectively made same-sex marriage legal in every state in the nation (Liptak 2015). The decision was widely celebrated for providing gay and lesbian couples with the same right to marriage that their heterosexual peers had enjoyed for centuries.

Our choices about dating, marriage, cohabitation, divorce, having children, or being child-free may seem highly personal and based on our desire for love and companionship, or adventure and freedom. Yet sociologists recognize that our choices are powerfully shaped by cultural beliefs and social structures such as laws. Whether, when, and under what conditions Edith and Thea could marry were dictated by law. Yet cultural factors, including social norms and subcultural or religious beliefs, also shape our decisions regarding our family lives as well as our attitudes toward others' families. For example, although the majority of Americans today support same-sex marriage, these attitudes vary widely by generation, religious views, and even geographic region. Nearly three-quarters of those born after 1981 support the legalization of gay marriage, whereas just 41 percent of those born between 1928 and 1945 do so. While just 35 percent of white evangelical Christians support same-sex marriage, 85 percent of persons who are unaffiliated with religious organizations are in favor (Pew Research Center 2017b). Understanding families in contemporary society requires a sociological imagination that takes into account both personal preferences and the powerful impact of social structures and cultural beliefs.

Sociological research on families typically involves descriptive or scientific studies that aim to solve some of the most interesting puzzles about marriage and families in the contemporary world. Taken together, the work of these researchers is fascinating and demonstrates that sociology can give us insights that are by no means obvious.

An extended Kazak family in Mongolia (left). Kazaks usually live in extended families and collectively herd their livestock. The youngest son will inherit the father's house, and the elder sons will build their own houses close by when they get married. In industrial societies, the nuclear family, which is made up of an adult or adult couple and their children, is the predominant family form (right). How is a nuclear family different from an extended family?

Basic Concepts

Before delving into questions about why and how people form their families, some basic concepts require review. A **family** is a group of persons directly linked by kin connections, the adult members of which assume responsibility for caring for children. **Kinship** refers to connections among individuals, typically established either through marriage or through the lines of descent that connect blood relatives (mothers, fathers, offspring, grandparents, etc.). **Marriage** can be defined as a socially and legally acknowledged and approved sexual union between two individuals. When two people marry, they become kin to each other; the marriage bond also, however, connects a wider range of kinspeople. Parents, brothers, sisters, and other blood relatives become relatives of the partner through marriage.

In virtually all societies, sociologists and anthropologists have documented the presence of the **nuclear family**, two adults living together in a household with biological or adopted children. In most traditional societies, the nuclear family was part of a larger kinship network of some type. When close relatives in addition to a married couple and children live either in the same household or in a close and continuous relationship with one another, we speak of an **extended family**. An extended family might, for example, include grandparents, brothers and their wives, sisters and their husbands, aunts, and nephews.

Families can also be divided into **families of orientation** and **families of procreation**. The first is the family into which a person is born or adopted; the second is the family into which one enters as an adult and within which a new generation of children is brought up. A further important distinction concerns place of residence. In the United States, when a couple forms a permanent union, they are usually expected to set up an independent household, separate from either partner's family of orientation. This may be in the same region in which one of the partner's parents live or a different town or city altogether.

nuclear family
A family group consisting of an adult or adult couple and their dependent children.

extended family
A family group consisting of more than two generations of relatives.

family of orientation
The family into which an individual is born or adopted.

family of procreation
The family an individual initiates through marriage or by having children.

monogamy

A form of marriage in which each married partner is allowed only one spouse at any given time.

polygamy

A form of marriage in which a person may have two or more spouses simultaneously.

polygyny

A form of marriage in which a man may have two or more wives simultaneously.

> Review the development of sociological thinking about families and family life.

polyandry

A form of marriage in which a woman may have two or more husbands simultaneously.

In some other societies, however, everyone who marries or forms a permanent partnership is expected to live close to or within the same dwelling as the parents of one of the two partners.

In Western societies, marriage, and therefore family, is associated with **monogamy**. It is illegal for a man or woman to be married to more than one individual at any one time. But in many parts of the world, monogamy is far less common than it is in Western nations. In his classic research, George Murdock (1967, 1981) compared several hundred societies from 1960 through 1980 and found that **polygamy**, a marriage that allows a husband or wife to have more than one spouse, was permitted in over 80 percent (see also Gray 1998). There are two types of polygamy: **polygyny**, in which a man may be married to more than one woman at the same time, and **polyandry**, much less common, in which a woman may have two or more husbands simultaneously. Of the 1,231 societies tracked, Murdock found that just 15 percent were monogamous, 37 percent had occasional polygyny, 48 percent had more frequent polygyny, and less than 1 percent had polyandry (Murdock 1981).

Recent work suggests that polygamy has grown less common over time, due to multiple social and economic conditions, including increasing levels of democracy, a declining acceptance of arranged marriage, an increase in marriages based on a desire for love and companionship, and strides in the education and human rights protections afforded to women. Polygyny is widely considered disadvantageous to women and, as such, has declined as women have gained more rights and power in many parts of the world (Bailey and Kaufman 2010).

How Do Sociological Theories Characterize Families?

Sociologists with diverse theoretical orientations have studied family life. Many of the perspectives that prevailed just a few decades ago now seem much less convincing in light of recent research and important changes in the social world. Nevertheless, it is valuable to briefly trace the evolution of sociological thinking before turning to contemporary approaches to studying families.

Functionalism

The functionalist perspective sees society as a set of social institutions that perform specific functions to ensure continuity and stability. According to this perspective, families perform important tasks that contribute to society's basic needs and help perpetuate the existence of major social institutions and practices. Sociologists working in the functionalist tradition have regarded the nuclear family as fulfilling certain specialized roles in modern societies. With the advent of industrialization, families became less important as a unit of economic production and more focused on reproduction, child rearing, and socialization.

According to the American sociologist Talcott Parsons, the family's two main functions are primary socialization and personality stabilization (Parsons and Bales 1955). **Primary socialization** is the process by which young children learn the cultural norms of the society into which they are born. **Personality stabilization** refers to the role that families play in assisting adult family members emotionally. Marriage is the arrangement through which adult personalities are supported and kept healthy. In industrial societies, families may play a critical role in stabilizing adult personalities. This is because the nuclear family is often distanced from its extended kin and is unable to draw on larger kinship ties as families could before industrialization.

Parsons regarded the nuclear family as the unit best equipped to handle the demands of industrial society. In the "conventional family," one spouse can work outside the home while the other spouse cares for the home and children. In practical terms, this specialization of roles within the nuclear family involved the husband adopting the "instrumental" role as breadwinner and the wife assuming the "affective," emotional role in domestic settings.

Parsons's view of families is now widely regarded by sociologists as inadequate and outdated. Functionalist theories of families have come under heavy criticism for justifying the division of household labor between men and women as something natural and unproblematic. Moreover, functionalist perspectives presume that a male–female married couple is essential for the successful rearing of children and for the efficient operation of households; Parsons failed to consider that same-sex and single-parent families may run efficiently and effectively parent and socialize children. He also failed to recognize that in many families, wives may be better suited to breadwinning and their husbands better suited to child rearing or that the two would share both tasks equally. Functionalist perspectives also overstate the importance of families in performing certain functions, especially socializing children, neglecting and minimizing the role that other social institutions—such as government, media, and schools—play.

Parsons's views reflect the historical period in which he was living and working. The immediate post–World War II years were a period when same-sex relationships were hidden, if not outright illegal in some parts of the United States. Divorce and single parenthood were relatively rare and often stigmatized. This era also saw women returning to their traditional domestic roles and men reassuming positions as sole breadwinners; this arrangement was rational for the family, as men typically earned far more than women (Becker 2009).

Symbolic Interactionist Approaches

Symbolic interactionist approaches to studying the family stand in stark contrast with functionalist perspectives. Whereas functionalist approaches emphasize stability and maintaining the current social order, symbolic interactionism emphasizes the contextual, subjective, and even ephemeral nature of family relationships (LaRossa and Reitzes 1993). Sociologist Ernest Burgess (1926) was one of the earliest scholars to apply symbolic interactionist approaches to the family, which he described as "a unity of interacting personalities" in which the behavior or identities of individual family members mutually shaped one another over time.

Symbolic interactionist approaches do not take power differentials for granted, nor do they assume that men have more power than women or that adults have more power than children. For example, Willard Waller (1938) developed the principle of least interest to show that the partner who is least committed to, or interested in, the romantic relationship has more power and might exploit that power. More contemporary work emphasizes

primary socialization

The process by which children learn the cultural norms of the society into which they are born. Primary socialization occurs largely in one's family.

---

personality stabilization

According to the theory of functionalism, the role families play in assisting adult members emotionally. Marriage between adults is the arrangement through which adult personalities are supported and kept healthy.

the ways that family members continually negotiate, define, and redefine their roles. Recall from Chapter 10 the concept of "doing gender" (West and Zimmerman 1987). Marriage and romantic relationships are a particularly important site for "doing gender." Studies have explored the ways that couples negotiate housework and how they do gender, even when no longer performing the household tasks typically associated with their sex. Emslie and colleagues (2009) studied the ways that colorectal cancer patients did gender when their illness prevented them from carrying out the gender-typed household roles they previously performed. The couples developed narratives to maintain their gendered identities, where women organized "cover" for housework and childcare when they were ill and men focused on making sure that their families were financially secure and spouses were "protected" from the stress of the men's cancer battles.

Symbolic interactionist approaches have been applied to parent–child relationships as well. Whereas scholarship on functionalist traditions presumed that parents taught and socialized their children, symbolic interactionist studies find that children often shape, influence, and guide their parents. Several studies of immigrant families, for instance, show that parents and children often must renegotiate their roles when they inhabit unfamiliar contexts (e.g., Katz 2014). Children may have relatively higher status than their parents, especially if they have a better understanding of the language and practices in the United States. This knowledge allows them to serve as the family's liaison to schoolteachers and health care providers.

Symbolic interactionism is critiqued for placing too much emphasis on cooperation and consensus and for being an overly descriptive approach. It tells us what is happening, but it does not tell us why. Some scholars, especially those working in the feminist tradition, find fault with the perspective's lack of attention to social structure and deeply embedded gender differences in social and interpersonal power.

Feminist Approaches

For many people, families provide a vital source of solace and comfort, love and companionship. Yet families can also be a site for exploitation, loneliness, and inequality. Feminist theories have challenged the vision of the family as harmonious and egalitarian. In the 1960s, American feminist Betty Friedan described in her landmark book *The Feminine Mystique* the isolation and boredom that gripped many suburban American housewives, who felt relegated to an endless cycle of childcare and housework.

During the 1970s and 1980s, feminist perspectives dominated most debates and research on families. If, previously, the sociology of families had focused on family structures, the historical development of the nuclear and extended family, and the importance of kinship ties, feminism redirected attention to the experiences of women in the domestic sphere. Many feminist writers have questioned the vision of families as cooperative units based on common interests and mutual support. They have sought to show that unequal power relationships mean that certain family members tend to benefit more than others.

Feminist writings emphasize a broad spectrum of topics, but three main themes are of particular importance. One is the domestic division of labor—the ways in which tasks such as childcare and housework are allocated among members of a household. Feminist sociologists have shown that women continue to bear the main responsibility for domestic tasks and enjoy less leisure time than men, despite the fact that more women are working in paid employment outside the home than ever before (Bianchi et al. 2007; Hochschild

Women's rights activist Betty Friedan participates in a New York City march commemorating the fiftieth anniversary of women's suffrage in August 1970.

and Machung 1989). In same-sex couples, partners tend to share housework more equally than heterosexual couples do, highlighting the complex ways that gender shapes household arrangements (Goldberg, Smith, and Perry-Jenkin 2012).

Second, feminists have drawn attention to the unequal power relationships within many families. Intimate partner violence (IPV), marital rape, incest, and the sexual victimization of children have received more public attention as a result of feminists' assertions that the violent and abusive sides of family life have long been ignored in both academic contexts and legal and policy circles. Feminist sociologists have sought to understand how families serve as an arena for gender oppression and even physical abuse. For example, through much of U.S. history, a husband had the legal right to engage his wife in coerced or forced sex. Owing in large part to efforts of feminist activists and scholars, marital rape became illegal in all fifty states in 1993 (Hines, Malley-Morrison, and Dutton 2012).

The study of caring activities is a third area in which feminists have made important contributions. Care work encompasses a variety of processes, from childcare to elder care. Sometimes caring means simply being attuned to someone else's psychological well-being—several feminist writers have been interested in "emotion work" within relationships. Not only do women tend to shoulder concrete tasks, such as cleaning and childcare, but they also invest large amounts of emotional labor in maintaining personal relationships (Duncombe and Marsden 1993). While caring work often is grounded in love and deep emotion, it is also a form of labor that requires an ability to listen, perceive, negotiate, and act creatively. Caring work also happens outside of one's family; thousands of women find work in jobs that require caring for others. Ironically, jobs that involve caring, such as childcare worker, nanny, or elderly companion, are among the lowest paid of all occupations, and such jobs are typically held by women of color and immigrants in the United States (Macdonald 2011; Rodriquez 2011).

CONCEPT CHECKS

1. According to the functionalist perspective, what are two main functions of families?

2. What themes guide symbolic interactionist approaches to the study of families?

3. According to feminist perspectives, what three aspects of family life are sources of concern? Why are these three aspects troubling to feminists?

How Have Families Changed over Time?

Understand how families have changed over the last 300 years. See that although a diversity of family forms exists in different societies today, widespread changes are occurring that relate to the spread of globalization.

Sociologists once thought that prior to the modern period, the extended or multigenerational family was the predominant family form in Western Europe. Research has shown this view to be mistaken. The nuclear family seems long to have been preeminent. Premodern household size was larger than it is today, but the difference is not especially great. In the United States, for example, throughout the seventeenth, eighteenth, and nineteenth centuries, the average household size was 4.75 persons. The current average is 2.5 (U.S. Bureau of the Census 2018b). This low number is partly due to the high proportion of Americans who live alone today, especially older widowed women and young professionals who maintain their own homes (Klinenberg 2012a). Since the earlier figure includes domestic servants, the absolute difference in family size is small.

Furthermore, more Americans are living in multigenerational households today than ever before. In 2016, a record 64 million Americans, or 20 percent of the total population, lived in a family household that contained at least two adult generations or a grandparent and at least one other generation (Cohn and Passel 2018). This pattern is caused partly by the recent economic recession and home foreclosures, which forced families to live together. Young college graduates, in particular, are moving back into their parents' homes in unprecedented numbers, as they face bleak job prospects. In 2016, 33 percent of adults between the ages of twenty-five and twenty-nine lived in multigenerational households (Cohn and Passel 2018). The large and growing immigrant population and the rising number of single or divorced parents who reside with their own parents have also contributed to this trend.

Finally, the number of grandparents living with and raising their grandchildren has increased steadily since the 1990s. These households are often referred to as "skip-generation" households because the "middle" generation (parents) is not present. Social problems such as HIV/AIDS, the heroin and opioid epidemics, and "three strikes" policies, which put a disproportionate number of young African American men in jail, created a context where the young children of deceased or imprisoned parents would go to live with their grandmothers (Baker, Silverstein, and Putney 2008).

"The Way We Never Were": Myths of the Traditional Family

Were families of the past as peaceful and harmonious as many people recall it, or is this simply a nostalgic fiction? As Stephanie Coontz points out in her book *The Way We Never Were* (1992), as with other visions of a golden age of the past, the rosy light shed on the "traditional family" dissolves when we look back to previous times to see what things were really like.

Popular lore depicts the family of colonial America as disciplined and stable; however, colonial families suffered from the same disintegrative forces as their counterparts in Europe. High death rates meant that the average length of marriage was less than twelve years, and more than half of all children saw the death of at least one parent by the time they were twenty-one. The admired discipline of the colonial family was rooted in the strict authority of parents over their children. The way in which this authority was exercised would be considered exceedingly harsh by today's standards.

In the Victorian period, wives were largely confined to the home. According to Victorian morality, women were supposed to be strictly virtuous, while men were sexually licentious: Many visited prostitutes and paid regular visits to brothels. Spouses often had little to do with each other, communicating only through their children. Moreover, domesticity wasn't even an option for poorer groups of this period. African American slaves in the southern United States lived and worked in what were frequently appalling conditions. In the factories and workshops of the North, families worked long hours with little time for home life. Children often labored long hours under dangerous conditions in factories until states started to pass compulsory schooling laws in the late nineteenth and early twentieth centuries. By 1918, every state required children to complete elementary school, which virtually eliminated the pool of child workers (Graham 1974).

Our most recent memory draws us to the 1950s as the time of the ideal American family. This was a period when large numbers of white middle-class women worked only in the home, while men were held responsible for earning a family wage. Yet large numbers of women didn't want to retreat to a purely domestic role and felt unfulfilled and trapped. Many women had held paid jobs during World War II as part of the war effort. They lost these jobs when men returned from the war. Moreover, men were still emotionally removed from their wives and often exercised a strong sexual double standard, seeking sexual adventures for themselves but setting strict codes for their spouses. Betty Friedan's best-selling book *The Feminine Mystique* first appeared in 1963, but its research referred to the 1950s. Friedan struck a chord in the hearts of thousands of women when she spoke of the "problem with no name": the oppressive nature of a domestic life bound up with childcare, domestic drudgery, and a husband who often prioritized work over his family life.

Changes in Family Patterns Worldwide

Family life across the globe has also been transformed over the past three centuries. In some areas, such as more remote regions in Asia, Africa, and the Pacific Rim, traditional family systems are little altered. In most countries in the Global South, however, widespread changes are occurring. The origins of these changes are complex, but several factors can be picked out as especially important. One is the spread of Western culture. Western ideals of romantic love, for example, have spread to societies in which they were previously unknown. One study found evidence of romantic love in nearly 147 of the 166 traditional societies studied in sub-Saharan Africa, East Eurasia, and elsewhere (Jankowiak and Fisher 1992).

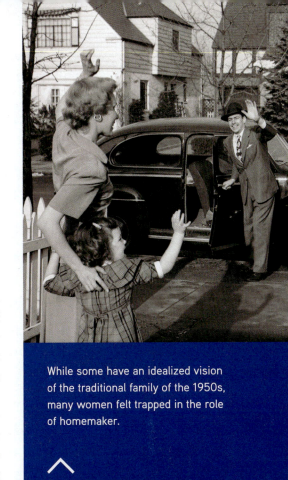

While some have an idealized vision of the traditional family of the 1950s, many women felt trapped in the role of homemaker.

Another factor is the development of centralized government in areas previously composed of autonomous smaller societies. People's lives are influenced by their involvement in a national political system; moreover, governments make active attempts to alter traditional ways of behaving. Because of rapidly expanding population growth, states frequently introduce programs advocating smaller families, for example, by promoting the use of contraception. One of the world's most effective population control programs was the one-child policy in China, implemented in 1978. Births subsequently dropped from 5 per woman in the 1970s to 3 in 1980 to an estimated 1.6 in 2015 (Central Intelligence Agency 2015).

In November 2015, the Chinese government relaxed this policy. Married couples are now allowed to have a second child. There are many reasons behind the government's decision to change the policy, including public opposition to a policy viewed as highly restrictive, widespread use of abortion when women became pregnant with a second child, and fears that the small cohorts of young people would not be sufficient to support much larger cohorts of older adults in China (Buckley 2015).

A further influence on family life is the large-scale migration from rural to urban areas. Often men go to work in towns or cities, leaving family members in the home village. Alternatively, a nuclear family group will move as a unit to the city. In both cases, traditional family forms and kinship systems may become weakened. Finally, and perhaps most important, employment opportunities away from the land and in such organizations as government bureaucracies, mines, plantations, and—where they exist—industrial firms tend to have disruptive consequences for family systems previously centered on agricultural production in the local community.

DIRECTIONS OF CHANGE

Families are being transformed throughout the globe today, with extended family systems giving way to the predominance of the nuclear family. This was first documented by William J. Goode in his book *World Revolution in Family Patterns* (1963), and subsequent

CONCEPT CHECKS

1. Briefly describe changes in family size over the past three centuries.

2. How has Stephanie Coontz dispelled the myth of the harmonious family believed to exist in past decades?

3. Give two examples of problems facing families in past centuries.

4. What are three conditions that have contributed to changing family forms throughout the world?

5. How has migration from rural to urban areas affected family life?

6. Name at least two recent shifts in family life that relate to globalization.

Learn about patterns of marriage, childbearing, divorce, remarriage, and child-free families. Analyze how different these patterns are today compared with other time periods.

research has shown that these changes continue. Building on Goode's work, sociologists have identified seven important changes that have characterized global family change over the past half-century:

1. Clans, or small family groups based on shared heredity, and other types of kin groups are declining in their influence.

2. There is a general trend toward the free choice of a spouse.

3. The rights of women are becoming more widely recognized, in respect to both the initiation of marriage and decision making within families.

4. Kin marriages are becoming less common.

5. Higher levels of sexual freedom are developing in societies that were very restrictive.

6. Birthrates are declining, meaning that women are giving birth to fewer babies.

7. There is a general trend toward the extension of children's rights.

In many countries, especially Western industrial societies, five additional trends have occurred within the past three decades:

1. An increase in the number of births that occur outside of marriage.

2. A liberalization of laws and norms regarding divorce.

3. An increase in nonmarital cohabitation among romantic partners.

4. An increasing age at first marriage and first birth.

5. A growing number of and acceptance for same-sex couples and their families.

Taken together, most industrial societies and an increasing number of nations in the Global South have witnessed a slow yet gradual decline of the nuclear family as the preeminent family form.

What Do Marriage and Family in the United States Look like Today?

The United States has long had high marriage rates; nearly 90 percent of adults in their mid-fifties today are or have been married (U.S. Bureau of the Census 2018b). However, recent evidence shows that the age at which Americans marry for the first time has risen sharply in recent decades. In 2017, the median age at first marriage in the United States was 27.4 for women and 29.5 for men; this marks a dramatic increase over 1960, when the median ages were 20.3 and 22.8 years for women and men, respectively (U.S. Bureau of the Census 2017d). Sociologists offer several explanations for this trend.

First, increases in cohabitation among younger people account for the decreases (or delays) in marriage among this group. In past decades, young people who wanted to live with their romantic partners typically married them, given the stigma of "living in sin" and having sexual relations with a person to whom one was not legally married. Within the past four decades, however, cohabitation has grown exponentially in popularity among young adults, alongside the disappearing stigma of premarital sexual relations. Second, increases in postsecondary school enrollment, especially among women, are partially responsible for delays in marriage. Most couples prefer to delay marriage until they have completed their formal schooling (Wang and Parker 2014).

Third, women's increased participation in the labor force often leads to delays in marriage as women work to establish their careers before marrying and starting a family (Goldstein and Kenney 2001). Labor force participation also increases economic independence among women. By earning their own income, many women no longer need a male breadwinner in their home. The flip side of the economic independence argument is the idea that the deterioration of men's economic position since the late 1980s has made them less attractive mates and less ready to marry (Sweeney 2002).

Finally, some researchers believe that modernization, changing gender roles, and a rise in attitudes that promote individualism make marriage less important than it once was. This sharp increase in age at first marriage has led researchers to debate whether marriage is simply being delayed or whether it is being forgone all together. According to Census Bureau predictions, a sizable proportion of today's young adults, especially African American women, will remain single for life (Parker and Stepler 2017).

As noted earlier, an extraordinary increase in the proportion of people living alone in the United States has also taken place in recent years—a phenomenon that partly reflects the high levels of marital separation and divorce. A record 28 percent of households in 2017 consisted of only one person, and the percentage of Americans who live alone has doubled over the last fifty years (U.S. Bureau of the Census 2017e; Klinenberg 2012a). There has been a particularly sharp rise in the proportion of individuals living alone in the forty-five to sixty-four age bracket. Older women are especially likely to live alone, as most will outlive their husbands and will not subsequently remarry.

Some people still suppose that the average American family is made up of a married couple and their children. This is very different from the real situation: Just one-fifth of U.S. households today are the "traditional" family, down from a quarter a decade ago and 43 percent in 1950 (Tavernise 2011). One reason is the rising rate of divorce: A substantial proportion of the population live either in single-parent households or in stepfamilies. The expectation that the "traditional" family includes a working husband and stay-at-home wife is also a thing of the past. Dual-career marriages and single-parent families are now the norm. Although men historically were the only breadwinners or the higher earners in the marriage, this pattern is rapidly changing; women outearn their husbands in roughly one in four households today (Livingston and Bialik 2018).

There are also significant differences in patterns of childbearing between parents in the 1950s and later generations. The birthrate rose sharply just after World War II and again during the 1950s. Women in the 1950s had their first children earlier than later generations did, and their subsequent children were born closer together. Since the late 1960s, the average age at which women have their first children has risen steadily. In 2016, the average age of first-time moms was a record-high 26.6 years—5 years older than in 1970 (Martin et al. 2018; Mathews and Hamilton 2016). This sharp rise in the mean age of first-time moms is due largely to a decrease in births to teen mothers. In addition, the proportion of women having their first children in their thirties and forties has also increased sharply (Mathews and Hamilton 2016).

Family structures and patterns are powerfully shaped by both structural and cultural factors. Structural factors—including shifts in educational attainment, economic prospects of young adults, and whether one has the legal right to marry—have a powerful influence on the ways families are formed. At the same time, cultural factors—ranging from attitudes toward marriage, sexuality, and cohabitation to beliefs about the appropriate context for raising children—shape family lives. For these reasons, American families vary widely based on factors such as social class, race, ethnicity, religion, and even the geographic region where one lives. We briefly focus on the ways that race and social class shape family life in the contemporary United States.

Race, Ethnicity, and American Families

Family sociologists have detected considerable variations in family structure across racial and ethnic groups. The most striking differences are between the family lives of whites and Blacks; Asian American families resemble white families in many ways, while Latino and Native American patterns are highly varied. Early studies suggested that these differences reflected cultural differences, including beliefs about the importance of marriage and of being economically self-sufficient (Lewis 1969). However, in recent decades, scholars have placed much greater emphasis on structural factors, recognizing that socioeconomic resources, including education, and the opportunity to work in jobs that provide a living wage are among the key factors that contribute to racial and ethnic differences in family structure—most notably, the greater tendency of whites and Asians (relative to Blacks, Latinos, and Native Americans) to marry, remain married, and have children within marriage. However, as we have seen elsewhere in this book, ethnicity and socioeconomic status are so closely intertwined in the United States that it is difficult to parse the distinctive effects of one over the other.

NATIVE AMERICAN FAMILIES

Kinship ties are very important in Native American families. As noted family demographer Andrew Cherlin (2005) observed, "Kinship networks constitute tribal organization; kinship ties confer identity" for Native Americans. However, for those who live in cities or away from reservations, kinship ties may be less prominent. Native Americans have higher rates of intermarriage than any other racial or ethnic group. According to 2013 American Community Survey data, fully 58 percent of all newly wed Native Americans don't marry other natives. By way of comparison, just 7 percent of whites, 19 percent of Blacks, and 28 percent of Asians who married in 2013 wed a partner of a different racial background (Wang 2015).

Native American women have a low overall birthrate. Compared to all U.S. women, a high proportion of Native American births are to women under age twenty (10 percent compared to 5 percent for women of all ethnicities), and Native American women have the youngest mean age of mother at first birth (23.2 years). Furthermore, more than two-thirds (68 percent) of all births to Native American women in 2016 were to unmarried women (Martin et al. 2018). These patterns are powerfully shaped by structural factors, including limited access to higher education, high unemployment rates, high levels of poverty, and high rates of mortality and imprisonment among young men, thus leaving young women without steady partners (Sandefur and Liebler 1997). Native Americans are also at particularly high risk of domestic violence; yet, as we will see later in this chapter, family violence can afflict persons of any ethnicity (Grossman and Lundy 2007).

LATINO FAMILIES

Latinos are heterogeneous when it comes to family patterns. Mexicans, Puerto Ricans, and Cubans are three of the largest Hispanic subgroups. In 2015, Mexicans constituted 63 percent of the Hispanic population, Puerto Ricans constituted 10 percent, and Cubans were just 3.7 percent; the rest of the Hispanic population was made up of much smaller groups from many Latin American nations (Flores 2017).

Mexican American families have a high birthrate and often live in multigenerational households. Economically, Mexican American families are more successful than Puerto Rican families but less so than Cuban families. Defying cultural stereotypes of a Mexican American home with a male breadwinner and female homemaker, more than half of all Mexican American women are in the labor force (U.S. Bureau of Labor Statistics 2017d). However, this is often out of necessity rather than desire. Many Mexican American families would prefer the breadwinner–homemaker model but are constrained by finances (Hurtado 1995).

Mexican family members often must grapple with separation from one another. Often, family members will migrate from Mexico to the United States sequentially, where one person (usually the father) secures a job and sends money or "remittances" back to his family. The plans for the rest of the family to move to the United States and reunite are often delayed or halted due to immigration laws (Smokowski and Bacallao 2011). Even after family members arrive, those who are undocumented may risk deportation, again causing family separation (Chang-Muy 2009).

The case is very different for Puerto Ricans, because Puerto Rico is a U.S. commonwealth. Because of their status as U.S. citizens, Puerto Ricans move freely between Puerto Rico and the mainland. When barriers to immigration are high, only the most able (physically, financially, and so on) members of a society can move to another country; but because Puerto Ricans face fewer barriers, even the least able can manage the migration process. Thus, they are the most economically disadvantaged of all the major Hispanic groups. Puerto Rican families have a higher percentage of children born to unmarried mothers than any other Hispanic group—64 percent in 2016 (Martin et al. 2018). Only African Americans (70 percent) and Native Americans (68 percent) had higher rates of births to unmarried women. However, consensual unions—cohabiting relationships in which couples consider themselves married but are not legally married—are often the context for births to unmarried mothers. Puerto Ricans may respond to tough economic times by forming consensual unions as the next best option to a more expensive legal marriage (Landale and Fennelly 1992).

Cuban American families are the most prosperous of all the Hispanic groups but less prosperous than whites (Brown and Patten 2013). Most Cuban Americans settled in the Miami area, forming enclaves in which they rely on other Cubans

Like many Mexican American families, the Camargo household contains multiple generations, including Beatriz Camargo and her husband and three kids, Beatriz's parents, two brothers, and her sister.

FIGURE 11.1

White and Black Households, 2017

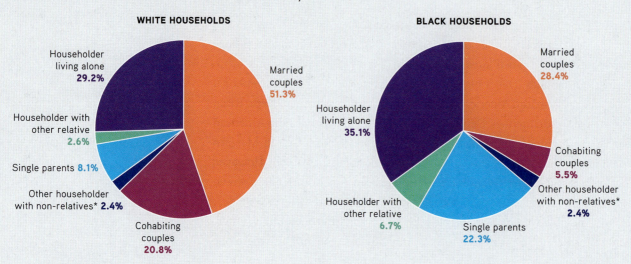

WHITE HOUSEHOLDS

Householder living alone **29.2%**

Householder with other relative **2.6%**

Single parents **8.1%**

Other householder with non-relatives* **2.4%**

Cohabiting couples **20.8%**

Married couples **51.3%**

BLACK HOUSEHOLDS

Married couples **28.4%**

Householder living alone **35.1%**

Cohabiting couples **5.5%**

Other householder with non-relatives* **2.4%**

Householder with other relative **6.7%**

Single parents **22.3%**

Non-relatives is a broad category that includes roommates, boarders, foster children, a paid live-in helper, and other persons not related by blood or law to the householder.

Source: U.S. Bureau of the Census 2017c.

for their business and social needs (such as banking, schools, and shopping). The relative wealth of Cuban Americans is driven largely by family business ownership. In terms of childbearing, Cuban Americans have lower levels of fertility than non-Hispanic whites and equally low levels of nonmarital fertility, suggesting that economic factors are equally if not more important than cultural factors in shaping family lives in the United States (López 2015).

AFRICAN AMERICAN FAMILIES

Black and white families differ dramatically in terms of family structure (Figure 11.1), although these differences are largely attributable to structural factors, including economic resources that facilitate marriage and marital stability. Blacks have higher rates of childbearing outside marriage, are less likely ever to marry, and are less likely to marry after having a nonmarital birth. These differences are of particular interest to sociologists, because single parenthood in the United States is both a cause and a consequence of poverty (Harknett and McLanahan 2004).

The contemporary state of Black families has deep historical roots. More than fifty years ago, Senator Daniel Patrick Moynihan (1965) described Black families as "disorganized" and caught up in a "tangle of pathology." Moynihan, among others, sought reasons for this in the history of the Black family. For one thing, the circumstances of slavery prevented Blacks from maintaining the cultural customs of their societies of origin; for example, members of similar tribal groups were dispersed to different plantations. Also, although some owners treated their slaves considerately and fostered their family lives, others regarded them as little better than livestock and inherently promiscuous and, therefore, unworthy of marriage. After emancipation, new cultural experiences and structural factors threatened Black families. Among these were new forms of discrimination against

African Americans; changes in the economy, such as the development of sharecropping in the South after the Civil War; and the migration of Black families to northern cities early in the twentieth century (Jones 1986).

A persistent puzzle facing researchers is the question of why Black and white family patterns have diverged even further since the 1960s, when Moynihan's study was published and a time when public benefits for poor families were expanded dramatically. Here we focus on contemporary factors that have contributed to the increasingly large gap in the structure of Black and white families. In 2017, married couples accounted for 51 percent of all white households yet just 28 percent of all Black households (U.S. Bureau of the Census 2018b). According to the U.S. Census, a household is composed of one or more people who occupy a housing unit. Not all households contain families; for instance, a household could include unmarried roommates. In 1960, 21 percent of African American families with children under eighteen were headed by females; the comparable rate among white families was 8 percent. By 2017, the proportion of Black families with children under eighteen that were female headed had risen to 53 percent, while the comparable proportion for white families was 21 percent (U.S. Bureau of the Census 2018b).

One social condition that contributes to high rates of nonmarital childbearing (and, consequently, female-headed households) is what sociologist William Julius Wilson (1987) calls a shortage of "marriageable" Black men. Marriage opportunities for heterosexual women are constrained if there are not enough men employed in the formal labor market. A woman will be less inclined to marry a man who is not earning a living wage and may instead opt to have and raise her child on her own rather than enter a marital union marked by financial instability. Contemporary research provides some support for Wilson's "marriageable male" hypothesis. Recent research confirms that one of the best predictors of whether parents marry after a nonmarital birth is the availability of eligible partners in a geographic area (Harknett and McLanahan 2004), demonstrating the continued importance of marriage markets even after the birth of a child. However, as we shall see in the section that follows, the patterns that Wilson documented are largely limited to lower-income African Americans; middle-class Blacks reveal family patterns that are very similar to whites' family patterns.

African Americans are often embedded in larger and more complex family networks than whites, but these ties are a source of both support and strain. In a now-classic study, white anthropologist Carol Stack (1997) lived in a Black ghetto community in Illinois to study the support systems that poor Black families formed. Getting to know the kinship system from the inside, she demonstrated that families adapted to poverty by forming large, complex support networks. Thus, a mother heading a one-parent family is likely to have a close and supportive network of relatives to depend on. Yet, these family ties often place demands—especially on older African American women, who are more likely than any other group to live with and raise their grandchildren. They step into this role when their own children, or the grandchildren's parents, can no longer adequately fulfill their parenting role. For many older Black women, family ties are a source of strain and demand as well as social support (Hughes et al. 2007).

ASIAN AMERICAN FAMILIES

Asian American families historically have been characterized by an interdependence among members of the extended family, a practice that some sociologists attribute to cultural beliefs related to filial piety, or respect for and a sense of responsibility for caring for one's elders (Bengtson et al. 2000). In 2016, 29 percent of Asians in the United States resided in multigenerational households, compared with just 16 percent of whites (Cohn and Passel 2018). Family members' interdependence also helps Asian Americans prosper financially. Asian American family and friend networks often pool money to help their members start a business or buy a house. This help is reciprocated as more of the recipients who prosper as a result then contribute to other family members. The result is a median family income for Asian Americans that is higher than the median family income for non-Hispanic whites.

As each generation of Asian Americans grows increasingly acculturated to life in the United States, scholars predict that they will come to resemble white families more and more (Pew Research Center for the People and the Press 2012b). This process may be further hastened by outmarriage. Asians are more likely than other racial groups to "marry out": 28 percent of Asians who wed in 2013 married someone of a different race, compared to just 7 percent of whites (Wang 2015).

Although there is less research on differences among various Asian American subgroups than among Hispanic subgroups, some fertility differences have been established. Chinese American and Japanese American women have much lower fertility rates than any other racial or ethnic group, due partly to their high levels of educational attainment. These differences in educational attainment reflect a range of economic factors, including the type of jobs their parents held, the conditions under which their families emigrated to the United States, and their language skills. As women remain in school and delay marriage and childbearing, they typically go on to have fewer children. Chinese, Japanese, and Filipino families have lower levels of nonmarital fertility than all other racial or ethnic groups, including non-Hispanic whites. Low levels of nonmarital fertility combined with low levels of divorce for most Asian American groups demonstrate the cultural emphasis on marriage as the appropriate forum for family formation and maintenance (Pew Research Center 2013c).

Social Class and the American Family

Sociologists studying racial and ethnic differences in American families are keenly aware of the role that social class plays. As we learned earlier in this book, whites, Blacks, Native Americans, Latinos, and Asians differ starkly with respect to their levels of education, the kinds of jobs they hold, their income, their savings, and whether they own homes. Economic and occupational stability is a powerful influence on families, where those with richer resources are more likely to marry and have children within (rather than outside of) marriage. Even studies that focus primarily on race are essentially studies of social class because race and class historically have been so closely intertwined. For example, while some early work (Lewis 1969; Stack 1997) attributed racial differences in the organization of the extended family to cultural or interpersonal factors, contemporary researchers have concluded that the differences between Black and white extended family relationships are mainly due to contemporary differences in social and economic class positions of group members (Sarkisian and Gerstel 2004). Most scholars today agree that cultural factors are less significant than structural obstacles.

This leads to a thought-provoking and policy-relevant question: Are racial differences in family formation due primarily to economic or to cultural factors? Recent studies show that race and class each have distinctive and often complicated effects on family behavior. For instance, while whites from working-class and poor backgrounds—often residing in the southern United States—report very strong ideological support for marriage and bearing

Dating and Mating Online

How did you meet your last romantic partner? Perhaps you met at a party or sat next to each other in your introduction to sociology course. Can you remember what it was that drew you to him or her? Was there something subtle or unexpected that signaled to you there might be an attraction, like a tone of voice, a wink, or a light touch to the shoulder? Or did you already have a clear-cut notion of the kind of person you wanted to date—maybe someone tall, or who shared your religious background, or who had professional goals similar to your own—and you carefully surveyed those whom you saw as an "appropriate" partner before making your move?

While popular music suggests that two strangers will lock eyes across a crowded room and true love will follow, in our current digital age, meetings often happen in a far less romantic and more strategic way. Dozens of smartphone apps, such as Tinder, Hinge, and Happn, allow people to search through endless photos of eligible partners and screen them, or "swipe right," based on personal preferences like education, occupation, age, height, body weight, gender, sexual orientation, and race. GPS functionality allows users to find like-minded people in their vicinity at any given time (Wortham 2013).

Although apps may take the romance and intrigue out of dating, they do fulfill a practical function. Young people can shop for a date in exactly the same way they would shop for a new car; they can specify precisely what they want and search for potential partners who possess those traits. From a sociological perspective, many apps provide strong evidence that norms of "homogamy," or dating and marrying a partner similar to oneself, are still pervasive in U.S. society. New studies also show that apps provide evidence of "hypergamy," or the preference for women, typically, to partner with a man with richer socio-economic resources than her own.

For instance, sociologist Kevin Lewis (2013) analyzed data from more than 126,000 dating site users and found that users tended to show the greatest interest in those of their same ethnic background. He analyzed only the first message sent and the first reply of each user. He found the tendency to initiate contact within one's own race to be strongest among East and South Asians and Indians and weakest among whites. He found that while users would respond to a message from someone from a different ethnic or racial group, this open-mindedness was relatively short-lived; most would promptly return to their old patterns of communicating only with members of their own group.

Lewis's analyses also uncovered evidence that users hold dating preferences consistent with "highly gendered status hierarchies." For instance, women tend to seek out men with more education and more income than they themselves have. Although men also sought educated partners, they tended to show greatest interest in women with a college education—"no more and no less." Racial hierarchies also emerged. White men, Lewis found, enjoyed a privileged position, receiving the most initial messages, while Black women received the fewest.

Thinking about sociological writings on family formation, what do you see as the pros and cons of such apps? Why do you think app users, and daters more generally, prefer to date someone of their own racial background? How would you explain women's preferences for men who are more educated and wealthy than themselves? What cultural and structural factors may underlie the patterns found in Lewis's study?

Dating apps like the League, which narrows the dating pool to Ivy League graduates, provide evidence of modern homogamy.

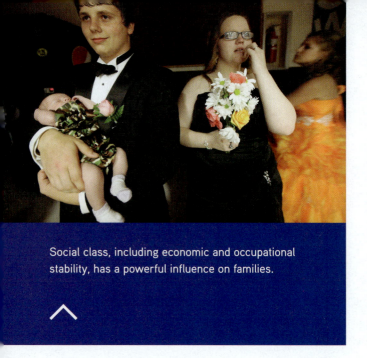

Social class, including economic and occupational stability, has a powerful influence on families.

children within marriage, their behaviors often depart from these conservative ideals. White young adults of lower- and working-class backgrounds are much more likely than their wealthier peers to get pregnant prior to marriage, to marry young, and, subsequently, to divorce (Cahn and Carbone 2010). Because they often do not attend college, they marry young and bear children young—often before they are financially or emotionally prepared. As a result, they often struggle unsuccessfully with the challenges of marriage and babies and ultimately divorce. In the past decade, white working-class families have been particularly vulnerable to the opioid crisis, where a young adult's addiction can derail his or her life chances and destabilize family well-being (Egan 2018). For young adult addicts with small children, addiction and death by overdose often mean that older white adults become custodial grandparents to their orphaned grandchildren (Whalen 2016).

Middle-class young adults, by contrast, show much more stable family formation patterns. They often cohabit while in school or working in their first jobs, so they marry later and bear children later. Their delay of marriage until they are emotionally and financially ready is one of the key reasons college-educated young whites have lower rates of divorce than their more economically disadvantaged counterparts (Cahn and Carbone 2010).

Research comparing middle-class Blacks with their less advantaged counterparts was very scarce until the past decade, when scholars began to explore in depth middle-class Black families (Lacy 2007; Landry and Marsh 2011; Pattillo 2013). Although middle-class Black families are more likely than their less economically advantaged peers to live in married-couple households, recent studies detect a new form of middle-class Black family, especially among young adults—the single-person household. Due in part to the shortage of marriageable men described earlier, college-educated Black women often live on their own without a romantic partner (Marsh et al. 2007). Studies of intersectionality, or the complex interplay between race and class, provide important insights into the ways both culture and structure shape family lives.

Divorce and Separation

While divorce rates—calculated by looking at the number of divorces per 1,000 married men or women per year—increased steadily through the 1970s and 1980s, they have leveled off, and even declined, since the 1990s (Figure 11.2). Attitudes have changed in tandem, with a stark decline in the proportion of Americans who disapprove of divorce.

Divorce has a substantial impact on children. In one calculation, about half of children born in 1980 became members of a one-parent family. Since two-thirds of women and three-fourths of men who are divorced eventually remarry, most of these children nonetheless grew up in a family environment, often acquiring new stepsiblings in the process. Remarriage rates are substantially lower for African Americans. Only 32 percent of Black women and 55 percent of Black men who divorce remarry within 10 years. White children are more than twice as likely as Black children (74 percent vs. 36 percent) to reside in a home with two married parents (U.S. Bureau of the Census 2018b).

The economic well-being of women and children declines in the immediate aftermath of divorce. According to one study, the living standards of divorced women and their children on average fell by 27 percent in the first year following the divorce settlement. By contrast, the average standard of living of divorced men rose by 10 percent. However, more in-depth investigations find that the economic toll on men varies widely based on the spouses' financial arrangements prior to divorce and custody arrangements following divorce. For instance, one study found that men who were contributing more than 80 percent to family income prior to divorce experienced an improvement in their living standards after divorce by approximately 10 percent at the median but that for men who contributed a smaller share to family income, divorce resulted in a reduction in living standards (McManus and DiPrete 2001).

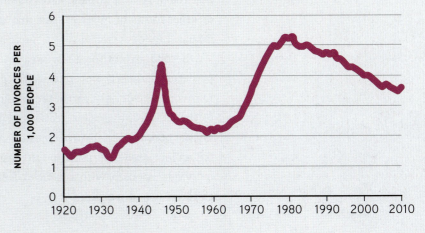

FIGURE 11.2

Divorce Rate in the United States

Source: Centers for Disease Control and Prevention 2013g.

Of special interest to family demographers is the fact that the women with the most resources—especially education—are increasingly following trajectories that provide their children with greater resources, primarily by delaying fertility and being more involved in the labor market. In contrast, women who have the fewest resources—low levels of educational attainment or few economic resources—are increasingly following a trajectory of early fertility and infrequent employment. These different trajectories are problematic because they lead to even higher levels of inequality in the future educational, economic, and health experiences of children (McLanahan 2004).

Young adults from financially disadvantaged homes are less likely to attend college and thus are more likely to marry younger—often right after high school. As a result, they tend to hold lower-paying jobs than their counterparts who graduate from college. Marrying young, before one is emotionally and financially ready, is considered one of the most powerful predictors of divorce (Elliott and Simmons 2011). In this way, early marriage and subsequent divorce play a role in perpetuating social class disadvantage across generations.

Other factors linked to the likelihood of divorce:

- Parental divorce (people whose parents divorce are more likely to divorce themselves)

- Premarital childbearing (people who marry after having children are more likely to divorce)

- Marriage at an early age (people who marry as teenagers have a higher divorce rate)

- Low incomes (divorce is more likely among couples with low incomes) (Amato 2010)

REASONS FOR DIVORCE

Why did divorce become increasingly common throughout the latter half of the twentieth century? There are several reasons, which involve wider changes in modern societies and social institutions. First, changes in the law made divorce easier. Today, individuals may file for a "no-fault" divorce, which doesn't require evidence that one spouse is at "fault." Through the first six decades of the twentieth century, by contrast, one party had to prove that the other was at fault, necessitating compelling evidence of abuse, neglect, infidelity, or other forms of mistreatment that could be difficult to prove from a legal standpoint (Marvell 1989).

Second, except for a few wealthy people, marriage no longer reflects the desire to perpetuate property and status across generations. The fact that little stigma now attaches to divorce is not only partly the result of these developments but also adds momentum to them. Third, there is a growing tendency to evaluate marriage in terms of personal satisfaction. Higher divorce rates do not indicate dissatisfaction with marriage as such but an increased determination to make it a rewarding and satisfying relationship for the individual spouses, with a reduced desire to stay (unhappily) married "for the sake of the children" (Cherlin 2010).

DIVORCE AND CHILDREN

The effects of divorce on children are difficult to gauge, and both public sentiment and scholarly research have evolved dramatically over the past four decades. How contentious the relationship is between the parents prior to separation, the age of a child at the time, whether there are siblings, the availability of grandparents and other relatives, the child's relationship with his or her individual parents, and how frequently the child continues to see both parents can all affect the process of adjustment. Because children whose parents are unhappy with each other but stay together may also be affected, assessing the consequences of divorce for children is doubly problematic.

Some of the earliest studies of divorce consequences were based on clinical samples, that is, populations of people seeking counseling; these found that children often suffer a period of marked emotional anxiety following the separation of their parents. Judith Wallerstein and Joan Kelly (1980) studied 131 children of 60 families in Marin County, California, following the separation of the parents. The researchers found evidence of both short- and long-term deleterious consequences of parental divorce. Almost all the children experienced intense emotional disturbance at the time of the divorce. Yet even ten or fifteen years later, nearly half the then–young-adult children reported difficulties in their romantic relationships, compromised self-esteem, and a sense of underachievement. However, these harmful effects partly reflect the fact that the study was based on a clinical sample; by definition, all had already been seeking professional help for their troubles prior to the start of the study.

In sharp contrast, more recent studies based on population-based samples find that the majority of people with divorced parents do not have serious mental health problems. Syntheses of decades of research have identified several common consequences of divorce for children (Amato 2001; Amato and Keith 1991):

- Almost all children experience an initial period of intense emotional upset after their parents separate.

- Most adjust without serious problems within two years of the separation.

- A minority of children experiences some long-term problems as a result of the breakup that may persist into adulthood.

Remarriage and Stepparenting

Before 1900, almost all marriages in the United States were first marriages. Most remarriages involved at least one widowed person. With the progressive rise in the divorce rate, the level of remarriage also began to climb, and in an increasing proportion of marriages at least one person was divorced. Today, fully 40 percent of marriages involve at least one previously married person, and 20 percent involve two previously married persons (Livingston 2014). For younger adults, remarriages typically follow divorce, whereas for older persons, the second walk down the aisle is taken by widows and widowers (Sweeney 2010). Increasingly, however, older adults are opting not to remarry after their marriages end, instead preferring an arrangement called "living apart together," where partners essentially "go steady" but maintain their own homes and separate bank accounts (Upton-Davis 2015).

The rise in rates of remarriage has been accompanied by rising numbers of stepfamilies. A **stepfamily** refers to a family in which at least one partner has children from a previous marriage. Many who remarry become stepparents of children who regularly visit rather than live in the same household. By this definition, the number of stepfamilies is much greater than official statistics indicate, because the statistics usually refer only to families with whom stepchildren live. Stepfamilies give rise to kin ties resembling those of some traditional societies in non-Western countries. Children may now have two "mothers" and two "fathers"—their biological parents and their stepparents. Some stepfamilies regard all the children and close relatives (including grandparents) from previous marriages as part of their family.

When two families merge and become a stepfamily, challenges may arise. First, there is usually a biological parent living elsewhere whose influence over the child or children remains powerful. Cooperative relations between divorced individuals often become strained when one or both remarry. Second, because most stepchildren belong to two households, the possibilities of clashes of habits and outlooks are considerable (Fine, Coleman, and Ganong 1998). There are few established norms defining the relationship between stepparent and stepchild. Should a child call a new stepparent by name, or is "Dad" or "Mom" more appropriate? Should the stepparent play the same part in disciplining the children as the natural parent? The difficulties in negotiating these decisions often intensify in later life as aging parents need to turn to their children for caregiving and stepchildren may not willingly oblige (Sherman, Webster, and Antonucci 2013).

stepfamily

A family in which at least one partner has children from a previous marriage.

Single-Parent Households

Single-parent households have become increasingly common. In 2017, nearly one-third (31 percent) of all families with children under eighteen were single-parent families, up from less than 13 percent in 1970 (U.S. Bureau of the Census 2017b). Black children are more than twice as likely as white children (53 percent vs. 20 percent) to reside in a single-parent household (U.S. Bureau of the Census 2017c). There are two main pathways to single-parent households: divorce and nonmarital childbearing. The vast majority of single-parent households are headed by women because unmarried women often do not maintain contact with the birth father of the child and may even prefer to raise a child on their own. Moreover, in the case of divorce, the mother usually obtains primary custody of the children. In 2017, of the 11.7 million single-parent families in the United States, 81 percent were headed by women (U.S. Bureau of the Census 2017c).

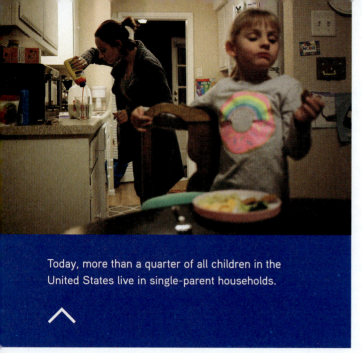

Today, more than a quarter of all children in the United States live in single-parent households.

^

CONCEPT CHECKS

1. Briefly describe changes in family structure in the United States since 1960.

2. Contrast both general and nonmarital fertility rates among whites, Blacks, Hispanics, Asians, and Native Americans in the United States.

3. What are the main reasons divorce rates increased sharply during the latter half of the twentieth century?

4. How does divorce affect the well-being of children?

In 1950, only 4 percent of all children in the United States were born to unmarried parents; by 2016, nearly 40 percent of all children were born outside of marriage (Martin et al. 2018). Although we often think of nonmarital births as births to teenagers—often abandoned by their male partners—the data show otherwise. An increasingly large number of nonmarital births are to men and women in their twenties who are delaying marriage but not delaying childbearing (Cherlin 2010). In fact, the majority of all nonmarital births today occur within cohabiting relationships (Curtin, Ventura, and Martinez 2014). Although the mother may be legally "unmarried," both she and her child may very well have a dedicated and involved male figure in their lives and in their home (Kennedy and Bumpass 2008).

A growing minority of Americans are choosing to become single parents, setting out to have a child or children without the support of a spouse or partner. "Single mothers by choice" is an apt description of some parents, normally those who possess sufficient resources to manage satisfactorily as a single-parent household. According to the National Center for Health Statistics, the most rapidly increasing rates of nonmarital births between the years 2007 and 2012 were to unmarried women ages thirty-five and older. During the same time period, rates of nonmarital births among younger women either declined or stayed stable. Women in their late thirties and forties today recognize that they can have a child on their own without facing the stigma that plagued single mothers in earlier generations. Many also have the financial means to support a child on their own (Curtin, Ventura, and Martinez 2014). For most unmarried or never-married mothers, however, the reality is different: There is a high correlation between the rate of births outside marriage and indicators of poverty and social deprivation. As we saw earlier, these influences are very important in explaining the high proportion of single-parent households among Black families in the United States.

Sociologists have long debated the impact on children of growing up with a single parent. The most exhaustive set of studies carried out to date, by Sara McLanahan and Gary Sandefur (1994), rejects the claim that children raised by only one parent do just as well as children raised by both parents. A large part of the reason is economic—the sudden drop in income associated with divorce. But about half of the disadvantage comes from inadequate parental attention and lack of social ties. Separation or divorce weakens the connection between child and father as well as the link between the child and the father's network of friends and acquaintances. On the basis of wide empirical research, the authors conclude that it is a myth that there are usually strong support networks or extended family ties available to single mothers (McLanahan and Sandefur 1994).

Others have been quick to point out that although children who grow up in a single-parent home are on average disadvantaged, it is better for children's mental health if parents in extremely high-conflict marriages divorce than if they stay together (Amato et al. 1995; Musick and Meier 2010). This suggests that divorce may benefit children growing up in high-conflict marriages but may harm children whose parents have relatively low levels of marital conflict before divorcing.

Why Does Family Violence Happen?

Learn about sexual abuse and violence within families.

Family relationships—between spouses, parents and children, siblings, or more distant relatives—can be warm and fulfilling. But they can also be full of extreme tension, driving people to despair or imbuing them with a deep sense of anxiety and guilt. Family discord can take many forms. The two broad categories of family violence are child abuse and intimate partner violence (IPV). Because of the sensitive and private nature of violence within families, it is difficult to obtain national data on levels of domestic violence. Data on child abuse are particularly sparse because of the issues of cognitive development and ethical concerns involved in studying child subjects.

Child Abuse

The most common definition of child abuse is serious physical harm (trauma, sexual abuse with injury, or willful malnutrition) with intent to injure. One national study of married or cohabiting adults indicated that about 3 percent of respondents abused their children, though cohabiting adults are no more or less likely to abuse their children than married couples (Brown 2004; Sedlak and Broadhurst 1996).

More recent statistics are based on national surveys of child welfare professionals. However, these surveys miss children who are not seen by professionals or reported to state agencies. Researchers estimate that as many as 50 percent to 60 percent of child deaths from abuse or neglect are not recorded (U.S. Department of Health & Human Services 2008). Statistics based on the National Child Abuse and Neglect Data System indicate that in 2016, an estimated 676,000 children were victims of abuse or neglect. Of these, 75 percent suffered neglect, 18 percent suffered physical abuse, and 8.5 percent were sexually abused; 91 percent of victims were maltreated by their parents. Accounting for 21 percent of all victims, African American children suffered a higher rate of abuse than white and Hispanic children, at 8.6 victims per 1,000 children (U.S. Department of Health & Human Services 2018).

The effects of child mistreatment can linger for years, if not decades, after the child escapes the abusive situation; recent studies show that adult men and women who suffered physical or sexual abuse in childhood are at elevated risk of multiple health conditions in midlife, including depression, chronic pain (Goldberg 1994), sleep problems (Greenfield et al. 2011), and metabolic syndrome (Lee, Tsenkova, and Carr 2014). Some evidence also suggests that victims of child abuse are more likely to commit abuse against partners and their own children, thus amplifying and perpetuating the cycle of abuse for future generations (Widom, Czaja, and DuMont 2015).

Intimate Partner Violence (IPV)

Abuse of one's spouse or romantic partner is widespread in the United States. A 1985 study by Murray Straus and his colleagues found that IPV had occurred at least once in the past year in 16 percent of all marriages and at some point in the marriage in 28 percent of all marriages. This does not, however, distinguish between severe acts, such as beating up and

Marriage and Family Therapist

If you ask Americans to name their main source of happiness, one answer consistently tops the list: their families. A recent Harris Poll found that 86 percent of Americans say they have positive relationships with their family (Sifferlin 2017). Yet anyone who has a family also knows that relationships are not always rosy. Siblings squabble. Romantic partners may argue, cheat, and even become physically aggressive with each other. Parents ideally provide their children with love and support, but some neglect their children or harm them physically and emotionally. Troubled families, as well as relatively happy families who want to make their relationships stronger or who are grappling with serious problems like a young child's terminal illness, a teenager's drug use, or a parent's job loss, may seek out help from a marriage and family therapist (MFT). A master's degree and license are typically required for this position, yet a strong undergraduate background in sociology is excellent preparation for the challenges and duties of a MFT (MFT-License.com 2013).

MFTs offer guidance to couples and families who are managing problems that may undermine their emotional health and well-being. Many of the issues that MFTs address are topics of great sociological concern, including marital conflict, intimate partner and family violence, divorce, child and teen behavioral problems, and family disagreements. MFTs also work with families from all walks of life, rich and poor, with and without children, same-sex and male–female partnerships, urban and rural. For this reason, they work in diverse settings, including social service agencies, family services, outpatient mental health and substance abuse centers, hospitals, government, schools, and private practices.

Several themes and concepts central to sociology guide the work of MFTs. Whereas some branches of therapy focus on an individual client, MFTs recognize that there are multiple perspectives on any problem. It is only when family members come to a shared

threatening with or using a gun or knife, and less severe acts, such as slapping, grabbing, or shoving. When the authors disaggregated this number, they found that approximately 3 percent of all husbands admitted perpetrating at least one act of severe violence on their spouses in the previous year, but this is likely an underestimate.

More recent data, based on the National Intimate Partner and Sexual Violence Survey, find even higher rates of severe violence among both married and nonmarried romantic partners (Breiding et al. 2014). For instance, these data from 2011 show that 16 percent of women and 10 percent of men had ever experienced sexual violence by an intimate partner during their lifetimes. Severe physical violence by an intimate partner (including acts such as being hit with something hard, being kicked or beaten, or being burned on purpose) was reported by 22 percent of women and 14 percent of men during their lifetimes. Relatively little is known about the victimization of persons in same-sex relationships, although most experts agree that this abuse is especially likely to go undetected, given fears of stigmatization of being "outed" if one reports the abuse (Shwayder 2013).

definition of the situation that they can understand and resolve their problems, underscoring a key theme of symbolic interactionism. Family relationships also are dynamic, shifting over time. Gender roles, the personalities of family members, and the relative power that any one family member holds can change over time and across contexts, as sociologists like Ernest Burgess (1926) have observed.

A strong grasp of sociological research approaches also can be very helpful. An important task of a MFT is observing how family members interact and using this information to draw educated conclusions about the sources of relationship problems. Understanding how to do a systematic evaluation, identify patterns, and recognize that one particular interaction may not be a "random sample" or accurate representation of the overall relationship helps MFTs to offer guidance and advice. For instance, an MFT might observe that a husband is aloof, quiet, and disengaged from what his wife and child are doing and saying during their session. Before drawing a strong conclusion, the therapist will want to observe the family on several occasions to make sure that the quietness isn't a fluke, perhaps due to illness on that day, but rather an accurate reflection of long-standing family dynamics.

MFTs also depend on academic research as they do their evaluations and offer guidance. An MFT who is not abreast of recent literature may believe, based on the early conclusions of Wallerstein and Kelly (1980), that parents should stay married at any cost. This therapist may dissuade feuding parents from getting a divorce, for fear that the couple's children will suffer in the long term. Yet a therapist well versed in sociological research would understand that these pessimistic conclusions are based on clinical populations of young people who had emotional problems even prior to the divorce. A sociologically informed MFT would know that newer research shows children fare just fine after divorce, and may even be hurt if parents remain in very contentious marriages (Amato 2001). Compassion and a desire to help others are essential to being a good MFT, yet a knowledge of sociological research and theory also is critical.

Despite 86 percent of Americans saying that they live in happy families, family relationships can be volatile. MFTs must be equipped to handle a myriad issues and a strong sociological base is key for an effective practice.

Trend data can tell us about the prevalence of IPV but not the nature of or context giving rise to abuse. Michael Johnson (1995) has identified two broad types of IPV: patriarchal terrorism, which is perpetuated by feelings of power and control, versus common couple violence, which generally relates to a specific incident and is not rooted in power or control. Sociological studies show that IPV is closely related to structural factors, including low levels of power among women, and cultural factors, including widespread acceptance of violence and beliefs that male power is equated with violence (Jewkes 2002). Violence is particularly likely to occur among couples whose relationships are marked by conflict, especially conflicts about finances, jealousy, substance use, and the husband's belief that his wife is violating traditional gender roles. Women who are more empowered educationally, economically, and socially are most protected from IPV (Jewkes 2002).

CONCEPT CHECKS

1. How do social scientists measure and track patterns of child abuse?

2. Describe structural and cultural factors that are linked with intimate partner violence.

How Do New Family Forms Affect Your Life?

>

Learn about some alternatives to traditional marriage and family patterns that are becoming more widespread.

cohabitation

Two people living together in a sexual relationship of some permanence without being married to each other.

Cohabitation

If current statistics are any indicator, more than half of all students reading this textbook will live with their romantic partners before marrying. **Cohabitation**—in which a couple lives together in a sexual relationship without being married—has become increasingly widespread in most Western societies. The proportion of young couples who cohabit has risen steeply, from 11 percent in the early 1970s to 44 percent in the early 1980s to roughly 50 percent today (Stepler 2017).

While for some, cohabitation may be a substitute for marriage, for many it is viewed as a stage in the process of relationship building that precedes marriage. Men and women differ in their reasons for cohabiting. One recent study conducted focus group interviews with cohabiting men and women and found that both genders said their primary motives for cohabiting included spending time together, sharing expenses, and evaluating compatibility. Men, however, were concerned about the loss of freedom marriage would entail, whereas women worried that cohabitation would further delay marriage (Huang et al. 2011).

For most young adults today, cohabitation does not end in marriage. Only about 40 percent of cohabiters married their partners within three years of starting to live together (Copen, Daniels, and Mosher 2013). Increasingly, evidence shows that rather than being a "stage in the process" between dating and marriage, cohabitation may be an end in itself. For a very small subset, cohabitation is preferable to remarriage. Longtime cohabiters with no plans to marry say that they prefer cohabitation due to their unease about the meanings associated with marriage and concerns about what marriage does to the relationship (Hatch 2015).

For some couples, cohabitation is "marriage-like" in that it is a context for bearing and rearing children. In the early 1980s, 29 percent of all nonmarital births were to those in cohabiting unions. In the 2010–2014 period, more than half of all nonmarital births were to cohabiting couples (Curtin, Ventura, and Martinez 2014). Part of the reason more babies are born into cohabiting unions is that pregnant women today don't feel the same social pressure to marry their partners as they would have in past years. In the past, unmarried couples who got pregnant might have had a "shotgun" wedding before the birth of the baby. Currently, however, only 11 percent of pregnant single women are married by the time the child is born. Fewer unmarried couples—whether cohabiting or not—are marrying before the birth of their child. There has, however, been an increase in the rate at which unmarried women begin cohabiting with their partners once they find out they are pregnant: Currently 11 percent of pregnant single women begin cohabiting by the time the baby is born—the same percentage who marry to "legitimize" a pregnancy.

The United States is certainly not alone in the increasing prevalence of cohabitation. Many European countries are experiencing similar, and in some cases much greater, proportions of unions beginning with cohabitation rather than marriage. The northern European countries of Denmark, Sweden, and Finland, along with France, show particularly high rates of cohabitation. However, unions in the southern European countries of Spain, Greece, and Italy—along with Ireland and Portugal—still largely begin with marriage.

Maternity Leave

Of the 185 countries included in a 2014 report by the International Labour Organization, all but two countries—the United States and Papua New Guinea—mandate paid leave for new mothers. In Eastern Europe and Central Asia, new mothers are given an average of almost 27 weeks of leave.

Length of maternity leave

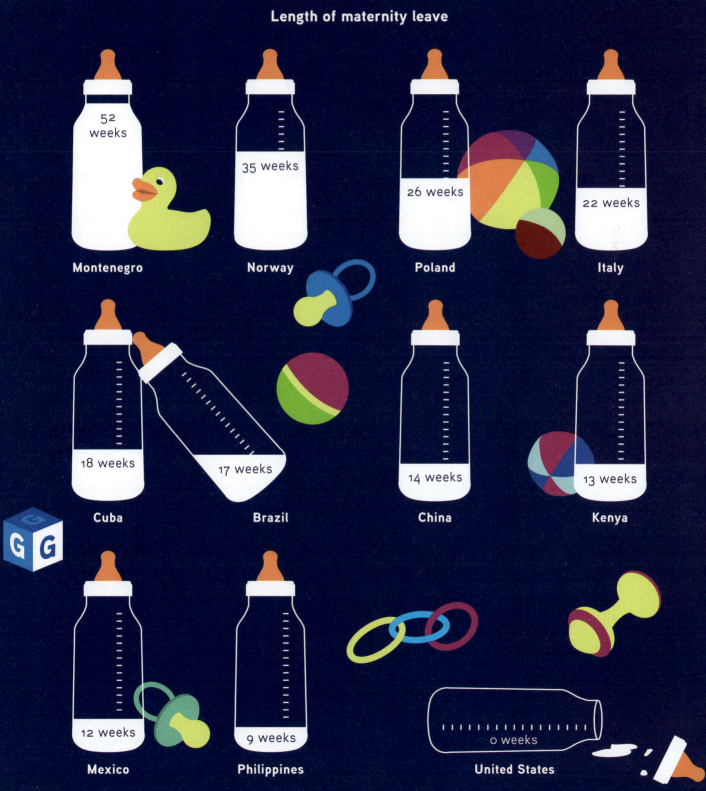

52 weeks — **Montenegro**

35 weeks — **Norway**

26 weeks — **Poland**

22 weeks — **Italy**

18 weeks — **Cuba**

17 weeks — **Brazil**

14 weeks — **China**

13 weeks — **Kenya**

12 weeks — **Mexico**

9 weeks — **Philippines**

0 weeks — **United States**

Note: All the countries featured pay 80%–100% of previous earnings for the entire period of leave.

Source: International Labour Organization 2014a.

DOES LIVING TOGETHER HELP REDUCE THE CHANCES FOR DIVORCE?

As we noted earlier, most readers of this textbook will cohabit before marrying. How will this experience of cohabitation affect their marriages? Many college students believe that by living with a boyfriend or girlfriend, they will learn whether they're right for each other and thus have more successful marriages. As early as the 1960s, anthropologist Margaret Mead (1966) predicted that living together would allow people to make better decisions about marriage. Yet some studies suggest that those who live with their partners before marrying them are slightly more likely to divorce than individuals who do not cohabit before marriage (Goodwin et al. 2010). How could this be so?

Sociologists offer two explanations: the "selection explanation" and the "experience of living together" explanation. The selection explanation proposes that people who live together would be more likely to divorce even if they hadn't ever lived together before marriage. That is, the very people who would choose to cohabit differ from those who don't, and the traits that distinguish the two groups are associated with their chances of divorcing, such as religiosity or education (Kamp Dush, Cohan, and Amato 2003).

Although the selection explanation suggests that there is nothing inherent about cohabitation that promotes divorce, the competing explanation suggests that living together is the kind of experience that erodes belief in the permanence of marriage. As people go through their twenties living with various partners, they develop a sense that relationships can be started and ended easily and that they have many options for intimate relations outside of marriage (Teachman 2003).

Which explanation is right? Early studies found that cohabitation slightly increased the probability of divorce, due to selection. However, recent evidence challenges the notion that cohabiters are more likely than others to divorce, and evidence of selection effects has virtually disappeared. Today, cohabitation is widespread and relatively "universal," with young people from all walks of life (not just disadvantaged backgrounds) opting for this living arrangement (Lu et al. 2012). Sociologists today generally agree that there is little evidence that living together "causes" divorce.

Same-Sex-Parent Families

LGBTQ college students today can look forward to a much more accepting social world than the one that greeted generations before them. Many gay, lesbian, and bisexual young adults live in stable relationships as couples, and this number is projected to increase dramatically with the 2015 Supreme Court decision guaranteeing the right to marriage for all.

Alongside the increase in same-sex marriage, same-sex couples are forming families with children in unprecedented numbers. Nearly one-fifth of same-sex-couple households include children, 72 percent of whom live with female same-sex couples and 28 percent of whom reside with male same-sex couples (Payne and Manning 2015). According to the Williams Institute, an estimated 37 percent of LGBTQ-identified persons have been a parent, and an estimated 6 million children in the United States have lived with a gay parent at some point in their lives (Gates 2013). Although lesbian couples may have a child by donor insemination and gay men may rely on a surrogate to carry a biological child, LGBTQ-identified persons are far more likely than heterosexuals to have a child through adoption. Same-sex couples raising children are four times more likely than their heterosexual counterparts to be raising an adopted child and more than six times as likely to be raising foster children (Gates 2013).

The legal, cultural, and technological landscapes facing gay parents and their children have changed dramatically in recent years. Increasingly tolerant attitudes toward LGBTQ persons have been accompanied by a growing tendency for courts to allocate custody of children to mothers living in lesbian relationships. All fifty states now allow LGBTQ individuals to adopt a child, although states vary in their policies regarding **second parent adoption** (where one partner adopts a child and the partner applies to be a second or co-parent) and **joint adoption** (where the partners adopt a child together) (Family Equality Council 2014). Popular media images in recent years, such as the TV series *The Fosters* and *Modern Family*, depict same-sex parents as providing the same love and guidance as heterosexual parents.

One reason for this growing acceptance of gay adoption and parenting is that widespread consensus has emerged among scholars that the ability to parent effectively is not related to sexual orientation. Although one recent and controversial study argued that children of parents who had ever had a same-sex relationship would go on to face greater adversity, this study has since been discredited for its serious methodological limitations (Regnerus 2012). The most persuasive and comprehensive reviews of scientific research to date, including the American Psychological Association's (2005) review of sixty studies and the American Academy of Pediatrics's (2013) seminal report, have concluded that "children growing up in households headed by gay men or lesbians are not disadvantaged in any significant respect relative to children of heterosexual parents" (Perrin, Cohen, and Caren 2013). The sixty studies considered a range of outcomes, including school performance, social adjustment, and emotional well-being, concluding that children's well-being is much more closely tied to their parents' "sense of competence and security"—and the "social and economic support" they provide their children—than sexual orientation.

For example, Rosenfeld (2010) finds that children of gay parents are just as likely as the children of straight parents to progress successfully throughout their school grades without being left back. Part of the reason children fare equally well regardless of parental sexual orientation is that sexual orientation has no bearing on one's capacity to be a loving parent. Moreover, children of same-sex couples usually share a common peer and school environment with children of heterosexual couples. As such, their experiences at school and with peers are very similar regardless of their parents' romantic preferences (Rosenfeld 2010). Most studies also show no differences in the psychological adjustment of children raised by same-sex or opposite-sex parents; for example, several studies of teenagers show no differences in their depressive symptoms, anxiety, self-esteem, or risk of attention-deficit hyperactivity disorder (Gatrell and Bos 2010; Lamb 2012).

Although children of gay or straight parents are neither significantly advantaged nor disadvantaged, some studies point out a small number of differences. For example, sociologists Judith Stacey and Timothy Biblarz (2001) reviewed twenty-one studies dating back to the 1980s and found that children in gay households are more likely to buck stereotypical male–female behavior. For example, boys raised by lesbians appear to be less aggressive and more nurturing than boys raised in heterosexual families. Daughters of lesbians are more likely to aspire to become doctors, lawyers, engineers, and astronauts. Heterosexual mothers tend to encourage sons to participate in historically "masculine" games and activities, such as Little League, and daughters in more "feminine" pursuits, such as ballet. In

second parent adoption

A family in which one partner adopts a child and the other partner applies to be a second parent or co-parent.

joint adoption

A family in which both partners adopt a child together.

The balance of studies shows that sexual orientation has no bearing on one's capacity to be a loving parent.

contrast, lesbian mothers had no such interest; their preferences for their children's play were gender neutral. The balance of evidence shows that children of gay parents are just as happy, healthy, and academically successful as their peers raised by heterosexual parents, and they may have more flexible views of gender and gender-typed behaviors.

Being Single

The broad category of "single" encompasses both people who have never married and those who have married but are now single due to divorce, separation, or widowhood. The number of people classified as single has increased dramatically in recent decades. Several factors have contributed to this trend. First, people are marrying later than ever. That means that more and more people in their twenties, thirties, and even forties are unmarried, either cohabiting with a nonmarital partner or waiting for the "right one" to come along. Second, the rise and stabilization of divorce rates over the past half-century mean that many more people are living on their own when their marriages end. Third, the "graying" of the U.S. population is accompanied by growing numbers of older adults whose partners have died and who now are technically "single" and live alone as widows and widowers. Fourth, the "stigma" of being single has diminished, due in part to television shows such as *Broad City* and *Girls*, which portray the active social lives and close friendships of unmarried women and men. As such, many more Americans are happily choosing to live their lives on their own (Byrne and Carr 2005; Klinenberg 2012a).

Yet are people really happy on their own, or are they better off being married? A large body of literature dating back to Émile Durkheim's classic *Suicide* (1966; orig. 1897) argues that social ties, especially marriage and parenthood, are essential to one's physical, social, and emotional well-being. Contemporary studies also show that divorced and widowed people report more sickness, depression, and anxiety compared with their married counterparts, although much of this disadvantage reflects the strains that precede a marital transition (such as a husband's illness or marital strife) as well as the strains that follow from the dissolution, such as financial worries or legal battles (Carr and Springer 2010).

But what about people who are long-term singles or who choose to live alone without a spouse or partner? To date, these individuals are relatively rare, as nearly 90 percent of American adults do ultimately marry. However, researchers have projected that as many as one in five Millennials (those born in the 1980s and thereafter) will never marry. Millennials are particularly likely to face financial obstacles to marriage, given their high levels of unemployment, school debt, and tendency to live with their parents rather than on their own during their twenties (Wang and Parker 2014).

Yet is lifelong singlehood a bad thing? Are their lives marked by loneliness and isolation? Research by sociologist Eric Klinenberg and others finds that living alone can "promote freedom, personal control, and self-realization—all prized aspects of contemporary life" (Klinenberg 2012b). After interviewing more than 300 people who live alone, he found that they had more, rather than less, social interaction than their married counterparts, and much of their social interactions were those they sought out by choice: encounters with friends, volunteering, arts events, classes, and other meetings that rounded out their lives. These patterns hold among older adults and younger adults alike. As Klinenberg observes, for many people, living alone and being able to choose how and with whom they spend their time are sought-after luxuries.

Taken together, research showing that people who live alone (by choice) are no better or worse off than their partnered peers underscores one of the core themes of the sociology of families: There is no one "best" or "typical" way in which Americans arrange their social lives. It is the freedom to choose one's relationships that is essential to one's happiness.

Being Child-Free

The number of Americans who do not have children increased steadily throughout the 1980s and 1990s yet has recently dipped (Livingston 2018). Calculating precisely who is childless is difficult; researchers historically have classified a woman as having no children if she has had zero children by age forty-four. Using this metric, the proportion of women ages forty to forty-four with no children climbed from 10 percent in 1986 to 15 percent in 1986 to 20 percent in 2006; the share had dropped to 14 percent by 2016.

White women are considerably more likely than Black, Asian, or Hispanic women to be child-free (17 percent vs. 15 percent, 13 percent, and 10 percent, respectively) (Livingston 2015). Women with a college degree or higher are more likely than high school graduates or dropouts to have no children. Most studies find that a relatively small fraction of these women are involuntarily childless; with advances in health and technology, the proportion who cannot physically bear children is modest. Rather, the reasons are often social and psychological, including not having a partner with whom one would want to have a child, a preference for a child-free lifestyle, concerns about the environment and bringing a child into an unsafe world, and concerns about whether one has the financial and emotional wherewithal to have a child (Connidis and McMullin 1996; Jacobson and Heaton 1991).

Childlessness was historically viewed as a stigmatized identity, a mark of a "barren" woman or a woman who was "selfish" and prioritized her own career pursuits over motherhood (May 1997). However, in recent decades, childlessness has been increasingly recognized as a status that is desirable and even preferable for many women and men. Those who do not have children have myriad opportunities to "give back" to the next generation by volunteering or caring for nieces or nephews, should they choose to do so (Sandler 2013).

Decisions about our family lives are among the most important ones we make. As we have seen in this chapter, marrying (or staying single), having (or not having) children—whether with a partner or on our own—and staying married (versus divorcing) are personal choices. Yet they are also powerfully shaped by other social factors, including our birth cohort, race, ethnicity, and social class. Although our personal preferences and values may shape our choices, the relationship between values and personal decision making is complex. Our personal decisions are also shaped by social institutions, including the law and the economy.

When Edith Windsor, whom we met in the chapter opener, was a young woman in the 1960s, she was unable to marry her longtime love, Thea Spyer. Edith and Thea lived together as a loving couple for more than forty years but were not allowed to marry legally until 2007. Their relationship ultimately changed history by leading to the repeal of DOMA in 2013 and the Supreme Court decision in 2015 guaranteeing individuals the right to same-sex marriage. Edith Windsor died in 2017 at age eighty-eight, yet her legacy is this historic verdict that ensures that all Americans have the right to marry, eliminating once and for all a structural barrier to marriage equality.

CONCEPT CHECKS

1. Why has cohabitation become so common in the United States and worldwide?

2. Does cohabitation lead to divorce? Why or why not?

3. What are two factors that have contributed to the rise in the number of people classified as single?

4. What are two main reasons for being child-free?

The Big Picture

Families and Intimate Relationships

Thinking Sociologically

1. Using this textbook's presentation, compare the characteristics of contemporary white non-Hispanic, Asian American, Latino, and African American families.

2. Increases in cohabitation and single-parent households suggest that marriage may be beginning to fall by the wayside in our contemporary society. However, this chapter claims that marriage and family remain firmly established institutions in our society. Explain the rising patterns of cohabitation and single-parent households and show how these seemingly paradoxical trends can be reconciled with the claims offered by this textbook.

Terms to Know

Concept Checks

family • kinship • marriage • nuclear family •
extended family • family of orientation •
family of procreation • monogamy • polygamy •
polygyny • polyandry

primary socialization • personality stabilization

1. According to the functionalist perspective, what are two main functions of families?
2. What themes guide symbolic interactionist approaches to the study of families?
3. According to feminist perspectives, what three aspects of family life are sources of concern? Why are these three aspects troubling to feminists?

1. Briefly describe changes in family size over the past three centuries.
2. How has Stephanie Coontz dispelled the myth of the harmonious family believed to exist in past decades?
3. Give two examples of problems facing families in past centuries.
4. What are three conditions that have contributed to changing family forms throughout the world?
5. How has migration from rural to urban areas affected family life?
6. Name at least two recent shifts in family life that relate to globalization.

stepfamily

1. Briefly describe changes in family structure in the United States since 1960.
2. Contrast both general and nonmarital fertility rates among whites, Blacks, Hispanics, Asians, and Native Americans in the United States.
3. What are the main reasons divorce rates increased sharply during the latter half of the twentieth century?
4. How does divorce affect the well-being of children?

1. How do social scientists measure and track patterns of child abuse?
2. Describe structural and cultural factors that are linked with intimate partner violence.

cohabitation • second parent adoption •
joint adoption

1. Why has cohabitation become so common in the United States and worldwide?
2. Does cohabitation lead to divorce? Why or why not?
3. What are two factors that have contributed to the rise in the number of people classified as single?
4. What are two main reasons for being child-free?

12

Education
and Religion

Religious Affiliation

p. 381

Malala Yousafzai has become an internationally recognized advocate for girls' education. At age fifteen, in 2012, Malala was shot and gravely wounded by members of the Taliban. Malala survived the injury and was awarded the Nobel Peace Prize in 2014.

THE BIG QUESTIONS

Why are education and literacy so important?

Know how and why systems of mass education emerged in the United States. Know some basic facts about the education systems and literacy rates of developing countries.

What is the link between education and inequality?

Become familiar with the most important research on whether education reduces or perpetuates inequality. Learn the social and cultural influences on educational achievement.

How do sociologists think about religion?

Learn the elements that make up religion. Know the sociological approaches to religion developed by Marx, Durkheim, and Weber as well as the religious economy approach.

How does religion affect life throughout the world?

Understand the various ways religious communities are organized and how they have become institutionalized. Recognize how the globalization of religion is reflected in religious activism in poor countries and the rise of religious nationalist movements.

How does religion affect your life in the United States?

Learn about the sociological dimensions of religion in the United States.

Most young American men and women take for granted that they will graduate high school and even go on to college or graduate school. Yet in some parts of the world, young women have to fight to receive even a middle school education. For the remarkable Malala Yousafzai, her desire to receive an education nearly cost the teenager her life.

In October 2012, when Malala was just fifteen years old, she was shot in the head and neck as she rode the bus home from her school in Mingora in the Swat district of Pakistan's Khyber Pakhtunkhwa province. The gunmen were members of the Taliban, an Islamic fundamentalist group that has long oppressed women. The Taliban had set an edict that girls in Mingora could not attend school after the age of fifteen. They had reportedly blown up more than 100 schools and threatened to blow up others. But why would they single out Malala for attack?

Several years earlier, Malala had maintained a blog and had spoken out publicly against the Taliban's mistreatment of girls and women. In one of her early public speeches, the bold teenager asked, "How dare the Taliban take away my basic right to education?" By challenging

At age seventeen, Malala became the youngest person to receive the Nobel Peace Prize. Recently, she cofounded the Malala Fund, organizing projects in six countries to advocate for girls' education and empower girls to become leaders.

the religious beliefs of the Taliban and advocating for the education of girls and women, Malala had made herself a target (Peer 2012; Yousafzai and Lamb 2013).

The shooting of Malala sent shock waves throughout the world. In the days after her attack, she lay unconscious at a local hospital in critical condition. As her condition stabilized, she was sent to a hospital in England for continued care. Her health improved, but she remained at risk; members of the Taliban publicly stated that they still intended to kill Malala and her father, a poet and social activist. Malala ultimately triumphed, however. She made a full recovery, wrote a book documenting her ordeal (Yousafzai and Lamb 2013), and became an internationally recognized heroine—an advocate for women's education worldwide. She received a litany of awards, including the 2014 Nobel Peace Prize. Malala also graced the cover of *Time* magazine and was named one of the 100 most influential people in the world. Her ordeal called attention to the state of girls' education in Pakistan, as well as other parts of the world where a high school diploma is not a taken-for-granted part of teenagers' lives. Just 2.6 percent of Pakistan's gross domestic product is dedicated to education. According to United Nations data, nearly 5 million children in Pakistan are not attending school; 62 percent of them are female. Due in part to this gender gap in education, a stark gender gap in literacy rates persists; 34 percent of young women ages fifteen to twenty-four (versus just 20 percent of their male peers) are illiterate, meaning that they cannot read or write (UNESCO 2018a).

Malala's story highlights many important themes at the core of the sociology of education and the sociology of religion. Education is a social institution that teaches individuals how to be active, engaged members of society. Through education, we become aware of the common characteristics we share with other members of society and gain at least some knowledge about our society's geographical and political position in the world and its history. The educational system both directly and indirectly exposes young people to the lessons that they will need to learn to become players in other major social institutions, such as the economy and the family. Yet education also gives us power; it provides us with the intellectual resources to scrutinize the world around us. As we saw in Chapter 7, formal education also provides the tools and credentials needed to seek gainful employment and, in some cases, to become upwardly mobile economically. It's mainly for these reasons that the Taliban so opposed Malala and other girls receiving an education: It would give them freedom, independence, and knowledge of a world beyond the confines of their insular Muslim community.

Like education, religion is an institution that exercises a socializing influence. However, while education is intended to be universalistic and to expose all young people to similar messages, religious institutions vary widely in the values, beliefs, and practices that they espouse. Some religions, for example, teach that all persons are created equal, whereas others are based on a foundation of oppression, where some groups are viewed as morally superior to and more worthy than others. Sociologists of religion try to assess under what conditions religion unites communities and under what conditions it divides them. The study of religion is a challenging enterprise that places special demands on the sociological imagination, as we must be sensitive to individual beliefs that may be rooted in faith more so than science.

This chapter focuses on the socializing processes of education and religion. To study these issues, we look at how present-day education developed and analyze its socializing influence. We also look at education in relation to social inequality and consider how far the education system exacerbates or reduces such inequality. Then we move to studying religion and the different forms that religious beliefs and practices take. We also analyze the various types of religious organizations and the effect of social change on the position of religion in the wider world.

Why Are Education and Literacy So Important?

Know how and why systems of mass education emerged in the United States. Know some basic facts about the education systems and literacy rates of developing countries.

The term *school* has its origins in a Greek word meaning "leisure" or "recreation." In pre-modern societies, schooling existed for the few who had the time and resources available to pursue the cultivation of the arts and philosophy. For some, engagement with schooling was like taking up a hobby. For others, such as religious leaders or priests, schooling was a way of gaining skills and thus increasing their ability to interpret sacred texts. But for the vast majority of people, growing up meant learning by example the same social habits and work skills as their elders. Learning was a family affair—there were no schools for the mass of the population. Since children often started to help with domestic duties and farming work at very young ages, they rapidly became full-fledged members of the community.

Education in its modern form, the instruction of pupils within specially constructed school premises, gradually emerged in the first few years of the nineteenth century, when primary schools began to be built in Europe and the United States. One main reason for the rise of large educational systems was the process of industrialization, with its ensuing expansion of cities.

Education and Industrialization

Until the first few decades of the nineteenth century, most of the world's population had no schooling whatsoever. But as the industrial economy rapidly expanded, there was a great demand for specialized schooling that could produce an educated, capable workforce. As occupations became more differentiated and were increasingly located away from the home, it was impossible for work skills to be passed on directly from parents to children.

As education systems became universal, more and more people were exposed to abstract learning (of subjects like math, science, history, and literature) rather than to the practical transmission of specific skills. In a modern society, people have to be furnished with basic skills—such as reading, writing, and calculating—and a general knowledge of their physical, social, and economic environments, but it is also important that they know how to learn so that they are able to master new, sometimes very technical forms of information. An advanced society also needs pure research and insights with no immediate practical value to push out the boundaries of knowledge. For example, developing complex reasoning skills, the ability to debate the merits of competing theories, and an understanding of philosophical and religious debates are three skills essential to a cultured and well-educated society.

In the modern age, education and other qualifications became an important stepping-stone into job opportunities and careers. For example, colleges and universities not only broaden people's minds and perspectives but are expected to prepare new generations of citizens for participation in economic life. Think about your own college education. Perhaps you're required to take certain general education courses to provide you with a broad base of knowledge; you might also take very specific courses in your major that help prepare you for your future career. It can be difficult to achieve the right balance between receiving a generalist education and mastering concepts and skills related to

With the spread of industrialization, the demand for educated workers increased. The newly expanded education systems emphasized basic skills like reading, writing, and mathematics instead of specific skills for work.

>

home schooling

The practice of parents or guardians educating their children at home, for religious, philosophical, or safety reasons.

one's chosen profession. Specialized forms of technical, vocational, and professional training often supplement pupils' liberal arts education and facilitate the transition from school to work. Internships, for example, allow young people to gain specific knowledge applicable to their future careers.

Although schools and universities seek above all to provide students with a well-rounded education, policymakers and employers are concerned with ensuring that education and training programs produce a stream of graduates who can meet a country's employment demands. Yet in times of rapid economic and technical change, the priorities of the education system don't always match up with the availability of professional opportunities. The rapid expansion of a country's health care system, for example, would dramatically increase the demand for trained health professionals, laboratory technicians, capable administrators, and computer systems analysts familiar with public health issues.

The complex relationship between the education system and the country's employment demands may be further complicated by an emerging trend: home schooling. Between 1999 and 2012, the number of students who were home schooled more than doubled; an estimated 1.8 million (or 3.4 percent of all) children are currently home schooled (Redford et al. 2017). **Home schooling** means that a child is taught by his or her parents, guardians, or a team of adults who oversee the child's educational development. The curriculum studied by home-schooled children varies widely from state to state; some states mandate quite strict curricula, while others are much more lax and provide the parent with great leeway.

A survey conducted by the U.S. Department of Education in 2012 queried parents about their motivation for home schooling their children. The most frequently cited reasons were a concern about the school environment (91 percent), a desire to provide moral instruction (77 percent), and dissatisfaction with the academic instruction at other schools (74 percent). It remains to be seen how well home schooling prepares young adults for the future challenges of college or employment in the United States. Very few empirical studies have been done on the effectiveness of home schooling on subsequent academic performance, and those that have been done typically examine a small number of cases. One study that examined how college students who were home schooled compare with their classmates who attended regular high schools found that home-schooled young adults enjoyed higher ACT scores, grade point averages, and graduation rates compared with other college students (Cogan 2010). Another study, based on self-reports, found that college students who had been home schooled had higher self-esteem, lower rates of depression, and greater academic success than those given a traditional education (Drenovsky and Cohen 2012). However, many sociologists would like to see these findings confirmed in other samples before concluding that home schooling provides the same benefits as traditional schools.

Sociological Theories of Education

Sociologists have debated why formal systems of schooling developed in modern societies through the three lenses we have used elsewhere in this textbook: functionalism, symbolic interactionism, and conflict theory.

FUNCTIONALISM

Functionalist theory focuses on the way schooling helps contribute to the stability of society. Robert K. Merton, one of the founders of functionalist theory, argued that "schools are of course the official agency for passing on [society's] prevailing values" (Merton, 1968: 191). This is no less true today than it was fifty years ago, achieving this through both their manifest and latent functions. The manifest functions (education's intended consequences) include socializing children into a society's norms and values, providing the knowledge and skills deemed to be culturally important (for example, reading, writing, mathematics, history, civics, arts, and humanities), preparing students for work and eventual careers, and developing a sense of national identity. Reciting the Pledge of Allegiance at the beginning of class or celebrating President's Day or Lincoln's birthday are among the ways that schools promote a sense of national identity and patriotic feelings among students. In 1983, under President Ronald Reagan, Martin Luther King Day was also made a national holiday, in honor of the civil rights leader who was assassinated in 1968; a national commitment to King's belief in the nonviolent pursuit of racial equality was thus acknowledged as an integral part of the American identity. According to this approach, the content of education is particularly important in creating a common culture.

But education has latent functions, or unintended consequences, that also serve to promote social stability, for example, learning to sit (relatively) quietly for long hours and to respect authority. This then serves as preparation for the desk-bound jobs many graduates will pursue. Other latent functions include keeping young people in classrooms for long hours rather than on the streets, where they are more likely to get into trouble.

SYMBOLIC INTERACTIONISM

Symbolic interactionism provides a way to better understand how education's manifest and latent functions play out in practice, because it provides a lens through which to examine how meaning is negotiated and constructed through the many interactions that occur in the school environment. For example, students develop a sense of self based on their

Schools promote a sense of national identity among students through customs like the Pledge of Allegiance.

interpretations of how they are viewed by their peers and their teachers. In a classic study, Robert Rosenthal and Lenore Jacobson (2003; orig. 1968) gave intelligence (IQ) tests to students in a California elementary school. One-fifth of the students were randomly selected and labeled "intellectual bloomers" with high IQs, even though they were in reality no different, on average, than other students. Teachers were then told who these allegedly gifted students were. Based on subsequent testing, the students identified as intellectual bloomers outperformed other students. Rosenthal and Jacobson concluded that because teachers had higher expectations for the bloomers, they consciously or unconsciously interacted with them in ways that reinforced the bloomers' sense of self as gifted. The intellectual bloomers were expected to perform well; they received special attention from their teachers; and they, on average, lived up to their teachers' expectations.

From a symbolic interactionist perspective, the intellectual bloomers' improved academic performance resulted from their heightened sense of themselves as achievers, as seen through their teachers' eyes. Although the Rosenthal and Jacobson study was done a half-century ago, subsequent studies have reinforced the symbolic interactionist insight that one's sense of self can be strongly shaped by interactions with others in educational settings at all levels, whether it be developing a professional identity (Heggen and Terum 2017) or learning how to improvise as a musician (Monk 2013; Isbell 2008). One detailed review (termed a "meta-analysis") of nearly three dozen studies of student–teacher interactions, such as praising or blaming, concluded that teachers initiated more interactions with male students than with female students—and that the interactions with males were more likely to be negative than those with females (no gender differences were observed in terms of positive interactions). Such interactions play an important part in shaping gender roles (Jones and Dindia 2004).

CONFLICT THEORY

Conflict theory is less concerned with the content of an official curriculum and focuses instead on the ways that schooling reproduces social inequality (we will further explore the link between education and inequality later in the chapter). Working within this perspective, Bowles and Gintis (1976) argue that the expansion of education was brought about by employers' need for certain personality characteristics in their workers—self-discipline, dependability, punctuality, obedience, and the like—which are all taught in schools. Another influential conflict perspective comes from the sociologist Randall Collins, who has argued that the primary social function of mass education derives from the need for diplomas and degrees to determine one's credentials for a job, even if the work involved has nothing to do with the education one has received. Over time, the practice of credentialism results in demands for higher credentials, which require higher levels of educational attainment. Jobs that thirty years ago would have required a high school diploma now require a college degree. Because educational attainment is closely related to class position, credentialism reinforces the class structure within a society (Collins 1971, 1979).

Education and Literacy in the Developing World

Literacy is the "baseline" of education. Without it, schooling cannot proceed. We take it for granted in the West that the majority of people are literate, but this is only a recent development in Western history. The rise of literacy in Europe was closely tied

literacy

The ability to read and write.

to sweeping social transformations, particularly the Protestant Reformation of the sixteenth century, when Martin Luther, John Calvin, and others challenged the traditional authority of the Catholic Church and the right of the pope in Rome to determine acceptable Christian practice. Spurred by the development of the printing press a century earlier, the Protestant Reformation brought about individual study of the Bible. The social and political upheavals that resulted from the Protestant Reformation coincided with the beginning of a revolution in science, which also questioned the authority of the Catholic Church to dictate the laws of nature and astronomy. Literacy spread during this period due largely to the development of printing from movable type. Compulsory schooling, established in Europe and the United States in the nineteenth century, was perhaps the most important influence on the high rates of literacy in the world today (Barton 2006).

During the period of colonialism, colonial governments regarded education with some trepidation. Until the twentieth century, most believed indigenous populations were too primitive to be worth educating. Later, education was seen as a way of making local elites acclimate to European ways of life. To some extent, this backfired: The majority of those who led anticolonial and nationalist movements were educated elites who had attended schools in Europe. They were able to compare firsthand the democratic institutions of the European countries with the absence of democracy in their lands of origin.

The education that the colonizers introduced usually focused on issues relevant to Europe, not the colonial areas themselves. Educated Africans in the British colonies knew about the kings and queens of England and read Shakespeare but knew next to nothing about their own countries' histories or cultural achievements. Policies of educational reform since the end of colonialism have not completely altered the situation even today. Partly as a result of the legacy of colonial education, which was not directed toward the majority of the population, the education systems in many developing countries are top heavy: Higher education is disproportionately developed, relative to primary and secondary education. The result is a correspondingly overqualified group that, having attended colleges and universities, cannot find white-collar or professional jobs. Given the low level of industrial development, most of the better-paid positions are in government, and there are not enough of those to go around.

Today, 37 percent of the population ages fifteen and older in the least-developed countries are illiterate. Literacy rates are lowest in sub-Saharan Africa (65 percent) and South and West Asia (72 percent). Nearly two-thirds of the global illiterate adult population are women (UNESCO 2018a). In recent years, some developing countries, recognizing the shortcomings of the curricula inherited from colonialism, have tried to redirect their education programs toward the rural poor. They have had limited success, because usually there is insufficient funding to pay for the scale of the necessary innovations. As a result, countries like India have begun programs of self-help education. Communities draw on existing resources without creating demands for high levels of financing. Those who can read and write and who perhaps possess job skills are encouraged to take others on as apprentices.

CONCEPT CHECKS

1. Why did schooling become widespread only after the Industrial Revolution?

2. What are three of the most frequently cited reasons for home schooling?

3. What are some of the functions of formal schooling?

4. What are some of the reasons there are many illiterate people in the developing world?

What Is the Link between Education and Inequality?

Become familiar with the most important research on whether education reduces or perpetuates inequality. Learn the social and cultural influences on educational achievement.

The expansion of education in both developing and wealthy nations has always been closely linked to the ideals of democracy. Reformers value education for its own sake—for the opportunity it provides for individuals to develop their capabilities. Yet education has also consistently been seen as a means of promoting equality. In their classic study of intergenerational mobility, which we discussed in Chapter 7, Peter Blau and Otis Dudley Duncan found that educational attainment was a key factor behind occupational success. Long-range intergenerational mobility was rare, they found, because a child's education is strongly influenced by family social status. Subsequent studies have confirmed the importance of education in status attainment, although how important it is relative to other factors, such as family background or prevailing economic conditions, is debated (Hauser 1980; Campbell et al. 2005; Schoon, 2008; Lui et al. 2013).

Would access to universal education help reduce disparities of wealth and power? Has education in fact proved to be a great equalizer? The answer depends, in part, on whether educational opportunities are truly equal for everyone. A large body of research suggests that they are not.

"Fire in the Ashes"

During the latter third of the twentieth century, after the 1954 *Brown v. Board of Education* decision outlawed segregation and southern schools were integrated—often in the face of violent resistance—it seemed possible that all students, regardless of race, might benefit equally from access to education. Sadly, this has not turned out to be the case.

Between the 1980s and the first decade of the twenty-first century, the journalist Jonathan Kozol studied schools in about thirty neighborhoods around the United States. Through this work, he has vividly shown how unequal schools are in the United States and that these inequalities powerfully influence students' lives as they enter adulthood. What startled him most was the segregation within these schools and the inequalities among them. Kozol brought these terrible conditions to the attention of the American people in his best-selling book *Savage Inequalities* (Kozol 1991).

In his passionate opening chapter, he first took readers to East St. Louis, Illinois, a city that has been roughly 98 percent Black for the past several decades. At the time of Kozol's research, the city had no regular trash collection and few jobs. Three-quarters of its residents were living on welfare at the time. City residents were forced to use their backyards as garbage dumps, which attracted a plague of flies and rats during the hot summer months. East St. Louis also had some of the sickest children in the United States, with extremely high rates of infant death and asthma, poor nutrition, and extremely low rates of immunization. Among the city's other social problems were crime, dilapidated housing, poor health care, and lack of education.

Kozol showed how the problems of the city affected the school on a daily basis. Teachers often had to hold classes without chalk or paper. One teacher commented on these conditions affecting her teaching: "I have no materials with the exception of a single textbook given to each child. If I bring in anything else—books or tapes or magazines—I bring it in myself. . . . The AV equipment in the school is so old that we are pressured not to use it." Comments from students reflected the same concerns. "I don't go to physics class because my lab has no equipment," said one student. Only 55 percent of the students in this high school ultimately graduated, about one-third of whom went on to college.

More than two decades later, Kozol went back and revisited the neighborhoods and children he studied to find out what had happened to them. His portraits are often depressing, with many of the children from the poorer districts growing up to be troubled adults. Their lives often were derailed by alcohol abuse, unwanted pregnancies, murders, prison time, and even suicide. Yet Kozol did find that a handful of the students succeeded even though the odds were stacked against them. Most of these resilient children had been fortunate to have especially devoted parents, support from their religious communities, or a serendipitous scholarship opportunity. As Kozol notes in his 2012 book *Fire in the Ashes: Twenty-Five Years among the Poorest Children in America*, "These children had unusual advantages: Someone intervened in every case." For example, one young girl named Pineapple, whom Kozol met when she was a kindergartner, went on to graduate college and become a social worker. Pineapple attended a school that Kozol described as "almost always in a state of chaos because so many teachers did not stay for long." A local minister helped her get scholarships to private schools. The daughter of Spanish-speaking immigrants, Pineapple had to work hard to overcome deficits in reading, writing, and basic study skills, but she and her older sister both were the first in their family to finish high school and go to college (Kozol 2012).

While the personal tales of Pineapple and her sister are inspiring, Kozol's analyses reveal that very little has improved in the past two decades. As with his earlier studies of East St. Louis, Kozol visited schools that were just a few miles apart geographically but that offered vastly different educational opportunities. While suburban white schools would offer advanced math, literature, and an array of arts electives, the nearby primarily Black school would offer classes like hairdressing, typing, and auto shop. There remain vast disparities in educational spending in largely Black and Latino central cities versus largely white well-to-do suburbs. Because school funding tends to come from local property taxes, wealthier areas generate more funding for schools, while poorer neighborhoods with few lavish private homes have far less money for schools. For example, in 2013, the Chicago Ridge School District, where two-thirds of students come from low-income families, spent $9,794 per child. Less than an hour north, in a wealthy Chicago suburb, the Rondout District spent $28,639 per student (Turner et al. 2016). According to the U.S. Department of Education (2015), school districts with high levels of poverty spend, on average, 16 percent less per student than districts with low levels of poverty; high-poverty districts also tend to be those with higher percentages of minority populations.

Kozol's poignant journalistic account of educational inequality has become part of our nation's conventional wisdom on the subject. But many sociologists have argued that although Kozol's book is a moving portrait, it provides an inaccurate and incomplete view of educational inequality. Why would Kozol's research not be compelling? There are several reasons, including the unsystematic way that he chose the schools that he studied.

Sociologists, however, have proposed a variety of theories and identified myriad factors that contribute to the inequality and differential outcomes that Kozol witnessed in the schools he visited.

Coleman's Study of "Between-School Effects" in American Education

Studies comparing how schools differ from one another (or "between-school effects" studies) have been the focus of sociological research on the education system for the past three decades. One of the classic investigations of educational inequality was undertaken in the United States in the 1960s. As part of the Civil Rights Act of 1964, the commissioner of education was required to prepare a report on educational inequalities resulting from differences of ethnic background, religion, and national origin. James Coleman, a sociologist, was director of the research program. The outcome was a study, published in 1966, based on one of the most extensive research projects ever carried out in sociology.

Information was collected on more than half a million pupils who were given a range of achievement tests assessing verbal and nonverbal abilities, reading levels, and mathematical skills. Sixty thousand teachers also completed forms providing data for about 4,000 schools. The report found that a large majority of children went to schools that effectively segregated Black from white. Almost 80 percent of schools attended by white students contained only 10 percent or less African American students. White and Asian American students scored higher on achievement tests than did Blacks and other ethnic-minority students. Coleman had supposed his results would also show schools that were mainly African American to have worse facilities, larger classes, and inferior buildings than schools that were predominantly white. But surprisingly, the results showed far fewer differences of this type than had been anticipated.

Coleman therefore concluded that the material resources provided in schools made little difference to educational performance; the decisive influence was the children's backgrounds. In Coleman's words, "Inequalities imposed on children by their home, neighborhood, and peer environment are carried along to become the inequalities with which they confront adult life at the end of school" (Coleman et al. 1966: 325). There was, however, some evidence that students from deprived backgrounds who formed close friendships with those from more favorable circumstances were likely to be more successful educationally. The findings of Coleman's study have been replicated many times over the past decades, most notably by Christopher Jencks and colleagues (Jencks et al. 1972; Schofield 1995).

The Resegregation of American Schools?

More than sixty years have passed since *Brown v. Board of Education* outlawed school segregation on the basis of race. What has been achieved? Early experiments with school busing as a way of achieving more integrated schools met with resistance, particularly on the part of white parents. In recent years, court decisions have weakened the ability of schools to bus students (Orfield et al. 2016), and the impetus to achieve mandated integration through busing has weakened.

It appears that racial segregation of schools remains strong and in fact may be increasing. One study of white, Black, and Hispanic students at more than 86,000 public schools found that the racial and ethnic makeup of schools tended to reflect that of their surrounding neighborhoods, defined in the study as census tracts with a roughly two-mile radius (Whitehurst et al. 2017). Because many neighborhoods remain racially segregated, such segregation is reflected in schools as well: White and Black students tend to go to different schools. On the other hand, an increase in the Hispanic population has resulted in an overall decline in school segregation within neighborhoods, as the number of Hispanics in neighborhood schools has increased: Between 1990 and 2013, the Black share of enrollment in public schools remained at roughly 15 percent, while the white share dropped from 69 percent to 50 percent and the Hispanic share increased from 11 percent to 25 percent (Orfield et al. 2016).

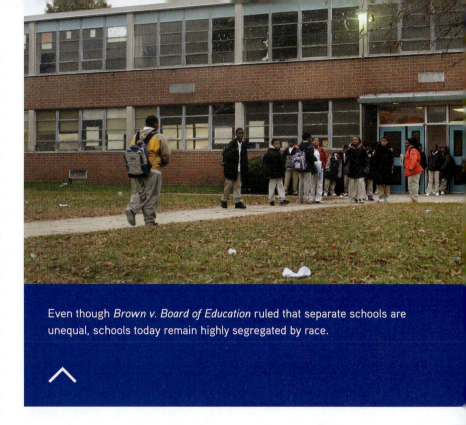

Even though *Brown v. Board of Education* ruled that separate schools are unequal, schools today remain highly segregated by race.

The Whitehurst study found that even though schools have become more diverse as a result of an influx of Hispanic and Asian students, white and Black students remain very separate. In fact, Black and white students are no more likely to be in the same classrooms than they were decades ago (Whitehurst et al. 2017: 6). The study also noted that racial segregation between school districts (as opposed to within districts) has actually increased, suggesting that school district boundaries may be geographically constructed to produce more racially homogenous schools. School segregation matters, the study concludes, because research shows that students in high-poverty schools have worse outcomes, on average, than students in more economically integrated schools. Because Black and Hispanic students are much more likely to be poor, they are also more likely to attend schools with a high proportion of poor students: 45 percent of Black and Hispanic students are in high-poverty schools, compared to just 8 percent of white students. In other words, "Black students are four times as likely to be in a high-poverty school as a low-poverty one; for whites the ratio is the other way around" (Whitehurst et al. 2017: 7).

Tracking and "Within-School Effects"

The practice of **tracking**—dividing students into groups that receive different instruction on the basis of assumed similarities in ability or attainment—is common in American schools. In some schools, students are tracked only for certain subjects, in others, for all subjects. Sociologists have long believed that tracking partly explains why schooling seems to have little effect on existing social inequalities, because being placed in a particular track labels a student as either able or otherwise. Children from more privileged backgrounds, in which academic work is encouraged, are likely to find themselves in the higher tracks early on—and by and large stay there.

In a classic study of school tracking, Jeannie Oakes (1985) studied twenty-five middle schools and high schools, both large and small and in both urban and rural areas, and concentrated on differences within schools rather than among them. She found that although several schools claimed they did not track students, virtually all of them had mechanisms

tracking

Dividing students into groups that receive different instruction on the basis of perceived similarities in ability.

for sorting students into groups on the basis of purported ability to make teaching easier. In other words, they employed tracking but did not choose to use the term *tracking* itself. Oakes found that tracking made both teachers and students label students based on their tracks—high ability, low achieving, slow, average, and so on. Individual students in these groups came to be defined by teachers, other students, and themselves in terms of such labels. A student in a "high-achieving" group was considered a high-achieving person—smart and quick. Pupils in a "low-achieving" group came to be seen as slow and below average.

What is the impact of tracking on students in the "low" group? A subsequent study by Oakes found that these students received a poorer education in terms of the quality of courses, teachers, and textbooks made available to them. Moreover, tracking had a negative impact primarily on students who were poor, and particularly on African American and Latino students (Oakes 1990). Being assigned to a low track (or failing to get into a high track) can have negative psychological effects as well: Just as the students labeled as "intellectual bloomers" in the Rosenthal and Jacobson (2003) study came to be seen (and therefore most likely came to see themselves) as gifted, students identified as unworthy of special attention may begin to take on that label themselves. Because students relegated to lower tracks are more likely to be poor and nonwhite, tracking may also contribute to a cycle of poverty (Chiu et al. 2008; Ansalone 2003).

Despite these negative consequences, school systems typically track students on the assumption that bright children learn more quickly and effectively in a group of others who are equally able and that clever students are held back if placed in mixed groups. This assumption was partially supported by a subsequent study by the sociologist Adam Gamoran. He and his colleagues agreed with Oakes's conclusions that tracking reinforces previously existing inequalities for average or poor students. However, they also found that tracking has positive benefits for some high-achieving students but little or no effect for low-achieving students (Gamoran et al. 1995; Oakes 1992; Winner 1997; Carbonaro 2005).

The Social Reproduction of Inequality

As we mentioned earlier, the education system provides more than formal instruction: It socializes children to get along with one another, teaches basic skills, and transmits elements of culture, such as language and values. Sociologists have looked at education as a form of social reproduction, a concept discussed in Chapter 1 and elsewhere. In the context of education, *social reproduction* refers to the ways in which schools help perpetuate social and economic inequalities across the generations. A number of sociologists have argued that the hidden curriculum is the mechanism through which social reproduction occurs. The **hidden curriculum** addresses the fact that much of what is learned in school has nothing directly to do with the formal content of lessons. The hidden curriculum teaches children that their role in life is "to know their place and to sit still with it" (Illich 1983). Children spend long hours in school and get an early taste of what the world of work will be like, learning that they are expected to be punctual and apply themselves diligently to the tasks that those in authority set for them.

In their classic study of social reproduction, Samuel Bowles and Herbert Gintis (1976) provide an example of how the hidden curriculum works. Modern education, they propose, is a response to the economic needs of industrial capitalism. Schools help provide the technical and social skills required by industrial enterprise, and they instill discipline

hidden curriculum

Traits of behavior or attitudes that are learned at school but not included within the formal curriculum, for example, gender differences.

and respect for authority in the future labor force. Authority relations in school, which are hierarchical and place strong emphasis on obedience, directly parallel those dominating the workplace. Under the current system, schools "are destined to legitimize inequality, limit personal development to forms compatible with submission to arbitrary authority, and aid in the process whereby youth are resigned to their fate" (266). If there were greater democracy in the workplace and more equality in society at large, Bowles and Gintis argue, a system of education could be developed that would provide for greater individual fulfillment.

Adherents of this perspective don't completely dismiss the content of the official curriculum. They accept that the development of mass education has had many beneficial consequences, including low illiteracy rates compared with premodern times. But because education has expanded mainly as a response to economic needs, schooling has not become the "great equalizer"; rather, within the current economic and political system, schooling reproduces social class stratification and merely produces for many the feelings of powerlessness that continue throughout their experience in industrial settings.

How well did you behave today?

In addition to the official curriculum of reading, writing, and calculating, there exists a hidden curriculum in schools whereby students are instructed on how to behave.

Intelligence and Inequality

Suppose differences in educational attainment, and in subsequent occupations and incomes, directly reflected differential intelligence. In such circumstances, it might be argued, there is in fact equality of opportunity in the school system for people to find a level equivalent to their innate potential. For years, psychologists, geneticists, statisticians, and others have debated whether there exists a single human capability that can be called **intelligence** and, if so, whether it rests on innately determined differences. Intelligence is difficult to define because, as the term is usually employed, it covers qualities that may be unrelated to one another. We might suppose, for example, that the "purest" form of intelligence is the ability to solve abstract mathematical puzzles. However, people who are very good at such puzzles sometimes show low capabilities in other areas.

Since the concept has proved so resistant to definition, some psychologists have proposed (and many educators have by default accepted) that intelligence should simply be regarded as "what **IQ (intelligence quotient)** tests measure." Most IQ tests consist of a mixture of conceptual and computational problems. The tests are constructed so that the average score is 100: Anyone scoring below is thus labeled "below-average intelligence," and anyone scoring above is "above-average intelligence." Despite the fundamental difficulty in measuring intelligence, IQ tests are widely used in research studies as well as in schools and businesses.

Scores on IQ tests do in fact correlate highly with academic performance (which is not surprising, because IQ tests were originally developed to predict success at school). They therefore also correlate closely with social, economic, and ethnic differences,

intelligence

Level of intellectual ability, particularly as measured by IQ (intelligence quotient) tests.

IQ (intelligence quotient)

A score attained on tests of symbolic or reasoning abilities.

because these are associated with variations in levels of educational attainment. White students score better, on average, than African Americans or members of other disadvantaged minorities, although the gap has narrowed over time (Dickens and Flynn 2006). The relationship between race and intelligence is best explained by social rather than biological causes, according to a team of Berkeley sociologists in their 1996 book *Inequality by Design: Cracking the Bell Curve Myth* (Fischer et al. 1996). The authors conducted this research as a way to rigorously evaluate the controversial claims made by Richard J. Herrnstein and Charles Murray in their book *The Bell Curve* (1994), which argued that the Black–white gap in IQ is due in part to genetic differences in intelligence. All societies have oppressed ethnic groups. Low status, often coupled with discrimination and mistreatment, leads to socioeconomic deprivation, group segregation, and a stigma of inferiority. The combination of these forces often prevents racial minorities from obtaining education, and consequently, their scores on standardized intelligence tests are lower.

The average lower IQ score of African Americans in the United States is remarkably similar to that of deprived ethnic minorities in other countries—such as the "untouchables" in India (who are at the very bottom of the caste system), the Maori in New Zealand, and the *burakumin* of Japan. Children in these groups score an average of 10 to 15 IQ points below children belonging to the ethnic majority. Such observations strongly suggest that the IQ variations between African Americans and whites in the United States result from social, cultural, and economic—rather than genetic—factors.

Educational Reform in the United States

Research done by sociologists has played a major role in reforming the educational system. The object of James Coleman's research, commissioned as part of the 1964 Civil Rights Act, was not solely academic; it was undertaken to influence policy. Education has long been a political battleground. In the 1960s, partly in response to Coleman's work, some politicians, educators, and community activists pushed for universal access to high-quality education through such initiatives as busing programs to mitigate racial segregation, bilingual education programs, multicultural education, open admissions to college, the establishment of ethnic studies programs on campuses, and more equitable funding schemes. Such initiatives were seen as supporting civil rights and equality. Educational policies in the twenty-first century have similarly intended to provide quality education to all children and close the achievement gap. However, scholars disagree about how successful recent policies have been in meeting this goal.

One important target of educational policy today is improving levels of **functional literacy** in the United States. Literacy is more than the ability to read and write; literacy is also the ability to process complex information in our increasingly technology-focused society. The National Center for Education Statistics breaks literacy into three components: prose literacy, document literacy, and quantitative literacy. *Prose literacy* means that a person can look at a short piece of text to get a small piece of uncomplicated information. *Document literacy* refers to a person's ability to locate and use information in forms, schedules, charts, graphs, and other informational tables. *Quantitative literacy* is the ability to do simple addition. In the United States today, only 13 percent of the population is proficient in these three areas (Kutner et al. 2007). Of course, the United States is a country of immigrants, who, when they arrive, may not be able to read and write and who may also have

functional literacy

Having reading and writing skills that are beyond a basic level and are sufficient to manage one's everyday activities and employment tasks.

trouble with English. However, this doesn't explain why America lags behind most other industrial countries in terms of its level of functional literacy.

Some policymakers believe that one of the most effective ways to enhance literacy and other academic outcomes is through formal testing and benchmarking of student progress. The No Child Left Behind (NCLB) Act, signed into law by President George W. Bush in 2002, aimed to improve academic outcomes for all children and to close achievement gaps. As we saw in Chapter 10, rates of high school graduation and college attendance vary dramatically by race, ethnicity, and socioeconomic background. At the top of the act's agenda is instituting **standardized testing**, in which all students in a state take the same test under the same conditions as a means of measuring students' academic performance. The act also provides a strong push for school choice; that is, in the spirit of competition, parents are to be given choices as to where they send their children to school. Low-performing schools, at risk of losing students, may jeopardize their funding and eventually be closed. Another significant implication of NCLB is that for the first time since 1968, states are not required to offer non-English-speaking students bilingual education. Instead, the act emphasizes learning English over using students' native language to support learning objectives and favors English-only program models. NCLB also provides support for a "zero tolerance" approach to school discipline that was first mandated in the 1990 Gun-Free School Act.

The NCLB was widely criticized, as teachers must "teach to the test." Critics argued that the emphasis on standardized testing as the means of assessment encourages teachers to teach a narrow set of skills that will increase students' test performance rather than helping them to acquire an in-depth understanding of important concepts and skills (Hursh 2007). Because teachers and principals at underperforming schools risk job loss, some critics described the program as a punitive model of school reform. They also noted that achievement gaps had not changed and that the policy neglected the important fact that the broader socioeconomic context affects school functioning.

In 2009, shortly after he took office, President Barack Obama—eager to leave his own stamp on education—implemented Race to the Top, a program that rewards states that demonstrate improvements in student outcomes, including closing achievement gaps, increasing graduation rates, and better preparing students for college. States competing for the more than $4 billion in grant money had to outline plans for developing and adopting common standards and assessments, building data systems that track student growth, recruiting and retaining high-quality teachers, and improving the lowest-achieving schools. Like NCLB, however, Race to the Top was roundly criticized for relying too heavily on high-stakes testing and also for failing to address the true causes of low student achievement, namely, poverty and lack of opportunity (Dillon 2010). In 2012, recognizing that the NCLB may not be effective for all school systems, Obama granted waivers from NCLB requirements to thirty-two states, allowing them to develop their own standards and exempting them from the 2014 targets set by NCLB. And in 2015, at the end of his presidency, Obama replaced NCLB entirely with a much weaker act, dubbed Every Student Succeeds, which retained standardized testing but ceded control to states and school districts.

The crisis in American schools won't be solved in the short term, and it won't be solved by educational reforms alone, no matter how well intended. In fact, a 2006 study by the U.S. Department of Education found that schools identified as most in need of improvement

standardized testing

A procedure whereby all students in a state take the same test under the same conditions.

In 1970, a U.S. judge in North Carolina ordered that Black students be bused to white schools and that white students be bused to Black schools in an attempt to end the de facto segregation of public schools caused by white students living in predominantly white neighborhoods and Black students living in predominantly Black neighborhoods.

were disproportionately urban, high-poverty schools and that school poverty and district size were more powerful predictors of school success than any policies actually implemented by the schools (U.S. Department of Education 2006). A further unintended consequence of the current emphasis on testing is that schools have narrowed their course offerings to focus much more heavily on tested subject areas while cutting time in science, social studies, music, art, and physical education (Center on Education Policy 2007).

The lesson of sociological research is that inequalities and barriers in educational opportunity reflect wider social divisions and tensions. While the United States remains wracked by racial tensions, and the polarization between decaying cities and affluent suburbs persists, the crisis in the school system is likely to prove difficult to turn around.

What is to be done? Some have proposed giving schools more control over their budgets (a reform that has been carried out in Britain). The idea is that more responsibility for and control over budgeting decisions will create a greater drive to improve the school. Further proposals include the refunding of federal programs, such as Head Start, to ensure healthy early-childhood development and thus save millions of dollars in later costs. In March 2017, President Trump signed bills that weakened federal regulation of local schools, overturning such Obama-era rules as requiring states to rate teacher-training programs on the basis of student performance on standardized tests (Brown 2017b). The clearly stated goal of educational reform under Trump, however, spearheaded by Education Secretary Betsy DeVos, is to greatly expand "school choice": the ability of parents to use public funds, in the form of vouchers or tax credits, to enroll their children in various alternatives to public schools. Such alternatives might include private schools, charter schools (public schools run by private companies), Catholic schools, or, ideally, any educational environment of their choice. Proponents of such reforms argue that this will enable parents, especially parents of low-income children, to get their children into far better schools than the ones they currently attend; they also argue that public schools will improve if forced to compete with alternatives (Coulson 2009; Jeynes 2012; edChoice 2018; Chingos and Peterson 2018). Opponents claim that it will greatly weaken the public school system by redirecting public funds to private schools and is far more likely to benefit middle- or upper-income families than poor families, because vouchers will be insufficient to make alternative schools affordable to the poor (Hopkinson 2011; Singer 2017; Klein 2017).

One thing seems clear: Continually changing efforts at educational reform have resulted in "reform fatigue" among teachers. A recent national survey of more than 500 K–12 teachers found that 58 percent reported they have "experienced too much or way too much reform in the past two years," and fully 68 percent are skeptical that "new" education reforms are truly new (Education Week Research Center 2017). Teachers are not only frustrated with the ever-changing national requirements; they are also discouraged by low pay. Nationally, teachers earn roughly three-quarters as much as other college graduates; in Arizona and Oklahoma, they earn only two-thirds as much (Allegretto 2018). In early 2018, teachers in West Virginia, Kentucky, and Oklahoma walked out in protest of low pay. Perhaps a good starting point for improving public education would be to provide teachers with more equitable (and competitive) salaries.

CONCEPT CHECKS

1. According to Kozol, has education become an equalizer in American society? Why or why not?

2. How do Coleman's findings differ from the results of Kozol's research? Whose theory, in your opinion, can better explain the racial gap in educational achievement?

3. What effect does tracking have on academic achievement?

4. How do schools perpetuate existing inequalities across generations?

5. Describe the components and critiques of the No Child Left Behind Act and Race to the Top.

How Do Sociologists Think about Religion?

Whereas modern education emerged in the nineteenth century, religion is one of the oldest human institutions. Cave drawings suggest that religious beliefs and practices existed more than 40,000 years ago. According to anthropologists, there have probably been about 100,000 religions throughout human history (Hadden 1997a). Sociologists define **religion** as a cultural system of commonly shared beliefs and rituals that provides a sense of meaning and purpose by creating an idea of reality that is sacred, all-encompassing, and supernatural (Berger 1967; Durkheim 1965, orig. 1912; Wuthnow 1988). There are three key elements in this definition:

1. **Religion is a form of culture.** You will recall from Chapter 2 that culture consists of the shared beliefs, values, norms, and material conditions that create a common identity among a group of people. Religion shares all of these characteristics.

2. **Religion involves beliefs that take the form of ritualized practices.** All religions have a behavioral aspect—special activities that identify believers as members of the religious community.

3. **Perhaps most important, religion provides a sense of purpose—a feeling that life is meaningful.** It does so by explaining what transcends or overshadows everyday life in ways that other aspects of culture (such as an education system or a belief in democracy) typically cannot (Geertz 1973; Wuthnow 1988).

What is absent from the sociological definition of religion is as important as what is included: Nowhere is there mention of God. We often think of **theism**—a belief in one or more supernatural deities (the term originates from the Greek word for God)—as basic to religion, but this is not necessarily the case. Some religions, such as Buddhism, believe in the existence of spiritual forces rather than in a particular God.

Four broad conditions set the stage for the sociological study of religion:

1. **Sociologists are not concerned with whether religious beliefs are true or false.** From a sociological perspective, religions are regarded not as being decreed by God but as being socially constructed by human beings. As a result, sociologists put aside their personal beliefs when they study religion. They are concerned with the human rather than the divine aspects of religion. Sociologists ask, How is the religion organized? How is it related to the larger society? What explains its success or failure in recruiting and retaining believers? The question of whether a particular belief is "good" or "true," however important it may be to the believers of the religion under study, is not something that sociologists are able to address as sociologists. (As individuals, they may have strong opinions, but one hopes that they can keep these opinions from biasing their research.)

2. **Sociologists are especially concerned with the social organization of religion.** Religions are among the most important institutions in society. They are

Learn the elements that make up religion. Know the sociological approaches to religion developed by Marx, Durkheim, and Weber as well as the religious economy approach.

religion

A set of beliefs adhered to by the members of a community, incorporating symbols regarded with a sense of awe or wonder together with ritual practices. Religions do not universally involve a belief in supernatural entities.

theism

A belief in one or more supernatural deities.

Surrounded by his parents and rabbi, a thirteen-year-old Jewish boy carries a Torah during his Bar Mitzvah, effectively becoming a man in the eyes of his synagogue (left). In a Catholic church in Brooklyn, an infant is baptized into the Christian faith when a priest ceremonially anoints him or her with holy water (right). What are some similarities and differences among the rituals of various religions?

a primary source of the most deeply seated norms and values. At the same time, religions are typically practiced through an enormous variety of social forms. The sociology of religion is concerned with how different religious institutions and organizations actually function. The earliest European religions were often indistinguishable from the larger society, as religious beliefs and practices were incorporated into daily life. This is still true in many parts of the world today. In modern industrial society, however, religions have become established in separate, often bureaucratic, organizations, and so sociologists focus on the organizations through which religions must operate to survive (Hammond 1992).

3. **Sociologists often view religions as a major source of social solidarity because religions provide their believers with a common set of norms and values.** Religious beliefs, rituals, and bonds help to create a "moral community" in which all members know how to behave toward one another (Wuthnow 1988). If a single religion dominates a society, the religion may be an important source of social stability. However, religion can also be oppressive if, like the religion practiced by the Taliban, it requires absolute conformity to a particular set of beliefs and punishes those who deviate from these beliefs. For example, the attempted murder of Malala Yousafzai, whom we met in the chapter's opener, was an effort by some Taliban leaders to punish a young girl who defied their beliefs.

4. **Sociologists tend to explain the appeal of religion in terms of social forces rather than purely personal, spiritual, or psychological factors.** For many people, religious beliefs are a deeply personal experience, involving a powerful sense of connection with forces that transcend everyday reality. Sociologists do not question the depth of such feelings and experiences, but they are unlikely to limit themselves to a purely spiritual explanation of religious commitment. Some

researchers argue that people often "get religion" when their fundamental sense of a social order is threatened by economic hardship, loneliness, loss or grief, physical suffering, or poor health (Berger 1967; Glock 1976; Schwartz 1970; Stark and Bainbridge 1980). In explaining the appeal of religious movements, sociologists are more likely to focus on the problems of the social order than on the psychological response of the individual.

Theories of Religion

Sociological approaches to religion are strongly influenced by the classical theories of Marx, Durkheim, and Weber. None of the three was religious himself, and they all believed that religion would become less and less significant in modern times. Each argued that religion was fundamentally an illusion: The very diversity of religions and their obvious connection to different societies and regions of the world made their advocates' claims inherently implausible. An individual born into an Australian society of hunters and gatherers would hold different religious beliefs from someone born into the caste system of India or into the Catholic Church of medieval Europe.

MARX: RELIGION AND INEQUALITY

Despite the influence of his views on the subject, Karl Marx never studied religion in any detail. His thinking on religion was derived mostly from the writings of Ludwig Feuerbach, who believed that through a process he called **alienation**, human beings tend to attribute their own culturally created values and norms to divine forces or gods because they do not understand their own history. Thus, the story of the Ten Commandments that God gave to Moses is a mythical version of the origins of the moral precepts that govern the lives of Jewish and Christian believers.

Marx accepted the view that religion represents human self-alienation. In a famous phrase, Marx declared that religion is the "opium of the people." Religion defers happiness and reward to the afterlife, he said, teaching the resigned acceptance of existing conditions in the earthly life. Attention is thus diverted from injustices in this world by the promise of what is to come in the next. Religious belief also can provide justifications for those in power. For example, "the meek shall inherit the earth" suggests attitudes of humility and nonresistance to oppression.

alienation

The sense that our own abilities as human beings are taken over by other entities. The term was originally used by Karl Marx to refer to the projection of human powers onto gods. Subsequently, he used the term to refer to the loss of workers' control over the nature and products of their labor.

DURKHEIM: RELIGION AND FUNCTIONALISM

In contrast to Marx, Émile Durkheim studied religion extensively and connected religion not with social inequalities or power but with the overall nature of a society's institutions. In *The Elementary Forms of the Religious Life* (1965; orig. 1912), Durkheim concentrated on totemism, the worship of objects, such as animals or plants, believed to embody mystical spirits. Durkheim studied totemism as practiced by Australian aboriginal societies, arguing that such totemism represented religion in its most "elementary" form (hence the title of his book). Durkheim sought to show that totems served a dual purpose: They are "at once the symbol of the god and the society," which raises a central question: "Is that not because the god and the society are only one?" In answering that question, Durkheim concludes that "the god of the clan, the totemic principle, can therefore be nothing else than the clan itself, personified and represented to the imagination under the visible form of the animal or vegetable which serves as totem" (Durkheim, 1965; orig. 1912: 206). Although Durkheim focused on Australian aboriginal societies, he argued that these elementary forms of religion provided insights into modern society as well. Religion, for Durkheim, was society writ large. Sociologist Reza Aslan (2017) extends this argument, drawing on archeological as well as contemporary evidence.

Durkheim defined religion in terms of a distinction between the sacred and the profane. **Sacred** objects and symbols, he held, are treated as apart from the routine aspects of day-to-day existence—the realm of the **profane**. A totem, Durkheim argued, is a sacred object, regarded with veneration and surrounded by ritual activities. These ceremonies and rituals, in Durkheim's view, are essential to unifying the members of groups.

Durkheim's theory of religion is a good example of the functionalist tradition in sociology. To analyze the function of a social behavior or social institution like religion is to study the contribution it makes to the continuation of a group, community, or society. According to Durkheim, religion has the function of uniting a society by ensuring that people meet regularly to affirm common beliefs and values.

WEBER: THE WORLD RELIGIONS AND SOCIAL CHANGE

Whereas Durkheim based his arguments on a restricted range of examples, Max Weber embarked on a massive study of religions worldwide. No scholar before or since has undertaken a task of this scope. Weber's writings on religion differ from those of Durkheim because they concentrate on the connection between religion and social change, something to which Durkheim gave little direct attention. They also contrast with those of Marx, because Weber argued that religion is not necessarily a conservative force; on the contrary, religiously inspired movements have often produced dramatic social transformations. Thus, Protestantism, particularly Puritanism, according to Weber, was the source of the capitalistic outlook found in the modern West. The early entrepreneurs were mostly Calvinists. Their drive to succeed, which helped initiate Western economic development, was originally prompted by a desire to serve God. Material success was a sign of divine favor. But because Calvinists also believed that one should not ostentatiously flaunt one's wealth, Calvinist entrepreneurs were likely to reinvest their wealth in their enterprises rather than spend it on personal consumption. Such "worldly asceticism," as Weber called it, resulted in capital accumulation—the hallmark of a successful capitalist system.

Weber conceived of his research on the world religions as a single project. His discussion of the impact of Protestantism on the development of the West was connected to a comprehensive attempt to understand the influence of religion on social and economic life in various cultures. After analyzing Eastern religions, Weber concluded that they provided barriers to the development of industrial capitalism like that which took place in the West. Eastern civilizations, he observed, were oriented toward different values, such as escape from the toils of the material world.

Weber regarded Christianity as a salvation religion. According to such religions, human beings can be "saved" if they are converted to the beliefs of the religion and follow its moral tenets. The notions of "sin" and of being rescued from sinfulness by God's grace are important. They generate a tension and an emotional dynamism essentially absent from the Eastern religions. Salvation religions have a "revolutionary" aspect. Whereas the religions of the East cultivate an attitude of passivity or acceptance within the believer, Christianity demands a constant struggle against sin and so can stimulate revolt against

---

sacred

Describing something that inspires awe or reverence among those who believe in a given set of religious ideas.

---

profane

That which belongs to the mundane, everyday world.

the existing order. Religious leaders—such as Luther or Calvin—have arisen who reinterpret existing doctrines in such a way as to challenge the extant power structure.

CRITICAL ASSESSMENT OF THE CLASSICAL VIEW

Marx, Durkheim, and Weber each identified some important general characteristics of religion, and in some ways, their views complement one another. Marx was correct to claim that religion often has ideological implications, serving to justify the interests of ruling groups at the expense of others. There are innumerable instances of this in history. For example, the European missionaries who sought to convert "heathen" peoples to Christian beliefs were no doubt sincere in their efforts. Yet their teachings contributed to the destruction of traditional cultures and the imposition of white domination. Almost all Christian denominations tolerated, or endorsed, slavery in the United States and other parts of the world into the nineteenth century. Doctrines were developed proclaiming slavery to be based on divine law, disobedient slaves being guilty of an offense against God as well as their masters (Stampp 1956).

Weber also emphasized the unsettling and often revolutionary impact of religious ideals on the established social order. Despite many churches' early support for slavery in the United States, church leaders later played a key role in fighting to abolish the institution. Religious beliefs have prompted social movements seeking to overthrow unjust systems of authority; for instance, religious sentiments played a prominent part in the civil rights movement of the 1960s.

These divisive influences of religion, so prominent in history, find little mention in Durkheim's work. Durkheim emphasized the role of religion in promoting social cohesion. Yet it is not difficult to redirect his ideas toward explaining religious division, conflict, and change as well as solidarity. After all, much of the strength of feeling that may be generated against other religious groups derives from the commitment to religious

secular thinking

Worldly thinking, particularly as seen in the rise of science, technology, and rational thought in general.

secularization

A process of decline in the influence of religion. Secularization can refer to levels of involvement with religious organizations, the social and material influence wielded by religious organizations, and the degree to which people hold religious beliefs.

values generated within each community of believers. Among the most valuable points of Durkheim's writings is his stress on ritual and ceremony. All religions comprise regular assemblies of believers, at which ritual prescriptions are observed. As Durkheim rightly points out, ritual activities also mark the major life stages—birth, the transition to adulthood (rituals associated with puberty are found in many cultures), marriage, and death (van Gennep 1977).

Finally, the theories of Marx, Durkheim, and Weber on religion were based on their studies of societies in which a single religion predominated. As a consequence, it seemed reasonable for them to examine the relationship between a predominant religion and the society as a whole. However, in the past fifty years, this classical view has been challenged by some U.S. sociologists. Because of their own experience in a society that is highly tolerant of religious diversity, these theorists have focused on religious pluralism rather than on religious domination. Not surprisingly, their conclusions differ substantially from those of Marx, Durkheim, and Weber, each of whom regarded religion as closely bound up with the larger society. Religion was believed to reflect and reinforce society's values, or at least the values of those who were most powerful; to provide an important source of solidarity and social stability; and to drive social change. According to this view, religion is threatened by the rise of **secular thinking**, particularly as seen in the rise of science, technology, and rational thought in general.

The classical theorists argued that the key problem facing religions in the modern world is **secularization**, or the process by which religious belief and involvement decline and thus result in a weakening of the social and political power of religious organizations. Peter Berger (1967) has described religion in premodern societies as a "sacred canopy" that covers all aspects of life and is therefore seldom questioned. In modern society, however, the sacred canopy is more like a quilt, a patchwork of different religious and secular belief systems. When multiple belief systems coexist, it becomes increasingly difficult to sustain the idea that there is any single true faith. According to this view, secularization is the likely result.

According to the religious economy approach, the presence of numerous religious groups increases participation.

CONTEMPORARY APPROACHES: "RELIGIOUS ECONOMY"

One of the most influential contemporary approaches to the sociology of religion is tailored to societies, such as the United States, that offer many different faiths from which to pick and choose. Sociologists who favor the **religious economy** approach argue that religions can be thought of as organizations in competition with one another for followers (Finke and Stark 1988, 1992; Hammond 1992; Moore 1994; Roof and McKinney 1990; Stark and Bainbridge 1987; Warner 1993).

Like contemporary economists who study businesses, these sociologists argue that competition is preferable to monopoly when it comes to ensuring religious vitality. This position is exactly opposite to that of the classical theorists. Marx, Durkheim, and Weber assumed that religion weakens when challenged by different religious or secular viewpoints, whereas the religious economists argue that competition increases the overall level of religious involvement in modern society. Religious economists believe that this is true for two reasons. First, competition makes each religious group try harder to win followers. Second, the presence of numerous religions means that there is likely to be something for just about everyone. In a culturally diverse society like the United States, a single religion will probably appeal to only a limited range of followers, whereas the presence of Indian gurus and fundamentalist preachers, in addition to mainline churches, is likely to encourage a high level of religious participation.

A criticism of the religious economy approach is that it overestimates the extent to which people rationally pick and choose among different religions, as if they were shopping for a new car or a pair of shoes. Among deeply committed believers, particularly in societies that lack religious pluralism, it is not obvious that religion is a matter of rational choice. Even when people are allowed to choose among different religions, most are likely to practice their childhood religion without ever questioning whether there are more appealing alternatives. Moreover, the spiritual aspects of religion may be overlooked if sociologists simply assume that religious buyers are always on spiritual shopping sprees. Wade Clark Roof's (1993) study of 1,400 baby boomers found that one-third had remained loyal to their childhood faith, while another third had continued to profess their childhood beliefs, although they no longer belonged to a religious organization. Only one-third were actively searching for a new religion, making the sorts of choices presumed by the religious economy approach (Pew Forum on Religion & Public Life 2008).

CONCEPT CHECKS

1. What are the three main components of religion as a social institution?

2. How do sociologists differ from other scholars in their approach to studying religion?

3. What are the differences between classical and contemporary approaches to understanding religion?

How Does Religion Affect Life throughout the World?

Religion is one of the most truly global of all social institutions, affecting almost all aspects of social life. In this section, we describe the way religion shapes life throughout the globe. We begin, however, by briefly describing the different ways that world religions are organized.

Types of Religious Organizations

Early theorists such as Max Weber (1963, orig. 1921), Ernst Troeltsch (1931), and Richard Niebuhr (1929) described religious organizations as falling along a continuum based on the degree to which they are well established and conventional: Churches lie at one end (they are conventional and well established), cults lie at the other (they are neither), and sects fall somewhere in the middle. These distinctions were based on the study of those religions that account for the majority of persons in Europe and the United States. There is much debate over how well they apply to the non-Christian world.

Understand the various ways religious communities are organized and how they have become institutionalized. Recognize how the globalization of religion is reflected in religious activism in poor countries and the rise of religious nationalist movements.

churches

Large, established religious bodies, normally having a formal, bureaucratic structure and a hierarchy of religious officials. The term is also used to refer to the place in which religious ceremonies are carried out.

sect

A religious movement that breaks away from orthodoxy and follows its own unique set of rules and principles.

Today, sociologists are aware that the terms *sect* and *cult* have negative connotations, something they wish to avoid. For this reason, contemporary sociologists of religion sometimes use the phrase *new religious movements* to characterize novel religious organizations that have not yet achieved the respectability that comes with being well established for a long period of time (Hadden 1997b; Hexham and Poewe 1997).

CHURCHES AND SECTS

Churches are large, established religious bodies; one example is the Roman Catholic Church. They normally have a formal, bureaucratic structure, with a hierarchy of religious officials. Churches often represent a traditional face of religion, because they are integrated within the existing institutional order. Most of their adherents are born into and grow up within the church.

A **sect** is typically described as a religious subgroup that breaks away from the larger organization and consequently follows its own unique set of rules and principles. Sects are smaller, less highly organized groups of committed believers, usually set up in protest against an established church. Sects aim to discover or follow the "true way" and either try to change the surrounding society or withdraw from it into communities of their own, a process known as revival. Many sects have few or no officials, and all members are regarded as equal participants. For the most part, people are not born into sects but actively join them to further commitments in which they believe.

DENOMINATIONS AND CULTS

A **denomination** is a sect that has cooled down and become an institutionalized body rather than an activist protest group. Sects that survive over any period of time inevitably become denominations. Denominations are recognized as legitimate by churches and exist alongside them, often cooperating harmoniously with them.

Cults, by contrast, are the most loosely knit and transient of all religious organizations. They are composed of individuals who reject what they see as the values of the outside society, unlike sects, which try to revive an established church. They are a form of religious innovation rather than revival. Their focus is on individual experience, bringing like-minded people together. Like sects, cults often form around the influence of an inspirational leader.

Similar to sects, cults flourish when there is a breakdown in well-established and widespread societal belief systems. This is happening throughout the world today, in places as diverse as Japan, India, and the United States. When such a breakdown occurs, cults may originate within a society, or they may be "imported" from outside. In the

Members of the Unification Church, also known as "Moonies," named for its founder Reverend Sun Myung Moon, participate in a mass wedding. The Holy Marriage Blessing Ceremony strengthens participants' dedication to the church.

United States, examples of homegrown, or indigenous, cults include New Age religions based on such things as spiritualism, astrology, and religious practices adapted from Asian or Native American cultures. One of the largest imported cults is the late Reverend Sun Myung Moon's Unification Church ("Moonies"), which originated in South Korea and is now led by the Reverend's wife.

Globalization and Religion

More than half of the world's population follow one of two faiths: Christianity (31 percent) or Islam (23 percent), religions that have long been unconstrained by national borders (Pew Research Center 2015g). The current globalization of religion is reflected in political activism among religious groups in poor countries and in the rise of religious nationalist movements in opposition to the modern secular state.

THE GLOBAL RISE OF RELIGIOUS NATIONALISM

One of the most important trends in global religion today is the rise of **religious nationalism**, the linking of strongly held religious convictions with beliefs about a people's social and political destiny. In countries around the world, religious nationalist movements reject the notion that religion, government, and politics should be separate and call instead for a revival of traditional religious beliefs that are directly embodied in the nation and its leadership (Beyer 1994). These nationalist movements represent a strong reaction against the impact of technological and economic modernization on local religious beliefs. In particular, religious nationalists oppose what they see as the destructive aspects of "Western" influence on local culture and religion, ranging from American television to the missionary efforts of foreign evangelicals.

Religious nationalist movements accept many aspects of modern life, including modern technology, politics, and economics. For example, Islamic fundamentalists use video and television to reach millions of Muslims worldwide. However, they also emphasize a strict interpretation of religious values and completely reject the notion of secularization (Juergensmeyer 1993, 2001). Nationalist movements do not simply revive ancient religious beliefs. Rather, nationalist movements partly "invent" the past, selectively drawing on different traditions and reinterpreting past events to serve their current beliefs and interests. Violent conflicts between religious groups sometimes result from their competing interpretations of the same historical event (Anderson 1991; Juergensmeyer 1993, 2001; van der Veer 1994).

Religious nationalism is on the rise throughout the world—perhaps because in times of rapid social change, unshakable ideas have strong appeal. An early example is the Islamic Republic of Iran, which took power through a revolution in 1979. The aim of the Islamic Republic was to organize government and society so that traditional Islamic teachings would dominate all spheres of life. The Guardian Council of religious leaders determines whether laws, policies, and candidates for parliament conform to an extremely strict interpretation of Islamic beliefs, even though Iran has a U.S.-style constitution providing for elected officials and the separation of powers. Although recent years have seen some hopeful signs that Iran is liberalizing, such hopes have proved to be short-lived in the past. The 2013 election saw a swing in the liberal direction with the election of Hassan Rouhani as president, who has sought to improve relations with the West. Rouhani easily

denomination

A religious sect that has lost its revivalist dynamism and become an institutionalized body, commanding the adherence of significant numbers of people.

cults

Fragmentary religious groupings to which individuals are loosely affiliated but that lack any permanent structure.

religious nationalism

The linking of strongly held religious convictions with beliefs about a people's social and political destiny.

won reelection in 2017, but since then, he has been confronted with widespread protests, as Iranians across the country vented their anger over economic stagnation and political repression. In siding with the protesters, Rouhani made it clear that he intended to pursue reforms, even in the face of opposition by hard-line religious leaders, who wield considerable power (Cunningham 2018). The success of Rouhani's reform efforts will likely be shaped by Iran's relationship with the United States. When Iran agreed to roll back its nuclear program in exchange for an easing of Western economic sanctions in 2015, there was hope not only that a military confrontation could be avoided but that Iran's citizens would benefit economically. With the Trump administration threatening to end the agreement, and the economic benefits yet to filter down to ordinary Iranians, tensions are again on the rise, and Rouhani's reforms may be in jeopardy.

Religious nationalist movements sometimes turn violent as they seek to impose their vision of the world on others. For most Americans, the most prominent example today is ISIS—the Islamic State of Iraq and Syria—which has used terror and extreme brutality in its effort to spread its power over parts of the Middle East. But as sociologist Mark Juergensmeyer has shown in *Terror in the Mind of God* (2017) and other books (Jerryson and Juergensmeyer 2010; Juergensmeyer 1993, 2005), extremist religious nationalism is not confined to any one religion. While Islamist extremism has been of greatest concern to Americans since 9/11, India has experienced violent Sikh nationalism in the past and Hindu nationalism today; Christian extremism in the United States has resulted in assassinations and the 1995 bombing of the Alfred F. Murrah Federal Building in Oklahoma City, which killed 168 people; Buddhist nationalism, led by militant Buddhist monks, has resulted in a genocidal ethnic cleansing of Myanmar's Rohingya Muslims; and Jewish nationalism was responsible for the 1995 assassination of Israel's president Yizhak Rabin following the Oslo Peace Accord between Israel and the Palestinian Liberation Organization. Juergensmeyer argues that "what makes religious violence particularly savage and relentless is that its perpetrators have placed such religious images of divine struggle—cosmic war—in the service of worldly political battles. For this reason, acts of religious terror serve not only as tactics in a political strategy but also as evocations of a much larger spiritual confrontation" (Juergensmeyer 2017: 184).

ACTIVIST RELIGION AND SOCIAL CHANGE THROUGHOUT THE WORLD

Religion has played a critical role in effecting positive social change over the past fifty-plus years. In Vietnam in the 1960s, Buddhist priests burned themselves alive to protest the policies of the South Vietnamese government. Their willingness to sacrifice their lives for their beliefs, seen on television sets around the world, contributed to growing U.S. opposition to the war. Buddhist monks in Thailand are currently protesting deforestation.

An activist form of Catholicism, termed **liberation theology**, combines Catholic beliefs with a passion for social justice for the poor, particularly in Central and South America and in Africa. Catholic priests and nuns organize farming cooperatives, build health clinics and schools, and challenge government policies that impoverish the peasantry. A similar role is played by Islamic socialists in Pakistan and Buddhist socialists in Sri Lanka (Berryman 1987; Juergensmeyer 1993; Sigmund 1990). In some Central and Eastern European countries once dominated by the former Soviet Union, long-suppressed religious organizations provided an important basis for overturning socialist regimes during the early 1990s. In

liberation theology

An activist Catholic religious movement that combines Catholic beliefs with a passion for social justice for the poor.

Religious Affiliation

More than eight in ten people in the world are affiliated with a religion. While Christianity is currently the largest religion in the world, Islam is growing at a faster rate. Consequently, it is projected that there will be nearly equal numbers of Muslims and Christians by mid-century.

Global religious affiliation*

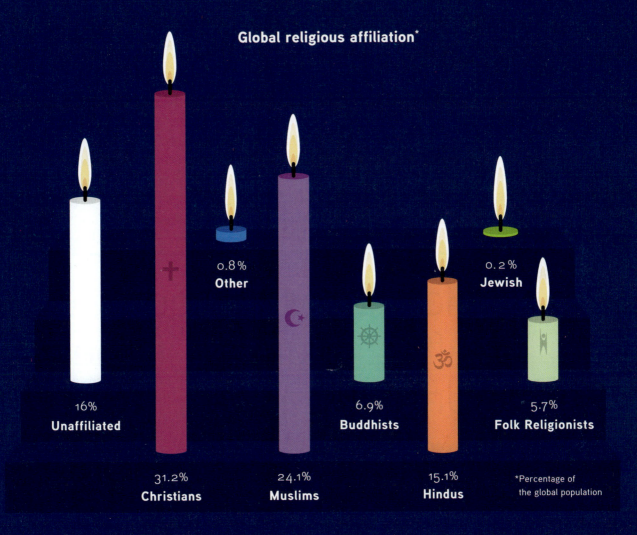

0.8%
Other

0.2%
Jewish

16%
Unaffiliated

6.9%
Buddhists

5.7%
Folk Religionists

31.2%
Christians

24.1%
Muslims

15.1%
Hindus

*Percentage of the global population

Religious affiliation in the United States

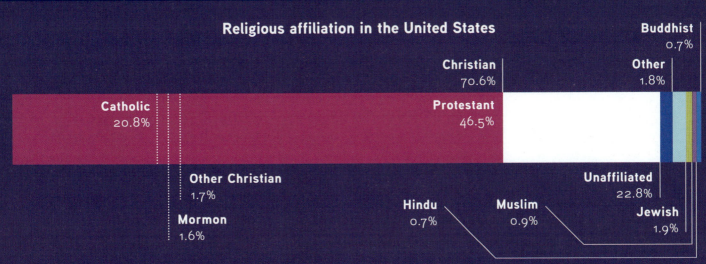

Buddhist
0.7%

Christian
70.6%

Other
1.8%

Catholic
20.8%

Protestant
46.5%

Other Christian
1.7%

Unaffiliated
22.8%

Mormon
1.6%

Hindu
0.7%

Muslim
0.9%

Jewish
1.9%

Source: Pew Research Center 2017c.

Poland, the Catholic Church was closely allied with the Solidarity movement, which toppled the socialist government in 1989. Many religious leaders have paid with their lives for their activism, which government and military leaders often regard as subversive.

Religious organizations also played a key role in the U.S. civil rights movement. The Southern Christian Leadership Conference, led by Reverend Martin Luther King Jr., organized key marches, boycotts, and other forms of nonviolent civil disobedience that resulted in the 1964 Civil Rights Act. Religious groups were also a major force in the Arab Spring that began with the self-immolation of Mohamed Bouazizi, a poor Tunisian street vendor who set himself on fire to protest harassment by corrupt local officials who had made it impossible for him to eke out a living. Bouazizi's sacrifice, seen around the world on YouTube and other social media, sparked revolutions in the Middle East and North Africa that overturned dictatorships in Egypt, Libya, and Yemen and sparked a civil war in Syria.

The Arab Spring proved fleeting: Syria, Yemen, and Libya were ravaged by internal strife and civil wars that had, at least in part, a religious dimension. Conflicts spilled over into Jordan and Lebanon as well as parts of North Africa. With the exception of Egypt, where it was repressed, there was a resurgence of Islamic fundamentalism throughout the region. The only country to remain democratic, among all the countries that had participated in the Arab Spring, was Tunisia—the country where it began.

CONCEPT CHECKS

1. Describe four types of religious organizations.

2. What is religious nationalism? Why can it be viewed as a reaction to economic modernization of local religious beliefs and Westernization?

>

Learn about the sociological dimensions of religion in the United States.

How Does Religion Affect Your Life in the United States?

Trends in Religious Affiliation

The United States is the most religiously diverse country in the world, with more than 1,500 distinct religions (Melton 1989). Yet the vast majority of people belong to a relatively small number of religious denominations: More than 70 percent of Americans identify as Christian (Pew Research Center 2015c).

In comparison with the citizens of other industrial nations, Americans are highly religious; nearly nine in ten U.S. adults believe in God, and one in two attend services at least monthly (Pew Research Center 2015c). However, levels of religious participation have declined since the mid-twentieth century. As measured by indicators such as belief in God, religious membership, and attendance at religious services, religiosity reached its highest levels in the 1950s and has been declining ever since—in part because post–World War II baby boomers have been less religious than their predecessors (Roof 1999). In one national survey, overwhelming majorities of Catholics, liberal Protestants, and conservative Protestants reported attending church on a weekly basis while they were children, but their attendance had dropped sharply by the time they reached their early twenties. Levels of participation, however, remain high among members of conservative Protestant groups.

Another survey of nearly 114,000 adults in 1990 and more than 50,000 adults in 2008 found that religious identification had declined sharply during the eighteen-year period. In 1990, 90 percent of all adults identified with some religious group; in 2008, the figure was less than 80 percent. The principal decline was among self-identified Christians (from 86 percent to 76 percent). By 2014, the Christian share of the U.S. population had dropped even further—to 71 percent (Table 12.1). This decline has been driven in large part by the growing number of adults who report no religious affiliation, referred to as the "rise of the nones." This rapid growth in religious "nones"—people who identify as atheists or agnostics or who say their religion is "nothing in particular"—was reflected in a 2014 survey conducted by the Pew Research Center, which found that about 23 percent of the U.S. adult population is religiously unaffiliated, up from 16 percent in 2007. Much of this growth can be tied to generational replacement: 35 percent of Millennials are religiously unaffiliated (Lipka 2015). Despite this rise of the nones, a significant majority—77 percent—of Americans identify with a religion. One reason so many Americans are religiously affiliated is that religious organizations are an important source of social ties and friendship networks. Churches, synagogues, and mosques are communities of people who share the same beliefs and values and who support one another during times of need. Religious communities thus often play a family-like role, offering help in times of emergency as well as more routine assistance, such as childcare.

The 2016 American Values Atlas Religion Report, conducted by the Public Religion Research Institute, surveyed more than 101,000 Americans in all fifty states and found that American religion is experiencing significant changes: White Christians, once the dominant religious group, now account for less than half of all U.S. adults, down from roughly 81 percent in 1976; less than half the states are majority white Christian (compared with nearly four out of five states only a decade ago). Only 17 percent of white Americans reported being Evangelical Protestants, compared with 23 percent a decade earlier; similar trends were noted for white mainline Protestants (18 percent to 13 percent) and white Catholics (16 percent to 11 percent). These trends partly reflect the growing nonwhite population in the United States, particularly the Hispanic population. One interesting finding is that the most youthful religious groups (those with large proportions of members who are under age thirty) are all non-Christian: 42 percent of Muslims are under age thirty, along with 36 percent of Hindus and 35 percent of Buddhists (Jones and Cox 2017).

PROTESTANTS: THE STRENGTH OF CONSERVATIVE DENOMINATIONS

A more detailed picture of recent trends in American religion can be obtained if we break down the large Protestant category into major subgroups. According to the Pew Research Center's 2014 U.S. Religious Landscape Study, the largest number of households

TABLE 12.1

Changes in Religious Affiliation in the United States

RELIGIOUS SELF-IDENTIFICATION	2007	2014
Evangelical Protestant	26.3%	25.4%
Catholic	23.9%	20.8%
Mainline Protestant	18.1%	14.7%
Jewish	1.7%	1.9%
Muslim	0.4%	0.9%
Buddhist	0.7%	0.7%
Hindu	0.4%	0.7%
Unaffiliated	16.1%	22.8%

Source: Pew Research Center 2015c.

From Pulpits to iPads?

The United States is one of the most religious countries in the world. Yet there are still millions of Americans for whom religion is not an important part of their everyday lives. For example, more than one in five Americans report that they have no religious affiliation; this proportion grows to more than one-third among Americans ages eighteen to twenty-nine (Pew Research Center 2015c). And while roughly 40 percent of Americans report that they "usually" attend religious services once a week, recent research based on daily diary data shows that the proportion of Americans who regularly attend services is as low as 24 percent, with rates even lower among young adults (Brenner 2011).

In the past decade, young adults and other Americans have found new ways to incorporate religion and spirituality into their lives, beyond the pews of their local churches and synagogues. The Internet and smartphones have allowed Americans to participate in religious activities on their own grounds and on their own schedules. For example, a spate of new smartphone apps allow users to download full texts of scriptures like the Bible, Book of Mormon, Koran, or Torah. Muslims can use apps to ascertain the time of day for their five daily prayers and to learn what direction to face when praying toward Mecca. Jews and Seventh-Day Adventists can use programs like the Sabbath App to calculate sunset times for Friday evening and Saturday evening each week, so they'll know exactly when the Sabbath begins and ends in their hometowns. Hindus can use their phones to present virtual offerings of incense and coconut to the god Ganesh.

Other apps allow users to type in prayers and send them off to God, to the Wailing Wall in Jerusalem, or simply into cyberspace (Wagner 2011). For those who believe that scriptures can be used to substantiate their political views, there are apps that help users quickly locate a biblical passage to support arguments for (or against) everything from abortion to same-sex marriage (Vitello 2011). Even those without religious views can use such apps to support their politics; new apps (like BibleThumper) "allow the atheist to keep the most funny and irrational Bible verses right in their pocket" (Vitello 2011).

Technology also keeps us connected to religious communities. Hundreds if not thousands of religious organizations allow people to "attend" religious services virtually. For example, many synagogues throughout the United States live-stream their services over the High Holidays of Rosh Hashanah and Yom Kippur. Advocates say that technology helps bring worship to people who don't have another way to participate in services and sermons, such as members of the military, the homebound, or Jews who live in areas without a local congregation (Mandel 2010). Similarly, websites like CyberChurch.com give users access to Christian services throughout the world.

Do you believe that technology will help people become more engaged in religion by enabling them to practice their faith where, when, and how they are comfortable? Or do you believe that these apps undermine some of the core aspects of religion, including interacting with a community of like-minded others or rituals like praying together? Do you think apps will ever replace in-person participation in religious services or activities? Why or why not?

Smartphone apps now allow users to participate virtually in religious services, make offerings to their gods, and read scriptures. Are these apps a reasonable substitute for traditional forms of worship?

were Baptist, accounting for 33 percent of all Protestants—more than three times the size of the second-largest group, Methodists (10 percent). There were far fewer Lutherans (8 percent), Presbyterians (5 percent), and Episcopalians (3 percent) (Pew Research Center 2015c).

These figures are important because they reveal the relative strength of conservative Protestants in the United States. Conservative Protestants, which include Pentecostals as well as evangelical wings of historically mainline Protestant churches, emphasize a literal interpretation of the Bible, morality in daily life, and conversion through evangelizing. A quarter of all U.S. adults today identify as evangelical. Conservative Protestants can be contrasted with the more historically established mainline and liberal Protestants, such as Episcopalians, Presbyterians, and Methodists, who tend to adopt a more flexible, humanistic approach to religious practice. While mainline Protestant churches have seen their numbers decline rapidly in recent years, conservative Protestant churches have been much more stable (Pew Research Center 2015c). Evangelical Protestants had an outsized effect on the 2016 presidential election, when four out of five voted for Trump—by far the highest of any religious group (Smith and Martínez 2016).

Evangelical Christians account for a majority of Protestants in the United States today. Many attend mega-churches, which attract as many as 30,000 congregants on any given Sunday.

CATHOLICISM

Catholics make up about one-fifth (21 percent) of the U.S. population, although only 16 percent of Millennials are Catholic (Pew Research Center 2015c). Currently, one-third of Catholics in the United States are Hispanic, and this proportion is likely to grow in the coming decades. While the Catholic share of the U.S. population has been relatively stable over the long term, the number of Catholics appears to be declining. Part of this decline is due to the fact that more people are leaving Catholicism for another faith than are joining the Church: Of the nearly one-third of Americans who were born Catholic, 41 percent no longer identify with the Catholic Church.

Church attendance declined sharply in the 1960s and 1970s, leveling off in the mid-1970s. While the reasons for this decline are unclear, one reason likely has to do with the papal encyclical of 1968 that reaffirmed the ban on the use of contraceptives by Catholics. The encyclical offered no leeway for people whose conscience allowed for the use of contraceptives. They were faced with disobeying the Church, and many Catholics did just that. According to a 2015 Pew survey, three-quarters (76 percent) of U.S. Catholics say the Church should allow the use of birth control. Church attendance has continued to decline in recent decades: In 1975, 47 percent of Catholics reported attending Mass at least once a week. By 2015, that figure had dropped to 39 percent (Pew Research Center 2015n).

OTHER RELIGIOUS GROUPS

Judaism in the United States has historically been divided into three major movements: Orthodox Judaism, which believes in the divine origins of the Jewish Bible (called by Christians the Old Testament) and follows highly traditional religious practices; Conservative Judaism, which is a blend of traditional and more contemporary beliefs and practices; and Reform Judaism, which rejects most traditional practices and is progressive in its ritual practices (services, for example, are more likely to be conducted in English than in Hebrew). Both Conservative and Reform Judaism reflect efforts by Jewish immigrants (or their descendants) to develop beliefs and rituals that turned away from "Old World" ones, developing forms more consistent with their new homeland.

Despite (or partly because of) these efforts to modernize Judaism, the number of Jews in the United States has declined as a result of low birthrates, intermarriage, and assimilation. Nearly 60 percent of Jews who have gotten married since 2000 have a non-Jewish spouse (Pew Research Center 2013b). Estimates of the number of Jewish Americans in the United States today vary, which may reflect precisely how Jews are identified and counted. Some Americans identify as Jewish if they have a Jewish mother—even if they have never practiced their religion. Yet other people may self-identify as Jewish only if they participate actively in the religion, whereas others may identify as "culturally Jewish," meaning they celebrate their heritage and culture but do not actively practice religion. In 2014, just 1.9 percent of the U.S. population identified as Jewish (Pew Research Center 2015c).

Among Muslims, growing emigration from Asia and Africa may change the U.S. religious profile. According to the Pew Research Center, Muslims in the United States will outnumber Jews by 2040, becoming the second-largest religious group after Christians (Mohamed 2018). In 2017, 1.1 percent of the U.S. population identified as Muslim, reflecting large gains since even just 2007; by 2050, that number is predicted to nearly double, to 2.1 percent. These figures may be underestimates, as many Muslims are reluctant to disclose their religious affiliation in the post-9/11 political climate—fewer than half of Americans held a favorable view of Islam in 2017 (Pew Research Center 2017a). The vast majority of Muslims in the United States are immigrants (58 percent), with many Muslims coming from the Asia-Pacific region, the Middle East, North Africa, and sub-Saharan Africa (Pew Research Center 2017d).

Religious Affiliation and Socioeconomic Status

The principal religious groupings in the United States vary substantially by region and socioeconomic status. Liberal Protestants tend to be well educated and have jobs and incomes that would classify them as middle or upper class. They are concentrated in the northeastern states and, to a small extent, in the West as well. Moderate Protestants fall at a somewhat lower level than liberal Protestants in terms of education and income. In fact, they are typical of the national average on these measures. They tend to live in the Midwest and, to some extent, in the West. Black Protestants are, on average, the least educated and poorest of any of the religious groups. Conservative Protestants have a similar profile, although they fall at a marginally higher level on all these measures (Pew Research Center 2015c). Catholics strongly resemble moderate Protestants (which is to say, average Americans) in terms of their socioeconomic profile. They are largely concentrated in the Northeast, although many live in the West and the Southwest as well.

Jews historically have had the most successful socioeconomic profile. Jews tend to be college graduates in middle- or upper-income categories. In 2014, 44 percent of Jews earned more than $100,000 a year, compared to 17 percent of Christians. Hindus—who also report high levels of educational attainment—have a similarly high socioeconomic profile, with 36 percent earning more than $100,000 (Pew Research Center 2015c). Whereas the large majority of Jews once lived in the northeastern states, today fewer than half do, as many have relocated throughout the United States. One recent study suggests that this high degree of geographical mobility is associated with lowered involvement in Jewish institutions. Jews who move across the country are less likely to belong to synagogues, have Jewish friends, or be married to Jewish spouses (Goldstein and Goldstein 1996).

There are political differences across religious groups as well. Jews tend to be the most heavily Democratic of any major religious groups, while fundamentalist and evangelical Christians are the most Republican. The more moderate Protestant denominations are somewhere in between (Jones and Cox 2017; Kosmin, Mayer, and Keysar 2001). As such, religious groups also differ widely regarding their views on major social issues in the United States, including abortion. On average, Jews and persons with no affiliation tend to hold the most liberal political views, meaning that they are likely to say that they believe women should have legal access to abortion. Fundamentalist and evangelical Christians are least likely to support these stances, while liberal Protestants, moderate Protestants, and Catholics sit toward the middle of the political continuum (Pew Research Center 2015c).

Religion has a subtle yet powerful influence on daily life in the United States and throughout the world. In analyzing religious practices and traditions, we must be sensitive to ideals that inspire profound conviction in believers, yet we must also take a balanced view of them. We must confront ideas that seek the eternal while recognizing that religious groups also promote mundane goals, such as earning money or attracting followers. We need to recognize not only the diversity of religious beliefs and models of conduct but also the nature of a global phenomenon.

We have also seen that education and religion are two social institutions that are powerful socializing agents. Religion and education teach young people the skills and beliefs that are an essential part of one's culture. However, the two institutions differ in a critical way: Education is intended to be universalistic and to expose all children to similar messages, whereas religious institutions vary widely in the values, beliefs, and practices that they impart. As we saw in our chapter opener, these institutions may occasionally collide; Malala's attempt to obtain an education was at odds with the fundamentalist religious beliefs that reigned in her village. However, both education systems and religious institutions are dynamic and may evolve as social contexts and policies change. The efforts of pioneering young women like Malala may be instrumental in helping to create a context where education and religious institutions meet the needs of all citizens, regardless of their gender, ethnicity, or social background.

CONCEPT CHECKS

1. What are the reasons so many Americans belong to religious organizations?

2. Describe the main differences between conservative and liberal Protestants.

3. Contrast the political views and socioeconomic statuses of major religious groups in the United States.

CHAPTER 12

The Big Picture

Education and Religion

Thinking Sociologically

1. From your reading of this chapter, describe what might be the principal advantages and disadvantages of having children go to private versus public schools in the United States at this time. Assess whether privatization of our public schools would help to improve them.

2. Karl Marx, Émile Durkheim, and Max Weber had different viewpoints on the nature of religion and its social significance. Briefly explain the viewpoints of each. Which theorist's views have the most to offer in explaining the rise of national and international fundamentalism today? Why?

Why Are Education and Literacy So Important?

p. 357

Know how and why systems of mass education emerged in the United States. Know some basic facts about the education system and literacy rates of developing countries.

What Is the Link between Education and Inequality?

p. 362

Become familiar with the most important research on whether education reduces or perpetuates inequality. Learn the social and cultural influences on educational achievement.

How Do Sociologists Think about Religion?

p. 371

Learn the elements that make up religion. Know the sociological approaches to religion developed by Marx, Durkheim, and Weber, as well as the religious economy approach.

How Does Religion Affect Life throughout the World?

p. 377

Understand the various ways religious communities are organized and how they have become institutionalized. Recognize how the globalization of religion is reflected in religious activism in poor countries and the rise of religious nationalist movements.

How Does Religion Affect Your Life in the United States?

p. 382

Learn about the sociological dimensions of religion in the United States.

1. Why did schooling become widespread only after the Industrial Revolution?
2. What are three of the most frequently cited reasons for home schooling?
3. What are some of the functions of formal schooling?
4. What are some of the reasons there are many illiterate people in the developing world?

home schooling • literacy

tracking • hidden curriculum • intelligence •
IQ (intelligence quotient) • functional literacy •
standardized testing

1. According to Kozol, has education become an equalizer in American society? Why or why not?
2. How do Coleman's findings differ from the results of Kozol's research? Whose theory, in your opinion, can better explain the racial gap in educational achievement?
3. What effect does tracking have on academic achievement?
4. How do schools perpetuate existing inequalities across generations?
5. Describe the components and critiques of the No Child Left Behind Act and Race to the Top.

religion • theism • alienation • sacred • profane •
secular thinking • secularization • religious
economy

1. What are the three main components of religion as a social institution?
2. How do sociologists differ from other scholars in their approach to studying religion?
3. What are the differences between classical and contemporary approaches to understanding religion?

churches • sect • denomination • cults •
religious nationalism • liberation theology

1. Describe four types of religious organizations.
2. What is religious nationalism? Why can it be viewed as a reaction to economic modernization of local religious beliefs and Westernization?

1. What are the reasons so many Americans belong to religious organizations?
2. Describe the main differences between conservative and liberal Protestants.
3. Contrast the political views and socioeconomic statuses of major religious groups in the United States.

13

Politics and Economic Life

Voter Turnout

p. 403

Workers gather at the Capitol in April 2015 to protest for a higher minimum wage. The current minimum—$7.25 per hour—hasn't changed since 2009.

THE BIG QUESTIONS

How did the state develop?

Learn the basic concepts underlying modern nation-states.

How do democracies function?

Learn about different types of democracy, how this form of government has spread around the world, key theories about power in a democracy, and some of the problems associated with modern-day democracy.

What is the social significance of work?

Assess the sociological ramifications of paid and unpaid work. Understand that modern economies are based on the division of labor and economic interdependence. Familiarize yourself with modern systems of economic production.

What are key elements of the modern economy?

See the importance of the rise of large corporations; consider particularly the global impact of transnational corporations.

How does work affect everyday life today?

Learn about the impact of global economic competition on employment. Consider how work will change over the coming years.

On April 15, 2015—the day income taxes are typically due—fast-food workers, home-care and day-care workers, adjunct faculty at colleges and universities, airport workers, and labor union members in 236 U.S. cities went on strike to raise the minimum wage to $15 an hour. Seattle and the District of Columbia had already done so; many other cities soon followed suit. By the end of 2015, fourteen other cities, counties, and states had called for immediate or phased-in minimum hourly wages of $15. The minimum wage became a national issue when Vermont senator Bernie Sanders made it a central part of his campaign platform. It is an issue that strongly divides the Democratic and Republican parties as well. The Democratic Party adopted the $15 minimum as a nonbinding part of its 2016 presidential election National Platform; the Republican Party, while not addressing the issue directly in its National Platform, regards any increases in the minimum wage as interfering with businesses and generally detrimental for economic growth.

FIGURE 13.1

Who Earns Less than $15 an Hour?

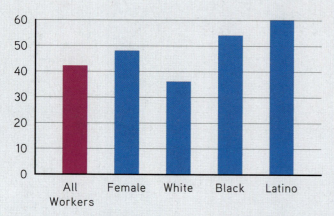

Note: Proportion of workers within each demographic group making less than $15 per hour.

Source: Tung, Lathrop, and Sonn 2015.

When the U.S. government established a national minimum wage in 1938, it argued that wages should be sufficient to provide "the minimum standard of living necessary for health, efficiency, and general well-being" (U.S. FLSA 2011). The current U.S. minimum wage, which hasn't changed since 2009, is $7.25 an hour. That works out to a yearly income of $15,131 for a full-time worker, which is below the official poverty level for a two-person household (roughly $16,000). This amount clearly fails to provide a "minimum standard of living." Raising the minimum hourly wage to $15—$31,305 in yearly income for full-time workers—would make a big difference in their lives, enabling them to cover basic living costs.

What proportion of Americans actually earn less than $15 an hour (Figure 13.1)? Opponents to raising the minimum wage claim that it is mainly high school students or young adults who are working part time in fast-food restaurants or as sales clerks, often living with their parents, who will eventually move on to higher-paid jobs. But one recent study found that more than two out of every five workers in the United States currently earn less than $15 an hour—a figure that rises to half of all African American workers and three out of every five Latino workers. Nor are low-wage jobs restricted to the very young: The study found that more than a quarter of all workers between the ages of thirty-five and forty-nine make less than $15 an hour (Tung, Lathrop, and Sonn 2015).

Some businesses have responded by increasing the minimum wage for their own workers. Walmart, McDonald's, and T.J. Maxx announced in early 2015 that they would raise their base-level wages to somewhere in the $9–$10 range. Costco, on the other hand, offers a starting wage of $11.50; its average wage is $21, an extremely high level for an industry in which four out of five workers earn less than $15 an hour. Costco also enjoys low worker turnover, high worker satisfaction, and sales per employee that far exceed its principal competitors. Facebook, Google, and Ben & Jerry's are among a small number of firms that adopted a $15 minimum wage in 2015, as have some universities (Tung, Lathrop, and Sonn 2015).

Will raising the minimum wage to $15 cause businesses to fail and their workers to lose their jobs, as some economists predict? Supporters of an increase point out that, in today's dollars, the minimum wage grew from roughly $5.00 an hour in 1940 to a high of $10.86 in 1968, without adversely affecting a period of rapid economic growth; since that time, it has actually lost ground in terms of actual purchasing power (Kurtz and Yellin 2015). Raising the minimum wage, proponents argue, will put more money into the pockets of workers, whose spending will stimulate economic growth. Recent research also finds that raising the minimum wage to $12 would reduce the cost of public assistance by $17 billion annually, resulting in savings in government expenditures (Cooper 2016).

Research on whether increases in the minimum wage will help or hurt workers has found mixed results (Neumark 2015). Some economists favor a smaller increase to $12, fearful that a $15 minimum would "put us in uncharted waters, and risk undesirable and unintended consequences," such as job loss (Krueger 2015). Perhaps, as former U.S. labor secretary Robert Reich (2015) has concluded, "maybe some jobs are worth risking if a strong moral case can be made for a $15 minimum. That moral case is that no one should be working full time and still remain in poverty." Reich regards an adequate wage as a human right—a topic to which we will return later, when we consider social rights as one of the rights of citizenship in modern states. Some observers have argued that the only way to ensure a reasonable standard of living for low-wage workers is for them to unionize (Eidelson 2013a).

As the growing movement for a $15 minimum wage reveals, the government, economics, and politics are closely intertwined. **Government** refers to the regular enactment of policies, decisions, and matters of state on the part of the officials within a political apparatus. A government often enacts policies, such as the federal minimum wage, that have sweeping economic consequences, whether for nations, states, cities, or even the individual lives of workers. **Politics** concerns the means by which power is used to affect the scope and content of governmental activities. But the sphere of the political is not limited only to those who work in government; it also involves the actions of others.

There are many ways in which people outside the political apparatus seek influence. The workers who went on strike across the country on April 15 attempted to exert power both on their employers and on public policies by protesting and walking off their jobs. A few major businesses responded by providing modest wage hikes for their workers; some state and local governments have raised their minimum wages or plan to do so; and the $15 minimum became a national issue during the 2016 presidential campaign. These actions show how politics are frequently intertwined with economics. The **economy** consists of institutions that provide for the production and distribution of goods and services, including jobs.

In this chapter, we study the main factors affecting political and economic life today. We begin with a discussion of politics and then turn to work and the economy. The sphere of government is the sphere of political power. All political life is about power: the people who hold it, how they achieve it, and what they do with it. As mentioned in Chapter 1, the study of power is of fundamental importance for sociology. **Power** is the ability of individuals or groups to make their own interests or concerns count, even when others resist. It sometimes involves the direct use of physical force, such as when the United States and its allies use military force to counter the growth of extremist groups, such as ISIS in the Middle East. At other times, it involves the use of threats, whether a threat to vote out of office a leader who does not meet the needs of the electorate or a threat to walk off the job. Power is an element in almost all social relationships, such as that between employer and employee. This chapter focuses on a narrower aspect of power: governmental power. In this form, it is almost always accompanied by ideologies, belief systems that are used to justify the actions of the powerful. For example, Democratic congresspersons who support the minimum wage hike tend to embrace an ideology that emphasizes social justice and workers' rights, whereas their Republican colleagues tend to subscribe to an ideology that emphasizes fiscal conservatism and free enterprise.

Authority is a government's legitimate (that is, lawful) use of power: Those who are subject to a government's legitimate authority consent to it. Power is thus different from authority: A government may rule by the use of power but lack legitimate authority in the eyes of its citizens. Contrary to what many believe, democracy—a system of government in which, as we discuss later, the citizens ultimately exert authority through their representatives—is not the only type of government people consider legitimate. Dictatorships, in which a single individual or group exercises virtually total authority, can have legitimacy as well, as can states governed by religious leaders. But as we shall see later, democracy is currently the most widespread form of government considered legitimate.

government

The enacting of policies, decisions, and matters of state on the part of officials within a political apparatus. In most modern societies, governments are run by officials who do not inherit their positions of power but are elected or appointed on the basis of qualifications.

politics

The means by which power is used to affect the nature and content of governmental activities.

economy

The system of production and exchange that provides for the material needs of individuals living in a given society. Economic institutions are of key importance in all social orders.

power

The ability of individuals or the members of a group to achieve aims or further the interests they hold. Power is a pervasive element in all human relationships.

authority

A government's legitimate use of power.

How Did the State Develop?

A **state** exists where there is a political apparatus of government (institutions like a parliament or congress, plus civil service officials) ruling over a given territory whose authority is backed by a legal system and by the capacity to use force to implement its policies. All modern states lay claim to specific territories, possess formalized codes of law, and are backed by the control of military force. **Nation-states** have come into existence at various times in different parts of the world (for example, the United States in 1776 and the Czech Republic in 1993). Their main characteristics, however, contrast rather sharply with those of states in traditional civilizations.

Characteristics of the State

SOVEREIGNTY

The territories ruled by traditional states were always poorly defined, the level of control wielded by the central government being quite weak. The notion of **sovereignty**—that a government possesses authority over an area with clear-cut borders, within which it is the supreme power—had little relevance. All modern nation-states, by contrast, are sovereign states.

CITIZENSHIP

In traditional states, most of the population ruled by a king or emperor showed little awareness of, or interest in, those who governed them. Nor did they have any political rights or influence. Normally only the dominant classes or more affluent groups felt a sense of belonging to an overall political community. In modern societies, by contrast, most people living within the borders of the political system are **citizens**, having common rights and duties and knowing themselves to be members of a national political community (Brubaker 1992). Although some people, such as political refugees, are "stateless," almost everyone in the world today sees himself or herself as a member of a definite national political community.

NATIONALISM

Nation-states are associated with the rise of **nationalism**, which can be defined as a set of symbols and beliefs providing the sense of being part of a single national political community. Thus, individuals feel a sense of pride and belonging in being American, Indian, or Chinese. Probably people have always felt some kind of identity with social groups of one form or another—their family, village, or religious community. Nationalism, however, made its appearance only with the development of the modern state. It is the main expression of feelings of identity with a distinct national political community.

Nationalistic loyalties do not always fit the physical borders marking the territories of states in the world today. Virtually all nation-states were built from communities of diverse backgrounds. As a result, **local nationalisms** have frequently arisen in opposition to those fostered by the states. Thus, in Canada, for instance, nationalist feelings among the French-speaking population in Quebec may present a challenge to the feeling

state

A political apparatus ruling over a given territorial order whose authority is backed by law and the ability to use force.

nation-state

A particular type of state, characteristic of the modern world, in which a government has sovereign power within a defined territorial area and the population comprises citizens who believe themselves to be part of a single nation or people.

sovereignty

The undisputed political rule of a state over a given territorial area.

citizens

Members of a political community, having both rights and duties associated with that membership.

nationalism

A set of symbols and beliefs expressing identification with a national community.

of "Canadianness," while such feelings among the Basque population of Spain may challenge the feeling of being a Spaniard. Yet, while the relation between the nation-state and nationalism is a complicated one, the two have come into being as part of the same process.

We can now offer a comprehensive definition of the nation-state: It is possessed of a government apparatus that is recognized to have sovereign rights within the borders of a territorial area, it is able to back its claims to sovereignty by the control of military power, and many of its citizens have positive feelings of commitment to its national identity.

CITIZENSHIP RIGHTS

Some modern nation-states first became centralized, effective political systems through the activities of monarchs, such as kings or queens, who successfully concentrated more and more power in their own hands. Citizenship did not originally carry rights of political participation in these states. Such rights were achieved largely through struggles that limited the power of monarchs, as in Britain, or actively overthrew them—sometimes by a process of revolution, as in the cases of the United States and France, followed by a period of negotiation between the new ruling elites and their subjects (Tilly 1996).

Three types of rights are associated with the growth of citizenship (Marshall 1973). **Civil rights** refer to the rights of the individual by law. These include privileges many of us take for granted today but that took a long time to achieve (and are by no means fully recognized in all countries). Examples are the freedom of individuals to live where they choose, freedom of speech and religion, the right to own property, the right to legally marry, and the right to equal justice before the law. These rights were not fully established in most European countries until the early nineteenth century. Although in 1789, the U.S. Constitution granted such rights to Americans well before most European states had them, African Americans were excluded. Even after the Civil War, when Blacks were formally given these rights by the Fourteenth Amendment, they were not able to exercise them. Women also were denied many civil rights; for example, at the turn of the nineteenth century in the United States, women had few rights independent of their husbands. They could not own property, write wills, collect an inheritance, or even earn a salary. Throughout the nineteenth century, states gradually began affording such rights to women regardless of their marital status (Speth 2011).

The second type of citizenship rights consists of **political rights**, especially the right to participate in elections and to run for public office. Again, these were not won easily or quickly. Except in the United States, the achievement of full voting rights even for all men is relatively recent and was gained only after a struggle in the face of governments reluctant to admit the principle of the universal vote. In most European countries, the vote was at first limited to male citizens owning a certain amount of property, which effectively limited voting rights to an affluent minority. Universal **franchise** for men was won in most Western nations by the early years of the twentieth century. Women had to wait longer; in most Western countries, the vote for women was achieved partly as a result of the efforts of women's movements and partly as a consequence of the mobilization of women into the formal economy during World War I. In the United States, women did not get the vote until the Nineteenth Amendment was ratified in 1920.

Feelings of nationalism are on full display at international events like the Olympic Games.

local nationalisms

The belief that communities that share a cultural identity should have political autonomy, even within smaller units of a nation-state.

civil rights

Legal rights held by all citizens in a given national community.

political rights

Rights of political participation, such as the right to vote in elections and to run for office, held by citizens of a national community.

franchise

The right to vote.

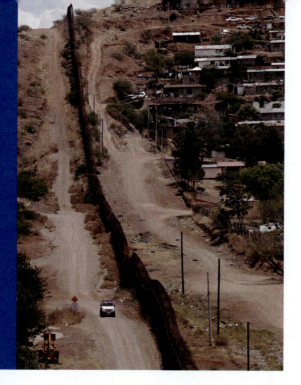

A U.S. Border Patrol agent drives along the wall that separates Nogales, Arizona, from Nogales, Sonora, Mexico, on the U.S.–Mexico border. What strategies has the United States used to achieve social closure?

The third type is **social rights**, the right of every individual to enjoy a certain minimum standard of economic welfare and security. Social rights include such entitlements as sickness benefits, benefits in case of unemployment, and, as we have seen, the guarantee of minimum levels of wages. In most societies, social rights have been the last to develop. This is because the establishment of civil and particularly political rights has usually been the basis of the fight for social rights. Social rights have been won largely as a result of the political strength that poorer groups have been able to develop after obtaining the vote.

The broadening of social rights is closely connected with the **welfare state**, which has been firmly established in Western societies only since World War II. A welfare state exists where government organizations provide material benefits for those who are unable to support themselves adequately through paid employment—the unemployed, the sick, the disabled, and the elderly. While all Western countries today provide extensive welfare benefits, these benefits are virtually nonexistent in many poorer countries.

Although an extensive welfare state was seen as the culmination of the development of citizenship rights, in recent years, welfare states have come under pressure from increasing global economic competition and the movement of people from poorer, often war-torn countries to richer ones. As a result, the United States and some European countries have sought to reduce benefits to noncitizens and to prevent new immigrants from entering the country. For example, for many years, the U.S. government has patrolled its border with Mexico and constructed walls of concrete and barbed wire in an attempt to keep illegal immigrants out of the country—an issue that resurfaced during the 2016 presidential campaign. In 2015, more than a million refugees from civil war and religious violence in Syria, Afghanistan, Iraq, and other countries fled their homelands and sought asylum in Europe, straining resources and provoking an anti-immigrant backlash among some segments of the European population. Citizenship, and the bundle of rights and privileges accompanying it, serve as a powerful instrument of social closure (Brubaker 1992), whereby prosperous nation-states have attempted to exclude the migrant poor from the status and benefits that citizenship confers. Concern over immigrants was one of the driving forces behind the 2016 British vote to leave the European Union (the so-called Brexit) and contributed to Donald Trump's electoral victory in 2016; during the campaign, one of Trump's promises was to "build a wall" between the United States and Mexico and make Mexico pay for it.

Having learned some of the important characteristics of modern states, we now consider the nature of democracy in modern societies.

How Do Democracies Function?

The word **democracy** has its roots in the Greek term *demokratia*—*demos* (people) and *kratos* (rule). Its basic meaning is therefore a political system in which the people, not monarchs or dictators, rule. What does it mean to be ruled by the people? The answer to that question has taken contrasting forms, at varying periods and in different societies. For example, "the people" have been variously understood as owners of property, white men, educated men, men, and adult men and women. In some societies, the officially accepted version of democracy is limited to the political sphere, whereas in others, it is extended to other areas of social life.

Participatory Democracy

In **participatory democracy**, decisions are made communally by those affected by them. This was the original type of democracy practiced in ancient Athens. Those who were citizens, a small minority of Athenian society, regularly assembled to consider policies and make major decisions. Participatory democracy is of limited importance in modern societies, where the vast majority of the population has political rights, rendering it impossible for everyone to participate actively in the making of all the decisions that affect them. In modern societies, direct democracy is a much more realistic approach to engaging citizens in decisions. A **direct democracy** is a form of participatory democracy in which citizens vote directly on laws and policies; however, they do not need to convene in one setting to do so. For example, Americans can visit voting booths in their hometowns to vote directly on legislation that affects their lives.

Yet some facets of participatory democracy do play a part in modern societies. The holding of a referendum, for example, whereby the majority express their views on a particular issue, is one form of participatory democracy. Direct consultation of large numbers of people is made possible by simplifying the issue to one or two questions to be answered. Referenda are employed frequently on a state level in the United States to decide controversial issues, such as the legalization of marijuana.

Monarchies and Liberal Democracies

Some modern states, including Britain and Belgium, still have monarchs, but these are few and far between. Where traditional rulers of this sort are still found, their real power is usually limited or nonexistent. In a tiny number of countries, such as Saudi Arabia and Jordan, monarchs continue to hold some degree of control over government, but in most cases, they are symbols of national identity rather than personages having any direct power in political life. The queen of England, the king of Sweden, and even the emperor of Japan are all **constitutional monarchs**: Their real power is severely restricted by the constitution, which vests authority in the elected representatives of the people. The vast majority of modern states are *republican*—there is no king or queen. Almost every modern state, including constitutional monarchies, professes adherence to democracy.

Learn about different types of democracy, how this form of government has spread around the world, key theories about power in a democracy, and some of the problems associated with modern-day democracy.

democracy
A political system that allows the citizens to participate in political decision making or to elect representatives to government bodies.

participatory democracy
A system of democracy in which all members of a group or community participate collectively in making major decisions.

direct democracy
A form of participatory democracy that allows citizens to vote directly on laws and policies.

constitutional monarchs
Kings or queens who are largely figureheads. Real power rests in the hands of other political leaders.

Countries in which voters can choose between two or more political parties and in which the majority of the adult population has the right to vote are usually called **liberal democracies**. The United States, the Western European countries, Japan, Australia, and New Zealand all fall into this category. Some developing countries, such as India, also have liberal democratic systems.

The Spread of Liberal Democracy

For much of the twentieth century, the political systems of the world were divided primarily between liberal democracy and communism, as found in the former Soviet Union (and which still exists in some form in China, Vietnam, Cuba, and North Korea). As we learned in Chapter 1, the philosophical roots of **communism** can be found in the writings of Karl Marx, who predicted that in the future, capitalism would be replaced by a society in which there were no classes—no divisions between rich and poor—and the economic system would come under communal ownership. Under these circumstances, Marx believed, a more equal society would be established. Marx's work had a far-reaching effect in the twentieth century. Through most of the century, until the fall of Soviet communism in the early 1990s, more than a third of the world population lived in societies whose governments claimed to derive their inspiration from Marx's ideas. In practice, however, communism often exists as a system of one-party rule. Voters are typically given a choice not between different parties but between different candidates of the same party—the Communist Party; sometimes only one candidate runs. The party controls the economy as well as the political system.

Since 1989, when the hold of the Soviet Union over Eastern Europe was broken, processes of democratization have swept across the world in a sort of chain reaction. The number of democratic nations almost doubled between 1989 and 2017, from 66 to 123 (Freedom House 2005, 2018). Freedom House classifies a nation as democratic if it maintains a competitive multiparty political system, all adults have a right to vote, election procedures are transparent, and major political parties have access to the general public via media and campaigning. Yet the trend toward democracy is hardly irreversible. Over the past decade, 113 countries have experienced declines in political rights and civil liberties, including the United States (Freedom House 2018).

THE INTERNET AND DEMOCRATIZATION

The Internet can be a powerful democratizing force. It transcends national and cultural borders, facilitates the spread of ideas around the globe, and allows like-minded people to find one another in the realm of cyberspace. More and more people in countries around the world access the Internet regularly and consider it to be important to their lives. One prominent example of the political role of the Internet is provided by MoveOn.org, a liberal organization that was originally created in 1998 by twenty-two-year-old activist Eli Pariser and software entrepreneurs Wes Boyd and Joan Blades to electronically mobilize opposition to the impeachment of President Bill Clinton. The organization now boasts 8 million members; it enables activists to start and circulate petitions, advocate for liberal causes, and reach millions of people.

Since MoveOn.org was founded, the Internet has flourished as a means of diffusing political information (and rhetoric), especially among young persons. All ends of the political spectrum, from MoveOn.org to the Tea Party, have used the Internet to recruit volunteers, share information on candidates, and organize rallies (Pilkington 2009). As we will see in Chapter 16, the Internet played an essential role in mobilizing protesters during the Arab Spring and Occupy Wall Street movements. In the United States, the Internet is replacing television and newspapers as the principal source for news for a growing number of people. As of 2017, two out of every three American adults reported getting at least some of their news on social media. Nearly half (45 percent) reported getting news from Facebook, 18 percent from YouTube, and 11 percent from Twitter (Shearer and Gottfried 2017).

While the Internet (especially social media) played a major role in the 2016 presidential election, television still proved to be a more important source of information. During the presidential primaries, 62 percent of Trump supporters reported that they got most of their election news from television, compared with only 28 percent who relied on news websites and social media. For those supporting Clinton, the corresponding figures were similar: 56 percent and 28 percent (Gottfried, Barthel, and Mitchell 2017). Both as candidate and now as president, Trump has used Twitter to mobilize his supporters. Twitter provides direct access to his strongest supporters, bypassing the more mainstream print and television news outlets. Because his tweets are often provocative and controversial, they are frequently covered by the mainstream media, enabling Trump to reach an even wider audience.

Does this easy access to the world's information result in greater open-mindedness—an ability to find information that challenges one's pet beliefs, thereby contributing to the free exchange of ideas ideally associated with democracy? This answer is far from clear-cut. While the Internet may democratize access to wide-ranging sources, most people get their news and information from Internet sources that reinforce their beliefs. On Facebook, for example, politically related postings are likely to come from like-minded friends (Pew Research Center 2015m).

Populist Authoritarianism

In recent years, there has been a turn toward what has been described as "populist authoritarianism" in many countries, including European countries and the United States. Populist authoritarianism is both a philosophy and a style of governance characterized by assertive leadership that values security over civil liberties. It is typically coupled with a strong nationalism that is anti-immigrant and—in its current form—strongly anti-globalization. As the term suggests, it combines two ideas: populism (the belief in "popular sovereignty and direct democracy at any cost, if necessary overriding minority rights, elite expertise, constitutional checks-and-balances, conventional practices, and decision-making by elected representatives . . . [in which leaders] maintain direct links with their followers, through public rallies, television studios, and social media") and authoritarianism ("policy positions which endorse the values of tough security against threats from outsiders, xenophobic nationalism, strict adherence to conventional moral norms, and intolerance of multiculturalism"). Populist authoritarianism can become a challenge to liberal democracy (Norris and Inglehart 2018).

Democracy in the United States

The United States is a representative democracy in which political participation is achieved through elected representatives. Political parties have come to play a key role in elections, while interest groups have significant (and, some would argue, growing) influence behind the scenes and on electoral politics.

Political Activist

When Barack Obama ran for his second term as president in 2012, he had a well-organized team of political activists who provided the grassroots support for his successful campaign. His campaign organization, "Obama for America," was based in his home city of Chicago; its national network of key activists, spread out across the country, included nearly a dozen who listed their undergraduate major as sociology (Obama for America 2012).

It is not surprising that young people, fresh from a major in sociology, might be attracted to political and social activism. As we have shown throughout this chapter, sociology looks at the intersection of politics and economics—the ways in which politics plays a role in shaping the world of work or our response to inequality (Manza and Sauder 2009). Obama's presidential victory in November 2008 inspired many young people who wanted change during a time of growing inequality. Sociology majors tend to be motivated in large part by idealism (Spalter-Roth and Van Vooren 2008b)—a belief that they can contribute to "change we can believe in" (Obama's campaign slogan) in the service of a more just and sustainable world.

But an activist's work doesn't stop once their candidate is elected. Political activists also are committed to larger social changes such as the fight for economic equality, which inspired the Occupy Wall Street movement. This movement first erupted in fall 2011 when a group of protesters created a tent city in New York City's financial district to protest the growing wealth and power of what they termed the "1 percent" (Gamson and Sifry 2013). The movement spread to other cities and helped to motivate young people to get involved in politics, including volunteering for Obama's campaign for his second term. At the other end of the political spectrum, the Trump campaign in 2016 also focused on declining prospects for the working

POLITICAL PARTIES

A political party is an organization of individuals with broadly similar political aims, oriented toward achieving legitimate control of government through an electoral process. Where elections are based on the principle of **winner take all**, as in the United States, two parties tend to dominate the political system. Where elections are based on different principles, as in **proportional representation**, five or six different parties, or even more, may be represented in the assembly. An advantage to the system of proportional representation is that minority political parties have a say. For example, in the United States, the Green Party has almost no presence at the national level. Yet in Germany, which abides by the proportional representation system, new and smaller political parties that are supported by even a small part of the electorate have a chance of being represented in parliament (Krennerich 2014). When they lack an overall majority, some of the parties have to form a coalition—an alliance to form a government—and government by coalition can lead to indecision and stalemate if compromises can't be worked out.

In the United States, the system has become effectively a two-party system comprising Republicans and Democrats, although no formal restriction is placed on the number

winner take all

An electoral system in which the seats in a representative assembly go to the candidate who receives the most votes in his or her electoral district (in the U.S. House of Representatives, for example, a candidate who gets 50 percent + 1 vote represents an entire congressional district, even if another candidate got 50 percent – 1 vote).

class, especially among white men in such industries as mining and manufacturing. Political activists also are running for office themselves. The November 2018 Senate races included candidates like Alexandria Ocasio-Cortez, a 29-year-old Democrat from New York who had previously fought human rights violations related to the Flint water crisis, the Dakota Access Pipeline, and other social problems affecting disadvantaged populations.

Inequality is a topic about which sociologists have a great deal to say. While inequality is often studied primarily in economic terms, sociologists also focus on social inequality— its causes and effects—in such intersecting sociocultural categories as class, race and ethnicity, gender, and age. Sociologists understand the structural sources of inequality— the ways in which changes in the economy, unequal educational opportunities, or racism and sexism contribute to different economic and social outcomes for different social groups (Keister and Southgate 2012; Grusky and Szelényi 2011). A strong grasp of sociology is critical to the work that political activists do, regardless of whether they are focusing on fighting racism, sexism, poverty, the exploitation of immigrant populations, or public policies that systematically discriminate against some social groups. Activists are motivated by a key theme of the sociological imagination: public issues require sweeping social, political, and economic changes rather than personal adaptations on the part of those individuals who are suffering.

Among prominent politicians, former sociology majors include both past and current members of Congress: senators Daniel Patrick Moynihan and Barbara Mikulski and representatives Shirley Chisholm, Tim Holden, and Maxine Waters. And as you learned in Chapter 1, former president Ronald Reagan majored in sociology and economics while an undergraduate at Eureka College and former First Lady Michelle Obama chose sociology at Princeton. Even if you don't end up serving in the White House, a degree in sociology can help you play a key role in determining who does—whether as a political activist, campaign adviser, or up-and-coming politician.

U.S. Congresswoman Maxine Waters got her bachelor's degree in sociology from California State University, Los Angeles, before launching her political career. She has been an outspoken critic of Trump and his policy of separating immigrant families at the border.

of political parties. The nation's founders made no mention of parties in the Constitution because they thought that party conflict might threaten the unity of the new republic. Building mass support for a party in the United States is difficult because the country is so large and includes so many different regional, cultural, and ethnic groups. Each party has tried to develop its electoral strength by forging broad regional bases of support and by campaigning for very general political ideals.

As measured by levels of membership, party identification, and voting support, both of the major American parties are in decline. In recent years, the proportion of registered voters who identify as Independents has grown, while the proportion identifying as Republicans has declined. In 2002, roughly equal proportions of registered voters identified as Democrats (34 percent) and Republicans (33 percent), with significantly fewer identifying as Independents (26 percent). By 2017, the proportion identifying as Democrats had stayed about the same (33 percent), the proportion identifying as Republicans had declined (26 percent), and the share identifying as Independents had grown significantly (37 percent). If one takes a longer view, however, in 1964 the majority of Americans (51 percent) identified as Democrats, compared with 25 percent as Republicans and 23 percent as

proportional representation

An electoral system in which seats in a representative assembly, often called a parliament, are allocated according to the proportions of the vote received; the head of state (called a prime minister) is the head of the party that has the largest number of seats.

Independents. The U.S. electorate is increasingly divided, with growing numbers refusing to identify with either party (Pew Research Center 2015f, 2018c).

Moreover, Democrats and Republicans have become increasingly polarized in the past few years. For example, Republicans are much more likely to be pro-business and believe in limited government, whereas Democrats are more likely to be critical of business while favoring stronger government support for the middle class and poor (Pew Research Center 2014a). As we saw earlier in this chapter, Democrats tend to be more supportive of raising the minimum wage to help low-wage workers secure an adequate standard of living, while their Republican counterparts worry that raising the minimum wage will be hurtful to business owners, especially small, independent business owners.

POLITICS AND VOTING

Since the early 1960s, the proportion of the population that turns out to vote in the United States has generally decreased, although recent elections suggest a possible reversal of that trend. In 1960, 64 percent of the eligible voting-age population turned out to vote; by 1996, that figure had dropped to 52 percent. The presidential election of 2008 bucked this declining trend with voter turnout levels of 62 percent of eligible voters, the highest since the Kennedy and Johnson elections in 1960 (64 percent) and 1964 (63 percent), respectively. The rate dipped slightly to 59 percent in 2012 and to 60 percent in 2016 (U.S. Elections Project 2018).

There are significant differences in voter turnout by race and ethnicity, age, educational attainment, and income. Political scientists have documented that voter turnout is highest historically among whites and lowest among Hispanics, with Blacks and Asian Americans in between. Highly educated persons and those with greater income also are more likely to vote than persons with fewer means. Generally, turnout increases directly with age. In the 2016 election, turnout was highest among non-Hispanic whites (65 percent), somewhat lower for Blacks (60 percent), and considerably lower for Hispanics (48 percent). Turnout also varied considerably with age, with older citizens voting at much higher rates than younger adults: 71 percent of eligible citizens 65 and older turned out to vote, compared with 67 percent of adults between the ages of 45 and 64, 59 percent of adults ages 30 to 44, and only 46 percent of adults ages 18 to 29. Education also influences voting behavior: In the 2016 election, 35 percent of persons who lacked a high school diploma turned out to vote, compared with 76 percent of those with a bachelor's degree or higher. Turnout was just 48 percent among voters whose family income was less than $30,000, rising to 78 percent among voters whose family income was more than $100,000 (U.S. Bureau of the Census 2017g).

Voter turnout in the United States is among the world's lowest. Many studies have found that countries with high rates of literacy, high average incomes, and well-established political freedoms and civil liberties have high voter turnout (International Institute for Democracy and Electoral Assistance 2004). Even though the United States ranks high on all these measures, it fails to motivate people to vote. The United States ranks thirty-first in voter turnout among the thirty-five advanced industrial countries that belong to the Organization for Economic Cooperation and Development (OECD), which includes the United States, Japan, Israel, most European countries, and such emerging economies as Chile, Turkey, and South Korea. Voting is compulsory for six OECD countries, but that alone cannot account for the United States' low ranking (twenty-five OECD countries that do not legally require people to vote still outrank the United States) (Desilver 2016).

Voter Turnout

Despite high literacy rates, high average incomes, and well-established political freedoms and civil rights, voter turnout in the United States is among the world's lowest.

Voter turnout in presidential elections*

◗ Compulsory voting

- 2016 **United States**** 56.0%
- 2017 **UK** 63.3%
- 2017 **France** 67.9%
- 2017 **Germany** 69.3%
- 2018 **Turkey** 89.0%
- 2018 **Mexico** 66.0%
- 2016 **Haiti** 17.8%
- 2018 **Egypt** 39.9%
- 2018 **Venezuela** 43.7%
- 2015 **Nigeria** 32.1%
- 2017 **Kenya** 61.4%
- 2016 **South Korea** 77.9%

*Voter turnout is defined as the percentage of the voting-age population that voted.

**Represents the percentage of entire voter-age population that voted in the 2016 election. Excluding non-citizens and those ineligible to vote due to past felony convictions, approximately 60.2% percent of the total voter-eligible population voted in 2016, according to the U.S. Elections Project.

Who voted for Trump?

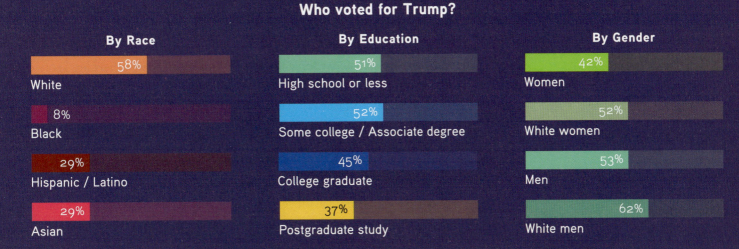

By Race
- White — 58%
- Black — 8%
- Hispanic / Latino — 29%
- Asian — 29%

By Education
- High school or less — 51%
- Some college / Associate degree — 52%
- College graduate — 45%
- Postgraduate study — 37%

By Gender
- Women — 42%
- White women — 52%
- Men — 53%
- White men — 62%

Sources: Desilver 2018; International Institute for Democracy and Electoral Assistance 2018; U.S. Elections Project 2016; Edison Research for the National Election Pool 2016.

Of those expressing continuing trust in government, most vote in presidential elections; of those who lack trust, most do not vote. As we have seen, younger people have traditionally had less interest in electoral politics than older generations have, although the young have a greater interest than their elders in issues like the environment. In the 2008 presidential election, Obama's call for "change we can believe in" apparently resonated with young voters: Turnout in the eighteen- to twenty-four-year-old age group reached 49 percent, a 2 percent increase over the previous presidential election, with two-thirds voting for Obama. Bernie Sanders's primary campaign in 2016, with its focus on ending inequality (and free higher education), similarly attracted large numbers of young voters. Yet 5 million fewer young voters bothered to turn out in the presidential election that year—most likely because they were dissatisfied with both Clinton and Trump, the two major party candidates (Purtill 2016).

Why is voter turnout so low in the United States? While there are no clear-cut answers, a number of factors undoubtedly play a role. First, in the United States—unlike in many other countries—voter registration is not automatic. Many people find the process of registering to vote burdensome and so don't bother. In some states, significant barriers have been raised to make it more difficult for some people to vote, including so-called voter ID laws, which require a driver's license, a birth certificate, or some other form of identification to vote. These laws are most likely to affect low-income and minority voters (Hajnal, Lajevardi, and Nielson 2017; Highton 2017).

Another possible reason is that since winner-take-all elections discourage the formation of third parties, voters may sometimes feel that they lack viable choices when it comes time to vote. A large (and growing) number of people clearly feel that the current system is unresponsive to their needs. This was an important factor in the rise of populism in the 2016 presidential campaigns, with Sanders and Trump—both rejecting "business as usual"—attracting the most enthusiastic support among dedicated followers. In many countries, including most European countries, some system of proportional representation is practiced, under which parties receive seats in proportion to the vote they get in electoral districts. Under this system, even small parties can often muster sufficient support to elect one or two representatives. When voters have a wider range of choices, they are more likely to vote.

INTEREST GROUPS

interest group

A group organized to pursue specific interests in the political arena, operating primarily by lobbying the members of legislative bodies.

Interest groups and lobbying play a distinctive part in American politics. An **interest group** is any organization that attempts to influence elected officials to consider its aims when deciding on legislation. The American Medical Association, the National Organization for Women, and the National Rifle Association are three examples. Interest groups vary in size; some are national, others statewide. Some are permanently organized; others are short-lived. "Lobbying" is the act of contacting influential officials to present arguments to convince them to vote in favor of a cause or otherwise lend support to the aims of an interest group. The word *lobby* originated in the British parliamentary system: In days past, members of parliament did not have offices, so their business was conducted in the lobbies of the parliament buildings.

To run as a candidate is enormously expensive, and interest groups provide much of the funding at all levels of political office. Donald Trump, the 2016 Republican candidate for president, raised roughly $398 million, while Hillary Clinton, the Democratic candidate, raised nearly twice as much ($768 million). On the other hand, Trump received far greater media coverage throughout the campaign; such "free media" exposure was estimated as worth more than $5.9 billion for Trump, twice the $2.8 billion in free media estimated for Clinton. When "outside money" is factored in, including money from political action committees (PACs) that are set up by interest groups to raise and distribute campaign funds, total spending on the 2016 presidential and congressional campaigns is estimated to have exceeded $6.4 billion: nearly $2.4 billion for the presidency and more than $4 billion for House and Senate seats (Sultan 2017).

Incumbents, or those already in office, have an enormous advantage in soliciting money. In the 2016 congressional elections, incumbents raised nearly three times more money than their challengers—more than $1 billion compared with $352 million (Center for Responsive Politics 2017). Incumbents are favored as fund-raisers partly because they can curry favor with special interests and other contributors, because they are in a position to ensure favorable votes on issues of importance to their funders and to obtain spending on pet projects and other "pork" for their districts. Incumbency also provides familiarity—a formidable (and costly) obstacle for most challengers to overcome. In 2016, 97 percent of incumbent House members and 93 percent of incumbent Senate members were reelected.

More than 40 percent of the funding in congressional elections comes from PACs, which are set up by interest groups to raise and distribute campaign funds. Paid lobbyists play a significant role in influencing the outcome of votes in Congress and decisions by the president. The Center for Responsive Politics (2018) reported that businesses, unions, and other advocacy groups spent some $3.4 billion in 2017, employing more than 11,500 lobbyists. The largest sector, health, was the target of $561 million in lobbying efforts, with the pharmaceutical industry accounting for more than half of that amount. The U.S. Chamber of Commerce, broadly representing business interests, was the top spender in 2017 ($82 million), followed by the National Association of Realtors ($54 million). All of the top twenty organizations in 2017 represented business interests.

The National Rifle Association is an interest group that advocates for gun rights. In 2015, the NRA spent more than $3 million to influence gun policy.

The Political Participation of Women

Voting has a special meaning for women, given their long struggle to obtain universal **suffrage**, or the legal right to vote. The members of the early women's movements saw the vote both as a symbol of political freedom and as the means of achieving greater economic and social equality. After what was often a long, hard fight, women now can vote in nearly all of the world's nations; however, this has not greatly altered the nature of politics. Women's voting patterns, like those of men, are shaped by party preferences, policy options, and the choice of available candidates. In the 2016 election, 54 percent of women voted for Clinton, compared to 42 percent for Trump. The reverse was true for men, with 53 percent voting for Trump and 41 percent for Clinton, suggesting a strong gender gap in the first U.S. election in which a major party presidential candidate was female (Tyson and Maniam 2016).

suffrage

A legal right to vote guaranteed by the Fifteenth Amendment to the U.S. Constitution; guaranteed to women by the Nineteenth Amendment.

A leader in the Democratic party, Elizabeth Warren became the first female senator of Massachusetts when she beat Republican Scott Brown in November 2012.

The influence of women on politics cannot be assessed solely through voting patterns, however. Feminist groups have made an impact on political life independently of the franchise, particularly in recent decades. Since the early 1960s, the National Organization for Women (NOW) and other women's groups in the United States have played a significant role in the passing of equal opportunity acts and have pressed for a range of issues directly affecting women to be placed on the political agenda. Such issues include equal rights at work, the availability of abortion, changes in family and divorce laws, and lesbian rights. In 1973, women achieved a legal victory when the Supreme Court ruled in *Roe v. Wade* that women had a legal right to abortion. The 1989 Court ruling in *Webster v. Reproductive Health Services*, which placed restrictions on that right, resulted in a resurgence of involvement in the women's movement.

Women such as Nancy Pelosi, Elizabeth Warren, Kristen Gillibrand, and Nikki Haley now play central roles in American politics. Still, women remain underrepresented in government. In 2018, there were only twenty-three women in the Senate and only eighty-four in the House of Representatives, representing 20 percent of all seats in Congress (Center for American Women and Politics 2018a). While these percentages may seem low, from a historical perspective, they represent a sea change in women's roles in politics. In 1970, there was only a single woman in the Senate, and there were just ten in the House. Despite the gender gap in Congress and other elected offices, both the Democratic and Republican parties today are nominally committed to securing equal opportunities for women and men. Since 1990, female candidates for political office have been successful when they have run for office. The critical factor seems to be that political parties (which are largely run by men) have not recruited as many women to run for office.

We now broaden our scope to look at some basic ideas of political power. First, we take up the issue of who actually holds the reins of power, drawing on comparative materials to help illuminate the discussion. We then consider whether democratic governments around the world are "in crisis."

Who Rules? Theories of Democracy

Classical sociology offers three different ideas of how modern democracies actually function: through rule by elites who possess the necessary expertise but are accountable to the electorate; through interest groups that compete for influence, providing a form of checks and balances against one another; and through an elite of the wealthy and powerful that operates in the background, thereby shaping policy in their interest.

DEMOCRATIC ELITISM

One of the most influential views of the nature and limits of modern democracy was set out by Max Weber and, in rather modified form, by the economist Joseph Schumpeter (1983, orig. 1942). The ideas they developed are sometimes referred to as the theory of **democratic elitism**.

Weber began from the assumption that direct democracy is impossible as a means of regular government in large-scale societies. This is not only for the obvious logistical reason that millions of people cannot meet to make political decisions but because running a complex society demands expertise. Participatory democracy, Weber believed, can

democratic elitism

A theory of the limits of democracy, which holds that in large-scale societies, democratic participation is necessarily limited to the regular election of political leaders.

succeed only in small organizations in which the work to be carried out is fairly simple and straightforward. When more complicated decisions have to be made, or policies worked out, even in modest-sized groups—such as a small business firm—specialized knowledge and skills are necessary. Experts have to carry out their jobs on a continuous basis; positions that require expertise cannot be subject to the regular election of people who may only have a vague knowledge of the necessary skills and information. While higher officials, responsible for overall policy decisions, are elected, there must be a large substratum of full-time bureaucratic officials who play an essential part in running a country (Weber 1979, orig. 1921).

Weber placed a great deal of emphasis on the importance of leadership in democracy—which is why his view is referred to as "democratic elitism." He argued that rule by elites is inevitable; the best we can hope for is that those elites effectively represent our interests and that they do so in an innovative and insightful fashion. Weber valued multiparty democracy more for the quality of leadership it generates than for the mass participation in politics it makes possible.

PLURALIST THEORIES

According to **pluralist theories of modern democracy**, government policies in a democracy are influenced by the continual processes of bargaining among numerous groups representing different interests—business organizations, trade unions, ethnic groups, environmental organizations, religious groups, and so forth. While pluralists accept that individual citizens can have little or no direct influence on political decision making, they argue that the presence of interest groups can limit the centralization of power in the hands of government officials. A democratic political order is one in which there is a balance among competing interests, all having some impact on policy but none dominating the actual mechanisms of government. Elections are also influenced by this situation; to achieve a broad enough base of support to lay claim to government, parties must be responsive to numerous diverse interest groups. The United States, it is held, is the most pluralistic of industrialized societies and, therefore, the most democratic. Competition among diverse interest groups occurs not only at the national level but within the states and in the politics of local communities.

pluralist theories of modern democracy

Theories that emphasize the role of diverse and potentially competing interest groups, none of which dominate the political process.

THE POWER ELITE

The view suggested by C. Wright Mills in his celebrated work *The Power Elite* (1956) is quite different from pluralist theories. Mills argues that during the course of the twentieth century, a process of institutional centralization occurred in the political order, the economy, and the sphere of the military. Not only did each of these spheres become more centralized, according to Mills, but each became increasingly merged with the others to form a unified system of power. Those who are in the highest positions in all three institutional areas come from similar social backgrounds, have parallel interests, and often know one another on a personal basis. By the mid-twentieth century, they had become a single **power elite** that ran, and continues to run, the country—and, given the international position of the United States, also influences a great deal of the rest of the world. Today, this elite group is sometimes called the "deep state"—entrenched government officials and bureaucrats, often linked to global business interests, who are unresponsive

power elite

Small networks of individuals who, according to C. Wright Mills, hold concentrated power in modern societies.

to the electorate, even in liberal democracies. Widespread dissatisfaction with governments considered unresponsive has fueled the rise of antigovernment movements on both the right and left throughout the world.

The power elite, in Mills's portrayal, is composed mainly of white Anglo-Saxon Protestants (WASPs). Many are from wealthy families, have been to the same prestigious universities, belong to the same clubs, and sit on government committees with one another. They have closely connected concerns. Business and political leaders work together, and both have close relationships with the military through weapons contracting and the supply of goods for the armed forces. There is a great deal of movement among top positions in the three spheres. Politicians have business interests; business leaders often run for public office; higher military personnel sit on the boards of the large companies.

Since Mills published his study, numerous other research investigations have analyzed the social background and interconnections of leading figures in the various spheres of American society (Dye 1986). All studies agree on the finding that the social backgrounds of those in leading positions are highly unrepresentative of the population as a whole (Domhoff 1971, 1979, 1983, 1998, 2013).

THE ROLE OF THE MILITARY

Mills's argument that the military plays a central role among the power elite was buttressed by a well-known warning from a former military hero and U.S. president, Dwight David Eisenhower. In his farewell presidential speech in 1961, Eisenhower—who was the supreme commander of the Allied forces in Europe in World War II—warned of the dangers of what he termed the "military–industrial complex." As Eisenhower bluntly put it, "The conjunction of an immense military establishment and a large arms industry is new in the American experience. In the councils of government, we must guard against the acquisition of unwarranted influence, whether sought or unsought, by the military–industrial complex. The potential for the disastrous rise of misplaced power exists and will persist" (Eisenhower Library 1961).

With the collapse of the Soviet Union in 1991, the United States has emerged as the world's unrivaled military superpower, accounting for 35 percent of total world military spending in 2017 (Stockholm International Peace Research Institute 2018). Eisenhower's dire warning seems no less apt today than when he uttered it more than fifty years ago.

Democracy in Trouble?

Democracy almost everywhere is in some difficulty today. Even in the United States, voter turnout is low, and many people tell pollsters that they don't trust politicians. In 1964, confidence in government was fairly high: Nearly four of five people answered "most of the time" or "just about always" when asked "How much of the time do you trust the government in Washington to do the right thing?" However, as we saw in Chapter 5, confidence in the U.S. government has neared historic lows in recent years. Following the terrorist attacks of 9/11, a solid majority (55 percent) of Americans reported that they trusted the government "most of the time" or "just about always." Recently, however, trust in the government has declined as a result of disillusionment over the wars in Iraq and Afghanistan, the economic collapse of 2008, the tepid recovery that followed, and government paralysis in Washington. By the end of 2017, just 18 percent of Americans said they could trust the government in Washington always or most of the time (Pew Research Center 2017c).

Roughly half of all Americans believe that government is too large, a perception that in part reflects this low level of trust (Pew Research Center 2017e). Yet among the twenty-seven industrial democracies in the OECD, the United States ranks twenty-fourth in terms of federal, state, and local government spending as a proportion of its total economy. Nine OECD countries spend more than half of their GDP on government at all levels, with France and Finland topping the list at 57 percent of GDP. The United States, by way of comparison, spends only 37 percent (OECD 2015).

The last few decades have also been a period in which, in several Western countries, the welfare state has come under attack. Rights and benefits, fought for over long periods, have been contested and cut back. One reason for this governmental retrenchment is the decline in revenues available to governments as a result of the general world recession that began in the early 1970s, the financial collapse of 2008, and the effects of globalization, which have shifted much economic growth (and therefore growth in tax revenues) from Europe, North America, and Japan to emerging economies, such as China. Yet an increasing skepticism also seems to have developed, shared not only by some governments but by many of their citizens, about the effectiveness of relying on the state for the provision of many essential goods and services. This skepticism is rooted in the belief that the welfare state is bureaucratic, alienating, and inefficient and that welfare benefits are all too often ineffective. In response to a 2011 survey asking "What is more important—that everyone be free to pursue their life's goals without government interference, or that the state play an active role in guaranteeing that nobody is in need?" 58 percent of American respondents favored the former, with only 25 percent favoring the latter. These percentages were reversed in many European countries, with roughly three out of every five respondents in Britain, France, and Germany believing that government has a role to play in helping the less fortunate (Stokes 2013).

Finally, for the past quarter-century, there has been a steady stream of well-financed criticism from conservative think tanks, news media, and politicians who share the belief that government is part of the problem rather than part of the solution. This has been highly effective in changing the national discourse on the role of government (Gonzales and Delgado 2006).

Why are so many Americans dissatisfied with democracy, a political system that not long ago seemed to be sweeping across the world? The answers, curiously, are bound up with the factors that have helped spread democracy—the impact of capitalism and the globalization of social life. While capitalist economies generate more wealth than any other type of economic system, that wealth is unevenly distributed, as we learned in Chapter 7. And economic inequalities influence who votes, joins parties, and gets elected. Wealthy individuals and corporations back interest groups that lobby for elected officials to support their aims when deciding on legislation. Not being subject to election, interest groups are not accountable to the majority of the electorate.

CONCEPT CHECKS

1. Why is it problematic for contemporary states to have participatory democracy?

2. What are two explanations for low voter turnout in the United States?

3. Describe the role interest groups play in American politics.

4. Compare and contrast pluralist theories of modern democracy and the power elite model.

What Is the Social Significance of Work?

Assess the sociological ramifications of paid and unpaid work. Understand that modern economies are based on the division of labor and economic interdependence. Familiarize yourself with modern systems of economic production.

Because politics is inextricably linked with economic life, we now turn our attention to the ways that work and the economy have changed. **Work** refers to carrying out tasks that require mental and physical effort, with the objective of the production of goods and services that cater to human needs. An **occupation**, or job, is work that is done in exchange for a regular wage or salary. In all cultures, work is the basis of the economic system.

The study of economic institutions is of major importance in sociology because the economy influences all segments of society and therefore social reproduction in general. Hunting and gathering, pastoralism, agriculture, industrialism—these different ways of gaining a livelihood have a fundamental influence on the lives people lead. The distribution of goods and variations in the economic position of those who produce them also strongly influence social inequalities of all kinds. Wealth and power do not inevitably go together, but in general, the privileged in terms of wealth are also among the more powerful groups in a society.

In the remainder of this chapter, we analyze the nature of work in modern societies and look at the major changes affecting economic life today. We investigate the changing nature of industrial production and of work itself. Modern industry differs in a fundamental way from premodern systems of production, which were based above all on agriculture. Most people worked in the fields or cared for livestock. In modern societies, by contrast, only a tiny fraction of the population works in agriculture, and farming itself has become industrialized—it is carried on largely by means of machines.

Modern industry is itself always changing—technological change is one of its main features. **Technology** involves the use of science and machinery to achieve greater productive efficiency. The nature of industrial production also changes in relation to wider social and economic influences. We focus on both technological and economic change, showing how these are transforming industry today. We will also see that globalization makes a great deal of difference to our working lives; the nature of the work we do is being changed by forces of global economic competition.

The Importance of Paid and Unpaid Work

We often associate the notion of work with drudgery—with a set of tasks that we want to minimize and, if possible, escape from altogether. Is this most people's attitude toward their work, and if so, why?

Work is more than just drudgery, or people would not feel so lost and disoriented when they become unemployed. How would you feel if you thought you would never get a job? In modern societies, having a job is important for maintaining a sense of purpose. Even where work conditions are relatively unpleasant and the tasks involved are dull, work tends to be a structuring element in people's psychological makeup and the cycle of their daily activities.

Work need not conform to the standard categories of paid employment. Nonpaid labor (such as repairing one's own car or doing one's own housework) is an important

---

work

Carrying out tasks that require mental and physical effort, with the objective of the production of goods and services that cater to human needs. Work should not be thought of exclusively as paid employment. In modern societies, there remain types of work that do not involve direct payment.

occupation

Any form of paid employment in which an individual regularly works.

technology

The application of knowledge of the material world to production; the creation of material instruments (such as machines) used in human interaction with nature.

Will a Robot Take Your Job?

The microchip has changed the way we live. Communication is easy and borderless. The world's information is at our fingertips. The goods and services we consume are cheaply sourced from around the planet. Cars are smarter (even if their drivers aren't). Are we heading for a utopian future in which "smart machines" do the drudge work, freeing up humans for more satisfying pursuits? Or will jobs disappear, resulting in massive unemployment?

Because the U.S. population is growing, more than a million new jobs must be created each year just to avoid an increase in the unemployment rate (Drum 2016). Yet, over the past decade, employment has failed to keep up. Many workers are so discouraged that they are no longer looking for work; while 66 percent of the adult population was in the labor force in 2006, eleven years later, in 2017, the percentage had dropped to 63 percent (U.S. Bureau of Labor Statistics 2018d).

These changes are due in part to technology that has led to the offshoring of manufacturing and service-sector jobs to countries where wages are lower. Those firms that remain in the United States often employ fewer workers, thanks also to advances in technology (Sherk 2010): automobile plants where robots assemble cars, banks that use ATMs, supermarkets with automatic checkout services. While much has been made of "twenty-first-century onshoring"—the return of some manufacturing jobs to the United States—these jobs often pay low wages and, thanks to automation, also require fewer workers (Semuels 2015).

While many lost jobs affect lower-skilled workers, rapid advances in software, in such areas as business intelligence, decision making, and "big data" analysis, hold the promise of automating occupations that currently require college degrees. The services provided by professionals such as lawyers, accountants, radiologists, and many middle managers can increasingly be performed by artificial intelligence software. One estimate is that as many as two-fifths of all jobs in the United States could be replaced by software (Brynjolfsson and McAfee 2014; Ford 2009, 2010). As one technology writer has noted, "the evidence is irrefutable that computerized automation, networks and artificial intelligence (AI)—including machine-learning, language-translation, and speech- and pattern-recognition software—are beginning to render many jobs simply obsolete" (*The Economist* 2011).

One technology that has the ability to radically change the nature of work is additive manufacturing, or 3D printing. This involves the ability, using computerized technology, to add layers of materials that result in physical objects such as fashion accessories, car parts, artificial organs, weapons, and even component parts for the Mars rover. Because it is software driven, 3D printing holds the promise of instantly delivering products that meet the exact requirements of individual consumers (Chowdry 2013; Peels 2012). Nike and Adidas, for example, are both developing the ability to scan a customer's foot and print out a custom-designed athletic shoe in the store—or eventually, as 3D printers come down in cost, perhaps even in the customer's home (Matisons 2015). If this indeed comes to pass, what will happen to the hundreds of thousands of jobs, around the globe, currently devoted to footwear manufacturing?

Will a robot take your job? Significantly, at least one robot—Siri—hasn't a clue.

Additive manufacturing, or 3D printing, which was used to create this robot, could radically change the nature of work by reducing the role of workers in creating complex products.

aspect of many people's lives. Much of the work done in the informal economy, for example, is not recorded in official employment statistics. The term **informal economy** refers to transactions outside the sphere of regular employment, sometimes involving the exchange of cash for services provided but also often involving the direct exchange of goods or services. Your child's babysitter might be paid in cash "off the books," or without any receipt being given or details of the job recorded.

The informal economy includes not only "hidden" cash transactions but many forms of self-provisioning that people carry on inside and outside the home. Do-it-yourself activities with household appliances and tools, for instance, provide goods and services that would otherwise have to be purchased (Gershuny and Miles 1983). Housework is usually unpaid, but it is work—often very hard and exhausting work—nevertheless. Volunteer work, for charities or other organizations, has an important social role. Having a paid job is important—but the category of "work" stretches more widely.

The Importance of the Division of Labor

The economic system of modern societies rests on a highly complex **division of labor**. Recall that in Chapter 1, we introduced this concept, which the nineteenth-century scholar Émile Durkheim viewed as the basis for the social cohesion that results when multiple parts of society function as an integrated whole. Under a division of labor, work is divided into an enormous number of different occupations in which people specialize. In traditional societies, nonagricultural work entailed the mastery of a specific skill. A worker typically learned craft skills through a lengthy period of apprenticeship and then carried out all aspects of the production process from beginning to end. For example, a metalworker making a plow would forge the iron, shape it, and assemble the implement itself. With the rise of modern industrial production, most traditional crafts have disappeared altogether, replaced by skills that form part of more large-scale production processes.

The contrast in the division of labor between traditional and modern societies is truly extraordinary. Even in the largest traditional societies, there usually existed no more than twenty or thirty major craft trades, together with such specialized pursuits as merchant, soldier, and priest. In a modern industrial system, there are literally thousands of distinct occupations. The U.S. Census Bureau lists more than 31,000 distinct jobs in the American economy. In traditional communities, most of the population worked on farms and were economically self-sufficient. They produced their own food, clothes, and other necessities of life. One of the main features of modern societies, by contrast, is an enormous expansion of **economic interdependence**. The vast majority of people in modern societies do not produce the food they eat or the material goods they consume. Moreover, the division of labor is now truly global, since the components of virtually all products are sourced from many factories in different countries.

Industrial Work

Writing more than two centuries ago, Adam Smith, one of the founders of modern economics, identified advantages that the division of labor provides in terms of increasing productivity. His most famous work, *The Wealth of Nations*, opens with a description of the division of labor in a pin factory. A person working alone could perhaps make twenty pins per day. By breaking down that worker's task into a number of simple operations, however, ten workers carrying out specialized jobs in collaboration with one another could collectively produce 48,000 pins per day. The rate of production per worker, in other words, is increased from 20 to 4,800 pins, each specialist operator producing 240 times more than when working alone.

More than a century later, these ideas reached their most developed expression in the writings of Frederick Winslow Taylor, an American management consultant. Taylor's approach to what he called "scientific management" involved the detailed study of industrial processes to break them down into simple operations that could be precisely timed and organized. Taylor's principles were appropriated by the industrialist Henry Ford. In 1908, Ford designed his first auto plant in Highland Park, Michigan, to manufacture only one product—the Model T Ford—

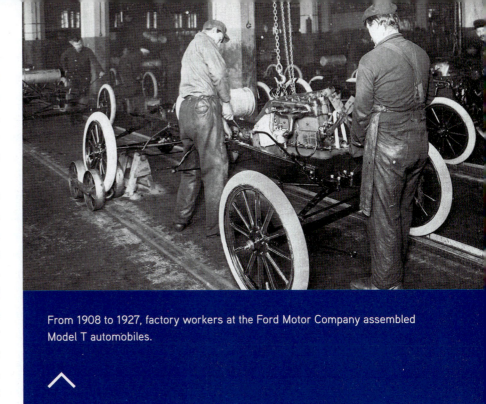

From 1908 to 1927, factory workers at the Ford Motor Company assembled Model T automobiles.

thereby allowing the introduction of specialized tools and machinery designed for speed, precision, and simplicity of operation. One of Ford's most significant innovations was the introduction of the assembly line, said to have been inspired by Chicago slaughterhouses, in which animals were disassembled section by section on a moving conveyor belt. Each worker on Ford's assembly line was assigned a specialized task, such as fitting the left-side door handles as the car bodies moved along the line. By 1929, when production of the Model T ceased, more than 15 million cars had been assembled.

WORK AND ALIENATION

Karl Marx was one of the first writers to grasp that the development of modern industry would reduce many people's work to dull, uninteresting tasks. According to Marx, the division of labor alienates human beings from their work. **Alienation** refers to feelings of indifference or hostility not only to work but to the overall framework of industrial production within a capitalist setting.

In traditional societies, Marx pointed out, work was often exhausting; peasant farmers sometimes toiled from dawn to dusk. Yet peasants had control over their work, which required much knowledge and skill. Many industrial workers, by contrast, have little control over their jobs, only contribute a fraction to the creation of the overall product, and have no influence over how or to whom it is eventually sold. Work thus appears as something alien, a task that the worker must carry out to earn an income but that is intrinsically unsatisfying.

INDUSTRIAL CONFLICT

There have long been conflicts between workers and those with economic and political authority over them. Riots against high taxes and food riots at periods of harvest failure were common in urban areas of Europe in the eighteenth century. These "premodern" forms of labor conflict continued up to the late nineteenth century in some countries. Such

alienation

The sense that our own abilities as human beings are taken over by others. Karl Marx used the term to refer to the loss of workers' control over the nature and products of their labor.

traditional forms of confrontation were not just sporadic, irrational outbursts of violence: The threat or use of violence had the effect of lowering the price of grain and other essential foodstuffs (Booth 1977; Rudé 1964; Thompson 1971).

Industrial conflict between workers and employers at first tended to follow these older patterns. In situations of confrontation, workers would quite often leave their places of employment and form crowds in the streets; they would make their grievances known through their unruly behavior or by engaging in acts of violence against the authorities. Workers in some parts of France in the late nineteenth century would threaten disliked employers with hanging (Holton 1978). Use of the strike as a weapon, today commonly associated with organized bargaining between workers and management, developed only slowly and sporadically.

A **strike** is a temporary stoppage of work by a group of employees to express a grievance or enforce a demand (Hyman 1984). As we saw earlier in this chapter, workers who were dissatisfied with their low wages and working conditions staged work stoppages throughout American cities in April 2015. Workers go on strike for many reasons. They may be seeking to gain higher wages, forestall a proposed reduction in their earnings, protest against technological changes that make their work more dull or lead to layoffs, or obtain greater job security. However, in all these circumstances, the strike is essentially a mechanism of power: a weapon of people who are relatively powerless in the workplace and whose working lives are affected by managerial decisions over which they have little control. Strikes typically occur when other negotiations have failed, because workers on strike either receive no income or depend on union funds, which might be limited. As Figure 13.3 shows, work stoppages in the United States dropped off considerably in the 1980s. This is due in large part to the fact that union membership decreased markedly at this time. In the next section, we explain the reasons behind this precipitous drop.

LABOR UNIONS

Although their levels of membership and the extent of their power vary widely, union organizations exist in all Western countries, which also all legally recognize the right of workers to strike in pursuit of economic objectives. In the early development of modern industry, workers in most countries had no political rights and little influence over their working conditions. **Unions** developed as a means of redressing the imbalance of power between workers and employers. As we saw earlier in this chapter, one tactic unions use is **collective bargaining**. This is the process of negotiations between employers and their workers; these negotiations are used to reach agreements about a broad range of working conditions, including pay scales, working hours, training, health and safety, and the right to file a grievance. Whereas workers may have limited power as individuals, through collective organization, their influence is considerably increased. An

strike

A temporary stoppage of work by a group of employees in order to express a grievance or enforce a demand.

——

unions

Organizations that advance and protect the interests of workers with respect to working conditions, wages, and benefits.

——

collective bargaining

The rights of employees and workers to negotiate with their employers for basic rights and benefits.

FIGURE 13.3

Work Stoppages,* 1947–2017

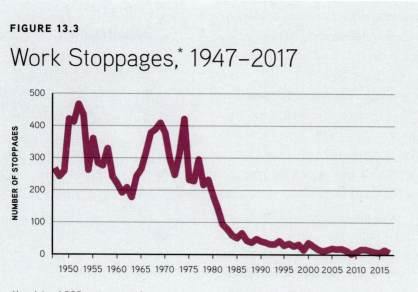

*Involving 1,000 or more workers.

Source: Bureau of Labor Statistics 2018i.

employer can do without the labor of any particular worker but not without that of all or most of the workers in a factory or plant.

After 1980, unions suffered declines across the advanced industrial countries. In the United States, the share of the workforce belonging to unions declined from 23 percent in 1980 to 11 percent in 2017. The decline has been steepest in the private sector: While just 6.5 percent of all private sector wage and salary workers were unionized in 2017, more than one-third (34 percent) of all government employees belonged to unions, reaching 40 percent for local government (Hirsch and Macpherson 2004; U.S. Bureau of Labor Statistics 2018h). Americans' perceptions of labor unions also took a hit following widespread protests by state workers in 2011. In Wisconsin and several other states, state and local workers—ranging from firefighters to teachers—protested because they feared the loss of their pensions, a reduction in their health benefits, and a loss of their right to collective bargaining. Many Americans, themselves facing higher health insurance costs and reduced pensions in their public-sector jobs, showed little empathy for the protesting state workers. However, since that time, attitudes toward unions have once again increased, with 60 percent having a favorable view, compared with 35 percent unfavorable (Maniam 2017).

There are several widely accepted explanations for the difficulties confronted by unions since 1980. One major factor is the loss of once-unionized manufacturing jobs to low-wage countries around the world, particularly in East Asia and most notably China—a country where independent labor unions are illegal. Such job loss, real or threatened, has greatly weakened the bargaining power of unions in the manufacturing sector and, as a result, has lowered their appeal to workers. Why join a union and pay union dues if the union cannot deliver wage increases or job security? Unionization efforts in the United States have also been hampered in recent years by decisions of the National Labor Relations Board (NLRB), the government agency responsible for protecting the right of workers to form unions and engage in collective bargaining. The NLRB has proven ineffective at protecting efforts to unionize workplaces, failing to take aggressive action when businesses harass or fire union organizers (Clawson and Clawson 1999; Estlund 2006).

capitalism

An economic system based on the private ownership of wealth.

CONCEPT CHECKS

1. Why is it important for sociologists to study economic institutions?

2. Define and provide an example of an informal economy.

3. Using the concept of division of labor, describe the key differences in the nature of work in traditional versus modern societies.

4. What is a labor union? Why have unions in the United States suffered from a decline in membership since the 1980s?

What Are Key Elements of the Modern Economy?

Modern societies are, in Marx's term, *capitalistic*. As we learned in Chapter 1, **capitalism** is a way of organizing economic life that is distinguished by the following important features: private ownership of the means of production; profit as incentive; competition for markets to sell goods, acquire cheap materials, and utilize cheap labor; and expansion and investment to accumulate capital. Capitalism, which began to spread with the growth of the Industrial Revolution in the early nineteenth century, is a vastly more dynamic economic system than any other that preceded it in history. Although the system has had many critics, such as Marx, it is now the most widespread form of economic organization in the world.

See the importance of the rise of large corporations; consider particularly the global impact of transnational corporations.

So far, we have been looking at industry mostly from the perspective of occupations and employees. But we also have to concern ourselves with the nature of the business firms in which the workforce is employed. (It should be recognized that many people today are employees of government organizations, although we will not consider these here.) What is happening to business corporations today, and how are they run?

Corporations and Corporate Power

The world economy is increasingly influenced by the rise of large business **corporations**. In 2017, the 2,000 largest corporations in the world had total assets valued at $170.5 trillion, yet just seventy-six of these corporations accounted for fully half of total assets. The largest twenty-four corporations, in terms of assets, are banks, claiming 30 percent of total assets. The largest bank in the world is the Industrial and Commercial Bank of China, with more than $4 trillion in assets (Forbes 2018).

Of course, there still exist thousands of smaller firms and enterprises within the American economy. In these companies, the image of the **entrepreneur**—the boss who owns and runs the firm—is by no means obsolete. The large corporations are a different matter. Ever since Adolf Berle and Gardiner Means published their celebrated study *The Modern Corporation and Private Property* more than eighty years ago, it has been accepted that most of the largest firms are not run by those who own them (Berle and Means 1982, orig. 1932). In theory, the large corporations are the property of their shareholders, who have the right to make all important decisions. But Berle and Means argued that since share ownership is so dispersed, actual control has passed into the hands of the managers who run firms on a day-to-day basis.

The power of the major corporations is very extensive. Corporations often cooperate in setting prices rather than freely competing with one another. Thus, the giant oil companies normally follow one another's lead in the price charged for gasoline. When one firm occupies a commanding position in a given industry, it is said to be in a **monopoly** position. More common is a situation of **oligopoly**, in which a small group of giant corporations predominates. In situations of oligopoly, firms are able more or less to dictate the terms on which they buy goods and services from the smaller firms that are their suppliers. The emergence of the global economy has contributed to a wave of corporate mergers and acquisitions on an unprecedented scale, totaling more than $3.5 trillion in 2017, creating oligopolies in industries such as health care and communications and media (Reuters 2017).

TYPES OF CORPORATE CAPITALISM

The development of business corporations has several general stages, although each overlaps with the others and all continue to coexist today. The first stage, characteristic of the nineteenth and early twentieth centuries, was dominated by **family capitalism**. Large firms were run either by individual entrepreneurs or by members of the same family and then passed on to their descendants. The famous corporate dynasties, such as the Rockefellers and Fords, belong in this category. These individuals and families did not just own a single large corporation but held a diversity of economic interests and stood at the apex of economic empires.

Most of the big firms founded by entrepreneurial families have since become public companies—that is, shares of their stock are traded on the open market—and

corporations

Business firms or companies.

entrepreneur

The owner or founder of a business firm.

monopoly

A situation in which a single firm dominates in a given industry.

oligopoly

The domination by a small number of firms in a given industry.

family capitalism

Capitalistic enterprise owned and administered by entrepreneurial families.

have passed into managerial control. In the large corporate sector, family capitalism was increasingly succeeded by **managerial capitalism**. As managers came to have more and more influence through the growth of very large firms, the entrepreneurial families were displaced. The result has been described as the replacement of the family in the company by the company itself (Allen 1981). Managerial capitalism has left an indelible imprint on modern society. The large corporation drives not only patterns of consumption but also the experience of employment in contemporary society. It is difficult to imagine how different the work lives of many Americans would be in the absence of large factories or corporate bureaucracies.

Sociologists have identified another area in which the large corporation has left a mark on modern institutions. **Welfare capitalism** refers to a practice that sought to make the corporation—rather than the state or trade unions—the primary shelter from the uncertainties of the market in modern industrial life. Beginning at the end of the nineteenth century, large firms began to provide certain services to their employees, including childcare, recreational facilities, profit-sharing plans, paid vacations, and group life and unemployment insurance. By the end of World War II, many corporations, as well as public employers such as governments and educational institutions, also began to offset much of the cost of purchasing private medical insurance for their employees. These programs often had a paternalistic bent, such as sponsoring "home visits" for the "moral education" of employees. Viewed in less benevolent terms, a major objective of welfare capitalism was coercion, as employers deployed all manner of tactics—including violence—to avoid unionization.

Despite the overwhelming importance of managerial capitalism in shaping the modern economy, many scholars now see the contours of a different phase in the evolution of the corporation emerging. They argue that managerial capitalism has today partly ceded place to **institutional capitalism**. This term refers to the emergence of a consolidated network of business leadership concerned not only with decision making within single firms but also with the development of corporate power beyond them. Institutional capitalism is based on the practice of corporations holding shares in other firms. **Interlocking directorates**—linkages among corporations created by individuals who sit on two or more corporate boards—exercise control over much of the corporate landscape. This reverses the process of increasing managerial control, since the managers' shareholdings are dwarfed by the large blocks of shares owned by other corporations. Rather than investing directly by buying shares in a business, individuals can now invest in money market, trust, insurance, and pension funds that are controlled by large financial organizations, which in turn invest these grouped savings in industrial corporations. However, in coming decades, Americans may be reluctant to put their resources into pension funds because of the recent economic crisis, a time when many people saw their investments decline, if not disappear entirely.

Finally, some sociologists argue that we have now entered a new stage in the development of business corporations—a form of institutional capitalism they see as **global capitalism**. According to this view, corporations are increasingly stateless: Giant transnational entities roam freely around the planet in search of lower costs and higher profits, loyal to no country regardless of where they might be headquartered. The major corporations today are global not only in the sense that they operate transnationally but

managerial capitalism

Capitalistic enterprises administered by managerial executives rather than by owners.

welfare capitalism

Practice in which large corporations protect their employees from the vicissitudes of the market.

institutional capitalism

Capitalistic enterprise organized on the basis of institutional shareholding.

interlocking directorates

Linkages among corporations created by individuals who sit on two or more corporate boards.

global capitalism

The current transnational phase of capitalism, characterized by global markets, production, and finances; a transnational capitalist class whose business concerns are global rather than national; and transnational systems of governance (such as the World Trade Organization) that promote global business interests.

Container ships are cargo ships that carry all of their load in truck-sized containers. As this technique greatly accelerates the speed at which goods can be transported to and from ports, these ships now carry the majority of the world's dry cargo.

also because their shareholders, directors, and top officers are drawn from many countries (Robinson 2004; Sklair 2000b).

TRANSNATIONAL CORPORATIONS

With the intensifying of globalization, most large corporations now operate in an international economic context. When they establish branches in two or more countries, they are referred to as multinational or **transnational corporations**, indicating that these companies operate across many different national boundaries. Swiss researchers identified more than 43,000 transnational corporations networked together in 2007. The top fifty firms were primarily financial institutions, including Barclays, JPMorgan Chase, and Merrill Lynch, which strongly suggests that the financial services industry has a great deal of power and influence in the global economy (Vitali, Glattfelder, and Battiston 2011). As we noted earlier, in 2017 the twenty-four largest corporations in the world, in terms of assets, were banks, reinforcing the conclusions of the earlier Swiss study.

The largest transnationals are gigantic; their wealth outstrips that of many countries. The scope of these companies' operations is staggering. The combined revenues of the world's 500 largest transnational corporations totaled $30 trillion in 2017 (Fortune 2018). To give an idea of the magnitude of that number, that same year, $75.8 trillion in goods and services were produced by the *entire world* (World Bank 2018f).

The United States is home to the largest number of firms among the top 500 transnational corporations, although the share of American-based companies has fallen sharply in recent years as the number of transnational corporations based in other countries—especially Asian countries such as South Korea and China—has increased. While U.S.- and European-based transnational corporations continue to dominate the global economy

transnational corporations

Business corporations located in two or more countries.

by a wide margin, China has begun to have a significant presence, surpassing Japan in terms of number of Global 500 corporations.

The reach of the transnationals over the past thirty years would not have been possible without advances in transportation and communications. Air travel now allows people to move around the world at a speed that would have seemed inconceivable even sixty years ago; container ships the size of small cities move hundreds of thousands of tons of goods across the Pacific Ocean in eleven days. Telecommunications technologies now permit more or less instantaneous communication from one part of the world to another.

CONCEPT CHECKS

1. What are the main features of capitalism?

2. Compare and contrast the different types of capitalism.

How Does Work Affect Everyday Life Today?

The globalization of economic production, together with the spread of information technology, is altering the nature of the jobs most people do. As discussed earlier, the proportion of people working in blue-collar jobs in industrial countries has progressively fallen. Fewer and fewer people work in factories. New jobs have been created in offices, in service centers such as superstores like Walmart, and in airports. Many of these new jobs are filled by women.

Learn about the impact of global economic competition on employment. Consider how work will change over the coming years.

Work and Technology

The relationship between technology and work has long been of interest to sociologists. How is our experience of work affected by the type of technology that is involved? As industrialization has progressed, technology has assumed an ever-greater role at the workplace—from factory automation to the computerization of office work. The current information technology revolution has attracted renewed interest in this question. Technology can lead to greater efficiency and productivity, but how does it affect the way work is experienced by those who carry it out? For sociologists, one of the main questions is how the move to more complex systems influences the nature of work and the institutions in which it is performed.

AUTOMATION AND THE SKILL DEBATE

The concept of **automation**, or programmable machinery, was introduced in the mid-1800s, when Christopher Spencer, an American, invented the Automat, a programmable lathe that made screws, nuts, and gears. Automation has thus far affected relatively few industries. The majority of the robots used in industry worldwide are found in automobile manufacture. For example, Ford used robots with lasers for eyes and suction cups for hands to manufacture its 2013 Escape (Nishimoto 2012). While the usefulness of robots is still somewhat limited, this is changing rapidly, and combined with breakthroughs in 3D printing, it is certain that automated production will spread rapidly in coming years. Robots are becoming more sophisticated, while their costs are decreasing.

The spread of automation has provoked a heated debate over the impact of the new technology on workers, their skills, and their level of commitment to their work. Harry Braverman, in his influential book *Labor and Monopoly Capital* (1974), argued that automation

automation

Production processes monitored and controlled by machines with only minimal supervision from people.

was part of the overall "deskilling" of the industrial labor force. In both industrial settings and modern offices, new technologies have reduced the need for creative human input. One function of automation, Braverman argued, is to increase control over workers; all that is required in a highly automated factory is an unthinking, unreflective body capable of endlessly carrying out the same unskilled task.

Although Braverman was primarily writing about the kind of assembly-line work that occurs in automobile-manufacturing facilities, his arguments apply with equal force to the giant electronics plants in China that assemble our smartphones and computers, the factories throughout the world that make our clothing, and the fast-food workers who serve up our orders in a matter of minutes. The introduction of computerized technology in the workplace has resulted in a two-tiered workforce composed of a small group of highly skilled professionals with a high degree of flexibility and autonomy in their jobs and a larger group of clerical, service, and production workers who lack autonomy in their jobs.

GLOBAL PRODUCTION

For much of the twentieth century, the most important business organizations were large manufacturing firms that controlled both the production and sale of goods. Giant automobile companies such as Ford and General Motors typify this approach, employing tens of thousands of factory workers and making everything from components to the final cars, which are then sold in the manufacturers' showrooms. Such manufacture-dominated production processes are organized as large bureaucracies, often controlled by a handful of large firms.

During the past quarter-century, however, another form of production has become important—one that is controlled by giant retailers. In retailer-dominated production, firms such as Walmart sell thousands of different brands of goods; the brands, in turn, arrange to have their products made by independently owned factories around the globe. Almost no major U.S. companies today make their own apparel or footwear, for example; rather, they outsource to independently owned factories that do the work for them. These factories range from tiny sweatshops to giant plants owned by transnational corporations. Most so-called garment manufacturers actually employ no garment workers at all. Instead, they rely on thousands of factories around the world to make their apparel, which they then sell in department stores and other retail outlets. Clothing manufacturers do not own any of these factories and are therefore free to use them or not, depending on their needs. While this provides the manufacturers with the flexibility previously discussed, it creates great uncertainty both for the factories, which must compete with one another for orders, and for the workers, who may lose their jobs if their factory loses business. Critics of this system argue that such competition has resulted in a global "race to the bottom," in which retailers and manufacturers will go anyplace on earth

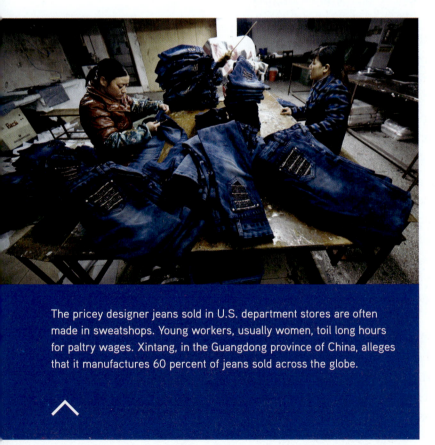

The pricey designer jeans sold in U.S. department stores are often made in sweatshops. Young workers, usually women, toil long hours for paltry wages. Xintang, in the Guangdong province of China, alleges that it manufactures 60 percent of jeans sold across the globe.

where they can pay the lowest wages possible (Bonacich and Appelbaum 2000). As we saw in Chapter 8, one result of globalization is that much of the clothing we buy today is made in sweatshops by young workers—most likely teenage girls—who get paid pennies for making clothing or pricey athletic shoes (Bonacich and Appelbaum 2000).

Trends in the Occupational Structure

The occupational structure in all industrialized countries has changed dramatically since the beginning of the twentieth century. In 1900, approximately three-quarters of the employed population was in manual work, either farming or blue-collar work such as manufacturing. White-collar professional and service jobs were much fewer in number. By 1960, however, more people worked in white-collar professional and service jobs than in manual labor. By 1993, the occupational system had nearly reversed its structure from 1900. Almost three-quarters of the employed population worked in white-collar professional and service jobs, while the rest worked in blue-collar and farming jobs. These trends have continued; in 2016, the manufacturing workforce was only 8 percent, while white-collar professional and service employment accounted for 80 percent (U.S. Bureau of Labor Statistics 2017c).

There are several reasons for the transformation of the occupational structure. One is the introduction of labor-saving machinery, culminating in the spread of information technology and computerization in industry in recent decades. Another is the rise of the manufacturing industry in other parts of the world, primarily Asia. The older industries in Western societies have experienced major job cutbacks because of their inability to compete with Asian producers, whose labor costs are lower. As we have seen, this global economic transformation has forced American companies to adopt new forms of production, which in turn has impelled employees to learn new skills and new occupations as manufacturing-related jobs move to other countries. A final important trend is the decline of full-time paid employment with the same employer over a long period of time. Not only has the transformation of the global economy affected the nature of day-to-day work but it has also changed the career patterns of many workers.

THE KNOWLEDGE ECONOMY

Taking these trends into account, some observers suggest that what is occurring today is a transition to a new type of society no longer based primarily on industrialism. We are entering, they claim, a phase of development beyond the industrial era altogether. A variety of terms have been coined to describe this new social order, such as the *postindustrial society*, the *information age*, and the *"new" economy*. The term that has come into most common use, however, is the **knowledge economy**.

A precise definition of the knowledge economy is difficult to formulate, but in general terms, it is an economy in which ideas, information, and forms of knowledge underpin innovation and economic growth. Knowledge-based industries include high technology, education and training, research and development, and the financial sector. Much of the workforce is involved not in the physical production or distribution of material goods but in their design, development, marketing, sale, and servicing. These employees can be termed *knowledge workers*. The knowledge economy is dominated by the constant flow of information and opinions and by the powerful potentials of science and technology.

The World Bank (2012a) recently developed the Knowledge Economy Index (KEI), which rates countries based on their overall preparedness to compete in the knowledge

knowledge economy

A society no longer based primarily on the production of material goods but based instead on the production of knowledge. Its emergence has been linked to the development of a broad base of consumers who are technologically literate and have made new advances in computing, entertainment, and telecommunications part of their lives.

economy. The factors contributing to this index include technological adoption and innovation, education, and the information and communications infrastructure. Scandinavian nations including Sweden, Finland, and Denmark topped the list. Sweden ranked number one because of its high levels of education and high levels of Internet penetration as well as a high number of patents for inventions.

THE CONTINGENT WORKFORCE

Another important employment trend of the past decade has been the replacement of full-time workers by part-time workers and contingency workers, or workers who are hired on a contract or "freelance" basis, often for a short-term task. Most temporary or contingent workers are hired for the least-skilled, lowest-paying jobs. More generally, part-time jobs do not provide the benefits associated with full-time work, such as medical insurance, paid vacation time, or retirement benefits. Because employers can save on the costs of wages and benefits, the use of part-time and contingent workers has become increasingly common. During economic recessions, in particular, cash-strapped employers may need to rely on this low-cost strategy.

Scholars have debated the psychological effects of part-time work on the workforce. Many temporary workers fulfill their assignments in a prompt and satisfactory manner, but others rebel against their tenuous positions by shirking their responsibilities or sabotaging their results. Some temporary workers have been observed trying to "look busy" or to work longer than necessary on rather simple tasks. However, some recent surveys of work indicate that part-time workers register higher levels of job satisfaction than those in full-time employment. This may be because most part-time workers are women, who find that part-time work is preferable to full-time employment when trying to juggle work and family demands. Yet men, too, may find that they are able to balance paid part-time work with other activities and enjoy a more varied life. Some people might choose to give full commitment to paid work from their youth to their middle years, then perhaps change to a second part-time career, which would open up new interests. However, workers who desire full-time employment but are able to secure only part-time work are often dissatisfied and anxious about their precarious financial status.

Unemployment

The experience of unemployment—being unable to find a job when one wants it—is a perennially important social problem. Yet some contemporary scholars argue that we should think about the relation between being "in work" and "out of work" in a completely different way from how we did in the recent past.

Unemployment rates fluctuated considerably over the course of the twentieth century. In Western countries, unemployment reached a peak in the Depression years of the early 1930s, when some 20 percent of the workforce was out of work in the United States. The economist John Maynard Keynes, who strongly influenced public policy in Europe and the United States during the post–World War II period, believed that unemployment results from consumers lacking sufficient resources to buy goods. Governments can intervene to increase the level of demand in an economy, leading to the creation of new jobs; the newly employed then have the income with which to buy more goods, thus creating yet more jobs for people who produce them. State management of economic life, most people came to believe, meant that high rates of unemployment belonged to the past. Commitment to full

unemployment rate

The proportion of the population sixteen and older that is actively seeking work but is unable to find employment.

employment became part of government policy in virtually all Western societies. Until the 1970s, these policies seemed successful, and economic growth was more or less continuous.

During the 1970s and 1980s, however, Keynesianism was largely abandoned. In the face of economic globalization, governments lost the ability to control economic life as they once had. One consequence was that unemployment rates shot up in many countries. Several factors explain the increase in unemployment levels in Western countries at that time. First is the rise of international competition in industries on which Western prosperity used to be founded. In 1947, 60 percent of steel production in the world was carried out in the United States. Today, the figure is only about 5 percent (World Steel Association 2017). Second is the economic recession that began in 2008. The U.S. unemployment rate was 5 percent in 2007; two years later, it had doubled to 10 percent. Although the economic recovery took many years, by 2015, unemployment was once again down to 5 percent. This recovery, however, was partly due to the fact that between 2006 and 2015, nearly four million workers gave up looking for work and so were no longer included among the unemployed (U.S. Bureau of Labor Statistics 2016). Third, the increasing use of microelectronics in industry has reduced the need for labor power. Finally, beginning in the 1970s, more women sought paid employment, meaning that more people were chasing a limited number of available jobs.

The Future of Work

Over the past twenty years, in all the industrialized countries except the United States, the average length of the working week has become shorter. Workers still undertake long stretches of overtime, but some governments are beginning to introduce new limits on permissible working hours. In France, for example, annual overtime is restricted to a maximum of 130 hours a year. In most countries, there is a general tendency toward shortening the average working career. More people would probably quit the labor force at age sixty or earlier if they could afford to do so.

If the amount of time devoted to paid employment continues to shrink, and the need to have a job becomes less central, the nature of working careers might be substantially reorganized. Job sharing or flexible working hours, which arose primarily as a result of the increasing numbers of working parents trying to balance the commitments of workplace and family, might become more common. Some work analysts have suggested that sabbaticals of the university type should be extended to workers in other spheres: People would be entitled to take a year off in order to study or pursue some form of self-improvement. Some might opt to work part time throughout their lives, rather than being forced to because of a lack of full-time employment opportunities.

The nature of the work most people do and the role of work in our lives, like so many other aspects of the societies in which we live, are undergoing major changes. As we will see in Chapter 16, the chief reasons are global economic competition, the widespread introduction of information technology and computerization, and the large-scale entry of women into the workforce. How will work change in the future? It appears very likely that people will take a more active look at their lives than in the past, moving in and out of paid work at different points. These are only positive options, however, when they are deliberately chosen. The reality for most is that regular paid work remains the key to day-to-day survival and that unemployment is experienced as a hardship rather than an opportunity.

CONCEPT CHECKS

1. Why does automation lead to worker alienation?

2. What are some of the changes that occurred in the occupational structure in the twentieth century? How can they be explained?

3. How did Keynes explain unemployment? What was his solution to high unemployment rates?

4. In your opinion, how will globalization change the nature of work?

CHAPTER 13

The Big Picture

Politics and Economic Life

Thinking Sociologically

1. Discuss the differences between the "pluralistic" and the "power elite" theories of democratic political processes. Which theory do you find most appropriate to describe U.S. politics in recent years?

2. Discuss some of the important ways that the nature of work will change for the contemporary worker as companies apply more automation and larger-scale production processes and as oligopolies become more pervasive. Explain each of these trends and how they affect workers, both now and in the future.

3. What was the main goal of the workers who protested their working conditions in April 2015? What kind of public policies might address their concerns? What does their battle reveal about the sociology of work and the sociology of politics?

Terms to Know

Concept Checks

government • politics •
economy • power • authority

state • nation-state • sovereignty • citizens •
nationalism • local nationalisms • civil rights •
political rights • franchise • social rights •
welfare state

1. Describe three main characteristics of the state.
2. What is a welfare state? Can the United States be classified as a welfare state? Why?

democracy • participatory democracy • direct
democracy • constitutional monarchs • liberal
democracies • communism • winner take all •
proportional representation • interest group •
suffrage • democratic elitism • pluralist theories
of modern democracy • power elite

1. Why is it problematic for contemporary states to have participatory democracy?
2. What are two explanations for low voter turnout in the United States?
3. Describe the role interest groups play in American politics.
4. Compare and contrast pluralist theories of modern democracy and the power elite model.

work • occupation • technology • informal
economy • division of labor • economic
interdependence • alienation • strike • unions •
collective bargaining

1. Why is it important for sociologists to study economic institutions?
2. Define and provide an example of an informal economy.
3. Using the concept of division of labor, describe the key differences in the nature of work in traditional versus modern societies.
4. What is a labor union? Why have unions in the United States suffered from a decline in membership since the 1980s?

capitalism • corporations • entrepreneur •
monopoly • oligopoly • family capitalism •
managerial capitalism • welfare capitalism •
institutional capitalism • interlocking
directorates • global capitalism • transnational
corporations

1. What are the main features of capitalism?
2. Compare and contrast the different types of capitalism.

automation • knowledge economy •
unemployment rate

1. Why does automation lead to worker alienation?
2. What are some of the changes that occurred in the occupational structure in the twentieth century? How can they be explained?
3. How did Keynes explain unemployment? What was his solution to high unemployment rates?
4. In your opinion, how will globalization change the nature of work?

14

The Sociology of the Body: Health, Illness, and Sexuality

THE BIG QUESTIONS

How do social contexts affect the human body?

Understand how social, cultural, and structural contexts shape attitudes toward "ideal" body forms and give rise to two body-related social problems in the United States: eating disorders and obesity.

How do sociologists understand health and illness?

Learn about functionalist and symbolic interactionist perspectives on physical and mental health and illness in contemporary society. Recognize the ways that disability challenges theoretical perspectives on health and illness. Understand the relationship between traditional medicine and complementary and alternative medicine (CAM).

How do social factors affect health and illness?

Recognize that health and illness are shaped by cultural, social, and economic factors. Learn about race, class, gender, and geographic differences in the distribution of disease.

How do social contexts shape sexual behavior?

Understand the diversity of sexual orientation today. Learn about the debate over the importance of biological versus social and cultural influences on human sexual behavior. Explore cultural differences in sexual behavior and patterns of sexual behavior today.

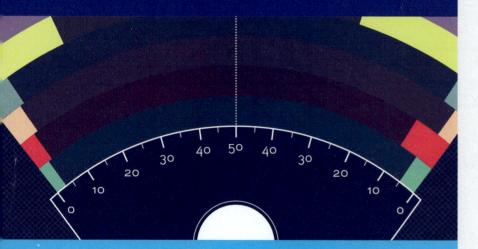

Obesity Rates

p. 433

Friends and family of Dr. Shalon Irving were shocked and saddened when the thirty-six-year-old collapsed and died from pregnancy-related complications just four months after giving birth to her daughter, Soleil. Deaths due to pregnancy-related complications are relatively rare in the United States, with 26 women dying per every 100,000 births each year. Yet that overall statistic belies a stark racial difference, where Black mothers are three to four times as likely as their white counterparts to die (Martin and Montagne 2017). Put differently, Black expectant and new mothers in the United States have death rates similar to women in much poorer nations, including Mexico and Uzbekistan (World Health Organization 2016a).

Shalon's death was particularly shocking because she was a well-respected scientist at the Centers for Disease Control and Prevention (CDC) who had dedicated her life to studying, understanding, and, ideally, eradicating racial disparities in health. After earning her PhD in sociology from Purdue University, Shalon continued on to the nation's top public health school, Johns Hopkins, where she earned her master's in public health. She then joined the ranks at the CDC, conducting important research on race disparities in health, intimate partner violence, and elder abuse. Dr. Irving lived the comfortable life of a highly educated professional in Atlanta, one of the nation's premier cities for health care. She had top-of-the-line health insurance, a lovely home, and a strong and close-knit community of friends and family to help her through her pregnancy. Yet despite her many accomplishments and rich support networks, Shalon carried with her the imprints of her lifelong experiences with racism, stress, high blood pressure, and early-life obesity, which placed her at high risk during pregnancy. These risks were amplified by the fact that some of her doctors failed to recognize that pregnancy-related health concerns do not end when a woman gives birth; rather, postpartum care is also essential, especially for those who had high-risk pregnancies (Martin and Montagne 2017).

The untimely deaths of Dr. Irving and others like her are an all-too-common outcome for African American women, who face multiple stressors throughout their lives that may cause "weathering." According to epidemiologist Arline Geronimus, weathering refers to the wearing down of one's body and one's health due to long-standing and far-reaching stressors, rendering its victims vulnerable to chronic diseases like diabetes and high blood pressure (Villarosa 2018). Weathering can affect men and women alike, although its physical toll is particularly acute for women during and shortly after pregnancy, a physiologically complex period in a woman's life (Martin and Montagne 2017).

In general, Americans are living longer lives than ever before. The average American can expect to live until age 79, with women surviving 81 years on average and men surviving 76. Yet, as we saw in the tragic case of Dr. Irving, simple statistical snapshots may conceal stark differences on the basis of race, socioeconomic status, and gender. For instance, although life expectancy has been rising steadily throughout the late twentieth and early twenty-first centuries, working-class whites have been dying younger and younger. Experts describe these premature deaths as "deaths of despair," a product of financial stress; obesity; and reliance on alcohol, smoking, unhealthy foods, or drugs like opioids to cope with financial difficulties (Case and Deaton 2017).

Another important reason for the stark socioeconomic divide in health is that not all Americans have access to health insurance. The proportion of U.S. adults who are uninsured decreased dramatically between 2013 and 2016, following the implementation of the Affordable Care Act (ACA) under President Barack Obama. While fully 20 percent of working-age Americans lacked health insurance in 2013, by 2016, that figure had dropped to just over 10 percent, although the proportions uninsured are considerably higher among Blacks and Latinos (Kaiser Family Foundation 2017). The future of the ACA is uncertain, but even with the dramatic expansion of health insurance that occurred under President Obama, the United States still lags behind other wealthy nations, including Canada, most European nations, Japan, Israel, New Zealand, Taiwan, and a growing number of nations in Latin America. In those countries, each and every individual has some form of health insurance through **universal health coverage**, which ensures that all people can obtain the health services they need without incurring financial hardship (World Health Organization 2010).

Health insurance is about more than just providing access to health care for people after they get sick. Many health insurance plans promote preventive health care and help people maintain healthy lifestyles and detect health problems early on rather than seeking medical care only after their conditions have advanced to a dangerous stage. For example, people may have access to covered services like depression screening, substance use disorder screening, blood pressure

universal health coverage

Public health care programs motivated by the goal of providing affordable health services to all members of a population.

screening, obesity counseling and screening, assistance with quitting smoking, vaccinations, and counseling for domestic abuse victims (Kaiser Family Foundation 2013). These services are particularly important for low-income and Black Americans, who are at a greater risk for nearly every major health condition relative to their wealthier white counterparts.

One of the most consistent patterns documented by sociologists of health and illness is the **social class gradient in health**. This gradient refers to the fact that socioeconomic resources are strongly linked to health, where those with higher levels of education, income, and assets are less likely than their disadvantaged counterparts to suffer from heart disease, diabetes, high blood pressure, early onset of dementia, physical disability, sleep problems, substance use problems, mental illness, and premature death (CDC 2011a). These patterns are so pronounced that one of the four main goals of Healthy People 2020, the federal government's health agenda, is to "achieve health equity, eliminate disparities, and improve the health of all groups," especially disparities on the basis of socioeconomic resources and race, as Irving's death revealed (U.S. Department of Health and Human Services 2010).

Judging by the chapter title, you might have expected to read about biology, or about the physical ways that our bodies function. You might have been surprised to read about something as seemingly far removed from our everyday lives as federal health care policy, or something as pervasive as stress and racism. Yet public policy and macrosocial factors are powerful influences on our health. The field known as **sociology of the body** investigates how and why our bodies are affected by our social experiences and the norms and values of the groups to which we belong. Using this framework, we begin our chapter by analyzing why obesity and an equally problematic phenomenon, eating disorders, have become so common in the Western world. We then describe the ways that sociologists theorize about health and medicine; discuss social dimensions of health and illness, with an emphasis on the ways that social class, race, and gender affect our health; and provide an overview of health issues that affect the lives of people in low-income nations. We conclude by examining social and cultural influences on our sexuality; as we will see, sexual orientation and behaviors, like health, are a product of biological, cultural, and social forces.

social class gradient in health

The strong inverse association between socioeconomic resources and risk of illness or death.

sociology of the body

Field that focuses on how our bodies are affected by our social experiences. Health and illness, for instance, are shaped by social, cultural, and economic influences.

How Do Social Contexts Affect the Human Body?

Understand how social, cultural, and structural contexts shape attitudes toward "ideal" body forms and give rise to two body-related social problems in the United States: eating disorders and obesity.

Social contexts affect our bodies in myriad ways. The types of jobs we hold, the neighborhoods in which we live, how much money we earn, the cultural practices we partake in, and our personal relationships all shape how long we live, the types of illnesses we suffer from, and even the shapes and sizes of our bodies. Later in this chapter, we will delve more fully into how key features of our social lives, including race and social class, affect our physical and mental health. In this section, we will focus on one specific aspect of our bodies to show the power of social, economic, and cultural contexts: our body weight. Whether we are slender or heavy is not just a consequence of our personal choices (such as what foods we eat) or our genes; rather, body weight is shaped by powerful social structures as well as cultural forces.

Let's take the case of eating disorders, such as anorexia nervosa or bulimia, and obesity, or excessive body weight. Both are important social problems in wealthy nations. Although both are conditions of the body, their causes reflect social factors more than physical or

biological factors. If both conditions reflected biology alone, we would expect that rates would be fairly constant across history—because human physiology has changed little throughout the millennia. However, both are very recent social problems. Both conditions are also highly stratified by social factors, such as gender, social class, race, and ethnicity. Women are far more likely than men to have an eating disorder, while economically disadvantaged persons are far more likely than their wealthier counterparts to struggle with obesity today.

Both conditions are also shaped by the cultural context. Fashion magazines regularly show images of models who are severely underweight yet uphold these women as paragons of beauty. The average fashion model today is 23 percent thinner than the average American woman, yet twenty-five years ago, that number was 8 percent (Derenne and Beresin 2006). At the same time, our culture also promotes excessive eating. A Big Mac is less expensive than a healthy salad in most parts of the country, perpetuating the social class gradient in obesity rates. By contrast, both eating disorders and obesity are virtually unknown in impoverished societies where food is scarce and cherished.

Both obesity and eating disorders also illustrate the ways that a "personal trouble" (for example, self-starvation or obesity-related complications, such as diabetes) reflects "public issues" (for example, a culture that promotes an unrealistic "thin ideal" for young women) or the ways that poverty makes it difficult for individuals to buy costly healthful foods or to reach public parks and other spaces for regular exercise.

Eating Disorders

Anorexia is related to the idea of dieting, and it reflects changing views of physical attractiveness in modern society. In most premodern societies, the ideal female shape was a fleshy one. Thinness was not desirable, partly because it was associated with hunger and poverty. The notion of slimness as the desirable feminine shape originated among some middle-class groups in the late nineteenth century, but it became generalized as an ideal for most women only recently. A historical examination of the physiques of Miss America winners between 1922 and 1999, for example, shows that for much of the twentieth century, pageant winners had a body weight that would be classified as "normal," yet in recent years, the majority of winners would be classified as "underweight" using medical guidelines (Rubinstein and Caballero 2000).

Anorexia was identified as a disorder in France in 1874, but it remained obscure until the past thirty or forty years (Brown and Jasper 1993). Since then, it has become increasingly common among young women. So has bulimia—bingeing on food, followed by self-induced vomiting. Anorexia and bulimia often occur in the same individual. An estimated 0.9 percent of females have suffered from anorexia at some time during their lives, and an estimated 0.5 percent of women have suffered from bulimia (Hudson et al. 2007). Nearly all (95 percent) of those who have eating disorders are between the ages of twelve and twenty-six (National Association of Anorexia Nervosa and Associated Disorders 2010).

Eating disorders have long been considered a health problem exclusive to young white women, yet recent data suggest that women of color and gay men also are vulnerable. Overall, women account for more than 90 percent of all persons with eating disorders, so it is difficult for researchers to estimate rates among men. But recent evidence suggests that boys and young men, especially athletes and gay and bisexual men, increasingly struggle to maintain a lean and muscular physique (Feldman and Meyer 2007; Field et al. 2014). New evidence also challenges the assumption that women of color are immune to eating disorders; national data show that white women are especially susceptible to anorexia nervosa, whereas Black and Latino women show higher rates of bulimia (Marques et al. 2011).

Once a young person starts to diet and exercise compulsively, he or she can become locked into a pattern of refusing food or vomiting up what he or she has eaten. As the body loses muscle mass, it loses heart muscle, so the heart gets smaller and weaker, which ultimately leads to heart failure. About half of all anorexics also have low white blood cell counts, and about one-third are anemic. Both conditions can lower the immune system's resistance to disease, leaving an anorexic vulnerable to infections. Anorexia has the highest mortality rate of any psychological disorder (Arcelus et al. 2011).

Why are rates of eating disorders higher among women (especially young women) and gay and bisexual men? Sociologists note that social norms stress the importance of physical attractiveness more for women than for men and that desirable body images of men differ from those of women. However, men are also less likely to seek treatment for eating disorders because they are considered to be female disorders; as a result, their illnesses are less likely to be reported and detected (National Association of Anorexia Nervosa and Associated Disorders 2010). Emerging research on gay and bisexual men highlights the importance of a lean, muscular physique for these men's self-concept but also emphasizes that experiences of discrimination, social rejection, or fear of rejection may make some young men especially vulnerable to negative body image and eating disorders (McClain and Peebles 2016).

This unhealthy cultural emphasis on a lean physique—and the resulting eating disorders—extends beyond the United States and Europe. As Western images of feminine beauty have spread to the rest of the world, so, too, have associated illnesses. Eating problems have also surfaced among young, primarily affluent women in Hong Kong and Singapore as well as in urban areas in Taiwan, China, the Philippines, India, and Pakistan (Pike and Dunn 2015). One famous study showed that in Fiji, a nation where larger bodies were long considered the cultural ideal, rates of eating disorders among young women increased markedly after American television shows like *Beverly Hills 90210* started to air there (Becker 2004).

The rise of eating disorders in Western societies coincides with the globalization of food production. Since the 1960s, supermarket shelves have been abundant with foods from all parts of the world. Most foods are available all the time, not just when they are in season locally. When all foods are available all the time, we must decide what to eat. First, we have to choose what to eat in relation to the new medical information with which science bombards us—for instance, that cholesterol levels contribute to heart disease. Second, we worry about calorie content. The fact that we have much more control over our own bodies than before presents us with positive possibilities as well as new anxieties and problems.

The Obesity Epidemic

obesity

Excessive body weight indicated by a body mass index (BMI) over 30.

Eating disorders are a major social problem in the United States and, increasingly, worldwide. Yet a very different weight-related health issue, obesity, is considered the top public health problem facing Americans today. **Obesity** is defined as a body mass index (BMI) of 30 or greater (CDC 2008a). Over the past two decades, obesity rates among adults in the United States have risen dramatically. In 2000, 31 percent of adults were obese; by 2016, that proportion had climbed steadily to nearly 40 percent. An even more troubling trend is that nearly 19 percent of children and adolescents are obese (Hales et al. 2017).

Obesity increases an individual's risk for a wide range of health problems, including heart disease, diabetes, sleep apnea, osteoarthritis, and some forms of cancer (Haslam and James 2005; Wang et al. 2011). Excessive body weight may also take a psychological toll. Overweight and obese Americans are more likely than their thinner peers to experience depression; strained family relationships; poorer-quality sex and dating lives; employment discrimination; discrimination by health care providers; and daily experiences of teasing, insults, and shame (Carr and Friedman 2005, 2006; Carr et al. 2007, 2013). Negative attitudes toward overweight and obese persons develop as early as elementary school (Puhl and Latner 2007). Sociologists are fascinated with the persistence of negative attitudes toward overweight and obese persons, especially because these individuals make up the statistical majority of all Americans.

The reasons behind the obesity crisis are widely debated. Some argue that the apparent increase in the overweight and obese population is a statistical artifact. The proportion of the U.S. population who are middle-aged has increased rapidly during the past two decades, with the aging of the large baby boom cohort. Middle-aged persons, due to slowing metabolism, are at greater risk of excessive body weight. Others attribute the pattern—especially the rise in childhood obesity—to shifts in the ethnic makeup of the U.S. population. The proportion of children today who are Black or Hispanic is

Obesity Rates

Obesity rates worldwide have more than tripled since 1975. In 2016, more than 650 million people—13 percent of all adults worldwide—were obese. Once considered a "first world" problem, rates of obesity have been rising in low- and middle-income countries.

Proportion of adults who are obese*

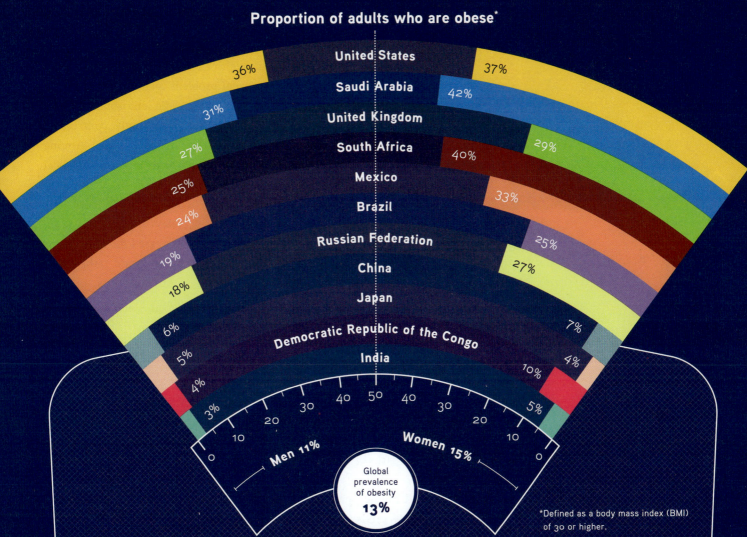

Country	Men	Women
United States	36%	37%
Saudi Arabia	31%	42%
United Kingdom	27%	29%
South Africa	25%	40%
Mexico	24%	33%
Brazil	19%	25%
Russian Federation	18%	27%
China	6%	7%
Japan	5%	4%
Democratic Republic of the Congo	4%	10%
India	3%	5%

Men 11% Women 15%

Global prevalence of obesity **13%**

*Defined as a body mass index (BMI) of 30 or higher.

Obesity in the United States

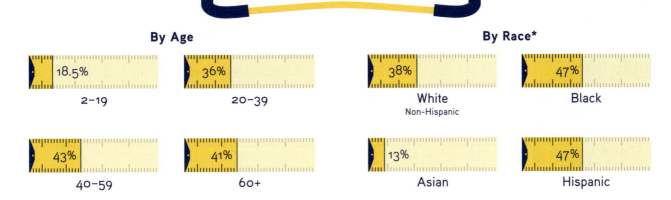

By Age

18.5% — 2–19	36% — 20–39
43% — 40–59	41% — 60+

By Race*

38% — White Non-Hispanic	47% — Black
13% — Asian	47% — Hispanic

*Among adults age 20 and over.

Sources: World Health Organization 2017a; Hales et al. 2017.

Americans live in what sociologists refer to as an "obesogenic environment," meaning an environment that contributes to weight gain.

food deserts

Geographic areas in which residents do not have easy access to high-quality affordable food. These regions are concentrated in rural areas and poor urban neighborhoods.

CONCEPT CHECKS

1. Why is anorexia more likely to strike young women than heterosexual young men?

2. What explanations are offered for the recent increase in obesity rates?

3. In what ways is the United States an "obesogenic environment"?

higher than in earlier decades, and these two ethnic groups are at a much greater risk for becoming overweight than their white peers. Still others argue that the measures used to count and classify obese persons have shifted, thus leading to an excessively high count. Finally, some social observers believe that public concern over obesity is blown out of proportion and reflects more of a "moral panic" than a "public health crisis" (Campos et al. 2006; Saguy 2012).

Most public health experts believe, however, that obesity is a very real problem caused by what Kelly Brownell calls the "obesogenic environment"—or a social environment that unwittingly contributes to weight gain (Brownell and Horgen 2004). Among adults, sedentary desk jobs have replaced physical jobs, such as farming. Children are more likely to spend their after-school hours sitting in front of a computer, smartphone, or television than playing tag or riding their bikes around the neighborhood. Parents are pressed for time, given their hectic work and family schedules, and turn to unhealthy fast food rather than home-cooked meals. Restaurants, eager to lure bargain-seeking patrons, provide enormous serving sizes at low prices. The social forces that promote high fat and sugar consumption and that restrict the opportunity to exercise are particularly acute for poor persons and ethnic minorities. Small grocery stores in poor neighborhoods rarely sell fresh or low-cost produce. Large grocery stores are scarce in poor inner-city neighborhoods and rural areas as well as in predominantly African American neighborhoods (Morland et al. 2002). Given the scarcity of high-quality healthy foods in poor neighborhoods, scholars have dubbed these areas **food deserts** (Walker, Keane, and Burke 2010). Additionally, high crime rates and high levels of traffic in inner-city neighborhoods make exercise in public parks or jogging on city streets potentially dangerous (Brownell and Horgen 2004).

Policymakers and public health professionals have proposed a broad range of solutions to the obesity crisis. Some have (unsuccessfully) proposed practices that place the burden directly on the individual. For example, some schools have considered having a "weight report card," where children and parents would be told the child's BMI, in an effort to trigger healthy behaviors at home. Yet most experts endorse solutions that attack the problem at a large-scale level, such as making healthy low-cost produce more widely available; providing safe public places to exercise, free or low-cost fitness classes, and classes in health and nutrition to low-income children and their families; and requiring restaurants and food manufacturers to clearly note the fat and calorie content of their products. Only in attacking the "public issue" of the obesogenic environment will the "private trouble" of excessive weight be resolved (Brownell and Horgen 2004).

How Do Sociologists Understand Health and Illness?

Sociologists of health and illness also are concerned with understanding the experience of illness—how individuals experience being sick, chronically ill, or disabled and how these experiences are shaped by one's social interactions with others. If you have ever been ill, even for a short period, you know that patterns of daily life are temporarily modified and your interactions with others change. This is because the normal functioning of the body is a vital, but often taken-for-granted, part of our daily lives. For most people, our sense of self is predicated on the expectation that our bodies will facilitate, not impede, our social interactions. One important exception is the experience of persons with a physical, sensory, or cognitive impairment. People with blindness, hearing impairments, or physical disabilities that may limit their movement adapt to these conditions and even base their identities and senses of self on their capacity to adapt and thrive (Darling 2003).

Illness has both personal and public dimensions. When we fall ill, others are affected as well. Our friends, families, and coworkers may extend sympathy, care, support, and assistance with practical tasks. They may struggle to understand our illness and its cause or to adjust the patterns of their own lives to accommodate it. Others' reactions to our illness, in turn, shape our own interpretations and can pose challenges to our senses of self. For instance, a longtime smoker who develops lung disease may be made to feel guilty by family members.

Two sociological perspectives on the experience of illness have been particularly influential. The first, associated with the functionalist school, proposes that "being sick" is a social role, just as "worker" or "mother" is a social role. As such, unhealthy persons are expected to comply with a widely agreed-upon set of behavioral expectations. The second view, favored by symbolic interactionists, explores how the meanings of illness are socially constructed and how these meanings influence people's behavior.

The Sick Role

The functionalist thinker Talcott Parsons (1951) developed the notion of the **sick role** to describe patterns of behavior that a sick person adopts to minimize the disruptive impact of illness or injury. Functionalist thought holds that society usually operates in a smooth and consensual manner. Illness is, therefore, seen as a dysfunction that can disrupt the flow of this normal state. An individual who has fallen ill, for example, might be unable to perform standard responsibilities or be less reliable and efficient than usual. Because sick people cannot carry out their normal roles, the lives of people around them are disrupted: Assignments at work go unfinished and cause stress for coworkers, responsibilities at home are not fulfilled, and so forth.

According to Parsons, people learn the sick role through socialization and enact it—with the cooperation of others—when they fall ill or suffer an injury. Sick persons face societal expectations for how to behave; at the same time, other members of society

Learn about functionalist and symbolic interactionist perspectives on physical and mental health and illness in contemporary society. Recognize the ways that disability challenges theoretical perspectives on health and illness. Understand the relationship between traditional medicine and complementary and alternative medicine (CAM).

sick role

A term Talcott Parsons used to describe the patterns of behavior that a sick person adopts to minimize the disruptive impact of his or her illness on others.

abide by a generally agreed-upon set of expectations for how they will treat the sick individual. The sick role is distinguished by three sets of normative expectations:

1. The sick person is not held personally responsible for his or her poor health.

2. The sick person is entitled to certain rights and privileges, including a release from normal responsibilities.

3. The sick person is expected to take sensible steps to regain his or her health, such as consulting a medical expert and agreeing to become a patient.

EVALUATION

Although the sick-role model reveals how the ill person is an integral part of a larger social context, a number of criticisms can be levied against it. Some argue that the sick-role formula does not adequately capture the "lived experience" of illness. Others point out that it cannot be applied across all contexts, cultures, and historical periods. For example, it does not account for instances in which doctors and patients disagree about a diagnosis or have opposing interests. It also fails to explain illnesses that do not necessarily lead to a suspension of normal activity, such as alcoholism, certain disabilities, and some chronic diseases. It also presumes a short-term condition and that people will return to normal functioning when the illness passes. This scenario does not apply to persons who have permanent or long-lasting disabilities yet adapt and thrive in their environments through the use, for example, of hearing aids or wheelchairs (Thomas 2007).

Furthermore, taking on the sick role is not always a straightforward process. Some individuals who suffer for years from chronic pain or from misdiagnosed symptoms are denied the sick role until they get a clear diagnosis. Other sick people, such as young women with autoimmune diseases, often appear physically healthy despite constant physical pain and exhaustion; because of their "healthy" outward appearance, they may not be readily granted sick-role status. In other cases, social factors like race, class, and gender can affect whether and how readily the sick role is granted. Single parents or people caring for ailing relatives may fail to acknowledge their own symptoms for fear that shirking their social roles will hurt their loved ones. The sick role cannot be divorced from the social, cultural, and economic influences that surround it.

The realities of life and illness are more complex than the sick role suggests. The leading causes of death today are heart disease and cancer, two diseases that are associated with unhealthy behaviors such as smoking, a high-fat diet, and a sedentary lifestyle. Given the emphasis on taking control over one's health and lifestyle, individuals bear ever-greater responsibility for their own well-being. This contradicts the first premise of the sick role—that the individual is not to blame for his or her illness. Moreover, in modern societies, the shift away from acute infectious disease toward chronic illness has made the sick role less applicable because there is no single formula for chronically ill or disabled people to follow.

Illness as "Lived Experience"

Symbolic interactionists study the ways people interpret the social world and the meanings they ascribe to it. Many sociologists have applied this approach to health and illness and view this perspective as a partial corrective to the limitations of functionalist approaches. Symbolic interactionists are not concerned with identifying risk factors for specific illnesses or conditions; rather, they address questions about the personal experience of illness: How do people react and adjust to news about a serious illness? How does illness shape individuals' daily lives? How does a chronic illness affect an individual's self-identity?

One theme that sociologists address is how chronically ill individuals cope with the practical and emotional implications of their illness. Certain illnesses require regular treatments that can affect daily routines. Undergoing dialysis or insulin injections or taking large numbers of pills requires individuals to adjust their schedules. Other illnesses have unpredictable effects, such as sudden loss of bowel or bladder control or violent nausea. People suffering from such conditions often develop strategies for managing their illness in daily life. These include practical considerations—such as noting the location of the restrooms when in an unfamiliar place—as well as skills for managing interpersonal relations, both intimate and commonplace. Although symptoms can be embarrassing and disruptive, people develop coping strategies to live as normally as possible (Kelly 1992).

At the same time, the experience of illness can pose challenges for individuals to manage their illnesses within the overall contexts of their lives (Jobling 1988; Williams 1993). Corbin and Strauss (1985) identified three types of "work" incorporated into the everyday strategies of the chronically ill. *Illness work* refers to activities involved in managing the condition, such as treating pain, doing diagnostic tests, or undergoing physical therapy. *Everyday work* pertains to the management of daily life—maintaining relationships with others, running household affairs, and pursuing professional or personal interests. *Biographical work* involves the process of incorporating the illness into one's life, making sense of it, and developing ways of explaining it to others. Such a process can help people with mental and physical illness restore meaning and order to their lives.

This is especially the case for those who have long-lasting or permanent physical disabilities. A flourishing body of research shows that persons with deafness and blindness, for instance, view these experiences as critical to their identity and belong to cultural communities with their own languages and practices. Rather than viewing their bodies as a deficit or "disorder," persons with disabilities view their bodies as simply another source of personal and cultural difference, just as race, ethnicity, or gender is a source of difference. For instance, many persons with deafness do not want to be "fixed" with hearing aids or cochlear implants and instead embrace their own culture and means of communication (Tucker 1998).

The process of adaptation may be particularly difficult for those who suffer from a stigmatized health condition, such as extreme obesity, alcoholism, schizophrenia, or HIV/AIDS. Sociologist Erving Goffman (1963) developed the concept of **stigma**, which refers to any personal characteristic that is devalued in a particular social context. Stigmatized individuals and groups are often treated with suspicion, hostility, or discrimination. Stigmas are rarely based on valid understandings or scientific data. They spring from stereotypes or perceptions that may be false or only partially correct. Furthermore, the nature of a stigma varies widely across sociocultural context: The extent to which a trait is devalued depends on the values and beliefs of those who do the stigmatizing. For instance, in the United States, obese persons are much more likely to be stigmatized by white upper-middle-class persons than they are to be stigmatized by African Americans or working-class whites (Carr and Friedman 2005). By contrast, other health conditions,

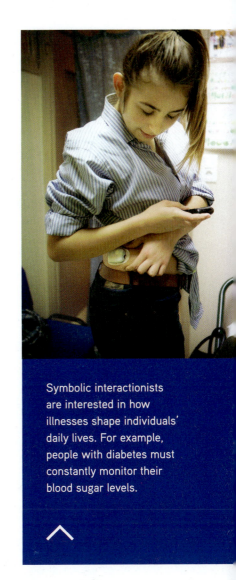

Symbolic interactionists are interested in how illnesses shape individuals' daily lives. For example, people with diabetes must constantly monitor their blood sugar levels.

stigma

Any physical or social characteristic that is labeled by society as undesirable.

including major mental illness and HIV/AIDS (as we will read about later), are much more widely stigmatized. One recent study of sixteen countries found that even in the most liberal, tolerant countries, the majority of the public held stigmatizing attitudes and a willingness to exclude people with schizophrenia from close, personal relationships and positions of authority, seeing them as unpredictable and potentially dangerous (Pescosolido et al. 2013).

Changing Conceptions of Health and Illness

A key theme in the sociological study of health is that cultures and societies differ in what they consider healthy and normal. All cultures have known concepts of physical health and illness, but most of what we now recognize as medicine is a consequence of developments in Western society over the past three centuries. In premodern cultures, the family was the main institution for coping with sickness or affliction. There have always been individuals who specialized as healers, using a mixture of physical and spiritual remedies, and many such traditional systems survive today. For instance, traditional Chinese medicine aims to restore harmony among aspects of the personality and bodily systems, involving the use of herbs and acupuncture for treatment.

Modern medicine sees the origins and treatment of disease as physical and explicable in scientific terms. The application of science to medical diagnosis and cure underlies the development of modern health care systems. Sociologists have argued that in contemporary Western societies, conditions that were previously viewed as having their roots in social, cultural, or religious causes are now "medicalized." Medicalization, according to sociologist Peter Conrad (2007), is the process by which some variations in human traits, behaviors, or conditions become defined as medical conditions that require treatment. For example, sociologist Allan Horwitz has argued that in the United States, the emotion of "sadness"—a normal response to stressors like loss, failure, and disappointment—has now been transformed into the medical disorder of "depression," which is believed to have its roots in biological causes, such as brain chemistry or genetics (Horwitz and Wakefield 2007). As such, depressed persons today are much more likely to be treated with medications, such as antidepressants, than "talk therapy," in which a therapist would focus on the social or emotional roots of the sad feelings.

This process through which human variation is medicalized and deviations from the norm are labeled medical disorders to be treated has been critiqued and contested by disability rights activists (Beaudry 2016). In response to the medicalization of disability, scholars and activists have called for a more critical perspective that views disability as a social phenomenon caused by social oppression and prejudices rather than by individual "impairments." The daily challenges that persons with disability often face are a function not of their eyes or ears or limbs but of social exclusion and society's failure to provide physical and social environments that foster inclusiveness (Oliver 2009).

In addition to medicalization, another important feature of modern medicine is the acceptance of the hospital as the setting within which to treat serious illnesses and the development of the medical profession as a body with codes of ethics and significant social power. The scientific view of disease is linked to the requirement that medical training be systematic and long term; self-taught healers are typically excluded. Although professional medical practice is not limited to hospitals, the hospital provides an environment in which doctors can treat and study large numbers of patients in circumstances permitting the concentration of medical technology.

COMPLEMENTARY AND ALTERNATIVE MEDICINE

Alternative therapies, such as herbal remedies, acupuncture, and chiropractic treatments, are being explored by a record high number of adults in the United States today and are slowly gaining acceptance by the mainstream medical community. Medical sociologists refer to such unorthodox medical practices as **complementary and alternative medicine (CAM)**. CAM encompasses a diverse set of approaches and therapies for treating illness and promoting well-being that generally fall outside standard medical practices. Alternative medicine is meant to be used *in place of* standard medical procedures, whereas complementary therapies are meant to be used *in conjunction with* medical procedures to increase their efficacy or reduce side effects (Saks 1992).

Industrialized countries have some of the best-developed, best-resourced medical facilities in the world. Why, then, are a growing number of people exploring treatments that have not yet proven effective in controlled clinical trials, such as aromatherapy and hypnotherapy? A 2012 survey conducted by the CDC found that 33 percent of American adults had used some form of CAM in the past year (Clarke et al. 2015). CAM use is more frequent among women and individuals with higher levels of educational attainment (Figure 14.1). Furthermore, whites are more likely to use CAM than their Black and Asian counterparts.

There are many reasons for seeking the services of an alternative medicine practitioner or pursuing CAM regimens on one's own. Some people perceive orthodox medicine to be deficient or ineffective in relieving chronic pain or symptoms of stress and anxiety. Others are dissatisfied with features of modern health care systems, such as long waits, referrals through chains of specialists, and financial restrictions. Connected to this are concerns about the harmful side effects of medication and the intrusiveness of surgery, both staples of modern Western medicine. The asymmetrical power relationship between doctors and patients also drives people to seek alternative medicine. They feel that the role of the passive patient does not grant them enough input into their own treatment. Finally, some individuals profess religious or philosophical objections to orthodox medicine, which treats the mind and body separately. They believe that orthodox medicine often overlooks the spiritual and psychological dimensions of health and illness. All these concerns are critiques of the **biomedical model of health**, which defines disease in objective terms and believes that scientifically based medical treatment can restore the body to health (Beyerstein 1999).

The growth of alternative medicine is a fascinating reflection of the transformations occurring within modern societies. We are living in an age where much more information is available. The proliferation of health-related websites such as WebMD and MedicineNet

complementary and alternative medicine (CAM)

A diverse set of approaches and therapies for treating illness and promoting well-being that generally fall outside of standard medical practices.

---

biomedical model of health

The set of principles underpinning Western medical systems and practices that defines diseases objectively and holds that the healthy body can be restored through scientifically based medical treatment.

FIGURE 14.1

Use of Complementary Medicine

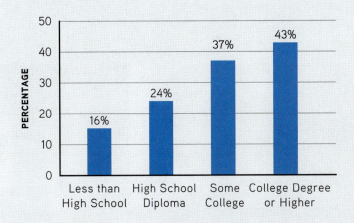

Source: Clarke et al. 2015.

Can Wearable Tech Keep You Healthy?

Until fairly recently, when a person felt sick, he or she would call a doctor to make an appointment. During this visit, the doctor would likely diagnose the patient's symptoms and perhaps prescribe medication to help treat the patient. Although many Americans, especially those with health insurance and access to providers, still see a health care professional on a regular basis, more and more Americans are trying to diagnose themselves, often with the assistance of health-related smartphone apps and fitness trackers. For the past decade or two, people have been visiting websites like WebMD to determine whether their headache is due to a head cold or is a sign of something more dire. More recently, smartphone owners have relied on apps and fitness trackers to do everything from take their pulses to chart their ovulation cycles to identify the best medication for depression.

Health-related apps and fitness trackers range from the very simple to the very complex. Basic fitness trackers keep users informed of steps taken and calories burned, while more expensive trackers keep tabs on users' heart rates and sleep patterns, even detailing how much time a user spends in light sleep versus deep sleep. Accompanying apps allow users to dig even deeper: Jawbone's UP Coffee app monitors caffeine consumption and then syncs with the user's fitness tracker to assess how caffeine is affecting sleep patterns.

Smartphones are particularly helpful in guiding us to make healthy food choices. For instance, with Fooducate, users swipe the bar codes of food items they're considering buying at the grocery store and are then given detailed information on the product's ingredients and nutritional value (Summers 2013).

Yet activity trackers and smartphones are increasingly being used for more serious health-related issues, like monitoring blood pressure, heart rate, and ovulation cycles and even assessing hearing and vision. For example, ECG Check allows patients to analyze their own heart rhythms, while apps like Glooko and Glucose Buddy help diabetics monitor their blood sugar levels. Fertility Friend helps women who are hoping to conceive by monitoring their menstrual cycles (Edney 2013). PsychDrugs helps people determine which antidepressant or antianxiety medication will best treat their symptoms.

It's not just patients who use apps to enhance their health; health care providers also rely on apps to help them deliver care. Apps like Epocrates help doctors review drug-prescribing and safety information, research potentially harmful drug interactions, and perform calculations like BMI and glomerular filtration rate, an indication of how well one's kidneys are functioning (Glenn 2013).

Many health care providers and patients are enthusiastic about the role of technology in helping to enhance medical care. Doctors believe that symptom-monitoring apps and fitness trackers encourage patients to be proactive and knowledgeable about their own health (Edney 2013). However, others counter that even the best app or activity tracker is not a substitute for a regular checkup. What do you think? Are health and wellness apps and fitness trackers a cost-effective and efficient way for people to look after their own health, or do they keep people from receiving potentially valuable professional care?

Fitness trackers and smartphone apps allow users to do everything from monitor their sleep patterns to track their caloric intake.

provides instant access to information on health symptoms and treatments, while some fitness trackers allow users to monitor their activity levels as well as their heart rates and sleep patterns (see Digital Life box). Thus, individuals are increasingly becoming health consumers, adopting an active stance toward their own health and well-being. Not only are we choosing the types of practitioners to consult but we are also demanding more involvement in our own care and treatment.

Physicians increasingly believe that such unorthodox therapies may be an important complement to (although not a substitute for) traditional Western medicine, provided they are held up to the same level of scientific scrutiny and rigorous scientific evaluation. Debates about CAM also shed light on the changing nature of health and illness over the past two centuries. Many conditions and illnesses for which individuals seek alternative medical treatment seem to be products of the modern age itself. Rates of insomnia, anxiety, stress, depression, fatigue, and chronic pain (caused by arthritis, cancer, and other diseases) are increasing in industrialized societies (Kessler and Üstün 2008). Although these conditions have long existed, they are causing greater distress and disruption to people's health than ever before. Ironically, these consequences of modernity are ones that orthodox medicine has difficulty addressing. Alternative medicine is unlikely to overtake mainstream health care altogether, but indications are that its role will continue to grow.

CONCEPT CHECKS

1. How do functionalist theorists and symbolic interactionists differ in their perspectives on health and illness?

2. What is stigma, and how does it pertain to health and illness?

3. What is the biomedical model of health?

4. How does disability pose a challenge to both functionalist and biomedical models of health?

5. Compare complementary medicine with alternative medicine.

How Do Social Factors Affect Health and Illness?

Recognize that health and illness are shaped by cultural, social, and economic factors. Learn about race, class, gender, and geographic differences in the distribution of disease.

The twentieth and early twenty-first centuries witnessed a significant increase in life expectancy for people living in industrialized countries. Infectious diseases such as polio, scarlet fever, and diphtheria have been all but eradicated. Infant mortality rates have dropped precipitously, leading to an increase in the average life span in the developed world. Compared with other parts of the world, standards of health and well-being are high. Many advances in public health have been attributed to the power of modern medicine. It is commonly assumed that medical research has been—and will continue to be—successful in uncovering the biological causes of disease and in developing effective treatments. At the same time, the proportion reporting mental health conditions, including depression and anxiety disorders, has increased steeply through the twentieth and twenty-first centuries, raising new questions about how we define, detect, and diagnose mental health conditions (Greenberg 2010; Horwitz 2013).

Sociologists recognize that whereas, on average, physical health and risk of mortality have improved while mental health has, by some accounts, declined in the past century, these patterns vary widely throughout the population both within the United States and across the globe. As we saw in the chapter opener, certain groups of people enjoy much better health than others. These health inequalities appear to reflect larger systems of social stratification, including race, gender, and social class stratification.

Social Class–Based Inequalities in Health

In Chapter 7, we defined social class as a concept that encompasses education, income, occupation, and assets. In U.S. society, people with better educations, higher incomes, and more prestigious occupations have better health. What is fascinating is that each of these dimensions of social class may be related to health and mortality for different reasons.

Income is the most obvious factor. In countries such as the United States, where medical care is expensive and the ACA is still in its infancy, those with more financial resources have better access to physicians and medicine. But inequalities in health also persist in countries such as Great Britain that have a long history of national health insurance. Differences in occupational status may lead to inequalities in health and illness even when medical care is fairly evenly distributed. One highly influential study of health inequalities in Great Britain, the *Black Report* (Townsend and Davidson 1982), found that manual workers had substantially higher mortality rates than professional workers, even though Britain had made great strides in equalizing the distribution of health care. Those who work in offices or in domestic settings face less risk of injury and exposure to hazardous materials.

Education is also a powerful predictor of health, where those with higher levels of education have longer life spans than those with fewer years of schooling. In recent years, the least-educated Americans—especially women—have actually experienced a reversal in life expectancy, while all other groups have experienced gains (Cockerham 2014). Numerous studies have found a positive correlation between education and a broad array of preventive health behaviors. Better-educated people are significantly more likely to engage in aerobic exercise and to know their blood pressure (Shea et al. 1991) and are less likely to smoke (Kenkel, Lillard, and Mathios 2006) or be overweight (Himes 1999). Highly educated smokers are also much more likely to quit smoking when faced with a new health threat, such as a heart attack (Wray et al. 1998). Poorly educated people engage in more cigarette smoking; they also have more problems with cholesterol and body weight (Winkleby et al. 1992).

Mental health is similarly affected by social class–based inequalities. In general, persons with lower levels of education and income fare worse along most mental health outcomes, including risk of depression, anxiety, and suicidal ideation. The stressors related to economic adversity, including unsatisfying jobs, strained marriages, and worries about money and one's personal safety, may overwhelm one's ability to cope. Depressive symptoms (feelings of profound sadness and hopelessness) and anxiety (nervousness about one's daily experiences) are emotional consequences of living under persistently stressful circumstances (Carr 2014).

Mental health and physical health can be closely intertwined. Where poor physical health compromises one's emotional well-being, one's mental health can undermine one's physical health, especially for those who try to soothe their feelings of sadness, anxiety, or alienation through behaviors like drinking excessively, smoking, or using drugs such as opioids. As we saw earlier in this chapter, growing numbers of Americans with low levels of education are dying in middle age rather than old age. Many are grappling with precarious employment, financial strain, and other stressors that may make them vulnerable to alcohol and drug abuse and, in the worst-case scenario, deaths from suicide or chronic liver disease. Health researchers describe these "deaths of despair" as a dramatic example of social class–based disparities in health (Case and Deaton 2017).

Race-Based Inequalities in Health

Blacks fare worse than whites in the United States on nearly all health indicators, ranging from body weight to mortality rates to risk of major illnesses like diabetes and cancer. In the United States, life expectancy at birth in 2015 was 84 for Hispanic females and about 81 for white females but 78 for Black females. Likewise, life expectancy at birth in 2015 was 79 for Hispanic males and 76 for white males yet 72 for Black males (CDC 2017a). An even more startling gap emerges when early-life mortality is considered: Black infants have more than twice the mortality rate of white infants, and as we saw earlier in this chapter, Black expectant and new mothers are nearly four times as likely as white mothers to die of pregnancy-related complications, as Irving's death revealed.

Racial differences in health reveal the complex interrelations among ethnicity, race, social class, and culture. A powerful example of the multiple ways that race affects health is the Hispanic health paradox: Although Hispanics in the United States have poorer socioeconomic resources than whites, on average, their health—and especially the health of their infants—is just as good as, if not better than, that of whites. Blacks, by contrast, face economic disadvantages that are similar to those of Hispanics, yet Blacks do not enjoy the same health benefits. Experts attribute Hispanics' relative health advantage not only to cultural factors such as social cohesion but also to methodological factors. Studies of Hispanic health in the United States focus on those who successfully migrated to the United States; as such, they are believed to be in better health, or more robust, than those Latinos who remained in their native countries (Perea 2012).

A close inspection of Blacks' health and mortality disadvantage further reveals the multiple ways that race matters for health. One of the main reasons for Blacks' health disadvantage is that Blacks as a group have less money than whites, as noted in Chapter 7. Yet Black–white disparities in health go beyond economic causes and reflect other important aspects of the social and cultural landscape. Recall that Irving was a highly educated doctor who earned a good living, yet she remained vulnerable to stressors that are pervasive for African Americans, including exposure to racism. To take another example, consider racial gaps in mortality due to homicide. The homicide rate for Black men is more than ten times higher than for their white peers (Murphy et al. 2017; Smith and Cooper 2013; Flaherty and Sethi 2010). This gap has been attributed to the violent crime that has accompanied the rise of widespread crack cocaine addiction, especially in the late 1980s and 1990s, mainly affecting poor African American neighborhoods plagued by high levels of unemployment (Wilson 1996).

Other race-based inequalities in health status, health behaviors, and health care are also stark. There is a higher prevalence of hypertension among Blacks than whites, especially among Black men (41 percent of Black men vs. 30 percent of white men in 2016)—a

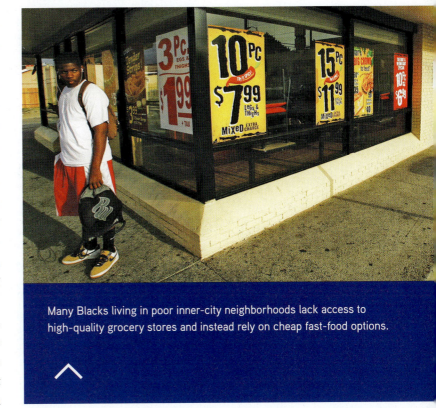

Many Blacks living in poor inner-city neighborhoods lack access to high-quality grocery stores and instead rely on cheap fast-food options.

difference that may be partly biological (Fryar et al. 2017). The pattern may also reflect Blacks' tendency to eat high-fat foods, a pattern encouraged by the fast-food industry's targeting of African Americans as a market (Henderson and Kelly 2005). Black women are also far less likely than white women to exercise regularly, a pattern that most social scientists attribute to their hectic schedules of juggling work and family and to the high costs of fitness programs and gym memberships (August and Sorkin 2010). Cumulative exposure to racism, whether institutional discrimination from employers or everyday microaggressions and unkind treatment, also get "under the skin" via the process of weathering. As we learned earlier, cumulative exposure to stress can lead to wear and tear on one's cardiovascular, metabolic, and immune systems, rendering the body vulnerable to disease and even to premature death (Villarosa 2018).

Racial differences in mental health are far less well understood than racial differences in physical health. In general, most studies show that Blacks report fewer symptoms of depression than whites, yet when socioeconomic factors are controlled for, Blacks actually report lower rates of depression than whites (Dunlop et al. 2003). At first blush, this finding is seen as surprising, given that Blacks are exposed to more stressful circumstances that may trigger depression, such as workplace discrimination, unsafe living environments, and financial troubles. However, some scholars also note that African Americans have important resources on which they can draw that mitigate against symptoms of depression, including social support from their religious community and a strong sense of racial identity and community (Oates and Goode 2012).

COUNTERING RACIAL INEQUALITIES IN HEALTH

Despite the persistence of inequalities in the health of Blacks and whites, some progress has been made in eradicating them. According to the CDC, racial differences in cigarette smoking have decreased. In 1965, half of white men and 60 percent of Black men age eighteen and older smoked cigarettes. By 2016, only 18 percent of white men and 20 percent of Black men smoked (Jamal et al. 2018). Hypertension among Blacks has also been greatly reduced. In the early 1970s, half of Black adults between the ages of twenty and seventy-four suffered from hypertension. By 2016, the proportion of Black adults over age eighteen suffering from hypertension had dropped to 40 percent (Fryar et al. 2017).

Patterns of physician visitation, hospitalization, and preventive medicine have also improved, yet racial equity still remains elusive. For example, Black women historically have been less likely than white women to receive mammograms. This gap has narrowed in recent years. Still, some studies suggest that Black women delay receiving mammograms and thus those with breast cancer have their condition detected at a later—and more dangerous—stage of the disease's progression (Smith-Bindman et al. 2006).

Gender-Based Inequalities in Health

Women in the United States generally live longer than men, and this gender gap increased steadily throughout the twentieth century. In 1900, there was only a two-year difference in female and male life expectancies. By 1940, this gap had increased to 4.4 years, and by 1970, to 7.7 years. Since reaching its peak in the 1970s, however, the gap has been decreasing. By 2016, the gender gap had fallen to slightly less than five years (Cleary 1987; Kochanek 2017).

Despite the female advantage in mortality, most large surveys show that women more often report poorer physical and mental health. Women have higher rates of illness from acute conditions and nonfatal chronic conditions, including arthritis and osteoporosis. They are slightly more likely to report their health as fair or poor and spend about 57 percent more days sick in bed each year. Women also report that their physical activities are either restricted or impossible about 50 percent more than men do. In addition, they make more physician visits each year and undergo twice the number of surgical procedures as men (CDC 2013c; National Center for Health Statistics 2003, 2011). Women are also twice as likely as men to report symptoms of depression and to be diagnosed with a major depressive disorder (Van de Velde, Bracke, and Leveque 2010).

There are two main explanations for the gender gap in health: (1) Greater life expectancy and age bring poorer health, and (2) women make greater use of medical services, including preventive care (CDC 2003). Men may experience as many, or more, health symptoms as women, but they may ignore their symptoms, underestimate the extent of their illness, or utilize preventive services less often (Waldron 1986). Furthermore, men who are socialized to believe that men should be "traditionally masculine," strong, and self-sufficient are less likely to seek out annual checkups (Springer and Mouzon 2011).

Explanations for the gender gap in depression are less clear, yet scholars typically attribute women's greater level of depression to measurement issues and differences in stress exposure. First, standard instruments used to measure depression typically emphasize symptoms that are more likely to be endorsed by women, leading some scholars to question whether depression in men goes underdetected. Second, given the strains that women increasingly face in juggling work and family roles—compounded by other stressors, such as single parenthood, that disproportionately strike women—women may be more likely than men to be depleted and depressed by the daunting demands they face (Nolen-Hoeksema 1993).

A major question for sociologists is whether the gender gap in mortality will continue to decline in coming years. Many researchers believe that it will, yet for an unfortunate reason: Women's life expectancies may erode and thus become more similar to men's. As men's and women's gender roles have converged over the past several decades, women have increasingly taken on unhealthy "male-typed" behaviors, such as smoking and alcohol use, as well as emotional and physical stress in the workplace. These patterns are particularly pronounced for women of low socioeconomic status. One recent study found that American women have lost ground with respect to life expectancy compared with women from other nations. In the early 1980s, the life expectancy of women in the United States ranked fourteenth in the world, yet by 2010, American women had fallen to forty-first place (Karas-Montez and Zajacova 2013). These disheartening findings reveal that gender differences in health and mortality are not a function of biology alone but of the social advantages and adversities experienced by men and women in particular sociohistorical contexts.

Disparities in Infectious Diseases Worldwide

Socioeconomic disparities in health exist not only in the United States but also worldwide. Lower-income nations have higher rates of illness from infectious disease, higher mortality rates, and lower life expectancies than wealthier nations. We briefly describe why and how infectious diseases, and HIV/AIDS, in particular, pose a threat to low-income nations and what public health practitioners and policymakers are doing to help fight these devastating diseases.

INFECTIOUS DISEASES TODAY

Although major strides have occurred in reducing, and in some cases eliminating, infectious diseases in the developing world, they remain far more common there than in the West. The most important example of a disease that has almost completely disappeared is smallpox, which, as recently as the 1960s, was a scourge of Europe as well as many other regions. Campaigns against malaria have been much less successful. When the insecticide DDT was first produced, it was hoped that the mosquito, the prime carrier of malaria, could be eradicated. At first, there was considerable progress, but this has slowed because some strains of mosquito have become resistant to DDT. Owing to the pesticide's declining benefits as well as increased awareness of its harmful effects on both the environment and humans, the Environmental Protection Agency banned the use of DDT in the United States in 1972. In 2006, however, the World Health Organization declared its support of the use of the pesticide in African countries with high rates of malaria (EPA 2015). In 2016, an estimated 216 million new cases of malaria were reported worldwide, and 445,000 people died from the disease; rates are highest in sub-Saharan Africa, and children are at a particularly high risk (World Health Organization 2018).

The health threats posed by infectious disease in low-income nations have a long history. During the colonial era, efforts to bring Western ideals to developing societies brought certain diseases into other parts of the world. Smallpox, measles, and typhus, among other major maladies, were unknown to the indigenous populations of Central and South America before the Spanish Conquest in the early sixteenth century. The English and French colonists brought the same diseases to North America (Dubos 1959). Some of these illnesses produced epidemics that ravaged or completely wiped out native populations, which had little or no resistance to them.

In Africa and subtropical parts of Asia, infectious diseases have been rife for a long time. Tropical and subtropical conditions are especially conducive to diseases such as malaria, carried by mosquitoes, and sleeping sickness, carried by the tsetse fly. Historians believe that risks from infectious diseases were lower in Africa and Asia prior to the time that Europeans tried to colonize these regions—as they often brought with them practices that negatively affected the health of local natives. The most significant consequence of the colonial system was its effect on nutrition and, therefore, on levels of resistance to illness as a result of the changed economic conditions involved in producing for world markets. In many parts of Africa, the nutritional quality of native diets became substantially depressed as cash-crop production supplanted the production of native foods.

HUMAN IMMUNODEFICIENCY VIRUS (HIV) AND ACQUIRED IMMUNE DEFICIENCY SYNDROME (AIDS)

HIV/AIDS is one infectious disease that has been cause for great concern in both low- and high-income nations, and it remains a global epidemic. Nearly 37 million people were living with HIV in 2017. In 2017 alone, 1.8 million people became newly infected with HIV, and nearly 1 million people died from AIDS-related illnesses. The majority of people affected in the world today are heterosexuals; about half are women (UNAIDS 2018).

In high-income countries, the rate of new infections has declined, yet the demographics are striking. In the United States, approximately 40,000 people become infected with HIV each year, and roughly 1.1 million people are living with HIV (CDC 2017b). The

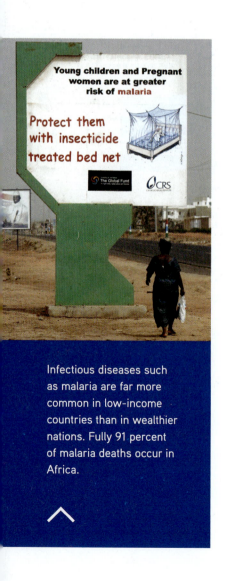

Young children and Pregnant women are at greater risk of malaria

Protect them with insecticide treated bed net

The Global Fund   CRS

Infectious diseases such as malaria are far more common in low-income countries than in wealthier nations. Fully 91 percent of malaria deaths occur in Africa.

incidence of infection, however, is not proportionately represented throughout the United States. Despite representing just 13 percent of the U.S. population, African Americans accounted for 44 percent of all new HIV diagnoses in 2016. Hispanics are also disproportionately affected: They account for about 18 percent of the population but 25 percent of HIV diagnoses in 2016 (CDC 2017b). Although there was a steep drop in AIDS-related deaths after the introduction of antiretroviral therapy, African Americans are less likely than whites to benefit from such life-prolonging treatments. African Americans have the highest death rate of people with HIV, seven times higher than that of their white counterparts and nearly four times as high as Hispanics (CDC 2017c).

The stigma that associates HIV-positive status with sexual promiscuity, men who have sex with men, and IV drug use results in avoidance of HIV/AIDS prevention and treatment programs. In the United States, one in seven people living with HIV/AIDS does not know he or she is infected (CDC 2017d). Part of the reason is the high level of fear and denial associated with being diagnosed as HIV positive. The stigma of having HIV and the discrimination against people living with these infections are major barriers to the treatment of the epidemic worldwide. A recent study of 1,450 HIV-positive patients seeking care in India found that two-thirds of them reported authoritarian behavior from doctors, and 55 percent felt they were not treated in a dignified manner (Mehta 2013).

Although the spread of AIDS has slowed in many low-income nations, the illness is still a source of crisis. Besides the devastation to individuals who suffer from it, the AIDS epidemic is creating severe social consequences, including sharply rising numbers of orphaned children. Frail older adults are increasingly called on to provide physical care to their adult children who suffer from AIDS or to care for their grandchildren who were orphaned by their parents' deaths from AIDS (Knodel 2006). The decimated population of working adults combined with the surging population of orphans set the stage for massive social instability; economies break down, and governments cannot provide for the social needs of orphans, who become targets for recruitment into gangs and armies.

CONCEPT CHECKS

1. How do social class and race affect health?

2. Name at least two explanations for the gender gap in health.

3. Identify at least two reasons why the gender gap in life expectancy may narrow in the future.

4. Why are infectious diseases more common in developing nations than in the United States today?

5. What are three social consequences of the AIDS epidemic in developing nations?

How Do Social Contexts Shape Sexual Behavior?

As with the study of health and illness, scholars disagree as to the importance of biological versus social and cultural influences on human sexual orientations and behaviors, important facets of the sociology of the body.

The Diversity of Human Sexuality

Human sexuality is fascinating, diverse, and dynamic. In Chapter 9, we discussed in great detail gender identity and emphasized that while most people are cisgender and identify as the sex they were born, growing numbers of people identify as noncisgender. Noncisgender persons challenge the gender binary of male–female, moving fluidly between the categories of male and female or rejecting the binary altogether (Padawer 2014; Schulman 2013).

Understand the diversity of sexual orientation today. Learn about the debate over the importance of biological versus social and cultural influences on human sexual behavior. Explore cultural differences in sexual behavior and patterns of sexual behavior today.

heteronormativity

The pervasive cultural belief that heterosexuality is the only normal and natural expression of human sexuality.

heterosexuality

Sexual or romantic attraction to persons of the opposite sex.

homosexuality

Sexual or romantic attraction to persons of one's own sex.

bisexuality

Sexual or romantic attraction to persons of one's own and the opposite sex.

sexual orientation

The direction of one's sexual or romantic attraction.

QUILTBAG

Acronym that captures the diversity of sexual orientations and gender identities, including queer/questioning, undecided, intersex, lesbian, trans, bisexual, asexual, and/or gay/genderqueer.

Similarly, growing numbers of persons are challenging the cultural norm of heteronormativity. **Heteronormativity** is the pervasive cultural belief that heterosexuality is the only normal and natural expression of human sexuality. **Heterosexuality**, or being sexually attracted to persons of the opposite sex (i.e., straight), historically has been considered the norm, whereas **homosexuality** (i.e., gay or lesbian), or being sexually attracted to persons of the same sex, and **bisexuality** (i.e., sexual attraction to both sexes) have been considered immoral, pathological, or a mental health disorder in need of a cure (Drescher 2015). Although heterosexuality may be the prevailing norm in most cultures, it is not "normal" in the sense of being dictated by some universal moral or religious standard. Like all behavior, heterosexual behavior is socially learned within a particular culture.

Heterosexuality, homosexuality, and bisexuality are examples of **sexual orientation**, or who we are attracted to sexually. It is important to note that *sexual orientation* is a more appropriate term when describing human sexuality than *sexual preference*. The latter is misleading because it implies that one's sexual or romantic attraction is entirely a matter of personal choice. As you will see, sexual orientation results from a complex interplay of biological and social factors not yet fully understood. It is difficult to document sexual orientation because of the lingering stigma attached to same-sex relationships, which may result in the underreporting of sexuality in demographic surveys. However, most estimates suggest that from 2 to 5 percent of all women and from 3 to 10 percent of all men in the United States are attracted to same-sex partners (Smith 2003; Stephens-Davidowitz 2013).

Early writings viewed sexual orientation, like gender identity, as a binary, where people could be categorized as either heterosexual (straight) or homosexual (gay). Yet, in recent decades, scholars have recognized that these two categories are far too simplistic to capture the nuances of human sexuality. Sociologist Judith Lorber (1994) identified as many as ten different sexual identities, including straight (heterosexual) woman, straight man, lesbian woman, gay man, bisexual woman, and bisexual man. More contemporary studies of human sexuality use the acronyms **QUILTBAG** or LGBTQ to capture even greater complexity. QUILTBAG is a short-hand label that encompasses queer/questioning, undecided, intersex, lesbian, trans, bisexual, asexual, and/or gay/genderqueer. LGBTQ, similarly, refers to lesbian, gay, bisexual, transgender, and queer persons. You may notice that some of these terms reflect one's gender identity, such as transgender, whereas others refer specifically to sexual orientation. Queer, gay, and genderqueer generally refer to persons who are sexually attracted to or prefer sexual relations with same-sex persons, whereas bisexual refers to those who are sexually attracted to both men and women. Other categories, still, acknowledge that some people have not decided on their sexual orientations (undecided) or have low or no desire for sexual relationships (asexual).

The specific language we use to talk about sexual behavior has changed dramatically over time, yet the notion that human sexuality is complex and constrained by societal norms has existed for centuries. In the late nineteenth and early twentieth centuries, Sigmund Freud argued that human beings are born with a wide range of sexual tastes that are ordinarily curbed through socialization—although some adults may follow these even when, in a given society, they are regarded as immoral or illegal. Freud began his research during the Victorian period, when many people were sexually prudish, yet his patients still revealed an amazing diversity of sexual pursuits.

In most societies, sexual norms encourage some practices and discourage or condemn others. Such norms, however, vary among cultures and often challenge the notion of heteronormativity. For example, the anthropologist Gilbert Herdt (1981, 1984, 1986) reported that among more than twenty tribes in Melanesia and New Guinea, ritually prescribed same-sex encounters among young men and boys were considered necessary for subsequent masculine virility (Herdt and Davidson 1988). Ritualized male–male sexual encounters also occurred among the Azande of Africa's Sudan and Congo (Evans-Pritchard 1970), Japanese samurai warriors in the nineteenth century (Leupp 1995), and highly educated Greek men and boys at the time of Plato (Rousselle 1999). These examples underscore the importance of social and historical contexts in shaping sexuality.

Cross-cultural variations have been detected for myriad aspects of human sexuality, including precisely what is included and valued in a sexual encounter, and the traits that one views as attractive in a potential sexual partner. The most comprehensive cross-cultural study of sexual practices was carried out by Clellan Ford and Frank Beach (1951), using anthropological evidence from more than 200 societies. Striking variations were found in what was regarded as "natural" sexual behavior and in norms of sexual attractiveness. For example, in some cultures, extended foreplay is desirable and even necessary before intercourse; in others, foreplay is nonexistent. In some societies, it is believed that overly frequent intercourse leads to physical debilitation or illness.

In most cultures, norms of sexual attractiveness (held by both cisgender females and cisgender males) focus more on physical looks for women than for men, a situation that may be changing in the West as women become active in spheres outside the home. The traits seen as most important in female beauty, however, differ greatly. In the modern West, a slim, small physique is admired, while in other cultures, a more generous shape is attractive. Sometimes the breasts are not considered a source of sexual stimulus, whereas some societies attach erotic significance to them. Some societies value the shape of the face, whereas others emphasize the shape and color of the eyes or the size and form of the nose and lips.

Sexuality in Western Culture: A Historical Overview

Western attitudes toward sexual behavior were for nearly 2,000 years molded primarily by Christianity, whose dominant view was that all sexual behavior was suspect except that needed for reproduction. During some periods, this view produced an extreme prudishness, but at other times, many people ignored the church's teachings and engaged in practices such as adultery. The idea that sexual fulfillment can and should be sought through marriage was rare.

In the nineteenth century, religious presumptions about sexuality were partly replaced by medical ones. Most early writings by doctors about sexual behavior, however, were as stern as the views of the church. Some argued that any type of sexual activity unconnected with reproduction would cause serious physical harm. Masturbation was said to cause blindness, insanity, and heart disease, while oral sex was claimed to cause cancer. In Victorian times, sexual hypocrisy abounded. Many Victorian men—who appeared to be sober, well-behaved citizens, devoted to their wives—regularly visited prostitutes or kept mistresses. Such behavior was accepted, whereas "respectable" women who took lovers were regarded as scandalous and were shunned in polite society. The differing attitudes toward the sexual activities of men and women formed a double standard, which persists today.

Currently, traditional attitudes exist alongside much more permissive attitudes, which developed widely in the 1960s. Some people, particularly those influenced by Christian teachings, believe that premarital sex is wrong; they frown on all forms of sexual behavior except heterosexual activity within marriage—although it is now more commonly accepted that sexual pleasure is an important feature of marriage. Sexual attitudes have undoubtedly become more permissive over recent decades in most Western countries, although some behaviors remain consistently more acceptable than others. For example, the proportion of Americans saying that premarital sex is "always wrong" dropped from 34 percent in 1972 to 26 percent in 2016 (Bowman 2018). However, attitudes toward premarital sex among young teens are far less permissive; more than three-quarters of Americans disapprove of sexual relations between unmarried teens ages fourteen to sixteen. Disapproval for extramarital sex also has remained consistently high and has even increased: The proportion of Americans saying that extramarital sex is "always wrong" increased from 71 to 81 percent between 1972 and 2016 (Labrecque and Whisman 2017; Smith and Son 2013).

Although attitudes toward sexual behavior have grown more permissive in recent decades, sexual behavior—especially among young women—is still highly regulated, monitored, and judged. Scholars, activists, and even celebrities like Lady Gaga and Amber Rose have called attention to practices like "slut-shaming," which maligns young women for being sexually active, and "prude-shaming," where young women are shamed or embarrassed for not being sexually active. Activists underscore how dangerous these judgments can be, pointing out that some young women who have been the victims of sexual assault are blamed or shamed for wearing miniskirts or low-cut tops, drinking, flirting, or kissing their assailants prior to their attacks (Nguyen 2013).

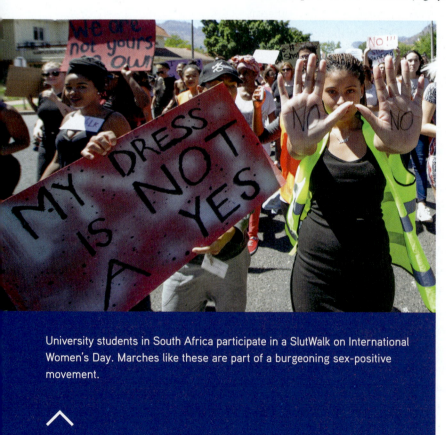

University students in South Africa participate in a SlutWalk on International Women's Day. Marches like these are part of a burgeoning sex-positive movement.

These public condemnations of judging others' sexuality are part of a larger movement called sex positivity. The sex-positive movement is a philosophy and a social movement that encourages and embraces diverse forms of sexuality and sexual expression, emphasizing the importance of safe, healthy, and consensual sex (Ivanski and Kohut 2017). Although the sex-positive movement has flourished in recent years, facilitated by social media and public events like SlutWalk protest marches, its core idea dates back to early-twentieth-century doctor and psychoanalyst Wilhelm Reich. Both Reich and current-day advocates of the sex-positive movement view sexuality as a matter of personal choice and avoid making moral judgments or distinctions. All forms of healthy, consensual sex, whether same-sex or opposite-sex relations, masturbation, polyamory, asexuality, or voluntary sadomasochism, should be respected and spared judgment. The movement's larger goal is to advocate for comprehensive sex education for all.

SEXUAL BEHAVIOR: KINSEY'S STUDY

We can speak more confidently about public values and attitudes concerning sexuality than we can about actual sexual behavior, because these deeply private practices have gone undocumented for much of history. Alfred Kinsey broke major ground when he initiated the first major investigation of sexual behavior in the United States in the 1940s and 1950s. Kinsey and his coresearchers (1948, 1953) faced condemnation from religious organizations, and his work was denounced as immoral in newspapers and in Congress. But he persisted, thus making his research the largest rigorous study of sexuality at that time, although his sample was not representative of the overall American population.

Kinsey's results were surprising because they revealed a tremendous discrepancy between prevailing public expectations of sexual behavior and actual sexual conduct. The gap between publicly accepted attitudes and actual behavior was probably especially pronounced just after World War II, the time of Kinsey's study. A phase of sexual liberalization had begun in the 1920s, when many younger people felt freed from the strict moral codes that had governed earlier generations. Sexual behavior probably changed, but issues concerning sexuality were not openly discussed. People participating in sexual activities that were still strongly disapproved of on a public level concealed them, not realizing that others were engaging in similar practices behind closed doors. The more permissive 1960s brought openly declared attitudes more into line with the realities of behavior.

SEXUAL BEHAVIOR SINCE KINSEY

In the 1960s, social movements that challenged the existing order, such as those associated with countercultural lifestyles, also broke with existing sexual norms. These movements preached sexual freedom, and the introduction of the contraceptive pill allowed sexual pleasure to be separated from reproduction. Women's groups also started pressing for greater independence from male sexual values, rejection of the double standard, and the need for women to achieve greater sexual satisfaction in their relationships—efforts that persist today, as part of the sex-positivity movement. Even so, until recently, it was unclear to what extent sexual behavior had changed since the time of Kinsey's research.

In the late 1980s, sociologist Lillian Rubin (1990) interviewed 1,000 Americans between the ages of thirteen and forty-eight to identify changes in sexual behavior and attitudes over the previous thirty years or so. Her findings indicated significant changes. Sexual activity begins at a younger age; moreover, teenagers' sexual practices are as varied and comprehensive as those of adults. There is still a double standard, but it is not as powerful. Contemporary scholarship confirms this. Studies of high school–age students find that sexual permissiveness is much greater today than it was in the 1970s. According to the CDC in 2017, 40 percent of all high school students reported having ever had sexual intercourse, and 10 percent reported having had four or more partners (Kann et al. 2018). Both figures represent declines from 1991, when more than 54 percent of high school students had had sex and nearly 19 percent had had four or more partners.

Recent research on the sexual lives of college students shows that a "hookup culture" is alive and well on campus, where both male and female students will have one-night stands, short-lived sexual relationships, or "friends with benefits" relationships in which friends will have sexual relations without the expectation that their friendship will transform into a full-blown romance (Garcia et al. 2012; Hamilton and Armstrong 2009). However, when a team of sociologists delved more closely into the sexual lives of college students, they found that while casual sexual encounters were relatively common, men and women were fairly selective in such encounters. Sociologist Paula England and colleagues interviewed more than 14,000 undergraduate students at nineteen universities and colleges about their romantic and sexual lives. Nearly three-quarters (72 percent) of both women and men said that they'd had at least one "hookup" during their senior year. But, for most, hookups were relatively rare. Of those students who said that they had ever hooked up, equal proportions said that they had fewer than three (40 percent) or between four and nine (40 percent) hookups. Just one in five reported ten or more hookups in their lifetimes. Moreover, not all of these hookups involved sexual intercourse. Fully 20 percent of college seniors reported never having had sexual intercourse (England, Shafer, and Fogarty 2012).

Studies of the sexual lives of adults beyond college age also reveal that Americans report relatively few sexual partners throughout their lives and less frequent sex than their counterparts in other nations. For example, in 1994, a team of researchers led by Edward Laumann published *The Social Organization of Sexuality: Sexual Practices in the United States*, the most comprehensive study of sexual behavior since Kinsey. Their findings reflect an essential sexual conservatism among Americans. For instance, 83 percent of their subjects had had only one partner (or no partner at all) in the preceding year, and among married people, the figure was 96 percent. Fidelity to one's spouse was also quite common: Only 10 percent of women and less than 25 percent of men reported having an extramarital affair during their lifetimes. More recent data reveal that little has changed; according to the CDC (2017e), in 2015, men reported an average of just 6.1 sexual partners in their lives, while women reported just 4.2 partners.

IS SEXUAL ORIENTATION INBORN OR LEARNED?

Most sociologists believe that sexual orientation—whether straight, gay, lesbian, bisexual, or asexual—results from a complex interplay between biological factors and social learning. Since heterosexuality is the norm for most people in U.S. culture, considerable research has focused on why some people prefer same-sex partners. Some scholars argue that biological influences predispose certain people to become gay from birth (Bell, Weinberg, and Hammersmith 1981; Green 1987). Biological explanations have included differences in brain characteristics of gay and straight men (LeVay 2011) and the effect on fetal development of the mother's *in utero* hormone production during pregnancy (Blanchard and Bogaert 1996; Manning, Koukouratis, and Brodie 1997; McFadden and Champlin 2000). Such studies, which are based on small numbers of cases, give highly inconclusive (and highly controversial) results (Healy 2001). It is virtually impossible to separate biological from early social influences in determining a person's sexual orientation (LeVay 2011).

Studies of twins may shed light on any genetic basis for homosexuality, since identical twins share identical genes. In two related studies, Bailey and Pillard (1991; Bailey et al. 1993) examined 167 pairs of brothers and 143 pairs of sisters, with each pair of siblings raised in the same family, in which at least one sibling defined himself or herself as gay or lesbian. Some of these pairs were identical twins (who share all genes), some were fraternal twins (who share some genes), and some were adoptive brothers or sisters (who share no genes).

The results offer some support that same-sex attraction, like opposite-sex sexual attraction, results from a combination of biological and social factors. Among the men and women studied, when one twin was gay, there was about a 50 percent chance that the other twin was gay. In other words, a woman or man is five times as likely to be gay or lesbian if his or her identical twin is gay than if his or her sibling is gay but related only through adoption. These results offer some support for the importance of biological factors, since the higher the percentage of shared genes, the greater the percentage of cases in which both siblings were gay. However, because approximately half of the identical twin brothers and sisters of individuals who identified as gay were not themselves gay, social learning must also be involved.

Even studies of identical twins cannot fully isolate biological from social factors. It is often the case that even in infancy, identical twins are treated more like each other by parents, peers, and teachers than are fraternal twins, who in turn are treated more like each other than are adoptive siblings. Thus, identical twins may have more than genes in common: They may also share a higher proportion of similar socializing experiences. Sociologist Peter Bearman (2002) has shown the intricate ways that genetics and social experience are intertwined. Bearman found that males with a female twin are twice as likely to report same-sex attractions. He theorized that parents of opposite-sex twins are more likely to give them unisex treatment, leading to a less traditionally masculine influence on the males. Having an older brother decreases the rate of homosexuality. Bearman hypothesized that an older brother establishes gender-socializing mechanisms for the younger brother to follow, which allows him to compensate for unisex treatment. Bearman's work is consistent with the statements offered by professional organizations such as the American Academy of Pediatrics (2004), which concludes that "sexual orientation probably is not determined by any one factor but by a combination of genetic, hormonal, and environmental influences."

HOMOPHOBIA AND HETEROSEXISM

Homophobia, a term coined in the late 1960s, refers to both attitudes and behaviors marked by an aversion to or hatred of gays and lesbians, their lifestyles, and their practices. It is a form of prejudice reflected not only in overt acts of hostility and violence toward lesbian, gay, and bisexual persons but also in forms of verbal abuse that are widespread in American culture—for example, using terms like *fag* or *homo* to insult heterosexual males or using female-related offensive terms such as *sissy* or *pansy* to insult gay men (Pascoe 2011). Similarly, **transphobia** refers to negative attitudes, feelings, or actions toward transgender and gender-nonconforming people, their lifestyles, and their practices.

Homophobia and transphobia generally encompass individual-level behavior, whereas more systematic oppression of LGBTQ persons falls under the umbrella of **heterosexism**, which refers to an ideological system that denies, denigrates, and stigmatizes any nonheterosexual form of behavior, identity, relationship, or community (Herek 2004). Like institutional racism, which we learned about in Chapter 10, heterosexism subtly or overtly underlies many societal customs and institutional practices. Examples of heterosexism in the United States include banning transgender persons from enlisting in the military; widespread lack of legal protection from antigay discrimination in employment, housing, and services; laws requiring people to use the bathrooms linked with the sex listed on their birth certificates rather than with the gender with which they currently identify; or being denied services because of one's sexual orientation or gender identity. The latter

homophobia

An irrational fear or disdain of homosexuality.

—

transphobia

Negative attitudes, feelings, or actions toward transgender and gender-nonconforming people, their lifestyles, and their practices.

—

heterosexism

An ideological system that denies, denigrates, and stigmatizes any nonheterosexual form of behavior, identity, relationship, or community.

Health Care Provider

Medicine is commonly lauded as a "noble profession." Doctors, nurses, and other health care providers dedicate their lives to diagnosing and treating health problems, and helping their patients live long and comfortable lives. A deep knowledge of biology, chemistry, and anatomy is critical to the medical professions, but a sociological imagination is also necessary. The study of sociology helps physicians understand why some people may not have access to the health care they need, how power dynamics in the clinician–patient encounter may affect the quality of care, why some patients ignore sound medical advice, why social and environmental factors like stress make us sick, and so on. In fact, a strong grasp of human behavior is so important to health care providers that in 2015 the Medical College Admissions Test (MCAT) introduced a new required module on Psychological, Social, and Biological Foundations of Behavior. The Association of American Medical Colleges (AAMC) has also underscored the importance of sociology to medical education, noting that "medicine now faces complex societal problems like addiction, obesity, violence, and end-of-life care, which require behavioral and social science research and interventions" (AAMC 2011: 5).

One area of particular concern to health care providers is cultural competency, or the skills and ability to interact effectively with patients from cultural backgrounds different from one's own. "Culture" refers to more than a patient's race, ethnicity, or national origin; it also encompasses characteristics such as age, gender, gender identity, sexual orientation, disability, religion, income level, education, geographical location, or profession (Substance Abuse and Mental Health Services Administration 2016). An understanding

is evidenced in myriad ways, ranging from bakers who refuse to bake cakes for same-sex weddings to health care providers or clergy referring LGBTQ youth to conversion therapy.

Heterosexism may take a profound toll on the health, well-being, and personal safety of LGBTQ persons. One survey of more than 7,500 high school students found that nearly 44 percent of gay male and 40 percent of lesbian teens said they had been bullied in the previous year, compared with just 26 and 15 percent of heterosexual boys and girls, respectively (Berlan et al. 2010). Another national study from 2015 found that 85 percent of LGBTQ youth reported that they had been verbally harassed at school, 27 percent had been physically harassed, and 49 percent had been victims of cyberbullying. More than half of LGBTQ students (58 percent) felt "unsafe" at school (Kosciw et al. 2016). Mistreatment of transgender youth is even more devastating, epitomized by the violent 2017 murders of seventeen-year-old Ally Lee Steinfeld in Missouri and eighteen-year-old Jaquarrius Holland in Louisiana.

This pervasive culture of fear, intimidation, and harassment can have dire consequences: LGBTQ youth have much higher rates of suicide, suicidal thoughts, depression, and substance use than straight youth, due in large part to the victimization and teasing they suffer at the hands of their classmates and, at times, their families and teachers, who fail to protect them (Espelage, Aragon, and Birkett 2008; Russell and Joyner 2001; Ryan et al. 2009). For many, families are a source of cruelty and victimization rather than support. An estimated 20 to 40 percent of all homeless youth identify as LGBTQ, many of whom have been put out on the streets by homophobic or transphobic parents (Durso and Gates 2012).

of sociology helps practitioners to be respectful, responsive, and sensitive to the cultural and linguistic needs as well as the health beliefs and practices of their patients. The classic book *The Spirit Catches You and You Fall Down* vividly portrayed the difficulties a Hmong family faced when trying to get medical treatment for their daughter Lia Lee, who suffered a rare and severe form of epilepsy. Lia's parents believed in spiritual approaches to medicine and refused to give their daughter certain medications, while their doctors didn't understand Hmong culture, couldn't establish a rapport or empathy with their patients, and created a context of distrust that impeded Lia's treatment and prognosis (Fadiman 1997).

Health care providers also need to recognize their own unconscious biases, which may affect how they interact with and treat their patients. Several studies have found that Black and Latino emergency room patients are much less likely than white patients with similar injuries to be prescribed the painkillers they need. One explanation is that health care providers hold assumptions that ethnic minorities are more likely to abuse drugs, or may believe that the patients are misrepresenting their conditions simply to secure drugs (Pletcher et al. 2008). Other studies have found that health care providers who do not understand the needs and experiences of gender-nonconforming and transgender patients may treat them insensitively, referring to them by the wrong gender, and may even neglect particular symptoms or fail to offer tests the patient might have needed (Sallans 2016). Medical practitioners who understand social, cultural, and interpersonal influences on health and health care seeking will be especially well-equipped to provide respectful and high-quality care to their increasingly diverse patients.

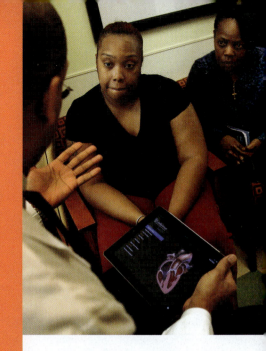

Understanding the nuances of a diverse body of patients can help health care workers provide care that considers many different experiences.

Despite the devastating statistics on bullying and homelessness among LGBTQ youth, data suggest that homophobia and transphobia in the United States are slowly starting to erode. The majority of Americans today view same-sex relationships as morally acceptable, signifying a marked increase from 2001, when just 40 percent of Americans agreed with the sentiment (Gallup 2013a). In May 2011, for the first time in its history, a Gallup poll found that the majority of Americans (53 percent) supported gay marriage (Gallup 2013b); by 2017, that proportion had risen to 64 percent (McCarthy 2017). Public policies both reflect and shape private attitudes; as we saw in Chapter 11, in June 2015, the U.S. Supreme Court legalized gay marriage in *Obergefell v. Hodges*, guaranteeing same-sex married couples the same rights as opposite-sex married couples.

THE MOVEMENT FOR LGBTQ CIVIL RIGHTS

Until recently, most LGBTQ persons hid their sexual orientation for fear that "coming out of the closet"—publicly revealing one's sexual orientation—would cost them their jobs, families, and friends and leave them open to verbal and physical abuse. Yet, since the late 1960s, many gays and lesbians have openly acknowledged their preference for same-sex sexual partners, and in some cities, the lives of lesbian and gay Americans have become quite normalized (Seidman, Meeks, and Traschen 1999). New York City, San Francisco, London, and other large metropolitan areas worldwide have thriving LGBTQ communities.

The Stonewall Inn nightclub raid in 1969 is regarded as the first shot fired in the battle for gay rights in the United States. The twenty-fifth anniversary of the event was commemorated in New York City with a variety of celebrations as well as discussions on the evolution and future of gay rights.

LGBTQ activists have achieved important milestones in fighting heterosexism and forging institutional changes, ranging from changing medical notions of sexual orientation to legalizing same-sex marriage. The movement for LGBTQ civil rights in the United States arguably began with the Stonewall riots in June 1969, when New York City's gay community—angered by continual police harassment—fought the New York Police Department for two days (D'Emilio 1983; Weeks 1977). The Stonewall riots became a symbol of gay pride. In May 2005, the International Day Against Homophobia was first celebrated, with events held in more than forty countries.

Activists and advocates have had a profound impact on the policies and practices that affect LGBTQ persons. For example, in 1869, the U.S. medical community began using the term *homosexual* to characterize what was then regarded as a personality disorder. Same-sex attraction was medicalized and viewed as a pathology that required medical or psychiatric treatment. The American Psychiatric Association did not remove homosexuality from its list of mental illnesses until 1973 or from its influential *Diagnostic and Statistical Manual of Mental Disorders* (*DSM*) until 1980. These long-overdue steps were taken only after prolonged lobbying and pressure by LGBTQ advocacy organizations. The medical community was belatedly forced to acknowledge that no scientific research had ever found gays and lesbians as a group to be psychologically unhealthier than heterosexuals (Burr 1993). However, the *DSM-5* continues to classify other aspects of sexuality as "disorders," medicalizing fairly common sexual problems such as disorders of sexual arousal (e.g., lubrication and erectile problems) and orgasmic disorders (American Psychiatric Association 2013).

In addition to policy changes, social movements and cultural shifts have contributed to the slow and gradual erosion of heterosexism. One such shift is the public coming-out of LGBTQ persons in the public eye. Coming-out may be important not only for the person who does so but also for others in the larger society: Previously closeted LGBTQ persons discover they are not alone, while heterosexuals recognize that people whom they admire and respect are LGBTQ. Famous actors, singers, and performers, such as Ellen DeGeneres and Elton John, have been out publicly for many years. And in 2018, Daniela Vega, the Oscar-nominated star of the Chilean film *A Fantastic Woman*, achieved a major milestone when she became the first openly transgender person to present at the Academy Awards.

LGBTQ persons in other professions, especially professional sports, have been more reticent about acknowledging their sexual orientation, perhaps out of fear of persecution. NBA basketball player Jason Collins made national news in April 2013 when he told reporters that he was gay. With his announcement, Collins became the first active player in one of the four major American professional team sports to announce that he was gay (ESPN 2013a). The next year, Michael Sam, who was drafted by the NFL's St. Louis Rams in 2014, came out as gay. Since that time, literally dozens of pro athletes have publicly identified as LGBTQ. The 2018 U.S. Winter Olympic Team boasted more than fourteen "out" athletes, including free-style skier Gus Kenworthy and outspoken figure skater Adam Rippon.

Social change is occurring globally, slowly but steadily, even in countries that historically have had cruel and oppressive policies toward gays and lesbians. For example, in 2014, the Constitutional Court in Uganda invalidated a previously passed "antigay" bill, which provided jail terms up to life for persons convicted of having gay sex and stipulating lengthy jail terms for persons convicted of "attempted homosexuality" or the "promotion of homosexuality" (Gettleman 2014). This marked a significant change in a nation where, just three years earlier, the outspoken gay rights activist David Kato was beaten to death with a hammer.

How Does the Social Context of Bodies, Sexuality, and Health Affect Your Life?

As we have seen in this section, our bodies, health, health behaviors, and sexual orientations and practices reflect a complex set of biological, social, cultural, and historical influences. For example, although most American young adults believe they have the freedom to choose whomever they like as their romantic partners (and turn up their noses at the idea of arranged marriage), the gender of whom we choose, what we deem attractive, and when and under what circumstances we engage in sexual relationships are powerfully shaped by laws, norms, and cultural practices.

Similarly, although most people believe that their body size and shape reflect their own personal efforts, such as going to the gym four times a week and counting calories, or biological factors (e.g., "good genes"), sociologists have documented that social factors such as race, class, gender, and region affect one's access to health-enhancing resources like healthy food, safe walking and running paths, and high-quality health care. Solutions to sweeping public health crises, like the obesity epidemic, often require strategies that alter both individual-level choices and behaviors and macrosocial structures. Public programs that target both macro and micro levels by encouraging healthier food choices and exercise among individuals and by making larger social changes—such as bringing grocery stores to inner-city neighborhoods and ensuring that major corporations that supply food to public schools abide by healthier food production guidelines—are likely to be more effective.

Yet further and dramatic social changes are still needed to eradicate persistent racial and socioeconomic disparities in health. The life and premature death of Dr. Irving underscores just how powerfully social inequalities affect our bodies. Persistent stressors like racism get under one's skin, wearing down one's body and one's health and thus rendering its victims vulnerable to chronic diseases like diabetes and high blood pressure. Economic strains and precarious employment may make some vulnerable to substance use, including use of opioids or excessive drinking, which have contributed to "deaths of despair" from suicide or chronic liver disease. Support for programs like early screening for high blood pressure, obesity, substance use, and depression may help to ensure that health problems are detected in their earliest stages and that timely treatment is sought. Through the use of these strategies, it is possible that the United States may ultimately reach the goal articulated by the federal government to "achieve health equity, eliminate disparities, and improve the health of all groups" (U.S. Department of Health and Human Services 2010).

CONCEPT CHECKS

1. Describe several changes in sexual practices over the past two centuries.

2. What are the most important contributions of Alfred Kinsey's research on sexuality?

3. Name at least three important findings about sexual behavior discovered since Kinsey.

4. What are several of the most important achievements of LGBTQ rights movements?

CHAPTER 14

The Big Picture

The Sociology of the Body: Health, Illness, and Sexuality

Thinking Sociologically

1. Obesity is a major health concern in the United States, especially among poor Americans, Blacks, and Latinos. What types of public programs do you believe will be most effective in fighting the obesity epidemic? Why do you think the programs you've proposed are necessary?

2. Statistical studies of our national health show a gap in life expectancies between the rich and the poor. Review all the major factors that would explain why rich people live longer than poor people.

3. This chapter discusses the biological and sociocultural factors associated with sexual orientation. Why are twin studies the most promising type of research on the genetic basis of sexual orientation? Summarize the analysis of these studies, and show whether it presently appears that sexual orientation results from genetic differences, sociocultural practices and experiences, or both.

Learning Objectives

How Do Social Contexts Affect the Human Body?

p. 429

Understand how social, cultural, and structural contexts shape attitudes toward "ideal" body forms and give rise to two body-related social problems in the United States: eating disorders and obesity.

How Do Sociologists Understand Health and Illness?

p. 435

Learn about functionalist and symbolic interactionist perspectives on physical and mental health and illness in contemporary society. Recognize the ways that disability challenges theoretical perspectives on health and illness. Understand the relationship between traditional medicine and complementary and alternative medicine (CAM).

How Do Social Factors Affect Health and Illness?

p. 441

Recognize that health and illness are shaped by cultural, social, and economic factors. Learn about race, class, gender, and geographic differences in the distribution of disease.

How Do Social Contexts Shape Sexual Behavior?

p. 447

Understand the diversity of sexual orientation today. Learn about the debate over the importance of biological versus social and cultural influences on human sexual behavior. Explore cultural differences in sexual behavior and patterns of sexual behavior today.

Terms to Know

Concept Checks

universal health coverage • social class gradient in health • sociology of the body

obesity • food deserts

1. Why is anorexia more likely to strike young women than heterosexual young men?
2. What explanations are offered for the recent increase in obesity rates?
3. In what ways is the United States an "obesogenic environment"?

sick role • stigma • complementary and alternative medicine (CAM) • biomedical model of health

1. How do functionalist theorists and symbolic interactionists differ in their perspectives on health and illness?
2. What is stigma, and how does it pertain to health and illness?
3. What is the biomedical model of health?
4. How does disability pose a challenge to both functionalist and biomedical models of health?
5. Compare complementary medicine with alternative medicine.

1. How do social class and race affect health?
2. Name at least two explanations for the gender gap in health.
3. Identify at least two reasons why the gender gap in life expectancy may narrow in the future.
4. Why are infectious diseases more common in developing nations than in the United States today?
5. What are three social consequences of the AIDS epidemic in developing nations?

heteronormativity • heterosexuality • homosexuality • bisexuality • sexual orientation • QUILTBAG • homophobia • transphobia • heterosexism

1. Describe several changes in sexual practices over the past two centuries.
2. What are the most important contributions of Alfred Kinsey's research on sexuality?
3. Name at least three important findings about sexual behavior discovered since Kinsey.
4. What are several of the most important achievements of LGBTQ rights movements?

15

Urbanization, Population, and the Environment

According to the World Health Organization (WHO), six of the ten most polluted cities in the world are in India, including New Delhi, pictured here. The WHO has called air pollution, which increases the risk of respiratory illnesses as well as heart disease and stroke, the greatest environmental risk to health.

THE BIG QUESTIONS

How do cities develop and evolve?

Learn how cities have changed as a result of industrialization and urbanization. Learn how theories of urbanism have placed increasing emphasis on the influence of socioeconomic factors on city life.

How do rural, suburban, and urban life differ in the United States?

Learn about key developments affecting American cities, suburbs, and rural communities in the last several decades: suburbanization, urban decay, gentrification, and population loss in rural areas.

How does urbanization affect life across the globe?

See that global economic competition has a profound impact on urbanization and urban life. Recognize the challenges of urbanization in the developing world.

What are the forces behind world population growth?

Learn why the world population has increased dramatically, and understand the main consequences of this growth.

How do environmental changes affect your life?

See that the environment is a sociological issue related to economic development and population growth.

Urbanization

p. 475

With a population of 1.4 billion, China is the most populous country in the world; India is a close second, with 1.3 billion, and is poised to overtake China within the next few years. Together, China and India account for nearly two-fifths of the world's total population. Their population growth has been accompanied by rapid economic growth. The Chinese economy grew at an annual rate of 9–10 percent between 1980 and 2010; India's only a few percent lower. Since that time, both have slowed to 7 percent. By way of comparison, the U.S. economy is currently growing at an annual rate of about 2 percent (World Bank 2018j).

In their rush to re-create the Industrial Revolution that made Western nations wealthy, Chinese and Indian manufacturers—spurred on by Western investors—have become the world's smokestack. Their rapid industrialization has lifted hundreds of millions of people out of poverty and into the middle class, but at a high environmental cost. Six of the ten most polluted cities in the world are in India (Miles 2018). Toxic chemical spills have threatened the water supply of millions of people, while the air in major cities has become so polluted that the ultramodern skyscrapers are often not visible.

For many years, China's booming economy depended exclusively on burning coal. Every week or so, a new coal-burning power plant was brought online, most often one using outmoded technology. The sulfur dioxide from these plants contributed to nearly a half-million deaths a year in China, while causing acid rain that poisoned lakes, rivers, and farmlands. Climate-changing smoke and soot from China's power plants have been detected across the Pacific Ocean in California. China has surpassed the United States as the world's leading emitter of greenhouse gases (although when adjusted for population size, the United States remains the world's worst offender). China's environmental threats became so acute that in 2011, the nation's environment minister, Zhou Shengxian, publicly announced that carbon dioxide pollution from coal-burning factories, along with the nation's high demand for resources—both consequences of prior economic growth—may, ironically, threaten future economic growth. Even the state-run newspaper *China Daily*, historically known for keeping the nation's problems under wraps, described the nation's major cities as "barely suitable for living," given persistent environmental threats (Wagstaff 2013).

During the past several years, public discontent over toxic air quality has spurred China toward a more "green" development path, both by reducing its reliance on export manufacturing and by switching to greener technologies such as solar, wind, and nuclear power. In 2013, the Chinese government acknowledged that it would need to spend more than $800 billion each year to fight air pollution (Yongqiang 2013). China's coal consumption has declined significantly in recent years (Carrington 2016), contributing to a global drop in greenhouse gas emissions in 2015 (Vaughan 2015).

In Delhi, it is estimated that between 25,000 and 50,000 people die each year because of air pollution; ten times that number die for the same reason throughout the country as a whole. Unlike with China, India's problem does not stem so much from smokestack industries as from motor vehicle exhaust and smoke from home cooking. The government of India has responded by shuttering some heavily polluting industries and power plants; requiring buses, rickshaws, and taxis to convert to natural gas; banning the burning of rubbish; and ending government subsidies for diesel-powered motor vehicles (*The Economist* 2016). Unlike in China, where the government exerts strong control over the economy and many aspects of private life, the Indian government is weak and corrupt.

The potentially dire future facing China and India reveals in vivid detail the ways that population growth, urbanization, industrialization, and the environment are intertwined. How exactly do these forces mutually influence one another? In the case of China, for example, the rush to develop and meet the demands of its burgeoning population has severely taxed and depleted the nation's natural resources. China has begun building a network of highways, much like the United States did during the 1950s and 1960s. In fact, the country's 45,000 miles of high-speed freeways, which will connect all major cities in China, are modeled after the U.S. interstate highway system. As was the case in the United States, the highways will generate much more extensive automobile use, projected to outstrip that of the United States by the middle of the century. For a country where, as recently as a generation ago, the bicycle and rickshaw were the principal means of transportation, this is an enormous transformation, and one that has already contributed

A coal miner emerges from a mine after a day's work in Shanxi Province, China.

to urban traffic congestion, multiday traffic jams, and increased levels of energy use and pollution. China is making the transition from rural to urban in record time, and the government is calling for the relocation of some 400 million people—more than the entire U.S. population—to newly built urban centers. China already has more than 100 cities with populations exceeding 1 million, including six "megacities" with more than 10 million people (*The Economist* 2015b).

In terms of industrialization, China's explosive economic growth, as previously noted, was based on enormous resource consumption and the burning of greenhouse gas–emitting coal and other fossil fuels. Yet China's environmental challenges are not wholly due to internal factors. After all, the growth of manufacturing in China has been the direct result of European and North American firms' decisions to relocate their manufacturing to China, where environmental restrictions have historically been weaker (thereby lowering production costs).

In this chapter, we examine the ways in which population growth, urbanization, and environmental change go hand in hand, against the backdrop of the rapid industrialization that is transforming many parts of the world. We begin by studying the origins of cities and the vast growth in the numbers of city dwellers that has occurred over the past century. From there, we review the most influential theories of urban life. We then move on to consider patterns of urban development in North America compared with cities in the developing world. Cities in the Global South are growing at an enormous rate. We consider why this is happening and, at the same time, look at changes now taking place in world population patterns. We conclude by assessing the connections among urbanization, world population growth, and environmental problems.

How Do Cities Develop and Evolve?

Cities in Traditional Societies

The world's first cities appeared about 3500 BCE, in the river valleys of the Nile in Egypt, the Tigris and Euphrates in what is now Iraq, and the Indus in what is today Pakistan. Cities in traditional societies were very small by modern standards. Babylon, for example, one of the largest ancient Middle Eastern cities, extended over an area of only 3.2 square miles and at its height, around 2000 BCE, probably numbered no more than 15,000 to 20,000 people. Rome under Emperor Augustus in the first century BCE was easily the largest premodern city outside China, with some 300,000 inhabitants—the population of Anchorage, Cincinnati, or Pittsburgh today.

Most cities of the ancient world shared certain features. They were usually surrounded by walls that served as a military defense and emphasized the separation of the urban community from the countryside. The central area of the city was almost always occupied by a religious temple, a royal palace, government and commercial buildings, and a public square. This ceremonial, commercial, and political center was sometimes enclosed within a second inner wall and was usually too small to hold more than a minority of the citizens. The dwellings of the ruling class or elite tended to be concentrated in or near the center. Less privileged groups lived toward the perimeter of the city or outside the walls, moving inside if the city came under attack. Different ethnic and religious communities

Learn how cities have changed as a result of industrialization and urbanization. Learn how theories of urbanism have placed increasing emphasis on the influence of socioeconomic factors on city life.

were often segregated in separate neighborhoods, where their members lived and worked. Communication among city dwellers was erratic. Lacking any form of printing press, and with very low literacy rates, public officials had to shout at the tops of their voices to deliver pronouncements. A few traditional civilizations boasted sophisticated road systems linking particular cities, but these existed mainly for military purposes, and transportation for the most part was slow and limited. Merchants and soldiers were the only people who regularly traveled over long distances.

Although cities were the main centers for science, the arts, and cosmopolitan culture, their influence in surrounding areas was always weak. No more than a tiny proportion of the population lived in the cities, and the division between cities and countryside was pronounced. By far the majority of people lived in small rural communities and rarely came into contact with more than the occasional state official or merchant from the towns.

Industrialization and Urbanization

The contrast in size between the largest modern cities today and those of premodern civilizations is extraordinary. The most populous cities in the industrialized countries—sometimes termed "megacities"—number more than 10 million inhabitants (see the "Globalization by the Numbers" infographic on p. 475). A **conurbation**—a cluster of cities and towns forming a continuous network—may include even larger numbers of people. The peak of urban life today is represented by what is called the **megalopolis**, the "city of cities." The term was originally coined in ancient Greece to refer to a city-state that was planned to be the envy of all civilizations. The current megalopolis, though, bears little relation to that utopia. The term was first applied in modern times to refer to the Northeast Corridor of the United States, an area covering some 450 miles from north of Boston to south of Washington, D.C. In this region, more than 50 million people live at an average density of more than 800 persons per square mile. Large, dense urban populations can also be found in the lower Great Lakes region surrounding Chicago and in the San Francisco–East Bay–San Jose–Silicon Valley region of California (Scommegna 2011).

Britain was the first society to undergo industrialization, beginning in the mid-eighteenth century. The process of industrialization generated increasing **urbanization**—the movement of the population into towns and cities, away from the land. In 1800, less than 20 percent of the British population lived in towns or cities with more than 10,000 inhabitants. By 1900, this proportion had risen to 74 percent. London held approximately 1.1 million people in 1800; by the beginning of the twentieth century, it had increased in size to a population of more than 7.8 million, at that date the largest city ever seen in the world. It was a vast manufacturing, commercial, and financial center at the heart of the still-expanding British Empire.

The urbanization of most other European countries and the United States took place somewhat later. In 1800, the United States was more of a rural society than were the leading European countries. Less than 10 percent of Americans lived in communities with populations of more than 2,500 people. Between 1800 and 1900, as industrialization grew in the United States, the population of New York City leaped from 60,000 to 4.8 million. Today, slightly more than 80 percent of Americans reside in metropolitan areas.

Urbanization in the twenty-first century is a global process into which the developing world is increasingly being drawn. From 1900 to 1950, world urbanization increased five

conurbation

A cluster of towns or cities forming an unbroken urban environment.

megalopolis

The "city of all cities" in ancient Greece—used in modern times to refer to very large conurbations.

urbanization

The movement of the population into towns and cities and away from the land.

times faster than world population growth, and the years since have seen an even greater acceleration in urbanization. In 1950, roughly 30 percent of the world's population lived in urban areas. By 2018, the figure had grown to more than half (55 percent), and it is projected to reach 68 percent by 2050 (UN Department of Economic and Social Affairs 2018).

Urbanization goes hand in hand with economic growth. In 2018, the least urbanized countries were also low-income countries (where only 32 percent lived in urban areas); next came the more urbanized middle-income countries (with 53 percent urban), while high-income countries were the most urbanized (82 percent). At more than 82 percent urban, North America is the most urbanized, followed by Latin America and the Caribbean (81 percent) and Europe (74 percent). The least urbanized countries are in Eastern Africa (28 percent urban), which is also one of the poorest regions in the world (UN Department of Economic and Social Affairs 2018).

China went from 12 percent urbanized in 1950 to 59 percent in 2018, an unprecedented movement of hundreds of millions of people from rural countryside to enormous industrial cities, initially on China's east coast but more recently to metropolitan areas throughout the country. This urbanization, as noted previously, was in large part by government design, as central state planning called for urban-based industrial production to replace rural agriculture as the driving force in the economy. India, on the other hand, remains largely rural, its urban population increasing from 17 percent in 1950 to only 34 percent in 2018 (UN Department of Economic and Social Affairs 2018).

Traffic outside of the Bank of England in the financial district of London in 1896. In only a century, the population of London grew from just over 1 million people to more than 7 million.

Theories of Urbanism

THE CHICAGO SCHOOL

Scholars associated with the University of Chicago from the 1920s to the 1940s—especially Robert Park, Ernest Burgess, and Louis Wirth—developed ideas that were for many years the chief basis of theory and research in urban sociology. Two concepts developed by the "Chicago School" are worthy of special attention. One is the so-called **ecological approach** to urban analysis, the other the characterization of urbanism as a *way of life*, developed by Wirth (Park 1952; Wirth 1938). It is important to understand these ideas as they were initially conceived by the Chicago School and to see how they have been revised and even replaced by sociologists in more recent decades.

Urban Ecology *Ecology*—the study of the adaptation of plant and animal organisms to their environment—is a term taken from the physical sciences. In the natural world, organisms tend to be distributed in systematic ways over the terrain, such that a balance or equilibrium among different species is achieved. The Chicago School believed that the locations of major urban settlements and the distribution of different types of neighborhoods within them could be understood in terms of similar principles. Cities do not grow

ecological approach

A perspective on urban analysis emphasizing the "natural" distribution of city neighborhoods into areas having contrasting characteristics.

up at random but in response to advantageous features of the environment. For example, large urban areas in modern societies tend to develop along the shores of rivers, in fertile plains, or at the intersections of trading routes or railways.

According to Park, cities become ordered into "natural areas" through processes of competition, invasion, and succession—all of which also occur in biological ecology. Patterns of location, movement, and relocation in cities, according to the ecological view, take a similar form. Different neighborhoods develop through the adjustments made by inhabitants as they struggle to gain their livelihoods. A city can be pictured as a map of areas with distinct and contrasting social characteristics, in concentric rings, broken up into segments. In the center are the **inner-city** areas, a mixture of big-business prosperity and decaying private homes. Beyond these are older established neighborhoods, housing workers employed in stable manual occupations. Farther out still are the suburbs, in which higher-income groups tend to live. Processes of invasion and succession occur within the segments of the concentric rings. Thus, as property decays in a central or near-central area, ethnic/minority groups might start to move into it. As they do so, more of the preexisting population starts to leave, precipitating movement to neighborhoods elsewhere in the city or out to the suburbs. However, as we will see later, these traditional patterns are starting to change: Wealthy persons and the young are flooding into urban areas, seeking amenities such as arts and entertainment, and suburban areas are becoming more desirable (and affordable) to poor and working-class persons.

Another aspect of the **urban ecology** approach emphasizes the *interdependence* of different city areas. Differentiation—the specialization of groups and occupational roles—is the main way human beings adapt to their environment. Groups on which many others depend will have a dominant role, often reflected in their central geographical position. Business groups, for example, such as large banks or insurance companies, provide key services for many in a community and hence are usually found in the central areas of settlements (Hawley 1950, 1968).

Urbanism as a Way of Life   Wirth's (1938) thesis of **urbanism** outlines the ways that life in cities is different from life elsewhere. In cities, large numbers of people live in close proximity without knowing most others personally—a fundamental contrast to small, traditional villages. Most contact between city dwellers is fleeting and partial. Interactions with sales clerks in stores, baristas at coffee shops, or passengers or ticket collectors on trains are passing encounters, entered into not for their own sake but as means to other aims.

Wirth was among the first to address the "urban interaction problem" (Duneier and Molotch 1999), the necessity for city dwellers to respect social boundaries when so many people are in close physical proximity all the time. Many people walk down the street in cities acting unconcerned about others near them, often talking on cell phones or listening to music that blocks out the sounds of urban life. Through such appearance of apathy, they can avoid unwanted transgression of social boundaries.

Wirth's ideas have deservedly enjoyed wide currency. However, in assessing these ideas, we should consider that neighborhoods marked by close kinship and personal ties are often actively created by city life; they are not just remnants of a preexisting way of life that survive for a period within the city. Claude Fischer (1984) has put forward an explanation for why large-scale urbanism helps promote diverse subcultures. Those who live

inner city

The areas composing the central neighborhoods of a city, as distinct from the suburbs. In many modern urban settings in industrialized nations, inner-city areas are subject to decay and dilapidation, the more affluent residents having moved to outlying areas.

urban ecology

An approach to the study of urban life based on an analogy with the adjustment of plants and organisms to the physical environment. According to ecological theorists, the various neighborhoods and zones within cities are formed as a result of natural processes of adjustment on the part of populations as they compete for resources.

urbanism

A term used by Louis Wirth to denote distinctive characteristics of urban social life, such as its impersonal or alienating nature.

in cities are able to collaborate with others of similar background and interests to develop local connections, and they can join distinctive religious, ethnic, political, and other subcultural groups. A small town or village does not allow for the development of such subcultural diversity. For example, some gay and lesbian young people may find more hospitable communities in cities, such as San Francisco, with large gay subcultures, compared with the small towns where they may have grown up. A large city is a world of strangers, yet it ultimately supports and creates personal relationships.

JANE JACOBS: "EYES AND EARS UPON THE STREET"

Like most sociologists in the twentieth century, the Chicago School researchers were professors who saw their mission as contributing to a scholarly literature and advancing the field of social science. Yet one of the most influential urban scholars of the twentieth century, Jane Jacobs, author of *The Death and Life of Great American Cities* (1961), was an architecture critic with a high school education. Through her own independent reading and research in the 1950s, she transformed herself into one of the most learned figures in the emerging field of urban studies.

Like sociologists before her, such as Wirth of the Chicago School, Jacobs noted that "cities are, by definition, full of strangers," some of whom are dangerous. She argued that cities are most habitable when they feature a diversity of uses, thereby ensuring that many people will be coming and going on the streets at any time. When enough people are out and about, Jacobs wrote, "respectable" eyes and ears dominate the street and are fixed on strangers, who will thus not get out of hand. The more people who are out, or who are looking from their windows at the people who are out, the more their gazes will safeguard the street.

The world has changed a great deal since Jacobs wrote *The Death and Life of Great American Cities*. Most of the people on the sidewalks Jacobs was writing about were more alike in many respects than they are today; now homeless people, drug users, panhandlers, and others representing economic inequalities, cultural differences, and extremes of behavior can make sidewalk life unpredictable (Duneier 1999). Under these conditions, strangers do not necessarily feel the kind of solidarity and mutual assurance she described. Sociologists today must ask, What happens to urban life when "the eyes and ears upon the street" represent vast inequalities and cultural differences? Do the assumptions Jacobs made still hold up? In many cases, the answer is yes, but in other cases, the answer is no. More than five decades after her book was published, Jacobs's ideas remain extremely influential.

URBANISM AND THE CREATED ENVIRONMENT

Whereas the earlier Chicago School of sociology emphasized that the distribution of people in cities occurs naturally, more recent theories of the city have stressed that urbanism is not a natural process but rather is shaped by political and economic forces. These theories focus on the **created environment**, or those constructions established by humans to serve their own needs, including roads, railways, factories, offices, private homes, and other buildings. Urbanism is a core aspect of the created environment. Cities and urban areas were "created" by the spread of industrial capitalism.

Cities can promote diverse subcultures. The Castro district of San Francisco, for example, has a thriving gay subculture.

created environment

Constructions established by human beings to serve their own needs, including roads, railways, factories, offices, homes, and other buildings.

According to this view, it is not the stranger on the sidewalk who is most threatening to many urban dwellers, especially the poor; instead, it is the stranger far away, working in a bank or real estate development company, who has the power to make decisions that transform whole blocks or neighborhoods (Logan and Molotch 1987). This focus on the political economy of cities, and on different kinds of strangers, represented a new and critical direction for urban sociology. It emphasizes the ways that everyday life in urban areas is shaped by macrosocial forces and institutions, including corporations and public policies.

According to social geographer David Harvey (1973, 1982, 1985), space is continually restructured in modern urbanism. The process is determined by where large firms choose to place their factories, research and development centers, and so forth; the controls that governments operate over both land and industrial production; and the activities of private investors, who are buying and selling houses and land. Businesses, for example, are constantly weighing the relative advantages of new locations against existing ones. As production becomes cheaper in one area, or as the firm moves from one product to another, offices and factories will be closed down in one place and opened up elsewhere. Thus, at one period, when there are considerable profits to be made, there may be a spate of office-block buildings in the center of large cities. Once the offices have been built and the central area redeveloped, investors look for the potential for further speculative building elsewhere. Often what is profitable in one period will not be so in another, when the financial climate changes.

The activities of private home buyers are strongly influenced by how far, and where, business interests buy up land, as well as by rates of loans and taxes fixed by local and central government. After World War II, for instance, there was vast expansion of suburban development outside major cities in the United States. This was partly due to ethnic discrimination and the tendency of whites to move away from inner-city areas. However, it was made possible, Harvey argues, only because of government decisions to provide tax breaks to home buyers and construction firms and the willingness of financial organizations to set up special credit arrangements. These provided the basis for the building and buying of new homes on the peripheries of cities and at the same time promoted demand for industrial products, such as the automobile (Harvey 1973, 1982, 1985).

Like Harvey, Manuel Castells (1977, 1983) emphasizes that the spatial form of a society is closely linked to the overall mechanisms of its development. However, the nature of the created environment is not just the result of the activities of wealthy and powerful people. Castells stresses the importance of the struggles of underprivileged groups to alter their living conditions. Urban problems stimulate a range of social movements, concerned with improving housing conditions, protesting against air pollution, defending parks, and combating building development that changes the nature of an area. For example, Castells has studied the gay movement in San Francisco, which succeeded in restructuring neighborhoods around its own cultural values—allowing many gay organizations, clubs, and bars to flourish—and gained a prominent position in local politics.

CONCEPT CHECKS

1. What are two characteristics of ancient cities?

2. What is urbanization? How is it related to globalization?

3. How does urban ecology use physical science analogies to explain life in modern cities?

4. What is the urban interaction problem?

5. According to Jane Jacobs, the more people there are on the streets, the more likely it is that street life will be orderly. Do you agree with Jacobs's hypothesis and her explanation for this pattern?

How Do Rural, Suburban, and Urban Life Differ in the United States?

What are the main trends that have affected city, suburban, and rural life in the United States over the past several decades? How can we explain patterns such as suburban sprawl, the disappearance of traditional rural life, and population declines in central cities and older suburbs? These are questions we will take up in the following sections. One of the major changes in population distribution in the period since World War II is the movement of large parts of city populations to newly constructed suburbs; this movement outward has been a particularly pronounced feature of American cities and is related directly to central-city decay. At the same time, rural populations have continued to decline as young people seek richer professional and personal opportunities in our nation's large and small cities. We therefore begin with a discussion of rural America and suburbia before moving on to look at the inner city.

The Decline of Rural America?

Rural life has long been the focus of romanticized images among Americans: close-knit communities and families, stretches of picturesque cornfields, and isolation from social problems like poverty. Yet these stereotypes stand in stark contrast to life in many parts of rural America today. Rural areas of the United States are defined by the Census Bureau as those areas located outside urbanized areas or urban clusters. Rural areas have fewer than 2,500 people, who typically live in open country. Rural America contains approximately 97 percent of the nation's land area yet holds only about 19 percent of the total U.S. population (U.S. Bureau of the Census 2016d).

For most of the twentieth century, rural communities experienced significant population losses, despite several modest short-term reversals in the 1970s and 1990s (U.S. Department of Agriculture 2013). This trend is continuing into the twenty-first century. Population losses in rural areas are attributed to declines in farming and other rural industries; high poverty rates; scarce economic opportunities or lifestyle amenities for young people; lack of government services; and—in some regions—a dearth of natural amenities, such as forests, lakes, or temperate winters. Population losses are compounded by the fact that most people leaving rural areas are young people, meaning that fewer babies are born to replace the aging population (Johnson 2006). Many rural areas have disproportionately high numbers of older adults because young persons seek opportunities elsewhere and leave the older persons behind. This phenomenon, called *aging in place*, explains the relatively old populations in rural areas in the Rust Belt and upper Midwest (McGranahan and Beale 2002).

Poverty rates have been higher in rural than urban areas since this measure was first recorded in 1960, although the lower cost of rural housing may partly offset this difference (U.S. Department of Agriculture 2017). One out of every four rural children lives in

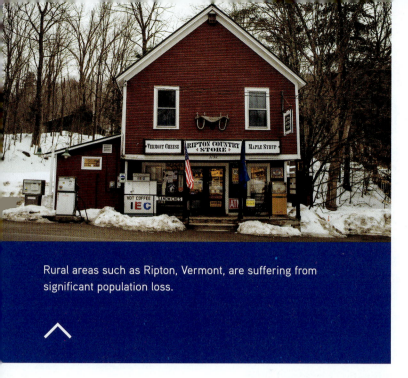

Rural areas such as Ripton, Vermont, are suffering from significant population loss.

poverty, significantly higher than the urban rate (one out of every five) (National Low Income Housing Coalition 2015). Not all areas are equally likely to be poverty stricken, however. Child poverty rates are highest in the most remote rural counties with the lowest population densities. The counties with persistent child poverty tend to cluster in Appalachia, along the Mississippi Delta, in the northern Great Plains, along the Texas–Mexico border, and in the Southwest (Schaefer, Mattingly, and Johnson 2016). Race also shapes rural poverty, just as it shapes urban poverty. Rural counties with the highest child poverty rates often are "majority minority" counties, where less than 50 percent of the population is non-Hispanic white. These areas include Black-majority counties in the Mississippi Delta and counties in the Midwest and West that have large Native American populations, often dwelling on Indian reservations (O'Hare and Mather 2008).

Despite the challenges facing rural America, many rural sociologists are guardedly optimistic about the future of nonmetropolitan life. Technological innovations in transportation and telecommunications give people the flexibility to work away from their urban offices. A number of government programs offer young people financial incentives to serve as teachers or health care professionals in remote areas. However, such programs are likely to be effective only in attracting workers and businesses to rural areas that have at least some natural or recreational amenities (Johnson 2006).

Suburbanization

suburbanization

The development of towns surrounding a city.

In the United States, **suburbanization**, the massive development and inhabiting of towns surrounding a city, rapidly increased during the 1950s and 1960s, a time of great economic growth. World War II had absorbed most industrial resources, and any development outside the war effort was restricted. But by the 1950s, war rationing had ended, and the postwar economic boom facilitated moving out of the city. The Federal Housing Administration (FHA) provided assistance in obtaining mortgage loans, making it possible in the early postwar period for families to buy housing in the suburbs for less than they would have paid for rent in the cities. The FHA did not offer financial assistance to improve older homes or to build new homes in the central areas of ethnically mixed cities; its large-scale aid went only to the builders and buyers of suburban housing.

Early in the 1950s, lobbies promoting highway construction launched Project Adequate Roads, aimed at convincing the federal government to support the building of highways. In 1956, the Highway Act was passed, authorizing $32 billion to be used for building such highways. The new highway program—funded in part by gasoline taxes, which grew rapidly as more and more people took to the wheel—led to the establishment of industries and services in suburban areas themselves. Consequently, the movement of businesses from the cities to the suburbs took jobs in the manufacturing and service industries with them. Many suburban towns became essentially separate cities, connected by rapid highways to the other suburbs around them. From the 1960s on, the proportion of people commuting between suburbs increased more steadily than the proportion commuting to cities.

Suburban Levittown, New York, in the 1950s (left). A housing development in the exurb of Highland, California (right).

An important change in suburbs today is that more and more members of racial and ethnic minorities are moving there. Blacks accounted for 7 percent of the suburban population in 1990, 9 percent in 2000, and 10 percent in 2010. Comparable increases for Latinos were even steeper, climbing from 8 to 12 to 17 percent during the same time period. Whites as a share of the suburban population declined steeply, from 81 percent in 1990 to 65 percent in 2010. The steady increase in minority suburban populations was concentrated in so-called melting-pot metros, or the metropolitan regions of New York, Los Angeles, Chicago, San Francisco, Miami, and other immigrant gateway cities (Frey 2011b). Members of minority groups move to the suburbs for reasons similar to those who preceded them: better housing, schools, and amenities. A notable exception to the growing diversity of suburban America is exurbs. **Exurban counties**, or low-density suburban counties on the periphery of large metro areas, grew rapidly from 2000 to 2010, with whites accounting for 73 percent of that growth (Frey 2011a).

While the last several decades saw a movement from the cities to the suburbs, they also witnessed a shift in the regional distribution of the U.S. population from North to South and from East to West. Between 2000 and 2010, regional growth was much more rapid in the South and West (14 percent) than it was in the Northeast (3 percent) and Midwest (4 percent) (Mackun and Wilson 2011).

Urban Problems

Inner-city decay is partially a consequence of the social and economic forces involved in the movement of businesses, jobs, and middle-class residents from major cities to the outlying suburbs, a trend that began in the 1950s. The manufacturing industries that provided employment for the urban blue-collar class largely vanished and were replaced by white-collar service industries. Millions of blue-collar jobs disappeared, affecting in particular the poorly educated, drawn mostly from minority groups. Although the overall educational levels of minority groups have improved since the mid-twentieth century, the improvement has not been sufficient to keep up with the demands of an information-based economy (Kasarda 1993). William Julius Wilson (1991, 1996) has argued that the problems of the urban underclass have grown out of this economic transformation (see Chapter 7).

exurban counties

Low-density suburban counties on the periphery of large metro areas.

These economic changes also contributed to increased residential segregation of racial and ethnic groups and social classes, as we saw in Chapter 10. Discriminatory practices by home sellers, real estate agents, and mortgage-lending institutions added to this pattern of segregation (Massey and Denton 1993). Currently the average white resident lives in a neighborhood that is 77 percent white; the average Black resident lives in a neighborhood that is 45 percent Black. While Hispanics constitute only 15 percent of the population, 45 percent of their neighbors are also Hispanic (Frey 2015). While some may argue that this is because most people prefer to live in racially homogeneous neighborhoods populated by people similar to themselves, it turns out that racial segregation today is in large part the result of government housing policies that began during the Great Depression of the 1930s and continued for several decades after World War II.

In *The Color of Law: A Forgotten History of How Our Government Segregated America*, Richard Rothstein (2017) argues that housing built under the federal Public Works Administration was deliberately segregated, with most going to whites. After World War II, the 1949 Housing Act provided Federal Housing Administration (FHA) guarantees for bank loans, enabling the construction of large-scale white-only housing developments, with deeds that prohibited resale to African Americans (so-called restrictive covenants). At the local level, African American neighborhoods were frequently rezoned to allow industrial (and often toxic) uses, effectively turning them into slums. These and other government policies enabled white working-class Americans to buy homes in decent neighborhoods at affordable prices, while consigning most African Americans to rental housing in substandard areas. The social isolation of minority groups, particularly those in the underclass or "ghetto poor," can escalate urban problems such as crime, lack of economic opportunities, poor health, and family breakdown (Massey 1996).

Adding to these difficulties is the fact that city governments today operate against a background of almost continual financial crisis. As businesses and middle-class residents moved to the suburbs, cities lost major sources of tax revenue. High rates of crime and unemployment in the city require it to spend more on welfare services, schools, police, and overall upkeep. Yet because of budget constraints, cities are forced to cut back many of these services. A cycle of deterioration develops in which the more suburbia expands, the greater the problems faced by city dwellers become. Problems of urban decline reached a pinnacle when the city of Detroit, Michigan, filed for bankruptcy. Over the course of several decades, both businesses and residents had left the city for neighboring suburbs, depleting the city's tax base. By the late 2000s, when the auto industry was in serious crisis, most of the major auto companies had already fled the city and moved their factories to the suburbs. After years of paying its bills with borrowed money, the city finally succumbed to bankruptcy in 2013 (Bomey, Snavely, and Priddle 2013).

Gentrification and Urban Renewal

Urban decay is not wholly a one-way process; it can stimulate countertrends, such as gentrification and urban renewal. More recently, there has been a revival of many central cities, as wealthier professionals in some regions have moved back into downtown areas. This revival has resulted in a process of **gentrification** whereby older, deteriorated housing and other buildings are renovated as more affluent groups move into an area. Such a renewal process is called gentrification because those areas or buildings become upgraded and return to the control of the urban "gentry"—high-income dwellers—rather than remaining in the hands of the poor.

gentrification

A process of urban renewal in which older, deteriorated housing is refurbished by affluent people moving into the area.

---

urban renewal

The process of renovating deteriorating neighborhoods by encouraging the renewal of old buildings and the construction of new ones.

In *Streetwise: Race, Class, and Change in an Urban Community* (1990), sociologist Elijah Anderson analyzed the effect of gentrification on cities. Although the renovation of a neighborhood generally increases its value, it rarely improves the living standards of its current low-income residents, who are usually forced to move out. The poor residents who continue to live in the neighborhood receive some benefits in the form of improved schools and police protection, but the resulting increases in taxes and rents often force them to leave for a more affordable neighborhood, most often deeper into the ghetto. The white newcomers come to the city in search of cheap "antique" housing, closer access to their city-based jobs, and a trendy urban lifestyle. They profess to be "open-minded" about racial and ethnic differences; in reality, however, little fraternizing takes place between the new and old residents unless they are of the same social class. Over time, the neighborhood is gradually transformed into a white middle-class enclave.

Often gentrification is a result not purely of market forces but of government policies that put public funds into poor areas. **Urban renewal** (or sometimes urban redevelopment) is the process of renovating deteriorating neighborhoods by using public funds to renew old buildings and construct new ones, often through large-scale demolition of slum housing. While the twin processes of gentrification and urban renewal are not new, they gained force after World War II, in large part when the 1949 and 1954 Federal Housing Acts pumped billions of federal dollars into urban renewal programs in run-down urban areas.

CONCEPT CHECKS

1. Describe at least two problems facing rural America today.

2. Why did so many Americans move to suburban areas in the 1950s and 1960s?

3. What are two unintended consequences of suburbanization? How do they deepen socioeconomic and racial inequalities?

How Does Urbanization Affect Life across the Globe?

In premodern times, cities were self-contained entities that stood apart from the predominantly rural areas in which they were located. Communication between cities was limited. The picture at the start of the twenty-first century could hardly be more different. Globalization has had a profound effect on cities, making them more interdependent and encouraging the proliferation of horizontal links between cities across national borders. Physical and virtual ties between cities now abound, and global networks of cities are emerging.

See that global economic competition has a profound impact on urbanization and urban life. Recognize the challenges of urbanization in the developing world.

Global Cities

The role of cities in the new global order has been attracting a great deal of attention from sociologists. Saskia Sassen has been one of the leading contributors to the debate on cities and globalization. She uses the term **global city** to refer to urban centers that are home to the headquarters of large transnational corporations and a superabundance of financial, technological, and consulting services. In *The Global City* (1991), Sassen based her work on the study of three such cities: New York, London, and Tokyo. The contemporary development of the world economy, she argues, has created a novel strategic role for major cities.

global city

A city—such as London, New York, or Tokyo—that has become an organizing center of the new global economy.

Most such cities have long been centers of international trade, but they now have four new traits:

1. They have developed into command posts—centers of direction and policy-making—for the global economy.

2. They are the key locations for financial and specialized service firms, which have become more important than manufacturing in influencing economic development.

3. They are the sites of production and innovation in these newly expanded industries.

4. They are markets on which the "products" of financial and service industries are bought, sold, or otherwise disposed of.

Within the highly dispersed world economy of today, cities like these provide for central control of crucial operations. Global cities are much more than simply places of coordination, however; they are also contexts of production. What is important here is not the production of material goods but the production of the specialized services required by business organizations for administering offices and factories scattered across the world and the production of financial innovations and markets. Services and financial goods are the "things" the global city makes.

Inequality and the Global City

The new global economy is highly problematic in many ways. Sassen argues that global cities have produced a new dynamic of inequality: It is no coincidence that the prosperous central business district adjoins impoverished inner-city areas in many global cities; business and poverty should be seen as interrelated phenomena. The growth sectors of the new economy—financial services, marketing, high technology—are reaping profits far greater than any found within traditional economic sectors. Those who work in finance and global services receive high salaries, and the areas where they live become gentrified. At the same time, manufacturing jobs are lost, and the very process of gentrification creates a vast supply of low-wage jobs—in restaurants, hotels, and boutiques. Affordable housing is scarce in gentrified areas, forcing an expansion of low-income neighborhoods. As the salaries and bonuses of the very affluent continue to climb, the wages of those employed to clean and guard their offices are dropping. Sassen (1998) argues that we are witnessing the "valorization" of work located at the forefront of the new global economy and the "devalorization" of work that occurs behind the scenes.

Within global cities, a geography of "centrality and marginality" is taking shape. Alongside resplendent affluence is acute poverty. These two worlds exist side by side, yet actual contact between them can be surprisingly minimal. As Mike Davis (1990) noted in his study of Los Angeles, there has been a "conscious 'hardening' of the city surface against the poor." Accessible public spaces have been replaced by walled compounds, neighborhoods guarded by electronic surveillance. Benches at bus stops are short or barrel shaped to prevent people from sleeping on them, the number of public toilets is fewer than in any other North American city, and sprinkler systems have been installed in many parks to deter the homeless from living in them. Police and city planners have attempted to contain the homeless population within certain regions of the city, but in periodically sweeping through and confiscating makeshift shelters, they have effectively created a population of "urban Bedouins."

Urbanization

Today 55 percent of the world's population resides in cities. This proportion is expected to rise to 68 percent by 2050, with China, India, and Nigeria alone accounting for more than a third of the projected growth of the world's urban population.

Largest cities in 2018

In millons

	2018	2035*
Tokyo, Japan	37.5	36.0
Delhi, India	28.5	43.3
Shanghai, China	25.6	34.3
São Paulo, Brazil	21.7	24.5
Mexico City, Mexico	21.6	25.4
Cairo, Egypt	20.1	28.5
Mumbai, India	20.0	27.3
Beijing, China	19.6	25.4
Dhaka, Bangladesh	19.6	31.2
Osaka, Japan	19.3	18.3
New York-Newark, U.S.	18.8	20.8
Karachi, Pakistan	15.4	23.1

*Projected.

Percentage of population residing in urban areas, 1950–2050

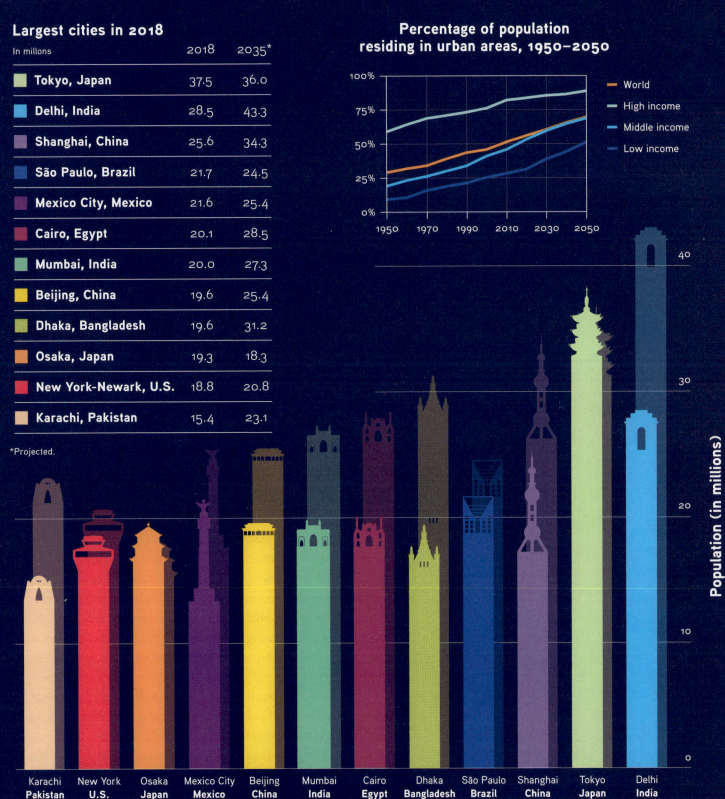

Legend:
- World
- High income
- Middle income
- Low income

Population (in millions)

Karachi **Pakistan** · New York **U.S.** · Osaka **Japan** · Mexico City **Mexico** · Beijing **China** · Mumbai **India** · Cairo **Egypt** · Dhaka **Bangladesh** · São Paulo **Brazil** · Shanghai **China** · Tokyo **Japan** · Delhi **India**

Source: United Nations Department of Economic and Social Affairs 2018.

Urbanization in the Global South

Most urban growth is now occurring in cities in the Global South. China, India, and Nigeria alone are projected to account for more than one-third of the increase in the global urban population from 2018 to 2050. New Delhi is projected to become the most populous city on the planet around 2028 (UN Department of Economic and Social Affairs 2018). Why will urban growth be limited largely to less industrialized regions in the coming decades? Apart from the obvious fact that the most economically developed nations are already highly urbanized, two factors in particular must be taken into account.

First, rates of population growth are higher in the Global South than they are in industrialized nations. Urban growth is fueled by high fertility rates among people already living in cities. Second, there is widespread *internal migration* from rural areas to urban ones. People are drawn to cities in the Global South either because their traditional systems of rural production have disintegrated or because the urban areas offer superior job opportunities. Rural poverty prompts many people to try their hand at city life. They may intend to migrate to the city for only a short time, aiming to return to their villages once they have earned enough money.

informal economy

Economic transactions carried on outside the sphere of formal paid employment.

CHALLENGES OF URBANIZATION IN THE GLOBAL SOUTH

Rapid urbanization in the Global South has brought with it many economic, environmental, and social challenges.

Economic Challenges   As a growing number of unskilled and agricultural workers migrate to urban centers, the formal economy often struggles to absorb the influx into the workforce. In most cities in the Global South, it is the **informal economy** that allows those who cannot find formal work to make ends meet. It is estimated that more than 60 percent of the employed population worldwide earn their living in the informal economy (International Labour Organization 2018). From casual work in manufacturing and construction to small-scale trading activities, the unregulated informal sector offers earning opportunities to poor or unskilled workers.

Informal economic opportunities are important in helping thousands of families (and women, especially) to survive in urban conditions, but they are also problematic. The informal economy is untaxed and unregulated. It is also less productive than the formal economy. Countries where economic activity is concentrated in this sector fail to collect much-needed revenue through taxation. The low level of productivity also hurts the general economy—the proportion of the GDP generated by informal economic activity is much lower than the percentage of the population involved in the sector. Some development analysts argue that attention should be paid to formalizing or regulating the large informal economy, where much of the excess workforce is likely to cluster in years to come.

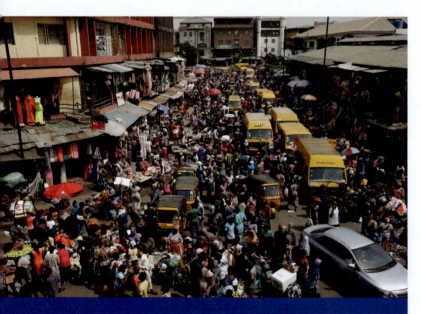

There are currently thirty-one megacities with 10 million or more inhabitants, twenty-four of which are in the Global South, including Lagos, Nigeria.

Residents of South New Delhi, India, collect water from a government water tank. Access to safe water supplies is a chronic problem in urban areas of the Global South.

Environmental Challenges The rapidly expanding urban areas in the Global South differ dramatically from cities in the industrialized world. Although cities everywhere are faced with environmental problems, those in the Global South are confronted by particularly severe risks. As we saw earlier in this chapter, China's leaders have called for a lower economic growth rate, with the explicit goal of capping the nation's energy use and reducing environmental degradation. Pollution, housing shortages, inadequate sanitation, and unsafe water supplies are chronic problems for cities in less industrialized countries.

Housing is one of the most acute problems in many urban areas. Cities such as Calcutta and São Paulo are massively congested. Housing shortages have resulted in rising prices for available housing, forcing the poor—and the migrants who flock to these cities in search of jobs—to crowd into squatters' zones that mushroom around the edges of cities. In urban areas in the West, newcomers are most likely to settle close to the central parts of the city, but the reverse tends to happen in the Global South, where migrants populate what has been called the "septic fringe" of the urban areas. Shanty dwellings made of concrete blocks—or, in worse cases, burlap or even cardboard—are set up around the edges of the city wherever there is a little space.

Congestion and overdevelopment in city centers lead to serious environmental problems in many urban areas. Mexico City is a prime example. About 94 percent of Mexico City consists of built-up areas, with only 6 percent of land left as open space. The level of green spaces—parks and open stretches of green land—is far below that found in even the most densely populated U.S. or European cities. Pollution is a major problem, coming mostly from the cars, buses, and trucks that pack the inadequate roads of the city, the rest deriving from industrial pollutants. It has been estimated that living in Mexico City is equivalent to smoking forty cigarettes a day.

Finally, global climate change, in combination with population growth and urbanization, can have deadly consequences. One study predicts that heat-related deaths in India will double by 2080, the direct result of global warming (*Economic Times* 2015); another study drew similar conclusions for New York City (Freedman 2013). In Pakistan, 1,250 people died during a heat wave in June 2015; during spring 2016, Karachi—Pakistan's largest city, with more than 16 million people—began to run out of water, with people queuing up for hours for potable water brought in on trucks (Awaz.TV 2016). Nor is Pakistan alone; nearly half the global population already resides in places that are potentially water scarce at least one month a year, three-quarters of whom are in Asia (UNESCO 2018b).

Social Challenges   Many urban areas in the Global South are overcrowded, and social programs are underresourced. Poverty is widespread, and existing social services cannot meet the demands for health care, family-planning advice, education, and training. The unbalanced age distribution in less industrialized countries adds to their social and economic difficulties. Compared with industrialized countries, a much larger proportion of the population in the Global South is under age fifteen. For example, in many African nations, nearly half of the population is under age fifteen. Niger is the youngest country in the world, with fully half of its national population under fifteen. This figure stands at 50 percent in Niger, 48 percent in Uganda and Mali, 47 percent in Chad and Angola, and 46 percent in Somalia and the Democratic Republic of the Congo. By way of contrast, in Japan and Germany—two of the world's oldest populations—just 13 percent of the national population is under age fifteen (World Bank 2018l).

A youthful population needs a good educational system, but many countries in the Global South lack the resources to provide universal education. When their families are poor, many children must work full time, and others have to eke out a living on the street, begging for whatever they can. Poor urban children fare worse than more well-off urban children and rural children in terms of health, are more likely to be underweight, and are less likely to receive important vaccinations. These disadvantages experienced by children in urban slums set off lifelong patterns of disadvantage. When the street children mature, most are unemployed, homeless, or both. Large numbers of unemployed or underemployed young people, especially young men, is often a prescription for unrest, providing a recruiting ground for organizations that advocate for radical (and sometimes violent) change (Goldstone 2002).

The Future of Urbanization in the Developing World

In considering the scope of the challenges facing urban areas in developing countries, it can be difficult to see prospects for change and development. Conditions of life in many of the world's largest cities seem likely to decline even further in the years to come. But the picture is not entirely negative.

First, although birthrates remain high in many countries, they are likely to drop in the years to come, as we will discuss in the next section. Second, globalization is presenting important opportunities for urban areas in the Global South. With economic integration, cities around the world are able to enter international markets, to promote themselves as locations for investment and development, and to create economic links across the borders of nation-states. Third, migrants to urban areas are often "positively selected" in terms of traits such as higher levels of educational attainment. Thus migration may be beneficial to those who find better work opportunities, and for their families, who benefit from *remittances*—the money that the migrant workers send back home.

CONCEPT CHECKS

1. Discuss the effects of globalization on cities.

2. What are the four main characteristics of global cities?

3. Urban growth in the developing world is much higher than elsewhere. Discuss several economic, environmental, and social consequences of such rapid expansion of cities in developing nations.

What Are the Forces behind World Population Growth?

In late October 2011, the United Nations Population Fund announced that the world population had reached 7 billion. This announcement invited a flurry of proclamations about where, exactly, "Baby 7 Billion" was born, with India, the Philippines, and other nations claiming this honor (Rauhala 2011). In the 1960s, Paul Ehrlich (Fremlin 1964, Ehrlich 1968) calculated that if the rate of population growth at that time continued, 900 years from now (not a long period in world history as a whole) there would be 60 quadrillion people on the face of the Earth. That is 100 people for every square yard of the Earth's surface, including both land and water.

Such a picture, of course, is nothing more than nightmarish fiction designed to drive home how cataclysmic the consequences of continued population growth would be. The real issue is what will happen over the next thirty or so years, by which time, if current trends are not reversed, the world's population will already have grown to unsustainable levels. Partly because governments and other agencies heeded the warnings of Ehrlich and others forty-plus years ago by introducing population-control programs, there are grounds for supposing that world population growth is beginning to trail off. Estimates calculated in the 1960s of the likely world population by the year 2000 turned out to be inaccurate. Nevertheless, considering that a century ago, there were only 1.5 billion people in the world, this still represents growth of staggering proportions. Moreover, the factors underlying population growth are by no means completely predictable, and all estimates have to be interpreted with caution.

Population Analysis: Demography

The study of population is called **demography**. The term was invented about a century and a half ago, at a time when nations were beginning to keep official statistics on the nature and distribution of their populations. Demography is concerned with measuring the size of populations, explaining their rise or decline, and documenting the distribution of such populations both within and across continents, nations, states, cities, and even neighborhoods. Population patterns are governed by three factors: births, deaths, and migrations. Demography is customarily treated as a branch of sociology because the factors that influence the level of births and deaths in a given group or society, as well as migrations of population, are largely social and cultural.

BASIC DEMOGRAPHIC CONCEPTS

Among the basic concepts used by demographers, the most important are crude birthrates, fertility, fecundity, and crude death rates. **Crude birthrates** are expressed as the number of live births per year per 1,000 persons in the population. They are called "crude" rates because of their very general character. Crude birthrates, for example, do not tell us what

Learn why the world population has increased dramatically, and understand the main consequences of this growth.

demography

The study of the size, distribution, and composition of populations.

crude birthrate

A statistical measure representing the number of births within a given population per year, normally calculated as the number of births per 1,000 members. Although the crude birthrate is a useful index, it is only a general measure, because it does not specify numbers of births in relation to age distribution.

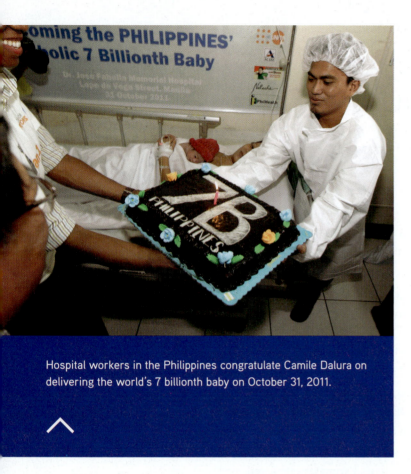

Hospital workers in the Philippines congratulate Camile Dalura on delivering the world's 7 billionth baby on October 31, 2011.

proportions of a population are male or female or what the age distribution of a population is (the relative proportions of young and old people in the population). Where statistics are collected that relate birth or death rates to such categories, demographers speak of "specific" rather than "crude" rates. For instance, an age-specific birthrate might specify the number of births per 1,000 women in the twenty-five- to thirty-four-year-old age group.

If we wish to understand population patterns in any detail, the information provided by specific birthrates is normally necessary. Crude birthrates, however, may be useful for making overall comparisons among different groups, societies, and regions. Thus the crude birthrate for the world as a whole in 2016 was 19 per 1,000. It was far lower in the United States (12 per 1,000), although other industrialized countries have lower rates: 8 per 1,000 in Portugal and Japan and some East Asian countries (Korea, Hong Kong) and 9 per 1,000 in many other European countries (for example, Greece, Germany, Spain). In many other parts of the world, crude birthrates are much higher. In Pakistan, for instance, the crude birthrate is 28 per 1,000. In many African nations, it is more than 40 per 1,000. For example, the crude birthrate in Niger is 48 per 1,000 (World Bank 2018l).

Among less industrialized countries, China's crude birthrate is the same as the United States (12 per 1,000)—a direct result of government-mandated family-planning programs introduced between 1978 and 1980 that enforced (sometimes through sanctions against violators) a one-child policy. The Chinese government instituted incentives (such as better housing and free health care and education) to promote single-child families, whereas families with more than one child face special hardships (wages are cut for those who have a third child). The Chinese government lifted the controversial ban in 2015 and now allows couples to have two children. This policy shift was triggered by several unintended and undesirable consequences of the one-child policy. Many Chinese couples were angry over the restrictive environment they were living in, while the Chinese public worried that generations of only children (nicknamed "little emperors") were becoming self-absorbed adults. Yet the most serious concern was that there would be insufficient numbers of young people to support China's rapidly aging population (Buckley 2013).

Birthrates are an expression of the fertility of women. **Fertility** refers to how many live-born children the average woman has. A fertility rate is usually calculated as the average number of live births per 1,000 women of childbearing age. Fertility is distinguished from **fecundity**, which refers to the number of children women are biologically capable of bearing. It is physically possible for a normal woman to bear a child every year during the period when she is capable of conception. There are variations in fecundity according to the age at which women reach puberty and menopause, both of which vary among countries as well as among individuals. Although there may be families in which a woman bears twenty or more children, fertility rates in practice are always much lower than fecundity

fertility

The average number of live-born children produced by women of childbearing age in a particular society.

fecundity

A measure of the number of children that it is biologically possible for a woman to produce.

rates, because social and cultural factors limit the actual number of children a woman gives birth to.

Crude death rates (also called "mortality rates") are calculated in the same way as birthrates—the number of deaths per 1,000 of population per year. Again, there are major variations among countries, but death rates in many societies in the Global South are falling to levels comparable to those of the West. The crude death rate for the world as a whole was 8 per 1,000 in 2016—a rate shared by many countries throughout the world, including the United States, Cuba, Argentina, Ghana, and Myanmar. Both India and China had slightly lower crude death rates (7 per 1,000). At the other extreme, Bulgaria, Latvia, and the Ukraine had the highest crude death rates, at 15 per 1,000. A high crude death rate can result from many factors, with poverty—and the poor health care that often goes along with it—being a major cause. But HIV/AIDS, warfare, and natural disasters also play a role.

Like crude birthrates, crude death rates provide only a very general index of **mortality** (the number of deaths in a population). Specific death rates give more precise information. A particularly important specific death rate is the **infant mortality rate**: the number of babies per 1,000 births in any year who die before reaching age one. One of the key factors underlying the population explosion has been reductions in infant mortality rates. As with crude birth and death rates, there is variation in the rate of infant mortality. For the world as a whole, in 2016, out of every 1,000 births, 31 babies died in infancy. Infant mortality ranged from a low of 2 per 1,000 in countries such as Norway, Sweden, and Finland, three Scandinavian countries with strong public health care systems, to highs in the 80s and 90s per 1,000 births in war-torn African countries that are also failed states, including Somalia, Sierra Leone, and the Central African Republic. The infant mortality rate in the United States—6 per 1,000—is at the low end, but it's not as low as Cuba's (4 per 1,000) (World Bank 2018l). Although Cuba is a poor (and undemocratic) country, its government has made preventive medicine, including a strong primary health care system, one of its top priorities (Medical Education Cooperation with Cuba 2016).

Declining rates of infant mortality are the most important influence on increasing **life expectancy**—that is, the number of years the average person can expect to live. In 1900, life expectancy at birth in the United States was about forty years. Today it has increased to nearly seventy-nine years. This does not mean, however, that most people at the turn of the century died when they were around forty. When there is a high infant mortality rate, as there is in many less industrialized nations, the average life expectancy—which is a statistical average—is brought down by deaths that occurred at age 0 or 0.5 years, for example. If we look at the life expectancy of only those people who survive the first year of life, we find that in 1900, the average person could expect to live to age fifty-eight. Illness, poor nutrition, and natural disasters are the other factors that influence life expectancy. Life expectancy has to be distinguished from **life span**, which is the maximum number of years that an individual could live. Although life expectancy has increased in most societies in the world over the past century, life span has remained unaltered.

Dynamics of Population Change

Rates of population growth or decline are measured by subtracting the yearly number of deaths per 1,000 from the number of births per 1,000. (Actual population growth or decline also requires taking into account the number of people who have migrated into the country as well as the number who have emigrated out of the country.) Some European

crude death rate

A statistical measure representing the number of deaths that occur annually in a given population per year, normally calculated as the number of deaths per 1,000 members. Crude death rates give a general indication of the mortality levels of a community or society but are limited in their usefulness because they do not take into account the age distribution.

mortality

The number of deaths in a population.

infant mortality rate

The number of infants who die during the first year of life, per 1,000 live births.

life expectancy

The number of years the average person can expect to live.

life span

The maximum length of life that is biologically possible for a member of a given species.

rates of population growth or decline

A measure of population change calculated by subtracting the yearly number of deaths per 1,000 from the number of births per 1,000.

Population Growth Rate, 2010–2015

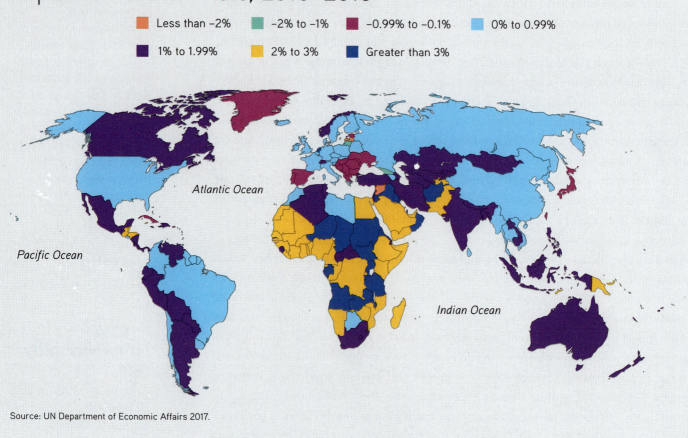

Less than –2%   –2% to –1%   –0.99% to –0.1%   0% to 0.99%

1% to 1.99%   2% to 3%   Greater than 3%

Atlantic Ocean

Pacific Ocean

Indian Ocean

Source: UN Department of Economic Affairs 2017.

exponential growth

A geometric, rather than linear, rate of increase. Populations tend to grow exponentially.

doubling time

The time it takes for a particular level of population to double.

countries have negative growth rates—in other words, their populations are declining. Virtually all the industrialized countries have growth rates of less than 0.5 percent. Rates of population growth were high in the eighteenth and nineteenth centuries in Europe and the United States but have since leveled off. Many countries in the Global South today have rates of between 2 and 3 percent (see Global Map 15.1). These may not seem very different from the rates of the industrialized countries, but in fact, the difference is enormous.

The reason is that growth in population is **exponential** rather than arithmetic. An ancient Persian myth helps illustrate this concept. A courtier asked a ruler to reward him for his services by giving him twice as many grains of rice for each service as he had the time before, starting with a single grain on the first square of a chessboard: that is, one grain on the first square, two on the second, four on the third, and so on. By the twenty-first square, over a million grains were needed, and more than a trillion (a million million) on the forty-first square (Meadows et al. 1972). This myth conveys an important mathematical principle: that starting with one item and doubling it, doubling the result, and so on, rapidly leads to huge numbers. Exactly the same principle applies to population growth. We can measure this effect by means of the **doubling time**, the period of time it takes for the population to double. The formula used to calculate doubling time is 70 divided by the current growth rate. For example, a population growth of 1 percent will produce a doubling of numbers in seventy years. At 2 percent growth, a population will double in thirty-five years.

Malthusianism

In premodern societies, birthrates were very high by the standards of the industrialized world today. Nonetheless, population growth remained low until the eighteenth century because there was a rough overall balance between births and deaths. Although there were sometimes periods of marked population increase, these were followed by increases in death rates. In medieval Europe, for example, when harvests were bad, marriages tended to be postponed and the number of conceptions fell, while deaths increased. These complementary trends reduced the number of mouths to be fed. No preindustrial society was able to escape from this self-regulating rhythm (Wrigley 1968).

During the rise of industrialism, many looked forward to a new age in which food scarcity would be a phenomenon of the past. The development of modern industry, it was widely supposed, would create a new era of abundance. In his celebrated work *Essay on the Principle of Population* (1798), Thomas Malthus criticized these ideas and initiated a debate about the connection between population and food resources that continues to this day. At the time Malthus wrote, the population in Europe was growing rapidly. Malthus pointed out that whereas population increase is exponential, food supply depends on fixed resources that can be expanded only by developing new land for cultivation. Population growth, therefore, tends to outstrip the means of support available. The inevitable outcome is famine, which, combined with the influence of war and plagues, acts as a natural limit to population increase. Malthus predicted that human beings would always live in circumstances of misery and starvation unless they practiced what he called "moral restraint." His cure for excessive population growth was for people to delay marriage and to strictly limit their frequency of sexual intercourse. (The use of contraception he proclaimed to be a "vice.")

For a while, **Malthusianism** was ignored. The population development of the Western countries followed a quite different pattern from that which he had anticipated. Rates of population growth trailed off in the nineteenth and twentieth centuries. In the 1930s, there were major worries about population decline in many industrialized countries, including the United States. Malthus also failed to anticipate the technological developments fostering increases in food production that would develop in the modern era. However, the upsurge in world population growth in the twentieth century again lent some credence to Malthus's views. Population expansion in the Global South seems to be outstripping the resources that those countries can generate to feed their citizenry.

The Demographic Transition

Demographers often refer to the changes in the ratio of births to deaths in the industrialized countries from the nineteenth century onward as the **demographic transition** (Figure 15.2). This theory was first developed by Warren S. Thompson (1929), who described a three-stage process in which one type of population stability would eventually be replaced by another as a society reached an advanced level of economic development.

Stage 1 refers to the conditions characteristic of most traditional societies, in which both birth and death rates are high and the infant mortality rate is especially high. Population grows little, if at all, as the high number of births is more or less balanced by the level of deaths. Stage 2, which began in Europe and the United States in the early part of the nineteenth century—with wide regional variations—occurs when death rates fall while fertility remains high. This is, therefore, a phase of marked population growth. It is

Malthusianism

A doctrine about population dynamics developed by Thomas Malthus, according to which population increase comes up against "natural limits," represented by famine and war.

demographic transition

The changes in the ratio of births to deaths in the industrialized countries from the nineteenth century onward.

FIGURE 15.2

Demographic Transition

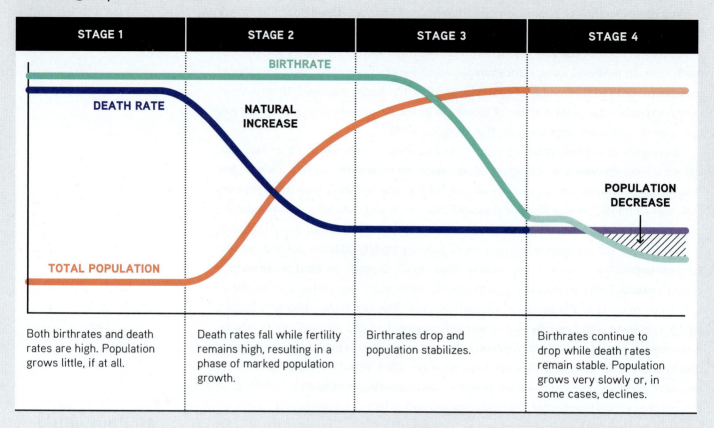

STAGE 1	STAGE 2	STAGE 3	STAGE 4
Both birthrates and death rates are high. Population grows little, if at all.	Death rates fall while fertility remains high, resulting in a phase of marked population growth.	Birthrates drop and population stabilizes.	Birthrates continue to drop while death rates remain stable. Population grows very slowly or, in some cases, declines.

subsequently replaced by stage 3, in which, with industrial development, birthrates drop to a level such that population is again fairly stable. Some societies in Europe have moved into a fourth stage, which we will discuss in the next section.

Industrial development was accompanied by a range of social changes that contribute to lower birthrates. As the economy transitioned from agricultural to manufacturing, parents no longer required many children to help maintain their farms. With the advent of compulsory schooling, children cost money rather than earning money for their families. Industrial development was also accompanied by technologies that allowed women to control their own fertility as well as a cultural change regarding people's views toward childbearing. How many children a woman would have was now viewed as under her own control rather than a "gift from God." In contemporary society, as women have achieved higher levels of education and higher earnings in the labor market, the incentive to have fewer children has increased. Higher education among both men and women is also linked to delayed marriage and, consequently, delayed (and thus diminished) childbearing (Caldwell et al. 2010).

If we look at the crude birth and death rates for some of the poorest countries in the world, we can get a better sense of how this is playing out in the world today. "Least developed countries" (LDCs) are low-income countries that the United Nations has determined suffer from "severe structural impediments to sustainable development," such as low per capita income, high levels of malnutrition and child mortality, and low levels of adult literacy.

According to the United Nations, forty-seven countries—mostly in Africa, with a few in Asia and Oceania—are currently characterized as LDCs (UN Department of Economic and Social Affairs 2017). In 1960, the average crude birthrate in the LDCs was 48 per 1,000; by 2015, it had dropped to 33 per 1,000, a decline of 15. During the same period, the average crude death rate dropped from 24 to 9 per 1,000, a decline of 15. The good news (from a population growth standpoint) is the large drop in the birthrate. While a decline in the death rate is obviously also good news, it also means that the difference between the number of births and the number of deaths (24 in both 1960 and 2015) remained unchanged over the period. This represents a 2.4 percent rate of annual population growth, which—were it to continue—would mean that the population of the poorest countries (more than 950 million people) would double every thirty years (UN Department of Economic and Social Affairs 2017).

The theories of demographic transition directly oppose the ideas of Malthus. Whereas for Malthus, increasing prosperity would automatically bring about population increase, the thesis of demographic transition emphasizes that economic development, generated by industrialism, would actually lead to a new equilibrium of population stability. We have already noted another flaw in Malthus's predictions: food production. Thanks to modern agricultural technologies, food production has thus far kept pace with population growth. Whether this will remain the case—or whether climate changes will result in an ecological limit to ever-increasing food production—remains a question. The distribution of food is another issue: Food is abundant among wealthier populations and scarce among those who live in poorer countries.

Prospects for Change

Fertility remains high in the developing world because traditional attitudes to family size have persisted. Having large numbers of children is often still regarded as desirable, providing a source of labor on family-run farms. Some religions either are opposed to birth

The Indian railroad, one of the world's largest rail networks, continues to be one of the only forms of affordable transportation available to the majority of Indians. India is projected to overtake China as the most populous country around 2024.

Demographer

You may notice changes happening in your hometown or the neighborhoods surrounding your college campus. Perhaps you see that a high-end grocery store or fitness studio is moving in. Or you may marvel at the construction site for a brand-new elementary school. You might have also noticed that the ATM at your local bank offers language options in English, Spanish, Mandarin, and Russian. How do businesses decide whether to locate their latest franchise in one neighborhood versus another? And how do planners decide whether their city or town needs a new school, senior center, or additional weekday routes on the commuter rail or bus schedule? How do companies decide which languages, if any, to translate their materials into?

Demographers hold the key to these and other questions that require a deep knowledge of the size, distribution, and composition of a population. Demography is a branch of sociology focused on the ways that populations grow and change. It uses complex statistical methods and population-based data, often obtained from large surveys or the U.S. Census. Given the statistical complexity of the work, most demographers have at least a master's degree in sociology or statistics, although a bachelor's degree in sociology may be sufficient for some entry-level positions.

Demographers make observations about the causes and consequences of population changes, such as increases in birthrates or immigration. Demographers also analyze data to identify current trends and predict future ones. These predictions can help governments, social service agencies, and private companies to plan ahead. For instance, if a demographer calculates that a particular state has seen a substantial increase in birthrates, then he or she might project that in twenty years that state may require more two- and four-year colleges to accommodate this large and growing number of young people.

Many demographers work for government agencies such as the Census Bureau or Bureau of Labor Statistics (BLS). For example, the BLS hires demographers to

control or affirm the desirability of having many children. Contraception is opposed by Islamic leaders in several countries and by the Catholic Church, whose influence is especially marked in South and Central America. Yet, a decline in fertility levels has at last occurred in some large countries in the Global South.

Some claim that the demographic changes that will continue to occur over the next century will be greater than any before in all of human history. It is difficult to predict with any precision the rate at which the world population will rise, but the United Nations has increased its estimates of future population growth, based on the most recent national population censuses and demographic and health surveys. World population—nearly 7.6 billion in 2017—is projected to reach 8.6 billion by 2030, 9.8 billion by 2050, and 11.2 billion by the end of the century. Africa alone is predicted to account for more than half of total world population growth by mid-century (UN Department of Economic and Social Affairs 2017). This overall population increase highlights two distinct trends. First, most countries in the Global South will undergo the process of demographic transition. This will result in research questions like, What subpopulations have especially high unemployment rates? We know that young Black men have especially high rates of unemployment, which has implications for their family lives and health. Likewise, demographers may find that women and men working in the same occupation have very different annual earnings, calling attention to the fact that some industries may be particularly guilty of sex discrimination. When analyzing and interpreting data, demographers require a sociological imagination. If one person is unemployed, that may be a personal issue reflecting his or her poor work ethic or bad luck. Yet if demographic analyses show clear patterns, where some groups are consistently more likely to face poor labor market prospects, then that may be an indication of a larger public issue such as racial discrimination or sexism in the labor market. By detecting trends and subgroup differences, demographers are identifying public issues that may require policy solutions.

Businesses, corporations, and market research firms also regularly employ demographers to help them make decisions about the kinds of goods and services they deliver. Let's say that a high-end fitness firm is debating whether to locate their latest studio in a particular neighborhood. They may want to know the average income of people living and working there. Demographers might also examine commuting or residential patterns. If urban workers leave the office at 5 p.m. and then return to their homes in the suburbs, the fitness studio would likely schedule classes at 5:30 p.m. rather than 8 p.m. so that members can work out before returning home for dinner.

Demographers also work in international organizations like the United Nations and World Bank. Demographic analyses can help us understand why birthrates are so high in places like sub-Saharan Africa, and why high birthrates are linked with undesirable outcomes like poverty and infant and child mortality. Demography is a fascinating profession that allows its workers to apply sophisticated statistical tools and sociological concepts to some of the most vexing social issues in the United States and worldwide.

People line up for a job fair in New York City. Demographers who work for the Bureau of Labor Statistics are concerned with issues such as unemployment and other labor market outcomes.

a substantial surge in the population as death rates fall. India is likely to see its population surpass 1.6 billion before growth levels off. Areas in Asia, Africa, and Latin America will similarly experience rapid growth before the population eventually stabilizes.

The second trend concerns the developed countries that have already undergone the first three stages of the demographic transition. These societies have moved into stage 4. In this stage, birthrates continue to drop while death rates remain steady, resulting in very slight population growth or even population decline as birthrates fall below replacement level. This has been the recent experience of many European countries, including Germany, the Netherlands, Belgium, Spain, Italy, and Sweden, as well as Japan and Singapore. A process of aging will occur in which the number of young people will decline in absolute terms and the older segment of the population will increase markedly. This will have widespread economic and social implications for developed countries. First, there will be an increase in the **dependency ratio**, the ratio of the number of economically dependent members of the population to the number of economically productive members. Economically dependent

dependency ratio

The ratio of people of dependent ages (children and the elderly) to people of economically active ages.

persons are those considered too young or too old to work, typically those under age fifteen and over age sixty-five. As the dependency ratio increases, pressure will mount on health care and social services.

What will be the consequences of these demographic changes? Some observers see the makings of widespread social upheaval—particularly in countries in the Global South undergoing demographic transition. The rapid growth of cities will likely lead to environmental damage, new public health risks, overloaded infrastructures, rising crime, and impoverished squatter settlements. Changes in the economy and labor markets may prompt widespread migration as people in rural areas search for work. Warfare and civic violence also cause people to migrate. Civil war in Syria, along with the rise of radical Islamist movements in the Middle East and North Africa, has laid waste to entire cities. Millions of refugees have fled war-torn countries as a result. Refugees, like all migrants, seek out major cities, where the possibility of finding work is the greatest. This enormous movement of people has placed significant strains on the resources of the destination countries. While Germany was initially among the most generous of countries in opening its doors to refugees, by 2016, even Germany began to close its borders. Other countries responded far more harshly: Those bordering on Greece—a destination of choice for many migrants, because the water crossing from Turkey to the Greek island of Samos is only five miles—closed their borders in 2015, stranding many migrants in one of Europe's poorest countries.

Famine and food shortages are another serious concern. About 815 million people—11 percent of the global population—suffer from hunger or undernourishment. In some parts of the world, more than one-third of the population is undernourished (Food and Agriculture Organization of the United Nations 2017). As the population rises, levels of food output will need to rise accordingly to avoid widespread scarcity. Yet many of the world's poorest areas are particularly affected by water shortages, shrinking farmland, and soil degradation—processes that reduce, rather than enhance, agricultural productivity. It is almost certain that food production will not occur at a level to ensure self-sufficiency. Large amounts of food and grain will need to be imported from areas where there are surpluses.

CONCEPT CHECKS

1. What is the difference between fertility and fecundity?

2. Explain Malthus's position on the relationship between population growth and the food supply.

3. Describe the four stages of the demographic transition.

4. How does the theory of demographic transition conflict with Malthus's ideas?

>

See that the environment is a sociological issue related to economic development and population growth.

How Do Environmental Changes Affect Your Life?

Today the human onslaught on the environment is so intense that few natural processes are uninfluenced by human activity. Nearly all cultivatable land is under agricultural production. What used to be almost inaccessible wildernesses are now often nature reserves, visited routinely by thousands of tourists. Modern industry, still expanding worldwide, has led to steeply climbing demands for sources of energy and raw materials. Yet the world's supply of such energy sources and raw materials is limited, and some key resources are bound to run out if global consumption is not restricted. Even the world's climate, as we shall see, has been affected by the global development of industry.

Global Environmental Threats

Global environmental threats are of several basic sorts. The ones that are most likely to be felt locally, at least in economically developed countries like the United States, are those associated with air and water pollution, the loss of forests and other wildlife habitats, the creation of waste that cannot be disposed of in the short term or recycled, and the depletion of resources that cannot be replenished. According to the International Union for Conservation of Nature (2017), the most widely accepted authoritative source, more than 25,000 species are currently threatened with extinction. The loss of biodiversity, in turn, means more to humans than merely the loss of natural habitat. Biodiversity also provides humans with new medicines and sources and varieties of food and plays a role in regulating atmospheric and oceanic chemistry.

The amount of domestic waste—what goes into our garbage cans—produced each day in the industrialized societies is staggering. Food is mostly bought in packages that are thrown away at the end of the day. The disposal of electronic waste—computers, cell phones, and the host of toys and gadgets that contain electronic circuits—is a growing problem. Discarded electronics, which contain toxins that cause cancer and other illnesses, are routinely "recycled" to landfills in poor countries, where there are few if any safeguards against contaminating local watersheds, farmlands, and communities.

Global green movements and political parties (such as Friends of the Earth, Greenpeace, or Conservation International) have developed in response to these new environmental hazards. Although green philosophies are varied, a common thread concerns taking action to protect the world's environment, conserve rather than exhaust its resources, and protect the remaining animal species.

Global Warming and Climate Change

Global warming occurs because carbon dioxide and other greenhouse gases are released into the atmosphere by the burning of fuels such as oil and coal in cars and power stations and gases released into the air by the use of such things as aerosol cans, material for insulation, and air-conditioning units. This buildup of greenhouse gases in the Earth's atmosphere functions like the glass of a greenhouse. The atmosphere allows the sun's rays to pass through but acts as a barrier to the rays passing back, causing the Earth to heat up. For this reason, global warming is sometimes termed the "greenhouse effect."

In 2007, the Intergovernmental Panel on Climate Change (IPCC), a blue-ribbon group of scientists created by the UN Environment Program and its World Meteorological Organization, took the planet's temperature and found it had risen steadily since the mid-twentieth century. Rising temperatures result in the rapid shrinking of arctic ice caps, along with mountain glaciers; long-term droughts in some regions, with greater rainfall in others; an increase in hurricane activity in the North Atlantic; and in general, more turbulence in global weather. Most significantly, the IPCC report stated unequivocally that human activity was the principal source of global warming, very likely causing most of the temperature increase over the last century. By 2014, when the IPCC issued its synthesis report based on the most recent research, the expert warnings were even more dire: The report warned of "severe, pervasive and irreversible impacts for people and ecosystems," including continued global warming to 2 degrees Celsius (3.6 degrees Fahrenheit) by the end of the century; more frequent and longer-lasting heat waves; and more extreme weather,

A worker at an e-waste recycling company in Bangalore, India, shows shredded pieces of printed circuit boards of obsolete electronic gadgets. E-waste is a growing environmental and public health concern.

Tracking Your Ecological Footprint

The United States makes up roughly 4 percent of the global population yet uses nearly one-third of the world's fossil fuel reserves (U.S. Energy Information Administration 2016). According to the Global Footprint Network, we would need 4.1 Earths to allow every person on the planet to live a typical American lifestyle (McDonald 2015). The many decisions we make each day—what kind of cars we drive, how much we fly, the size of our homes, and even how much time we spend in the shower each morning—contribute to the United States' high levels of resource consumption.

Many Americans, however, especially young adults, are taking conscious steps to monitor their energy use by biking to school or work, recycling and composting, bringing old cell phones to e-waste facilities, and using refillable water bottles. But how do we know whether these efforts are enough, or what their impact is? A number of websites and apps have been developed that enable users to estimate their "ecological footprints," the amount of energy they require given the details of their daily living. One of the most popular is the Ecological Footprint Quiz (http://www.earthday.org/take-action/footprint-calculator/), which provides an animated tour through one's lifestyle. While perhaps not perfectly scientifically accurate, it will give you a rough idea of how many planet Earths would be required if everyone on the planet were to consume the land and other resources required to enjoy your lifestyle.

You begin by creating (and styling) your avatar and choosing your country. You next enter information about the food you eat (are you a vegan? meat eater? is your food mainly local? organic?), the amount of clothing and other goods you have, the amount of trash you generate, the size and "greenness" of your home, how much you drive (and what gas mileage you get), whether you use public transportation, and how much you fly. As you chart your lifestyle, the many things you have come to enjoy surround your avatar.

If you are a struggling student at a U.S. college or university who lives in a dorm, is a vegan who eats only locally grown vegetables, gets around on a bicycle, travels long distances by bus, and never flies, you will require "only" 3.3 planets! That's because unless you are completely off the grid, you are still tied into a range of services that require energy, and in the United States, those services are typically not energy efficient. Still, you are likely doing less ecological damage than the authors of this textbook. When one of the authors took the quiz, nearly 7 Earths were required: Living in a full-sized house, driving a car—even one that is energy-efficient—and flying around the world to conferences clearly take a toll on the planet. On the other hand, if you lived in Switzerland and lived an ecologically conscious lifestyle, according to the quiz, you would require only 2.4 planets.

The lesson derived from this exercise is simple: Your individual lifestyle can, in fact, make a difference. Still, a significant reduction in the global ecological footprint will require the countries of the world to adopt policies that will move them toward much greater energy efficiency. The Paris agreement provides a hopeful direction.

How many planet Earths would it take if everyone on the planet were to consume the resources required to enjoy your lifestyle?

such as tropical storms, cyclones, and hurricanes (Intergovernmental Panel on Climate Change 2015).

The predicted increase in average global temperature is sufficient, in the eyes of some scientists, to destabilize the ice sheets that cover Greenland and the western part of Antarctica; were the ice sheets to melt in their entirety, sea levels would rise an estimated 30 feet or more. While such a catastrophic event is unlikely, during the past century, sea levels have risen faster than at any time during the past several millennia; the consensus has been that it will likely rise three to four feet by the end of the century. However, even that consensus may be too conservative: The most recent studies have found the West Antarctic ice sheet to be far less stable than scientists once thought, meaning sea levels may rise even higher—faster (Leber 2014; Tollefson 2016; von Kaenel 2016). Rising sea levels will likely have devastating consequences for cities near the coasts or in low-lying areas.

The most vulnerable people around the world would be the most severely affected by climate change, because they would lack access to the resources that might enable them to adjust. Apart from severe droughts— which will turn once-fertile lands into deserts and are already affecting major parts of the U.S. Southwest—global warming will threaten the water supplies of hundreds of millions of people, increase the danger of flooding for others, adversely affect agriculture in parts of the world, and further reduce planetary biodiversity. The 2014 IPCC report suggests ways that the worst consequences of global warming can be mitigated. The first and most important step would be for the countries of the world to adopt conservation measures that would limit global warming to no more than 2 degrees Celsius. This would require national policies that reduce green-

Kids play on a merry-go-round near an oil refinery at the Carver Terrace housing project playground in west Port Arthur, Texas. Port Arthur sits squarely on a two-state corridor routinely ranked as one of the country's most polluted regions.

house gas emissions, which in turn would require massive conversion from economies based on fossil fuels to economies based on clean energy. Because some amount of global warming seems inevitable, the recent IPCC report also emphasized adaptation measures that might be taken to offset some of the predicted consequences. These included recommendations for such things as early warning systems, flood and cyclone shelters, seawalls and levees, desalinization plants to convert seawater to drinking water, and programs to educate people about the dangers of global warming.

In December 2015, 195 nations met in Paris and reached an agreement to limit temperature increase to no more than 1.5 degrees Celsius (2.7 degrees Fahrenheit). As part of the agreement, each participating country submitted a climate plan. China, for example, pledged to lower emissions per unit of GDP by 60 to 65 percent (which would mean total emissions would continue to rise at least until 2030, because China hopes to continue building its rapidly growing GDP). While the Paris agreement is historically significant and is seen by many as a hopeful sign, it remains to be seen whether it will be honored. There are no enforcement mechanisms, apart from regular public reports that may shame countries into fulfilling their commitments. The agreement was further weakened in 2017, when President Trump announced that the United States would no longer honor it. It now remains up to China, and the other world signatories, to be leaders in addressing the principal causes of global climate change.

Rather than calling for a reining in of economic growth, some policy recommendations are focused on sustainable development. **Sustainable development** is a term that was first used in a report by the World Commission on Environment and Development (1987), popularly referred to as the Brundtland Report after former Norwegian prime minister Gro Harlem Brundtland, who chaired the commission. The report defined sustainable development simply as "development that meets the needs of the present without compromising the ability of future generations to meet their own needs" (p. 326). This definition sought to reconcile two seemingly intractably opposed communities: environmentalists, who were often seen as antigrowth, and businesspeople, who were often seen as anti-environment. Environmentalists could now argue that at least in wealthy industrial nations, economic development should be limited if it is environmentally harmful, while conceding that economic growth might be necessary to lift people out of poverty in the Global South.

The notion of sustainable development, while popular, remains unclear: What are the needs of the present? How much development can occur without compromising the future, particularly because we don't know what the future effects of technological change may be (Giddens 2009)? However imprecise the term, "sustainable development" is generally taken to mean that growth should, at least minimally, be carried on in such a way as to preserve and recycle physical resources rather than deplete them, maintain biodiversity, and keep pollution to a minimum by protecting clean air, water, and land.

Critics see the notion of sustainable development as neglecting the specific needs of poorer countries. The idea of sustainable development, according to critics, tends to focus attention only on the needs of richer countries; it does not consider the ways in which the high levels of consumption in more affluent countries are satisfied at the expense of other people. For instance, demands on Indonesia to conserve its rain forests could be seen as unfair because Indonesia has a greater need than industrialized countries for the revenue it must forgo by accepting conservation. Acknowledging this concern, the Paris meeting adopted a "loss and damage" principle, under which wealthy countries agreed to create a fund to help poorer countries cope with the effects of global warming (UNFCCC 2015). And there is some hopeful evidence that economies can grow even without increasing greenhouse gases: Even though global GDP grew by 3 percent during the two-year period of 2014–2015, carbon emissions remained flat (International Energy Agency 2016).

A New Ecological Paradigm?

Environmental sociology is concerned with the interactions between humans and the natural environment—especially the social forces that create environmental problems, and their possible solutions. Classical sociology tended to minimize, if not completely ignore, the importance of human impacts on the environment. This was partly because the negative effects were less pervasive (recall that by 1900, there were still fewer than 2 billion people in the world), but primarily because sociology generally took human domination of nature for granted (Leiss 1994). As we saw in Chapter 1, sociology emerged during the late nineteenth and early twentieth centuries, a period that reflected what has been termed "optimistic ethnocentrism" during an "age of exuberance" (Catton and Dunlop 1980). This refers to the belief that science, technology, and industrial development, fueled by vast lands and resources in North and South America, as well as a seemingly endless supply of fossil fuel, would provide for limitless opportunity and endless progress. This view has been termed the *human exceptionalism paradigm* (HEP), reflecting the belief that humans are unique among all creatures: We dominate all species and can remake the world to serve our needs, because human destiny is shaped by social and cultural factors and not the physical environment (Catton and Dunlap 1980).

This view held sway until the early 1970s, when it was seriously questioned by a growing number of sociologists and ecologists. This was a period during which social movements were already challenging prevailing worldviews (see Chapter 16), including the taken-for-granted assumption that human life could somehow be divorced from the natural world. The HEP was further challenged when an offshore oil drilling platform in the Santa Barbara, California, channel blew out in late 1969, covering miles of pristine coastline with oil, killing thousands of birds, and creating the largest ecological disaster in U.S. history. The oil spill received widespread media attention: Even President Nixon paid a visit. The spill inspired the first Earth Day as well as one of the country's first environmental studies programs, at UC Santa Barbara. The oil spill also highlighted the tensions between the petroleum industry and local residents, raising questions about the nature of power in America (Molotch 1970). Popular books, such as Rachel Carson's *Silent Spring* (1962), which exposed the effects of pesticides on water, raised public awareness about environmental issues.

The result was labeled the new ecological paradigm (NEP), a framework that emphasized the complex interactions involved in global ecosystems. Humans, it argued, are not somehow exempt from the "web of nature," and our biophysical environment is not limitless. On the contrary, the carrying capacity of our planet is limited, and we must come to understand and respect these limits, or we will pay a price (Catton and Dunlap 1980; see also Buttel 1987; Dunlap 2002; Freudenberg, Frickel, and Gramling 1995; Schnaiberg 1980). The challenge is similar to what the ecologist Garret Hardin (1968), drawing on a pamphlet written by a nineteenth-century English economist, called the "tragedy of the commons": When individuals pursue their own self-interest, which requires using common resources without thinking of others, those common resources are quickly depleted. The solution, Hardin argued, was to give up individual freedom so that shared resources are preserved. Hardin, echoing Malthus, believed that the principal problem was population growth; he suggested that people must be educated to give up what they see as their right to have as many children as they choose. While education and family planning have in fact contributed to a global decline in birthrates, future population growth has yet to be stemmed.

At the turn of the twenty-first century, Paul Crutzen, an atmospheric chemist and recipient of the Nobel Prize in Chemistry, popularized the term **Anthropocene** ("human epoch") to characterize the current geological period, a time when human activities have become the main agent of change in our planetary ecosystem. This label has caught on: The term has appeared in hundreds of scientific publications, and the International Union of Geological Sciences, the organization charged with identifying geological periods, has convened scholars to decide whether the Holocene (the period that began after the last ice age, some 12,000 years ago) has been officially superseded by the Anthropocene (Stromberg 2013).

The once radical ideas of the early environmental sociologists have by now become mainstream, acknowledged by the consensus of IPCC scientists, reflected in the Paris agreement, and perhaps even enshrined in a new—if problematic—geological era.

Anthropocene

A term used to denote the current geological epoch, in which many geologically significant conditions and processes are profoundly altered by human activities.

CONCEPT CHECKS

1. Describe the basic processes that give rise to global warming.

2. Define sustainable development, and provide at least one critique of the concept.

3. What is meant by the human exceptionalism paradigm? What has replaced it?

CHAPTER 15

The Big Picture

Urbanization, Population, and the Environment

Thinking Sociologically

1. Explain what makes the urbanization now occurring in developing countries different from and more problematic than the urbanization that took place a century ago in New York, London, Tokyo, and Berlin.

2. Following analysis presented in this chapter, concisely explain how the expanded quest for cheap energy and raw materials and present-day dangers of environmental pollution and resource depletion threaten not only the survival of people in developed countries but also that of people in less developed countries.

Learning Objectives

How Do Cities Develop and Evolve?

p. 463

Learn how cities have changed as a result of industrialization and urbanization. Learn how theories of urbanism have placed increasing emphasis on the influence of socioeconomic factors on city life.

How Do Rural, Suburban, and Urban Life Differ in the United States?

p. 469

Learn about key developments affecting American cities, suburbs, and rural communities in the last several decades: suburbanization, urban decay, gentrification, and population loss in rural areas.

How Does Urbanization Affect Life across the Globe?

p. 473

See that global economic competition has a profound impact on urbanization and urban life. Recognize the challenges of urbanization in the developing world.

What Are the Forces behind World Population Growth?

p. 479

Learn why the world population has increased dramatically, and understand the main consequences of this growth.

How Do Environmental Changes Affect Your Life?

p. 488

See that the environment is a sociological issue related to economic development and population growth.

Terms to Know

Concept Checks

conurbation • megalopolis • urbanization • ecological approach • inner city • urban ecology • urbanism • created environment

1. What are two characteristics of ancient cities?
2. What is urbanization? How is it related to globalization?
3. How does urban ecology use physical science analogies to explain life in modern cities?
4. What is the urban interaction problem?
5. According to Jane Jacobs, the more people are on the streets, the more likely it is that street life will be orderly. Do you agree with Jacobs's hypothesis and her explanation for this pattern?

suburbanization • exurban counties • gentrification • urban renewal

1. Describe at least two problems facing rural America today.
2. Why did so many Americans move to suburban areas in the 1950s and 1960s?
3. What are two unintended consequences of urbanization? How do they deepen socioeconomic and racial inequalities?

global city • informal economy

1. Discuss the effects of globalization on cities.
2. What are the four main characteristics of global cities?
3. Urban growth in the developing world is much higher than elsewhere. Discuss several economic, environmental, and social consequences of such rapid expansion of cities in developing nations.

demography • crude birthrate • fertility • fecundity • crude death rate • mortality • infant mortality rate • life expectancy • life span • rates of population growth or decline • exponential growth • doubling time • Malthusianism • demographic transition • dependency ratio

1. What is the difference between fertility and fecundity?
2. Explain Malthus's position on the relationship between population growth and the food supply.
3. Describe the four stages of the demographic transition.
4. How does the theory of demographic transition conflict with Malthus's ideas?

sustainable development • Anthropocene

1. Describe the basic processes that give rise to global warming.
2. Define sustainable development, and provide at least one critique of the concept.
3. What is meant by the human exceptionalism paradigm? What has replaced it?

16

Globalization in a Changing World

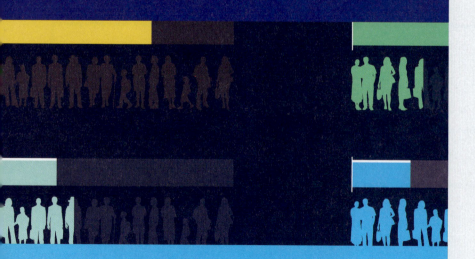

Global Wealth

p. 527

Syria descended into civil war in 2011 after pro-democracy protests turned violent. Since then, the Islamic State and other jihadist organizations have entered the fight, as have the United States and Russia. The conflict, which had claimed half a million lives as of early 2018 and resulted in nearly 12 million displaced people, has left much of the country in ruins.

THE BIG QUESTIONS

How does globalization affect social change?

Recognize that a number of factors influence social change, including the physical environment, political organization, culture, economics, and technology.

Why does terrorism seem to be on the rise in the world today?

Understand the relationship between globalization and terrorism.

What are social movements?

Understand what social movements are, why they occur, and how they affect society.

What factors contribute to globalization?

Recognize the importance of information flows, political changes, and transnational corporations.

How does globalization affect your life?

Recognize the ways that large global systems affect local contexts and personal experiences.

On December 17, 2010, Tunisian street vendor Mohamed Bouazizi set himself on fire in protest of the local police's confiscation of his wares and the harassment and humiliation that he experienced at the hands of a local government bureaucrat. This act of frustration and defiance, many people believe, was the initial catalyst for the demonstrations and riots that erupted in protest of widespread corruption and inequality in the country. In the months following the Tunisian Revolution, protests spread like wildfire throughout Jordan, Egypt, Libya, Yemen, and elsewhere in the Middle East in the spring of 2011—deemed the Arab Spring.

The nature and causes of the protests varied across countries and over time, yet most were led by educated but discontented young people who sought to fight against dictatorships, human rights violations, government corruption, economic declines, unemployment, extreme poverty, and persistent inequalities between the haves and have-nots. These revolutions eventually led to the resignation or overthrow of five heads of state: Tunisian president Zine

El Abidine Ben Ali, Egyptian president Hosni Mubarak, President Ali Abdullah Saleh of Yemen, Libyan leader Muammar al-Gaddafi, and Ukrainian president Viktor Yanukovych. In Ukraine, more than 300,000 young people took to the streets in 2014, protesting corruption, economic stagnation, and Yanukovych's decision to sign a treaty and loan agreement with Russia, instead of an agreement that would have fostered closer ties with the European Union. When Ukraine's security forces responded with force, the demonstrations mounted and Yanukovych was driven from office (Black, Pearson, and Butenko 2014).

Unlike revolutions and protests at earlier points in history, all of these events were facilitated by the Internet. Through the use of Twitter, Facebook, YouTube, online chatrooms, and other forms of social media, protesters and refugees—as well as jihadists—could report in "real time" what they did and saw; these messages were transmitted not only to their peers and fellow protesters but to viewers worldwide, inspiring large protests across the globe (Kulish 2011). In India, hundreds of thousands of disillusioned young people supported rural activist Kisan Baburao "Anna" Hazare in his hunger strike. Hazare starved himself for twelve days, until the Indian Parliament met some of his demands to implement an anticorruption measure. In Israel, an estimated 430,000 people gathered in Tel Aviv, Jerusalem, and Haifa to protest high unemployment, high costs of living, and other social injustices. In London, violence erupted at a protest march organized by the Trades Union Congress; an estimated 250,000 to 500,000 people marched from the Thames Embankment to the Houses of Parliament to Hyde Park to show their opposition to planned public spending cuts. In cities across the United States, protests have been organized over income inequality (Occupy Wall Street), police brutality and the mass incarceration of Blacks (Black Lives Matter), and a higher minimum wage (Fight for 15), among other causes.

Yet protests do not always result in positive social change. While the overthrow of Egyptian president Mubarak did pave the way for the first democratic election in Egyptian history, the winner of the election—Mohamed Morsi, a leading member of the Muslim Brotherhood—soon issued a constitutional declaration giving himself virtually unlimited power. This triggered another wave of popular protests, and the Egyptian military staged a coup. Morsi was imprisoned and eventually sentenced to death. Thousands of his followers were killed or imprisoned by the Egyptian military, and Egypt returned to the autocratic rule that had prompted the uprisings in the first place. Yemen and Libya descended into civil war, with armed factions—often supported by outside powers—laying waste to cities and villages.

Protests in Syria were brutally repressed by the Assad regime, resulting in a civil war that, as of early 2018, had claimed half a million lives, injured countless others, and displaced nearly 12 million people—half internally and half as refugees who have fled to other countries (Specia 2018; CNN 2018). This, in turn, has resulted in a rise of anti-immigrant political movements throughout Europe and has threatened the European Union's open-border policy—one of the key components of European unity. Social media, which had prompted protests and calls for democracy across the world in 2011, was effectively used by Islamic extremists to recruit jihadists and suicide bombers from Europe and North America, facilitating the creation of the so-called Islamic State across a vast swath of the Middle East.

While the Arab Spring had many causes, globalization played a central role. **Globalization** refers to the fact that we all live in one world, so that individuals, groups, and nations become more interdependent. Such interdependence is increasingly at a global scale—that is, what happens halfway across the globe is more likely than ever before to have enormous consequences for our daily lives. For example, the global economic downturn that began in 2007 contributed to rising global food prices, increased unemployment, and greater inequality in a region that in recent decades

globalization

The development of social, cultural, political, and economic relationships stretching worldwide. In current times, we are all influenced by organizations and social networks located thousands of miles away.

had suffered from economic problems. Beginning in the 1950s and 1960s, countries throughout the region adopted semi-socialist economic policies that promised public-sector jobs and social services, such as health care and education, with the expectation that in exchange, citizens would consent to authoritarian governments. But in the mid-1980s, economic reforms spread throughout the region, spurred by the International Monetary Fund and the World Bank. The reforms required "free market" economic liberalization: privatizing state-owned industries, reducing government subsidies for food and fuel, and economic austerity programs that severely cut government spending. While these reforms were good for business and initially resulted in economic growth, the benefits did not filter down to the middle class or the poor. At the same time, a global commitment to mass education (expressed in the United Nations Millennium Development Goals) meant that there were many well-educated youth—young people who had little or no economic prospects, blamed corrupt and repressive governments for their problems, and were well aware of the existence of democratic alternatives. Since fifteen- to thirty-year-olds make up nearly a third of the population in the region, there was no shortage of angry young people who believed they had little to lose and much to gain by taking to the streets (Winckler 2013; Ansani and Daniele 2012; Kirk 2016; Lynch 2011).

A key part of the study of globalization is the emergence of a world system—for some purposes, we need to regard the world as forming a single social order. As we see in the case of the Arab Spring protests that surged throughout the world—and the often violent response to the push for greater democracy—we are all global citizens whose lives are increasingly interdependent. In this chapter, we examine these global processes and see what leading sociologists and other social scientists have had to say about them. Some of these ideas will already be familiar to you, since much of this book has been about the consequences of globalization. In this chapter, we go beyond our earlier discussions, considering why the modern period is associated with especially profound and rapid social change. We examine how globalization has contributed to such rapid social change and offer some thoughts on what the future is likely to bring.

social change

Alteration in basic structures of a social group or society. Social change is an ever-present phenomenon in social life but has become especially intense in the modern era. The origins of modern sociology can be traced to attempts to understand the dramatic changes shattering the traditional world and promoting new forms of social order.

Recognize that a number of factors influence social change, including the physical environment, political organization, culture, economics, and technology.

How Does Globalization Affect Social Change?

The ways of living and the social institutions characteristic of the modern world are radically different from those of even the recent past. During a period of only two or three centuries—a small sliver of time in the context of human history—human social life has been wrenched away from the types of social order in which people lived for thousands of years. **Social change** can be defined as the transformation over time of the institutions and culture of a society. Globalization has accelerated the pace of social change, bringing virtually all of humanity into the same turbulent seas. As a result, far more than any generations before us, we face an uncertain future. To be sure, conditions of life for previous generations were always insecure: People were at the mercy of natural disasters, plagues, and famines. Yet, although these problems still trouble much of the world, today we must also deal with the social forces that we ourselves have unleashed.

Social theorists have tried for the past two centuries to develop a single grand theory that explains the nature of social change. Marx, for example, emphasized the importance of economic factors in shaping all aspects of social life, including politics and culture. But no single-factor theory can adequately account for the diversity of human social development from hunting-and-gathering and pastoral societies to traditional civilizations and finally to the highly complex social systems of today. In analyzing social change, we can at most accomplish two tasks: We can identify major factors that have consistently influenced social change, such as the physical environment, political organization, culture, economics, and technology, and we can also develop theories that account for particular periods of change, such as modern times.

The Physical Environment

The physical environment often has an effect on the development of human social organization. This is clearest in extreme environmental conditions, where people must organize their ways of life in relation to the weather. People who live in Alaska, where the winters are long and cold and the days very short, tend to follow different patterns of social life from people who live in the much warmer American South. Most Alaskans spend more of their lives indoors and, except for the summer months, plan outdoor activities carefully, given the frequently inhospitable environment in which they live.

Less extreme physical conditions can also affect society. The native population of Australia has never stopped being hunters and gatherers, since the continent contained hardly any indigenous plants suitable for regular cultivation or animals that could be domesticated. Ease of communication across land and the availability of sea routes are also important: Societies cut off from others by mountain ranges, impassable jungles, or deserts often remain relatively unchanged over long periods of time.

A strong case for the importance of environment is made by Jared Diamond (2005) in his widely acclaimed book *Collapse: How Societies Choose to Fail or Succeed*. Diamond—a physiologist, biologist, and geographer—examines more than a dozen past and present societies, some of which collapsed (past examples include Easter Island and the Anasazi of the southwestern United States; more recent candidates include Rwanda and Haiti) and some of which overcame serious challenges to succeed. Diamond identifies five sets of factors that can contribute to a society's collapse: the presence of hostile neighbors, the absence (or collapse of) trading partners for essential goods, climate change, environmental problems, and an inadequate response to environmental problems. Three of these five factors have to do with environmental conditions. The first four factors are often outside of a society's control and need not always result in collapse. The final factor, however, is always crucial: As the subtitle of his book suggests, success or failure depends on the choices made by a society and its leaders.

The collapse of Rwanda, for example, is typically attributed to ethnic rivalries between Hutu and Tutsi, fueled by Rwanda's colonial past. According to some explanations of the genocide that left more than 800,000 Tutsi dead in the span of a few horrific months in 1994, a large part of the cause lay in the legacy of colonialism. During the first part of the twentieth century, Belgium ran Rwanda through Tutsi administrators because, according to the prevailing European racial theories of the time, the Tutsi—who tended on average to be somewhat taller and lighter skinned than the Hutu and, therefore, closer in resemblance to Europeans—were believed by the Belgians to be more civilized. This led to resentments and hatred, which boiled over in 1994, fueled by Hutu demagogues urging the killing of all Tutsi.

Rwandan refugees try to reach the United Nations camp in Tanzania. More than 800,000 Tutsi and moderate Hutu were killed during a period of 100 days in 1994. Hundreds of thousands of Rwandans fled to neighboring countries to escape the bloodshed.

Diamond does not reject this explanation but shows that it is only part of the story and by itself cannot account for the depth of the violence. Instead, through careful analysis of patterns of landholding, population, and killing, he argues that the root causes are found in overpopulation and resulting environmental destruction. Rwanda, he shows, had one of the fastest-growing populations in the world, with disastrous consequences for its land as well as its people, who had become some of the most impoverished on the planet. Faced with starvation and the absence of land to share among the growing number of (male) children, Rwanda was ripe for violence and collapse. Although ethnic rivalries may have fueled the fires of rage, Diamond shows that in some hard-hit provinces, Hutu killed other Hutu, as young men sought to acquire scarce farmland by any means.

Some have criticized Diamond for overemphasizing the importance of the environment at the expense of other factors. The environment alone does not necessarily determine how a society develops. Today especially, when humans can exert a high degree of control over their immediate living conditions, environment would seem to be less important: Modern cities have sprung up in the arctic cold and the harshest deserts.

Since global warming is resulting in extreme climate-related events, such as hurricanes, droughts, and wildfires, the physical environment in many parts of the world will change dramatically. Populations will be forced to either move due to food scarcity and land loss or somehow adapt to the changing climate conditions. Examples are large coastal populations affected by rising sea levels, from Boston to Bangladesh, or residents of drought-stricken areas, from the U.S. Southwest to sub-Saharan Africa. The poorest people—and the poorest countries—are the most vulnerable (EPA 2017). Such changes, and any required adaptations, will prove costly to governments: According to one study, for each degree increase in average global temperature, the cost to the United States alone will be 2 percent of GDP, or roughly $380–$390 billion (based on 2017 GDP). The study notes that since its estimates are for the United States alone, these costs would likely be higher: Many of our trading partners would likely suffer worse effects, impacting the U.S. economy, while migration would increase as climate change devastates poorer countries (Hsiang et al. 2017).

Political Organization

A second factor strongly influencing social change is the type of political organization that operates in a society. In hunter-gatherer societies, this influence is minimal, since there are no political authorities capable of mobilizing the community. In all other types of society, however, the existence of distinct political agencies—chiefs, lords, monarchs, and governments—strongly affects the course of development a society takes.

How a society and its leaders respond to a crisis can play a decisive role in whether they thrive or fail. A leader capable of pursuing dynamic policies and generating a mass following or radically altering preexisting modes of thought can overturn a previously established order. However, individuals can reach positions of leadership and become effective only if favorable social conditions exist. Mahatma Gandhi, the famous pacifist leader in India, effectively secured his country's independence from Britain because World War II and other events had unsettled the existing colonial institutions there.

The most important political factor that has helped speed up patterns of change in the modern era is the emergence of the modern state, which has proved a vastly more efficient mechanism of government than the types that existed in premodern societies. Globalization today may be challenging the ability of national governments to effectively exert leadership. Sociologist William Robinson (2001, 2004, 2014), for one, claims that as economic power has become increasingly deterritorialized, so, too, has political power: Just as transnational corporations operate across borders, with little or no national allegiance, transnational political organizations are becoming stronger even as national governments are becoming weaker. The World Trade Organization (WTO), for example, has the power to punish countries that violate its principles of free trade (Conti 2011).

Culture

The third main influence on social change is culture, including communication systems, religious and other belief systems, and popular culture. Communication systems, in particular, affect the character and pace of social change. The invention of writing, for instance, allowed for effective recordkeeping, making possible the development of large-scale organizations. In addition, writing altered people's perception of the relation among past, present, and future. Societies that write keep a record of past events, which then enables them to develop a sense of the society's overall line of evolution. The existence of a written constitution and laws makes it possible for a country to have a legal system based on the interpretation of specific legal precedents—just as written scripture enables religious leaders to justify their beliefs by citing religious texts like the Bible or the Qur'an.

Religion, as we have seen, may be either a conservative or an innovative force in social life. Some forms of religious belief and practice have acted as a brake on change, emphasizing above all the need to adhere to traditional values and rituals. Yet, as Max Weber emphasized, religious convictions frequently play a mobilizing role in pressures for social change. For instance, throughout history, many American church leaders have promoted attempts to lessen poverty or diminish inequalities in society. Religious leaders, such as Dr. Martin Luther King Jr., were at the forefront of the American civil rights movement.

Yet, at the same time, religion today has become one of the driving forces against many of the cultural aspects of globalization. Islamic fundamentalists, fundamentalist

Christians, and ultra-Orthodox Jewish *haredim* all reject what they regard as the corrupting influences of modern secular culture, now rapidly spreading throughout the world thanks to mass media and the Internet (Juergensmeyer 2003, 2009; Juergensmeyer, Griego, and Soboslai 2015). Fundamentalist Islamists call this "westoxification"—literally, getting drunk on the temptations of modern Western culture. Although such religious communities are usually willing to embrace modern technology, which they often use effectively to disseminate their ideas, they reject what they view as the "McWorld" corruptions that go along with it.

Juergensmeyer (1994) predicted that in the twenty-first century, the principal cultural clashes would not be between so-called civilizations but rather between those who believe that truthful understanding is derived from religious faith and those who argue that such understanding is grounded in science, critical thinking, and secular thought. Although it is too early to fully evaluate the accuracy of Juergensmeyer's prediction, we do see compelling evidence already. For example, in the United States, political debates rage between religious conservatives who promote teaching creationism (versus evolution) in schools and liberals who believe policies should be guided by scientific evidence and preservation of civil rights. Not surprisingly, creationism is much more likely to be taught in public school systems in politically conservative districts in the South than in more liberal regions of the North (Kirk 2014).

Economic Factors

Of economic influences, the farthest reaching is global capitalism. Capitalism differs in a fundamental way from preexisting production systems because it involves the constant expansion of production and the ever-increasing accumulation of wealth. In traditional production systems, levels of production were fairly stable, since they were geared to habitual, customary needs. Capitalism requires the constant revision of the technology of production, a process into which science is increasingly drawn. The rate of technological innovation fostered in modern industry is vastly greater than in any previous type of economic order. And such technological innovation, as we have seen, has helped create a truly global economy—one whose production lines draw on a worldwide workforce. Thanks to modern technology, firms today are able to produce their goods and services by hiring factories and service centers around the world, providing jobs for people in emerging economies, but often under harsh and unsafe conditions.

Economic changes help shape other changes as well. Science and technology, for example, are driven in part (often in large part) by economic factors. Governments often get into the act, spending far more money than individual businesses can afford in an effort to ensure that their countries don't fall behind technologically, militarily, or economically. For instance, when the Soviet Union launched the world's first satellite (Sputnik) into space in 1957, the United States responded with a massive and costly space program, inspired by fear that the Russians were winning the space race. Even as recently as 2013, President Barack Obama proposed boosting funding for the Energy Department to modernize the United States' existing nuclear weapons with the goal of maintaining "a safe, secure and effective nuclear deterrent" (Guarino 2013). In each of these historical cases, the arms race—fueled by government contracts with corporations—provides major economic support for scientific research as well as more general support for the U.S. economy.

Technology

Technology has influenced social change since humans evolved from our prehuman primate ancestors. For example, the invention of primitive tools enabled early humans to hunt animals and gather food sources more efficiently; harnessing fire hundreds of thousands of years ago changed the nature of food and provided warmth, which in turn facilitated the spread of humans around the globe; the development of bronze (adding tin to copper) 4,000 years ago made for stronger metals, improved tools, and (not surprisingly) better weapons; paper, invented in China two millennia ago, transformed writing and reading (although an early use was for toilet paper!); gunpowder, a Chinese invention 1,000 years after paper, forever changed the nature of warfare; and the printing press, which was first developed in Europe in the fifteenth century, enabled the mass production of pamphlets and books, allowing the spread of ideas that resulted in the Protestant Revolution.

We have seen in this and previous chapters how the Internet and the proliferation of smartphones have transformed our personal relationships, the nature of politics and social movements, our forms of recreation, and the ways in which we learn and work—in fact, almost every aspect of modern life. These changes have been among the most rapid in human history, resulting in what geographer David Harvey (1989) has aptly referred to as the "time-space compression"—technological changes that cause the relative distances between places (measured in terms of travel time or cost) to shrink, effectively making such places grow "closer." Until the steam locomotive was developed in the mid-nineteenth century, a person traveling by horse-drawn carriage would be fortunate to average a mile per hour on rough roads; a 100-mile trip might take many days. The steam locomotive, which could reach sixty miles an hour, reduced the same trip to a couple hours; the propeller airplane in the 1950s to fifteen minutes; and the jet by half that time. Today, we can virtually travel anywhere in the world instantaneously, whether it be through video-conferencing software, Skype, FaceTime, Google Hangouts, or other easily available apps.

The rapid technological advances that have occurred in most of your lifetimes ushered in a social change as significant as the agricultural revolution 12,000 years ago or the eighteenth-century Industrial Revolution—the information revolution. The result is said to be a **postindustrial society**—one in which knowledge and information expand the importance of services in the economy, while the physical production of goods declines in significance (Bell 1976; Touraine 1974). The blue-collar worker, employed in a factory or workshop, is no longer the most essential type of employee. White-collar (clerical and professional) workers outnumber blue-collar workers, with professional and technical occupations growing

postindustrial society

A society based on the production of knowledge and information rather than material goods, resulting in the rise of an economic service sector and the decline of the manufacturing sector.

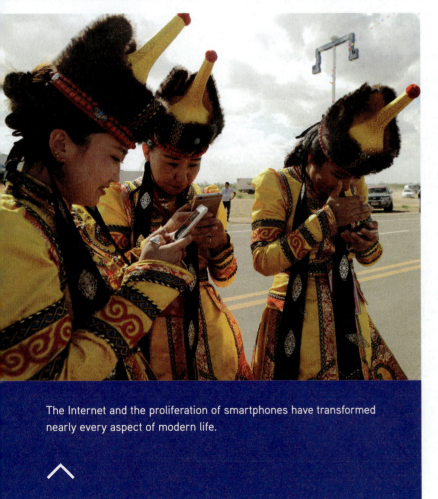

The Internet and the proliferation of smartphones have transformed nearly every aspect of modern life.

fastest of all. The shift from producing goods to providing services in the U.S. economy does not mean that factory production has ceased to be important: Manufacturing jobs have simply been offshored to take advantage of lower wages and weaker environmental protections in poor countries (see Chapters 8 and 13 for a more extensive discussion).

These are all advances that stem from the invention of the integrated circuit—a tiny microchip containing numerous electronic circuits that permit vastly increased speeds for data processing. While integrated circuits were first developed in the 1950s, the first commercially useful microprocessors, involving the use of numerous chips as the central processing unit (CPU) for computers, were developed by Intel in the 1970s. The world that we know today didn't exist even a generation ago: our smartphones that harness more computing power than that possessed by entire governments when most of your grandparents were your age; the Internet and cloud computing, with their access to global information; entirely new sources of information that didn't even exist two decades ago, including Google (which was invented in 1998), Facebook (2005), Reddit (2005), Twitter (2006), Instagram (2010), Snapchat (2011), and, in China at least, WeChat (2011). As we have previously discussed, these new technologies played a central role in the Arab Spring. Throughout this chapter, we will return to the role of technology, examining its role not only in social movements but in other changes that stem from globalization in the world today.

CONCEPT CHECKS

1. Name three examples of cultural factors that may influence social change.

2. What are the most important political factors that influence social change?

3. How does technology affect social change?

4. What is a postindustrial society?

Why Does Terrorism Seem to Be on the Rise in the World Today?

Understand the relationship between globalization and terrorism.

terrorism
A public act of violence meant to be intimidating.

Terrorism has seized national headlines in recent decades, yet defining *terrorism* can be a complex and nuanced process (Turk 2004). Terrorism broadly refers to "any action [by a nonstate organization] . . . that is intended to cause death or serious bodily harm to civilians or noncombatants, when the purpose of such an act, by its nature or context, is to intimidate a population, or to compel a Government or an international organization to do or to abstain from doing any act" (Panyarachun et al. 2004). In simpler terms, according to sociologist Mark Juergensmeyer (2015), terrorism is "a public act of violence meant to be intimidating." Note that this definition leaves out the notion that terrorism is limited to nonstate actors; states, in this view, can commit terrorist acts as well.

Terrorism today is most famously associated with loosely connected extremist groups that claim to embrace a form of Islam they believe to be true to its original teachings. Organizations such as al Qaeda, ISIS (which had hoped to create an Islamic State), and Boko Haram engage in very public acts of extreme violence aimed largely at civilian populations. Beheadings, suicide bombings in markets and other public places, and public executions are widely circulated through social media in the hope of intimidating opponents and attracting recruits. Although they frame their acts in terms of Islam, they no more reflect Islamic beliefs than acts of Christian terrorists reflect the beliefs of Christianity.

In fact, before 9/11, the worst act of terrorism in the United States was the bombing of the Alfred P. Murrah Federal Building in Oklahoma City in 1995, which claimed 168 lives (including 19 children in a day-care center), while injuring more than 800 others. That act was carried out by Timothy McVeigh, a former U.S. serviceman and white supremacist strongly influenced by the Christian Identity movement. There are currently more than 950 active hate groups in the United States, most espousing racist ideologies that draw on a vision of a supposedly pure (and white) Christian culture (Southern Poverty Law Center 2017). In fact, as Juergensmeyer has shown, when it comes to terrorism, no religion has a monopoly on the use of violence to intimate nonbelievers: Muslims, Christians, Jews, Sikhs, Hindus, and even Buddhists have all spawned terrorist offshoots in recent years (Juergensmeyer 2003, 2009; Jerryson and Juergensmeyer 2010).

The fact that terrorism today often grows out of religious beliefs would seem to be puzzling: After all, when sociology emerged in the late nineteenth and early twentieth centuries, its founders generally assumed that with the rise of science and secular thinking, religion would decline in importance (see Chapter 12). But in the twenty-first century, as globalization contributed to the spread of Western secular beliefs throughout the world, traditional beliefs that once provided a sense of identity, accountability, and security have been undermined. The loss of personal and spiritual identity provides fertile ground for leaders who promise the absolute certainty that comes with a commitment to unquestioned religious beliefs. In the view of religious fanatics, terrorism is a political weapon to be used in what they see as a "cosmic war" between good and evil: It shows that their beliefs are powerful enough to provoke and frighten even the most powerful countries (Juergensmeyer 2009).

CONCEPT CHECKS

1. How do we define terrorism?

2. What are some of the causes of religious terrorism in the world today?

> Understand what social movements are, why they occur, and how they affect society.

social movement

Large groups of people who seek to accomplish, or to block, a process of social change. Social movements normally exist in conflict with organizations whose objectives and outlook they oppose. However, movements that successfully challenge power can develop into organizations.

What Are Social Movements?

In addition to economics, technology, politics, and culture, one of the most common ways social change occurs is through social movements. As we saw earlier in this chapter, **social movements** are collective attempts to further a common interest or secure a common goal (such as forging social change) through action outside the sphere of established institutions. A wide variety of social movements, some enduring, some transient, have existed in modern societies. They are a vital feature of the contemporary world, as are the formal, bureaucratic organizations they often oppose.

Social movements come in all shapes and sizes. Some are very small, numbering fewer than a dozen members; others include thousands or even millions of people. Some social movements carry on their activities within the laws of the society, such as those carrying out peaceful protests in public squares, while others operate as illegal or underground groups, perhaps by breaking into a nuclear power plant to protest its operations. However, social movements operate near the margins of what is defined as legally permissible by governments at any particular time or place. Increasingly, many are also global in scope and rely heavily on the use of information technology to link local social movement participants to global issues.

Social movements often arise with the aim of bringing about a major change, such as expanding civil rights for a segment of the population. In response, countermovements sometimes arise in defense of the status quo. The campaign for women's right to legal abortion, for example, has been vociferously challenged by anti-abortion ("pro-life") activists, who believe that abortion should be illegal. Similarly, protests calling for the rights of transgender individuals have often been met by counterprotests from religious conservatives; most recently, the countermovement against transgender equality has taken the form of "bathroom bills" that prevent people from using a bathroom that doesn't correspond to the person's biological sex.

Often, laws or policies are altered as a result of the action of social movements. These changes in legislation can have far-ranging effects. For example, it used to be illegal for groups of workers to call their members out on strike, and striking was punished with varying degrees of severity in different countries. Eventually, however, the laws were amended, making the strike a permissible tactic of industrial conflict.

Classical Theories of Social Movements

Sociology arose in the late nineteenth century as part of an effort to come to grips with the massive political and economic transformations that Europe experienced on its way from the preindustrial to the modern world (Moore 1966). Perhaps because sociology was founded in this context, sociologists have never lost their fascination with these transformations.

Since mass social movements have been so important in world history over the past two centuries, it is not surprising that a range of theories exist to try to account for them. Some theories were formulated early in the history of the social sciences; the most important was that of Karl Marx. Marx, who lived well before any of the social movements undertaken in the name of his ideas took place, intended his views to be taken not just as an analysis of the conditions that fostered revolutionary change but as a means of furthering such change. Whatever their intrinsic validity, Marx's ideas had an immense practical impact on twentieth-century social change.

We shall look at three classical frameworks for the study of social movements: economic deprivation, resource mobilization, and structural strain.

ECONOMIC DEPRIVATION

Marx's view of social movements is based on his general interpretation of human history (see Chapter 1). According to Marx, the development of societies is marked by periodic class conflicts that, when they become acute, tend to end in a process of revolutionary change. Class struggles derive from the unresolvable tensions (he termed them "contradictions") in societies. The main sources of tension can be traced to economic changes, or changes in the *forces of production*. In any stable society, there is a balance among the economic structure, social relationships, and the political system. As the forces of production alter, contradictions are intensified, leading to open clashes between classes— and ultimately, Marx predicted, to revolution. In Marx's view, revolutionary social movements emerge from below: An increasingly discontented working class eventually organizes, rises up, and overthrows the system that oppressed them. For Marx, revolutionary social movements were a legitimate (indeed, inevitable) response to an inequitable and self-destructive capitalist economic system.

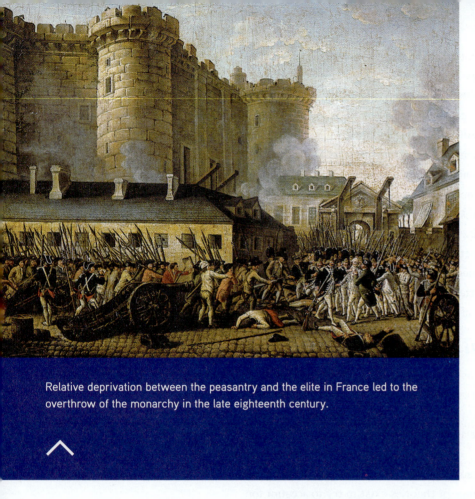

Relative deprivation between the peasantry and the elite in France led to the overthrow of the monarchy in the late eighteenth century.

Marx's views were not shared by Max Weber, who rejected the notion that revolutionary change necessarily rises up from impoverished groups that are marginalized by an oppressive economic system (Collins 2001). Weber's study of revolutions in ancient Rome and Greece and in medieval Italian city-states, as well as upheavals in Russia a decade before the communist revolution of 1917, led him to conclude that revolutions (and the social movements that swirled around them) were far more complex than Marx envisioned: They often resulted from conflict between different status groups contending for power, the breakdown of state power, and geopolitical strain (for example, defeat in war). Weber also questioned whether it was legitimate for revolutionary movements to seek to topple governments, since the state (even if oppressive), he argued, is the "human community that (successfully) claims the *monopoly of the legitimate use of physical force* within a given territory" (Weber, 1946: orig. 1921: emphasis in original).

Contrary to Marx's predictions, and perhaps more in keeping with Weber's theories, revolutions failed to occur in the advanced industrialized societies of the West. Why? More than a half-century ago, the sociologist James Davies (1962), a critic of Marx, pointed to periods of history when people lived in dire poverty but did not rise up in protest. Davies argued that social protest, and ultimately revolution, is more likely to occur when there is an improvement in people's living conditions that raises their level of expectations. When those expectations are not met, people experience *relative deprivation*, a discrepancy between the lives they are forced to lead and what they think could realistically be achieved; they then rise up in protest. While a number of recent studies have found some support for the theory of relative deprivation, it is clear that individual frustration is not sufficient to explain the rise of social movements (Walker and Smith 2002; Abrams and Grant 2012; Smith et al. 2012). Sociological, as opposed to more psychological, factors are also at work, including the ability of social movement organizations to mobilize necessary resources as well as structural strains in society that result in social conflict.

RESOURCE MOBILIZATION

Resource mobilization theory emerged during the 1970s, a period of heightened activism over the rights of women and minorities, the Vietnam War, and the environment. Sociologists questioned whether self-interest alone (such as that implied by the notion of relative deprivation) was sufficient to explain this rise in social movements, which often involved young, relatively affluent white students.

Resource mobilization theory emphasizes the interaction between a group's ability to mobilize the necessary resources to be effective, its internal organization, the degree to which the group is able to form coalitions with other groups, and the political (or other) circumstances that created opportunities for action (Tilly 1978). A key study of the civil rights movement drew on this framework to explain its rise and subsequent decline, examining such factors as changing relationships among activists and an overall decline in funding for key organizations (McAdam 1982). Resource mobilization theory also emphasizes the entrepreneurial aspects of social movements: how they are best able to amass needed money, facilities, and dedicated participants, including the tactics used to garner broader support, such as different forms of communication and access to existing supportive networks (McCarthy and Zald 1977, Zald and McCarthy 1979). By focusing on the concrete environments in which social movements exist, this theory highlights the importance of what the French sociologist Alain Touraine (1977, 1981) has termed *fields of action*—the arenas within which social movements interact with established organizations; social movements engage in conflict with the very organizations and networks on which they often draw for support.

STRUCTURAL STRAIN

Neil Smelser's (1963) theory of structural strain emphasizes the importance of structural conditions (rather than the conscious actions of social movement activists) in shaping social movements and collective action. His "value-added" theory underscores the cumulative effect of six conditions necessary to bring about social change:

1.   *Structural conduciveness* refers to the general social conditions promoting or inhibiting the formation of social movements of different types.

2.   Just because the conditions are conducive to the development of a social movement does not mean those conditions will bring it into being. There must be **structural strain**, or tensions that produce conflicting interests within societies. Uncertainties, anxieties, ambiguities, or direct clashes of goals are expressions of such strains.

3.   *Generalized beliefs and ideologies* crystallize grievances and suggest courses of action that might be pursued to remedy them.

4.   *Precipitating factors* are events or incidents that actually trigger direct action by those who become involved in the movement.

5.   The first four conditions combined might precede minor protests, but they do not lead to the development of social movements unless there is a coordinated group that becomes mobilized for action. *Leadership* and some means of regular *communication* among participants, together with funding and material resources, are necessary for a social movement to exist.

6.   The development of a social movement is strongly influenced by *social control forces*. A harsh reaction by governing authorities might encourage further protest and help solidify the movement, whereas divisions within the military can be crucial in deciding the outcome of confrontations with revolutionary movements.

Smelser's model is useful for analyzing the sequences in the development of social movements, and collective action in general. But his theory—along with resource mobilization theory—can be seen as overly deterministic, treating social movements as

structural strain

Tensions that produce conflicting interests within societies.

responses to situations rather than allowing that their members might take conscious action to achieve desired social changes.

Globalization and Social Movements

Since the emergence of classical social movement theory in the 1960s and 1970s, new theories have emerged that emphasize the quality of private life rather than the broader economic or political concerns that tended to be a key focus of the classical theories. Research has also focused on the changing role of technology in shaping social movements, particularly the role of the Internet and social media. These developments have occurred within the context of globalization, which shapes not only social movements but also our understanding of them.

NEW SOCIAL MOVEMENTS

Until the 1964 Civil Rights Act was passed—the result of a decade of action by members of the civil rights movement—African Americans in the South were barred from using "white only" restrooms and public facilities, such as swimming pools, or from eating with whites in some restaurants. In the 1960s and 1970s, the women's movement played a similar role in the fight for women's rights. For example, prior to 1971, Idaho law gave men preference over women when it came to managing the estate of a deceased, holding that "males must be preferred to females" (Napikoski 2017). A feminist attorney, and now Supreme Court justice, Ruth Bader Ginsburg, successfully argued the position of the women's movement before the U.S. Supreme Court, which then overturned the law.

The civil rights and women's movements of the 1960s and 1970s; the gay rights campaign of the 1990s; and the transgender rights, environmental, gun control, Black Lives Matter, #MeToo, alt-right, and antifa (anti-fascist) movements today are often referred to as *new social movements*. This description seeks to differentiate contemporary social movements from those that preceded them in earlier decades. They are often concerned with aspects of private life as much as with political and economic issues, calling for large-scale changes in the way people think and act. In other words, what makes new social movements "new" is that—unlike conventional social movements—they are not based on single-issue objectives that typically involve changes in the distribution of economic resources or power. Rather, they involve the creation of collective identities based around entire lifestyles, often calling for sweeping cultural changes. New social movements have emerged in recent years around issues such as ecology, peace, gender and sexual identity, gay and lesbian rights, women's rights, and opposition to globalization. Although the triggers of the Arab Spring protests differed across nations, the catalysts for many of the uprisings in Northern African and the Persian Gulf countries included discontent over the persistent concentration of wealth in the hands of autocrats in power, insufficient transparency in government policies and practices, corruption, and the refusal of the youth to accept the status quo (Khalidi 2011).

Participation in new social movements is often viewed as a moral obligation (and sometimes even a pleasure) as much as a calculated effort to achieve some specific goal. Moreover, the kinds of protest chosen by new social movements are a form of "expressive logic" whereby participants make a statement about who they are: Protest is an end in itself, a way of affirming one's identity, and a means to achieving concrete objectives (Polletta and Jasper 2001). The Women's March in January 2017, a worldwide protest in support of women's

rights and other human rights issues, attracted an estimated 3–5 million participants—men as well as women. The marches and demonstrations were not directed at immediate actions or changes but rather were intended to show solidarity among those committed to feminist ideals and opposed to perceived threats to gains that had been made. The marches held one year later, which reportedly attracted some 2 million participants, were more closely tied to effecting change in the 2018 midterm elections (Lopez 2018).

The rise of new social movements in recent years is a reflection of the changing risks facing human societies. The conditions are ripe for social movements: Increasingly, traditional political institutions are unable to cope with the challenges before them. Existing democratic political institutions cannot hope to fix sweeping problems like climate change and the dangers of nuclear energy. As a result, these unfolding challenges are frequently ignored or avoided until it is too late and a full-blown crisis is at hand. The cumulative effect of these new challenges and risks is a sense that people are losing control of their lives in the midst of rapid change. Individuals feel less secure and more isolated—a combination that leads to a sense of powerlessness. By contrast, corporations, governments, and the media appear to be dominating more and more aspects of people's lives, heightening the sensation of a runaway world. There is a growing sense that, left to its own logic, globalization will present ever greater risks to citizens' lives.

Although faith in traditional politics seems to be waning, the growth of new social movements is evidence that citizens today are not apathetic or uninterested in politics, as is sometimes claimed. Rather, there is a belief that direct action and participation are more useful than reliance on politicians and political systems. New social movements are helping to revitalize democracy in many countries. They are at the heart of a strong civic culture or **civil society**—the sphere between the state and the marketplace occupied by family, community associations, and other noneconomic institutions.

Technology and Social Movements

In recent years, two of the most influential forces in late modern societies—information technology and social movements—have come together with astonishing results. In our current information age, social movements around the globe are able to join together in

civil society

The sphere of activity that lies between the state and the marketplace, including the family, schools, community associations, and other noneconomic institutions. Civil society, or civic culture, is essential to vibrant democratic societies.

Online Activism Trends Upward

For young adults today, political protests have been reinvented by digital media technologies. As we have seen in this chapter, technology has played a critical role in mobilizing both social movements and public protests. The Arab Spring (as well as the civil wars and jihadist movements that followed), Occupy Wall Street, and Black Lives Matter were all fueled by social media.

Black Lives Matter, which began as an online protest movement against police killings of African Americans, has grown into a social movement that "affirms the lives of Black queer and trans folks, disabled folks, Black-undocumented folks, folks with records, women and all Black lives along the gender spectrum. It centers those that have been marginalized within Black liberation movements. It is a tactic to (re)build the Black liberation movement" (BlackLivesMatter 2016). It played a role in the 2015 presidential primaries, using social media to disrupt campaign rallies as well as forcing the resignations of the president of the University of Missouri and a Yale professor (Foran 2015).

To take another example, Twitter was an essential player in the antigovernment protests in Turkey in 2013. Since the local media did not adequately cover the protests, many Turks would have had little knowledge of what was happening without Twitter. Over the course of three days, an estimated 10 million tweets were posted by protesters and observers using the most popular hashtags, according to researchers from New York University. Unlike other recent protests, however, these tweets came from people on the front lines rather than people sharing their views from outside the nation. Researchers have documented that roughly 90 percent of geotagged tweets were coming from inside the country, with half from Istanbul—the epicenter of the protests (Fitzpatrick 2013).

Dissatisfied with the local mainstream media's coverage of the uprisings, young Turkish protesters began live tweeting about their actions and using their smartphones to live-stream video of the daily events. These tweets and videos, along with articles in the Western news media, became the major source of information about the movement. Protesters even urged their fellow Turks to turn off their televisions in protest of the lack of coverage by the local mainstream media, using the hashtag #BugünTelevizyonlarıKapat (literally, "turn off the TVs today"). Instead, they directed people to turn to the Internet to find out what was really happening (Fitzpatrick 2013).

Electronic media were also critical players in the January 2014 protests in the Ukraine. For example, early tweets by journalists and activists were considered the primary trigger that brought hundreds of thousands of Ukrainians into the streets on the eve of November 21, 2013. Even before dedicated Twitter feeds and Facebook pages were created, protesters tracked the events using hashtags. Very early on, #Euromaidan emerged as the main hashtag used for protest-related tweets. Shortly thereafter, an official Euromaidan Facebook page was created. Its popularity set a record in Ukraine, attracting 76,000 "likes" in its first week. The page was used to provide real-time updates as well as information on activists' future plans and advice on how to deal with potentially aggressive police officers. The speed and reach of such digital messages were remarkable and unprecedented (Arndt 2014).

Have you ever used a Facebook site, followed tweets, or used digital media to participate in or spread news about a political issue or event? What do you see as the pros and cons? What can digital media achieve that old 1960s-style protests could not?

The January 25, 2011, protest in Egypt relied so heavily on social media for its organization that pundits refer to the day as the "Facebook Revolution."

huge regional and international networks comprising nongovernmental organizations, religious and humanitarian groups, human rights associations, consumer protection advocates, environmental activists, and others who campaign in the public interest. These electronic networks now have the unprecedented ability to respond immediately to events as they occur; gain access to and share sources of information; and put pressure on corporations, governments, and international bodies as part of their campaigning strategies. For example, crowdsourcing websites like Rally and ActBlue allow like-minded individuals to make contributions to the political causes and candidates they support.

The Internet and social media have facilitated the work of social movement activists; with the click of a finger, local stories are disseminated internationally. The ability of citizens to coordinate international protests is highly worrisome for governments. For example, in response to massive protests against government corruption in January 2011, the Egyptian government blocked social media sites and mobile phone networks before ultimately pulling the plug on Egypt's access to the Internet. In an effort to control information flows, the Chinese government created the "Great Firewall," which makes it virtually impossible for Chinese citizens to access Google or Facebook.

From global protests in favor of canceling developing nations' debt to promoting fair trade to ending international corruption, the Internet has the potential to unite campaigners across national and cultural borders. Some observers argue that the information age is witnessing a migration of power away from nation-states into new nongovernmental alliances and coalitions. Such global social movements (Bennett 2012; Ghimire 2005; Milani and Laniado 2007) are examples of cross-border networks of activists who join together in pursuit of common goals—a task greatly facilitated by the Internet. For example, the International Campaign to Abolish Nuclear Weapons (ICAN), which received the Nobel Peace Prize in 2017, is a coalition of 468 nongovernmental organizations in more than 100 countries; its mission is to promote implementation of the UN's treaty on the prohibition of nuclear weapons (ICAN 2018). Transnational feminist networks (Moghadam 2005) unite groups fighting for women's rights around the world, including the Association for Women's Rights in Development (2018), "an international, feminist, membership organization committed to achieving gender equality, sustainable development and women's human rights," and the Sisterhood Is Global Institute (2018), whose global communication network provides "urgent action alerts" to organizations around the world. Networked organizations such as these depend on websites and social media to get the word out and coordinate activities among their members.

CONCEPT CHECKS

1. Compare and contrast the three classical frameworks for studying social movements.

2. Provide an example of a new social movement.

3. What distinguishes new social movements from their precursors?

What Factors Contribute to Globalization?

Recognize the importance of information flows, political changes, and transnational corporations.

Globalization is often portrayed solely as an economic phenomenon. Some make much of the role of transnational corporations whose massive operations stretch across national borders, influencing global production processes and the international distribution of labor. Others point to the electronic integration of global financial markets and the enormous volume of global capital flows. Still others focus on the unprecedented scope of world trade, which involves a much broader range of goods and services than ever before.

Although economic forces are an integral part of globalization, it would be wrong to suggest that they alone produce it. Globalization is created by the coming together of technological, political, and economic factors. It has been driven forward above all by the development of information and communications technologies that have intensified the speed and scope of interaction between people all over the world.

Information Flows

The explosion in global communications has been facilitated by some important advances in technology and the world's telecommunications infrastructure. In the post–World War II era, there has been a profound transformation in the scope and intensity of telecommunications flows. Traditional telephone communication, which depended on analog signals sent through wires and cables, has been replaced by integrated systems in which vast amounts of information are compressed and transferred digitally. Cable technology and the spread of communications satellites, beginning in the 1960s, have been integral in expanding international communications.

The impact of these communications systems has been staggering. In countries with highly developed telecommunications infrastructures, homes and offices now have multiple links to the outside world, including telephones (both landlines and cell phones), digital and cable television, and the Internet. The Internet has emerged as the fastest-growing communication tool ever developed. More than 3.5 billion people worldwide (nearly half of the world's population) were estimated to be using the Internet in 2017—representing nearly 250 percent growth in usage since 2005 (International Telecommunication Union 2017).

As we noted earlier, these forms of technology facilitate the compression of time and space: Two individuals located on opposite sides of the planet can hold a conversation in real time and send documents and images or tweet their ideas to each other with the help of satellite technology. Widespread use of the Internet and smartphones is spurring on and accelerating processes of globalization; more and more people are becoming interconnected through the use of these technologies and are doing so in places that have previously been isolated or poorly served by traditional communications. Although the telecommunications infrastructure is not evenly developed around the world, a growing number of countries now have access to international communications networks in a way that was previously impossible.

Globalization is also being driven forward by the electronic integration of the world economy.

At a call center in Gurgaon, India, employees field questions and concerns from people in the United States and elsewhere, representing a global flow of information.

The global economy is increasingly dominated by activity that has been described as "weightless" and "intangible" (Quah 1999). This so-called weightless economy is one in which products have their base in information, as is the case with computer software, media and entertainment products, and Internet-based services. The emergence of such an economy has been linked to the development of a broad base of consumers who are technologically literate and who eagerly integrate new advances in computing, entertainment, and telecommunications into their everyday lives.

While the idea of a "weightless economy" is useful in pointing to the growing economic importance of information, it obscures the fact that we cannot live on information alone: Someone is ultimately making the goods we consume, writing the code that powers our software, staffing call centers around the world (Aneesh 2015), and generally providing other supposedly "weightless" services. All too often, the people who do so are underpaid and work in unsafe and unhealthy conditions. Since the weightless economy increasingly drives production and consumption, it also contributes to resource depletion and climate change. Cloud computing now represents an estimated 1.8 percent of total energy consumption in the United States. Amazon alone required almost as much electricity to power its servers as its hometown city of Seattle in 2015; its annual global carbon dioxide emissions are estimated to be 3.3 million tons and growing—roughly the same as 650,000 passenger cars (Ryan 2018; EPA 2018).

The very operation of the global economy reflects the changes that have occurred in the information age. Many aspects of the economy now work through networks that cross national boundaries rather than stopping at them (Castells 1996). To be competitive in globalizing conditions, businesses and corporations have restructured themselves to be more flexible and less hierarchical. Production practices and organizational patterns have become more flexible, partnering arrangements with other firms have become commonplace, and participation in worldwide distribution networks has become essential for doing business in a rapidly changing global market.

Whether a job is in a factory or a call center, it can be done more cheaply in China, India, or some other developing country than in developed countries like the United States. This is increasingly true for the work of software engineers, graphic designers, and financial consultants. Of course, to the extent that global competition for labor reduces the cost of goods and services, it also provides for a wealth of cheaper products (Roach 2005). As consumers, we all benefit from low-cost flat-panel TVs made in China and inexpensive computer games programmed in India. It is an open question, however, whether the declining cost of consumption will balance out wage and job losses due to globalization.

Political Changes

A number of political changes are driving forces behind contemporary globalization. One of the most significant is the collapse of Soviet-style communism, which occurred in a series of dramatic revolutions in Eastern Europe in 1989 and culminated in the dissolution of the Soviet Union itself in 1991. Since the fall of Soviet-style communism, countries in the former Soviet bloc—including Russia, Ukraine, Poland, Hungary, the Czech Republic, the Baltic states, the states of the Caucasus and Central Asia, and many others—have moved, unevenly, toward Western-style political and economic systems. They are no longer isolated from the global community but are becoming integrated within it. The collapse of communism has hastened processes of globalization but should also be seen

In a June 2016 referendum popularly referred to as "Brexit," 52 percent of British voters voted to exit the European Union.

as a result of globalization itself. The centrally planned communist economies and the ideological and cultural control of communist political authority were ultimately unable to survive in an era of global media and an electronically integrated world economy.

A second important political factor leading to intensifying globalization is the growth of international and regional mechanisms of government. The United Nations and the European Union (EU) are the two most prominent examples of international organizations that bring together nation-states in a common political forum. Whereas the UN does this as an association of individual nation-states, the EU is a more pioneering form of transnational governance in which a certain degree of national sovereignty is relinquished by its member states. The governments of individual EU states are bound by directives, regulations, and court judgments from common EU bodies, but they also reap economic, social, and political benefits from their participation in the regional union.

Yet both the UN and the EU have been challenged in recent years. The UN, unfortunately, has proven to be a weak actor. One of the reasons is that significant UN actions require the consent of its Security Council, which in turn requires the agreement of at least nine of its fifteen members, including all five of its permanent members (the United States, France, England, Russia, and China). Another reason is that member nations are not willing to give up their sovereignty to the UN, which consequently lacks the means to enforce its actions. The EU has had difficulty managing the economic slowdown of its member nations, including the near-insolvency of debt-ridden countries such as Greece. The influx of more than a million refugees from war-torn Syria and other countries has created seemingly insurmountable challenges, particularly since once they are in any European country, migrants can freely cross borders into any other. Anti-migrant sentiments have grown, leading some to question the "open borders" policies that have thus far created a strongly unified Europe.

One result was the so-called Brexit vote in Britain, a June 2016 referendum in which slightly more than half of all voters (52 percent) called for Britain to withdraw from the EU. The vote passed both because of voters' concerns about immigration and the belief that Britain was surrendering too much national sovereignty to the EU governance system in Brussels. While opponents of Brexit argued that such concerns were greatly overblown, they could not assuage the fears of a majority of voters. The Brexit vote sent shock waves throughout the EU, since it raised fears that other countries may eventually follow suit.

A third important political factor is the growing importance of international governmental organizations (IGOs) and international nongovernmental organizations (INGOs). An IGO is a body that is established by participating governments and given responsibility for regulating or overseeing a particular domain of activity that is transnational in scope. The first such body, the International Telegraph Union, was founded in 1865. Since that time, a great number of similar bodies have been created to regulate a range of business activities, including civil aviation, broadcasting, and the disposal of hazardous waste. Prominent examples include the International Monetary Fund, the World Bank, and the WTO.

As the name suggests, INGOs differ from IGOs in that INGOs are not affiliated with government institutions. Rather, they are independent organizations that work alongside governmental bodies in making policy decisions and addressing international issues. Some of the best-known INGOs—Greenpeace, Médecins sans frontières (Doctors without Borders), the Red Cross, and Amnesty International—are involved in environmental protection and humanitarian relief efforts.

Finally, the spread of information technology has expanded the possibilities for contact among people around the globe. Every day, the global media bring news, images, and information into people's homes, linking them directly and continuously to the outside world. Some of the most gripping events of the past three decades—such as the fall of the Berlin Wall; the violent crackdown on democratic protesters in Beijing's Tiananmen Square; the terrorist attacks in New York City in 2001, Paris in 2015, and Brussels in 2016; and the protests of the Arab Spring—have unfolded through the media before a truly global audience. Such events, along with thousands of less dramatic ones, have resulted in a reorientation in people's thinking from the level of the nation-state to the global stage. Individuals are now more aware of their interconnectedness with others and more likely to identify with global issues and processes than in times past.

This shift to a global outlook has two significant dimensions. First, as members of a global community, people increasingly perceive that social responsibility does not stop at national borders but instead extends beyond them. There is a growing assumption that the international community has an obligation to act in crisis situations to protect the physical well-being or human rights of people whose lives are under threat. In the case of natural disasters, such interventions take the form of humanitarian relief and technical assistance. In recent years, earthquakes in Haiti and Japan, floods in Mozambique, famine in Africa, hurricanes in Central America, the tsunami that hit Asia and Africa, and the typhoon that struck the Philippines have been rallying points for global assistance. Today, with a growing awareness of the scientific consensus on the ramifications of global warming, environmental movements—united by social media—have mushroomed from the Marshall Islands to Miami.

Second, a global outlook means that people are increasingly looking to sources other than the nation-state in formulating their sense of identity. Local cultural identities in various parts of the world are experiencing powerful revivals at a time when the traditional hold of the nation-state is undergoing profound transformation. In Europe, for example, inhabitants of Scotland and the Basque region of Spain might be more likely to identify as Scottish or Basque—or simply as Europeans—rather than as British or Spanish. The nation-state as a source of identity is waning in many areas as political shifts at the regional and global levels loosen people's orientations toward the states in which they live. A form of nationalism based on ethnicity, religion, or culture—rather than nation-state—is reflected in growing persecution, and sometimes outright violence, in many countries against those perceived as nonnative, such as immigrants or members of religious minorities.

Economic Changes

Among the many economic factors driving globalization, the role of transnational corporations is particularly important. **Transnational corporations** are companies that produce goods or market services in more than one country. These may be relatively small firms with one or two factories outside the country in which they are based or gigantic international ventures whose operations crisscross the globe.

Transnational corporations account for some two-thirds of all world trade, they are instrumental in the diffusion of new technology around the globe, and they are major actors in international financial markets. As we noted in Chapter 13, a Swiss study of more

transnational corporations

Business corporations located in two or more countries.

Transnational corporations such as Coca-Cola are eager to tap growing markets in countries like China and India. Here, corporate leaders break ground on a new plant in the Gansu province of China. The plant is one of thirty-five bottling plants Coca-Cola has opened in mainland China since it reentered the country in 1979.

than 43,000 transnational corporations found that a mere 737 firms—less than 2 percent of the total—accounted for four-fifths of their combined monetary value. The financial services industry is a power player in the global economy: The top fifty firms were primarily financial organizations such as banks and giant investment firms (Vitali, Glattfelder, and Battiston 2011). The world's 500 largest transnational corporations had combined revenues of nearly $28 trillion in 2016 (Silva 2017); in the same year, $75.8 trillion in goods and services were produced by the entire world (World Bank 2018j). While the United States remains home to the largest number of giant transnational corporations, its share has slipped considerably in recent years, particularly with the rise of Asian countries, such as Japan, South Korea, and especially China.

The "electronic economy" is another factor that underpins economic globalization. Banks, corporations, fund managers, and individual investors are able to shift funds internationally with the click of a mouse or a tap on a smartphone. This new ability to instantaneously move "electronic money" carries with it great risks, however. Transfers of vast amounts of capital can destabilize economies, triggering international financial crises. As the global economy becomes increasingly integrated, a financial collapse in one part of the world can have an enormous effect on distant economies. This became painfully evident when the once-venerable financial services firm Lehman Brothers collapsed in 2008, sending financial shock waves throughout the United States and global economies. The Dow Jones dropped by more than 4 percentage points immediately following Lehman's filing for Chapter 11 bankruptcy, making it the largest single drop since the 9/11 attacks in 2001. Banks and insurers throughout the world, from Scotland to Japan, registered devastating losses as a result (Council on Foreign Relations 2013).

The political, economic, social, and technological factors we have described are joining together to produce a phenomenon that lacks any earlier parallel in terms of its intensity and scope. The consequences of globalization are many and far-reaching, as we will see later in this chapter. But first we turn our attention to the main views of globalization.

The Globalization Debate

In recent years, globalization has become a hotly debated topic. Most people accept that important transformations are occurring around us, but the extent to which it is valid to explain these as "globalization" is contested. As an unpredictable and turbulent process, globalization is seen and understood very differently by observers. David Held and his colleagues (1999) have surveyed the controversy and divided its participants into three schools of thought: "skeptics," "hyperglobalizers," and "transformationalists." These three tendencies within the globalization debate are summarized in Table 16.1.

TABLE 16.1

Conceptualizing Globalization: Three Tendencies

CHARACTERISTIC	SKEPTICS	TRANSFORMATIONALISTS	HYPERGLOBALIZERS
What's new?	Trading blocs, weaker geogovernance	Historically unprecedented levels of global interconnectedness	A global age
Dominant features	World less interdependent than in 1890s	"Thick" (intensive and extensive) globalization	Global capitalism, global governance, global civil society
Power of national governments	Reinforced or enhanced	Reconstituted, restructured	Declining or eroding
Driving forces of globalization	Governments and markets	Combined forces of modernity	Capitalism and technology
Pattern of stratification	Increased marginalization of Global South	New architecture of world order	Erosion of old hierarchies
Dominant motif	National interest	Transformation of political community	McDonald's, Beyoncé, etc.
Conceptualization of globalization	As internationalization and regionalization	As the reordering of interregional relations and action at a distance	As a reordering of the framework of human action
Historical trajectory	Regional blocs/clash of civilizations	Indeterminate: global integration and fragmentation	Global civilization
Summary argument	Internationalization depends on government acquiescence and support.	Globalization is transforming government power and world politics.	Globalization means the end of the nation-state.

Source: Adapted from Held et al. 1999.

THE SKEPTICS

Some thinkers argue that the idea of globalization is overrated—that the debate over globalization is a lot of talk about something that is not new. The skeptics in the globalization controversy believe that current levels of economic interdependence are not unprecedented. Pointing to nineteenth-century statistics on world trade and investment, they contend that modern globalization differs from the past only in the intensity of interaction between nations. The skeptics agree that there may now be more contact between countries than in previous eras, but in their eyes, the current world economy is not sufficiently integrated to constitute a truly globalized economy. This is because the bulk of trade occurs within three regional groups: Europe, Asia-Pacific, and North America (Hirst 1997).

Many skeptics focus on processes of regionalization within the world economy, such as the emergence of major financial and trading blocs. To skeptics, the growth of regionalization is evidence that the world economy has become less integrated rather than more so

(Boyer and Drache 1996; Hirst and Thompson 1999). Compared with the patterns of trade that prevailed a century ago, they argue, the world economy is less global in its geographical scope and more concentrated on intense pockets of activity.

According to the skeptics, national governments continue to be key players because of their involvement in regulating and coordinating economic activity. For example, skeptics point out that national governments are the driving force behind many trade agreements and policies of economic liberalization.

THE HYPERGLOBALIZERS

The hyperglobalizers take an opposing position to that of the skeptics (Ohmae 1990, 1995; Friedman 2000, 2005). They argue that globalization is a very real phenomenon, the consequences of which can be felt almost everywhere. They see globalization as a process that is indifferent to national borders. It is producing a new global order, swept along by powerful flows of cross-border trade and production. Much of the analysis of globalization offered by hyperglobalizers focuses on the changing role of the nation-state. It is argued that individual countries no longer control their economies because of the vast growth in world trade. Some hyperglobalizers believe that the power of national governments is also being challenged from above—by new regional and international institutions, such as the EU. Taken together, these shifts signal to the hyperglobalizers the dawning of a global age (Albrow 1997) in which national governments decline in importance and influence.

Social scientists endorsing what might be termed a "strong globalization" position include sociologists such as William Robinson (2001, 2004, 2005a, 2005b, 2014), Leslie Sklair (2002a, 2002b, 2003), and Saskia Sassen (1996, 2005). While these scholars do not see themselves as hyperglobalists, they argue that transnational economic actors and political institutions are challenging the dominance of national ones. Robinson, one of the strongest proponents of this position, has studied these changes throughout the world, with a special focus on Latin America. He argues that the most powerful economic actors on the world scene today are not bound by national boundaries; they are, instead, transnational in nature. For example, he argues that nation-states are being transformed into "component elements" of a transnational state—exemplified by the WTO, whose purpose is to serve the interests of global businesses as a whole by ensuring that individual countries adhere to the principles of free trade. Robinson (2001) concludes that "the nation-state is a historically-specific form of world social organization in the process of becoming transcended by globalization."

THE TRANSFORMATIONALISTS

The transformationalists take more of a middle position. Writers such as David Held (Held et al. 1999) and one of the authors of this textbook, Anthony Giddens (1990), see globalization as the central force behind a broad spectrum of changes that are currently shaping modern societies. In this view, the global order is being transformed, but many of the old patterns remain. Governments, for instance, retain a good deal of power despite the advance of global interdependence. These transformations are not restricted to economics alone but are equally prominent within the realms of politics, culture, and personal life. Transformationalists contend that the current level of globalization is breaking down established boundaries between internal and external, international and domestic. In trying to adjust to this new order, societies, institutions, and individuals are being forced to navigate contexts where previous structures have been shaken up.

Unlike hyperglobalizers, transformationalists see globalization as a dynamic and open process that is subject to influence and change. Globalization is not a one-way process, as some claim, but a two-way flow of images, information, and influences. Global migration, media, and telecommunications are contributing to the diffusion of cultural influences. The world's vibrant "global cities" are thoroughly multicultural, with ethnic groups and cultures intersecting and living side by side. According to transformationalists, globalization is a decentered and self-aware process characterized by links and cultural flows that work in a multidirectional way. Because globalization is the product of numerous intertwined global networks, it cannot be seen as being driven from one particular part of the world.

Rather than losing sovereignty, as the hyperglobalizers argue, countries are seen by transformationalists as restructuring in response to new forms of economic and social organization that are nonterritorial in basis (e.g., corporations, social movements, and international bodies). They argue that we are no longer living in a state-centric world; governments are being forced to adopt a more active, outward-looking stance toward governance under the complex conditions of globalization (Rosenau 1997).

Whose view is most nearly correct? There are elements of truth in all three views, although the view of the transformationalists is perhaps the most balanced. The skeptics underestimate how far the world is changing; world finance markets, for example, are organized on a global level much more than they ever were before. Yet, at the same time, the world has undergone periods of intense globalization before, only to withdraw into periods in which countries sought to protect their markets and closed their borders to trade. While the march of globalization today often seems inevitable, it is by no means certain that it will continue unabated: Countries that find themselves losing out may attempt to stem the tide.

On the one hand, the hyperglobalizers are correct in pointing to the current strength of globalization as dissolving many national barriers, changing the nature of state power, and creating new and powerful transnational social classes. On the other hand, they often see globalization too much in economic terms and as too much of a one-way process. In reality, globalization is much more complex. National governments will neither dissolve under the weight of a globalized economy (as some hyperglobalizers argue) nor reassert themselves as the dominant political force (as some skeptics argue) but rather will seek to steer global capitalism to their own advantage. The world economy of the future may be much more globalized than today's, with multinational corporations and global institutions like the WTO playing increasingly important roles. But some countries in the world economy may still be more powerful than even the most powerful transnational actors.

CONCEPT CHECKS

1. What is an example of a political change driving contemporary globalization?

2. How are transnational corporations contributing to globalization?

3. Compare and contrast how skeptics, hyperglobalizers, and transformationalists explain the phenomenon of globalization.

4. How might skeptics, hyperglobalizers, and transformationalists differently interpret the growing global prominence of China?

How Does Globalization Affect Your Life?

<

Recognize the ways that large global systems affect local contexts and personal experiences.

Although globalization is often associated with changes within big systems—such as the world financial markets, production and trade, and telecommunications—the effects of globalization are felt equally strongly in the private realm. Globalization is fundamentally changing the nature of our everyday experiences. As societies undergo profound transformations, the established institutions that used to underpin them have become outmoded. This is forcing a redefinition of intimate and personal aspects of our lives, such as the family, gender roles, sexuality, personal identity, interactions with others, and relationships to work.

The Rise of Individualism

In our current age, individuals have much more opportunity to shape their own lives than once was the case. At one time, tradition and custom exercised a very strong influence on the path of people's lives. Factors such as social class, gender, ethnicity, and even religious affiliation could close off certain avenues for individuals or open up others. In times past, individuals' personal identities were formed in the context of the community into which they were born. The values, lifestyles, and ethics prevailing in that community provided relatively fixed guidelines according to which people lived their lives.

Under conditions of globalization, however, we are faced with a move toward a new individualism in which people are more easily able to actively construct their own identities. Thanks to information technology and social media, a globalized production system with factories and service providers dispersed around the world, and the global spread of consumer culture more generally, the weight of tradition and established values is diminishing in many places as people around the world are exposed to new ideas. Even the small choices we make in our daily lives—what we wear, how we spend our leisure time, and how we take care of our health—are part of an ongoing process of creating and re-creating our self-identities.

One study of the degree to which societies have become more individualistic examined changes in values and behaviors in seventy-seven countries between 1960 and 2010. Values were measured by survey questions asking how respondents felt about such things as the importance of family and friends, teaching children to be independent, and preference for self-expression; behaviors were determined by census data, including household size, number of people living alone, and divorce rates. The study found significant increases in individualistic values in nearly three out of four countries, while more than four out of five showed significant increases in individualistic behavior (Santos, Varnum, and Grossman 2017). There were, however, some significant exceptions: China, for example, actually showed a slight decline in individualism.

Work Patterns

Globalization has unleashed profound transformations within the world of work. New patterns of international trade and the shift to a knowledge economy have had a significant impact on long-standing employment patterns. Many traditional industries have been made obsolete by new technological advances or are losing their share of the market to competitors abroad whose labor costs are lower. Global trade and new technology have had a strong effect on traditional manufacturing communities, where industrial workers in advanced economies have been left unemployed and without the types of skills needed to enter the new knowledge-based economy. These communities are facing a new set of social problems, including long-term unemployment and rising crime rates, as a result of economic globalization. Yet, at the same time, globalization has provided job opportunities for workers (especially women) in emerging economies.

While these jobs are typically poorly paid and frequently unsafe, they also provide a degree of financial independence that women in more traditional societies often lack.

If at one time people's working lives were dominated by employment with one employer over the course of several decades—the so-called job-for-life framework—today many more individuals create their own career paths, pursuing individual goals and exercising choice in attaining them. Often this involves changing jobs several times over the course of a career, building up new skills and abilities, and transferring them to diverse work contexts. Standard patterns of full-time work are being dissolved into more flexible arrangements: working from home with the help of information technology, job sharing, gig work, short-term consulting projects, and so forth. While this affords new opportunities for some, for most it means far greater uncertainty. Job security—and the health and retirement benefits that went with it—have largely become things of the past.

Popular Culture

The cultural effects of globalization have received much attention. Images, ideas, goods, and styles are now disseminated around the world more rapidly than ever before. Trade, information technologies, the international media, and global migration have all contributed to the free movement of culture across national borders. Many people believe that we now live in a single information order—a massive global network where information is shared quickly and in great volumes. Films like *Avengers: Infinity War* and *Star Wars: The Force Awakens* have enjoyed worldwide popularity. The film *Avatar*, the most popular film of all time, has grossed nearly $2.8 billion in fifty-five countries since its release in 2009—two-thirds of it outside the United States.

Some people worry that globalization is leading to the creation of a global culture in which the values of the most powerful and affluent—in this instance, Hollywood filmmakers—overwhelm the strength of local customs and tradition. According to this view, globalization is a form of **cultural imperialism** in which the values, styles, and outlooks of the Western world are being spread so aggressively that they smother individual national cultures.

Others, by contrast, claim that global society is now characterized by an enormous diversity of cultures existing side by side. Local traditions are joined by a host of additional cultural forms from abroad, presenting people with a bewildering array of lifestyle options from which to choose. Rather than a unified global culture, what we are witnessing is the fragmentation of cultural forms (Baudrillard 1988). Established identities and ways of life grounded in local communities and cultures are giving way to **hybridization**—the process by which new forms of hybrid identity are created out of elements from contrasting cultural sources (Hall 1992). For example, while *bhangra* melodies hail from the Punjab region of India, U.S. music fans may recognize *bhangra* harmonies and rhythms from hip-hop artists such as Beyoncé and Jay-Z.

American films such as *Star Wars: The Force Awakens* dominate the global box office. Does this amount to cultural imperialism?

cultural imperialism

When the values, styles, and outlooks of the world are being spread so aggressively that they smother individual national cultures.

hybridization

The process by which new forms of hybrid identity are created out of elements from contrasting cultural sources.

Globalization and Risk

The consequences of globalization are far-reaching, affecting virtually all aspects of the social world. Yet because globalization is an open-ended and internally contradictory process, it produces outcomes that are difficult to predict and control. Another way of thinking of this dynamic is in terms of risk. Many of the changes wrought by globalization are presenting us with new forms of risk that differ greatly from those that existed in previous eras. Unlike risks from the past, which had established causes and known effects, today's risks are incalculable in origin and indeterminate in their consequences.

THE SPREAD OF "MANUFACTURED RISK"

Humans have always had to face risks of one kind or another, but today's risks are qualitatively different from those of earlier times. Until quite recently, human societies were threatened by **external risk**—dangers such as drought, earthquakes, famines, and storms that spring from the natural world and are unrelated to the actions of humans. Today, however, we are increasingly confronted with various types of **manufactured risk**—risks that are created by the impact of our own knowledge and technology on the natural world. As we shall see, many environmental and health risks facing contemporary societies are instances of manufactured risk; they are the outcomes of our own interventions into nature.

One of the clearest illustrations of manufactured risk are threats to the natural environment (see Chapter 15). One of the consequences of accelerating industrial and technological development is that few aspects of the natural world remain untouched by humans. Urbanization, industrial production and pollution, large-scale agricultural projects, the construction of dams and hydroelectric plants, and nuclear power are just some of the ways in which human beings have had an impact on their natural surroundings. The collective outcome of such processes has been widespread environmental destruction whose precise cause is indeterminate and whose consequences are similarly difficult to calculate.

In our globalizing world, ecological risk confronts us in many guises. Concern over global warming has been mounting in the scientific community for some years. Most scientists now accept that Earth's temperature has been increasing due to a rising concentration of greenhouse gases—a by-product of man-made processes, such as deforestation and the burning of fossil fuels. The potential consequences of global warming are devastating: If polar ice caps continue to melt at the current rate, sea levels will rise and may threaten low-lying landmasses and their human populations. Changes in climate patterns have been cited as possible causes of the record number of hurricanes that swept through the Atlantic and the Gulf of Mexico in the fall of 2005 and of Hurricane Katrina, which devastated New Orleans, and Hurricane Sandy, which leveled entire neighborhoods in New Jersey and New York in 2012. Hurricanes Irma, Harvey, and Maria—three Category 5 hurricanes with sustained winds of over 150 miles per hour—swept through the Caribbean and southeastern United States within a month of one another in 2017. Harvey left large parts of Houston underwater, becoming one of the costliest hurricanes in U.S. history. Maria destroyed much of Puerto Rico, knocking out its electrical system and leaving 1.5 million homes and businesses without power; nearly two years later, the island's electrical grid remains unstable and vulnerable to the next storm (Resnick-Ault and Brown 2018).

In the past decade, the dangers posed to human health by manufactured risks have attracted great attention. For example, sun exposure has been linked to a heightened risk

---

external risk

Dangers that spring from the natural world and are unrelated to the actions of humans. Examples include droughts, earthquakes, famines, and storms.

---

manufactured risk

Dangers that are created by the impact of human knowledge and technology on the natural world. Examples include global warming and genetically modified foods.

of skin cancer in many parts of the world. This is thought to be related to the depletion of the ozone layer—the layer of Earth's atmosphere that normally filters out ultraviolet light. There is, however, some good news: According to a 2014 report, the ozone layer is on track to make a substantial recovery by mid-century, largely because of the 1987 Montreal Protocol, which banned ozone-depleting chemicals like chlorofluorocarbons (CFCs) (UNEP/WMO 2014).

Many examples of manufactured risk are linked to food. For example, chemical pesticides and herbicides are widely used in commercial agriculture, and many animals (such as chickens and pigs) are pumped full of hormones and antibiotics. Some scientists have suggested that farming techniques such as these, and widespread production of genetically modified foods, could compromise food safety and have an adverse effect on humans.

THE GLOBAL "RISK SOCIETY"

Manufactured risks have presented individuals with new choices and challenges in their everyday lives. Because there is no road map to these new dangers, individuals, countries, and transnational organizations must negotiate risks as they make choices about how lives are to be lived. The German sociologist Ulrich Beck (1992) sees these risks contributing to the formation of a global "risk society." As technological change progresses more and more rapidly and produces new forms of risk, we must constantly respond and adjust to these changes. The risk society, he argues, is not limited to environmental and health risks; it includes a whole series of interrelated changes within contemporary social life: shifting employment patterns, heightened job insecurity, the erosion of traditional family patterns, and the democratization of personal relationships. Because personal futures are much less fixed than they were in traditional societies, decisions of all kinds present risks for individuals. According to Beck (1995), an important aspect of the risk society is that its hazards are not restricted spatially, temporally, or socially. Today's risks affect all countries and all social classes; they have global, not merely personal, consequences.

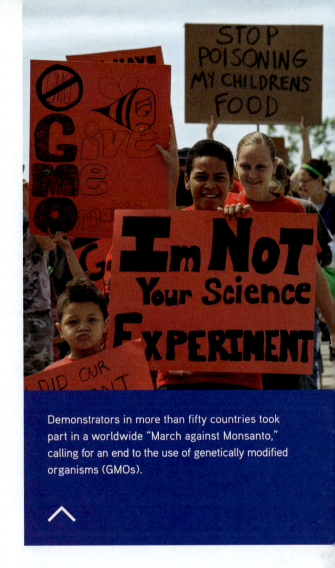

Demonstrators in more than fifty countries took part in a worldwide "March against Monsanto," calling for an end to the use of genetically modified organisms (GMOs).

Globalization and Inequality

While Beck and other scholars have drawn attention to risk as one of the main outcomes of globalization and technological advance, it is important to recognize that not all people and societies are affected equally. New forms of risk present complex challenges for both individuals and whole societies that are forced to navigate unknown terrain. Globalization is proceeding in an uneven way: Next to mounting ecological problems, the expansion of inequalities within and between societies is one of the most serious challenges facing the world today.

INEQUALITY AND GLOBAL DIVISIONS

As we learned in our discussions of types of societies (Chapter 2) and of global inequality (Chapter 8), the vast majority of the world's wealth is concentrated in the industrialized countries of the Global North, whereas countries in the Global South suffer from widespread poverty, overpopulation, inadequate educational and health care systems, and crippling foreign debt. The disparity between the industrial North and the Global South widened steadily over the course of the twentieth century and is now the largest it has ever been. A recent report on global wealth shows that global inequality is at extreme levels. The richest 9 percent of the global population now owns 86 percent of the world's wealth, and the richest 1 percent alone holds half of all global wealth. In sharp contrast, the bottom 70 percent of the global population owns less than 3 percent of total wealth (Credit Suisse 2017).

These vast disparities in economic well-being are all the more jarring when daily income is considered. Recent data from the World Bank (2017b) show that approximately 770 million people are living in extreme poverty; that is, they live on less than $1.90 per day. Although the proportion of persons who live under such dire circumstances has decreased markedly over the last three decades—from 44 percent in 1981 to 11 percent in 2017—the absolute numbers living in abject poverty remain high, because the populations in these poor nations are so large. Extreme poverty is clustered in sub-Saharan Africa, which accounts for half of the world's extreme poor.

In much of the Global South, levels of economic growth and output over the past century have not kept up with the rate of population growth, whereas the level of economic development in industrialized countries has far outpaced it. These opposing tendencies have led to a marked divergence between the richest and poorest countries of the world. China, for example, accounts for 22 percent of the global adult population but only 10 percent of global wealth (see the "Globalization by the Numbers" infographic on p. 527). The gap between population and wealth is even more extreme in India, which accounts for 17 percent of the global adult population but only 1.8 percent of global wealth (Credit Suisse 2017).

Globalization is exacerbating these trends by further concentrating income, wealth, and resources within a small core of countries. As we have seen in this chapter, the global economy is growing and integrating at an extremely rapid rate. The expansion of global trade has been central to this process. Global trade in goods and services has increased by nearly 60 percent in the last decade, from $13 trillion in 2005 to nearly $21 trillion in 2016. The volume of merchandise exports in 2016 exceeded $16 trillion—up from $10.6 trillion in 2005. The volume of service exports in 2016 was nearly $5 trillion—up from $2.6 trillion in 2005 (World Bank 2018i). And yet only a handful of developing countries have managed to benefit from that rapid growth, and the process of integration into the global economy has been uneven. Some countries—such as the East Asian economies, Chile, India, and Poland—have fared well, with significant growth in exports. Other countries—such as Russia, Venezuela, Algeria, and most of sub-Saharan Africa—have seen few benefits from expanding trade and globalization (UN Development Programme 2006). There is a danger that many of the countries most in need of economic growth will be left even further behind as globalization progresses (World Bank 2000).

Free trade is seen by many as the key to economic development and poverty relief. Organizations such as the WTO work to liberalize trade regulations and to reduce barriers to trade among the countries of the world. Free trade across borders is viewed as a win-win proposition for countries in the Global North and South. While the industrialized economies are able to export their products to markets around the world, it is claimed

Global Wealth

In 2017 global wealth reached $280 trillion (U.S. dollars), 27 percent higher than a decade ago at the onset of the financial crisis. Currently, the wealthiest 1 percent of the population owns 50.1 percent of all global household wealth. While emerging economies such as China and India are growing at a rapid clip, much of this wealth is still concentrated in Europe and the United States.

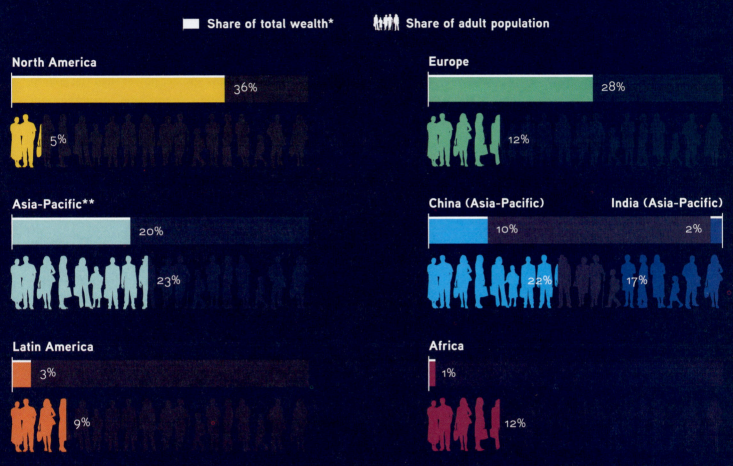

■ **Share of total wealth*** ♟♟♟ **Share of adult population**

North America
36%
5%

Europe
28%
12%

Asia-Pacific**
20%
23%

China (Asia-Pacific)
10%
22%

India (Asia-Pacific)
2%
17%

Latin America
3%
9%

Africa
1%
12%

*Global household wealth as defined by Credit Suisse as the marketable value of financial assets plus non-financial assets (principally housing and land) minus debts.
**Excludes China and India.

Real annual wealth growth rates (%), 2000–2015

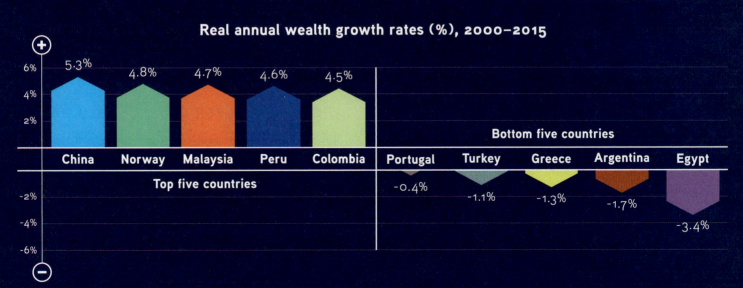

Top five countries					Bottom five countries				
China	Norway	Malaysia	Peru	Colombia	Portugal	Turkey	Greece	Argentina	Egypt
5.3%	4.8%	4.7%	4.6%	4.5%	-0.4%	-1.1%	-1.3%	-1.7%	-3.4%

Source: Credit Suisse 2017.

that countries in the Global South also benefit by gaining access to world markets. This, in turn, is supposed to improve their prospects for integration into the global economy.

But such integration has often come with significant human and ecological costs. Low-income countries are seldom able to compete in world markets; rather, they are far more likely to provide cheap labor and natural resources for the North American, European, and East Asian firms that produce the goods and services we all consume. As a result, low-income countries typically suffer from low wages and unsafe, polluting factories (Appelbaum and Lichtenstein 2017). These countries are also in a poor position to address these problems: If their governments try to raise wages or enforce strict environmental regulations, they run the risk of losing foreign investments (and the low-wage jobs they bring) to other countries that are more compliant. When it comes to hiring factories or purchasing services, transnational corporations have greater power than the governments of most low-income countries. And all too often, governmental corruption reinforces this power imbalance (Appelbaum et al. 2017; Robinson 2014).

Many low-income countries have also incurred substantial debt to international and regional banks and other lenders. According to the International Monetary Fund (2018), by 2017, two out of every five low-income developing countries were believed to be at some degree of risk of defaulting on their loans, twice the number from only four years earlier. Nearly a third of these—primarily countries in sub-Saharan Africa—were judged to be of high risk. In many of these countries, interest payments on borrowed money make up a large and growing part of government spending. If a country is judged to be in serious danger of default, the IMF may step in, provide loans to bail out the troubled government, and then require tax increases or cutbacks in government spending (and sometimes both) to enable the government to function without running a deficit that is funded by private borrowing. Not surprisingly, such "austerity measures" have not been popular in low-income countries, prompting anti-austerity protests.

In Tunisia, for example, where, as we saw at the beginning of this chapter, the Arab Spring first began, anti-austerity protests swept through the country in 2018. The IMF lent Tunisia $2.9 billion in 2015 to help the country balance its budget; two years later, it told Tunisia that "urgent action" was needed to reduce the country's deficit. The government's response—raising the price of gasoline and food and increasing taxes on a variety of goods and services—sparked uprisings that quickly turned violent (BBC 2018; DW 2018).

THE CAMPAIGN FOR GLOBAL JUSTICE

Not everyone agrees that free trade is the solution to poverty and global inequality. In fact, many critics argue that free trade is a rather one-sided affair that benefits those who are already well-off, leads to massive job loss of industrial workers in the advanced economies, and exacerbates existing patterns of poverty and dependency within the Global South. During the 2016 presidential

In the birthplace of the Arab Spring, people storm the streets to protest price hikes and other "austerity" measures implemented by the Tunisian government.

primary campaigns, Democrat Bernie Sanders and Republican Donald Trump seemed to agree on just one thing: Free trade agreements such as the proposed Trans-Pacific Partnership (TPP) were costing Americans their jobs. President Trump has doubled down on his campaign pledge to rethink all trade agreements: He pulled the United States out of the TPP and renegotiated the North American Free Trade Agreement (NAFTA) with Mexico and Canada.

Recently, much of this criticism of free trade has focused on the activities and policies of the WTO, which is at the forefront of efforts to increase global trade. Critics of the WTO argue that free trade and economic globalization succeed in further concentrating wealth in the hands of a few, while increasing poverty for the majority of the world's population. Most of these activists agree that global trade is necessary and potentially beneficial for national economies, but they claim that it needs to be regulated by different rules from those favored by the WTO. They argue that trade rules should be focused, first and foremost, on protecting human rights, the environment, labor rights, and local economies—not on ensuring larger profits for already rich corporations.

Critics also claim that the WTO is an undemocratic organization that is dominated by the interests of the world's richest nations—particularly the United States. Such imbalances have very real consequences. For example, although the WTO has insisted that countries in the Global South open their markets to imports from industrialized countries, it has allowed industrialized countries to maintain high barriers to agricultural imports and provide vast subsidies for their domestic agriculture production to protect their own agricultural sectors. Between 1995 and 2016, the U.S. government spent more than $350 billion on subsidies to boost the income of crop and livestock farmers (Environmental Working Group 2018). For certain crops, like sugar and rice, agricultural subsidies amount to as much as 80 percent of farm income (Stiglitz 2007). The EU spends $69 billion each year on its farmers, and the farm budget takes up to 40 percent of the EU's yearly expenditure (Reuters 2013). This has meant that the world's poorest countries, many of which remain predominantly agricultural, do not have access to the large markets for agricultural goods in industrialized countries.

Protesters against the WTO and other international financial institutions like the World Bank and the IMF argue that exuberance over global economic integration and free trade is forcing people to live in an economy rather than a society. Many are convinced that such moves will further weaken the economic position of poor societies by allowing transnational corporations to operate with few or no safety and environmental regulations. Commercial interests, they claim, are increasingly taking precedence over concern for human well-being. Not only within developing nations but in industrialized ones as well, there needs to be more investment in "human capital"—public health, education, and training—if global divisions are not to deepen even further. The issues raised in the WTO protests were, in many ways, echoed in the protests that sprung up throughout dozens of nations in 2011. The key challenge for the twenty-first century is to ensure that globalization works for people everywhere, not only for those who are already well placed to benefit from it.

CONCEPT CHECKS

1. What effects does globalization have on our everyday lives?

2. Why is globalization associated with new forms of risks? What are they?

3. What are two criticisms of free trade?

The Big Picture

Globalization in a Changing World

Thinking Sociologically

1. Discuss the many influences on social change: environmental, political, and cultural factors. Summarize how each element can contribute to social change.

2. According to this chapter, we now live in a society where we are increasingly confronted by various types of manufactured risks. Briefly explain what these risks consist of. Do you think the last decade has brought us any closer to or farther away from confronting the challenges of manufactured risks? Explain.

3. Provide an evaluation of how well the classical theories of social movements explain the Arab Spring uprisings. Which theory is most effective? Why?

How Does Globalization Affect Social Change?

p. 499

Recognize that a number of factors influence social change, including the physical environment, political organization, culture, economics, and technology.

Why Does Terrorism Seem to Be on the Rise in the World Today?

p. 505

Understand the relationship between globalization and terrorism.

What Are Social Movements?

p. 506

Understand what social movements are, why they occur, and how they affect society.

What Factors Contribute to Globalization?

p. 513

Recognize the importance of information flows, political changes, and transnational corporations.

How Does Globalization Affect Your Life?

p. 521

Recognize the ways that large global systems affect local contexts and personal experiences.

Terms to Know

Concept Checks

globalization

social change • postindustrial society

1. Name three examples of cultural factors that may influence social change.
2. What are the most important political factors that influence social change?
3. How does technology affect social change?
4. What is a postindustrial society?

terrorism

1. How do we define terrorism?
2. What are some of the causes of religious terrorism in the world today?

social movement • structural strain • civil society

1. Compare and contrast the three classical frameworks for studying social movements.
2. Provide an example of a new social movement.
3. What distinguishes new social movements from their precursors?

transnational corporations

1. What is an example of a political change driving contemporary globalization?
2. How are transnational corporations contributing to globalization?
3. Compare and contrast how skeptics, hyperglobalizers, and transformationalists explain the phenomenon of globalization.
4. How might skeptics, hyperglobalizers, and transformationalists differently interpret the growing global prominence of China?

cultural imperialism • hybridization • external risk • manufactured risk

1. What effects does globalization have on our everyday lives?
2. Why is globalization associated with new forms of risks? What are they?
3. What are two criticisms of free trade?

Glossary

A

absolute poverty The minimal requirements necessary to sustain a healthy existence.

activity theory A functionalist theory of aging that maintains that busy, engaged people are more likely to lead fulfilling and productive lives.

ageism Discrimination or prejudice against a person on the basis of age.

agency The ability to think, act, and make choices independently.

agents of socialization Groups or social contexts within which processes of socialization take place.

aging The combination of biological, psychological, and social processes that affects people as they grow older.

agrarian societies Societies whose means of subsistence are based on agricultural production (crop growing).

alienation The sense that our own abilities as human beings are taken over by other entities. The term was originally used by Karl Marx to refer to the projection of human powers onto gods. Subsequently, he used the term to refer to the loss of workers' control over the nature and products of their labor.

anomie A concept first brought into wide usage in sociology by Émile Durkheim, referring to a situation in which social norms lose their hold over individual behavior.

Anthropocene A term used to denote the current geological epoch, in which many geologically significant conditions and processes are profoundly altered by human activities.

assimilation The acceptance of a minority group by a majority population, in which the new group takes on the values and norms of the dominant culture.

authority A government's legitimate use of power.

automation Production processes monitored and controlled by machines with only minimal supervision from people.

B

biological determinism The belief that differences we observe between groups of people, such as men and women, are explained wholly by biological causes.

biological essentialism The view that differences between men and women are natural and inevitable consequences of the intrinsic biological natures of men and women.

biomedical model of health The set of principles underpinning Western medical systems and practices that defines diseases objectively and holds that the healthy body can be restored through scientifically based medical treatment.

bisexuality Sexual or romantic attraction to persons of one's own and the opposite sex.

Black feminism A strand of feminist theory that highlights the multiple disadvantages of gender, class, and race that shape the experiences of nonwhite women. Black feminists reject the idea of a single, unified gender oppression that is experienced evenly by all women and argue that early feminist analysis reflected the specific concerns of white, middle-class women.

blue- and pink-collar jobs Jobs that typically pay low wages and often involve manual or low-skill labor. Blue-collar jobs typically are held by men (e.g., factory worker), whereas pink-collar jobs are typically held by women (e.g., clerical assistant).

bourgeoisie People who own companies, land, or stocks (shares) and use these to generate economic returns, according to Marx.

broken windows theory A theory proposing that even small acts of crime, disorder, and vandalism can threaten a neighborhood and render it unsafe.

bureaucracy A type of organization marked by a clear hierarchy of authority and the existence of written rules of procedure and staffed by full-time, salaried officials.

C

capitalism An economic system based on the private ownership of wealth, which is invested and reinvested to produce profit.

caste system A social system in which one's social status is determined at birth and set for life.

churches Large, established religious bodies, normally having a formal, bureaucratic structure and a hierarchy of religious officials. The term is also used to refer to the place in which religious ceremonies are carried out.

citizens Members of a political community, having both rights and duties associated with that membership.

civil inattention The process whereby individuals in the same physical setting demonstrate to each other that they are aware of the other's presence.

civil rights Legal rights held by all citizens in a given national community.

civil society The sphere of activity that lies between the state and the marketplace, including the family, schools, community associations, and other noneconomic institutions. Civil society, or civic culture, is essential to vibrant democratic societies.

class Although it is one of the most frequently used concepts in sociology, there is no clear agreement about how the term should be defined. Most sociologists use the term to refer to socioeconomic variations among groups of individuals that create variations in their material prosperity and power.

clock time Time as measured by the clock, in terms of hours, minutes, and seconds. Before the invention of clocks, time reckoning was based on events in the natural world, such as the rising and setting of the sun.

cognition Human thought processes involving perception, reasoning, and remembering.

cohabitation Two people living together in a sexual relationship of some permanence without being married to each other.

collective bargaining The rights of employees and workers to negotiate with their employers for basic rights and benefits.

colonialism The process whereby Western nations established their rule in parts of the world away from their home territories.

communism A set of political ideas associated with Marx, as developed particularly by Lenin and institutionalized in the Soviet Union, Eastern Europe, and some developing countries.

community policing A renewed emphasis on crime prevention rather than law enforcement to reintegrate policing within the community.

comparative questions Questions concerned with drawing comparisons among different human societies.

comparative research Research that compares one set of findings on one society with the same type of findings on other societies.

complementary and alternative medicine (CAM) A diverse set of approaches and therapies for treating illness and promoting well-being that generally fall outside of standard medical practices.

compulsion of proximity People's need to interact with others in their presence.

concrete operational stage The stage of human cognitive development, as formulated by Jean Piaget, in which the child's thinking is based primarily on physical perception of the world. In this phase, the child is not yet capable of dealing with abstract concepts or hypothetical situations.

conflict theories A sociological perspective that emphasizes the role of political and economic power and oppression as contributing to the existing social order.

constitutional monarchs Kings or queens who are largely figureheads. Real power rests in the hands of other political leaders.

continuity theory Theoretical perspective on aging that specifies that older adults fare best when they participate in activities consistent with their personality, preferences, and activities from earlier in life.

control theory A theory that views crime as the outcome of an imbalance between impulses toward criminal activity and controls that deter it. Control theorists hold that criminals are rational beings who will act to maximize their own reward unless they are rendered unable to do so through either social or physical controls.

conurbation A cluster of towns or cities forming an unbroken urban environment.

conversation analysis The empirical study of conversations, employing techniques drawn from ethnomethodology. Conversation analysis examines details of naturally occurring conversations to reveal the organizational principles of talk and its role in the production and reproduction of social order.

core countries According to world-systems theory, the most advanced industrial countries, which take the lion's share of profits in the world economic system.

corporate crime Offenses committed by large corporations in society, including pollution, false advertising, and violations of health and safety regulations.

corporate culture An organizational culture involving rituals, events, or traditions that are unique to a specific company.

corporations Business firms or companies.

correlation coefficient A measure of the degree of correlation between variables.

countercultures Cultural groups within a wider society that largely reject the values and norms of the majority.

created environment Constructions established by human beings to serve their own needs, including roads, railways, factories, offices, homes, and other buildings.

crimes Any actions that contravene the laws established by a political authority.

crude birthrate A statistical measure representing the number of births within a given population per year, normally calculated as the number of births per 1,000 members. Although the crude birthrate is a useful index, it is only a general measure, because it does not specify numbers of births in relation to age distribution.

crude death rate A statistical measure representing the number of deaths that occur annually in a given population per year, normally calculated as the number of deaths per 1,000 members. Crude death rates give a general indication of the mortality levels of a community or society but are limited in their usefulness because they do not take into account the age distribution.

cults Fragmentary religious groupings to which individuals are loosely affiliated but that lack any permanent structure.

cultural appropriation When members of one cultural group borrow elements of another group's culture.

cultural capital Noneconomic or cultural resources that parents pass down to their children, such as language or knowledge. These resources contribute to the process of social reproduction, according to Bourdieu.

cultural imperialism When the values, styles, and outlooks of the world are being spread so aggressively that they smother individual national cultures.

cultural relativism The practice of judging a society by its own standards.

cultural universals Values or modes of behavior shared by all human cultures.

culture The values, norms, and material goods characteristic of a given group. The notion of culture is widely used in sociology and the other social sciences (particularly anthropology). Culture is one of the most distinctive properties of human social association.

culture of poverty The thesis, popularized by Oscar Lewis, that poverty is not a result of individual inadequacies but is instead the outcome of a larger social and cultural atmosphere into which successive generations of children are socialized. The culture of poverty refers to the values, beliefs, lifestyles, habits, and traditions that are common among people living under conditions of material deprivation.

D

data Factual information used as a basis for reasoning, discussion, or calculation. Social science data often refer to individuals' responses to survey questions.

debriefing Following a research study, informing study participants about the true purpose of the study and revealing any deception that happened during the study.

degree of dispersal The range or distribution of a set of figures.

democracy A political system that allows the citizens to participate in political decision making or to elect representatives to government bodies.

democratic elitism A theory of the limits of democracy, which holds that in large-scale societies democratic participation is necessarily limited to the regular election of political leaders.

demographic transition The changes in the ratio of births to deaths in the industrialized countries from the nineteenth century onward.

demography The study of the size, distribution, and composition of populations.

denomination A religious sect that has lost its revivalist dynamism and become an institutionalized body, commanding the adherence of significant numbers of people.

dependency culture A term popularized by Charles Murray to describe individuals who rely on state welfare provision rather than entering the labor market. The dependency culture is seen as the outcome of the "paternalistic" welfare state that undermines individual ambition and people's capacity for self-help.

dependency ratio The ratio of people of dependent ages (children and the elderly) to people of economically active ages.

dependency theories Marxist theories of economic development arguing that the poverty of low-income countries stems directly from their exploitation by wealthy countries and the multinational corporations that are based in wealthy countries.

developing world The less-developed societies, in which industrial production is either virtually nonexistent or only developed to a limited degree. The majority of the world's population lives in less-developed countries.

developmental questions Questions that sociologists pose when looking at the origins and path of development of social institutions.

deviance Modes of action that do not conform to the norms or values held by most members of a group or society who can enforce their definitions. What is regarded as deviant is as variable as the norms and values that distinguish different cultures and subcultures from one another.

deviant subculture A subculture whose members hold values that differ substantially from those of the majority.

diaspora The dispersal of an ethnic population from an original homeland into foreign areas, often in a forced manner or under traumatic circumstances.

differential association An interpretation of the development of criminal behavior proposed by Edwin H. Sutherland, according to whom criminal behavior is learned through association with others who regularly engage in crime.

direct democracy A form of participatory democracy that allows citizens to vote directly on laws and policies.

discrimination Behavior that denies to the members of a particular group resources or rewards that can be obtained by others. Discrimination must be distinguished from prejudice: Individuals who are prejudiced against others may not engage in discriminatory practices; conversely, people may act in a discriminatory fashion toward a group even though they are not prejudiced against that group.

disengagement theory A functionalist theory of aging that holds that it is functional for society to remove people from their traditional roles when they become elderly, thereby freeing up those roles for others.

displacement The transferring of ideas or emotions from their true source to another object.

division of labor The specialization of work tasks, by means of which different occupations are combined within a production system. All societies have at least some rudimentary form of division of labor, especially between the tasks allocated to men and those performed by women.

dominant group The group that possesses more wealth, power, and prestige in a society.

doubling time The time it takes for a particular level of population to double.

downward mobility Social mobility in which individuals' wealth, income, or status is lower than what they or their parents once had.

dyad A group consisting of two persons.

E

ecological approach A perspective on urban analysis emphasizing the "natural" distribution of city neighborhoods into areas having contrasting characteristics.

economic interdependence The fact that in the division of labor, individuals depend on others to produce many or most of the goods they need to sustain their lives.

economy The system of production and exchange that provides for the material needs of individuals living in a given society. Economic institutions are of key importance in all social orders.

egocentric According to Jean Piaget, the characteristic quality of a child during the early years of life. Egocentric thinking involves understanding objects and events in the environment solely in terms of the child's own position.

emerging economies Developing countries that, over the past two or three decades, have begun to develop a strong industrial base, such as Singapore and Hong Kong.

emigration The movement of people out of one country to settle in another.

empirical investigation Factual inquiry carried out in any area of sociological study.

encounter A meeting between two or more people in a situation of face-to-face interaction. Our daily lives can be seen in a series of different encounters strung out across the course of the day. In modern societies, many of these encounters are with strangers rather than people we know.

endogamy The forbidding of marriage or sexual relations outside one's social group.

entrepreneur The owner or founder of a business firm.

ethnicity Cultural values and norms that distinguish the members of a given group from others. An ethnic group is one whose members share a distinct awareness of a common cultural identity, separating them from other groups.

ethnocentrism The tendency to look at other cultures through the eyes of one's own culture and thereby misrepresent them.

ethnography The firsthand study of people using observation, in-depth interviewing, or both. Also called "fieldwork."

ethnomethodology The study of how people make sense of what others say and do in the course of day-to-day social interaction. Ethnomethodology is concerned with the "ethnomethods" by which people sustain meaningful exchanges with one another.

experiment A research method in which variables can be analyzed in a controlled and systematic way, either in an artificial situation constructed by the researcher or in naturally occurring settings.

exponential growth A geometric, rather than linear, rate of increase. Populations tend to grow exponentially.

extended family A family group consisting of more than two generations of relatives.

external risk Dangers that spring from the natural world and are unrelated to the actions of humans. Examples include droughts, earthquakes, famines, and storms.

exurban counties Low-density suburban counties on the periphery of large metro areas.

F

factual questions Questions that raise issues concerning matters of fact (rather than theoretical or moral issues).

family A group of individuals related to one another by blood ties, marriage, or adoption, who form an economic unit, the adult members of which are often responsible for the upbringing of children.

family capitalism Capitalistic enterprise owned and administered by entrepreneurial families.

family of orientation The family into which an individual is born or adopted.

family of procreation The family an individual initiates through marriage or by having children.

fecundity A measure of the number of children that it is biologically possible for a woman to produce.

feminism Advocacy of the rights of women to be equal with men in all spheres of life. Feminism dates from the late eighteenth century in Europe, and feminist movements exist in most countries today.

feminist theory A sociological perspective that emphasizes the centrality of gender in analyzing the social world and particularly the experiences of women. There are many strands of feminist theory, but they all share the intention to explain gender inequalities in society and to work to overcome them.

feminization of poverty An increase in the proportion of the poor who are female.

fertility The average number of live-born children produced by women of childbearing age in a particular society.

focused interaction Interaction between individuals engaged in a common activity or in direct conversation with each other.

folkways Norms that guide casual or everyday interactions. Violations are sanctioned subtly or not at all.

food deserts Geographic areas in which residents do not have easy access to high-quality affordable food. These regions are concentrated in rural areas and poor urban neighborhoods.

formal operational stage According to Jean Piaget, the stage of human cognitive development at which the growing child becomes capable of handling abstract concepts and hypothetical situations.

formal organization Means by which a group is rationally designed to achieve its objectives, often using explicit rules, regulations, and procedures.

formal relations Relations that exist in groups and organizations, laid down by the norms, or rules, of the official system of authority.

franchise The right to vote.

functional literacy Having reading and writing skills that are beyond a basic level and are sufficient to manage one's everyday activities and employment tasks.

functionalism A theoretical perspective based on the notion that social events can best be explained in terms of the functions they perform, that is, the contributions they make to the continuity of a society.

G

gender Social expectations about behavior regarded as appropriate for the members of each sex. Gender refers not to the physical attributes distinguishing men and women but to socially formed traits of masculinity and femininity.

gender binary The classification of sex and gender into two discrete, opposite, and nonoverlapping forms of masculine and feminine.

gender inequality The inequality between men and women in terms of wealth, income, and status.

gender socialization The learning of gender roles through social factors such as schooling, peers, the media, and family.

gender typing Designation of occupations as male or female, with "women's" occupations, such as secretarial and retail positions, having lower status and pay and "men's" occupations, such as managerial and professional positions, having higher status and pay.

generalized other A concept in the theory of George Herbert Mead, according to which the individual takes over the general values of a given group or society during the socialization process.

genocide The systematic, planned destruction of a racial, ethnic, religious, political, or cultural group.

gentrification A process of urban renewal in which older, deteriorated housing is refurbished by affluent people moving into the area.

glass ceiling A promotion barrier that prevents a woman's upward mobility within an organization.

global capitalism The current transnational phase of capitalism, characterized by global markets, production, and finances; a transnational capitalist class whose business concerns are global rather than national; and transnational systems of governance (such as the World Trade Organization) that promote global business interests.

global city A city—such as London, New York, or Tokyo—that has become an organizing center of the new global economy.

global commodity chains Worldwide networks of labor and production processes yielding a finished product.

global inequality The systematic differences in wealth and power among countries.

globalization The development of social, cultural, political, and economic relationships stretching worldwide. In current times, we are all influenced by organizations and social networks located thousands of miles away. A key part of study of globalization is the emergence of a world system—for some purposes, we need to regard the world as forming a single social order.

government The enacting of policies, decisions, and matters of state on the part of officials within a political apparatus. In most modern societies, governments are run by officials who do not inherit their positions of power but are elected or appointed on the basis of qualifications.

groupthink A process by which the members of a group ignore ways of thinking and plans of action that go against the group consensus.

H

hate crime A criminal act by an offender who is motivated by some bias, such as racism, sexism, or homophobia.

hegemonic masculinity Social norms dictating that men should be strong, self-reliant, and unemotional.

heteronormativity The pervasive cultural belief that heterosexuality is the only normal and natural expression of human sexuality.

heterosexism An ideological system that denies, denigrates, and stigmatizes any nonheterosexual form of behavior, identity, relationship, or community.

heterosexuality Sexual or romantic attraction to persons of the opposite sex.

hidden curriculum Traits of behavior or attitudes that are learned at school but not included within the formal curriculum, for example, gender differences.

home schooling The practice of parents or guardians educating their children at home, for religious, philosophical, or safety reasons.

homophobia An irrational fear or disdain of homosexuality.

homosexuality Sexual or romantic attraction to persons of one's own sex.

housework Unpaid work carried on in the home, usually by women; domestic chores such as cooking, cleaning, and shopping. Also called "domestic labor."

human resource management A style of management that regards a company's workforce as vital to its economic competitiveness.

hybridization The process by which new forms of hybrid identity are created out of elements from contrasting cultural sources.

hypothesis An idea or a guess about a given state of affairs, put forward as a basis for empirical testing.

I

ideal type A "pure type," constructed by emphasizing certain traits of a social item that do not necessarily exist in reality. An example is Max Weber's ideal type of bureaucratic organization.

ideology Shared ideas or beliefs that serve to justify the interests of dominant groups. Ideologies are found in all societies in which there are systematic and ingrained inequalities among groups. The concept of ideology connects closely with that of power.

immigration The movement of people into one country from another for the purpose of settlement.

impression management Preparing for the presentation of one's social role.

in-groups Groups toward which one feels particular loyalty and respect—the groups to which "we" belong.

income Payment, usually derived from wages, salaries, or investments.

industrialization The emergence of machine production, based on the use of inanimate power resources (such as steam or electricity).

industrialized societies Highly developed nation-states in which the majority of the population work in factories or offices rather than in agriculture and in which most people live in urban areas.

infant mortality rate The number of infants who die during the first year of life, per 1,000 live births.

infanticide The intentional killing of a newborn. Female babies are more likely than male babies to be murdered in cultures that devalue women.

informal economy Economic transactions carried on outside the sphere of formal paid employment.

informal networks Relations that exist in groups and organizations developed on the basis of personal connections; ways of doing things that depart from formally recognized modes of procedure.

information society A society no longer based primarily on the production of material goods but on the production of knowledge. The notion of the information society is closely bound up with the rise of information technology.

information technology Forms of technology based on information processing and requiring microelectronic circuitry.

informed consent The process whereby the study investigator informs potential participants about the risks and benefits involved in the research study. Informed consent must be obtained before an individual participates in a study.

inner city The areas composing the central neighborhoods of a city, as distinct from the suburbs. In many modern urban settings in industrialized nations, inner-city areas are subject to decay and dilapidation, the more affluent residents having moved to outlying areas.

instinct A fixed pattern of behavior that has genetic origins and that appears in all normal animals within a given species.

institutional capitalism Capitalistic enterprise organized on the basis of institutional shareholding.

institutional racism Patterns of discrimination based on ethnicity that have become structured into existing social institutions.

intelligence Level of intellectual ability, particularly as measured by IQ (intelligence quotient) tests.

interactional vandalism The deliberate subversion of the tacit rules of conversation.

interest group A group organized to pursue specific interests in the political arena, operating primarily by lobbying the members of legislative bodies.

intergenerational mobility Movement up or down a social stratification hierarchy from one generation to another.

interlocking directorates Linkages among corporations created by individuals who sit on two or more corporate boards.

intersectionality A sociological perspective that holds that our multiple group memberships affect our lives in ways that are distinct from single group membership. For example, the experience of a Black female may be distinct from that of a white female or a Black male.

intersex An individual possessing both male and female genitalia. Although statistically rare, this subpopulation is of great interest to gender scholars.

intragenerational mobility Movement up or down a social stratification hierarchy within the course of a personal career.

IQ (intelligence quotient) A score attained on tests of symbolic or reasoning abilities.

iron law of oligarchy A term coined by Weber's student Robert Michels meaning that large organizations tend toward centralization of power, making democracy difficult.

J

joint adoption A family in which both partners adopt a child together.

K

kinship A relation that links individuals through blood ties, marriage, or adoption.

knowledge economy A society no longer based primarily on the production of material goods but based instead on the production of knowledge. Its emergence has been linked to the development of a broad base of consumers who are technologically literate and have made new advances in computing, entertainment, and telecommunications part of their lives.

knowledge society Another common term for information society—a society based on the production and consumption of knowledge and information.

L

labeling theory An approach to the study of deviance that suggests that people become "deviant" because certain labels are attached to their behavior by political authorities and others.

language A system of symbols that represent objects and abstract thoughts; the primary vehicle of meaning and communication in a society.

latent functions Functional consequences that are not intended or recognized by the members of a social system in which they occur.

laws Rules of behavior established by a political authority and backed by state power.

leader A person who is able to influence the behavior of other members of a group.

liberal democracies Representative democracies in which elected representatives hold power.

liberal feminism Form of feminist theory that believes that gender inequality is produced by unequal access to civil rights and certain social resources, such as education and employment, based on sex. Liberal feminists tend to seek solutions through changes in legislation that ensure that the rights of individuals are protected.

liberation theology An activist Catholic religious movement that combines Catholic beliefs with a passion for social justice for the poor.

life chances A term introduced by Max Weber to signify a person's opportunities for achieving economic prosperity.

life course The various transitions and stages people experience during their lives.

life course theory A perspective based on the assumptions that the aging process is shaped by historical time and place; individuals make choices that reflect both opportunities and constraints; aging is a lifelong process; and the relationships, events, and experiences of early life have consequences for later life.

life expectancy The number of years the average person can expect to live.

life span The maximum length of life that is biologically possible for a member of a given species.

linguistic relativity hypothesis A hypothesis, based on the theories of Edward Sapir and Benjamin Lee Whorf, that perceptions are relative to language; also referred to as the Sapir-Whorf hypothesis.

literacy The ability to read and write.

local nationalisms The beliefs that communities that share a cultural identity should have political autonomy, even within smaller units of a nation-state.

looking-glass self A theory developed by Charles Horton Cooley that proposes that the reactions we elicit in social situations create a mirror in which we see ourselves.

lower class A social class composed of those who work part time or not at all and whose household income is typically low.

M

macrosociology The study of large-scale groups, organizations, or social systems.

Malthusianism A doctrine about population dynamics developed by Thomas Malthus, according to which population increase comes up against "natural limits," represented by famine and war.

managerial capitalism Capitalistic enterprises administered by managerial executives rather than by owners.

manifest functions The functions of a particular social activity that are known to and intended by the individuals involved in the activity.

manufactured risk Dangers that are created by the impact of human knowledge and technology on the natural world. Examples include global warming and genetically modified foods.

market-oriented theories Theories about economic development that assume that the best possible economic consequences will result if individuals are free to make their own economic decisions, uninhibited by governmental constraint.

marriage A socially approved sexual relationship between two individuals. Marriage normally forms the basis of a family of procreation; that is, it is expected that the married couple will produce and raise children.

Marxism A body of thought deriving its main elements from Karl Marx's ideas.

material culture The physical objects that society creates that influence the ways in which people live.

materialist conception of history The view developed by Karl Marx according to which material, or economic, factors have a prime role in determining historical change.

mean A statistical measure of central tendency, or average, based on dividing a total by the number of individual cases.

means of production The means whereby the production of material goods is carried on in a society, including not just technology but the social relations among producers.

measures of central tendency The ways of calculating averages.

median The number that falls halfway in a range of numbers.

Medicare A program under the U.S. Social Security Administration that reimburses hospitals and physicians for medical care provided to qualifying people over sixty-five years old.

megalopolis The "city of all cities" in ancient Greece—used in modern times to refer to very large conurbations.

melting pot The idea that ethnic differences can be combined to create new patterns of behavior drawing on diverse cultural sources.

microsociology The study of human behavior in contexts of face-to-face interaction.

middle class A social class composed broadly of those working in white-collar and highly skilled blue-collar jobs.

minority group A group of people who, because of their distinct physical or cultural characteristics, find themselves in situations of inequality within that society.

mode The number that appears most often in a data set.

modernization theory A version of market-oriented development theory that argues that low-income societies develop economically only if they give up their traditional ways and adopt modern economic institutions, technologies, and cultural values that emphasize savings and productive investment.

monogamy A form of marriage in which each married partner is allowed only one spouse at any given time.

monopoly A situation in which a single firm dominates in a given industry.

mores Norms that are widely adhered to and have great moral or social significance. Violations are generally sanctioned strongly.

mortality The number of deaths in a population.

multiculturalism The viewpoint according to which ethnic groups can exist separately and share equally in economic and political life.

N

nation-state A particular type of state, characteristic of the modern world, in which a government has sovereign power within a defined territorial area and the population comprises citizens who believe themselves to be part of a single nation or people.

nationalism A set of beliefs and symbols expressing identification with a national community.

neoliberalism The economic belief that free-market forces, achieved by minimizing or, ideally, eliminating government restrictions on business, provide the only route to economic growth.

network A set of informal and formal social ties that links people to one another.

nonbinary A gender identity that does not fit squarely into the male-female gender binary classification.

nonverbal communication Communication between individuals based on facial expression or bodily gestures rather than on language.

norms Rules of conduct that specify appropriate behavior in a given range of social situations. A norm either prescribes a given type of behavior or forbids it. All human groups follow norms, which are always backed by sanctions of one kind or another—varying from informal disapproval to physical punishment.

nuclear family A family group consisting of an adult or adult couple and their dependent children.

O

obesity Excessive body weight, indicated by a body mass index (BMI) over 30.

occupation Any form of paid employment in which an individual regularly works.

old old Sociological term for persons between the ages of seventy-five and eighty-four.

oldest old Sociological term for persons age eighty-five and older.

oligarchy Rule by a small minority within an organization or society.

oligopoly The domination by a small number of firms in a given industry.

oral history Interviews with people about events they witnessed earlier in their lives.

organic solidarity According to Émile Durkheim, the social cohesion that results from the various parts of a society functioning as an integrated whole.

organization A large group of individuals with a definite set of authority relations. Many types of organizations exist in industrialized societies, influencing most aspects of our lives. While not all organizations are bureaucratic, there are close links between the development of organizations and bureaucratic tendencies.

organized crime Criminal activities carried out by organizations established as businesses.

out-groups Groups toward which one feels antagonism and contempt—"those people."

P

pariah groups Groups that suffer from negative status discrimination—they are looked down on by most other members of society.

participant observation A method of research widely used in sociology and anthropology in which the researcher takes part in the activities of the group or community being studied.

participatory democracy A system of democracy in which all members of a group or community participate collectively in making major decisions.

pastoral societies Societies whose subsistence derives from the rearing of domesticated animals.

patriarchy The dominance of and privilege afforded to men over women. All known societies are patriarchal, although there are variations in the degree and nature of the power men exercise and are bestowed relative to women.

peer group A friendship group composed of individuals of similar age and social status.

peripheral countries Countries that have a marginal role in the world economy and are thus dependent on the core producing societies for their trading relationships.

personal space The physical space individuals maintain between themselves and others.

personal troubles Difficulties that are located in individual biographies and their immediate milieu, a seemingly private experience.

personality stabilization According to the theory of functionalism, the role families play in assisting adult members emotionally. Marriage between adults is the arrangement through which adult personalities are supported and kept healthy.

pilot study A trial run in survey research.

pluralism A model for ethnic relations in which all ethnic groups in a society retain their independent and separate identities yet share equally in the rights and powers of citizenship.

pluralist theories of modern democracy Theories that emphasize the role of diverse and potentially competing interest groups, none of which dominate the political process.

political rights Rights of political participation, such as the right to vote in elections and to run for office, held by citizens of a national community.

politics The means by which power is used to affect the nature and content of governmental activities.

polyandry A form of marriage in which a woman may have two or more husbands simultaneously.

polygamy A form of marriage in which a person may have two or more spouses simultaneously.

polygyny A form of marriage in which a man may have two or more wives simultaneously.

postindustrial society A society based on the production of knowledge and information rather than material goods, resulting in the rise of an economic service sector and the decline of the manufacturing sector.

postmodernism The belief that society is no longer governed by history or progress. Postmodern society is highly pluralistic and diverse, with no "grand narrative" guiding its development.

poverty line An official government measure to define those living in poverty in the United States.

power The ability of individuals or the members of a group to achieve aims or further the interests they hold. Power is a pervasive element in all human relationships. Many conflicts in society are struggles over power, because how much power an individual or group is able to obtain governs how far they are able to put their wishes into practice.

power elite Small networks of individuals who, according to C. Wright Mills, hold concentrated power in modern societies.

prejudice The holding of preconceived ideas about an individual or group, ideas that are resistant to change even in the face of new information. Prejudice may be either positive or negative.

preoperational stage According to Jean Piaget, the second stage of human cognitive development, in which a child has advanced sufficiently to master basic modes of logical thought.

primary deviance According to Edwin Lemert, the actions that cause others to label one as a deviant.

primary group A group that is characterized by intense emotional ties, face-to-face interaction, intimacy, and a strong, enduring sense of commitment.

primary socialization The process by which children learn the cultural norms of the society into which they are born. Primary socialization occurs largely in one's family.

profane That which belongs to the mundane, everyday world.

proletariat People who sell their labor for wages, according to Marx.

proportional representation An electoral system in which seats in a representative assembly, often called a parliament, are allocated according to the proportions of the vote received; the head of state (called a prime minister) is the head of the party that has the largest number of seats.

psychopath A specific personality type; such individuals lack the moral sense and concern for others held by most normal people.

public issues Difficulties or problems that are linked to the institutional and historical possibilities of social structure.

Q

qualitative methods Approaches to sociological research that often rely on personal and/or collective interviews, accounts, or observations of a person or situation.

quantitative methods Approaches to sociological research that draw on objective and statistical data and often focus on documenting trends, comparing subgroups, or exploring correlations.

QUILTBAG Acronym that captures the diversity of sexual orientations and gender identities, including queer/questioning, undecided, intersex, lesbian, trans, bisexual, asexual, and/or gay/genderqueer.

R

race Differences in human physical characteristics used to categorize large numbers of individuals.

race socialization The specific verbal and nonverbal messages that older generations transmit to younger generations regarding the meaning and significance of race.

racial microaggressions Small slights, indignities, or acts of disrespect that are hurtful to people of color even though they are often perpetuated by well-meaning whites.

racialization The process by which understandings of race are used to classify individuals or groups of people.

racism The attribution of characteristics of superiority or inferiority to a population sharing certain physically inherited characteristics.

radical feminism Form of feminist theory that believes that gender inequality is the result of male domination in all aspects of social and economic life.

random sampling Sampling method in which a sample is chosen so that every member of the population has the same probability of being included.

rape The forcing of nonconsensual vaginal, oral, or anal intercourse.

rape culture Social context in which attitudes and norms perpetuate the treatment of women as sexual objects and instill in men a sense of sexual entitlement.

rates of population growth or decline A measure of population change calculated by subtracting the yearly number of deaths per 1,000 from the number of births per 1,000.

reference group A group that provides a standard for judging one's attitudes or behaviors.

refugees People who have fled their homes due to a political, economic, or natural crisis.

regionalization The division of social life into different regional settings or zones.

relative deprivation The recognition that one has less than his or her peers.

relative poverty Poverty defined according to the living standards of the majority in any given society.

religion A set of beliefs adhered to by the members of a community, incorporating symbols regarded with a sense of awe or wonder together with ritual practices. Religions do not universally involve a belief in supernatural entities.

religious economy A theoretical framework within the sociology of religion that argues that religions can be fruitfully understood as organizations in competition with one another for followers.

religious nationalism The linking of strongly held religious convictions with beliefs about a people's social and political destiny.

representative sample A sample from a larger population that is statistically typical of that population.

resocialization The process of learning new norms, values, and behaviors when one joins a new group or takes on a new social role or when one's life circumstances change dramatically.

response cries Seemingly involuntary exclamations individuals make when, for example, being taken by surprise, dropping something inadvertently, or expressing pleasure.

S

sacred Describing something that inspires awe or reverence among those who believe in a given set of religious ideas.

sample A small proportion of a larger population.

sampling Studying a proportion of individuals or cases from a larger population as representative of that population as a whole.

sanction A mode of reward or punishment that reinforces socially expected forms of behavior.

scapegoats Individuals or groups blamed for wrongs that were not of their doing.

science The disciplined marshaling of empirical data, combined with theoretical approaches and theories that illuminate or explain those data. Scientific activity combines the creation of new modes of thought with the careful testing of hypotheses and ideas. One major feature that helps distinguish science from other idea systems (such as religion) is the assumption that all scientific ideas are open to criticism and revision.

scientific racism The use of scientific research or data to justify or reify beliefs about the superiority or inferiority of particular racial groups. Much of the "data" used to justify such claims are flawed or biased.

second parent adoption A family in which one partner adopts a child and the other partner applies to be a second parent or co-parent.

second shift The excessive work hours borne by women relative to men; these hours are typically spent on domestic chores following the end of a day of work outside the home.

secondary deviance According to Edwin Lemert, following the act of primary deviance, secondary deviation occurs when an individual accepts the label of deviant and acts accordingly.

secondary group A group characterized by its large size and by impersonal, fleeting relationships.

sect A religious movement that breaks away from orthodoxy and follows its own unique set of rules and principles.

secular thinking Worldly thinking, particularly as seen in the rise of science, technology, and rational thought in general.

secularization A process of decline in the influence of religion. Secularization can refer to levels of involvement with religious organizations (such as rates of church attendance), the social and material influence wielded by religious organizations, and the degree to which people hold religious beliefs.

segregation The practices of keeping racial and ethnic groups physically separate.

self-consciousness Awareness of one's distinct social identity as a person separate from others. Human beings are not born with self-consciousness but acquire an awareness of self as a result of early socialization.

self-identity The ongoing process of self-development and definition of our personal identity through which we formulate a unique sense of ourselves and our relationship to the world around us.

semiperipheral countries Countries that supply sources of labor and raw materials to the core industrial countries and the world economy but are not themselves fully industrialized societies.

sensorimotor stage According to Jean Piaget, the first stage of human cognitive development, in which a child's awareness of his or her environment is dominated by perception and touch.

service society A social order distinguished by the growth of service occupations at the expense of industrial jobs that produce material goods.

sex The biological and anatomical differences distinguishing females from males.

sex segregation The concentration of men and women in different jobs. These differences are believed to contribute to the gender pay gap.

sexual harassment The making of unwanted sexual advances by one individual toward another, in which the first person persists even though it is clear that the other party is resistant.

sexual orientation The direction of one's sexual or romantic attraction.

shaming A way of punishing criminal and deviant behavior based on rituals of public disapproval rather than incarceration. The goal of shaming is to maintain the ties of the offender to the community.

short-range downward mobility Social mobility that occurs when an individual moves from one position in the class structure to another of nearly equal status.

sick role A term Talcott Parsons used to describe the patterns of behavior that a sick person adopts in order to minimize the disruptive impact of his or her illness on others.

signifier Any vehicle of meaning and communication.

slavery A form of social stratification in which some people are owned by others as their property.

social aggregate A collection of people who happen to be together in a particular place but do not significantly interact or identify with one another.

social capital The social knowledge and connections that enable people to accomplish their goals and extend their influence.

social category People who share a common characteristic (such as gender or occupation) but do not necessarily interact or identify with one another.

social change Alteration in basic structures of a social group or society. Social change is an ever-present phenomenon in social life but has become especially intense in the modern era. The origins of modern sociology can be traced to attempts to understand the dramatic changes shattering the traditional world and promoting new forms of social order.

social class gradient in health The strong inverse association between socioeconomic resources and risk of illness or death.

social conflict theories of aging Arguments that emphasize the ways in which the larger social structure helps to shape the opportunities available to older adults. Unequal opportunities are seen as creating the potential for conflict.

social constraint The conditioning influence on our behavior by the groups and societies of which we are members. Social constraint was regarded by Émile Durkheim as one of the distinctive properties of social facts.

social construction of gender A perspective holding that gender differences are a product of social and cultural norms and expectations rather than biology.

social exclusion The outcome of multiple deprivations that prevent individuals or groups from participating fully in the economic, social, and political life of the society in which they live.

social facts According to Émile Durkheim, the aspects of social life that shape our actions as individuals. Durkheim believed that social facts could be studied scientifically.

social gerontologists Social scientists who study older adults and life course influences on aging processes.

social group A collection of people who regularly interact with one another on the basis of shared expectations concerning behavior and who share a sense of common identity.

social identity The characteristics that other people attribute to an individual.

social interaction The process by which we act with and react to those around us.

social mobility Upward or downward movement of individuals or groups among different social positions.

social movement Large groups of people who seek to accomplish, or to block, a process of social change. Social movements normally exist in conflict with organizations whose objectives and outlook they oppose. However, movements that successfully challenge power can develop into organizations.

social position The social identity an individual has in a given group or society. Social positions may be general in nature (those associated with gender roles) or may be more specific (occupational positions).

social reproduction The process whereby societies have structural continuity over time. Social reproduction is an important pathway through which parents transmit or produce values, norms, and social practices among their children.

social rights Rights of social and welfare provision held by all citizens in a national community.

social roles Socially defined expectations of an individual in a given status or occupying a particular social position. In every society, individuals play a number of social roles, such as teenager, parent, worker, or political leader.

Social Security A government program that provides economic assistance to persons faced with unemployment, disability, or old age.

social self According to the theory of George Herbert Mead, the identity conferred upon an individual by the reactions of others. A person achieves self-consciousness by becoming aware of this social identity.

social stratification The existence of structured inequalities among groups in society in terms of their access to material or symbolic rewards.

socialization The social processes through which we develop an awareness of social norms and values and achieve a distinct sense of self.

socialization of nature The process by which phenomena regarded as "natural" have now become social.

society A group of people who live in a particular territory, are subject to a common system of political authority, and are aware of having a distinct identity from other groups. Some societies, like hunting-and-gathering societies, are small, numbering no more than a few dozen people. Others are large, numbering millions—modern Chinese society, for instance, has a population of more than a billion people.

sociobiology An approach that attempts to explain the behavior of both animals and human beings in terms of biological principles.

sociological imagination The application of imaginative thought to the asking and answering of sociological questions. Someone using the sociological imagination "thinks himself away" from the familiar routines of daily life.

sociology The study of human groups and societies, giving particular emphasis to analysis of the industrialized world. Sociology is one of a group of social sciences that includes anthropology, economics, political science, and human geography. The divisions among the various social sciences are not clear-cut, and all share a certain range of common interests, concepts, and methods.

sociology of the body Field that focuses on how our bodies are affected by our social experiences. Health and illness, for instance, are shaped by social, cultural, and economic influences.

sovereignty The undisputed political rule of a state over a given territorial area.

standard deviation A way of calculating the spread of a group of figures.

standardized testing A procedure whereby all students in a state take the same test under the same conditions.

state A political apparatus (government institutions plus civil service officials) ruling over a given territorial order whose authority is backed by law and the ability to use force.

state-centered theories Development theories that argue that appropriate government policies do not interfere with economic development, but rather can play a key role in bringing it about.

status The social honor or prestige that a particular group is accorded by other members of a society. Status groups normally display distinct styles of life—patterns of behavior that the members of a group follow. Status privilege may be positive or negative.

stepfamily A family in which at least one partner has children from a previous marriage.

stereotype A fixed and inflexible category.

stigma Any physical or social characteristic that is labeled by society as undesirable.

strike A temporary stoppage of work by a group of employees in order to express a grievance or enforce a demand.

structural strain Tensions that produce conflicting interests within societies.

structuration The two-way process by which we shape our social world through our individual actions and by which we are reshaped by society.

structure The recurrent patterned arrangements and hierarchies that influence or limit the choices and opportunities available to us.

subcultures Cultural groups within a wider society that hold values and norms distinct from those of the majority.

suburbanization The development of towns surrounding a city.

suffrage A legal right to vote guaranteed by the Fifteenth Amendment to the U.S. Constitution; guaranteed to women by the Nineteenth Amendment.

suffragettes Members of early women's movements who pressed for equal voting rights for women and men.

surplus value In Marxist theory, the value of a worker's labor power left over when an employer has repaid the cost of hiring the worker.

survey A method of sociological research in which questionnaires are administered to the population being studied.

sustainable development Development that meets the needs of the present without compromising the ability of future generations to meet their own needs.

symbol One item used to stand for or represent another—as in the case of a flag, which symbolizes a nation.

symbolic interactionism A theoretical approach in sociology developed by George Herbert Mead that emphasizes the role of symbols and language as core elements of all human interaction.

T

target hardening Practical measures used to limit a criminal's ability to commit crime, such as community policing and use of house alarms.

technology The application of knowledge of the material world to production; the creation of material instruments (such as machines) used in human interaction with nature.

terrorism A public act of violence meant to be intimidating.

theism A belief in one or more supernatural deities.

theoretical questions Questions posed by sociologists when seeking to explain a particular range of observed events. The asking of theoretical questions is crucial to allowing us to generalize about the nature of social life.

theory of racial formation The process by which social, economic, and political forces determine the content and importance of racial categories.

time-space When and where events occur.

toxic masculinity A cluster of potentially destructive values or behaviors that historically have been part of boys' socialization, such as the devaluation of and aggression toward women.

tracking Dividing students into groups that receive different instruction on the basis of perceived similarities in ability.

transactional leaders Leaders who are concerned with accomplishing the group's tasks, getting group members to do their jobs, and making certain that the group achieves its goals.

transformational leaders Leaders who are able to instill in the members of a group a sense of mission or higher purpose, thereby changing the nature of the group itself.

transgender A person who identifies as or expresses a gender identity that differs from their sex at birth. Transgender persons differ from nonbinary persons, who may have a fluid identity that shifts between male and female or who may identify as neither male nor female.

transnational corporations Business corporations located in two or more countries.

transnational feminism A branch of feminist theory that highlights the way that global processes—including colonialism, racism, and imperialism—shape gender relations and hierarchies.

transphobia Negative attitudes, feelings, or actions toward transgender and gender-nonconforming people, their lifestyles, and their practices.

triad A group consisting of three persons.

triangulation The use of multiple research methods as a way of producing more reliable empirical data than are available from any single method.

U

underclass A class of individuals situated at the bottom of the class system, often composed of people from ethnic minority backgrounds.

unemployment rate The proportion of the population sixteen and older that is actively seeking work but is unable to find employment.

unfocused interaction Interaction occurring among people present in a particular setting but not engaged in direct face-to-face communication.

Uniform Crime Reports (UCR) Documents that contain official data on crime that is reported to law enforcement agencies that then provide the data to the FBI.

unions Organizations that advance and protect the interests of workers with respect to working conditions, wages, and benefits.

universal health coverage Public health care programs motivated by the goal of providing affordable health services to all members of a population.

upper class A social class broadly composed of the more affluent members of society, especially those who have inherited wealth, own businesses, or hold large numbers of stocks (shares).

urban ecology An approach to the study of urban life based on an analogy with the adjustment of plants and organisms to the physical environment. According to ecological theorists, the various neighborhoods and zones within cities are formed as a result of natural processes of adjustment on the part of populations as they compete for resources.

urban renewal The process of renovating deteriorating neighborhoods by encouraging the renewal of old buildings and the construction of new ones.

urbanism A term used by Louis Wirth to denote distinctive characteristics of urban social life, such as its impersonal or alienating nature.

urbanization The movement of the population into towns and cities and away from the land.

V

values Ideas held by individuals or groups about what is desirable, proper, good, and bad. What individuals value is strongly influenced by the specific culture in which they happen to live.

W

wealth Money and material possessions held by an individual or group.

welfare capitalism Practice in which large corporations protect their employees from the vicissitudes of the market.

welfare state A political system that provides a wide range of welfare benefits for its citizens.

white-collar crime Criminal activities carried out by those in white-collar, or professional, jobs.

white privilege The unacknowledged and unearned assets that benefit whites in their everyday lives.

winner take all An electoral system in which the seats in a representative assembly go to the candidate who receives the most votes in his or her electoral district (in the U.S. House of Representatives, for example, a candidate who gets 50 percent + 1 vote represents an entire congressional district, even if another candidate got 50 percent −1 vote).

work Carrying out tasks that require mental and physical effort, with the objective of the production of goods and services that cater to human needs. Work should not be thought of exclusively as paid employment. In modern societies, there remain types of work that do not involved direct payment.

working class A social class broadly composed of people working in blue-collar, or manual, occupations.

working poor People who work but whose earnings are not enough to lift them above the poverty line.

world-systems theory Pioneered by Immanuel Wallerstein, a theory that emphasizes the interconnections among countries based on the expansion of a capitalist world economy. This economy is made up of core countries, semiperipheral countries, and peripheral countries.

Y

young old Sociological term for persons between the ages of sixty-five and seventy-four.

Bibliography

Abeles, R. P., & Riley, M. W. (1987). Longevity, social structure, and cognitive aging. In C. Schooler & K. Warner Schaie (Eds.), *Cognitive functioning and social structure over the life course* (pp. 161–175). Norwood, NJ: Ablex.

Abrams, D., & Grant, P. R. (2012). Testing the social identity relative deprivation (SIRD) model of social change: The political rise of Scottish nationalism. *British Journal of Social Psychology, 51,* 674–689.

Accad, E. (1991). Conflicts and contradictions for contemporary women in the Middle East. In C. T. Mohanty, A. Russo, & L. Torres (Eds.), *Third world women and the politics of feminism* (pp. 237–251). Bloomington, IN: Indiana University Press.

Acs, G. (2011). *Downward mobility from the middle class: Waking up from the American Dream.* Washington, DC: Pew Charitable Trusts.

Administration on Aging. (2015). *A profile of older Americans: 2015.* Retrieved from http://www.aoa.acl.gov/Aging_Statistics/Profile/2015/docs/2015-Profile.pdf

Administration on Aging. (2018). *A profile of older Americans: 2017.* Retrieved from https://www.acl.gov/index.php/aging-and-disability-in-america/data-and-research/profile-older-americans

Adorno, T. W., Frenkel-Brunswik, E., Levinson, D. J., & Sanford, R. N. (1950). *The authoritarian personality.* New York: Harper and Row.

Afshar, H., & Dennis, C. (1992). *Women and adjustment policies in the third world.* New York, NY: St. Martin's Press.

Ahmed, J., Homa, D. M., O'Connor, E., Babb, S. D., Caraballo, R. S., Tushar, S., . . . King, B. A. (2015). Current cigarette smoking among adults—United States, 2005–2014. *Morbidity and Mortality Weekly Report, 64*(44), 1233–1240. Retrieved from http://www.cdc.gov/mmwr/pdf/wk/mm6444.pdf

Ahrnsbrak, R., Bose, J., Hedden, S. L., Lipari, R. N., & Park-Lee, E. (2017). *Key substance use and mental health indicators in the United States: Results from the 2016 National Health Survey on Drug Use and Health.* Substance Abuse and Mental Health Services Administration. Retrieved from https://www.samhsa.gov/data/sites/default/files/NSDUH-FFR1-2016/NSDUH-FFR1-2016.pdf

Ahtone, T. (2011, March 31). Native American intermarriage puts benefits at risk. *National Public Radio.* Retrieved from http://www.npr.org/2011/03/31/134421470/native-american-intermarriage-puts-benefits-at-risk

Akechi, H., Senju, A., Uibo, H., Kikuchi, Y., Hasegawa, T., & Hietanen, J. K. (2013). Attention to eye contact in the West and East: Autonomic responses and evaluative ratings. *PLoS ONE, 8*(3), e59312.

Albrow, M. (1997). *The global age: State and society beyond modernity.* Stanford, CA: Stanford University Press.

Aldrich, H. E., & Marsden, P. (1988). Environments and organizations. In N. J. Smelser (Ed.), *Handbook of sociology* (pp. 361–392). Newbury Park, CA: Sage.

Alexander, M. (2012). *The new Jim Crow.* New York, NY: The New Press.

Al Jazeera. (2017, August 17). Charlottesville attack: What, where and who? Retrieved from http://www.aljazeera.com/news/2017/08/charlottesville-attack-170813081045115.html

Allegretto, S. (2018, April 4). *Teachers across the country have finally had enough of the teacher pay penalty.* Economic Policy Institute. Retrieved from https://www.epi.org/publication/teachers-across-the-country-have-finally-had-enough-of-the-teacher-pay-penalty/

Allen, M. P. (1981). Managerial power and tenure in the large corporation. *Social Forces, 60,* 482–494.

Almeida, J. et al. (2009). Emotional distress among LGBT Youth: The influence of perceived discrimination based on sexual orientation. *Journal of Youth and Adolescence, 38*(7): 1001–1014.

Alter, C. (2018, January 18). A year ago, they marched. Now a record number of women are running for office. *Time.* Retrieved from http://time.com/5107499/record-number-of-women-are-running-for-office/

Altman, A. (2015, December 21). Black Lives Matter: A new civil rights movement is turning a protest cry into a political force. *Time.* Retrieved from http://time.com/time-person-of-the-year-2015-runner-up-black-lives-matter/

Alvarez, L. (2013, September 6). A university band, chastened by hazing, makes its return. *New York Times.* Retrieved from http://www.nytimes.com/2013/09/07/us/a-university-band-chastened-by-hazing-makes-its-return.html?pagewanted=all

Alvarez, R., Robin, L., Tuan, M., & Huang, A. S.-I. (1996). Women in the professions: Assessing progress. In P. J. Dubeck & K. Borman (Eds.), *Women and work: A handbook* (pp. 118–122). New York, NY: Garland.

Amato, P. R. (2001). Children of divorce in the 1990s: An update of the Amato and Keith (1991) meta-analysis. *Journal of Family Psychology, 15*(3), 355.

Amato, P. R. (2010). Research on divorce: Continuing trends and new developments. *Journal of Marriage and Family, 72,* 650–666.

Amato, P. R., & Keith, B. (1991). Parental divorce and the well-being of children: A meta-analysis. *Psychological Bulletin, 110*(1), 26.

Amato, P. R., Loomis, L. S., & Booth, A. (1995). Parental divorce, marital conflict, and offspring well-being during early adulthood. *Social Forces, 73,* 895–915.

American Academy of Pediatrics. (2004). Sexual orientation and adolescents. *Pediatrics, 113*(6), 1827–1832.

American Association of Retired Persons. (2012). *Loneliness among older adults: A national survey of adults 45+.* Retrieved from http://www.aarp.org/content/dam/aarp/research/surveys_statistics/general/2012/loneliness_2010.pdf

American Association of University Women. (1992). *How schools shortchange girls*. Washington, DC: American Association of University Women Educational Foundation.

American Civil Liberties Union. (2015). ACLU-NJ launches mobile justice smartphone app. Retrieved from https://www.aclu-nj.org/news/2015/11/13/aclu-nj-launches-mobile-justice-smartphone-app

American Council on Education. (2001). *The American freshman: National norms for fall 2000*. Los Angeles, CA: UCLA Higher Education Research Institute and ACE.

American Medical Association. (1999). State-specific maternal mortality among black and white women: United States, 1987–1996. *Journal of the American Medical Association, 282*(13), 1220–1222.

American Psychiatric Association. (2013). *Diagnostic and statistical manual of mental disorders* (5th ed.). Arlington, VA: American Psychiatric Association.

American Psychological Association. (2005). *Lesbian and gay parenting*. Washington, DC: American Psychological Association.

American Sociological Association. (2015). *Jobs, careers and sociological skills: The early employment experiences of 2012 sociology majors*. Washington, DC: American Sociological Association.

American Sociological Association. (2017, November 2). Big data may amplify existing police surveillance practices. Retrieved from http://www.asanet.org/news-events/asa-news/big-data-may-amplify-existing-police-surveillance-practices

Amin, S. (1974). *Accumulation on a world scale*. New York, NY: Monthly Review Press.

Ammons, S., & Markham, W. (2004). Working at home: Experiences of skilled white collar workers. *Sociological Spectrum, 24*(2), 191–238.

Amsden, A. H. (1989). *Asia's next giant: South Korea and late industrialization*. New York, NY: Oxford University Press.

Amsden, A. H. (1994). Why isn't the whole world experimenting with the East Asian model to develop?: A review of the East Asian miracle." *World Development, 22*(4): 627–633.

Amsden, A., Kochanowicz, J., & Taylor, L. (1994). *The market meets its match: Restructuring the economies of Eastern Europe*. Cambridge, MA: Harvard University Press.

Andazola Marquez, V. (2017, October 24). I accidentally turned my dad in to Immigration Services. *New York Times*. Retrieved from https://www.nytimes.com/2017/10/24/opinion/ice-detained-father-yale.html

Anderson, E. (1990). *Streetwise: Race, class, and change in an urban community*. Chicago, IL: University of Chicago Press.

Anderson, B. (1991). *Imagined communities: Reflections on the origin and spread of nationalism* (Rev. ed.). New York, NY: Routledge.

Anderson, E. (2011). *The cosmopolitan canopy: Race and civility in everyday life*. New York, NY: Norton.

Anderson, P. B., & Struckman-Johnson, C. (Eds.). (1998). *Sexually aggressive women: Current perspectives and controversies*. New York, NY: Guilford Press.

Aneesh, A. (2015). *Neutral accent: How language, labor, and life become global*. Raleigh, NC: Duke University Press.

Angell, M., & Kassirer, J. P. (1998). Alternative medicine—the risks of untested and unregulated remedies. *New England Journal of Medicine, 339*, 839.

Angier, N. (1995, June 11). If you're really ancient, you may be better off. *New York Times*. Retrieved from http://www.nytimes.com/1995/06/11/weekinreview/the-nation-if-you-re-really-ancient-you-may-be-better-off.html

Annie E. Casey Foundation. (2017). *2017 Kids Count data book: State trends in child well-being*. Retrieved from https://www.aecf.org/m/resourcedoc/aecf-2017kidscountdatabook.pdf

Ansalone, G. (2003). Poverty, tracking, and the social construction of failure: International perspectives on tracking. *Journal of Children and Poverty, 9*(1), 3–20.

Ansani, A., & Daniele, V. (2012). About a revolution: The economic motivations of the Arab Spring. *International Journal of Development and Conflict, 3*(3), 1–29.

Anyon, J. (2005). *Radical possibilities: Public policy, urban education, and a new social movement*. New York, NY: Taylor & Francis.

Anzaldua, G. (1990). *Making face, making soul: Haciendo caras: Creative and cultural perspectives by feminists of color*. San Francisco, CA: Aunt Lute Foundation.

Appadurai, A. (1986). Introduction: Commodities and the politics of value. In A. Appadurai (Ed.), *The social life of things* (pp. 3–63). Cambridge, UK: Cambridge University Press.

Appelbaum, R. P., & Christerson, B. (1997). Cheap labor strategies and export-oriented industrialization: Some lessons from the East Asia/Los Angeles apparel connection. *International Journal of Urban and Regional Research, 21*(2), 202–217.

Appelbaum, R., & Lichtenstein, N. (Eds.). (2017). *Achieving workers' rights in the global economy*. Ithaca, NY: Cornell University Press.

Appelbaum, R. P., Cong, C., Han, X., Parker, R., & Simon, D. (2018). *Innovation in China: Challenging the global science and technology system*. London, UK: Polity Press.

Arcelus, J., Mitchell, A. J., Wales, J., & Nielsen, S. (2011). Mortality rates in patients with anorexia nervosa and other eating disorders. *Archives of General Psychiatry, 68*, 724–731.

Arias, E. (2016). *Changes in life expectancy by race and Hispanic origin in the United States* (NCHS Data Brief No. 244). National Center for Health Statistics. Retrieved from http://www.cdc.gov/nchs/data/databriefs/db244.pdf

Ariès, P. (1965). *Centuries of childhood*. New York, NY: Random House.

Arndt, F. (2014, January 9). Social media in Ukraine's #Euromaidan Protests [Blog]. *Epoch Times*. Retrieved from http://www.theepochtimes.com/n3/blog/social-media-in-ukraines-euromaidan-protests/

Asch, S. (1952). *Social psychology*. Englewood Cliffs, NJ: Prentice Hall.

Ashworth, A. E. (1980). *Trench warfare: 1914–1918*. London, UK: Macmillan.

Aslan, R. (2017). *God: A human history*. New York, NY: Random House.

Association for Women's Rights in Development. (2018). About us. Retrieved from https://www.awid.org/

Association of American Medical Colleges. (2011). *Behavioral and social science foundations for future physicians. Report of the behavioral and social science expert panel*. Retrieved from https://www.aamc.org/download/271020/data/behavioralandsocialsciencefoundationsforfuturephysicians.pdf

Astor, M., Caron, C., & Victor, D. (2017, August 13). A guide to the Charlottesville aftermath. *New York Times*. Retrieved from https://www.nytimes.com/2017/08/13/us/charlottesville-virginia-overview.html

Atchley, R. C. (1989). A continuity theory of normal aging. *Gerontologist, 29*, 183–190.

Atchley, R. C. (2000). *Social forces and aging: An introduction to social gerontology* (9th ed.). Belmont, CA: Wadsworth.

Atique, A. (2015, August 18). The water crisis in Pakistan may be a bigger threat than militancy. *Muftah*. Retrieved from http://muftah.org/the-water-crisis-in-pakistan-may-be-a-bigger-threat-than-militancy/#.VwCkWXAmiHl

Attaran, M. (2004). Exploring the relationship between information technology and business process reengineering. *Information & Management, 41*(5), 585–596.

August, K. J., & Sorkin, D. H. (2010). Racial and ethnic disparities in indicators of physical health status: Do they still exist throughout late life? *Journal of the American Geriatrics Society, 58*, 2009–2015.

Avert.org. (2013). *Worldwide HIV & AIDS statistics*. Retrieved from http://avert.org/worldstats.htm

Avery, R., & Canner, G. (2005, Summer). New information reported under HMDA and its application in fair lending enforcement. *Federal Reserve Bulletin*. Retrieved from http://www.federalreserve.gov/pubs/bulletin/2005/3-05hmda.pdf

Awaz TV. (2016, April 2). Pakistan's largest city is running out water. Retrieved from http://www.awaztoday.tv/News-Talk-Shows/113790/Pakistans-largest-city-is-running-out-of-water-Roshan-Pakistan.aspx

Bachmann, H. (2012, December 20). The Swiss difference: A gun culture that works. *Time*. Retrieved from http://world.time.com/2012/12/20/the-swiss-difference-a-gun-culture-that-works/

Bailey, M., & Kaufman, A. J. (2010). *Polygamy in the monogamous world: Multicultural challenges for western law and policy*. New York, NY: Praeger.

Bailey, J., & Pillard, R. (1991). A genetic study of male sexual orientation. *Archives of General Psychiatry, 48*, 1089–1096.

Bailey, J. M., Pillard, R. C., Neale, M. C., & Agyei, Y. (1993). Heritable factors influence sexual orientation in women. *Archives of General Psychiatry, 50*, 217–223.

Bair, J. (2009). *Frontiers of commodity chain research.* Stanford, CA: Stanford University Press.

Baker, J. & Cangemi, J. (2016). Why are there so few women CEOs and senior leaders in corporate America? *Organizational Development Journal, 34*(2).

Baker, L. A., Silverstein, M., & Putney, N. M. (2008). Grandparents raising grandchildren in the United States: Changing family forms, stagnant social policies. *Journal of Societal & Social Policy, 7,* 53.

Bales, R. F. (1953). The egalitarian problem in small groups. In T. Parsons (Ed.), *Working papers in the theory of action* (pp. 111–161). Glencoe, IL: Free Press.

Bales, R. F. (1970). *Personality and interpersonal behavior.* New York, NY: Holt, Rinehart, and Winston.

Ballesteros, C. (2017, November 7). Racism might have spared black and Latino communities from opioid epidemic, drug abuse expert says. *Newsweek.* Retrieved from http://www.newsweek.com/racism-opiod-epidemic-blacks-latinos-trump-704370

Banda, P. S., & Ricciardi, N. (2013, August 23). Coy Mathis case: Colorado Civil Rights Division rules in favor of transgender 6-year-old in bathroom dispute. *Huffington Post.* Retrieved from http://www.huffingtonpost.com/2013/06/24/coy-mathis_n_3488306.html

Barak, A., Boniel-Nissim, M., & Suler, J. (2008). Fostering empowerment in online support groups. *Computers in Human Behavior, 25,* 1867–1883. doi:10.1016/j.chb.2008.02.004

Barnett, J. C., & Berchick, E. R. (2017). *Health insurance coverage in the United States: 2016.* Washington, DC: U.S. Government Printing Office.

Barry, A. (2015, September 16). Bridging apps reviewed app: Clevermind. BridgingApps. Retrieved from http://bridgingapps.org/2015/09/bridgingapps-reviewed-app-clevermind/

Barton, D. (2006). *Literacy: An introduction to the ecology of written language* (2nd ed.). Malden, MA: Wiley Blackwell.

Barzilai-Nahon, K., & Barzilai, G. (2005). Cultured technology: Internet and religious fundamentalism. *Information Society, 21*(1), 25–40.

Basu, S. (2014). Sex, money, and brutality. *Contexts, 13,* 17–18.

Baudrillard, J. (1988). *Jean Baudrillard: Selected writings.* Stanford, CA: Stanford University Press.

Bauerlein, M., & Jeffery, C. (2010). Too big to jail? *Mother Jones.* Retrieved from http://www.motherjones.com/politics/2010/01/too-big-jail

Baumann, L. (2015, April 21). Gov. Bullock signs Montana anti-bullying bill into law. *Great Falls Tribune.* Retrieved from http://www.greatfallstribune.com/story/news/local/2015/04/21/gov-bullock-signs-montana-anti-bullying-bill-law/26145567

Baxter, S. (2011, March 29). New Santa Cruz police smartphone application includes police scanner, alerts, tip submissions. *San Jose Mercury News.* Retrieved from http://www.mercurynews.com/ci_17728800

Bazelon, E. (2015, May 29). Have we learned anything from the Columbia rape case? *New York Times.* Retrieved from http://www.nytimes.com/2015/05/29/magazine/have-we-learned-anything-from-the-columbia-rape-case.html

BBC. (2016, March 4). Migrant crisis: Migration to Europe explained in seven charts. Retrieved from http://www.bbc.com/news/world-europe-34131911

BBC. (2018, January 11). Tunisia hit by new anti-austerity protests. Retrieved from http://www.bbc.com/news/world-africa-42644326

Bearman, P. (2002). Opposite-sex twins and adolescent same-sex attraction. *American Journal of Sociology, 107,* 1179–1205.

Beaudry, J.-S. (2016). Beyond (models of) disability? *Journal of Medicine and Philosophy: A Forum for Bioethics and Philosophy of Medicine, 41,* 210–228.

Beck, U. (1992). *Risk society.* London, UK: Sage.

Beck, U. (1995). *Ecological politics in an age of risk.* Cambridge, UK: Polity Press.

Becker, H. S. (1963). *Outsiders: Studies in the sociology of deviance.* New York, NY: Macmillan.

Becker A. (2004). Television, disordered eating, and young women in Fiji: Negotiating body image and identity during rapid social change. *Culture, Medicine and Psychiatry, 28,* 533–559.

Becker, G. S. (2009). *A Treatise on the family.* Cambridge, MA: Harvard University Press.

Bell, D. (1976). *The coming of post-industrial society: A venture in social forecasting.* New York, NY: Basic Books.

Bell, A., Weinberg, M. S., & Hammersmith, S. K. (1981). *Sexual preference: Its development in men and women.* Bloomington, IN: Indiana University Press.

Bellah, R.N., Madsen, R., Sullivan, W. M., Swidler, A., & Tipton, S. M. (1985). *Habits of the heart: Individualism and commitment in American life.* New York, NY: Harper & Row.

Bem, S. L. (1993). *The lenses of gender: Transforming the debate on sexual inequality.* New Haven, CT: Yale University Press.

Bengtson, V., Kim, K.-D., Myers, G. C., & Eun, K.-S. (2000). *Aging east and west: Families, states and the elderly.* New York, NY: Springer.

Bennett, J. W. (1976). *The ecological transition: Cultural anthropology and human adaptation.* New York, NY: Pergamon Press.

Bennett, E. A. (2012). Global social movements in global governance. *Globalizations, 9,* 799–813.

Berger, P. L. (1967). *The sacred canopy: Elements of a sociological theory of religion.* Garden City, NY: Anchor Books.

Berger, P. L. (1986). *The capitalist revolution: Fifty propositions about prosperity, equality, and liberty.* New York, NY: Basic Books.

Berger, P. L., & Hsiao, H. (1988). *In search of an East Asian development model.* New Brunswick, NJ: Transaction.

Berlan, E. D., Corliss, H. L., Field, A. E., Goodman, E., & Austin, S. B. (2010). Sexual orientation and bullying among adolescents in the Growing Up Today Study. *Journal of Adolescent Health, 46*(4), 366–371.

Berle, A., & Means, G. (1982). *The modern corporation and private property.* Buffalo, NY: Heim. (Original work published 1932)

Berryman, P. (1987). *Liberation theology: Essential facts about the revolutionary movement in Central America and beyond.* Philadelphia, PA: Temple University Press.

Beyer, P. (1994). *Religion and globalization.* Thousand Oaks, CA: Sage.

Beyerstein, B. L. (1999, Fall/Winter). Psychology and "alternative medicine": Social and judgmental biases that make inert treatments seem to work. *Scientific Review of Alternative Medicine, 3*(2), 20–33.

Bialik, K., & Matsa, K. E. (2017, October 4). *Key trends in social and digital news media.* Pew Research Center. Retrieved from http://www.pewresearch.org/fact-tank/2017/10/04/key-trends-in-social-and-digital-news-media/

Bianchi, S. M., Robinson, J. P., & Milkie, M. A. (2007). *Changing rhythms of American family life.* New York, NY: Russell Sage.

Birren, J., & Bengston, V. (Eds.). (1988). *Emerging theories of aging.* New York, NY: Springer.

Bjerga, A. (2017, October 18). Poor nations have hardest time ramping up food production. *Bloomberg.* Retrieved from https://www.bloomberg.com/news/articles/2017-10-18/poor-nations-have-hardest-time-ramping-up-food-production

Bjorkqvist, K. (1994). Sex differences in physical, verbal, and indirect aggression: A review of recent research. *Sex Roles, 30*(3–4), 177–188.

Bjorkqvist, K., Lagerspetz, K., & Osterman, K. (2006). Sex differences in covert aggression. *Aggressive Behavior, 202,* 27–33.

Black, P., Pearson, M., & Butenko, V. (2014, February 19). Ukraine protesters stand ground as European, U.S. leaders ramp up pressure. *CNN.* Retrieved from http://www.cnn.com/2014/02/19/world/europe/ukraine-protests/

Black Lives Matter. (2016). Home page. Retrieved from http://blacklivesmatter.com/

Blanchard, R., & Bogaert, A. (1996). Homosexuality in men and number of older brothers. *American Journal of Psychiatry, 153,* 27–31.

Blau, P. M., & Duncan, O. D. (1967). *The American occupational structure.* New York, NY: Wiley.

Blauner, R. (1964). *Alienation and freedom.* Chicago, IL: University of Chicago Press.

Blauner, R. (1972). *Racial oppression in America.* New York, NY: Harper & Row.

Blum, L. M. (1991). *Between feminism and labor: The significance of the comparable worth movement.* Berkeley, CA: University of California Press.

Bochenek, M., & Knight, K. (2012). Establishing a third gender category in Nepal. *Emory International Law Review, 26,* 11–41.

Boden, D., & Molotch, H. L. (1994). The compulsion of proximity. In D. Boden & R. Friedland (Eds.), *Now here: Space, time and modernity* (pp. 257–286). Berkeley, CA: University of California Press.

Bohrnstedt, G., Kitmitto, S., Ogut, B., Sherman, D., & Chan, D. (2015). *School composition and the black-white achievement gap* (NCES 2015-018). National Center for Education Statistics. Retrieved from https://nces.ed.gov/nationsreportcard/subject/studies/pdf/school_composition_and_the_bw_achievement_gap_2015.pdf

Bomey, N., Snavely, B., & Priddle, A. (2013, December 3). Judge rules Detroit eligible for historic Chapter 9 bankruptcy, says pensions can be cut. *Detroit Free Press*. Retrieved from http://www.freep.com/article/20131203/NEWS01/312030084/Detroit-bankruptcy-eligibility-Steven-Rhodes-Chapter-9-Kevyn-Orr

Bonacich, E., & Appelbaum, R. (2000). *Behind the label: Inequality in the Los Angeles garment industry*. Berkeley, CA: University of California Press.

Bonilla-Silva, E. (2006). *Racism without racists: Color-blind racism and the persistence of racial inequality in the United States*. Lanham, MD: Rowman & Littlefield Publishers, Inc.

Bonilla-Silva, E. (2015). The structure of racism in color-blind, "post-racial" America. *American Behavioral Scientist, 59*(11), 1358–1376. doi:10.1177/0002764215586826

Bonilla-Silva, E., & Baiocchi, G. (2008). Anything but racism: How sociologists limit the significance of racism. In T. Zuberi & E. Bonilla-Silva (Eds.), *White logic, white methods: Racism and methodology* (pp. 137–152). Lanham, MD: Rowman & Littlefield.

Bonnington, C. (2012, April 12). Are men and women using mobile apps differently? *Wired*. Retrieved from http://www.wired.com/gadgetlab/2013/04/men-women-app-usage/

Booth, A. (1977). Food riots in the northwest of England, 1770–1801. *Past & Present, 77*, 84–107.

Bositis, D. (2001). *Black elected officials: A statistical summary, 2001*. Joint Center for Political and Economic Studies. Retrieved from http://www.jointcenter.org/sites/default/files/upload/research/files/Black%20Elected%20Officials%20A%20Statistical%20Summary%202001.pdf

Bouma, H., et al. (2004). *CHI '04 extended abstracts on human factors in computing systems*. Conference on Human Factors in Computing Systems. New York, NY: ACM Press.

Bourdieu, P. (1984). *Distinction: A social critique of judgement of taste*. Cambridge, MA: Harvard University Press.

Bourdieu, P. (1988). *Language and symbolic power*. Cambridge, UK: Polity Press.

Bourdieu, P. (1990). *The logic of practice*. Palo Alto, CA: Stanford University Press.

Bowles, S., & Gintis, H. (1976). *Schooling in capitalist America*. New York, NY: Basic Books.

Bowman, C. (2018, January 3). Is premarital sex wicked? Changing attitudes about morality. Forbes. Retrieved from https://www.forbes.com/sites/bowmanmarsico/2018/01/03/is-premarital-sex-wicked-changing-attitudes-about-morality/#7bd73d2a18ac

Bowman, Q., & Amico, C. (2010, November 5). Congress loses hundreds of years of experience—but majority of incumbents stick around. *PBS Newshour*. Retrieved from http://www.pbs.org/newshour/rundown/congress-loses-hundreds-of-years-of-experience-but-vast-majority-of-incumbents-stick-around/

boyd, d. (2014). *It's complicated: The social lives of networked teens*. New Haven, CT: Yale University Press.

Boyer, R., & Drache, D. (Eds.). (1996). *States against markets: The limits of globalization*. New York, NY: Routledge.

Braithwaite, J. (1996). Crime, shame, and reintegration. In P. Cordella & L. Siegal (Eds.), *Readings in contemporary criminological theory* (pp. 33–41). Boston, MA: Northeastern University Press.

Brass, D. J. (1985). Men's and women's networks: A study of interaction patterns and influence in an organization. *Academy of Management Journal, 28*, 327–343.

Braverman, H. (1974). *Labor and monopoly capital*. New York, NY: Monthly Review Press.

Brayne, S. (2017). Big data surveillance: The case of policing. *American Sociological Review, 82*(5), 977–1008. doi:10.1177/0003122417725865

Breiding, M., Smith, S., Basile, K., Walters, M. I., Chen, J., & Merrick, M. (2014). Prevalence and characteristics of sexual violence, stalking, and intimate partner violence victimization—National Intimate Partner and Sexual Violence Survey, United States, 2011. *MMWR Surveillance Summaries, 63*(8). Retrieved from http://www.cdc.gov/mmwr/pdf/ss/ss6308.pdf

Brenner, P. S. (2011). Exceptional behavior or exceptional identity? Overreporting of church attendance in the U.S. *Public Opinion Quarterly, 75*(1), 19–41.

Bresnahan, T. F., Brynjolfsson, E., & Hitt, L. (2002). Information technology, workplace organization, and the demand for skilled labor: Firm-level evidence. *Quarterly Journal of Economics, 117*(1), 33976.

Brewer, R. M. (1993). Theorizing race, class, and gender: The new scholarship of black feminist intellectuals and black women's labor. In S. M. James & A. P. A. Busia (Eds.), *Theorizing black feminisms: The visionary pragmatism of black women* (13–30). New York, NY: Routledge.

Bricker, J., Kennickell, A. B., Moore, K. B., & Sabelhaus, J. (2012). Changes in U.S. family finances from 2007 to 2010: Evidence from the Survey of Consumer Finances. *Federal Reserve Bulletin, 98*(2), 1–80. Retrieved from http://www.federalreserve.gov/pubs/bulletin/2012/pdf/scf12.pdf

Bricout, J. C. (2004). Using telework to enhance return to work outcomes for individuals with spinal cord injuries. *Neurorehabilitation, 19*(2), 147–159.

Brizendine, L. (2006). *The Female Brain*. New York: Morgan Road Books.

Brizendine, L. (2010). *The Male Brain*. New York: Broadway Books.

Bronner, S. (2015, September 18). The steps one homeless teen took to get a full ride at Yale. *Huffpost College*. Retrieved from http://www.huffingtonpost.com/entry/the-steps-one-homeless-teen-took-to-get-full-ride-at-yale_55dc926ce4b0a40aa3ac3bae

Brown, J. K. (1977). A note on the division of labor by sex. In N. Glazer & H. Y. Waehrer (Eds.), *Woman in a man-made world* (2nd ed.). Chicago, IL: Rand McNally.

Brown, D. E. (1991). *Human universals*. New York, NY: McGraw-Hill.

Brown, S. L. (2004). Family structure and child well-being: The significance of parental cohabitation. *Journal of Marriage and Family, 66*, 351–367.

Brown, A. (2017a). *The data on women leaders*. Pew Research Center. Retrieved from http://www.pewsocialtrends.org/2017/03/17/the-data-on-women-leaders/

Brown, E. (2017b, March 27). Trump signs bills overturning Obama era educational regulations. *Washington Post*. Retrieved from https://www.washingtonpost.com/news/education/wp/2017/03/27/trump-signs-bills-overturning-obama-era-education-regulations/?utm_term=.7ac39aa310e5

Brown, C., & Jasper, K. (Eds.). (1993). *Consuming passions: Feminist approaches to eating disorders and weight preoccupations*. Toronto, ON: Second Story Press.

Brown, A., & Patten, E. (2013). Hispanics of Cuban origin in the United States, 2011. Pew Research Center. Retrieved from http://www.pewhispanic.org/2013/06/19/hispanics-of-cuban-origin-in-the-united-states-2011/

Brown, A. & Stepler, R. (2016, April). *Foreign-born population in the United States statistical portrait*. Pew Research Center. Retrieved from http://www.pewhispanic.org/2016/04/19/2014-statistical-information-on-immigrants-in-united-states/

Brown-Dean, K., Hajnal, Z., Rivers, C., & White, I. (2015). *50 years of the Voting Rights Act: The state of race in politics*. Joint Center for Political and Economic Studies. Retrieved from http://jointcenter.org/sites/default/files/VRA%20report%2C%203.5.15%20%281130%20am%29%28updated%29.pdf

Brownell, K., & K. Horgen. (2004). *Food fight: The inside story of the food industry, America's obesity crisis, and what we can do about it*. New York, NY: McGraw-Hill.

Brownmiller, S. (1986). *Against our will: Men, women, and rape* (Rev. ed.). New York, NY: Bantam.

Brubaker, R. (1992). *The politics of citizenship*. Cambridge, MA: Harvard University Press.

Brynjolfsson, E., & McAfee, A. (2014). *The second machine age: Work, progress, and prosperity in a time of brilliant technologies*. New York, NY: Norton.

Buckley, C. (2013, November 16). China to ease longtime policy of 1-child limit. *New York Times*. Retrieved from http://www.nytimes.com/2013/11/16/world/asia/china-to-loosen-its-one-child-policy.html?_r=0

Buckley, C. (2015, October 29). China ends one child policy, allowing families two children. *New York Times*. Retrieved from http://www.nytimes.com/2015/10/30/world/asia/china-end-one-child-policy.html

Budig, M., Misra, J., & Boeckmann, I. The motherhood pay gap cross-nationally: How work-family policies and cultural attitudes intersect. *Social Politics, 19*(2), 163–193.

Bull, P. (1983). *Body movement and interpersonal communication*. New York, NY: Wiley.

Burgess, E. W. (1926). The family as a unity of interacting personalities. *Families in Society: The Journal of Contemporary Social Services, 7*(1), 3–9.

Burman, D. D., Bitan, T., & Booth, J. R. Sex differences in neural processing of language among children. *Neuropsychologia, 46*(5): 1349–1362.

Burns, J. M. (1978). *Leadership*. New York, NY: Harper & Row.

Burr, C. (1993, March). Homosexuality and biology. *Atlantic Monthly*. Retrieved from https://www.theatlantic.com/magazine/archive/1993/03/homosexuality-and-biology/304683/

Burris, B. H. (1993). *Technocracy at work*. Albany, NY: State University of New York Press.

Burris, B. H. (1998). Computerization of the workplace. *Annual Review of Sociology, 24*, 141–157.

Busby, M. (2017, August 31). Canada introduces gender-neutral "X" option on passports. *Guardian*. Retrieved from https://www.theguardian.com/world/2017/aug/31/canada-introduces-gender-neutral-x-option-on-passports

Butler, J. (1989). *Gender trouble: Feminism and the subversion of identity*. New York, NY: Routledge.

Butler, B. (2014, October 29). The story behind that "10 hours of walking in NYC" viral street harassment video [Blog]. *Washington Post*. Retrieved from https://www.washingtonpost.com/blogs/she-the-people/wp/2014/10/29/the-story-behind-that-10-hours-of-walking-in-nyc-viral-street-harassment-video/

Buttel, F. H. (1987). New directions in environmental sociology. *Annual Review of Sociology, 13*, 465–488.

Byrne, A., & Carr, D. (2005). Caught in the cultural lag: The stigma of singlehood. *Psychological Inquiry, 16*, 84–90.

Caciopppo, J., Hawkley, L., Crawford, L., Ernst, J., Burleson, M., Kowalewski, R., . . . Berntson, G. G. (2002). Loneliness and health: Potential mechanisms. *Psychosomatic Medicine, 64*, 407–417.

Cahn, N., & Carbone, J. (2010). *Red families v. blue families: Legal polarization and the creation of culture*. New York, NY: Oxford University Press.

Caixin Media. (2016, March 24). China's rural youngsters drop-out of school at an alarming rate, researchers find. Retrieved from http://reap.fsi.stanford.edu/news/caixin-media-chinas-rural-youngsters-drop-out-school-alarming-rate-researchers-find

Caldwell, J., Caldwell, B. K., Caldwell, P., McDonald, P. F., & Schindlmayr, T. (2010). *Demographic transition theory*. New York, NY: Springer.

Calvert, S., & Bauerlein, V. (2015). Viral videos shape view of police conduct. *Wall Street Journal*. Retrieved from http://www.wsj.com/articles/viral-videos-shape-views-of-police-conduct-1451512011

Calvin, K., Bond-Lamberty, B., Clarke, L., Edmonds, J., Eom, J., Hartin, C., . . . Wise, M. (2016). The SSP4: A world of deepening inequality. *Global Environmental Change, 42*, 284–296.

Campbell, M., Haveman, R., Sandefur, G., & Wolfe, B. (2005). Economic inequality and educational attainment across a generation. *Focus, 23*(3), 11–15.

Campos, P., Saguy, A., Ernsberger, P., Oliver, E., & Gaesser, G. (2006). The epidemiology of overweight and obesity: Public health crisis or moral panic? *International Journal of Epidemiology, 35*, 55–60.

Carbonaro, W. (2005). Tracking, students' effort, and academic achievement. *Sociology of Education, 78*(1), 27–49.

Cardoso, F. H., & Faletto, E. (1979). *Dependency and development in Latin America*. Berkeley, CA: University of California Press.

Carnegie Council on Adolescent Development. (1989). *Turning points: Preparing American youth for the 21st century*. Washington, DC: Task Force on Education of Young Adolescents.

Carneiro, R. L., & Perrin, R. G. (2002). Herbert Spencer's *Principles of sociology*: A centennial retrospective and appraisal. *Annals of Science, 59*(3), 221–261.

Carnevale, A. P., Jayasundera, T., & Gulish, A. (2016). *America's divided recovery: College haves and have-nots*. Georgetown University Center on Education and the Workforce. Retrieved from https://cew.georgetown.edu/cew-reports/americas-divided-recovery/#full-report

Carnevale, A. P., Strohl, J., & Melton, M. (2011). *What's it worth: The economic value of college majors*. Washington, DC: Georgetown University Center on Education and the Workforce.

Caron, C. (2017, October 19). Californians will soon have nonbinary as a gender option on birth certificates. *New York Times*. Retrieved from https://www.nytimes.com/2017/10/19/us/birth-certificate-nonbinary-gender-california.html

Carr, D. (2010). Golden years? Poverty among older adults. *Contexts, 9*(1), 62–63.

Carr, D. (2014). *Worried sick: Why stress hurts and what to do about it* (Pinpoint Series). New Brunswick, NJ: Rutgers University Press.

Carr, D., & Friedman, M. (2005). Is obesity stigmatizing? Body weight, perceived discrimination and psychological well-being in the United States. *Journal of Health and Social Behavior, 46*, 244–259.

Carr, D., & Friedman, M. (2006). Body weight and interpersonal relationships. *Social Psychology Quarterly, 69*, 127–149.

Carr, D., & Moorman, S. (2011). Social relations and aging. In R. A. Settersten & J. L. Angel (Eds.), *Handbook of sociology of aging* (pp. 145–160). New York, NY: Springer.

Carr, D., & Springer, K. (2010). Advances in families and health research in the 21st century. *Journal of Marriage and Family, 72*(3), 743–761.

Carr, D., Friedman, M. A., & Jaffe, K. (2007). Understanding the relationship between obesity and positive and negative affect: The role of psychosocial mechanisms. *Body Image, 4*(2), 165–177.

Carr, D., Murphy, L. F., Batson, H. D., & Springer, K. W. (2013). Bigger isn't always better: The effect of obesity on the sexual well-being of adult men in the U.S. *Men and Masculinities, 16*, 452–477.

Carrington, D. (2016, January 19). China's coal-burning in significant decline, figures show. *Guardian*. Retrieved from http://www.theguardian.com/environment/2016/jan/19/chinas-coal-burning-in-significant-decline-figures-show

Carson, R. (1962). *Silent spring*. New York, NY: Houghton Mifflin.

Carson, E. A. (2018). *Prisoners in 2016*. Bureau of Justice Statistics. Retrieved from https://www.bjs.gov/index.cfm?ty=pbdetail&iid=6187

Case, A., & Deaton, A. (2017). Mortality and morbidity in the 21st century. *Brookings Papers on Economic Activity*, 397.

Castells, M. (1977). *The urban question: A Marxist approach*. Cambridge, MA: MIT Press.

Castells, M. (1983). *The city and the grass roots: A cross-cultural theory of urban social movements*. Berkeley, CA: University of California Press.

Castells, M. (1992). Four Asian tigers with a dragon head: A comparative analysis of the state, economy, and society in the Asian Pacific Rim. In R. P. Appelbaum & J. Henderson (Eds.), *States and development in the Asian Pacific Rim* (pp. 33–70). Newbury Park, CA: Sage.

Castells, M. (1996). *The rise of the network society*. Malden, MA: Blackwell.

Castells, M. (1998). *End of millennium*. Malden, MA: Blackwell.

Castells, M. (2000). *The rise of the network society*. Oxford, UK: Oxford University Press.

Castells, M. (2001). *The internet galaxy*. Oxford, UK: Oxford University Press.

Castles, S., & Miller, M. J. (1993). *The age of migration: International population movements in the modern world*. London, UK: Macmillan.

Castles, S., & Miller. M. J. (2009). *The age of migration: International population movements in the modern world* (4th ed.). London, UK: Guilford Press.

Catalano, S. M. (2005). *Criminal victimization, 2004*. Table 2 (National Crime Victimization Survey, NCJ210674). Bureau of Justice Statistics. Retrieved from http://www.bjs.gov/content/pub/pdf/cv04.pdf

Catalyst. (2017). *Women in S&P 500 finance*. Retrieved from http://www.catalyst.org/knowledge/women-sp-500-finance

Catalyst. (2018a). *Women in S&P 500 companies*. Retrieved from http://www.catalyst.org/knowledge/women-sp-500-companies

Catalyst. (2018b). *Women CEOs of the S&P 500*. Retrieved from http://www.catalyst.org/knowledge/women-ceos-sp-500

Catton, W. R., Jr., & Dunlap, R. E. (1980). A new ecological paradigm for post-exuberant sociology. *American Behavioral Scientist, 24*(1), 15–47.

Center for American Progress. (2016, March 29). *Sexual orientation and gender identity data collection in the Behavioral Risk Factor surveillance system*. Retrieved from https://www.americanprogress.org/issues/lgbt/report/2016/03/29/134182/sexual-orientation-and-gender-identity-data-collection-in-the-behavioral-risk-factor-surveillance-system/

Center for American Women and Politics. (2018a). *Women in elective office 2018*. Retrieved from http://www.cawp.rutgers.edu/women-elective-office-2018

Center for American Women and Politics. (2018b). *Women in the U.S. Congress: 2018*. Retrieved from http://www.cawp.rutgers.edu/women-us-congress-2018

Center for Climate and Security (2018, Feb. 26). *A responsibility to prepare: Strengthening national and homeland security in the face of a changing climate.* Retrieved from https://climateandsecurity.files.wordpress.com/2018/02/climate-and-security-advisory-group_a-responsibility-to-prepare_2018_02.pdf

Center for Responsive Politics. (2017). *Incumbent advantage.* Retrieved from https://www.opensecrets.org/overview/incumbs.php

Center for Responsive Politics. (2018). *Lobbying database.* Retrieved from https://www.opensecrets.org/lobby/

Center on Education Policy. (2007). *Choices, changes, and challenges: Curriculum and instruction in the NCLB era.* Retrieved from http://www.cep-dc.org/publications/index.cfm?selectedYear=2007

Centers for Disease Control and Prevention. (2003). *National Ambulatory Care Survey, 2001 summary* (Advanced Data from Vital and Health Statistics, No. 337). Retrieved from http://www.cdc.gov/nchs/data/ad/ad337.pdf

Centers for Disease Control and Prevention. (2008a). *Childhood overweight and obesity.* Retrieved from http://www.cdc.gov/obesity/childhood

Centers for Disease Control and Prevention. (2011a). *CDC health disparities and inequalities report—United States, 2011.* Retrieved from http://origin.glb.cdc.gov/mmwr/pdf/other/su6001.pdf

Centers for Disease Control and Prevention. (2011b). Rates of diagnoses of HIV infection among adults and adolescents, by area of residence, 2011—United States and 6 dependent areas. *HIV Surveillance Report, 23,* 5–84.

Centers for Disease Control and Prevention. (2013a). Births: Final data for 2011. *National Vital Statistics Reports, 62*(1), 1–70.

Centers for Disease Control and Prevention. (2013b). Deaths: Final data for 2010. *National Vital Statistics Reports, 61*(4), 1–118.

Centers for Disease Control and Prevention. (2013c). *Health, United States, 2012: With special feature on emergency care.* Table 98. Retrieved from http://www.cdc.gov/nchs/data/hus/hus12.pdf

Centers for Disease Control and Prevention. (2014). *Prevalence and characteristics of sexual violence, stalking, and intimate partner violence victimization—National Intimate Partner and Sexual Violence Survey, United States, 2011.* Retrieved from http://www.cdc.gov/mmwr/preview/mmwrhtml/ss6308a1.htm?s_cid=ss6308a1_e

Centers for Disease Control and Prevention. (2017a). *Health, United States, 2016: With chartbook on long-term trends in health.* Retrieved from https://www.cdc.gov/nchs/data/hus/hus16.pdf#015

Centers for Disease Control and Prevention. (2017b). *HIV in the United States: At A Glance.* Retrieved from https://www.cdc.gov/hiv/statistics/overview/ataglance.htm

Centers for Disease Control and Prevention. (2017c). *HIV Surveillance Report, 2016,* Vol. 28, Nov. 2017, Retrieved from https://www.cdc.gov/hiv/pdf/library/reports/surveillance/cdc-hiv-surveillance-report-2016-vol-28.pdf

Centers for Disease Control and Prevention. (2017d). *HIV/AIDS: Basic Statistics.* Retrieved from https://www.cdc.gov/hiv/basics/statistics.html

Centers for Disease Control and Prevention. (2017e). Key statistics from the National Survey of Family Growth: Number of sexual partners in lifetime. Retrieved from https://www.cdc.gov/nchs/nsfg/key_statistics/n.htm#numberlifetime

Centers for Disease Control and Prevention. (2018). Current cigarette smoking among adults—United States, 2016. *Morbidity and Mortality Weekly Report, 67*(2), 53–59.

Central Intelligence Agency. (2000). *The world factbook.* Retrieved from http://www.cia.gov/cia/publications/factbook/geos/rs.html#Econ

Central Intelligence Agency. (2013a). China. *The world factbook.* Retrieved from https://www.cia.gov/library/publications/the-world-factbook/geos/ch.html

Central Intelligence Agency. (2013d). Country comparison: Exports. *The world factbook.* Retrieved from https://www.cia.gov/library/publications/the-world-factbook/rankorder/2078rank.html

Central Intelligence Agency. (2013e). *The world factbook.* Retrieved from https://www.cia.gov/library/publications/the-world-factbook/rankorder/2066rank.html

Central Intelligence Agency. (2014). *The world factbook.* Retrieved from https://www.cia.gov/library/publications/the-world-factbook/fields/2177.html

Central Intelligence Agency. (2015). China. *The world factbook.* Retrieved from https://www.cia.gov/library/publications/the-world-factbook/geos/ch.html

Central Intelligence Agency. (2016a). Country comparison: Exports. *The world factbook.* Retrieved from https://www.cia.gov/library/publications/the-world-factbook/rankorder/2078rank.html

Central Intelligence Agency. (2017). *The world factbook: Ethnic groups.* Retrieved from https://www.cia.gov/library/publications/the-world-factbook/fields/2075.html#71

Cerulo, K. A. (2009). Nonhumans in social interaction. *Annual Review of Sociology, 35,* 531–552. Retrieved from http://www.jstor.org/stable/27800090

Chafetz, J. S. (1990). *Gender equity: An integrated theory of stability and change.* Newbury Park, CA: Sage.

Chambliss, W. J. (1988). *On the take: From petty crooks to presidents.* Bloomington, IN: Indiana University Press.

Chan, J. (2017). #iSLAVEat10. Made in China: A Quarterly on Chinese Labour, Civil Society, and Rights, 2(3), 20–23. Retrieved from http://www.chinoiresie.info/PDF/Made-in-China_3_2017.pdf

Chan, M. (2017, September 14). How the kindness of strangers is helping hurricane victims rebuild their lives. *Time.* Retrieved from http://time.com/4941015/kindness-of-strangers-hurricane-victims/

Chang-Muy, F. (2009). Legal classifications of immigrants. In F. Chang-Muy & E. P. Congress (Eds.), *Social work with immigrants and refugees: Legal issues, clinical skills, and advocacy* (pp. 39–62). New York, NY: Springer.

Chase-Dunn, C. (1989). *Global formation: Structures of the world economy.* Cambridge, MA: Basil Blackwell.

Cheng, C.-Y., & Lee, F. (2009). Multiracial identity integration: Perceptions of conflict and distance among multiracial individuals. *Journal of Social Issues, 65*(1): 51–68.

Cherlin, A. (2005). American marriage in the early twenty-first century. *The Future of Children, 15*(2), 33–55.

Cherlin, A. (2010). *The marriage-go-round: The state of marriage and the family in America today.* New York, NY: Vintage.

Chetty, R., Grusky, D., Hell, M., Hendren, N., Manduca, R. & Narang, J. (2017). The fading American dream: Trends in absolute income mobility since 1940, *Science, 356*(6336), 398–406.

Child Trends. (2015). Child Trends databank: Family structure. Retrieved from http://www.childtrends.org/?indicators=family-structure

Chingos, M., & Peterson, P. E. (2018). *The effects of school vouchers on college enrollment: Experimental evidence from New York City.* Program on Education Policy and Governance, Harvard Kennedy School. Retrieved from http://www.hks.harvard.edu/pepg/PDF/Impacts_of_School_Vouchers_FINAL.pdf

Chira, S., & Einhorn, C. (2017, December 19). How tough is it to change a culture of harassment? Ask women at Ford. *New York Times.* Retrieved from https://www.nytimes.com/interactive/2017/12/19/us/ford-chicago-sexual-harassment.html

Chishti, M., & Hipsman, F. (2015, May 21). *In historic shift, new migration flows from Mexico fall below those from China and India.* Migration Policy Institute. Retrieved from https://www.migrationpolicy.org/article/historic-shift-new-migration-flows-mexico-fall-below-those-china-and-india

Chiu, D., Beru, Y., Watley, E., Wubu, S., Simson, E., Kessinger, . . . Wigfield, A. (2008). Influences of math tracking on seventh-grade students' self-beliefs and social comparisons. *Journal of Educational Research, 102,* 125–136.

Choo, H. Y., & Ferree, M. M. (2010). Practicing intersectionality in sociological research: A critical analysis of inclusions, interactions, and institutions in the study of inequalities. *Sociological Theory, 28,* 129–149.

Chowdry, A. (2013, October 8). What can 3D printing do? Here are 6 creative examples. *Forbes.* Retrieved from http://www.forbes.com/sites/amitchowdhry/2013/10/08/what-can-3d-printing-do-here-are-6-creative-examples/#774877e261b0

Christakis, N. A., & Fowler, J. H. (2009). *Connected: The surprising power of our social networks and how they shape our lives.* New York, NY: Little, Brown.

Cioffi, J. W. (2007). Review: *A theory of global capitalism: Production, class, and state in a transnational world* by William I. Robinson. *Journal of Politics, 69,* 880–882.

Clarke, T. C., Black, L. I., Stussman, B. J., Barnes, P. M., & Nahin, R. L. (2015). *Trends in the use of complementary health approaches among adults: United States, 2002–2012* (National Health Statistics Reports, No. 79). Hyattsville, MD: National Center for Health Statistics.

Clawson, D., & Clawson, M. A. (1999). What has happened to the U.S. labor movement? Union decline and renewal. *Annual Review of Sociology, 25,* 95–119.

Cleary, P. D. (1987). Gender differences in stress-related disorders. In R. C. Barnett (Ed.), *Gender and stress* (pp. 39–72). New York, NY: Free Press.

Cliff, G., & Wall-Parker, A. (2017, April). Statistical analysis of white-collar crime. In H. N. Pontell (Editor in Chief), *Oxford Research Encyclopedia of Criminology* (pp. 1–24). Oxford, UK: Oxford University Press. doi:10.1093/acrefore/9780190264079.013.267

Cloward, R., & Ohlin, L. E. (1960). *Delinquency and opportunity.* New York, NY: Free Press.

CNN. (2002, April 29). Is teaching abstinence only the best approach? Retrieved from http://edition.cnn.com/2002/ALLPOLITICS/04/24/cf.crossfire/index.html

CNN. (2013, May 30). The CNN Freedom Project: Ending modern-day slavery [Blog]. Retrieved from http://thecnnfreedomproject.blogs.cnn.com/

CNN. (2016). How a smartphone app can cure blindness in poor countries. Retrieved from http://www.cnn.com/2016/03/31/africa/peek-eye-app/

CNN. (2018, May 3). Syrian Civil War fast facts. Retrieved from https://www.cnn.com/2013/08/27/world/meast/syria-civil-war-fast-facts/index.html

Coate, J. (1994). Cyberspace innkeeping: Building online community. Retrieved from www.well.com:70/0/Community/innkeeping

Cockerham, W. C. (2014). The emerging crisis in American female longevity. *Social Currents, 1,* 220–227.

Cogan, M. F. (2010). Exploring academic outcomes of homeschooled Students. *Journal of College Admission, 208,* 18–25.

Cohan, W. (2016, January 6) Putting the heat on Yik Yak after a killing on campus. *New York Times.* Retrieved from http://www.nytimes.com/2016/01/07/business/dealbook/07db-streetscene.html

Cohen, A. (1955). *Delinquent boys: The culture of the gang.* Glencoe, IL: Free Press.

Cohen, P. (2012). *In our prime: The invention of middle age.* New York, NY: Scribner.

Cohen, L., Broschak, J. P., & Haveman, H. A. (1998). And then there were more? The effect of organizational sex composition on the hiring and promotion of managers. *American Sociological Review, 63*(5), 711–727.

Cohn, D., & Passel, J. (2016, August 11). *A record 60.6 million Americans live in multigenerational households.* Pew Research Center. Retrieved from http://www.pewresearch.org/fact-tank/2016/08/11/a-record-60-6-million-americans-live-in-multigenerational-households/

Cohn, D., & Passel, J. S. (2018, April 5). *A record 64 million Americans live in multigenerational households.* Pew Research Center. Retrieved from http://www.pewresearch.org/fact-tank/2018/04/05/a-record-64-million-americans-live-in-multigenerational-households/

Coker, A. L., Cook-Craig, P. G., Williams, C. M., Fisher, B. S., Clear, E. R., Garcia, L. S., & Hegge, L. M. (2011). Evaluation of Green Dot: An active bystander intervention to reduce sexual violence on college campuses. *Violence against Women, 17,* 777–796.

Colapinto, J. (2001). *As nature made him: The boy who was raised as a girl.* New York: HarperCollins.

Coleman, J. S. (1988). Social capital in the creation of human capital. *American Journal of Sociology, 94,* S95–S120.

Coleman, J. S. (1990). *The foundations of social theory.* Cambridge, MA: Harvard University Press.

Coleman, J. S., et al. (1966). *Equality of educational opportunity.* Washington, DC: U.S. Government Printing Office.

Collins, R. (1971). Functional and conflict theories of educational stratification. *American Sociological Review, 36,* 1002–1019.

Collins, R. (1979). *The credential society: An historical sociology of education.* New York, NY: Academic Press.

Collins, R. (2001). Weber and the sociology of revolution. *Journal of Classical Sociology, 1,* 171–194.

Coltrane, S. (1992). The micropolitics of gender in nonindustrial societies. *Gender & Society, 6*(1), 86–107.

Committee on Understanding and Eliminating Racial and Ethnic Disparities in Health Care, Board on Health Sciences Policy, Institute of Medicine. (2003). *Unequal treatment: Confronting racial and ethnic disparities in healthcare* (B. Smedley, A. Y. Stith, & A. R. Nelson, Eds.). Washington, DC: National Academies. Retrieved from https://www.ncbi.nlm.nih.gov/pubmed/25032386

Congressional Research Service. (2016). *Membership of the 114th Congress: A profile.* Retrieved from https://www.fas.org/sgp/crs/misc/R43869.pdf

Conley, D. (1999). *Being black, living in the red: Race, wealth, and social policy in America.* Berkeley, CA: University of California Press.

Connell, R. W. (1987). *Gender and power: Society, the person, and sexual politics.* Boston, MA: Allen & Unwin.

Connell, R., & Messerschmidt, J. W. (2005). Hegemonic masculinity: Rethinking the concept. *Gender and Society, 19,* 829–859.

Connidis, I. A., & McMullin, J. A. (1996). Reasons for and perceptions of childlessness among older persons: Exploring the impact of marital status and gender. *Journal of Aging Studies, 10*(3), 205–222.

Conrad, P. (2007). *The medicalization of society: On the transformation of human conditions into medical disorders.* Baltimore, MD: Johns Hopkins University Press.

Conti, J. (2011). *Between law and diplomacy: The social contexts of disputing at the World Trade Organization.* Palo Alto, CA: Stanford University Press.

Conway, M. (2004). Women's political participation at the state and local level in the United States. *Political Science & Politics, 37*(1), 60–61.

Cooley, C. H. (1964). *Human nature and the social order.* New York, NY: Schocken Books. (Original work published 1902)

Cooney, S. (2016). Shark Tank funds fewer women than men, with less money. *Mashable.* Retrieved from http://mashable.com/2016/01/15/shark-tank-women-entrepreneurs/#3Af08×WvxkqI

Coontz, S. (1992). *The way we never were: American families and the nostalgia trap.* New York, NY: Basic Books.

Cooper, D. (2016, February 3). *Balancing paychecks and public assistance* (EPI Briefing Paper No. 418). Economic Policy Institute. Retrieved from http://www.epi.org/files/2015/wages-and-transfers.pdf

Copen, C. E., Daniels, K., & Mosher, W. D. (2013). *First premarital cohabitation in the United States: 2006–2010.* (National Health Statistics Reports, No. 64). Hyattsville, MD: National Center for Health Statistics.

Coplan, J. H. (2011, Fall). In the fight for marriage equality, it's Edith Windsor vs. the United States of America. *New York University Alumni Magazine.* Retrieved from http://www.nyu.edu/alumni.magazine/issue17/17_FEA_DOMA.html

Corbin, J., & Strauss, A. (1985). Managing chronic illness at home: Three lines of work. *Qualitative Sociology, 8,* 224–247.

Correll, S. J., Benard, S., & Paik, I. (2007). Getting a job: Is there a motherhood penalty? *American Journal of Sociology, 112,* 1297–1338.

Corsaro, W. (1997). *The sociology of childhood.* Thousand Oaks, CA: Pine Forge Press.

Cosmides, L., & Tooby, J. (1997). *Evolutionary psychology: A primer.* University of California at Santa Barbara, Institute for Social, Behavioral, and Economic Research, Center for Evolutionary Psychology. Retrieved from http://www.cep.ucsb.edu/primer.html

Coulson, A. J. (2009). Comparing public, private, and market schools: The international evidence. *Journal of School Choice, 3,* 31–54.

Council on Foreign Relations. (2013, September 13). *Reflecting on Lehman's global legacy.* Retrieved from http://www.cfr.org/economics/reflecting-lehmans-global-legacy/p31391

Coursaris, C. K., & Liu, M. (2009). An analysis of social support exchanges in online HIV/AIDS self-help groups. *Computers in Human Behavior, 25,* 911–918. doi:10.1016/j.chb.2009.03.006

Credit Suisse. (2015, October). *Global wealth report: 2015.* Retrieved from https://publications.credit-suisse.com/tasks/render/file/?fileID=F2425415-DCA7-80B8-EAD989AF9341D47E

Credit Suisse. (2017, November 14). *Global wealth report 2017: Where are we ten years after the crisis?* Retrieved from https://www.credit-suisse.com/corporate/en/articles/news-and-expertise/global-wealth-report-2017-201711.html/

Creswell, J., & Thomas, L. (2009, January 24). The talented Mr. Madoff. *New York Times.* Retrieved from http://www.nytimes.com/2009/01/25/business/25bernie.html?pagewanted=1

Crossman, A. (2017, March 18). How sociology can prepare you for a career in the business world. *ThoughtCo.* Retrieved from https://www.thoughtco.com/sociology-and-business-3026175

Crowdsourcing.org. (2015). Global crowdfunding market to reach $34.4b in 2015, predicts Massolution's 2015cf industry report. Retrieved from http://www.crowdsourcing.org/editorial/global-crowdfunding-market-to-reach-344b-in-2015-predicts-massolutions-2015cf-industry-report/45376

Cumings, B. (1987). The origins and development of the northeast Asian political economy: Industrial sectors, product cycles, and political consequences. In F. C. Deyo (Ed.), *The political economy of the new Asian industrialism* (pp. 44–83). Ithaca, NY: Cornell University Press.

Cumings, B. (1997). *Korea's place in the sun: A modern history.* New York, NY: Norton.

Cumming, E. (1963). Further thoughts on the theory of disengagement. *International Social Science Journal, 15,* 377–393.

Cumming, E. (1975). Engagement with an old theory. *International Journal of Aging and Human Development, 6,* 187–191.

Cumming, E., & Henry, W. E. (1961). *Growing old: The process of disengagement.* New York, NY: Basic.

Cunningham, L. (2013, September 18). Hiring more women seen as answer to economic malaise: "Womenomics" pushed as fix for population woes. *Washington Post.* Retrieved from http://www.japantimes.co.jp/news/2013/09/18/national/hiring-more-women-seen-as-answer-to-economic-malaise/#.UwYs0YU2XO0

Cunningham, E. (2018, February 24). Anti-government protests now look like an opportunity for Iran's president. *Washington Post.* Retrieved from https://www.washingtonpost.com/world/anti-government-protests-now-look-like-an-opportunity-for-irans-president/2018/02/22/17f33326-0b4c-11e8-998c-96deb18cca19_story.html?utm_term=.42d01bc06f67

Curtin, S. C., Ventura, S. J., & Martinez, G. M. (2014). *Recent declines in nonmarital childbearing in the United States* (NCHS Data Brief No. 162). Hyattsville, MD: National Center for Health Statistics.

Cutrone, C., & Nisen, M. (2012, December 18). 30 people with "soft" college majors who became extremely successful. *Business Insider.* Retrieved from http://www.businessinsider.com/successful-liberal-arts-majors-2012-12

Damaske, S., & Frech, A. (2016). Women's work pathways across the life course. *Demography, 53,* 365–391.

Daniels, J. (2009). *Cyber racism: White supremacy online and the new attack on civil rights.* Lanham, MD: Rowman & Littlefield.

Darling, R. B. (2003). Toward a model of changing disability identities: A proposed typology and research agenda. *Disability & Society, 18,* 881–895.

Darwin, H. (2017). Doing gender beyond the binary: A virtual ethnography. *Symbolic Interaction, 40*(3):317–334.

Davies, J. (1962). Toward a theory of revolution. *American Sociological Review, 27,* 5–19.

Davis, K. (1937). The sociology of prostitution. *American Sociological Review, 11,* 744–755.

Davis, M. (1990). *City of quartz: Excavating the future in Los Angeles.* New York, NY: Verso.

Davis, G. (2015a). *Contesting intersex: The dubious diagnosis.* New York, NY: NYU Press.

Davis, L. C. (2015b, October 7). The flight from conversation. *The Atlantic.* Retrieved from http://www.theatlantic.com/technology/archive/2015/10/reclaiming-conversation-sherry-turkle/409273/

Davis, L., & James, S. D. (2011). Canadian mother raising "genderless" baby, storm, defends her family's decision. *ABC News.* Retrieved from http://abcnews.go.com/Health/genderless-baby-controversy-mom-defends-choice-reveal-sex/story?id=13718047

Davis, K., & Moore, W. E. (1945). Some principles of stratification. *American Sociological Review, 10,* 242–249.

Davis, D. D., & Polonko, K. A. (2001). *Telework in the United States: Telework American Research Study 2001.* Washington, DC: International Telework Association & Council.

Davis, E., & Snell, T. L. (2018, April 30). *Capital punishment, 2016—statistical brief.* Bureau of Justice Statistics. Retrieved from https://www.bjs.gov/index.cfm?ty=pbdetail&iid=6246

Dawsey, J. (2018, January 12). Trump derides protections for immigrants from "shithole" countries. *Washington Post.* Retrieved from https://www.washingtonpost.com/politics/trump-attacks-protections-for-immigrants-from-shithole-countries-in-oval-office-meeting/2018/01/11/bfc0725c-f711-11e7-91af-31ac729add94_story.html?tid=pm_politics_pop&utm_term=.28e249c023f4

Deacon, T. (1998). *The symbolic species: The co-evolution of language and the brain.* New York, NY: Norton.

Death Penalty Information Center. (2013). *The death penalty in 2013: A year end report.* Retrieved from http://deathpenaltyinfo.org/documents/YearEnd2013.pdf

de Jong Gierveld,[CE1] J., & Havens, B. (2004). Cross-national comparisons of social isolation and loneliness: Introduction and overview. *Canadian Journal on Aging, 23,* 109–113.

de Jong Gierveld, J., van Groenou, M. B., Hoogendoorn, A. W., & Smit, J. H. (2009). Quality of marriages in later life and emotional and social loneliness. *Journals of Gerontology. Series B, Psychological Sciences and Social Sciences, 64*(4), 497–506.

D'Emilio, J. (1983). *Sexual politics, sexual communities: The making of a homosexual minority in the United States, 1940–1970.* Chicago, IL: University of Chicago Press.

Demos. (2010). *At what cost? How student debt reduces lifetime wealth.* Retrieved from http://www.demos.org/sites/default/files/imce/AtWhatCostFinal.pdf

DeNavas-Walt, C., & Proctor, B. (2015). *Income and poverty in the United States: 2014* (Current Population Reports, P60-252). U.S. Bureau of the Census. Retrieved from http://www.census.gov/content/dam/Census/library/publications/2015/demo/p60-252.pdf

DeNavas-Walt, C., Proctor, B. D., & Lee, C. H. (2005). *Income, poverty, and health insurance coverage in the United States: 2004* (Current Population Reports, P60-229). U.S. Bureau of the Census. Retrieved from https://www.census.gov/prod/2005pubs/p60-229.pdf

Derenne, J. L., & Beresin, E. V. (2006). Body image, media, and eating disorders. *Academic Psychiatry, 30,* 257–261.

Desilver, D. (2016, August 2). *U.S. voter turnout trails most developed countries.* Pew Research Center. Retrieved from http://www.pewresearch.org/fact-tank/2016/08/02/u-s-voter-turnout-trails-most-developed-countries/

Desilver, D. (2018, May 21). *U.S. trails most developed countries in voter turnout.* Pew Research Center. Retrieved from http://www.pewresearch.org/fact-tank/2018/05/21/u-s-voter-turnout-trails-most-developed-countries/

Desmond, M. (2016). *Evicted: Poverty and profit in the American city.* New York, NY: Crown Books.

Desmond, M., & Emirbayer, M. (2016). *Race in America.* New York, NY: Norton.

Dettling, L. J., Hsu, J. W., Jacobs, L., Moore, K. B., & Thompson, J. P. (2017, September 27). Recent trends in wealth-holding by race and ethnicity: Evidence from the Survey of Consumer Finances. *FEDS Notes.* Retrieved from https://doi.org/10.17016/2380-7172.2083

Dewey, C. (2015, November 11). What is Yik Yak, the app that fielded racist threats at University of Missouri? *Washington Post.* Retrieved from https://www.washingtonpost.com/news/the-intersect/wp/2015/11/11/what-is-yik-yak-the-app-that-fielded-racist-threats-at-university-of-missouri/

Deyo, F. (1987). *The political economy of the new Asian industrialism.* Ithaca, NY: Cornell University Press.

Diamond, J. (2005). *Collapse: How societies choose to fail or succeed.* New York, NY: Penguin.

Dickens, W. T., & Flynn, J. R. (2006). Black Americans reduce the racial IQ gap: Evidence from standardization samples. *Psychological Science, 17,* 913–920.

Dillon, S. (2010, August 31). Formula to grade teachers' skill gains acceptance, and critics. *New York Times.* Retrieved from http://www.nytimes.com/2010/09/01/education/01teacher.html

Dimitrova, D. (2003). Controlling teleworkers: Supervision and flexibility revisited. *New Technology Work and Employment, 18*(3), 181–195.

DiSesa, N. (2008). *Seducing the boys club: Uncensored tactics from a woman at the top.* New York, NY: Ballantine.

Dolan, K. (2016, October 4). Inside the 2016 Forbes 400: Facts and figures about America's richest people. *Forbes.* Retrieved from http://www.forbes.com/sites/kerryadolan/2016/10/04/inside-the-2016-forbes-400-facts-and-figures-about-americas-richest-people/?ss=forbes400#41dc464d3973

Dolbeare, C. (1995). *Out of reach: Why everyday people can't find affordable housing.* Washington, DC: Low Income Housing Information.

Domhoff, G. W. (1971). *The higher circles: The governing class in America.* New York, NY: Vintage Books.

Domhoff, G. W. (1979). *The powers that be: Processes of ruling class domination in America*. New York, NY: Vintage Books.

Domhoff, G. W. (1983). *Who rules America now? A view for the '80s*. New York, NY: Prentice Hall.

Domhoff, G. W. (1998). *Who rules America?: Power and politics in the year 2000*. Belmont, CA: Mayfield.

Domhoff, G. W. (2013). *Who rules America? The triumph of the corporate rich*. New York, NY: McGraw-Hill.

Donadio, R. (2013, June 30). When Italians chat, hands and fingers do the talking. *New York Times*. Retrieved from http://www.nytimes.com/2013/07/01/world/europe/when-italians-chat-hands-and-fingers-do-the-talking.html?_r=0

Donkor, K. (1997). *Structural adjustment and mass poverty in Ghana*. Aldershot, Hants, England; Brookfield, VT: Ashgate.

Dowd, M. (2017, October 21). She's 26, and brought down Uber's C.E.O. What's next? *New York Times*. Retrieved from https://www.nytimes.com/2017/10/21/style/susan-fowler-uber.html

Drake, St. C., & Cayton, H. A. (1945). *Black Metropolis: A Study of Negro Life in a Northern City*. New York: Harcourt, Brace, and Jovanovich, Inc.

Dreier, P., & Appelbaum, R. (1992, Spring/Summer). The housing crisis enters the 1990s. *New England Journal of Public Policy*, 8, 1.

Drenovsky, C. K., & Cohen, I. (2012). The impact of homeschooling on the adjustment of college students. *International Social Science Review*, 87(1–2), 19–34.

Drescher, J. (2015). Out of DSM: Depathologizing homosexuality. *Behavioral Sciences, 5*, 565–575.

Drum, K. (2016). Chart of the day: Net new jobs in December. *Mother Jones*. Retrieved from http://www.motherjones.com/kevin-drum/2016/01/chart-day-net-new-jobs-december

Du Bois, W. E. B. (1903). *The souls of black folk*. Chicago, IL: A. C. McClurg.

Du Bois, W. E. B. (1925, November 17). [Letter to Rollo Ogden]. W. E. B. Du Bois Papers (microfilm reel no. 15, frame no. 1195), University of Massachusetts, Amherst.

Dubos, R. (1959). *Mirage of health*. New York, NY: Doubleday/Anchor.

Duggan, M. (2014, October 22). *Online harassment*. Pew Research Center. Retrieved from http://www.pewinternet.org/2014/10/22/online-harassment/

Duggan, M. (2017, July 11). *Online harassment 2017*. Pew Research Center. Retrieved from http://assets.pewresearch.org/wp-content/uploads/sites/14/2017/07/10151519/PI_2017.07.11_Online-Harassment_FINAL.pdf

Duignan, P., & Gann, L. H. (Eds.). (1998). *The debate in the United States over immigration*. Stanford, CA: Hoover Institution Press.

Duncan, G. J., Brooks-Gunn, J., Yeung, W. J., & Smith, J. R. (1998). How much does childhood poverty affect the life chances of children? *American Sociological Review*, 63(3), 406–423.coU.

Duncombe, J., & Marsden, D. (1993). Love and intimacy: The gender division of emotion and emotion work: A neglected aspect of sociological discussion of heterosexual relationships. *Sociology*, 27(2), 221–241.

Duneier, M. (1999). *Sidewalk*. New York, NY: Farrar, Straus and Giroux.

Duneier, M. (2016). *Ghetto: The invention of a place, the history of an idea*. New York, NY: Farrar, Straus and Giroux.

Duneier, M., & Molotch, H. (1999). Talking city trouble: Interactional vandalism, social inequality, and the urban interaction problem. *American Journal of Sociology*, 104(5), 1263–1295.

Dunlap, R. (2002). Environmental sociology: A personal perspective on the first quarter century. *Organization and Environment*, 15(1), 10–29.

Dunlop et al. (2003). Racial/ethnic differences of depression among preretirement adults. *American Journal of Public Health*, 93(11): 1945–1952.

Durkheim, É. (1964). *The division of labor in society*. New York, NY: Free Press. (Original work published 1893)

Durkheim, É. (1965). *The elementary forms of the religious life*. New York, NY: Free Press. (Original work published 1912)

Durkheim, É. (1966). *Suicide*. New York, NY: Free Press. (Original work published 1897)

Durso, L. E., & Gates, G. J. (2012). *Serving our youth: Findings from a national survey of service providers working with lesbian, gay, bisexual, and transgender youth who are homeless or at risk of becoming homeless*. Los Angeles, CA: Williams Institute with True Colors Fund and the Palette Fund.

DW. (2018, September 1). Tunisia anti-austerity protests turn deadly. Retrieved from http://www.dw.com/en/tunisia-anti-austerity-protests-turn-deadly/a-42079677

Dworkin, A. (1981). *Pornography: Men possessing women*. New York, NY: Pedigree.

Dworkin, A. (1987). *Intercourse*. New York, NY: Free Press.

Dye, T. (1986). *Who's running America?* (4th ed.). Englewood Cliffs, NJ: Prentice Hall.

Eagan, K., Stolzenberg, E. B., Ramirez, J. J., Aragon, M. C., Suchard, M. R., & Hurtado, S. (2014). *The American freshman: National norms fall 2014*. Los Angeles, CA: Higher Education Research Institute, UCLA.

Eagan, K., Stolzenberg, E. B., Bates, A. K., Aragon, M. C., Suchard, M. R., & Rios-Aguilar, C. (2015). *The American freshman: National norms fall 2015*. Los Angeles, CA: Higher Education Research Institute, UCLA.

Eagan, M. K., Stolzenberg, E. B., Zimmerman, H. B., Aragon, M. C., Whang Sayson, H., & Rios-Aguilar, C. (2017). *The American freshman: National norms fall 2016*. Los Angeles, CA: Higher Education Research Institute, UCLA.

Easterly, W. (2006). *The white man's burden: Why the West's efforts to aid the rest have done so much ill and so little good*. New York, NY: Penguin.

Eating Disorder Coalition. (2003). *Statistics*. Retrieved from www.eatingdisorderscoalition.org/reports/statistics.html

Eaton, J., & Pell, M. B. (2010, February 24). *Lobbyists swarm capitol to influence health care reform*. Center for Public Integrity. Retrieved from https://www.publicintegrity.org/articles/entry/1953/

Ebomoyi, E. (1987). The prevalence of female circumcision in two Nigerian communities. *Sex Roles*, 17(3–4), 139–151.

Economic Policy Institute. (2018). *State of Working America Data Library, college wage premium*. Retrieved from https://www.epi.org/data/#?subject=wagegap-coll

Economic Times. (2015). Heat-related deaths will double in urban India by 2080. Retrieved from http://articles.economictimes.indiatimes.com/2015-07-06/news/64143024_1_indian-institute-climate-change-mortality

EdChoice. (2018). *The ABCs of school choice: 2018 edition*. Retrieved from https://www.edchoice.org/?smd_process_download=1&download_id=8602

Edin, K., & Kefalas, M. J. (2005). *Promises I can keep: Why poor women put motherhood before marriage*. Berkeley, CA: University of California Press.

Edison Research for the National Election Pool. (2016). Exit Polls. Retrieved from https://www.cnn.com/election/2016/results/exit-polls

Edney, A. (2013, November 1). Medical advice just a touch away with smartphone apps. *Bloomberg News*. Retrieved from http://www.bloomberg.com/news/2013-11-01/medical-advice-just-a-touch-away-with-smartphone-apps.html

Education Week Research Center. (2017). *Teachers and education reform: Results from a national survey*. Retrieved from https://www.edweek.org/media/teachers-and-education-reform-report-education-week.pdf

Efron, S. (1997, October 18). Eating disorders go global. *Los Angeles Times*, p. A1.

Egan, J. (2018, May 9). Children of the opioid epidemic. *New York Times*. Retrieved from https://www.nytimes.com/2018/05/09/magazine/children-of-the-opioid-epidemic.html

Eggen, D. (2009, July 21). Lobbyists spend millions to influence health care reform. *Washington Post*. Retrieved from http://voices.washingtonpost.com/health-care-reform/2009/07/health_care_continues_its_inte.html

Ehrlich, P. (1968). *The population bomb*. New York, NY: Ballantine Books.

Eibl-Eibesfeldt, I. (1972). Similarities and differences between cultures in expressive movements. In R. A. Hinde (Ed.), *Nonverbal communication*. New York, NY: Cambridge University Press.

Eickmeyer, K. J. (2015). *Divorce rate in the U.S.: Geographic variation, 2014*. National Center for Family and Marriage Research. Retrieved from https://www.bgsu.edu/ncfmr/resources/data/family-profiles.html

Eidelson, J. (2013a, December 5). Biggest-ever fast food strike today! Thousands to walk out across 100 cities. *Salon*. Retrieved from http://www.salon.com/2013/12/05/biggest_ever_fast_food_strike_today_thousands_walk_out_across_100_cities/

Eidelson, J. (2013b, March 4). McDonald's to SEC: Strikes hurt, and we might have to hike pay. *Salon*. Retrieved from http://www.salon.com/2014/03/04/mcdonalds_to_sec_strikes_hurt_and_we_might_have_hike_pay/

Eisenhower Library. (1961). Farewell address. Abilene, KS: The Dwight D. Eisenhower Presidential Library. Retrieved from http://www.eisenhower.archives.gov/research/online_documents/farewell_address.html

Ekman, P., & Friesen, W. V. (1978). *Facial action coding system.* New York, NY: Consulting Psychologists Press.

El Dareer, A. (1982). *Woman, why do you weep? Circumcision and its consequences.* Westport, CT: Zed.

f, N. (1987). *Involvement and detachment.* Oxford, UK: Oxford University Press.

Elias, N., & Dunning, E. (1987). *Quest for excitement: Sport and leisure in the civilizing process.* Oxford, UK: Blackwell.

Elliott, S. (2013, May 31). Vitriol online for Cheerios ad with interracial family. *New York Times.* Retrieved from http://www.nytimes.com/2013/06/01/business/media/cheerios-ad-with-interracial-family-brings-out-internet-hate.html?_r=0

Elliott, D. B., & Simmons, T. (2011). *Marital events of Americans: 2009.* U.S. Bureau of the Census. Retrieved from http://www.census.gov/prod/2011pubs/acs-13.pdf

Elshtain, J. (1981). *Public man: Private woman.* Princeton, NJ: Princeton University Press.

Elwell, C. K. (2014, March 10). *The distribution of household income and the middle class.* Congressional Reference Service. Retrieved from http://fas.org/sgp/crs/misc/RS20811.pdf

Embrick, D. G., Domínguez, S., & Karsak, B. (2017, April). More than just insults: Rethinking sociology's contribution to scholarship on racial microaggressions. *Sociological Inquiry, 87,* 193–206.

Emirbayer, M., & Mische, A. (1998). What is agency? *American Journal of Sociology, 103,* 962–1023.

Emmanuel, A. (1972). *Unequal exchange: A Study of the imperialism of trade.* New York: Monthly Review Press.

Emslie, C., Browne, S., MacLeod, U., Rozmovits, L., Mitchell, E., & Ziebland, S. (2009). "Getting through" not "going under": A qualitative study of gender and spousal support after diagnosis with colorectal cancer. *Social Science & Medicine, 68,* 1169–1175.

Engelhardt, C. R., B. D. Bartholow, C. T. Kerr, and B. J. Bushman. (2011). This is your brain on violent video games: neural desensitization to violence predicts increased aggression following violent video game exposure. *Journal of Experimental Social Psychology, 47*(5), 1033–1036.

England, P., Shafer, E. F., & Fogarty, A. C. K. (2012). Hooking up and forming romantic relationships on today's college campuses. In M. Kimmel & A. Aronson (Eds.), *The gendered society reader* (5th ed., pp. 559–572). New York, NY: Oxford University Press.

Ennis, S. R., Rios-Vargas, M., & Albert, N. (2011). *The Hispanic population: 2010.* U.S. Bureau of the Census. Retrieved from http://www.census.gov/prod/cen2010/briefs/c2010br-04.pdf

Entertainment Software Association. (2011). Essential facts about the computer and video game industry. Retrieved from www.theesa.com/facts/pdfs/ESA_EF_2011.pdf.

Environmental Protection Agency. (2015). *DDT—A brief history and status.* Retrieved from https://www.epa.gov/ingredients-used-pesticide-products/ddt-brief-history-and-status

Environmental Working Group. (2018). *The United States farm subsidy information.* Retrieved from https://farm.ewg.org/region.php?fips=00000&statename=UnitedStates

EPA. (2017, January 19). *International climate impacts.* Retrieved from https://19january2017snapshot.epa.gov/climate-impacts/international-climate-impacts_.html

EPA. (2018). *Greenhouse gas emissions from a typical passenger vehicle.* Retrieved from https://www.epa.gov/greenvehicles/greenhouse-gas-emissions-typical-passenger-vehicle

Erdely, S. R. (2013, October 23). About a girl: Coy Mathis' fight to change gender. *Rolling Stone.* Retrieved from http://www.rollingstone.com/culture/news/about-a-girl-coy-mathis-fight-to-change-change-gender-20131028#ixzz3sRQOAiyb

Ericson, R. V., & K. D. Haggerty. (1997). *Policing the risk society.* Toronto, ON: University of Toronto Press.

Erlanger, S. (2011, April 11). France enforces ban on full-face veils in public. *New York Times.* Retrieved from http://www.nytimes.com/2011/04/12/world/europe/12france.html

Espelage, D. L., Aragon, S. R., & Birkett, M. (2008). Homophobic teasing, psychological outcomes, and sexual orientation among high school students: What influence do parents and schools have? *School Psychology Review, 37,* 202–216.

Espenshade, T. J., & Radford, A. W. (2009). *No longer separate, not yet equal: Race and class in elite college admission and campus life.* Princeton, NJ: Princeton University Press.

ESPN. (2013a, April 30). Jason Collins says he's gay. Retrieved from http://espn.go.com/nba/story/_/id/9223657/jason-collins-first-openly-gay-active-player

ESPN. (2013b, January 17). Story of Manti Te'o girlfriend a hoax. Retrieved from http://espn.go.com/college-football/story/_/id/8851033/story-manti-teo-girlfriend-death-apparently-hoax

ESPN. (2015, July 16). Caitlyn Jenner vows to "reshape the landscape" in ESPYS speech. Retrieved from http://espn.go.com/espys/2015/story/_/id/13264599/caitlyn-jenner-accepts-arthur-ashe-courage-award-espys-ashe2015

Estes, C. (1986). The politics of ageing in America. *Ageing and Society, 6*(2), 121–134.

Estes, C. (1991). The Reagan legacy: Privatization, the welfare state, and aging. In J. Myles & J. Quadagno (Eds.), *States, labor markets, and the future of old age policy.* Philadelphia, PA: Temple University Press.

Estes, C., E. A. Binney, & R. A. Culbertson. (1992). The gerontological imagination: Social influences on the development of gerontology, 1945–present. *International Journal of Aging & Human Development, 35*(1), 49–65.

Estlund, C. L. (2006) The death of labor law? *Annual Review of Law and Social Science, 2,* 105–123.

European Parliament. (2013). 40% of seats on company boards for women. Retrieved from http://www.europarl.europa.eu/news/en/news-room/content/20131118IPR25532/html/40-of-seats-on-company-boards-for-women

European Union. (2017). *2017 report on equality between women and men in the EU.* Brussels, Belgium: Author.

Evans, P. (1987). Class, state, and dependence in East Asia: Some lessons for Latin Americanists. In F. C. Deyo (Ed.), *The political economy of the new Asian industrialism.* Ithaca, NY: Cornell University Press.

Evans, P. (1995). *Embedded autonomy: States and industrial transformation.* Princeton, NJ: Princeton University Press.

Evans-Pritchard, E. (1970). Sexual inversion among the Azande. *American Anthropologist, 72*(6), 1428–1434.

Fadiman, A. (1997). *The spirit catches you and you fall down.* New York, NY: Farrar, Straus and Giroux.

Fairlie, R. W., & Sundstrom, W. A. (1999). The emergence, persistence, and recent widening of the racial unemployment gap. *Industrial and Labor Relations Review, 52*(2), 252–270.

Family Equality Council. (2014). *Equality maps.* Retrieved from http://www.familyequality.org/get_informed/equality_maps/joint_adoption_laws/

Fausto-Sterling, A. (2000). *Sexing the body: Gender politics and the construction of sexuality.* New York, NY: Basic Books.

Feagin, J. R., & McKinney, K. (2003). *The many costs of racism.* Lanham, MD: Rowman & Littlefield.

Federal Bureau of Investigation. (2011). *Hate crime statistics, 2011.* Retrieved from http://www.fbi.gov/news/stories/2012/december/annual-hate-crimes-report-released/annual-hate-crimes-report-released

Federal Bureau of Investigation. (2012a). Persons arrested. *Crime in the United States: 2011.* Retrieved from http://www.fbi.gov/about-us/cjis/ucr/crime-in-the-u.s/2011/crime-in-the-u.s.-2011/persons-arrested

Federal Bureau of Investigation. (2012b). Expanded homicide data tables. *Crime in the United States: 2012.* Retrieved from http://www.fbi.gov/about-us/cjis/ucr/crime-in-the-u.s/2012/crime-in-the-u.s.-2012/offenses-known-to-law-enforcement/expanded-homicide/expanded_homicide_data_table_6_murder_race_and_sex_of_vicitm_by_race_and_sex_of_offender_2012.xls

Federal Bureau of Investigation. (2012c). *Crime in the United States by volume and rate per 100,000 inhabitants, 1992–2011.* Retrieved from http://www.fbi.gov/about-us/cjis/ucr/crime-in-the-u.s/2011/crime-in-the-u.s.-2011/tables/table-1

Federal Bureau of Investigation. (2013a). *Crime in the United States by volume and rate per 100,000 inhabitants, 1992–2012.* Retrieved from http://www.fbi.gov/about-us/cjis/ucr/crime-in-the-u.s/2012/crime-in-the-u.s.-2012/tables/1tabledatadecoverviewpdf/table_1_crime_in_the_united_states_by_volume_and_rate_per_100000_inhabitants_1993-2012.xls

Federal Bureau of Investigation. (2013b). Persons arrested. *Crime in the United States: 2012.* Retrieved from http://www.fbi.gov/about-us/

cjis/ucr/crime-in-the.u.s/2012/crime-in-the-u.s.-2012/persons-arrested/persons-arrested

Federal Bureau of Investigation. (2015a). *2014 hate crime statistics*. Retrieved from https://www.fbi.gov/about-us/cjis/ucr/hate-crime/2014

Federal Bureau of Investigation. (2015b). Murder. *Crime in the United States: 2014*. Retrieved from https://www.fbi.gov/about-us/cjis/ucr/crime-in-the.u.s/2014/crime-in-the-u.s.-2014/cius-home

Federal Bureau of Investigation. (2015c). Persons arrested. *Crime in the United States: 2014*. Retrieved from https://ucr.fbi.gov/crime-in-the.u.s/2014/crime-in-the-u.s.-2014/persons-arrested/main

Federal Bureau of Investigation. (2015d). Table 42. *Crime in the United States: 2014*. Retrieved from https://www.fbi.gov/about-us/cjis/ucr/crime-in-the.u.s/2014/crime-in-the-u.s.-2014/tables/table-42

Federal Bureau of Investigation. (2015e). *Crime in the United States: 2014*. Retrieved from https://ucr.fbi.gov/crime-in-the.u.s/2014/crime-in-the-u.s.-2014/cius-home

Federal Bureau of Investigation. (2017a, November 13). *2016 hate crime statistics released*. Retrieved from https://www.fbi.gov/news/stories/2016-hate-crime-statistics

Federal Bureau of Investigation. (2017b, September 25). *Crime in the United States: 2016*. Retrieved from https://ucr.fbi.gov/crime-in-the.u.s/2016/crime-in-the-u.s.-2016/cius-2016

Federal Interagency Forum on Aging-Related Statistics. (2010). *Older Americans 2010: Key indicators of well-being*. Washington, DC: U.S. Government Printing Office.

Federal Interagency Forum on Aging-Related Statistics. (2013). *Older Americans 2012: Key indicators of well-being*. Washington, DC: U.S. Government Printing Office.

Federal Interagency Forum on Aging-Related Statistics. (2016). *Older Americans 2016: Key indicators of well-being*. Washington, DC: U.S. Government Printing Office.

Feldman, M. B., & Meyer, I. H. (2007). Eating disorders in diverse lesbian, gay, and bisexual populations. *International Journal of Eating Disorders, 40*, 218–226.

Fenner, L. (2012, January 27). Apps4Africa announces winners, more to come. *IIP Digital*. Retrieved from http://iipdigital.usembassy.gov/st/english/article/2012/01/201201271737116esiuolo.3623316.html#axzz2iNapYA39

Fenton, M. V., & Morris, D. L. (2003). The integration of holistic nursing practices and complementary and alternative modalities into curricula of schools of nursing. *Alternative Therapies in Health and Medicine, 9*(4), 62–67.

Field, A. E., Sonneville, K. R., & Crosby, R. D. (2014). Prospective associations of concerns about physique and the development of obesity, binge drinking, and drug use among adolescent boys and young adult men. *JAMA Pediatrics, 168*, 34–39.

Fields, R. (2017, November 15). New York City launches committee to review maternal deaths. *ProPublica*. Retrieved from https://www.propublica.org/article/new-york-city-launches-committee-to-review-maternal-deaths

File, T. (2013). *The diversifying electorate—voting rates by race and Hispanic origin in 2012*. U.S. Bureau of the Census. Retrieved from http://www.census.gov/content/dam/Census/library/publications/2013/demo/p20-568.pdf

File, T. (2017). Voting in America: A look at the 2016 presidential election [Blog]. U.S. Bureau of the Census. Retrieved from https://www.census.gov/newsroom/blogs/random-samplings/2017/05/voting_in_america.html

Fine, M. A., Coleman, M., & Ganong, L. H. (1998). Consistency in perceptions of the step-parent role among step-parents, parents and stepchildren. *Journal of Social and Personal Relationships, 15*(6), 810–828.

Finke, R. & Stark, R. (1988). Religious economies and sacred canopies: Religious mobilization in American cities, 1906. *American Sociological Review, 53*(1): 41–49.

Finke, R., & Stark, R. (1992). *The churching of America, 1776–1990: Winners and losers in our religious economy*. New Brunswick, NJ: Rutgers University Press.

Finley, N. J., Roberts, M. D., & Banahan, B. F. (1988). Motivators and inhibitors of attitudes of filial obligation toward aging parents. *Gerontologist, 28*, 73–78.

Fischer, C. (1984). *The urban experience* (2nd ed.). New York, NY: Harcourt Brace Jovanovich.

Fischer, C. S., Hout, M., Jankowski, M. S., Lucas, S. R., Swidler, A., & Voss, K. (1996). *Inequality by design: Cracking the bell curve myth*. Princeton, NJ: Princeton University Press.

Fisher, B. S., Cullen, F. T., & Turner, M. G. (2000, December). *The sexual victimization of college women*. Washington, DC: U.S. Department of Justice, National Institute of Justice, Bureau of Justice Statistics.

Fitzpatrick, A. (2013, June 3). Turkey protesters take to Twitter as local media turns a blind eye. *Mashable*. Retrieved from http://mashable.com/2013/06/03/twitter-turkey-protests/

Flaherty, B., & Sethi, R. (2010). Homicide in black and white. *Journal of Urban Economics, 68*(3), 215–230.

Flores, A. (2017, September 18). *How the U.S. Hispanic population is changing*. Pew Research Center. Retrieved from http://www.pewresearch.org/fact-tank/2017/09/18/how-the-u-s-hispanic-population-is-changing/

Flores, A., López, G., & Radford, J. (2017, September 18). *Facts on U.S. Latinos, 2015*. Pew Research Center. Retrieved from http://www.pewhispanic.org/2017/09/18/facts-on-u-s-latinos-current-data/

Food and Agriculture Organization. (2016). *The state of food and agriculture: Climate change, agriculture, and food security*. Retrieved from http://www.fao.org/3/a-i6030e.pdf

Food and Agriculture Organization of the United Nations. (2017). *The state of food security and nutrition in the world 2017*. Retrieved from http://www.fao.org/3/a-I7695e.pdf

Foran, C. (2015, December 31). A year of Black Lives Matter. *The Atlantic*. Retrieved from http://www.theatlantic.com/politics/archive/2015/12/black-lives-matter/421839/

Forbes. (2017, March 8). Forbes releases world's self-made women billionaires list. Retrieved from https://www.forbes.com/sites/forbespr/2017/03/08/forbes-releases-worlds-self-made-women-billionaires-list/#2044aad0203f

Forbes. (2018, June 6). Global 2000: The world's largest public companies. Retrieved from https://www.forbes.com/global2000/#36c3cc8c335d

Ford, M. (2009). *The lights in the tunnel: Automation, accelerating technology and the economy of the future*. Amazon CreateSpace Independent Publishing Platform.

Ford, M. (2010, April 8). Your job in 2020. *Forbes*. Retrieved from http://www.forbes.com/2010/04/08/unemployment-google-2020-technology-data-companies-10-economy.html

Ford, C. S., & Beach, F. A. (1951). *Patterns of sexual behavior*. New York, NY: Harper & Row.

Fortune. (2013). Fortune Global 500. Retrieved from http://money.cnn.com/magazines/fortune/global500/2013/full_list/

Fortune. (2016). Global 500. Retrieved from http://beta.fortune.com/global500

Fortune. (2017). Visualize the Global 500. Retrieved from http://fortune.com/global500/visualizations/?iid=recirc_g500landing-zone1

Foucault, M. (1979). *Discipline and punish: The birth of the prison*. New York, NY: Random House.

Foucault, M. (1988). Technologies of the self. In L. H. Martin, H. Gutman, & P. H. Hutton (Eds.), *Technologies of the self: A seminar with Michel Foucault* (pp. 16–49). Amherst, MA: University of Massachusetts Press.

Fox, M. (2014, November 24). A tale of two outbreaks: Why Congo conquered Ebola. *NBC News*. Retrieved from http://www.nbcnews.com/storyline/ebola-virus-outbreak/tale-two-outbreaks-why-congo-conquered-ebola-n253911

Frank, A. (1966). The development of underdevelopment. *Monthly Review, 18*(4), 17–31.

Frank, A. (1969a). *Latin America: Underdevelopment or revolution*. New York, NY: Monthly Review Press.

Frank, A. (1969b). *Capitalism and underdevelopment in Latin America: Historical studies of Chile and Brazil*. New York, NY: Monthly Review Press.

Frank, A. (1979). *Dependent accumulation and underdevelopment*. London, UK: Macmillan.

Fredrickson, G. (2002). *Racism: A brief history*. Princeton, NJ: Princeton University Press.

Freedman, A. (2013, May 19). Study projects steep increase in NYC heat-related deaths. *Climate Central*. Retrieved from http://www.climatecentral.org/news/study-projects-steep-increase-in-heat-related-deaths-in-new-york-16012

Freedom House. (2005). *Electoral democracies*. Retrieved from http://www.freedomhouse.org/template.cfm?page=205&year=200

Freedom House. (2016). *Freedom on the net 2016*. Retrieved from https://freedomhouse.org/report/freedom-net/freedom-net-2016

Freedom House. (2018). *Freedom in the world 2018: Democracy in crisis*. Retrieved from https://freedomhouse.org/report/freedom-world/freedom-world-2018

Fremlin, J. (1964). How many people can the world support? *New Scientist, 24*(415), 285–287.

French, H. (2001, January 1). Diploma at hand, Japanese women find glass ceiling reinforced with iron. *New York Times*, p. A1.

Freudenberg, W. R., Frickel, S., & Gramling, R. (1995). Beyond the nature/society divide: Learning to think about a mountain. *Sociological Forum, 10*(3), 361–392.

Frey, W. H. (2011a). *Census data: Blacks and Hispanics take different segregation paths.* Brookings Institution. Retrieved from http://www.brookings.edu/opinions/2010/1216_census_frey.aspx

Frey, W. H. (2011b). *Melting pot cities and suburbs: Racial and ethnic change in metro America in the 2000s.* Brookings Institution. Retrieved from http://www.brookings.edu/~/media/research/files/papers/2011/5/04%20census%20ethnicity%20frey/0504_census_ethnicity_frey.pdf

Frey, W. (2011c, May). *Melting pot cities and suburbs: Racial and ethnic change in metro America in the 2000s.* Brookings Institution. Retrieved from https://www.brookings.edu/wp-content/uploads/2016/06/0504_census_ethnicity_frey.pdf

Frey, W. (2015, June). A snapshot of race in America's neighborhoods [Blog]. Brookings Institution. Retrieved from https://www.brookings.edu/blog/the-avenue/2015/06/11/a-snapshot-of-race-in-americas-neighborhoods/

Frey, W. H., & Liaw, K.-L. (1998). The impact of recent immigration on population redistribution in the United States. In J. Smith & B. Edmonston (Eds.), *The immigration debate* (pp. 388–448). Washington, DC: National Academy Press.

Friedan, B. (1963). *The feminine mystique.* New York, NY: Norton.

Friedman, T. (2000). *The Lexus and the olive tree: Understanding globalization.* New York, NY: Anchor.

Friedman, T. (2005). *The world is flat: A brief history of the twenty-first century.* New York, NY: Farrar, Straus and Giroux.

Friedman, R. A., & Currall, S. C. (2003). Conflict escalation: Dispute exacerbating elements of e-mail communication. *Human Relations, 56*(11), 1325–1347.

Fry, R., & Passel, J. S. (2014). *In post-recession era, young adults drive continuing rise in multi-generational living.* Pew Research Center. Retrieved from http://www.pewsocialtrends.org/files/2014/07/ST-2014-07-17-multigen-households-report.pdf

Fryar, C. D., Carroll, M. D., & Ogden, C. L. (2016). *Prevalence of overweight, obesity, and extreme obesity among adults aged 20 and over: United States, 1960–1962 through 2013–2014.* National Center for Health Statistics. Retrieved from http://www.cdc.gov/nchs/data/hestat/obesity_adult_13_14/obesity_adult_13_14.pdf

Fryar, C. D., Ostchega, Y., Hales, C. M., Zhang, G., & Kruszon-Moran, D. (2017). *Hypertension prevalence and control among adults: United States, 2015–2016* (NCHS Data Brief No. 289). Hyattsville, MD: National Center for Health Statistics.

Fulwood, S., III. (2010, September 21). *Race and beyond: Majority-minority conflicts: Nationwide lessons from the D.C. mayoral election.* Center for American Politics. Retrieved from http://www.americanprogress.org/issues/2010/09/rab_092110.html

Furstenberg, F., & Kennedy, S. (2013, April). *The changing transition to adulthood in the U.S.: Trends in demographic role transitions and age norms since 2000.* Paper presented at the annual meeting of Population Association of America, New Orleans, LA.

Furstenberg, F., Kennedy, S., Mcloyd, V. C., Rumbaut, R. G., & Settersten, R. A. (2004). Growing up is harder to do. *Contexts, 3*, 33–41.

Gabbatt, A. (2013, June 26). Edith Windsor and Thea Spyer: A love affair that just kept on and on and on. *Guardian*. Retrieved from http://www.theguardian.com/world/2013/jun/26/edith-windsor-thea-spyer-doma

Gallagher, J. (2013, August 14). Optician's clinic that fits a pocket. *BBC News*. Retrieved from http://www.bbc.co.uk/news/health-22553730

Gallup. (2013a). In U.S., record-high say gay, lesbian relations morally OK. Retrieved from http://www.gallup.com/poll/162689/record-high-say-gay-lesbian-relations-morally.aspx

Gallup. (2013b). Same-sex marriage support solidifies above 50% in U.S. Retrieved from http://www.gallup.com/poll/162398/sex-marriage-support-solidifies-above.aspx

Gallup. (2017). *State of the American workplace.* Retrieved from https://news.gallup.com/reports/199961/7.aspx?g_source=link_wwwv9&g_campaign=item_236282&g_medium=copy

Gamoran, A., Nystrand, M., Berends, M., & LePore, P. C. (1995). An organizational analysis of the effects of ability grouping. *American Educational Research Journal, 32*(4), 687–715.

Gamson, W. A., & Sifry, M. (Eds.). (2013). The #Occupy movement. *Sociological Quarterly 54*(2), 159–334 [special section with 13 articles on OWS]. Retrieved from https://onlinelibrary.wiley.com/toc/15338525/54/2

Gans, D., & Silverstein, M. (2006). Norms of filial responsibility for aging parents across time and generations. *Journal of Marriage and the Family, 68*(4), 961–976.

Garcia, J. R., Reiber, C., Massey, S. G., & Merriwether, A. M. (2012). Sexual hookup culture: A review. *Review of General Psychology, 16*, 161–176.

Gardner, C. (1995). *Passing by: Gender and public harassment.* Berkeley, CA: University of California Press.

Garfinkel, H. (1963). A conception of, and experiments with, "trust" as a condition of stable concerted actions. In O. J. Harvey (Ed.), *Motivation and social interaction.* New York, NY: Ronald Press.

Garland, D. (2010). *Peculiar institution: America's death penalty in an age of abolition.* New York, NY: Oxford University Press.

Gates, G. J. (2013). *LGBT parenting in the United States.* Los Angeles, CA: The Williams Institute. Retrieved from http://williamsinstitute.law.ucla.edu/wp-content/uploads/LGBT-Parenting.pdf

Gatrell, N., & Bos, H. (2010). U.S. longitudinal lesbian family study. *Pediatrics, 126*(1), 28–36.

Gault, M. (2018, July 17). "Star citizen" court documents reveal the messy reality of crowdfunding a $200 million game. *Motherboard*. Retrieved from https://motherboard.vice.com/en_us/article/ne5n7b/star-citizen-court-documents-reveal-the-messy-reality-of-crowdfunding-a-dollar200-million-game

Geertz, C. (1973). *The interpretation of cultures.* New York, NY: Basic Books.

Gelb, I. (1952). *A study of writing.* Chicago, IL: University of Chicago Press.

Genworth. (2011). *Executive summary: Genworth 2011 cost of care survey.* Retrieved from http://www.genworth.com/content/etc/medialib/genworth_v2/pdf/ltc_cost_of_care.Par.85518.File.dat/ExecutiveSummary_gnw.pdf

Genworth. (2017). *2017 cost of care.* Retrieved from https://pro.genworth.com/riiproweb/productinfo/pdf/179703.pdf

Gereffi, G. (1995). Contending paradigms for cross-regional comparison: Development strategies and commodity chains in East Asia and Latin America. In P. H. Smith (Ed.), *Latin America in comparative perspective: New approaches to methods and analysis* (pp. 33–58). Boulder, CO: Westview Press.

Gereffi, G. (1996). Commodity chains and regional divisions of labor in East Asia. *Journal of Asian Business, 12*(1), 75–112.

Gershuny, J., & Miles, I. (1983). *The new service economy: The transformation of employment in industrial societies.* London, UK: Francis Pinter.

Gettleman, J. (2014, August 1). Uganda anti-gay law struck down by court. *New York Times*. Retrieved from http://www.nytimes.com/2014/08/02/world/africa/uganda-anti-gay-law-struck-down-by-court.html

Ghimire, K. B. (2005). *The contemporary global social movements: Emergent proposals, connectivity and development implications.* United Nations Research Institute for Social Development. Retrieved from http://www.unrisd.org/80256B3C005BCCF9/(httpAuxPages)/F0F8C2DF84C2FB2DC1257088002BFBD9/$file/ghimire.pdf

Gibbs, L. (2017, August 16). *The powerful story behind the NCAA's new sexual violence policy.* Think Progress. Retrieved from https://thinkprogress.org/ncaa-sexual-violence-brenda-tracy-fb7283b94a11/

Giddens, A. (1984). *The constitution of society.* Cambridge, UK: Polity Press.

Giddens, A. (1990). *The consequences of modernity.* Cambridge, UK: Polity Press.

Giddens, A. (2009). *The politics of climate change.* Cambridge, UK: Polity Press.

Gilens, M., & Page, B. I. (2014). Testing theories of American politics: Elites, interest groups, and average citizens, *Perspectives on Politics, 12*(3), 564–581.

Giuffre, P. A., & Williams, C. L. (1994). Boundary lines: Labeling sexual harassment in restaurants. *Gender & Society, 8*, 378–401.

Giuletti, C., Tonin, M., & Vlassopoulos, M. (2017). Racial discrimination in local public services: A field experiment in the US. *Journal of the European Economic Association*, 142–154. Retrieved from https://doi.org/10.1093/jeea/jvx045

Glenn, B. (2013, June 13). Physicians' top 5 most-used medical apps for smartphones and tablets. *Medical Economics*. Retrieved from http://medicaleconomics.modernmedicine.com/medical-economics/news/physicians-top-5-most-used-medical-apps-smartphones-and-tablets#sthash.O7srHRzZ.dpuf

Glieck, P. H. (2014). Water, drought, climate change, and conflict in Syria. *Weather, Climate, and Society*, 6(3), 331–340.

Global Workplace Analytics. (2015). *Latest telecommuting statistics*. Retrieved from http://globalworkplaceanalytics.com/telecommuting-statistics

Global Workplace Analytics. (2017). *2017 state of telecommuting in the U.S.* Retrieved from http://globalworkplaceanalytics.com/telecommuting-statistics

Glock, C. (1976). On the origin and evolution of religious groups. In C. Y. Glock & R. N. Bellah (Eds.), *The new religious consciousness*. Berkeley, CA: University of California Press.

Glueck, S. S., & Glueck, E. T. (1956). *Physique and delinquency*. New York, NY: Harper & Row.

Goel, V. (2014, June 29). Facebook tinkers with users' emotions in news feed experiment, stirring outcry. *New York Times*. Retrieved from http://www.nytimes.com/2014/06/30/technology/facebook-tinkers-with-users-emotions-in-news-feed-experiment-stirring-outcry.html

Goffman, E. (1963). *Stigma: Notes on the management of spoiled identity*. Englewood Cliffs, NJ: Prentice Hall.

Goffman, E. (1967). *Interaction ritual*. New York, NY: Doubleday/Anchor.

Goffman, E. (1971). *Relations in public: Microstudies of the public order*. New York, NY: Basic Books.

Goffman, E. (1973). *The presentation of self in everyday life*. New York, NY: Overlook Press.

Goffman, E. (1981). *Forms of talk*. Philadelphia, PA: University of Pennsylvania Press.

Gold, T. (1986). *State and society in the Taiwan miracle*. Armonk, NY: M. E. Sharpe.

Goldberg, R. T. (1994). Childhood abuse, depression, and chronic pain. *Clinical Journal of Pain*, 10(4), 277–281.

Goldberg, C. (1997, January 30). Hispanic households struggle amid broad decline in income. *New York Times*, pp. A1, A16.

Goldberg, A. E., Smith, J. Z., & Perry-Jenkins, M. (2012). The division of labor in lesbian, gay, and heterosexual new adoptive parents. *Journal of Marriage and Family*, 74, 812–828.

Goldscheider, F. (1990). The aging of the gender revolution: What do we know and what do we need to know? *Research on Aging*, 12(4), 531–545.

Goldscheider, F., & Goldscheider, C. (1999). *The changing transition to adulthood: Leaving and returning home*. Thousand Oaks, CA: Sage.

Goldstein, J. R., & Kenney, C. T. (2001). Marriage delayed or marriage forgone? New cohort forecasts of first marriage for U.S. women. *American Sociological Review*, 66, 506–519.

Goldstein, S., & Goldstein, A. (1996). *Jews on the move: Implications for Jewish identity*. Albany, NY: State University of New York Press.

Goldstone, J. A. (2002). Population and security: How demographic change can lead to violent conflict. *Journal of International Affairs*, 56(1), 3–21.

Gonzales, M. G., & Delgado, R. (2006). *Politics of fear: How republicans use money, race and the media to win*. Abingdon, UK: Routledge.

Gonzalez-Barrera, A. (2015). More Mexicans leaving than coming to the U.S. Pew Research Center. Retrieved from http://www.pewhispanic.org/2015/11/19/more-mexicans-leaving-than-coming-to-the-u-s/

Goode, W. (1963). *World revolution in family patterns*. New York, NY: Free Press.

Goodwin, P. Y., Mosher, W. D., & Chandra, A. (2010). Marriage and cohabitation in the United States: A statistical portrait based on cycle 6 (2002) of the National Survey of Family Growth. *Vital Health Statistics*, 23(28), 1–45.

Gottfredson, M. R., & Hirschi, T. (1990). *A general theory of crime*. Stanford, CA: Stanford University Press.

Gottfried, J., Barthel, M., & Mitchell, A. (2017, Jan. 18). *Trump, Clinton voters divided in their main source for election news*. Pew Research Center. Retrieved from http://www.journalism.org/wp-content/uploads/sites/8/2017/01/PJ_2017.01.18_Election-News-Sources_FINAL.pdf

Goudreau, J. (2012, May 21). A new obstacle for professional women: The glass escalator. *Forbes*. Retrieved from https://www.forbes.com/sites/jennagoudreau/2012/05/21/a-new-obstacle-for-professional-women-the-glass-escalator/#3607e871159d

Gove, W. R., Hughes, M., & Geerken, M. R. (1980). Playing dumb: A form of impression management with undesirable side effects. *Social Psychology Quarterly*, 43, 89–102.

Goyette, B. (2013). Cheerios commercial featuring mixed race family gets racist backlash. *Huffington Post*. Retrieved from http://www.huffingtonpost.com/2013/05/31/cheerios-commercial-racist-backlash_n_3363507.html

Grabe, S., Ward, L. M., & Hyde, J. S. (2008). The role of the media in body image concerns among women: A meta-analysis of experimental and correlational studies. *Psychological Bulletin*, 134, 460–476.

Graham, P. A. (1974). *Community and class in American education, 1865–1918*. New York, NY: Wiley.

Gramlich, J. (2016, December 28). *U.S. ends year with fewest executions since 1991*. Pew Research Center. Retrieved from http://www.pewresearch.org/fact-tank/2016/12/28/u-s-ends-year-with-fewest-executions-since-1991/

Granic, I., Lobel, A., & Engels, R. C. (2014). The benefits of playing video games. *American Psychologist*, 69(1), 66–78.

Granovetter, M. (1973). The strength of weak ties. *American Journal of Sociology*, 78(6), 1360–1380.

Grant Thornton. (2018). *Women in business: Beyond policy to practice*. Retrieved from https://www.grantthornton.global/globalassets/1.-member-firms/global/insights/women-in-business/grant-thornton-women-in-business-2018-report.pdf

Gray, J. (1998). Ethnographic atlas codebook. *World Cultures*, 10(1), 86–136.

Gray, J. (2003). *Al Qaeda and what it means to be modern*. Chatham, UK: Faber & Faber.

Green, F. (1987). *The "sissy boy" syndrome and the development of homosexuality*. New Haven, CT: Yale University Press.

Green, J. (2004). *The American religious landscape and political attitudes: A baseline for 2004*. Pew Forum on Religion & Public Life. Retrieved from http://pewforum.org/publications/surveys/green-full.pdf

Greenberg, G. (2010, December 27). Inside the battle to define mental illness. *Wired*. Retrieved from http://www.wired.com/2010/12/ff_dsmv/

Greenfield, E. A., Lee, C., Friedman, E. L., & Springer, K. W. (2011). Childhood abuse as a risk factor for sleep problems in adulthood: Evidence from a U.S. national study. *Annals of Behavioral Medicine*, 42(2), 245–256.

Greenwood, S., Perrin, A., & Duggan, M. (2016, November 11). *Social media update 2016*. Pew Research Center. Retrieved from http://www.pewinternet.org/2016/11/11/social-media-update-2016/

Griswold, A. (2014, September 10). The American concept of "prestige" has barely changed in 37 years [Blog]. *Slate*. Retrieved from http://www.slate.com/blogs/moneybox/2014/09/10/most_prestigious_jobs_in_america_the_short_list_has_barely_changed_in_37.html

Grossman A. H., & D'Augelli, A. R. (2007). Transgender youth and life-threatening behaviors. *Suicide & Life-Threatening Behavior*, 37, 527–537.

Grossman, S. F., & Lundy, M. (2007). Domestic violence across race and ethnicity: Implications for social work practice and policy. *Violence Against Women*, 13(10), 1029–1052.

Grusky, D., & Szelényi, S. (2011). *The inequality reader: Contemporary and foundational readings on race, class, and gender*. Boulder, CO: Westview Press.

Guarino, D. P. (2013, April 10). Obama seeks boost in DOE nuclear weapons spending, cut to nonproliferation. *Global Security Newswire*. Retrieved from http://www.nti.org/gsn/article/obama-seeks-boost-doe-nuclear-weapons-spending-cut-nonproliferation/

Gudmundsen, J. (2017, June 18). These kids' apps highlight diversity. *USA Today*. Retrieved from https://www.usatoday.com/story/tech/columnist/2017/06/18/these-kids-apps-highlight-diversity/102858966/

Haas, A. P., Rodgers, P. L., & Herman, J. L. (2014). *Suicide attempts among transgender and gender nonconforming adults: Findings of the National Transgender Discrimination Survey*. American Foundation for Suicide Prevention. Retrieved from http://williamsinstitute.law.ucla.edu/

wp-content/uploads/AFSP-Williams-Suicide-Report-Final.pdf

Hadden, J. (1997a). The concepts "cult" and "sect" in scholarly research and public discourse. Retrieved from http://religiousmovements.lib.virginia.edu/cultsect/concult.htm

Hadden, J. (1997b). New religious movements mission statement. Retrieved from http://religiousmovements.lib.virginia.edu/welcome/mission.htm

Haggard, S. (1990). Pathways from the periphery: The politics of growth in newly industrializing countries. Ithaca, NY: Cornell University Press.

Haig, M. (2011). Brand failures: The truth about the 100 biggest branding mistakes of all times (2nd ed.). London, UK: Kogan.

Hajnal, Z., Lajevardi, N., & Nielson, L. (2017). Voter identification laws and the suppression of minority votes. The Journal of Politics, 79: 363–379.

Hales, C. M., Carroll, M. D., Fryar, C., & Ogden, C. (2017). Prevalence of obesity among adults and youth: United States, 2015–2016. Retrieved from https://www.cdc.gov/nchs/data/databriefs/db288.pdf

Hall, E. (1969). The hidden dimension. New York, NY: Doubleday.

Hall, E. (1973). The silent language. New York, NY: Doubleday.

Hall, S. (1992). The question of cultural identity. In S. Hall, D. Held, & A. McGrew (Eds.), Modernity and its futures (pp. 273–326). Cambridge, UK: Polity Press.

Hall, A. K., Stellefson M., & Bernhardt, J. M. (2012). Healthy aging 2.0: The potential of new media and technology (Preventing Chronic Disease, Vol. 9). Retrieved from http://www.cdc.gov/pcd/issues/2012/pdf/11_0241.pdf

Hamamura, T. (2012). Are cultures becoming individualistic? A cross-temporal comparison of individualism–collectivism in the United States and Japan. Personality and Social Psychology Review, 16(1), 3–24.

Hamilton, L., & Armstrong, E. A. (2009). Double binds and flawed options: Gendered sexuality in early adulthood. Gender & Sexuality, 23, 589–616.

Hamilton, B. E., Martin, J. A., & Ventura, S. J. (2012). Births: Preliminary data for 2011. National Vital Statistics Reports, 61(5), 1–20. Retrieved from http://www.cdc.gov/nchs/data/nvsr/nvsr61/nvsr61_05.pdf

Hamilton, B. E., Martin, J., Osterman, M., Curtin, S., & Mathews, T. J. (2015). Births: Final data for 2014. National Vital Statistics Reports, 64(12), 1–63. Retrieved from https://www.cdc.gov/nchs/data/nvsr/nvsr64/nvsr64_12.pdf

Hammond, P. (1992). Religion and personal autonomy: The third disestablishment in America. Columbia, SC: University of South Carolina Press.

Hampton, K. N., Sessions Goulet, L., Rainie, L., & Purcell, K. (2011). Social networking sites and our lives. Washington, DC: Pew Internet and American Life Project.

Hardin, G. (1968). The tragedy of the commons. Science, 162, 1243–1248.

Hare, A. P., Borgatta, E. F., & Bales, R. F. (1965). Small groups: Studies in social interaction. New York, NY: Knopf.

Harknett, K., & McLanahan, S. S. (2004). Racial and ethnic differences in marriage after the birth of a child. American Sociological Review, 69, 790–811.

Harris, M. (1975). Cows, pigs, wars, and riches: The riddles of culture. New York, NY: Random House.

Harris, M. (1978). Cannibals and kings: The origins of cultures. New York, NY: Random House.

Harris, M. (1980). Cultural materialism: The struggle for a science of culture. New York, NY: Vintage Books.

Harris, J. (1998). The nurture assumption: Why children turn out the way they do. New York, NY: Free Press.

Harris, D. (2003). Racial classification and the 2000 census. Commissioned paper, Panel to Review the 2000 Census. Ann Arbor, MI: Committee on National Statistics, University of Michigan.

Harris, J. (2008). Review: A theory of global capitalism: Production, class, and state in a transnational world by William I. Robinson. Science & Society, 72, 113–115.

Harris, E. (2017, August 29). Family of boy who wears dresses sues Education Department. New York Times. Retrieved from https://www.nytimes.com/2017/08/29/nyregion/family-of-boy-who-wears-dresses-sues-education-department.html

Harris, D., & Sim, J. (2000). An empirical look at the social construction of race: The case of mixed-race adolescents (Population Studies Center Research Report 00-452). Ann Arbor, MI: University of Michigan.

Harris Poll. (2014, September 10). Doctors, military officers, firefighters, and scientists seen as America's most prestigious occupations. Retrieved from http://www.theharrispoll.com/politics/Doctors_Military_Officers_Firefighters_and_Scientists_Seen_as_Among_America_s_Most_Prestigious_Occupations.html

Hartig, T., Johansson, G., & Kylin, C. (2003). Residence in the social ecology of stress and restoration. Journal of Social Issues, 59(3), 611–636.

Harvard University Joint Center for Housing Studies. (2014, June 26). State of the nation's housing 2014: Key facts. Retrieved from http://www.jchs.harvard.edu/sites/jchs.harvard.edu/files/son_2014_key_facts.pdf

Harvey, D. (1973). Social justice and the city. Oxford, UK: Blackwell.

Harvey, D. (1982). The limits to capital. Oxford, UK: Blackwell.

Harvey, D. (1985). Consciousness and the urban experience: Studies in the history and theory of capitalist urbanization. Oxford, UK: Blackwell.

Harvey, D. (1989). The condition of postmodernity: An enquiry into the origins of cultural change. Cambridge, MA: Blackwell.

Haskell, W. (2014, April 28). A gossip app brought my high school to a halt. New York Magazine. Retrieved from http://nymag.com/thecut/2014/04/gossip-app-brought-my-high-school-to-a-halt.html

Haslam, D. W., & James, W. P. (2005). Obesity. Lancet, 366(9492), 1197–1209.

Haslett, B., Geis, F. L., & Carter, M. R. (1992). The organizational woman: Power and paradox. Norwood, NJ: Ablex.

Hatch, A. (2015). Saying "I don't" to matrimony: An investigation of why long-term heterosexual cohabitors choose not to marry. Journal of Family Issues, 38(12), 1651–1674.

Hathaway, A. D. (1997). Marijuana and tolerance: Revisiting Becker's sources of control. Deviant Behavior, 18(2), 103–124.

Hatuka, T., & Toch, E. (2016). The emergence of portable private-personal territory: Smartphones, social conduct, and public spaces. Urban Studies, 53, 2192–2208. doi:10.1177/0042098014524608

Haugen, E. (1977). Linguistic relativity: Myths and methods. In W. C. McCormack & S. A. Wurm (Eds.), Language and thought: Anthropological issues (pp. 11–28). The Hague, Netherlands: Mouton.

Hauser, R. M. (1980). Some exploratory methods for modeling mobility tables and other cross-classified data. Sociological Methodology, 11: 413–458.

Hawkes, T. (1977). Structuralism and semiotics. Berkeley, CA: University of California Press.

Hawley, A. (1950). Human ecology: A theory of community structure. New York, NY: Ronald Press.

Hawley, A. (1968). Human ecology. In International encyclopedia of social science (Vol. 4), 328–337. New York, NY: Free Press.

HBO. (2015). Heroin: Cape Cod, USA. Retrieved from https://www.hbo.com/documentaries/heroin-cape-cod-usa/heroin-cape-cod-usa

Healy, M. (2001, May 21). Pieces of the puzzle. Los Angeles Times. Retrieved from http://pqasb.pqarchiver.com/latimes/results.html?RQT=511&sid=1&firstIndex=460&PQACnt=1

Heffernan, M. (2013, August 7). What happened after the Foxconn suicides. CBS Interactive: MoneyWatch. Retrieved from https://www.cbsnews.com/news/what-happened-after-the-foxconn-suicides/

Heggen, K., & Terum, L. I. (2017). The impact of education on professional identity. In B. Blum, L. Evertsson, & M. Perlinski (Eds.), Social and caring professions in European welfare states (pp. 21–35). Bristol, UK: Policy Press.

Heine, F. (2013, August 13). M, F or blank: "Third gender" official in Germany from November. Der Spiegel. Retrieved from http://www.spiegel.de/international/germany/third-gender-option-to-become-available-on-german-birth-certificates-a-916940.html

Held, D., McGrew, A., Goldblatt, D., & Perraton, J. (1999). Global transformations: Politics, economics, and culture. Cambridge, UK: Polity Press.

Helm, L. (1992, November 21). Debt puts squeeze on Japanese. Los Angeles Times. Retrieved from http://articles.latimes.com/1992-11-21/news/mn-733_1_debt-problem

Hemez, P. (2017). Divorce rate in the U.S.: Geographic variation, 2016 (Family Profiles No. FP-17-24). Bowling Green, OH: National Center for Family & Marriage Research.

Henderson, D. B. (1989). The origin of strategy." Harvard Business Review. Retrieved from https://hbr.org/1989/11/the-origin-of-strategy

Henderson, J., & Appelbaum, R. P. (1992). Situating the state in the Asian development process. In R. P. Appelbaum & J. Henderson (Eds.), States and development in the Asian Pacific rim (pp. 1–26). Newbury Park, CA: Sage.

Henderson, V., & Kelly, B. (2005). Food advertising in the age of obesity: Content analysis of food advertising on general market and African American television. *Journal of Nutrition Education and Behavior, 37*, 191–196.

Hendricks, J. (1992). Generation and the generation of theory in social gerontology. *International Journal of Aging and Human Development, 35*(1), 31–47.

Hendricks, J., & Hatch, L. R. (1993). Federal policy and family life of older Americans. In J. Hendricks & C. J. Rosenthal (Eds.), *The remainder of their days: Impact of public policy on older families.* New York, NY: Greenwood.

Hendricks, J., & Hendricks, C. D. (1986). *Aging in mass society: Myths and realities.* Boston, MA: Little, Brown.

Henry, W. (1965). *Growing older: The process of disengagement.* New York, NY: Basic Books.

Hentges, B., & Case, K. (2013). Gender representations on Disney Channel, Cartoon Network, and Nickelodeon broadcasts in the United States. *Journal of Children and Media, 7*, 319–333.

Herdt, G. (1981). *Guardians of the flutes: Idioms of masculinity.* New York, NY: McGraw-Hill.

Herdt, G. (1984). *Ritualized homosexuality in Melanesia.* Berkeley, CA: University of California Press.

Herdt, G. (1986). *The Sambia: Ritual and gender in New Guinea.* New York, NY: Holt, Rinehart and Winston.

Herdt, G., & Davidson, J. (1988). The Sambia "turnim-man": Sociocultural and clinical aspects of gender formation in male pseudohermaphrodites with 5-alpha-reductase deficiency in Papua, New Guinea. *Archives of Sexual Behavior, 17*(1), 33–56.

Herek, G. M. (2004). Beyond "homophobia": Thinking about sexual prejudice and stigma in the twenty-first century. *Sexuality Research & Social Policy, 1*(2), 6–24.

Heritage, J. (1985). *Garfinkel and ethnomethodology.* New York, NY: Basil Blackwell.

Hernández, J. C. (2017, October 31). Xi Jinping vows no poverty in China by 2020. That could be hard. *New York Times.* Retrieved from https://www.nytimes.com/2017/10/31/world/asia/xi-jinping-poverty-china.html

Hernández, J. C. (2018, January 13). "Frost Boy" in China warms up the Internet, and stirs poverty debate. *New York Times.* Retrieved from https://www.nytimes.com/2018/01/13/world/asia/frozen-boy-china-poverty.html

Herrnstein, R. J., & Murray, C. (1994). *The bell curve: Intelligence and class structure in American Life.* New York, NY: Free Press.

Hershbein, B., & Kearney, M. S. (2014). *Major decisions: What graduates earn over their lifetimes.* Brookings Institution. Retrieved from http://www.hamiltonproject.org/papers/major_decisions_what_graduates_earn_over_their_lifetimes/

Herszenhorn, D. (2014, January 22). Unrest deepens in Ukraine as protests turn deadly. *New York Times.* Retrieved from http://www.nytimes.com/2014/01/23/world/europe/ukraine-protests.html

Hertz, T. (2006, April 26). *Understanding mobility in America.* Center for American Progress. Retrieved from http://cdn.americanprogress.org/wp-content/uploads/issues/2006/04/Hertz_MobilityAnalysis.pdf

Hess, A. (2015, October 28). Don't ban Yik Yak. *Slate.* Retrieved from http://www.slate.com/articles/technology/users/2015/10/yik_yak_is_good_for_university_students.html

Hexham, I., & Poewe, K. (1997). *New religions as global cultures.* Boulder, CO: Westview Press.

Higher Education Research Institute. (2012). *The American freshman: National norms fall 2012.* Los Angeles, CA: Higher Education Research Institute, UCLA.

Highton, B. (2017). Voter identification laws and turnout in the United States. *Annual Review of Political Science, 20*: 149–167.

Himes, C. (1999). Racial differences in education, obesity, and health in later life. *Annals of the New York Academy of Sciences, 896*(1), 370–372.

Hines, D. A., Malley-Morrison, K., & Dutton, L. B. (2012). *Family violence in the United States: Defining, understanding, and combating abuse.* New York, NY: Sage.

Hinrichs, P. (2014). Affirmative action bans and college graduation rates. *Economics of Education Review, 42*, 43–52. doi:10.1016/j.econedurev.2014.06.005

Hirsch, B. T., & Macpherson, D. A. (2004). Wages, sorting on skill, and the racial composition of jobs. *Journal of Labor Economics, 22*(1), 189–210.

Hirschi, T. (1969). *Causes of delinquency.* Berkeley, CA: University of California Press.

Hirst, P. (1997). The global economy: Myths and realities. *International Affairs, 73*(3), 409–425.

Hirst, P., & Thompson, G. (1992). The problem of "globalization": International economic relations, national economic management, and the formation of trading blocs. *Economy and Society, 24*(4), 357–396.

Hirst, P., & Thompson, G. (1999). *Globalization in question: The international economy and the possibilities of governance* (Rev. ed.). Cambridge, UK: Polity Press.

Hochschild, A. (1975). Disengagement theory: A critique and proposal. *American Sociological Review, 40*(5), 553–569.

Hochschild, A. R. (1983). *The managed heart: Commercialization of human feeling.* Berkeley, CA: University of California Press.

Hochschild, A., & Machung, A. (1989). *The second shift: Working parents and the revolution at home.* New York, NY: Viking.

Hoffman, J. (2010, December 4). As bullies go digital, parents play catch up. *New York Times.* Retrieved from http://www.nytimes.com/2010/12/05/us/05bully.html?pagewanted=all

Hoffman, K. M., Trawalter, S., Axt, J. R., & Oliver, M. N. (2016). Racial bias in pain assessment and treatment recommendations, and false beliefs about biological differences between blacks and whites. *Proceedings of the National Academy of Sciences USA, 113*, 4296–4301.

Hofstede, G. (1997). *Cultures and organizations: Software of the mind.* New York, NY: McGraw-Hill.

Hollaback! & ILR School. (2015, June 1). ILR and Hollaback! release largest analysis of street harassment to date. ILR Worker Institute. Retrieved from https://www.ilr.cornell.edu/worker-institute/news/ilr-and-hollaback-release-largest-analysis-street-harassment-date

Holton, R. (1978). The crowd in history: Some problems of theory and method. *Social History, 3*(2), 219–233.

Homans, G. (1950). *The human group.* New York, NY: Harcourt, Brace.

Hopkins, T. K., & Wallerstein, I. M. (1996). *The age of transition: Trajectory of the world-system, 1945–2025.* London, UK: Zed Books.

Hopkinson, N. (2011, December 4). Why school choice fails. *New York Times.* Retrieved from https://www.nytimes.com/2011/12/05/opinion/why-school-choice-fails.html

Horwitz, A. (2013). *Anxiety: A short history* (Johns Hopkins Biographies of Disease). Baltimore, MD: Johns Hopkins University Press.

Horwitz. A., & Wakefield, J. (2007). *The loss of sadness: How psychiatry has transformed normal sadness into depressive disorder.* New York, NY: Oxford University Press.

Howard, P. (2011, February 23). The Arab Spring's cascading effects. *Miller-McCune.* Retrieved from http://www.miller-mccune.com/politics/the-cascading-effects-of-the-arab-spring-28575/

Hsiang, S., Kopp, R., Jina, A., Rising, J. Delgado, M., Mohan S., . . . Houser, T. (2017). Estimating economic damage from climate change in the United States. Science, 365, 1362–1369.

Huang, P. M., Smock, P. J., Manning, W. D., & Bergstrom-Lynch, C. A. (2011). He says, she says: Gender and cohabitation. *Journal of Family Issues, 32*(7), 876–905.

Hudson, J. I., Hiripi, E., Pope, H. G., & Kessler, R. C. (2007). The prevalence and correlates of eating disorders in the National Comorbidity Survey Replication. *Biological Psychiatry, 61*, 348–358.

Huffington Post. (2012, June 21). Social media by gender: Women dominate Pinterest, Twitter, men dominate Reddit, YouTube (Infographic). *Huffington Post.* Retrieved from http://www.huffingtonpost.com/2012/06/20/social-media-by-gender-women-pinterest-men-reddit-infographic_n_1613812.html

Hughes, M. E., Waite, L. J., LaPierre, T. A., & Luo, Y. (2007). All in the family: The impact of caring for grandchildren on grandparents' health. *Journals of Gerontology. Series B, Psychological Sciences and Social Sciences, 62*(2), S108–S119.

Human Rights Watch. (1995). *The global report on women's human rights.* Retrieved from http://www.hrw.org/about/projects/womrep/

Hume, T. (2016, January 5). Deaths of migrants on the Mediterranean hit record numbers in 2015. *CNN.* Retrieved from http://www.cnn.com/2016/01/05/europe/mediterranean-record-migrants-deaths/

Humes, K. R., Jones, N. A., & Ramirez, R. R. (2010, March). *Overview of race and Hispanic origin: 2010.* U.S. Bureau of the Census. Retrieved from http://www.census.gov/prod/cen2010/ briefs/c2010br-02.pdf

Humphreys, L. (1970). *Tearoom trade: Impersonal sex in public places*. Chicago, IL: Aldine.

Hunt, K. S., & Dumville, R. (2016). *Recidivism among federal offenders: A comprehensive overview*. United States Sentencing Commission. Retrieved from https://www.ussc.gov/research/research-reports/recidivism-among-federal-offenders-comprehensive-overview

Hursh, D. (2007). Assessing No Child Left Behind and the rise of neoliberal education policies. *American Educational Research Journal, 44*, 493–518.

Hurtado, A. (1995). Variation, combinations, and evolutions: Latino families in the United States. In R. Zambrana (Ed.), *Understanding Latino families* (pp. 40–61). Thousand Oaks, CA: Sage.

Hyman, R. (1984). *Strikes* (2nd ed.). London, UK: Fontana.

Hyman, H. H., & Singer, E. (1968). *Readings in reference group theory and research*. New York, NY: Free Press.

Illegems, V., & Verbeke, A. (2004). Telework: What does it mean for management? *Long Range Planning, 37*(4), 319–334.

Illich, I. (1983). *Deschooling society*. New York, NY: Harper & Row.

Interactive Diversity Solutions. (2018). *Who am I? Race awareness game*. Retrieved from https://www.interactivediversitysolutions.com/

Intergovernmental Panel on Climate Change. (2007). *Climate change 2007: Synthesis report, summary for policymakers*. Retrieved from http://www.ipcc.ch/pdf/assessment-report/ar4/syr/ar4_syr_spm.pdf

Intergovernmental Panel on Climate Change. (2015). *Climate change 2014: Synthesis report*. Retrieved from http://www.ipcc.ch/pdf/assessment-report/ar5/syr/SYR_AR5_FINAL_full_wcover.pdf

International Campaign to Abolish Nuclear Weapons. (2018). *Campaign overview*. Retrieved from http://www.icanw.org/campaign/campaign-overview/

International Centre for Prison Studies. (2016). *World prison population list*. Retrieved from http://www.prisonstudies.org/sites/default/files/resources/downloads/world_prison_population_list_11th_edition.pdf

International Energy Agency. (2016, March 16). Decoupling of global emissions and economic growth confirmed. Retrieved from http://www.iea.org/newsroomandevents/pressreleases/2016/march/decoupling-of-global-emissions-and-economic-growth-confirmed.html?referrer=justicewire

International Institute for Democracy and Electoral Assistance (IIDEA). (2004). *Voter turnout in Western Europe since 1945*. Stockholm, Sweden: Author.

International Institute for Democracy and Electoral Assistance. (2009). *Voter turnout 2009*. Retrieved from http://www.idea.int/vt/view_data.cfm

International Institute for Democracy and Electoral Assistance. (2014). *Voter turnout database*. Retrieved from http://www.idea.int/vt

International Institute for Democracy and Electoral Assistance. (2018). *Voter turnout database*. Retrieved from https://www.idea.int/data-tools/data/voter-turnout

International Labour Organization. (2004a). *Breaking the glass ceiling: Women in management*. Retrieved from http://www.ilo.org/dyn/gender/docs/RES/292/F267981337/BreakingGlassPDFEnglish.pdf

International Labour Organization. (2004b). More women are entering the global labor force than ever before, but job equality, poverty reduction remain elusive. Retrieved from http://www.ilo.org/global/about-the-ilo/newsroom/news/WCMS_005243/lang—en/index.htm

International Labour Organization. (2011). *Female future: Turning the tide*. Retrieved from http://www.ilo.org/global/publications/magazines-and-journals/world-of-work-magazine/articles/WCMS_165993/lang—en/index.htm

International Labour Organization. (2013). *Global child labour trends 2008 to 2012*. Retrieved from http://www.ilo.org/ipec/Informationresources/WCMS_IPEC_PUB_23015/lang—en/index.htm

International Labour Organization. (2014). *Global employment trends for women: 2012*. Retrieved from http://www.ilo.org/wcmsp5/groups/public/—dgreports/—dcomm/documents/publication/wcms_195447.pdf

International Labour Organization. (2014a). *Maternity and paternity at work: Law and practice across the world*. Retrieved from https://www.ilo.org/global/topics/equality-and-discrimination/maternity-protection/publications/maternity-paternity-at-work-2014/lang--en/index.htm

International Labour Organization. (2016). *Women at Work: Trends 2016*. Geneva, Switzerland: International Labour Office.

International Labour Organization. (2017). *Global estimates of child labour: Results and trends, 2012–2016*. Retrieved from http://www.ilo.org/wcmsp5/groups/public/—dgreports/—dcomm/documents/publication/wcms_575499.pdf

International Labour Organization. (2018). *Women and men in the informal economy: A statistical picture* (3rd ed.). Retrieved from http://www.ilo.org/wcmsp5/groups/public/—dgreports/—dcomm/documents/publication/wcms_626831.pdf

International Monetary Fund. (2013). World economic outlook database. Retrieved from http://www.imf.org/external/pubs/ft/weo/2013/02/weodata/index.aspx

International Monetary Fund. (2018, March 22). *Macroeconomic developments and prospects in low-income developing countries*. Retrieved from https://www.imf.org/~/media/Files/Publications/PP/2018/pp021518-macroeconomic-developments-and-prospects-in-low-income-developing-countries.ashx

International Telecommunication Union. (2016). *ICT facts and figures 2016*. Retrieved from http://www.itu.int/en/ITU-D/Statistics/Pages/facts/default.aspx

International Telecommunication Union. (2017). *ICT facts and figures 2017: Global ICT development*. Retrieved from https://www.itu.int/en/ITU-D/Statistics/Pages/facts/default.aspx

International Telework Association and Council. (2004). *Telework facts and figures*. Retrieved from http://www.telecommute.org/ resources/abouttelework.htm

International Union for Conservation of Nature. (2017, March). *The IUCN red list of threatened species: Summary statistics*. Retrieved from http://www.iucnredlist.org/about/summary-statistics#Tables_1_2

Internet World Stats. (2012a). *Internet users—Top 20 countries*. Retrieved from http://www.internetworldstats.com/top20.htm

Internet World Stats. (2012b). *North American Internet usage statistics, population and telecommunications reports*. Retrieved from http://www.internetworldstats.com/stats14.htm

Internet World Stats. (2013). *Internet usage statistics*. Retrieved from http://www.internetworldstats.com/stats.htm

Internet World Stats. (2016a). *Internet usage statistics*. Retrieved from http://www.internetworldstats.com/stats.htm

Internet World Stats. (2016b). *Internet world users by language*. Retrieved from http://www.internetworldstats.com/stats7.htm

Internet World Stats. (2017a). *Internet users in the world by regions: December 31, 2017*. Retrieved from https://www.internetworldstats.com/stats.htm

Internet World Stats. (2017b). *Internet world users by language: Top 10 languages*. Retrieved from https://www.internetworldstats.com/stats7.htm

Inter-Parliamentary Union. (2016). *Women in national parliaments*. Retrieved from http://www.ipu.org/wmn-e/world.htm

Inter-Parliamentary Union. (2018). *Women in national parliaments*. Retrieved from http://archive.ipu.org/wmn-e/world.htm

Isbell, D. S. (2008). Musicians and teachers: The socialization and occupational identity of preservice music teachers. *Journal of Research in Music Education, 56*, 162–178.

Ivanski, C., & Kohut, T. (2017). Exploring definitions of sex positivity through thematic analysis. *Canadian Journal of Human Sexuality, 26*, 216–225.

Jacobs, J. (1961). *The death and life of great American cities*. New York, NY: Random House.

Jacobson, C. K., & Heaton, T. B. (1991). Voluntary childlessness among American men and women in the late 1980s. *Social Biology, 38*(1–2), 79–93.

Jaher, F. C. (Ed.). (1973). *The rich, the well born, and the powerful*. Urbana, IL: University of Illinois Press.

Jamal, A., Phillips, E., Gentzke, A. S., Homa, D. M., Babb, S. D., King, B. A., & Neff, L. J. (2018). Current cigarette smoking among adults—United States, 2016. *MMWR Morbidity and Mortality Weekly Report, 67*(2), 53–59.

Janis, I. (1972). *Victims of groupthink*. Boston, MA: Houghton Mifflin.

Janis, I. (1989). *Crucial decisions: Leadership in policy making and crisis management*. New York, NY: Free Press.

Janis, I., & Mann, L. (1977). *Decision making: A psychological analysis of conflict, choice, and commitment*. New York, NY: Free Press.

Jankowiak, W., & Fisher, E. 1992). Romantic love: A cross-cultural perspective. *Ethnology*, 149–156.

Jencks, C., Smith, M., Acland, H., Bane, M. J., Cohen, D., Gintis, H., . . . Michelson, S. (1972). *Inequality: A reassessment of the effects of family and school in America*. New York, NY: Basic Books.

Jerolmack, C. (2013). *The global pigeon*. Chicago, IL: University of Chicago Press.

Jerryson, M., & Juergensmeyer, M. (2010). *Buddhist warfare*. New York, NY: Oxford University Press.

Jewkes, R. (2002). Intimate partner violence: Causes and prevention. *Lancet*, 359(9315), 1423–1429.

Jeynes, W. H. (2012). A meta-analysis on the effects and contributions of public, public charter, and religious schools on student outcomes. *Peabody Journal of Education*, 87, 305–335.

Jiménez, T. R. (2010). *Replenished ethnicity: Mexican Americans, immigration, and identity*. Berkeley, CA: University of California Press.

Jobling, R. (1988). The experience of psoriasis under treatment. In M. Bury & R. Anderson (Eds.), *Living with chronic illness: The experience of patients and their families* (pp. 225–244). London, UK: Unwin Hyman.

Johnson, K. (2006). *Demographic trends in rural and small town America*. Carsey Institute Reports on Rural America. Retrieved from http://scholars.unh.edu/cgi/viewcontent.cgi?article=1004&context=carsey

Johnson, M. (1995). Patriarchal terrorism and common couple violence: Two forms of violence against women in U.S. families. *Journal of Marriage and the Family*, 57(2), 283–294.

Johnson, M., & Morton, J. (1991). *Biology and cognitive development: The case of face recognition*. Oxford, UK: Blackwell.

Johnson-Odim, C. (1991). Common themes, different contexts: Third world women and feminism. In C. T. Mohanty, A. Russo, & L. Torres (Eds.), *Third world women and the politics of feminism* (pp. 314–327). Bloomington, IN: Indiana University Press.

Joint Center for Housing Studies of Harvard University. (2011). *Rental market stresses: Impacts of the Great Recession on affordability and multifamily lending*. Retrieved from http://www.jchs.harvard.edu/publications/rental/jchs_what_works_rental_market_stresses.pdf

Joint Center for Housing Studies of Harvard University. (2017). *The state of the nation's housing*. Retrieved from http://www.jchs.harvard.edu/sites/default/files/harvard_jchs_state_of_the_nations_housing_2017_0.pdf

Joint Center for Political and Economic Studies. (2000). *Politics: Black elected officials*. Retrieved from http://www.pbs.org/fmc/book/10politics4.htm

Joint Center for Political and Economic Studies. (2011). *National roster of black elected officials*. Retrieved from http://www.jointcenter.org/research/national-roster-of-black-elected-officials

Jones, J. (1986). *Labor of love, labor of sorrow: Black women, work, and the family from slavery to the present*. New York, NY: Random House.

Jones, S. (1995). Understanding community in the information age. In S. G. Jones (Ed.), *CyberSociety: Computer-mediated communication and community* (pp. 10–35). Thousand Oaks, CA: Sage.

Jones, N. (2009). "I was aggressive for the streets, pretty for the pictures": Gender, difference, and the inner-city girl. *Gender and Society*, 23, 89–93.

Jones, J. (2018, January 8). Americans' identification as independents back up in 2017. *Gallup.*, Retrieved from http://news.gallup.com/poll/225056/americans-identification-independents-back-2017.aspx?g_source=link_NEWSV9&g_medium=tile_7&g_campaign=item_15370&g_content=Americans%27%2520Identification%2520as%2520Independents%2520Back%2520Up%2520in%25202017

Jones, N. A. & Bullock, J. (2012). *The two or more races population: 2010*. U. S. Census Bureau. Retrieved from https://www.census.gov/prod/cen2010/briefs/c2010br-13.pdf

Jones, R., & Cox, D. (2017). *America's changing religious identity: Findings from the 2016 American values atlas*. Public Religion Research Institute. Retrieved from https://www.prri.org/wp-content/uploads/2017/09/PRRI-Religion-Report.pdf

Jones, S. M., & Dindia, K. (2004). A meta-analytic perspective on sex equity in the classroom. *Review of Educational Research, 74*, 443–471.

Jones, R. K., & Dreweke, J. (2011). *Countering conventional wisdom: New evidence on religion and contraceptive use*. Guttmacher Institute. Retrieved from http://www.guttmacher.org/pubs/Religion-and-Contraceptive-Use.pdf

Jones, L. M., Mitchell, K. J., & Finkelhor, D. (2012). Trends in youth Internet victimization: Findings from three youth Internet safety surveys 2000–2010. *Journal of Adolescent Health, 50*(2), 179–186.

Journal of Blacks in Higher Education. (2007). Black student college graduation rates inch higher but a large racial gap persists. Retrieved from www.jbhe.com/preview/winter07preview.html

Juergensmeyer, M. (1993). *The new cold war? Religious nationalism confronts the secular state* (Comparative Studies in Religion and Society). Berkeley, CA: University of California Press.

Juergensmeyer, M. (2003). *Terror in the mind of God: The global rise of religious violence*. Berkeley, CA: University of California Press.

Jergensmeyer, M. (2005). *Religion in Global Civil Society*. New York: Oxford University Press.

Juergensmeyer, M. (2009). *Global rebellion: Religious challenges to the secular state, from Christian militias to al Qæda*, Berkeley, CA: University of California Press.

Juergensmeyer, M. (2015, June 20). The myth of the lone wolf terrorist [Blog]. Retrieved from http://juergensmeyer.org/the-myth-of-the-lone-wolf-terrorist/

Juergensmeyer, M. (2017). *Terror in the mind of God: The global rise of religious violence*. Oakland, CA: University of California Press.

Juergensmeyer, M., Griego, D., & Soboslai, J. (2015). *God in the tumult of the global square: Religion in global civil society*. Berkeley, CA: University of California Press.

Kaeble, D., & Cowhig, M. (2018, April). *Correctional populations in the United States*. Bureau of Justice Statistics. Retrieved from https://www.bjs.gov/content/pub/pdf/cpus16.pdf

Kaiser Family Foundation. (2010). *Generation M2: Media in the lives of 8- to 18-year-olds*. Retrieved from http://kaiserfamilyfoundation.files.wordpress.com/2013/01/8010.pdf

Kaiser Family Foundation. (2013). *Focus on health reform: Summary of the Affordable Care Act*. Retrieved from http://kaiserfamilyfoundation.files.wordpress.com/2011/04/8061-021.pdf

Kaiser Family Foundation. (2017, November 29). *Key facts about the uninsured population*. Retrieved from https://www.kff.org/uninsured/fact-sheet/key-facts-about-the-uninsured-population/

Kamp Dush, C. M., Cohan, C. L., & Amato, P. R. (2003). The relationship between cohabitation and marital quality and stability: Change across cohorts? *Journal of Marriage and Family*, 65, 539–549.

Kann, L., McManus, T., Harris, W. A., Shanklin, S. L., Flint, K. H., Hawkins, J., . . . Zaza, S. (2016). Youth risk behavior surveillance—United States, 2015. *MMWR Surveillance Summaries*, 65(6). Retrieved from https://www.cdc.gov/healthyyouth/data/yrbs/pdf/2015/ss6506_updated.pdf

Kann, L. et al. (2018). *Youth risk behavior surveillance-United States, 2017*. Centers for Disease Control and Prevention, 67(8): 1–114. Retrieved from https://www.cdc.gov/mmwr/volumes/67/ss/ss6708a1.htm

Kanter, R. (1977). *Men and women of the corporation*. New York, NY: Basic Books.

Kanter, R. (1983). *The change masters: Innovation for productivity in the American corporation*. New York, NY: Simon & Schuster.

Kanter, R. (1991). The future of bureaucracy and hierarchy in organizational theory. In P. Bourdieu & J. Coleman (Eds.), *Social theory for a changing society* (pp. 63–87). Boulder, CO: Westview.

Karas, D. (2012, June 22). Petition objecting to "whistling" Princeton MarketFair billboard leads to its removal. *The Times*. Retrieved from http://www.nj.com/mercer/index.ssf/2012/06/online_chain_reaction_gets_pri.html

Karas-Montez, J., & Zajacova, A. (2013). Explaining the widening education gap in mortality risk among U.S. white women. *Journal of Health and Social Behavior*, 54(2), 165–181.

Kasarda, J. (1993). Urban industrial transition and the underclass. In W. J. Wilson (Ed.), *The ghetto underclass* (pp. 43–64). Newbury Park, CA: Sage.

Kassie, E. (2015, Jan. 27). Male victims of campus sexual assault speak out: "We're up against a system that's not designed to help us." *Huffington Post*. Retrieved from https://www.huffingtonpost.com/2015/01/27/male-victims-sexual-assault_n_6535730.html

Katz, V. (2014). Children as brokers of their immigrant families' health-care connections. *Social Problems*, 61, 194–215.

Kautsky, J. (1982). *The politics of aristocratic empires.* Chapel Hill, NC: University of North Carolina Press.

Keister, L., & Southgate, D. (2012). *Inequality: A contemporary approach to race, class, and gender.* New York, NY: Cambridge University Press.

Kelley, J., & Evans, M. D. R. (1995). Class and class conflict in six western nations. *American Sociological Review, 60*(2), 157–178.

Kelling, G. L., & Coles, C. M. (1997). *Fixing broken windows: Restoring order and reducing crime in our communities.* New York, NY: Free Press.

Kelly, L. (1987). The continuum of sexual violence. In J. Hanmer & M. Maynard (Eds.), *Women, violence, and social control* (pp. 46–60). Atlantic Highlands, NJ: Humanities Press.

Kelly, M. (1992). *Colitis: The experience of illness.* London, UK: Routledge.

Kenkel, D. S., Lillard, D. R., & Mathios, A. D. (2006). The roles of high school completion and GED receipt in smoking and obesity. *Journal of Labor Economics, 24*(3), 635–660.

Kennedy, S., & Bumpass, L. (2008). Cohabitation and children's living arrangements: New estimates from the United States. *Demographic Research, 19*(47), 1663–1692.

Kernaghan, C. (2012). *Chinese sweatshop in Bangladesh.* Institute for Global Labour and Human Rights. Retrieved from http://www.globallabourrights.org/admin/reports/files/1203-Chinese-Sweatshop-in-Bangladesh.pdf

Kessler, R. C., & Üstün, T. B. (Eds.). (2008). *The WHO world mental health surveys: Global perspectives on the epidemiology of mental disorders.* New York, NY: Cambridge University Press.

Khalidi, R. (2011, March 3). The Arab Spring. *The Nation.* Retrieved from http://www.thenation.com/article/158991/arab-spring

Kim, E. C. (2012). Nonsocial transient behavior: Social disengagement on the Greyhound bus. *Symbolic Interaction, 35,* 267–283. Retrieved from http://www.jstor.org/stable/symbinte.35.3.267

Kimmel, M. S. (2008). *Guyland: The perilous world where boys become men.* New York: HarperCollins Publishers.

Kimmel, M. S. (2010). *Misframing men: The politics of contemporary masculinities.* New Brunswick, NJ: Rutgers University Press.

King, N. (1984). Exploitation and abuse of older family members: An overview of the problem. In J. J. Cosa (Ed.), *Abuse of the elderly.* Lexington, MA: Lexington Books.

Kingkade, T. (2014, October 8). Texas Tech frat loses charter following "no means yes, yes means anal" display. *Huffington Post.* Retrieved from https://www.huffingtonpost.com/2014/10/08/texas-tech-frat-no-means-yes_n_5953302.html

Kingston, P. (2001). *The classless society.* Palo Alto, CA: Stanford University Press.

Kinsey, A. C., Pomeroy, W. B., & penn, C. E. (1948). *Sexual behavior in the human male.* Philadelphia, PA: Saunders.

Kinsey, A., Pomeroy, W. B., Martin, C. E., Gebhard, P. H., Brown, J. M., Christenson, C. V., . . . Roehr, E. L. (1953). *Sexual behavior in the human female.* Philadelphia, PA: Saunders.

Kirk, C. (2014, January 26). Map: Publicly funded schools that are allowed to teach creationism. *Slate.* Retrieved from http://www.slate.com/articles/health_and_science/science/2014/01/creationism_in_public_schools_mapped_where_tax_money_supports_alternatives.html

Kirk, M. (2016). The role of global capitalism in the Arab uprisings and Taiwan's sunflower movement. *Taiwan in Comparative Perspective, 6,* 17–30.

Kjekshus, H. (1977). *Ecology, control, and economic development in East African history.* Berkeley, CA: University of California Press.

Klein, A. (2017, December 12). Many educators skeptical of school choice, including conservatives, survey shows. *Education Week.* Retrieved from https://www.edweek.org/ew/articles/2017/12/13/many-educators-skeptical-of-school-choice-including.html

Kliff, S. (2013, March 2). An average ER visit costs more than an average month's rent [Blog]. *Washington Post.* Retrieved from http://www.washingtonpost.com/blogs/wonkblog/wp/2013/03/02/an-average-er-visit-costs-more-than-an-average-months-rent/

Klinenberg, E. (2012a). *Going solo: The extraordinary rise and surprising appeal of living alone.* New York, NY: Penguin.

Klinenberg, E. (2012b, February 4). One's a crowd. *New York Times.* Retrieved from http://www.nytimes.com/2012/02/05/opinion/sunday/living-alone-means-being-social.html?_r=0

Kling, R. (1996). Computerization at work. In R. Kling (Ed.), *Computers and controversy* (2nd ed.). New York, NY: Academic Press.

Kluckhohn, C. (1949). *Mirror for man.* Tucson, AZ: University of Arizona Press.

Knodel, J. (2006). Parents of persons with AIDS: Unrecognized contributions and unmet needs. *Journal of Global Ageing, 4,* 46–55.

Knoke, D. (1990). *Political networks: The structural perspective.* New York, NY: Cambridge University Press.

Knorr-Cetina, K., & Cicourel, A. V. (Eds.). (1981). *Advances in social theory and methodology: Towards an integration of micro- and macro-sociologies.* Boston, MA: Routledge & Kegan Paul.

Kobrin, S. (1997). Electronic cash and the end of national markets. *Foreign Policy, 107,* 65–77.

Kochanek, K.D., Arias, E., & Anderson, R.N. (2015, Nov.). *Leading Causes of Death Contributing to Decrease in Life Expectancy Gap Between Black and White Populations: United States, 1999–2013.* Atlanta, GA: National Center for Health Statistics. Data brief, 218.

Kochanek, K. D., Murphy, S. L., Xu, J. Q., & Arias, E. (2017). *Mortality in the United States, 2016* (NCHS Data Brief No. 293). Atlanta, GA: National Center for Health Statistics.

Kochhar, R., & Cilluffo, A. (2018). *Income inequality in the U.S. is rising most rapidly among Asians,* Pew Research Center, Retrieved from http://www.pewsocialtrends.org/2018/07/12/income-inequality-in-the-u-s-is-rising-most-rapidly-among-asians/#income-gaps-across-racial-and-ethnic-groups-persist-and-in-some-cases-are-wider-than-in-1970

Kochhar, R., & Morin, R. (2014). *Despite recovery, fewer Americans identify as middle class.* Pew Research Center. Retrieved from http://www.pewresearch.org/fact-tank/2014/01/27/despite-recovery-fewer-americans-identify-as-middle-class/

Kochhar, R., Fry, R., & Taylor, P. (2011). *Twenty-to-one wealth gaps rise to record highs between whites, blacks and Hispanics.* Pew Research Center. Retrieved from http://pewsocialtrends.org/files/2011/07/SDT-Wealth-Report_7-26-11_FINAL.pdf

Kohn, M. (1977). *Class and conformity* (2nd ed.). Homewood, IL: Dorsey Press.

Kollock, P., & Smith. M. (1996). Managing the virtual commons: Cooperation and conflict in computer communities. In S. Herring (Ed.), *Computer-mediated communication* (pp. 109–128). Amsterdam: John Benjamins.

Kosciw, J. G., Greytak, E. A., Palmer, N. A., & Boesen, M. J. (2014). *The 2013 National School Climate Study: The experiences of lesbian, gay, bisexual, and transgender youth in our nation's schools.* Retrieved from https://www.glsen.org/sites/default/files/2013%20National%20School%20Climate%20Survey%20Full%20Report_0.pdf

Kosciw, J. G., Greytak, E. A., Giga, N. M., Villenas, C., & Danischewski, D. J. (2016). *The 2015 National School Climate Survey: The experiences of lesbian, gay, bisexual, transgender, and queer youth in our nation's schools.* New York, NY: GLSEN.

Kosmin, B., Mayer, E., & Keysar, A. (2001, December 19). *American Religious Identification Survey (ARIS).* CUNY Graduate Center. Retrieved from http://www.gc.cuny.edu/CUNY_GC/media/CUNY-Graduate-Center/PDF/ARIS/ARIS-PDF-version.pdf

Kozol, J. (1991). *Savage inequalities: Children in America's schools.* New York, NY: Crown.

Kozol, J. (2012). *Fire in the ashes: Twenty-five years among the poorest children in America.* New York, NY: Crown.

Kramer, A. D., Guillory, J. E., & Hancock, J. T. (2014). Experimental evidence of massive-scale emotional contagion through social networks. *Proceedings of the National Academy of Sciences USA, 111*(24), 8788–8790.

Kramer, K. Z., Kelly, E. L., & McCulloch, J. B. (2015). Stay-at-home fathers: Definition and characteristics based on 34 years of CPS data. *Journal of Family Issues, 36,* 1651–1673.

Krebs, C. P., Lindquist, C. H., Warner, T. D., Fisher, B. S., & Martin, S. L. (2007). *The campus sexual assault study.* Bureau of Justice Statistics, National Institute of Justice. Retrieved from https://www.ncjrs.gov/pdffiles1/nij/grants/221153.pdf

Krennerich, M. (2014). *Germany: The original mixed member proportional system.* The Electoral Project Network. Retrieved from http://aceproject.org/regions-en/countries-and-territories/DE/case-studies/germany-the-original-mixed-member-proportional-system/view

Kristof, N., & WuDunn, S. (2009). *Half the sky: Turning oppression into opportunity for women worldwide.* New York, NY: Knopf.

Krogstad, J. (2015). *Puerto Ricans leave in record numbers for mainland U.S.* Pew Research Center. Retrieved from http://www.pewresearch.org/fact-tank/2015/10/14/puerto-ricans-leave-in-record-numbers-for-mainland-u-s/

Krogstad, J. M., & Passel, J. S. (2015). *5 facts about illegal immigration in the U.S.* Pew Research Center. Retrieved from http://www.pewresearch.org/fact-tank/2015/11/19/5-facts-about-illegal-immigration-in-the-u-s/

Kroll, L. (2011, May 4). The world's richest self-made women. *Forbes.* Retrieved from http://www.forbes.com/forbes/2011/0523/focus-winfrey-fisher-hendricks-whitman-wynn-self-made.html

Kroll, L., & Dolan, K. (2018, March 6). Meet the members of the three-comma club. *Forbes.* Retrieved from https://bohemiarealtygroup.com/view-ad/?id=90061

Krueger, A. B. (2015, October 9). The minimum wage: How much is too much? *New York Times.* Retrieved from http://www.nytimes.com/2015/10/11/opinion/sunday/the-minimum-wage-how-much-is-too-much.html?_r=0

Kulish, N. (2011, September 27). As scorn for vote grows, protests surge around the globe. *New York Times.* Retrieved from http://www.nytimes.com/2011/09/28/world/as-scorn-for-vote-grows-protests-surge-around-globe.html?pagewanted=all

Kuo, L. (2015). A mobile app could make childbirth safer in Ethiopia, one of the deadliest countries to have a baby. *Quartz.* Retrieved from http://qz.com/548047/a-mobile-app-could-make-childbirth-safer-in-ethiopia-one-of-the-deadliest-countries-to-have-a-baby/

Kuo, L. (2016). A Kenyan smartphone app is being used to help prevent blindness in kids. *Quartz.* Retrieved from http://qz.com/629270/a-kenyan-smartphone-app-is-preventing-blindness-in-kids/

Kupers, T. A. (2005). Toxic masculinity as a barrier to mental health treatment in prison. *Journal of Clinical Psychology, 61,* 713–724.

Kurtz, A., & Yellin, T. (2015). Minimum wage since 1938. *CNN Money.* Retrieved from http://money.cnn.com/interactive/economy/minimum-wage-since-1938/

Kutner, M., Greenberg, E., Jin, Y., Boyle, B., Hsu, Y.-C., & Dunleavy, E. (2007). *Literacy in everyday life: Results from the 2003 National Assessment of Adult Literacy* (NCES 2007–480). Washington, DC: National Center for Education Statistics.

Kwabena, D. (1997). *Structural adjustment and mass poverty in Ghana.* Brookfield, VT: Ashgate.

Kweifio-Okai, C. (2015, December 25). Want to know how to deliver a baby? There's an app for that. *Guardian.* Retrieved from https://www.theguardian.com/global-development/2015/dec/25/safe-delivery-app-maternal-deaths-newborns-baby-pregnancy

Labrecque, L. T., & Whisman, M. A. (2017). Attitudes toward and prevalence of extramarital sex and descriptions of extramarital partners in the 21st century. *Journal of Family Psychology, 31,* 952–957.

Lachman, M. (2001). *Handbook of midlife development.* New York: Wiley.

Lacy, K. R. (2007). *Blue-chip black: Race, class, and status in the new black middle class.* Berkeley: CA: University of California Press.

Lajeunesse, S. (2013, September 17). Older adults learn to Skype with help from Penn State students. *Penn State News.* Retrieved from http://news.psu.edu/story/288102/2013/09/17/academics/older-adults-learn-skype-help-penn-state-students

Lamb, M. E. (2012). Mothers, fathers, families, and circumstances: Factors affecting children's adjustment. *Applied Developmental Science, 16*(2), 98–111.

Landale, N. S., & Fennelly, K. (1992). Informal unions among mainland Puerto Ricans: Cohabitation or an alternative to legal marriage? *Journal of Marriage and Family, 54*(2): 269–280.

Landale, N. S., Oropesa, R. S., & Bradatan, C. (2006). Hispanic families in the United States: Family structure and process in an era of family change. In M. Tienda & F. Mitchell (Eds.), *Hispanics and the future of America* (pp. 138–178). Washington, DC: National Academy Press.

Landry, B., & Marsh, K. (2011). The evolution of the new black middle class. *Annual Review of Sociology, 37,* 373–394.

Langone, A. (2018, January 18). These are the biggest donors to Hollywood's record-breaking #TimesUp campaign. *Money.* Retrieved from http://time.com/money/5107657/times-up-go-fund-me-donations/

LaPorte, N. (2013, April 13). Medical care: Aided by the crowd. *New York Times.* Retrieved from http://www.nytimes.com/2013/04/14/business/watsi-a-crowdfunding-site-offers-help-with-medical-care.html?pagewanted%253Dall&_r=0

Lareau, A. (2011). *Unequal childhoods: Class, race, and family life.* Berkeley, CA: University of California Press.

LaRossa, R., & Reitzes, D. C. (1993). Symbolic interactionism and family studies. In P. G. Boss, W. J. Doherty, R. LaRossa, W. R. Schumm, & S. K. Steinmetz (Eds.), *Sourcebook of family theories and methods: A contextual approach* (pp. 135–163). New York, NY: Plenum Press.

Latour, F. (2011, December 9). Ready. Set. Race [Blog]. Boston.com. Retrieved from http://www.boston.com/community/blogs/hyphenated_life/2011/12/bake_me_a_race_as_fast_as_you.html

Laumann, E. O., Gagnon, J. H., Michael, R. T., & Michaels, S. (1994). *The social organization of sexuality: Sexual practices in the United States.* Chicago, IL: University of Chicago Press.

Laumann, E. O., S. A. Leitsch, and L. J. Waite. (2008). Elder mistreatment in the United States: prevalence estimates from a nationally representative study. *Journal of Gerontology: Social Sciences, 63:* 248–54.

Lauzen, M. M. (2017). *The celluloid ceiling: Behind-the-scenes employment of women on the top 100, 250, and 500 films of 2016.* Center for the Study of Women in Television & Film. Retrieved from http://womenintvfilm.sdsu.edu/wpcontent/uploads/2017/01/2016_Celluloid_Ceiling_Report.pdf

Lazer, D., & Radford, J. (2017). Data ex machina: Introduction to big data. *Annual Review of Sociology, 43,* 19–39. doi:10.1146/annurev-soc-060116-053457

Le, V. (2015). The world's largest media companies of 2015. *Forbes.* Retrieved from http://www.forbes.com/sites/vannale/2015/05/22/the-worlds-largest-media-companies-of-2015/#7ffeeabc2b64

Leber, R. (2014, December 21). This is what our hellish world will look like after we hit the global warming tipping point. *New Republic.* Retrieved from https://newrepublic.com/article/120578/global-warming-threshold-what-2-degrees-celsius-36-f-looks

Lee, J. (2012, January 18). Stereotype promise [Blog]. Russell Sage Foundation. Retrieved from http://www.russellsage.org/blog/j-lee/stereotype-promise

Lee, J., & Zhou, M. (2014). The success frame and achievement paradox: The costs and consequences for Asian Americans. *Race and Social Problems, 6,* 38–55. doi:10.1007/s12552-014-9112-7

Lee, J., & Zhou, M. (2015). *The Asian American achievement paradox.* New York, NY: Russell Sage Foundation.

Lee, C., Tsenkova, V., & Carr, D. (2014). Childhood trauma and metabolic syndrome in men and women. *Social Science & Medicine, 105,* 122–130.

Leiss, W. (1994). *The Domination of Nature.* Montreal, Canada: McGill-Queen's University Press.

Leland, J. (2008, October 6). In "sweetie" and "dear," a hurt for the elderly. *New York Times.* Retrieved from http://www.nytimes.com/2008/10/07/us/07aging.html

Lemert, E. (1972). *Human deviance, social problems, and social control.* Englewood Cliffs, NJ: Prentice Hall.

Lenhart, A. (2007). *Cyberbullying.* Pew Internet & American Life Project. Retrieved from http://www.pewinternet.org/Reports/2007/Cyberbullying.aspx

Lenhart, A. (2015). *Teens, technology, and friendships.* Pew Research Center. Retrieved from http://www.pewinternet.org/2015/08/06/teens-technology-and-friendships/

Lenhart, A., Madden, M., Smith, A., Purcell, K., Zickuhr, K., & Rainie, L. (2011, November 9). *Teens, kindness and cruelty on social network sites.* Pew Research Internet Project. Retrieved from http://www.pewinternet.org/2011/11/09/teens-kindness-and-cruelty-on-social-network-sites/

Lenzer, R. (2011, November 20). The top 0.1% of the nation earn half of all capital gains. *Forbes.* Retrieved from http://www.forbes.com/sites/robertlenzner/2011/11/20/the-top-0-1-of-the-nation-earn-half-of-all-capital-gains/

Lesane-Brown, C. (2006). A review of race socialization within black families. *Developmental Review, 26,* 400–426.

Lesane-Brown, C. L., T. N. Brown, C. H. Caldwell, and R. M. Sellers. (2005). The comprehensive race socialization inventory. *Journal of Black Studies, 36*(2), 163–190.

Let's Move. (2017). *America's move to raise a healthier generation of kids.* Retrieved from https://letsmove.obamawhitehouse.archives.gov/

Leupp, G. (1995). *Male colors: The construction of homosexuality in Tokugawa Japan.* Berkeley, CA: University of California Press.

LeVay, Simon. (2011). *Gay, straight, and the reason why: The science of sexual orientation.* New York, NY: Oxford.

Lewin, T. (2012, November 19). College of future could be come one, come all. *New York Times.* Retrieved from http://www.nytimes.com/2012/11/20/education/colleges-turn-to-crowd-sourcing-courses.html?pagewanted=all

Lewis, O. (1969). The culture of poverty. In D. P. Moynihan (Ed.), *On understanding poverty: Perspectives from the social sciences.* New York, NY: Basic Books.

Lewis, K. (2013). The limits of racial prejudice. *Proceedings of the National Academy of Sciences USA, 110*(47), 18814–18819.

Li, V. (2014, May 7). Global 2000: The world's largest media companies of 2014. *Forbes.* Retrieved from http://www.forbes.com/sites/vannale/2014/05/07/global-2000-the-worlds-largest-media-companies-of-2014/

Li, Y., & Li, X. (2018). Freezing cold turns primary school student's hair white. *China Daily.* Retrieved from http://www.chinadaily.com.cn/a/201801/09/WS5a54cf27a3102e5b17371ae9_3.html

Liebow, E. (1967). *Tally's corner: A study of Negro streetcorner men.* New York, NY: Rowman & Littlefield.

Lieff, Cabraser, Heimann, & Bernstein, LLP, & Outten & Golden, LLP. (2013). Bank of America and Merrill Lynch sex discrimination lawsuit. Retrieved from http://www.bofagenderlawsuit.com/

Lightfoot-Klein, H. (1989). *Prisoners of ritual: An odyssey into female genital circumcision in Africa.* New York, NY: Haworth.

Lin, H. L. (2012, September 4). How your cell phone hurts your relationships. *Scientific American.* Retrieved from http://www.scientificamerican.com/article/how-your-cell-phone-hurts-your-relationships/

Lin, G., & Rogerson, P. A. (1995). Elderly parents and the geographic availability of their adult children. *Research on Aging, 17,* 303–331.

Lipka, M. (2015). *A closer look at America's rapidly growing religious "nones."* Pew Research Center. Retrieved from http://www.pewresearch.org/fact-tank/2015/05/13/a-closer-look-at-americas-rapidly-growing-religious-nones/

Lipka, M. (2016). *Muslims and Islam: Key findings in the U.S. and around the world.* Pew Research Center. Retrieved from http://www.pewresearch.org/fact-tank/2016/07/22/muslims-and-islam-key-findings-in-the-u-s-and-around-the-world/

Liptak, A. (2015, June 26). Supreme Court ruling makes same-sex marriage a right nationwide. *New York Times.* Retrieved from http://www.nytimes.com/2015/06/27/us/supreme-court-same-sex-marriage.html

Livingston, G. (2014). *Four-in-ten couples are saying "I do," again.* Pew Research Center. Retrieved from http://www.pewsocialtrends.org/files/2014/11/2014-11-14_remarriage-final.pdf

Livingston, G. (2015). *Childlessness falls, family size grows among highly educated women.* Pew Research Center. Retrieved from http://www.pewsocialtrends.org/files/2015/05/2015-05-07_children-ever-born_FINAL.pdf

Livingston, G. (2017, June 6). *The rise of multiracial and multiethnic babies in the U.S.* Pew Research Center. Retrieved from http://www.pewresearch.org/fact-tank/2017/06/06/the-rise-of-multiracial-and-multiethnic-babies-in-the-u-s/

Livingston, G., & Bialik, K. (2018, May 10). *7 facts about U.S. moms.* Pew Research Center. Retrieved from http://www.pewresearch.org/fact-tank/2018/05/10/facts-about-u-s-mothers/

Livingston, G., & Brown, A. (2017, May 18). *Intermarriage in the U.S. 50 years after Loving v. Virginia.* Pew Research Center. Retrieved from http://www.pewsocialtrends.org/2017/05/18/intermarriage-in-the-u-s-50-years-after-loving-v-virginia/

Lofquist, D. (2011). *Same-sex couple households* (American Community Survey Briefs). U.S. Bureau of the Census. Retrieved from https://www.census.gov/prod/2011pubs/acsbr10-03.pdf

Logan, J. R., & Molotch, H. (1987). *Urban fortunes: The political economy of place.* Berkeley, CA: University of California Press.

López, G. (2015, September 15). *Hispanics of Cuban origin in the United States, 2013.* Pew Research Center. Retrieved from http://www.pewhispanic.org/2015/09/15/hispanics-of-cuban-origin-in-the-united-states-2013/

Lopez, G. (2017, December 22). The opioid epidemic has now reached black America. *Vox.* Retrieved from https://www.vox.com/science-and-health/2017/12/22/16808490/opioid-epidemic-black-white

Lopez, G. (2018, January 23). A year after the first Women's March, millions are still actively protesting Trump. *Vox.* Retrieved from https://www.vox.com/policy-and-politics/2018/1/23/16922884/womens-march-attendance

López, G., & Bialik, K. (2017, May 3). *Key findings about U.S. immigrants.* Pew Research Center. Retrieved from http://www.pewresearch.org/fact-tank/2017/05/03/key-findings-about-u-s-immigrants/

Lopez, G., & Patten, E. (2015). *The impact of slowing immigration: Foreign-born share falls among 14 largest U.S. Hispanic origin groups.* Pew Research Center. Retrieved from http://www.pewhispanic.org/2015/09/15/the-impact-of-slowing-immigration-foreign-born-share-falls-among-14-largest-us-hispanic-origin-groups/

López, G., Ruiz, N. G., & Patten, E. (2017, September 8). *Key facts about Asian Americans, a diverse and growing population.* Pew Research Center. Retrieved from http://www.pewresearch.org/fact-tank/2017/09/08/key-facts-about-asian-americans/

Lorber, J. (1994). *Paradoxes of gender.* New Haven, CT: Yale University Press.

Lorber, J. (1996). Beyond the binaries: Depolarizing the categories of sex, sexuality, and gender. *Sociological Inquiry, 66,* 143–160.

Lorber, J. (2010, July 21). Republicans form caucus for Tea Party in the house. *New York Times.* Retrieved from http://www.nytimes.com/2010/07/22/us/politics/22tea.html?_r=0

Lu, B., Qian, Z., Cunningham, A., & Li, C. L. (2012). Estimating the effect of premarital cohabitation on timing of marital disruption using propensity score matching in event history analysis. *Sociological Methods & Research, 41*(3), 440–466.

Luckerson, V. (2015, January 29). Target will stop separating "girls" toys from "boys" toys in stores. *Time.* Retrieved from http://time.com/3989850/target-gender-signs/

Lui, C. K., Chung, P. J., Wallace, S. P., & Aneshensel, C. S. (2013). Social status attainment during the transition to adulthood. *Journal of Youth and Adolescence, 43,* 1134–1150.

Lynch, M. (2011, November 28). The big think behind the Arab Spring. *Foreign Policy.* Retrieved from http://foreignpolicy.com/2011/11/28/the-big-think-behind-the-arab-spring/

Lyotard, J. (1985). *The post-modern condition: A report on knowledge.* Minneapolis, MN: University of Minnesota Press.

Lytton, H., & Romney, D. M. (1991). Parents' differential socialization of boys and girls: A meta-analysis. *Psychological Bulletin, 109,* 267–296.

Macdonald, C. (2011). *Shadow mothers: Nannies, au pairs, and the micropolitics of mothering.* Berkeley, CA: University of California Press.

Mackun, P., & Wilson, S. (2011). *Population distribution and change: 2000 to 2010.* U.S. Bureau of the Census. Retrieved from http://www.census.gov/prod/cen2010/briefs/c2010br-01.pdf

Maddox, G. (1965). Fact and artifact: Evidence bearing on disengagement from the Duke Geriatrics Project. *Human Development, 8,* 117–130.

Maddox, G. (1970). Themes and issues in sociological theories of human aging. *Human Development, 13,* 17–27.

Maharidge, D. (1996). *The coming white minority.* New York, NY: Times Books.

Mahler, J. (2015, March 8). Who spewed that abuse? Anonymous Yik Yak isn't saying. *New York Times.* Retrieved from http://www.nytimes.com/2015/03/09/technology/popular-yik-yak-app-confers-anonymity-and-delivers-abuse.html?_r=1

Mallicoat, S. L. (2017). Gendered justice: Attributional differences between males and females in the juvenile courts. *Feminist Criminology, 2*(1), 4–30. doi:10.1177/1557085106296349

Malotki, E. (1983). *Hopi time: A linguistic analysis of the temporal concepts in the Hopi language.* Berlin, Germany: Mouton.

Malthus, T. (2003). *Essay on the principle of population* (Norton Critical Edition, 2nd ed., P. Appleman, Ed.). New York, NY: Norton. (Original work published 1798)

Mandel, J. (2010, September 7). Rosh Hashanah online: Yom Kippur services go mobile. *Jerusalem Post.* Retrieved from http://www.jpost.com/Arts-and-Culture/Entertainment/Rosh-Hashana-online-Yom-Kippur-services-go-mobile

Maniam, S. (2017, January 30). *Most Americans see labor unions, corporations favorably.* Pew Research Center. Retrieved from http://www.pewresearch.

org/fact-tank/2017/01/30/most-americans-see-labor-unions-corporations-favorably/

Manik, J. A., & Yardley, J. (2013, April 24). Building collapse in Bangladesh leaves scores dead. *New York Times*. Retrieved from http://www.nytimes.com/2013/04/25/world/asia/bangladesh-building-collapse.html?pagewanted=all

Manjoo, F. (2008). *True enough: Learning to live in a post-fact society*. New York, NY: John Wiley.

Manning, W. D., Brown, S. L., & Stykes, B. (2015). *Trends in births to single and cohabiting mothers, 1980–2013*. National Center for Family & Marriage Research. Retrieved from https://www.bgsu.edu/content/dam/BGSU/college-of-arts-and-sciences/NCFMR/documents/FP/FP-15-03-birth-trends-single-cohabiting-moms.pdf

Manning, J. T., Koukourakis, K., & Brodie, D. A. (1997). Fluctuating asymmetry, metabolic rate and sexual selection in human males. *Evolution and Human Behavior*, 18(1), 15–21.

Manning, W. D., and S. L. Brown. (2011). The demography of unions among older Americans, 1980–Present: A family change approach. In R. A. Settersten Jr., and J. L. Angel (Eds.) *Handbook of Sociology of Aging* (pp. 193–212). New York: Springer.

Manpower Inc. (2011). About ManPower Group. Retrieved from http://www.manpowergroup.com/about/about.cfm

Manza, J., & Sauder, M. (2009). *Inequality and society: Social science perspectives on social stratification*. New York, NY: Norton.

Mare, R. (1991). Five decades of educational assortative mating. *American Sociological Review*, 56(1), 15–32.

Marques, L., Alegria, M., Becker, A. E., Chen, C.-N., Fang, A., Chosak, A., & Diniz, B. J. (2011). Comparative prevalence, correlates of impairment, and service utilization for eating disorders across US ethnic groups: Implications for reducing ethnic disparities in health care access for eating disorders. *International Journal of Eating Disorders*, 44, 412–420.

Marquez, V. A. (2014, May 9). Essay for Yale University. *New York Times*. Retrieved from http://www.nytimes.com/interactive/2014/05/09/business/student-essays-your-money.html?_r=0

Marsden, P. V., & Lin, N. (1982). *Social structure and network analysis*. Beverly Hills, CA: Sage.

Marsh, J. (2013, July 25). Merrill bias suit: Women employees claim they were given book urging them to "stroke men's egos" to advance. *New York Post*. Retrieved from http://nypost.com/2013/07/25/merrill-bias-suit-women-employees-claim-they-were-given-book-urging-them-to-stroke-mens-egos-to-advance/

Marsh, K. W., Darity, A., Jr., Cohen, P. N., Casper, L. M., & Salters, D. (2007). The emerging black middle class: Single and living alone. *Social Forces*, 86, 735–762.

Marshall, T. (1973). *Class, citizenship, and social development: Essays by T. H. Marshall*. Westport, CT: Greenwood Press.

Martin, M. (2008). The intergenerational correlation in weight: How genetic resemblance reveals the social role of families. *American Journal of Sociology*, 114(S1), 67–105.

Martin, J. A., Hamilton, B. E., Ventura, S. J., Osterman, M., Wilson, E., & Mathews, T. J. (2012). Births: Final data for 2010. *National Vital Statistics Reports*, 61(1), 1–72. Retrieved from http://www.cdc.gov/nchs/data/nvsr/nvsr61/nvsr61_01.pdf

Martin, J. A., Hamilton, B. E., Osterman, M. J. K., Driscoll, A. K., & Drake, P. (2018). Births: Final data for 2016. *National Vital Statistics Reports*, 67(1), 1–55. Retrieved from https://www.cdc.gov/nchs/data/nvsr/nvsr67/nvsr67_01.pdf

Martin, N., & Montagne, R. (2017, December 7). Nothing protects black women from dying in pregnancy and childbirth. *ProPublica*. Retrieved from https://www.propublica.org/article/nothing-protects-black-women-from-dying-in-pregnancy-and-childbirth

Martineau, H. (2009). *Society in America* (3 vols.). Cambridge, UK: Cambridge University Press. (Original work published 1837)

Marvell, T. B. (1989). Divorce rates and the fault requirement. *Law and Society Review*, 23(4), 543–567.

Marwick, A. (2013). *Status update: Celebrity, publicity, and branding in the social media age*. New Haven, CT: Yale University Press.

Marx, K. (1977). *Capital: A critique of political economy* (Vol. 1). New York, NY: Random House. (Original work published 1864)

Massey, D. (1996). The age of extremes: Concentrated affluence and poverty in the twenty-first century. *Demography*, 33(4), 395–412.

Massey, D. (2012, Fall). The great decline in American immigration? *Pathways*. Retrieved from https://inequality.stanford.edu/sites/default/files/media/_media/pdf/pathways/fall_2012/Pathways_Fall_2012%20_Massey.pdf

Massey, D., & Denton, N. A. (1993). *American apartheid: Segregation and the making of the underclass*. Cambridge, MA: Harvard University Press.

Matchett, J. (2009). The undergraduate sociology degree's real-world application. *ASA Footnotes*, 37(5). Retrieved from http://www.asanet.org/sites/default/files/savvy/footnotes/mayjun09/undergrad_0509.html

Mather, M. (2008). *Population losses mount in U.S. rural areas*. Washington, DC: Population Reference Bureau.

Mathews, T. J., & Hamilton, B. E. (2014). *First births to older women continue to rise* (NCHS Data Brief No 152). National Center for Health Statistics. Retrieved from http://www.cdc.gov/nchs/data/databriefs/db152.pdf

Mathews, T. J., & Hamilton, B. E. (2016). *Mean age of mothers is on the rise: United States, 2000–2014* (NCHS Data Brief No. 232). Hyattsville, MD: National Center for Health Statistics.

Matisons, M. (2015). 3D printing shoes from home "not that far away," Nike COO predicts. Retrieved from 3DPrint.com website: https://3dprint.com/99927/nike-coo-3d-printing-not-far/

Matos, K. (2015). *Modern families: Same- and different-sex couples negotiating at home*. Families and Work Institute. Retrieved from http://www.familiesandwork.org/downloads/modern-families.pdf

Matsueda, R. (1992). Reflected appraisals, parental labeling, and delinquency: Specifying a symbolic interactionist theory. *American Journal of Sociology*, 97, 1577–1611.

Matthews, D. (2011, December 22). Racial identity becomes a guessing game—literally [Blog]. *CNN*. Retrieved from http://inamerica.blogs.cnn.com/2011/12/22/racial-identity-becomes-a-guessing-game-literally/

May, E. T. (1997). *Barren in the promised land: Childless Americans and the pursuit of happiness*. New Haven, CT: Harvard University Press.

McAdam, D. (1982). *Political process and the development of Black insurgency, 1930–1970*. Chicago, IL: University of Chicago Press.

McCabe, J., Fairchild, E., Grauerholz, L., Pescosolido, B. A., & Tope, D. (2011). Gender in twentieth-century children's books: Patterns of disparity in titles and central characters. *Gender & Society*, 25(2), 197–226.

McCarthy, M. M. (2015, October). Sex differences in the brain: How male and female brains diverge is a hotly debated topic, but the study of model organisms points to differences that cannot be ignored. *The Scientist*. Retrieved from http://www.the-scientist.com/?articles.view/articleNo/44096/title/Sex-Differences-in-the-Brain/

McCarthy, J. (2017, May 15). *U.S. support for gay marriage edges to new high*. Retrieved from http://news.gallup.com/poll/210566/support-gay-marriage-edges-new-high.aspx

McCarthy, J. D. & Zald, M. N. (1977). Resource mobilization and social movements: A partial theory." *American Journal of Sociology*, 82(6): 1212–1241.

McClain, Z., & Peebles, R. (2016). Body image and eating disorders among lesbian, gay, bisexual, and transgender youth. *Pediatric Clinics of North America*, 63, 1079–1090.

McDonald, C. (2015, June 16). How many Earths do we need? *BBC*. Retrieved from http://www.bbc.com/news/magazine-33133712

McFadden, D., & Champlin, C. A. (2000). Comparison of auditory evoked potentials in heterosexual, homosexual, and bisexual males and females. *Journal of the Association for Research in Otolaryngology*, 1, 89–99.

McFarland, J., Hussar, B., de Brey, C., Snyder, T., Wang, X., Wilkinson-Flicker, S., . . . Hinz, S. (2017). *The condition of education 2017* (NCES 2017-144). National Center for Education Statistics. Retrieved from https://nces.ed.gov/pubsearch/pubsinfo.asp?pubid=2017144

McFarland, J., Hussar, B., Wang, X., Zhang, J., Wang, K., Rathbun, A., . . . Bullock Mann, F. (2018). *The condition of education 2018* (NCES 2018-144). National Center for Education Statistics. Retrieved from https://nces.ed.gov/pubs2018/2018144.pdf

McGeehan, P. (2013, September 6). Bank of America to pay $39 million in gender bias case. *New York Times*. Retrieved from https://dealbook.nytimes.com/2013/09/06/bank-of-america-to-pay-39-million-in-gender-bias-case/

McGranahan, D. A. (2015). *Understanding the geography of growth in rural child poverty*.

Economic Research Service, USDA. Retrieved from http://www.ers.usda.gov/amber-waves/2015-july/understanding-the-geography-of-growth-in-rural-child-poverty.aspx#.V_7ti7uAOkq

McGranahan, D. A., & Beale, C. L. (2002). Understanding rural population loss. *Rural America, 17*(4), 2–11.

McIntosh, P. (1988). *White privilege and male privilege: A personal account of coming to see correspondences through work in women's studies* (Working paper). Wellesley, MA: Wellesley College Center for Research on Women.

McKenzie, L. (2017, September 12). The next Yik Yak. *Inside Higher Ed.* Retrieved from https://www.insidehighered.com/news/2017/09/12/could-college-messaging-app-islands-be-new-yik-yak

McLanahan, S. (2004). Diverging destinies: How children are faring under the second demographic transition. *Demography, 41,* 607–627.

McLanahan, S., & Sandefur, G. (1994). *Growing up with a single parent: What hurts, what helps.* Cambridge, MA: Harvard University Press.

McLaughlin, H., Uggen, C., & Blackstone, A. (2017). The economic and career effects of sexual harassment on working women. *Gender & Society, 31,* 333–358.

McManus, P. A., & DiPrete, T. A. (2001). Losers and winners: The financial consequences of separation and divorce for men. *American Sociological Review, 66*(2): 246–268.

Mead, M. (1963). *Sex and temperament in three primitive societies.* New York, NY: William Morrow. (Original work published 1935)

Mead, M. (1966). Marriage in two steps. *Redbook, 127*(3), 48–49, 84–86.

Mead, M. (1972). *Blackberry winter: My earlier years.* New York, NY: William Morrow.

Meadows, D. L., Meadows, D. H., Randers, J., & Behrens, W. W., III. (1972). *The limits to growth.* New York, NY: Universe Books.

Medical Education Cooperation with Cuba. (2016). About MEDICC. Retrieved from http://medicc.org/ns/

Meerwijk, E. L., & Sevelius, J. M. (2017, March 6). Transgender population size in the United States: A meta-regression of population-based probability samples [Online first]. *American Journal of Public Health, 107*(2), e1–e8.

Mehta, S. (2013, November 30). Treatment of HIV-AIDS still poses sociopsychological issues: Study. *Times of India.* Retrieved from http://articles.timesofindia.indiatimes.com/2013-11-30/visakhapatnam/44595899_1_hiv-aids-patients-treatment-study

Melton, J. (1989). *The encyclopedia of American religions* (3rd ed.). Detroit, MI: Gale Research Co.

Menasians, C. (2016, April 18). Where are the iPhone, iPad and Mac designed, made and assembled? A comprehensive breakdown of Apple's product supply chain. *Macworld.* Retrieved from http://www.macworld.co.uk/feature/apple/are-apple-products-truly-designed-in-california-made-in-china-iphonese-3633832/

Merica, D. (2017, August 16). Trump says both sides to blame amid Charlottesville backlash.

CNN Politics. Retrieved from http://www.cnn.com/2017/08/15/politics/trump-charlottesville-delay/index.html

Merton, R. (1938). Social structure and anomie. *American Sociological Review, 3,* 672–682.

Merton, R. (1957). *Social theory and social structure* (Rev. ed.). New York, NY: Free Press.

Merton, R. K. (1968). *Social theory and social structure.* New York, NY: Free Press.

Messerschmidt, J. W., & Messner, M. A. (2018). *Gender reckonings: New social theory and research* (pp. 35–36). New York, NY: NYU Press.

Meyer, J. W., & Rowan, B. (1977). Institutionalized organizations: Formal structure as myth and ceremony. *American Journal of Sociology, 83*(2), 340–363.

MFT-License.com. (2013). *Undergraduate options for students considering careers in marriage and family therapy.* Retrieved from https://www.mft-license.com/education/undergraduate-mft-preparation.html#context/api/listings/prefilter

Michels, R. (1967). *Political parties.* New York, NY: Free Press. (Original work published 1911)

Michigan Department of Community Health. (2010). *Watch out for date rape drugs.* Retrieved from http://www.michigan.gov/documents/publications_date_rape_drugs_8886_7.pdf

Migration Policy Institute. (2018). *Regions of birth for immigrants in the United States, 1960–present.* Retrieved from https://www.migrationpolicy.org/programs/data-hub/charts/regions-immigrant-birth-1960-present?width=900&height=850&iframe=true

Milani, C. R. S., & Laniado, R. N. (2007). Transnational social movements and the globalization agenda: A methodological approach based on the analysis of the World Social Forum. *Brazilian Political Science Review, 1*(2), 10–39.

Miles, R. (1993). *Racism after "race relations."* London, UK: Routledge.

Miles, T. (2018, May 3). *These are the world's most polluted cities.* World Economic Forum. Retrieved from https://www.weforum.org/agenda/2018/05/these-are-the-worlds-most-polluted-cities

Milgram, S. (1963). Behavioral studies in obedience. *Journal of Abnormal and Social Psychology, 67*(4), 371–378.

Milkman, R. (2017). A new political generation: Millennials and the post-2008 wave of protest. *American Sociological Review, 82,* 1–31. doi:10.1177/0003122416681031

Miller, M. (2016). *How app makers are pioneering gender-fluid design for kids.* Retrieved from Fast Company website: https://www.fastcodesign.com/3056346/how-app-makers-are-pioneering-gender-fluid-design-for-kids

Mills, C. W. (1956). *The power elite.* New York, NY: Oxford University Press.

Mills, C. W. (1959). *The sociological imagination.* New York, NY: Oxford University Press.

Mills, T. (1967). *The sociology of small groups.* Englewood Cliffs, NJ: Prentice Hall.

Mirza, H. (1986). *Multinationals and the growth of the Singapore economy.* New York, NY: St. Martin's Press.

Mody, A., Bartz, S., Hornik, C. P., Kiyimba, T., Bain, J., Muehlbauer, M., . . . Freemark, M. (2014). Effects of HIV infection on the metabolic and hormonal status of children with severe acute malnutrition. *PLoS ONE, 9*(7), e102233. Retrieved from http://www.plosone.org/article/info%3Adoi%2F10.1371%2Fjournal.pone.0102233

Moffitt, T. (1996). The neuropsychology of conduct disorder. In P. Cordella & L. Siegel (Eds.), *Readings in contemporary criminological theory* (pp. 85–106). Boston, MA: Northeastern University Press.

Moghadam, V. M. (2005). *Globalizing women: Transnational feminist networks.* Baltimore, MD: Johns Hopkins University Press.

Mohamed, B. (2018, January 3). *New estimates show U.S. Muslim population continues to grow.* Pew Research Center. Retrieved from http://www.pewresearch.org/fact-tank/2018/01/03/new-estimates-show-u-s-muslim-population-continues-to-grow/

Mohanty, C. T. (1991). Under Western eyes: Feminist scholarship and colonial discourse. In C. T. Mohanty, A. Russo, & L. Torres (Eds.), *Third world women and the politics of feminism* (pp. 51–80). Bloomington, IN: Indiana University Press.

Mohanty, C. T. (2003). *Feminism without borders.* Durham, NC: Duke University Press.

Mohanty, C. T. (2013). Transnational feminist crossings: On neoliberalism and radical critique. *Signs, 38*(4), 967–991.

Molotch, H. (1970). Oil in Santa Barbara and power in America. *Sociological Inquiry, 40*(1), 131–144.

Monk, A. (2013). Symbolic interactionism in music education: Eight strategies for collaborative improvisation. *Music Educators Journal, 99*(3), 76–81.

Monk, E. P., Jr. (2015). The cost of color: Skin color, discrimination, and health among African-Americans. *American Journal of Sociology, 121,* 396–444.

Monk, E., Jr. (2016). The consequences of "race and color" in Brazil. *Social Problems, 63,* 413–430. doi:10.1093/socpro/spw014

Monti, M. (2018, March 6). Former Italian prime minister: The election result will be a nightmare for the EU. *Washington Post.* Retrieved from https://www.washingtonpost.com/news/theworldpost/wp/2018/03/06/italian-election/

Moore, B., Jr., (1966). *Social origins of dictatorship and democracy: Lord and peasant in the making of the modern world.* Boston, MA: Beacon Press.

Moore, L. (1994). *Selling God: American religion in the marketplace of culture.* New York, NY: Oxford University Press.

Morello, C. (2012, April 26). Number of biracial babies soars over past decade. *Washington Post.* Retrieved from http://www.mixedracestudies.org/wordpress/?tag=carol-morello

Morgan, R. E., & Kena, G. (2017). *Criminal victimization, 2016* (NCJ 251150). Washington, DC: U.S. Department of Justice, Bureau of Justice Statistics. Retrieved from https://www.bjs.gov/content/pub/pdf/cv16.pdf

Morgan, S. L., Gelbgiser, D., & Weeden, K. A. (2013). Feeding the pipeline: Gender, occupational plans, and college major selection. *Social Science Research, 42*, 989–1005.

Morland, K., Wing, S., Diez Roux, & Poole, C. (2002). Neighborhood characteristics associated with the location of food stores and food service places. *American Journal of Preventive Medicine, 22*(1), 23–29.

Morris, A. (2015). *The scholar denied: WEB Du Bois and the birth of modern sociology*. Berkeley, CA: University of California Press.

Moynihan, D. P. (1965). *The negro family: The case for national action*. Washington, DC: Office of Policy Planning and Research, U.S. Department of Labor.

Mumford, L. (1973). *Interpretations and forecasts*. New York, NY: Harcourt Brace Jovanovich.

Muncie, J. (1999). *Youth and crime: A critical introduction*. London, UK: Sage.

Murdock, G. P. (1967). *Ethnographic atlas*. Pittsburgh, PA: Pittsburgh University Press.

Murdock, G. P. (1981). *Atlas of world cultures*. Pittsburgh, PA: University of Pittsburgh Press.

Murphy, S. L., Xu, J. Q., Kochanek, K. D., Curtin, S. C., & Arias, E. (2017). Deaths: Final data for 2015. *National Vital Statistics Reports, 66*(6).

Murray, C. (1984). *Losing ground: American social policy, 1950–1980*. New York, NY: Basic Books.

Musick, K., & Meier, A. (2010). Are both parents always better than one? Parental conflict and young adult well-being. *Social Science Research, 39*, 814–830.

Musu-Gillette, L., de Brey, C., McFarland, J., Hussar, W., Sonnenberg, W., & Wilkinson-Flicker, S. (2017). *Status and trends in the education of racial and ethnic groups 2017*. U.S. Department of Education, National Center for Education Statistics. Retrieved from https://nces.ed.gov/pubs2017/2017051.pdf

Mwizabi, G. (2013, May 7). Using cell-phones to fight HIV/AIDS. *Zambia Times*. Retrieved from http://www.times.co.zm/?p=9537

Myrdal, G. (1963). *Challenge to affluence*. New York, NY: Random House.

Nadal, K. L., Griffin, K. E., Wong, Y., Hamit, S., & Rasmus, M. (2014). The impact of racial microaggressions on mental health: Counseling implications for clients of color. *Journal of Counseling & Development, 92*(1), 57–66. doi:10.1002/j.1556-6676.2014.00130.x

Najman, J. (1993). Health and poverty: Past, present, and prospects for the future. *Social Science and Medicine, 36*(2), 157–166.

Napikoski, L. (2017, September 30). *Reed v. Reed*: Striking down sex discrimination. *ThoughtCo*. Retrieved from https://www.thoughtco.com/reed-v-reed-3529467

Narayan, D. (1999, December). *Can anyone hear us? Voices from 47 countries*. Washington, DC: World Bank Poverty Group, PREM.

Nasser, Haya-El. (2010, December 20). Census data show "surprising" segregation. *USA Today*. Retrieved from http://usatoday30.usatoday.com/news/nation/census/2010-12-14-segregation_N.ht

National Association of Anorexia Nervosa and Associated Disorders. (2010). *Eating disorder statistics*. Retrieved from http://www.anad.org/get-information/about-eating-disorders/eating-disorders-statistics/

National Association of State Budget Officers. (2012). *Examining fiscal 2010–2012 state spending*. Retrieved from http://www.nasbo.org/sites/default/files/State%20Expenditure%20Report_1.pdf

National Association of State Budget Officers. (2018). *State expenditure report: Examining fiscal 2015-2017*. Retrieved from https://www.nasbo.org/mainsite/reports-data/state-expenditure-report

National Center for Education Statistics. (2013). *The condition of education: The status of rural education*. Retrieved from https://nces.ed.gov/programs/coe/indicator_tla.asp

National Center for Education Statistics. (2016). Total fall enrollment in degree-granting postsecondary institutions, by level of enrollment, sex, attendance status, and race/ethnicity of student: Selected years, 1976 through 2015. *Digest of Education Statistics*. Retrieved from https://nces.ed.gov/programs/digest/d16/tables/dt16_306.10.asp?current=yes

National Center for Family and Marriage Research. (2013). *Divorce rate in the U.S., 2011*. Bowling Green, OH: National Center for Family & Marriage Research.

National Center for Health Statistics. (2003). *Women's health*. Retrieved from http://www.cdc.gov/nchs/fastats/womens_health.htm

National Center for Health Statistics. (2010). *Marriage and cohabitation in the United States*. Retrieved from http://www.cdc.gov/nchs/data/series/sr_23/sr23_028.pdf

National Center for Health Statistics. (2011). *Summary health statistics for U.S. adults: National Health Interview Survey, 2011*. Retrieved from http://www.cdc.gov/nchs/data/series/sr_10/sr10_256.pdf

National Center on Family Homelessness. (2011). *The characteristics and needs of families experiencing homelessness*. Retrieved from http://www.familyhomelessness.org/media/306.pdf

National Coalition of Homeless Veterans. (2011). *FAQ about homeless veterans*. Retrieved from http://www.nchv.org/background.cfm

National Immigration Forum. (2006). *Facts on immigration*. Retrieved from http://www.immigrationforum.org/DesktopDefault.aspx?tabid=790

National Immigration Law Center. (2013). *Basic facts about in-state tuition for undocumented immigrant students*. Retrieved from http://www.nilc.org/Basic-Facts-Instate.Html

National Low Income Housing Coalition. (2000). *Out of reach: The growing gap between housing costs and income of poor people in the United States*. Washington, DC: National Low Income Housing Coalition/Low Income Housing Information Service.

National Low Income Housing Coalition. (2015). *Child poverty rises in rural America*. Retrieved from http://nlihc.org/article/child-poverty-rises-rural-america

National Opinion Research Center. (2001). *The Paycheck Fairness Act: The next step in the fight for fair pay*. Retrieved from http://www.now.org/issues/economic/022709pfa.html

National Sleep Foundation. (2013). *International bedroom poll: Summary of findings*. Retrieved from http://www.sleepfoundation.org/sites/default/files/RPT495a.pdf

National Women's Law Center. (2016). *NWLC resources on poverty, income, and health insurance in 2016*. Retrieved from https://nwlc.org/resources/nwlc-resources-on-poverty-income-and-health-insurance-in-2016/

Neate, R. (2013, July 29). Apple investigates new claims of China factory staff mistreatment. *Guardian*. Retrieved from http://www.theguardian.com/technology/2013/jul/29/apple-investigates-claims-china-factory

Netherland, J., & Hansen, H. (2017). White opioids: Pharmaceutical race and the war on drugs that wasn't. *Biosocieties, 12*, 217–238.

Neumark, D. (2015, December 21). *The effects of minimum wage on employment* (FRBSF Economic Letter, No. 2015-37). Federal Reserve Bank of San Francisco. Retrieved from http://www.frbsf.org/economic-research/publications/economic-letter/2015/december/effects-of-minimum-wage-on-employment/el2015-37.pdf

Newman, K. (2000). *No shame in my game: The working poor in the inner city*. New York, NY: Vintage.

Newport, F. (2013, July 25). In U.S., 87% approve of black–white marriage, vs. 4% in 1958. *Gallup*. Retrieved from http://news.gallup.com/poll/163697/approve-marriage-Blacks-whites.aspx

Newport, F. (2015, April 28). Fewer Americans identify as middle class. *Gallup*. Retrieved from http://www.gallup.com/poll/182918/fewer-americans-identify-middle-class-recent-years.aspx

Newport, F. (2017, June 21). Middle-class identification in U.S. at pre-Recession levels. *Gallup*. Retrieved from https://news.gallup.com/poll/212660/middle-class-identification-pre-recession-levels.aspx

New York City Department of Health and Mental Hygiene. (2016). *Severe maternal morbidity in New York City, 2008–2012*. Retrieved from https://www1.nyc.gov/assets/doh/downloads/pdf/data/maternal-morbidity-report-08-12.pdf

New York Times. (2014, November 21). Ebola facts: How many Ebola patients have been treated outside of West Africa? Retrieved from http://www.nytimes.com/interactive/2014/07/31/world/africa/ebola-virus-outbreak-qa.html?_r=0#outside-africa

Nguyen, T. (2013). From SlutWalks to SuicideGirls: Feminist resistance in the third wave and postfeminist era. *Women's Studies Quarterly, 41*, 157–172.

Niahh, S. S. (2010). *Dancehall: From slave ship to ghetto*. Ottawa, ON: University of Ottawa Press.

Nibley, L. (2011). Two spirits. *PBS*. Retrieved from http://www.pbs.org/independentlens/two-spirits/resources/two-spirits-discussion.pdf

Nicodemo, A., & Petronio, L. (2018, February 26). Schools are safer than they were in the 90s, and school shootings are not more common than they used to be, researchers say. *News @ Northeastern.* Retrieved from https://news.northeastern.edu/2018/02/26/schools-are-still-one-of-the-safest-places-for-children-researcher-says/

Niebuhr, R. (1929). *The social sources of denominationalism.* New York, NY: Holt.

Nielsen, F. (1994). Income inequality and industrial development: Dualism revisited. *American Sociological Review, 59,* 654–677.

Nisbet, R. (2010). *Intelligence and how to get it.* New York, NY: Norton.

Nishimoto, A. (2012, April 12). Robots with laser eyes help manufacture 2013 Ford Escape. *Motor Trend.* Retrieved from http://wot.motortrend.com/robots-with-laser-eyes-help-manufacture-2013-ford-escape-191735.html#ixzz2wFvdD26i

Noble, S. U. (2018). *Algorithms of oppression.* New York, NY: New York University Press.

Nolen-Hoeksema, S. (1993). *Sex differences in depression.* Stanford, CA: Stanford University Press.

Nonprofit Voter Engagement Network. (2013). *America goes to the polls 2012: A report on voter turnout in the 2012 election.* Retrieved from http://www.nonprofitvote.org/download-document/america-goes-to-the-polls-2012.html

Noonan, M. C., & Glass, J. L. (2012, June). *The hard truth about telecommuting.* Bureau of Labor Statistics. Retrieved from https://www.bls.gov/opub/mlr/2012/06/art3full.pdf

Nordberg, J. (2010, September 21). Afghan boys are prized, so girls live the part. *New York Times.* Retrieved from http://www.nytimes.com/2010/09/21/world/asia/21gender.html

Norris, P., & Inglehart, R. (2018). *Cultural backlash: The rise of populist authoritarianism.* Electoral Integrity Project. Retrieved from https://www.electoralintegrityproject.com/populistauthoritarianism/

NPD. (2009). Total game console sales, May 2009 [Blog]. Retrieved from http://www.digital-digest.com/blog/DVDGuy/2009/06/13/game-consoles-may-2009-npd-sales-figure-analysis

NPR. (2014, October 24). Why do Ebola mortality rates vary so widely? Retrieved from http://www.npr.org/2014/10/23/358363535/why-do-ebola-mortality-rates-vary-so-widely

Nuasoft. (2000). *Irish Internet usage statistics.* Retrieved from http://www.nua.ie/surveys/how_many_online/index.html

Nuwer, H. (2013). Chronology of deaths among U.S. college students as a result of hazing, initiation, and pledging-related accidents. Retrieved from http://hazing.hanknuwer.com/listoflists.html

O'Hare, W., & Mather. M. (2008). *Child poverty is highest in rural counties in U.S.* Population Reference Bureau. Retrieved from http://www.prb.org/Publications/Articles/2008/childpoverty.aspx

Oakes, J. (1985). *Keeping track: How schools structure inequality.* New Haven, CT: Yale University Press.

Oakes, J. (1990). *Multiplying inequalities: The effects of race, social class, and tracking on opportunities to learn mathematics and science.* Santa Monica, CA: Rand.

Oakes, J. (1992). Can tracking research inform practice? Technical, normative, and political considerations. *Sociology of Education, 21*(4), 12–21.

Oates, G., & Goode. J. (2012). Racial differences in effects of religiosity and mastery on psychological distress: Evidence from national longitudinal data. *Society & Mental Health, 3,* 40–58.

Obama for America. (2012). Key people—President Barack Obama. Retrieved from http://www.p2012.org/candidates/obamaorg.html

Ogden, C. L., Carroll, M. D., Fryar, C. D., & Flegal, K. M. (2015). *Prevalence of obesity among adults and youth: United States, 2011–2014* (NCHS Data Brief No. 219). Hyattsville, MD: National Center for Health Statistics.

Ohmae, K. (1990). *The borderless world: Power and strategy in the industrial economy.* New York, NY: HarperCollins.

Ohmae, K. (1995). *The end of the nation state: How region states harness the prosperity of the global economy.* New York, NY: Free Press.

Oliphant, B. (2016, September 29). *Support for death penalty lowest in more than four decades.* Pew Research Center. Retrieved from http://www.pewresearch.org/fact-tank/2016/09/29/support-for-death-penalty-lowest-in-more-than-four-decades/

Oliver, M. (2009). *Understanding disability: From theory to practice* (2nd ed.). New York, NY: Palgrave Macmillan.

Oliver, M., & Shapiro, T. (1995). *Black wealth/white wealth: A new perspective on racial inequality.* New York, NY: Routledge.

Olson-Kennedy, J., Cohen-Kettenis, P. T., Kreukels, B. P., Meyer-Bahlburg, H. F., Garofalo, R., Meyer, W., & Rosenthal, S. M. (2016). Research priorities for gender nonconforming/transgender youth: Gender identity development and biopsychosocial outcomes. *Current Opinion in Endocrinology, Diabetes, and Obesity, 23,* 172–179.

Omi, M., & Winant, H. (1994). *Racial formation in the United States: From the 1960s to the 1990s* (2nd ed.). New York, NY: Routledge.

Ong, A. D., Burrow, A. L., Fuller-Rowell, T. E., Ja, N. M., & Sue, D. W. (2013). Racial microaggressions and daily well-being among Asian Americans. *Journal of Counseling Psychology, 60,* 188–199. doi:10.1037/a0031736

Oppel, R. (2011, May 23). Steady decline in major crime baffles experts. *New York Times.* Retrieved from http://www.nytimes.com/2011/05/24/us/24crime.html?_r=1

Oppenheimer, V. (1970). *The female labor force in the United States.* Westport, CT: Greenwood Press.

Orfield, G., & Frankenberg, E. (2014, May 15). *Brown at 60: Great progress, a long retreat and an uncertain future.* Civil Rights Project. Retrieved from https://www.civilrightsproject.ucla.edu/research/k-12-education/integration-and-diversity/brown-at-60-great-progress-a-long-retreat-and-an-uncertain-future/Brown-at-60-051814.pdf

Orfield, G., Ee, J., Frankenberg, E., & Siegel-Hawley, G. (2016, May 16). *Brown at 62: School segregation by race, poverty and state.* Civil Rights Project. Retrieved from https://www.civilrightsproject.ucla.edu/research/k-12-education/integration-and-diversity/brown-at-62-school-segregation-by-race-poverty-and-state/Brown-at-62-final-corrected-2.pdf

Organization for Economic Co-operation and Development. (2010). *OECD factbook 2010: Economic, environmental and social statistics.* Paris, France: OECD Publishing.

Organization for Economic Co-operation and Development. (2013). *Education at a glance: OECD indicators 2012: Japan.* Retrieved from http://www.oecd.org/education/EAG2012%20-%20Country%20note%20-%20Japan.pdf

Organization for Economic Co-operation and Development. (2015). *General government spending."* Retrieved from https://data.oecd.org/gga/general-government-spending.htm

Organization for Economic Co-operation and Development. (2016). *Population with tertiary education (indicator).* Retrieved from https://data.oecd.org/eduatt/population-with-tertiary-education.htm#indicator-chart

Owens, E. (2007). Nonbiological objects as actors. *Symbolic Interaction, 30,* 567–548.

Padavic, I., & Reskin, B. F. (2002). *Women and men at work* (2nd ed.). Thousand Oaks, CA: Pine Forge Press.

Padawer, R. (2014, October 15). When women become men at Wellesley. *New York Times Magazine.* Retrieved from https://www.nytimes.com/2014/10/19/magazine/when-women-become-men-at-wellesley-college.html

Pager, D. (2003). The mark of a criminal record. *American Journal of **Sociology,** 108,* 937–975.

Pager, D. (2007). *Marked: Race, crime, and finding work in an era of mass incarceration.* Chicago, IL: University of Chicago Press.

Pager, D., & Shepard, H. (2008). The sociology of discrimination: Racial discrimination in employment, housing, credit and consumer markets. *Annual Review of Sociology, 34,* 181–209.

Palantir Technologies. (2018). *Fighting crime in Salt Lake City. Impact study.* Retrieved from https://www.palantir.com/wp-assets/wp-content/uploads/2012/06/ImpactStudy_SLCPD.pdf

Palazzolo, J. (2013, March 15). Cost of housing federal prisoners continues to rise [Blog]. *Wall Street Journal.* Retrieved from http://blogs.wsj.com/law/2013/03/15/cost-to-house-federal-prisoners-continues-to-rise/

Palmore, E. (2015). Ageism comes of age. *Journals of Gerontology. Series B, Psychological Sciences and Social Sciences, 70*(6), 873–875.

Paludi, M. A., & Barickman, R. B. (1991). *Academic and workplace sexual harassment: A resource manual.* Albany, NY: State University of New York Press.

Panyarachun, M. A., et al. (2004). *A more secure world: Our shared responsibility: Report of the high-level panel on threats, challenges and change.*

United Nations. Retrieved from http://www. un.org/secureworld

Paoletti, J. (2012). *Pink and blue: Telling the girls from the boys in America*. Bloomington, IN: Indiana University Press.

Park, R. (1952). *Human communities: The city and human ecology*. New York, NY: Free Press.

Parker, K., & Stepler, R. (2017, September 14). As U.S. marriage rate hovers at 50%, education gap in marital status widens. Pew Research Center. Retrieved from http://www.pewresearch.org/fact-tank/2017/09/14/as-u-s-marriage-rate-hovers-at-50-education-gap-in-marital-status-widens/

Parsons, T. (1951). *The social system*. Glencoe, IL: Free Press.

Parsons, T. (1960). Toward a healthy maturity. *Journal of Health and Human Behavior*, *1*(3), 163–173.

Parsons, T. (1964). *The social system*. New York, NY: Free Press.

Parsons, T., & Bales, R. F. (1955). *Family, socialization, and interaction process*. Glencoe, IL: Free Press.

Pascoe, E. (2000, March 11). Can a sense of community flourish in cyberspace? *Guardian*. Retrieved from https://www.theguardian.com/theguardian/2000/mar/11/debate

Pascoe, C. J. (2011). *Dude, You're a fag: Masculinity and sexuality in high school*. Berkeley, CA: University of California Press.

Passel, J. S., Wang, W., & Taylor, P. (2010). *Marrying out*. Pew Research Center. Retrieved from http://pewresearch.org/pubs/1616/american-marriage-interracial-interethnic

Patillo, M. (2013). *Black picket fences: Privileges and peril among the black middle class* (2nd ed.). Chicago, IL: University of Chicago Press.

Payne, K. K., & Manning, W. D. (2015). *Number of children living in same-sex couple households: 2013*. National Center for Family & Marriage Research. Retrieved from https://www.bgsu.edu/ncfmr/resources/data/family-profiles/number-of-children-living-in-same-sex-couple-households—2013.html

Pearce, F. (1976). *Crimes of the powerful: Marxism, crime, and deviance*. London, UK: Pluto Press.

Pedulla, D. S., & Thébaud, S. (2015). Can we finish the revolution? Gender, work-family ideals, and institutional constraint. *American Sociological Review*, *80*(1), 116–139.

Peels, J. (2012). Why is 3D printing said to be the third industrial revolution? *Quora*. Retrieved from https://www.quora.com/Why-is-3D-printing-said-to-be-the-third-industrial-revolution

Peer, B. (2012, October 10). The girl who wanted to go to school. *New Yorker*. Retrieved from http://www.newyorker.com/online/bbgs/newdesk/2012/10/the-girl-who-wanted-to-go-to-school.html

Penner, A. M. (2008). Gender differences in extreme mathematical achievement: An international perspective on biological and social factors. *American Journal of Sociology*, *114*(S1), 138–170.

Perea, F. C. (2012). Hispanic health paradox. In S. Loue & M. Sajatovic (Eds.), *Encyclopedia of Immigrant Health* (pp. 828–830). New York, NY: Springer.

Perrin, A. (2018, June 28). *10 facts about smartphones as the iPhone turns 10*. Pew Research Center. Retrieved from http://www.pewresearch.org/fact-tank/2017/06/28/10-facts-about-smartphones/

Perrin, A., & Duggan, M. (2015). *Americans' Internet access: 2000–2015*. Pew Research Center. Retrieved from http://www.pewinternet.org/2015/06/26/americans-internet-access-2000-2015/

Perrin A. J., Cohen, P. N., & Caren, N. (2013). Are children of parents who had same-sex relationships disadvantaged? A scientific evaluation of the no-differences hypothesis. *Journal of Gay and Lesbian Mental Health*, *17*(3), 327–336.

Perrin, E. C., Siegel, B. S., & Committee on Psychosocial Aspects of Child and Family Health. (2013). Promoting the well-being of children whose parents are gay or lesbian. *Pediatrics*, *131*, 1374–1383.

Pescosolido, B. A., Perry, B. L., Long, J. S., Martin, J. K., Nurnberger, J. I., Jr., & Hesselbrock, V. (2008). Under the influence of genetics: How transdisciplinarity leads us to rethink social pathways to illness. *American Journal of Sociology*, *114*(S1), 171–201.thathavioral circumstances: whilg from genetics (pport system can es: whike a specific gene has been identified as increasing th

Pescosolido, B. A., Medina, T. R., Martin, J. K., & Long, J. S. (2013). The "backbone" of stigma: Identifying the global core of public prejudice associated with mental illness. *American Journal of Public Health*, *103*, 853–860.

Peterson, R. (1996). A re-evaluation of the economic consequences of divorce. *American Sociological Review*, *61*(3), 528–536.

Peterson-Withorn, C. (2015). Forbes billionaires: Full list of the 500 richest people in the world 2015. *Forbes*. Retrieved from http://www.forbes.com/sites/chasewithorn/2015/03/02/forbes-billionaires-full-list-of-the-500-richest-people-in-the-world-2015/#2010865b16e3

Pew Center on the States. (2008). *One in 100: Behind bars in America 2008*. Washington, DC: Pew Charitable Trusts.

Pew Forum on Religion & Public Life. (2008). *U.S. religious landscape survey: Religious affiliation: Diverse and dynamic*. Retrieved from http://religions.pewforum.org/pdf/report-religious-landscape-study-full.pdf

Pew Forum on Religion & Public Life. (2012). *Global religious landscape: Religious composition by country, in numbers*. Retrieved from http://features.pewforum.org/grl/population-number.php

Pew Forum on Religion & Public Life. (2013). *U.S. religious landscape survey*. Retrieved from http://religions.pewforum.org/

Pew Forum on Religion & Public Life. (2014). *The global religious landscape*. Retrieved from http://www.pewforum.org/2012/12/18/global-religious-landscape-exec/

Pew Internet & American Life Project. (2012). *Older adults and Internet use*. Retrieved from http://www.pewinternet.org/~/media//Files/Reports/2012/PIP_Older_adults_and_internet_use.pdf

Pew Internet & American Life Project. (2013). *Coming and going on Facebook*. Retrieved from http://pewinternet.org/Reports/2013/Coming-and-going-on-facebook.aspx

Pew Research Center. (2008). *Young voters in the 2008 election*. Retrieved from http://www.pewresearch.org/2008/11/13/young-voters-in-the-2008-election/www.pewglobal.org

Pew Research Center. (2010). *Distrust, discontent, anger and partisan rancor*. Retrieved from http://pewresearch.org/pubs/1569/trust-in-government-distrust-discontent-anger-partisan-rancor

Pew Research Center. (2012a). *The rise of Asian Americans*. Retrieved from http://www.pewsocialtrends.org/files/2013/04/Asian-Americans-new-full-report-04-2013.pdf

Pew Research Center. (2012b). *The rise of intermarriage: Rates, characteristics vary by race and gender*. Retrieved from http://www.pewsocialtrends.org/files/2012/02/SDT-Intermarriage-II.pdf

Pew Research Center. (2013a). *Modern parenthood: Roles of moms and dads converge as they balance work and family*. Retrieved from http://www.pewsocialtrends.org/files/2013/03/FINAL_modern_parenthood_03-2013.pdf

Pew Research Center. (2013b). *A portrait of Jewish Americans*. Retrieved from http://www.pewforum.org/2013/10/01/jewish-american-beliefs-attitudes-culture-survey/

Pew Research Center. (2013c). *The rise of Asian Americans*. Retrieved from http://www.pewsocialtrends.org/files/2013/04/Asian-Americans-new-full-report-04-2013.pdf

Pew Research Center. (2014a, June 26). *Beyond red and blue: The political typology*. Retrieved from http://www.people-press.org/2014/06/26/the-political-typology-beyond-red-vs-blue/

Pew Research Center. (2014b, February 11). *The rising cost of not going to college*. Retrieved from http://www.pewsocialtrends.org/2014/02/11/the-rising-cost-of-not-going-to-college/

Pew Research Center. (2015c). *America's changing religious landscape*. Retrieved from http://www.pewforum.org/2015/05/12/americas-changing-religious-landscape/

Pew Research Center. (2015d). *Beyond distrust: How Americans view their government*. Retrieved from http://www.people-press.org/2015/11/23/beyond-distrust-how-americans-view-their-government/

Pew Research Center. (2015e). *Changing attitudes on gay marriage*. Retrieved from http://www.pewforum.org/2015/07/29/graphics-slideshow-changing-attitudes-on-gay-marriage/

Pew Research Center. (2015f). *A deep dive into party affiliation*. Retrieved from http://www.people-press.org/2015/04/07/a-deep-dive-into-party-affiliation/

Pew Research Center. (2015g). *The future of world religions*. Retrieved from http://www.pewforum.org/2015/04/02/religious-projections-2010-2050/

Pew Research Center. (2015h). *Less support for death penalty, especially among Democrats.* Retrieved from http://www.people-press.org/2015/04/16/less-support-for-death-penalty-especially-among-democrats/

Pew Research Center. (2015i). *Mixed views of impact of long-term decline in union membership.* Retrieved from http://www.people-press.org/2015/04/27/mixed-views-of-impact-of-long-term-decline-in-union-membership/

Pew Research Center. (2015j). *Modern immigration wave brings 59 million to U.S., driving population growth and change through 2065.* Retrieved from http://www.pewhispanic.org/2015/09/28/modern-immigration-wave-brings-59-million-to-u-s-driving-population-growth-and-change-through-2065/

Pew Research Center. (2015k). *Most say government policies since recession have done little to help middle class, poor.* Retrieved from http://www.people-press.org/files/2015/03/03-04-15-Economy-release.pdf

Pew Research Center. (2015l). *Multiracial in America.* Retrieved from http://www.pewsocialtrends.org/2015/06/11/multiracial-in-america/

Pew Research Center. (2015m). *State of the news media 2015.* Retrieved from http://www.journalism.org/files/2015/04/final-state-of-the-news-media1.pdf

Pew Research Center. (2015n). *U.S. Catholics open to non-traditional families.* Retrieved from http://www.pewforum.org/2015/09/02/u-s-catholics-open-to-non-traditional-families/

Pew Research Center. (2016). *Changing attitudes on gay marriage.* Retrieved from http://www.pewforum.org/2016/05/12/changing-attitudes-on-gay-marriage/

Pew Research Center. (2017a, February 15). *Americans express increasingly warm feelings toward religious groups.* Retrieved from http://www.pewforum.org/2017/02/15/americans-express-increasingly-warm-feelings-toward-religious-groups/

Pew Research Center. (2017b, June 26). *Changing attitudes on gay marriage.* Retrieved from http://www.pewforum.org/fact-sheet/changing-attitudes-on-gay-marriage/

Pew Research Center. (2017c, December 14). *Public trust in government: 1958-2017.* Retrieved from http://www.people-press.org/2017/12/14/public-trust-in-government-1958-2017/

Pew Research Center. (2017d, July 26). *U.S. Muslims concerned about their place in society, but continue to believe in the American dream.* Retrieved from http://www.pewforum.org/2017/07/26/findings-from-pew-research-centers-2017-survey-of-us-muslims/

Pew Research Center. (2017e, April 24). *With budget debate looming, growing share of public prefers bigger government.* Retrieved from http://www.people-press.org/2017/04/24/with-budget-debate-looming-growing-share-of-public-prefers-bigger-government/

Pew Research Center. (2018a). *Internet/broadband fact sheet.* Retrieved from http://www.pewinternet.org/fact-sheet/internet-broadband/

Pew Research Center. (2018b, February 5). *Social media factsheet.* Retrieved from http://www.pewinternet.org/fact-sheet/social-media/

Pew Research Center. (2018c, March 20). *Wide gender gap, growing educational divide in voters' party identification.* Retrieved from http://www.people-press.org/2018/03/20/1-trends-in-party-affiliation-among-demographic-groups/

Pew Research Center. (2018d, Feb. 5). *Mobile Fact Sheet.* Retrieved from http://www.pewinternet.org/fact-sheet/mobile/

Pew Research Center for the People and the Press. (2010). *Public remains conflicted over Islam.* Retrieved from http://www.people-press.org/2010/08/24/public-remains-conflicted-over-islam/

Pew Research Center for the People and the Press. (2012a). *Pew Research Center for the People & the Press values survey, April.* Retrieved from http://www.people-press.org/question-search/?qid=1811708&pid=51&ccid=51#top

Pew Research Center for the People and the Press. (2012b). *Trend in party identification: 1939–2012.* Retrieved from http://www.people-press.org/2012/06/01/trend-in-party-identification-1939-2012/

Pew Research Center Global Indicators Database. (2017). *U.S. Image: Opinion of the United States.* Retrieved from http://www.pewglobal.org/database/indicator/1/

Pew Research Hispanic Trends Project. (2013). *A demographic portrait of Mexican-origin Hispanics in the United States.* Retrieved from http://www.pewhispanic.org/2013/05/01/a-demographic-portrait-of-mexican-origin-hispanics-in-the-united-states

Pew Research Hispanic Trends Project. (2014). *Statistical portrait of Hispanics in the United States, 2012.* Retrieved from http://www.pewhispanic.org/2014/04/29/statistical-portrait-of-hispanics-in-the-united-states-2012/

Pew Social & Demographic Trends. (2010). *The return of the multigenerational family household.* Retrieved from http://pewsocialtrends.org/files/2010/10/752-multi-generational-families.pdf

Pew Social & Demographic Trends. (2011). *Is college worth it?* Retrieved from http://pewsocialtrends.org/files/2011/05/Is-College-Worth-It.pdf

Pierce, C., & Dimsdale, J. (1986). Suppressed anger and blood pressure: The effects of race, sex, social class, obesity, and age. *Psychomatic Medicine, 48,* 430–436.

Pike, K. M., & Dunne, P. E. (2015). The rise of eating disorders in Asia: A review. *Journal of Eating Disorders, 3,* Article 70. Retrieved from https://www.ncbi.nlm.nih.gov/pmc/articles/PMC4574181/pdf/40337_2015_Article_70.pdf

Piketty, T., & Zucman, G. (2014). Capital is back: Wealth-income ratios in rich countries, 1700–2010. *Quarterly Journal of Economics, 129,* 1255–1310.

Pilkington, E. (2009, September 18). Republicans steal Barack Obama's Internet campaigning tricks: Since their election disaster, the right has used new media to gather strength, culminating in last weekend's huge protest. *Guardian.* Retrieved from http://www.guardian.co.uk/world/2009/sep/18/republicans-internet-barack-obama/print

Pillemer, K. (1985). The dangers of dependency: New findings in domestic violence against the elderly. *Social Problems, 33*(2), 146–158.

Pintor, R. L., & Gratschew, M. (2002). *Voter turnout since 1945: A global report.* International Institute for Democracy and Electoral Assistance, Stockholm. Retrieved from http://www.idea.int/publications/turnout/ VT_screenopt_2002.pdf

Pletcher, M. J., Kertesz, S. G., Kohn, M. A., & Gonzales. R. (2008). Trends in opioid prescribing by race/ethnicity for patients seeking care in US emergency departments. *Journal of the American Medical Association, 299*(1), 70–78. doi:10.1001/jama.2007.64

Pollak, O. (1950). *The criminality of women.* Philadelphia, PA: University of Pennsylvania Press.

Pollard, K. (2011). *The gender gap in college enrollment and graduation.* Washington, DC: Population Reference Bureau.

Polletta, F., & Jasper, J. M. (2001). Collective identity and social movements. *Annual Review of Sociology, 27,* 283–305.

Pomerleau, K. (2013). *Summary of latest federal income tax data.* Table 7. Tax Foundation. Retrieved from http://taxfoundation.org/article/summary-latest-federal-income-tax-data

Porter, S. R., & Umbach, P. D. (2006). College major choice: An analysis of student-environment fit. *Research in Higher Education, 47*(4), 429–449.

Potter, H. (2014, June 26). *What can we learn from states that ban affirmative action?* Century Foundation. Retrieved from https://tcf.org/content/commentary/what-can-we-learn-from-states-that-ban-affirmative-action/

Poushter, J. (2016). *Smartphone ownership and Internet usage continues to climb in emerging economies.* Pew Research Center. Retrieved from http://www.pewglobal.org/files/2016/02/pew_research_center_global_technology_report_final_february_22_2016.pdf

Poushter, J., & Bialik, K. (2017, June 26). *Around the world, favorability of the U.S. and confidence in its president decline.* Pew Research Center. Retrieved from http://www.pewresearch.org/fact-tank/2017/06/26/around-the-world-favorability-of-u-s-and-confidence-in-its-president-decline/

Prebisch, R. (1967). *Hacia una dinamica del desarrollo Latinoamericano.* Montevideo, Uruguay: Ediciones de la Banda Oriental.

Prebisch, R. (1971). *Change and development—Latin America's great task: Report submitted to the Inter-American Bank.* New York, NY: Praeger.

Prescott, S. (2015). Star Citizen has now attracted $100 million in crowdfunding. *PC Gamer.* Retrieved from http://www.pcgamer.com/star-citizen-has-now-attracted-100-million-in-crowdfunding/

President's Commission on Organized Crime. (1986). *Records of hearings, June 24–26, 1985.* Washington, DC: U.S. Government Printing Office.

Price, R. (2015). NZ introduces "gender diverse" option. *Stuff.* Retrieved from http://www.stuff.co.nz/national/70335912/nz-introduces-gender-diverse-option

Priest, N., Walton, J., White, F., Kowal, E., Baker, A., & Paradies, Y. (2014). Understanding the complexities of ethnic-racial socialization processes for both minority and majority groups: A 30-year review. *International Journal of Intercultural Relations, 43*, 139–155.

Prisons Bureau. (2017, July 19). Annual determination of average cost of incarceration. *Federal Register*. Retrieved from https://www.federalregister.gov/documents/2016/07/19/2016-17040/annual-determination-of-average-cost-of-incarceration

Puhl, R. M., & Latner, J. D. (2007). Stigma, obesity, and the health of the nation's children. *Psychological Bulletin, 133*, 557–580.

Purcell, K. (2011). *Half of adult cell phone owners have apps on their phones*. Pew Research Internet Project. Retrieved from http://www.pewinternet.org/2011/11/02/half-of-adult-cell-phone-owners-have-apps-on-their-phones/

Purtill, J. (2016, November 10). How one million young people staying home elected Donald Trump. *Triple J*. Retrieved from http://www.abc.net.au/triplej/programs/hack/one-million-young-people-staying-home-elected-donald-trump/8014712

Putnam, R. (1993). The prosperous community: Social capital and public life. *American Prospect, 13*, 35–42.

Putnam, R. (1995). Bowling alone: America's declining social capital. *Journal of Democracy, 6*(1), 65–78.

Putnam, R. (2000). *Bowling alone: The collapse and revival of American community*. New York, NY: Simon & Schuster.

Qian, Z.-C., & Lichter, D. T. (2011). Changing patterns of interracial marriage in a multiracial society. *Journal of Marriage and Family, 73*, 1065–1084.

Quah, D. (1999). *The weightless economy in economic development*. London, UK: Centre for Economic Performance.

Raghuram, S., & Wiesenfeld, B. (2004). Work-nonwork conflict and job stress among virtual workers. *Human Resource Management, 43*(2–3), 259–277.

Rainie, L. (2015). *Digital divides 2015*. Pew Research Center. Retrieved from http://www.pewinternet.org/2015/09/22/digital-divides-2015/

Rainie, L., & Zickuhr, K. (2015). *Americans' views on social etiquette*. Pew Research Center. Retrieved from http://www.pewinternet.org/files/2015/08/2015-08-26_mobile-etiquette_FINAL.pdf

Ramirez, F. O., & Boli, J. (1987). The political construction of mass schooling: European origins and worldwide institutionalism. *Sociology of Education, 60*(1), 2–17.

Rampell, C. (2009, February 5). As layoffs surge, women may pass men in job force. *New York Times*. Retrieved from http://www.nytimes.com/2009/02/06/business/06women.html?

Ranis, G. (1996). *Will Latin America now put a stop to "stop-and-go"?* New Haven, CT: Yale University, Economic Growth Center.

Ranis, G., & Mahmood, S. (1992). *The political economy of development policy change*. Cambridge, MA: Blackwell.

Rastogi, S., Johnson, T. D., Hoeffel, E. M., & Drewery, M. P., Jr. (2011, September). *The black population*. U.S. Bureau of the Census. Retrieved from https://www.census.gov/prod/cen2010/briefs/c2010br-06.pdf

Rauhala, E. (2011). The world welcomes "Baby 7 Billion," but what does her future hold? *Time*. Retrieved from http://world.time.com/2011/10/31/the-world-welcomes-baby-7-billion%E2%80%94what-does-her-future-hold/

Redding, S. (1990). *The spirit of Chinese capitalism*. Berlin, Germany: De Gruyter.

Kann, L., McManus, T., Harris, W. A., Shanklin, S.L., Flint, K.H., Queen, B. ...Ethier, K.A. (2018, June 15). Youth risk behavior surveillance—United States, 2017. *MMWR Surveillance Summaries, 67*(8). Retrieved from https://www.cdc.gov/healthyyouth/data/yrbs/pdf/2017/ss6708.pdf

Redford, J., Battle, D., & Bielick, S. (2017). *Homeschooling in the United States: 2012* (Report No. 2016-096.REV). Washington, DC: National Center for Education Statistics, Institute of Education Sciences, U.S. Department of Education.

Reeves, R. V., & Guyot, K. (2017, December 4). *Black women are earning more college degrees, but that alone won't close race gaps*. Washington, DC: Brookings Institution.

Regnerus, M. (2012). How different are the adult children of parents who have same-sex relationships? Findings from the New Family Structures Study. *Social Science Research, 41*(4), 752–770.

Reich, R. (2015, October 20). A $15 minimum wage is the only moral choice. *Salon*. Retrieved from http://www.salon.com/2015/10/20/robert_reich_a_15_minimum_wage_is_the_only_moral_choice_partner/

Rentz, C., & Donovan, D. (2015, June 27). After Freddie Gray death, cop-watchers film police to prevent misconduct. *Baltimore Sun*. Retrieved from http://www.baltimoresun.com/news/maryland/bs-md-copwatch-20150627-story.html

Renzetti, C. M., & Curran, D. J. (2003). *Women, men, and society* (4th ed.). Needham Heights, MA: Allyn and Bacon.

Reskin, B. F., & I. Padavic. (1994). *Women and men at work*. Thousand Oaks, CA: Pine Forge Press.

Resnick-Ault, J., & Brown, N. (2018, May 18). Puerto Rico power grid braces for hurricane season. *Reuters*. Retrieved from https://www.reuters.com/article/us-puertorico-power-army-corps/puerto-rico-power-grid-braces-for-hurricane-season-idUSKCN1IJ0DL

Reuters. (2013). Farm subsidies still get top share of EU austerity budget. Retrieved from http://www.reuters.com/article/eu-budget-agriculture-idUSL5N0B82UW20130208

Reuters. (2017). Mergers & acquisitions review: Full year 2017. Retrieved from https://www.thomsonreuters.co.jp/content/dam/openweb/documents/pdf/japan/market-review/2017/ma-4q-2017-e.pdf

Richardson, S. A., Goodman, N., Hastorf, A. H., & Dornbusch, S. M. (1961). Cultural uniformity in reaction to physical disabilities. *American Sociological Review, 26*, 241–247.

Rideout, V. J., Foehr, U. G., & Roberts, D. F. (2010). *Media in the lives of 8- to 18-year-olds*. Henry J. Kaiser Family Foundation. Retrieved from https://kaiserfamilyfoundation.files.wordpress.com/2013/04/8010.pdf

Rieff, D. (1991). *Los Angeles: Capital of the third world*. New York, NY: Simon & Schuster.

Rios, V. (2011). *Punished: Policing the lives of black and Latino boys*. New York, NY: NYU Press.

Ritzer, G. (1993). *The McDonaldization of society*. Newbury Park, CA: Pine Forge Press.

Roach, S. (2005, June 6). The new macro of globalization. *Global: Daily Economic Comment*. Morgan Stanley Global Economic Forum.

Rob Bliss Creative. (2014, October 28). 10 hours of walking in NYC as a woman [YouTube video]. Retrieved from https://www.youtube.com/watch?v=b1XGPvbWno A

Roberts, S. (2010, April 28). Listening to (and saving) the world's languages. *New York Times*. Retrieved from http://www.nytimes.com/2010/04/29/nyregion/29lost.html?hpw&_r=0

Robinson, W. (2001). Social theory and globalization: The rise of a transnational state. *Theory and Society, 30*(2), 157–200.

Robinson, W. (2004). *A theory of global capitalism: Production, class, and state in a transnational world*. Baltimore, MD: Johns Hopkins University Press.

Robinson, W. (2005a). Global capitalism: The new transnationalism and the folly of conventional thinking. *Science and Society, 69*(3), 316–328.

Robinson, W. (2005b). Gramsci and globalisation: From nation-state to transnational hegemony. *Critical Review of International Social and Political Philosophy, 8*(4), 1–16.

Robinson, W. I. (2007). *Global capitalism: Its fall and rise in the twentieth century*. New York, NY: Norton.

Robinson, W. (2014). *Global capitalism and the crisis of humanity*. New York, NY: Cambridge University Press.

Robinson, W. I. (2017). *Global capitalism: Reflections on a brave new world*. Great Transition Initiative. Retrieved from http://greattransition.org/publication/global-capitalism

Robles, F. (2017, June 11). 23% of Puerto Ricans vote in referendum, 97% of them for statehood. *New York Times*. Retrieved from https://www.nytimes.com/2017/06/11/us/puerto-ricans-vote-on-the-question-of-statehood.html

Rodriquez, J. (2011). "It's a dignity thing": Nursing home care workers' use of emotions. *Sociological Forum, 26*(2), 265–286.

Rogers, R. (2016). Mergers & acquisitions review. *Thomson Reuters*. Retrieved from http://share.thomsonreuters.com/general/PR/MA-4Q15-(E).pdf

Roof, W. (1993). *A generation of seekers: The spiritual journeys of the baby boom generation*. San Francisco, CA: Harper.

Roof, W. (1999). *Spiritual marketplace: Baby boomers and the remaking of American religion*. Princeton, NJ: Princeton University Press.

Roof, W., & McKinney, W. (1990). *American mainline religion: Its changing shape and future prospects.* New Brunswick, NJ: Rutgers University Press.

Rosenau, J. (1997). *Along the domestic-foreign frontier: Exploring governance in a turbulent world.* Cambridge, UK: Cambridge University Press.

Rosenfeld, M. J. (2010). Nontraditional families and childhood progress through school. *Demography, 47,* 755–775.

Rosenthal, E. (2013, October 12). The soaring cost of a simple breath. *New York Times.* Retrieved from http://www.nytimes.com/2013/10/13/us/the-soaring-cost-of-a-simple-breath.html

Rosenthal, R., & Jacobson, L. (2003). *Pygmalion in the classroom: Teacher expectation and pupils' intellectual development.* Norwalk, CT: Crown House. (Original work published 1968)

Rosin, H. (2014, October 29). The problem with that catcalling video [Blog]. *Slate.* Retrieved from http://www.slate.com/blogs/xx_factor/2014/10/29/catcalling_video_hollaback_s_look_at_street_harassment_in_nyc_edited_out.html

Rossi, A. (1973). The first woman sociologist: Harriett Martineau. In *The feminist papers: From Adams to de Beauvoir.* New York, NY: Columbia University Press.

Rostow, W. (1961). *The stages of economic growth.* Cambridge, UK: Cambridge University Press.

Rothstein, R. (2017). *The color of law: A forgotten history of how our government segregated America.* New York, NY: Liveright.

Rousselle, R. (1999). Defining ancient Greek sexuality. *Digital Archives of Psychohistory, 26* (4). Retrieved from http://www.geocities.ws/kidhistory/ja/defining.htm

Rowe, J. W., & Kahn, R. L. (1987). Human aging: Usual and successful. *Science, 237*(4811), 143–149.

Rubin, L. (1990). *Erotic wars: What happened to the sexual revolution?* New York, NY: Farrar, Straus and Giroux.

Rubin, J., Provenzano, F., & Luria, Z. (1974). The eye of the beholder: Parents' views on sex of newborns. *American Journal of Orthopsychiatry, 44,* 512–519.

Rubinstein, W. (1986). *Wealth and inequality in Britain.* Winchester, MA: Faber & Faber.

Rubinstein, S., & Caballero, B. (2000). Is Miss America an undernourished role model? *Journal of the American Medical Association, 283*(12), 1569.

Rudé, G. (1964). *The crowd in history: A study of popular disturbances in France and England, 1730–1848.* New York, NY: Wiley.

Rugh, J. S., & Massey, D. S. (2010). Racial segregation and the American foreclosure crisis. *American Sociological Review, 75,* 629–651. doi:10.1177/0003122410380868

Russell, S. T., & Joyner, K. (2001). Adolescent sexual orientation and suicide risk: Evidence from a national study. *American Journal of Public Health, 91,* 1276–1281.

Rutter, M., & Giller, H. (1984). *Juvenile delinquency: Trends and perspectives.* New York, NY: Guilford Press.

Ryan, T. (1985). The roots of masculinity. In A. Metcalf & M. Humphries (Eds.), *Sexuality of men* (pp. 26–27). London: Pluto.

Ryan, J. (2018, February 12). What Amazon doesn't want you to know about the cloud. *Northwest Public Broadcasting.* Retrieved from https://www.nwpb.org/2018/02/12/amazon-doesnt-want-know-cloud/

Ryan, C. L., & Bauman, K. (2016). *Educational attainment in the United States: 2015* (Current Population Reports, P20-578). U.S. Bureau of the Census. Retrieved from http://www.census.gov/content/dam/Census/library/publications/2016/demo/p20-578.pdf

Ryan, C., Huebner, D., Diaz, R. M., & Sanchez, J. (2009). Family rejection as a predictor of negative health outcomes in white and Latino lesbian, gay, and bisexual young adults. *Pediatrics, 123,* 346–352.

Saad, N. (2017, October 31). Harvey Weinstein's accusers: Full list includes fledgling actresses and Hollywood royalty. *Los Angeles Times.* Retrieved from http://www.latimes.com/entertainment/la-et-weinstein-accusers-list-20171011-htmlstory.html

Sachs, J. (2000, June 22). A new map of the world. *The Economist.* Retrieved from https://www.economist.com/unknown/2000/06/22/a-new-map-of-the-world

Sadker, M., & Sadker, D. (1994). *Failing at fairness.* New York, NY: Scribner.

Saez, E., & Zucman, G. (2014). Wealth inequality in the United States since 1913: Evidence from capitalized income tax data. *Quarterly Journal of Economics, 131*(2), 519–578.

Sagan, A. (2013, April 25). Anti-bullying apps will fail unless society takes action, critics say. *CBC News.* Retrieved from http://www.cbc.ca/news/canada/story/2013/04/18/f-anti-bullying-apps.html

Saguy, A. (2012). *What's wrong with fat?* New York, NY: Oxford University Press.

Sahn, D. E., Dorosh, P. A., & Younger, D. D. (1997). *Structural adjustment reconsidered: Economic policy and poverty in Africa.* New York, NY: Cambridge University Press.

Saks, M. (Ed.). (1992). *Alternative medicine in Britain.* Oxford, UK: Clarendon.

Sallans, R. K. (2016). Lessons from a transgender patient for health care professionals. *AMA Journal of Ethics, 18,* 1139–1146.

SAMHSA. (2010). *Violent behaviors among adolescent females.* Substance Abuse and Mental Health Services Administration. Retrieved from https://archive.samhsa.gov/data/2k9/171/171FemaleViolence.htm

Sampson, R. J., & Cohen, J. (1988). Deterrent effects of the police on crime: A replication and theoretical extension. *Law and Society Review, 22*(1), 163–189.

Sanburn, J. (2014, September 25). NYPD confrontation with pregnant woman is latest police video to go viral. *Time.* Retrieved from http://time.com/3426859/nypd-pregnant-woman-brooklyn-video/

Sandefur, G. D., & Liebler, C. A. (1997). The demography of American Indian families. *Population Research and Policy Review, 16*(12), 95–114.

Sandler, L. (2013, August 12). The childfree life: When having it all means not having children. *Time,* pp. 38–45.

Santos, H. C., Varnum, M. E., & Grossman, I. (2017). Global increases in individualism. *Psychological Science, 28,* 1228–1239.

Sarkar, M., & Torre, I. (2015, February 10). Same-sex marriage: Where is it legal? *CNN.* Retrieved from http://www.cnn.com/2015/02/10/world/gay-marriage-world/

Sarkisian, N., & Gerstel, N. (2004). Kin support among blacks and whites: Race and family organization. *American Sociological Review, 69,* 812–837.

Sartre, J. (1965). *Anti-Semite and Jew.* New York, NY: Schocken Books. (Original work published 1948)

Sassen, S. (1991). *The global city: New York, London, Tokyo.* Princeton, NJ: Princeton University Press.

Sassen, S. (1996). *Losing control: Sovereignty in the age of globalization.* New York, NY: Columbia University Press.

Sassen, S. (1998). *Globalization and its discontents.* New York, NY: New Press.

Sassen, S. (2005). *Denationalization: Territory, authority and rights.* Princeton, NJ: Princeton University Press.

Saulny, S. (2011, January 29). Black? White? Asian? More young Americans choose all of the above. *New York Times.* Retrieved from http://www.nytimes.com/2011/01/30/us/30mixed.html

Schaefer, A., Mattingly, M. J., & Johnson, K. M. (2016). *Child poverty higher and more persistent in rural America.* Retrieved from University of New Hampshire Scholars Repository website: https://scholars.unh.edu/cgi/viewcontent.cgi?article=1265&context=carsey

Schaffner, B. F., MacWilliams, M., & Nteta, T. (2017). *Explaining white polarization in the 2016 vote for president: The sobering role of racism and sexism.* Retrieved from UMassAmherst People website: https://people.umass.edu/schaffne/schaffner_et_al_IDC_conference.pdf

Schaie, K. (1983). *Longitudinal studies of adult psychological development.* New York, NY: Guilford Press.

Scheff, T. (1966). *Being mentally ill.* Chicago, IL: Aldine.

Schilt, K. (2010). *Just one of the guys? Transgender men and the persistence of gender inequality.* Chicago, IL: University of Chicago Press.

Schmitz, R. (2017, October 19). Xi Jinping's war on poverty moves millions of Chinese off the farm. *NPR.* Retrieved from https://www.npr.org/sections/parallels/2017/10/19/558677808/xi-jinping-s-war-on-poverty-moves-millions-of-chinese-off-the-farm

Schnaiberg, A. (1980). *The environment: From surplus to scarcity.* New York, NY: Oxford University Press.

Schofield, J. (1995). Review for research on school desegregation's impact on elementary and secondary school students. In J. A. Banks & C. A. M. Banks (Eds.), *Handbook on research on multicultural education* (pp. 597–616). New York, NY: Simon & Schuster.

Schoon, I. (2008). A transgenerational model of status attainment: The potential mediating role of school motivation and education. *National Institute Economic Review, 205,* 72–82.

Schulman, M. (2013, January 9). Generation LGBTQIA. *New York Times*. Retrieved from http://www.nytimes.com/2013/01/10/fashion/generation-lgbtqia.html?pagewanted=all

Schumpeter, J. (1983). *Capitalism, socialism, and democracy*. Magnolia, MA: Peter Smith. (Original work published 1942)

Schwartz, G. (1970). *Sect ideologies and social status*. Chicago, IL: University of Chicago Press.

Scommegna, P. (2011, November). *U.S. megalopolises 50 years later*. Population Reference Bureau. Retrieved from http://www.prb.org/Publications/Articles/2011/us-megalopolises-50-years.aspx

Scott, W. R. (2004). Reflections on a half-century of organizational sociology. *Annual Review of Sociology, 30*, 1–21.

Scott, S., & Morgan, D. (1993). Bodies in a social landscape. In S. Scott & D. Morgan (Eds.), *Body matters: Essays on the sociology of the body* (pp. 1–21). Washington, DC: Falmer Press.

Sedlak, A., & Broadhurst, D. D. (1996). *Third national incidence study of child abuse and neglect*. Washington, DC: U.S. Department of Health and Human Services.

Segarra, L. M. (2017, September 30). Donald Trump: Puerto Rico wants "everything to be done for them." *Time*. Retrieved from http://time.com/4963903/donald-trump-puerto-rican-leaders-want-everything-to-be-done-for-them/

Seidenberg, P., Nicholson, S., Schaefer, M., Semrau, K., Bweupe, M., Masese, N., . . . Thea, D. M. (2012). Early infant diagnosis of HIV infection in Zambia through mobile phone texting of blood test results. *Bulletin of the World Health Organization, 90*, 348–356.

Seidman, S., Meeks, C., & Traschen, F. (1999). Beyond the closet? The changing social meaning of homosexuality in the United States. *Sexualities, 2*(1), 9–34.

Semega, J. L., Fontenot, K. R, & Kollar, M. A. (2017). *Income and poverty in the United States: 2016*. U.S. Bureau of the Census. Retrieved from https://www.census.gov/content/dam/Census/library/publications/2017/demo/P60-259.pdf

Semuels, A. (2015, October 26). "Good" jobs aren't coming back. *The Atlantic*. Retrieved from http://www.theatlantic.com/business/archive/2015/10/onshoring-jobs/412201/

Sepúlveda, A. R., & Calado, M. (2012). Westernization: The role of mass media on body image and eating disorders. In I. J. Lobera (Ed.), *Relevant topics in eating disorders* (pp. 47–64). InTech. Retrieved from http://cdn.intechopen.com/pdfs/29049/InTech-Westernization_the_role_of_mass_media_on_body_image_and_eating_disorders.pdf

Seville Statement on Violence. (1990). *American Psychologist, 45*(10), 1167–1168. Retrieved from http://www.lrainc.com/swtaboo/taboos/seville1.html

Sewell, W., & Hauser, R. (1980). The Wisconsin longitudinal study of social and psychological factors in aspirations and achievements. In A. C. Kerckhoff (Ed.), *Research in Sociology and Education* (Vol. 1, pp. 59–99). Greenwich, CT: JAI Press.

Sharkey, P. (2018). *Uneasy peace: The great crime decline, the renewal of city life, and the next war on violence*. New York, NY: Norton.

Sharp, G. (2012, June 26). Working-class masculinity and street harassment. *Sociological Images*. Retrieved from http://thesocietypages.org/socimages/2012/06/

Sharp, E. A., Weiser, D. A., Lavigne, D. E., & Corby Kelly, R. (2017). From furious to fearless: Faculty action and feminist praxis in response to rape culture on college campuses. *Family Relations, 66*, 75–88.

Shea, S., Stein, A. D., Basch, C. E., Lantigua, R., Maylahn, C., Strogatz, S., & Novick, L. (1991). Independent associations of educational attainment and ethnicity with behavioral risk factors for cardiovascular disease. *American Journal of Epidemiology, 134*(6), 567–582.

Shearer, E., & Gottfried, J. (2017, September 7). *News use across social media platforms 2017*. Pew Research Center. Retrieved from http://www.journalism.org/2017/09/07/news-use-across-social-media-platforms-2017/

Sheldon, W. H. (1949). *Varieties of delinquent youth. An introduction to constitutional psychiatry*. New York, NY: Harper and Brothers.

Shenglin, B., Simonelli, F., Ruidong, Z., Bosc, R., & Li, W. (2018). *Digital infrastructure: Overcoming the digital divide in emerging economies*. G20 Insights. Retrieved from http://www.g20-insights.org/policy_briefs/digital-infrastructure-overcoming-digital-divide-emerging-economies/

Sherk, J. (2010, October 10). *Technology explains drop in manufacturing jobs*. Heritage Foundation. Retrieved from http://www.heritage.org/research/reports/2010/10/technology-explains-drop-in-manufacturing-jobs

Sherman, C. W., Webster, N. J., & Antonucci, T. C. (2013). Dementia caregiving in the context of late-life remarriage: Support networks, relationship quality, and well-being. *Journal of Marriage and Family, 75*(5), 1149–1163.

Sherwell, P. (2013, July 25). Merrill Lynch female trainees "given Seducing the Boys Club book." *Telegraph*. Retrieved from http://www.telegraph.co.uk/news/worldnews/northamerica/usa/10203266/Merrill-Lynch-female-trainees-given-Seducing-the-Boys-Club-book.htm

Sherwood, H. (2011, September 4). Israeli protests: 430,000 take to streets to demand social justice. *Guardian*. Retrieved from http://www.guardian.co.uk/world/2011/sep/04/israel-protests-social-justice

Shollenberger, T. L. (2014). Racial disparities in school suspension and subsequent outcomes. In D. J. Losen (Ed.), *Closing the school discipline gap: Equitable remedies for excessive exclusion* (pp. 31–43). New York, NY: Teachers College Press.

Shostak, M. (1981). *Nisa: The life and words of a !Kung woman*. Cambridge, MA: Harvard University Press.

Shwayder, M. (2013, November 5). A same-sex domestic violence epidemic is silent. *The Atlantic*. Retrieved from https://www.theatlantic.com/health/archive/2013/11/a-same-sex-domestic-violence-epidemic-is-silent/281131/

Sifferlin, A. (2017, July 26). Here's how happy Americans are right now. *Time*. Retrieved from http://time.com/4871720/how-happy-are-americans/

Sigmund, P. (1990). *Liberation theology at the crossroads: Democracy or revolution?* New York, NY: Oxford University Press.

Silva, B. (2017). Fortune Global 500 2017: These are the companies shaping the world. *Fortune*. Retrieved from http://fortune.com/global500/

Simmel, G. (1955). *Conflict and the web of group affiliations* (K. Wolff, Trans.). Glencoe, IL: Free Press.

Simms, J. (2013, February 27). Asia's women in the mix: Sakie Fukushima on gender adversity in Japan. *Forbes Asia*. Retrieved from http://www.forbes.com/sites/forbesasia/2013/02/27/asias-women-in-the-mix-sakie-fukushima-on-gender-adversity-in-japan/

Simpson, I. H., Stark, D., & Jackson, R. A. (1988). Class identification processes of married, working men and women. *American Sociological Review, 53*, 284–293.

Siner, E. (2014, January 10). Bitcoin takes stage in Texas senate campaign [Blog]. *National Public Radio*. Retrieved from http://www.npr.org/blogs/itsallpolitics/2014/01/10/260572933/bitcoin-takes-stage-in-texas-campaign

Singer, S. (2017, February 19). Top 10 reasons school choice is no choice. *Huffington Post*. Retrieved from https://www.huffingtonpost.com/entry/top-10-reasons-school-choice-is-no-choice_us_58a8d52fe4b0b0e1e0e20be3

Singer, H. (2018). Our leadership: Global executive team. Retrieved from Ashoka website: https://www.ashoka.org/en/leadership

Singhal, A., Tien, Y., & Hsia, R. Y. (2016, August). Racial-ethnic disparities in opioid prescriptions at emergency department visits for conditions commonly associated with prescription drug abuse. *PLoS ONE, 11*(8), e0159224.

Sisterhood Is Global Institute. (2018). Home page. Retrieved from https://sigi.org/

Sklair, L. (2000a). *The transnational capitalist class*. Malden, MA: Blackwell.

Sklair, L. (2000b). The transnational capitalist class and the discourse of globalisation. *Cambridge Review of International Affairs, 14*(1), 67–85.

Sklair, L. (2002a). Democracy and the transnational capitalist class. *Annals of the American Academy of Political and Social Science, 581*, 144–157.

Sklair, L. (2002b). *Globalization: Capitalism and its alternatives* (3rd ed.). New York, NY: Oxford University Press.

Sklair, L. (2003). Transnational practices and the analysis of the global system. In A. Hulsemeyer (Ed.), *Globalization in the twenty-first century: Convergence or divergence?* (pp. 15–32). New York, NY: Palgrave Macmillan.

Slapper, G., & Tombs, S. (1999). *Corporate crime*. Essex, UK: Longman.

Slater, D., & Tonkiss, F. (2013). *Market society: Markets and modern social theory*. New York, NY: Wiley.

Smedley, A. (1993). *Race in North America: Origin and evolution of a world view*. Boulder, CO: Westview Press.

Smelser, N. (1963). *Theory of collective behavior*. New York, NY: Free Press.

Smith, D. (2003). *The older population in the United States: March 2002* (Current Population Reports, P20-546). U.S. Bureau of the Census. Retrieved from https://www.census.gov/prod/2003pubs/p20-546.pdf

Smith, M. (2012, December 17). Gender-neutral Easy-Bake Oven announced by Hasbro following 13-year-old's petition. *Huffington Post*. Retrieved from http://www.huffingtonpost.com/2012/12/17/gender-neutral-easy-bake-oven_n_2318521.html

Smith, A. (2015a). *U.S. smartphone use in 2015, Chapter one: A portrait of smartphone use*. Pew Research Center. Retrieved from http://www.pewinternet.org/2015/04/01/chapter-one-a-portrait-of-smartphone-ownership/

Smith, A. (2015b). *Searching for work in the digital era*. Pew Research Center. Retrieved from http://www.pewinternet.org/2015/11/19/searching-for-work-in-the-digital-era/

Smith, S. (2017). *Why people are rich and poor: Republicans and Democrats have very different views*. Pew Research Center. Retrieved from http://www.pewresearch.org/fact-tank/2017/05/02/why-people-are-rich-and-poor-republicans-and-democrats-have-very-different-views/

Smith, E. L., & Cooper, A. (2013). *Homicide in the U.S. known to law enforcement, 2011*. Retrieved from https://www.bjs.gov/content/pub/pdf/hus11.pdf

Smith, G., & Martinez, J. (2016, November 9). *How the faithful voted: A preliminary 2016 analysis*. Pew Research Center. Retrieved from http://www.pewresearch.org/fact-tank/2016/11/09/how-the-faithful-voted-a-preliminary-2016-analysis/

Smith, J., & Medalia, C. (2015). *Health insurance coverage in the United States: 2014*. U.S. Bureau of the Census. Retrieved from http://www.census.gov/content/dam/Census/library/publications/2015/demo/p60-253.pdf

Smith, T. W., & Son, J. (2013). *Final report: Trends in public attitudes about sexual morality*. Chicago, IL: National Opinion Research Council. Retrieved from http://www.norc.org/PDFs/sexmoralfinal_06-21_FINAL.PDF

Smith, T. W., & Son, J. (2014). Measuring Occupational Prestige on the 2012 General Social Survey. *General Social Survey Methodological Report, 122*. Retrieved from http://gss.norc.org/Documents/reports/methodological-reports/MR122%20Occupational%20Prestige.pdf

Smith, W. A., Allen, W. R., & Danley, L. L. (2007, December). "Assume the position . . . you fit the description": Psychosocial experiences and racial battle fatigue among African American male college students. *American Behavioral Scientist, 52*, 5511–5578. doi:10.1177/0002764207307742

Smith, H. J., Pettigrew, T. F., Pippin, G. M., & Bialosiewicz, S. (2012). Relative deprivation: A theoretical and meta-analytic review. *Personality and Social Psychology Review, 16*, 203–232.

Smith, S. G., Chen, J., Basile, K. C., Gilbert, L. K., Merrick, M. T., Patel, N., . . . Jain, A. (2017). *The National Intimate Partner and Sexual Violence Survey (NISVS): 2010–2012 state report*. Atlanta, GA: National Center for Injury Prevention and Control, Centers for Disease Control and Prevention.

Smith-Bindman, R., Miglioretti, D. L., Lurie, N., Abraham, L., Barbash, R. B., Strzelczyk, J., . . . Kerlikowske K. (2006). Does utilization of screening mammography explain racial and ethnic differences in breast cancer? *Annals of Internal Medicine, 144*(8), 541–553.

Smokowski, P. R., & Bacallao, M. (2011). *Becoming bicultural: Risk, resilience, and Latino youth*. New York, NY: New York University Press.

Snyder, T. D., & Dillow, S. A. (2010). *Digest of education statistics 2009* (NCES 2010-013). National Center for Education Statistics, Institute of Education Sciences. Washington, DC: U.S. Department of Education.

Social Security Administration. (2012). *Income of the population 55 or older, 2010*. Retrieved from http://www.ssa.gov/policy/docs/statcomps/income_pop55/2010/incpop10.pdf

Social Security Administration. (2016). *December 2015 fact sheet*. Retrieved from https://www.ssa.gov/policy/docs/quickfacts/stat_snapshot/2015-12.pdf

Social Security Administration. (2018). *Annual statistical supplement, 2017*. Retrieved from https://www.ssa.gov/policy/docs/statcomps/supplement/2017/highlights.pdf

Sohn, E. (2015, May 28). Smartphones are so smart they can now test your vision." *NPR*. Retrieved from https://www.npr.org/sections/goatsandsoda/2015/05/28/409731415/smartphones-are-so-smart-they-can-now-test-your-vision

Southern Poverty Law Center. (2015b). *Lone wolf report, February 11*. Retrieved from https://www.splcenter.org/20150212/lone-wolf-report

Southern Poverty Law Center. (2017). *Hate map*. Retrieved from https://www.splcenter.org/hate-map

Southwick, S. (1996). Liszt: Searchable directory of e-mail discussion groups. Retrieved from http://www.liszt.com

Spalter-Roth, R., & Van Vooren, N. (2008a, January). *What are they doing with a bachelor's degree in sociology? Data brief on current jobs* (ASA Department of Research and Development Research Briefs). Retrieved from http://www.asanet.org/sites/default/files/savvy/research/BachelorsinSociology.pdf

Spalter-Roth, R., & Van Vooren, N. (2008b, April). *Pathways to job satisfaction: What happened to the class of 2005?* (ASA Department of Research and Development Research Briefs). Retrieved from http://www.asanet.org/sites/default/files/savvy/research/PathJobSatisfaction.pdf

Specia, M. (2018, April 13). How Syria's death toll is lost in the fog of war. *New York Times*. Retrieved from https://www.nytimes.com/2018/04/13/world/middleeast/syria-death-toll.html

Spectrem Group. (2018). New Spectrem Group market insights report reveals significant growth in U.S. household wealth in 2017. Retrieved from https://spectrem.com/Content/press-release-new-spectrem-group-market-insights-report-reveals-significant-growth-in-US-household-wealth-in-2017.aspx

Spencer, H. (1873). *The study of sociology*. New York, NY: Appleton.

Spencer, S. (2014). *Race and ethnicity*. Abington, UK: Routledge.

Speth, L. E. (2011). The married women's property acts, 1839–1865: Reform, reaction, or revolution? In J. R. Lindgren, N. Taub, B. A. Wolfson, & C. M. Palumbo (Eds.), *The law of sex discrimination* (4th ed., pp. 12–15). New York, NY: Wadsworth.

Spinks, W. A., & Wood, J. (1996). Office-based telecommuting: An international comparison of satellite offices in Japan and North America. In M. Igbaria (Chair) *SIGCPR '96 Proceedings of the 1996 ACM SIGCPR/SIGMIS conference on computer personnel research* (pp. 338–350). New York, NY: ACM.

Springer, K. W., & Mouzon, D. M. (2011). "Macho men" and preventive health care: Implications for older men in different social classes. *Journal of Health and Social Behavior, 52*, 212–227.

Stacey, J., & Biblarz, T. (2001). (How) does the sexual orientation of parents matter? *American Sociological Review, 66*(2), 159–183.

Stack, C. (1997). *All our kin: Strategies for survival in a black community*. New York, NY: HarperCollins.

Stalker, P. (2011). Stalker's guide to international migration—Map of migration flows. Retrieved from http://www.pstalker.com/migration/mg_map.htm

Stampp, K. (1956). *The peculiar institution*. New York, NY: Knopf.

Stark, R., & Bainbridge, W. S. (1980). Towards a theory of religion: Religious commitment. *Journal for the Scientific Study of Religion, 19*(2), 114–128.

Stark, R., & Bainbridge, W. S. (1987). *A theory of religion*. New Brunswick, NJ: Rutgers University Press.

Statista. (2016). *Number of Internet users worldwide from 2000 to 2015 (in millions)*. Retrieved from http://www.statista.com/statistics/273018/number-of-internet-users-worldwide/

Statista. (2017). *Distribution of game developers worldwide from 2014 to 2016, by gender*. Retrieved from https://www.statista.com/statistics/453634/game-developer-gender-distribution-worldwide/

St. Clair, D., & Cayton, H. R. (1945). *Black metropolis: A study of Negro life in a northern city*. Chicago, IL: University of Chicago Press.

Steffensmeier, D. J., Schwartz, J., & Roche, M. (2013). Gender and twenty-first-century corporate crime: Female involvement and the gender gap in Enron-era corporate frauds. *American Sociological Review, 78*, 448–476. doi:10.1177/0003122413484150

Steinmetz, S. (1983). Family violence toward elders. In S. Saunders, A. M. Anderson, C. A. Hart, & G. M. Rubenstein (Eds.), *Violent individuals and families: A handbook for practitioners*. Springfield, IL: Charles C. Thomas.

Stephens-Davidowitz, S. (2013, December 8). How many American men are gay? *New York Times*. Retrieved from http://www.nytimes.com/2013/12/08/opinion/sunday/how-many-american-men-are-gay.html?ref=opinion

Stepler, R. (2017, April 6). *Number of U.S. adults cohabiting with a partner continues to rise, especially among those 50 and older*. Pew Research Center. Retrieved from http://www.pewresearch.org/fact-tank/2017/04/06/number-of-u-s-adults-cohabiting-with-a-partner-continues-to-rise-especially-among-those-50-and-older/

Stepler, R., & Brown, A. (2016). *Statistical portrait of Hispanics in the United States*. Pew Research Center. Retrieved from http://www.pewhispanic.org/2016/04/19/statistical-portrait-of-hispanics-in-the-united-states/

Stevens, T., Morash, M., & Chesney-Lind, M. (2011). Are girls getting tougher, or are we tougher on girls? Probability of arrest and juvenile court oversight in 1980 and 2000. *Justice Quarterly, 28*, 719–744.

Stewart, J. B. (2013, March 15). Looking for a lesson in Google's perks. *New York Times*. Retrieved from http://www.nytimes.com/2013/03/16/business/at-google-a-place-to-work-and-play.html?pagewanted=all&_r=1&

Stiglitz, J. (2007). *Making globalization work*. New York, NY: Norton.

Stiglitz, J. (2017). *Globalization and its discontents revisited: Anti-globalization in the era of Trump*. New York, NY: W. W. Norton.

Stockholm International Peace Research Institute. (2018). *Trends in world military expenditure, 2017*. Retrieved from https://www.sipri.org/sites/default/files/2018-04/sipri_fs_1805_milex_2017.pdf

Stokes, B. (2013). *Public attitudes toward the next social contract*. Pew Research Center. Retrieved from http://www.pewglobal.org/files/pdf/Stokes_Bruce_NAF_Public_Attitudes_1_2013.pdf

Stokes, B. (2017, Feb. 1). *What it takes to truly be "one of us."* Pew Research Center. Retrieved from http://www.pewglobal.org/2017/02/01/what-it-takes-to-truly-be-one-of-us/

Stranges, S., Tigbe, W. Gomez-Olive, F. Z. Thorogood, M., & Kaldala, N.-B. (2012). Sleep problems: An emerging global epidemic? Findings from the in-depth WHO-SAGE study among more than 40,000 older adults from 8 countries. *Sleep, 35*(8), 1173–1181.

Straus, M. A., & Gelles, R. J. (1986). Societal change and change in family violence from 1975 to 1985 as revealed by two national surveys. *Journal of Marriage and the Family, 48*, 465–479.

Stromberg, J. (2013). What is the Anthropocene and are we in it? *Smithsonian*. Retrieved from http://www.smithsonianmag.com/science-nature/what-is-the-anthropocene-and-are-we-in-it-164801414/?no-ist

Stuart, F. (2016). *Down, out, and under arrest: Policing and everyday life in Skid Row*. Chicago, IL: University of Chicago Press.

Substance Abuse and Mental Health Services Administration. (2015). *Behavioral health trends in the United States: Results from the 2014 National Survey on Drug Use and Health*. Retrieved from http://www.samhsa.gov/data/sites/default/files/NSDUH-FRR1-2014/NSDUH-FRR1-2014.pdf

Substance Abuse and Mental Health Services Administration (SAMHSA) (2016, November 10). *Cultural competence*. Retrieved from https://www.samhsa.gov/capt/applying-strategic-prevention/cultural-competence

Sue, D. W. (2010). *Microaggressions in everyday life: Race, gender, and sexual orientation*. New York, NY: John Wiley.

Suitor, J. J., Sechrist, J., Gilligan, M., & Pillemer, K. (2011). Intergenerational relations in later-life families. In R. Settersten & J. Angel (Eds.), *Handbook of sociology of aging* (pp. 161–178). New York, NY: Springer.

Sultan, N. (2017, April 13). Election 2016: Trump's free media helped keep cost down, but fewer donors provided more of the cash. Center for Responsive Politics. Retrieved from https://www.opensecrets.org/news/2017/04/election-2016-trump-fewer-donors-provided-more-of-the-cash/

Summers, N. (2013). *The future of health apps: Personalized advice combining your diet, sleep pattern and fitness regime*. Retrieved from TNW website: http://thenextweb.com/apps/2013/08/07/the-future-of-health-apps-smart-recommendations-for-your-diet-sleep-pattern-and-fitness-regime/#!pn9Hz

Sunstein, C. (2012). *Republic.com 2.0*. Princeton, NJ: Princeton University Press.

Sutherland, E. (1949). *Principles of criminology*. Chicago, IL: Lippincott.

Sweeney, M. (2002). Two decades of family change: The shifting economic foundations of marriage. *American Sociological Review, 67*(1), 132–147.

Sweeney, M. (2010). Remarriage and stepfamilies: Strategic sites for family scholarship in the 21st century. *Journal of Marriage and Family, 72*, 667–684.

Swidler, A. (1986). Culture in action: Symbols and strategies. *American Sociological Review, 51*(2), 273–286.

Symantec Corporation. (2013). *Internet security threat report 2013: Vol. 18*. Retrieved from http://www.symantec.com/content/en/us/enterprise/other_resources/b-istr_main_report_v18_2012_21291018.en-us.pdf

Symantec Corporation. (2016, April). *2016 Internet security threat report*. Retrieved from https://www.symantec.com/content/dam/symantec/docs/reports/istr-21-2016-en.pdf?aid=elq_&om_sem_kw=elq_16285871&om_ext_cid=biz_email_elq_&elqTrackId=283a3acdb3ff42f4a70ab5a9f236eb71&elqaid=2902&elqat=2

Taggart, T., Grewe, M. E., Conserve, D. F., Gliwa, C., & Roman Isler, M. (2015). Social media and HIV: A systematic review of uses of social media in HIV communication. *Journal of Medical Internet Research, 17*(11), e248. doi:10.2196/jmir.4387

Tang, S., & Zuo, J. (2000). Dating attitudes and behaviors of American and Chinese college students. *Social Science Journal, 37*(1), 67–78.

Tavernise, S. (2011, August 16). More unwed parents live together, report finds. *New York Times*. Retrieved from http://www.nytimes.com/2011/08/17/us/17cohabitation.html

Taylor, C. (2016, March 23). Not-so-noble crowdfunding sparks backlash. *Reuters*. Retrieved from http://www.reuters.com/article/us-money-charity-crowdfunding-idUSKCN0WP1RO

Teachman, J. (2003). Premarital sex, premarital cohabitation, and the risk of subsequent marital dissolution among women. *Journal of Marriage and the Family, 65*, 444–455.

Television Bureau of Advertising. (2010). *TVB 2010 Media comparisons study*. Retrieved from http://www.tvb.org/media/file/TVB_PB_Media_Comparisons_2010_PERSONS.pdf

Telles, E., & Ortiz, V. (2009). *Generations of exclusion: Mexican Americans, assimilation, and race*. New York, NY: Russell Sage Foundation.

Thayer, M. (2010). *Making transnational feminism: Rural women, NGO activists, and northern donors in Brazil*. New York, NY: Routledge.

Thébaud, S. (2015). Status beliefs and the spirit of capitalism: Accounting for gender biases in entrepreneurship and innovation. *Social Forces, 94*, 61–86.

The Economist. (2003, May 6). A nation apart. Retrieved from http://www.economist.com/node/2172066

The Economist. (2011, November 4). Difference engine: Luddite legacy [Blog]. Retrieved from http://www.economist.com/blogs/babbage/2011/11/artificial-intelligence

The Economist. (2015a). How three teenagers invented an app to police the cops. Retrieved from http://www.economist.com/news/united-states/21684687-high-school-students-want-citizens-rate-their-interactions-officers-how-three

The Economist. (2015b). The great sprawl of China. Retrieved from http://www.economist.com/news/china/21640396-how-fix-chinese-cities-great-sprawl-china

The Economist. (2016). Urban population in India: Particular about particulates. Retrieved from http://www.economist.com/news/asia/21688447-bold-experiment-has-improved-delhis-air-indians-want-more-particular-about-particulates

Thomas, C.(2007). *Sociologies of disability and illness: Contested ideas in disability studies and medical sociology*. New York, NY: Palgrave Macmillan.

Thompson, W. (1929). Population. *American Journal of Sociology, 34*(6), 959–975.

Thompson, E. P. (1971). The moral economy of the English crowd in the eighteenth century. *Past & Present, 50*, 76–136.

Thornton, G. (2017). *Women in business New perspectives on risk and reward*. Retrieved from Grant Thornton website: https://www.grantthornton.global/globalassets/1.-member-firms/global/insights/article-pdfs/2017/grant-thornton_women-in-business_2017-report.pdf

Tian, N., Fleurant, A., Kuimova, A., Wezeman, P. D., & Wezeman, S. T. (2018, May). *Trends in world military expenditure, 2017*. Stockholm

International Peace Research Institute. Retrieved from https://www.sipri.org/sites/default/files/2018-04/sipri_fs_1805_milex_2017.pdf

Tilly, C. (1978). *From mobilization to revolution.* Reading, MA: Addison-Wesley.

Tilly, C. (1996). The emergence of citizenship in France and elsewhere. In C. Tilly (Ed.), *Citizenship, identity, and social history.* Cambridge, UK: Cambridge University Press.

Time. (2014, October 13). The 25 most influential teens of 2014. Retrieved from http://time.com/3486048/most-influential-teens-2014/

Tollefson, J. (2016, March 30). Antarctic model raises prospect of unstoppable ice collapse. *Nature.* Retrieved from http://www.nature.com/news/antarctic-model-raises-prospect-of-unstoppable-ice-collapse-1.19638

Totti, X. (1987, Fall). The making of a Latino ethnic identity. *Dissent, 34*(1), 537–542.

Toufexis, A. (1993, May 24). Sex has many accents. *Time.* Retrieved from http://content.time.com/time/magazine/article/0,9171,978575,00.html

Touraine, A. (1974). *The post-industrial society.* London, UK: Wildwood.

Touraine, A. (1977). *The self-production of society.* Chicago, IL: University of Chicago Press.

Touraine, A. (1981). *The voice and the eye: An analysis of social movements.* New York, NY: Cambridge University Press.

Townsend, P., & Davidson, N. (Eds.) (1982). *Inequalities in health: The Black report.* Harmondsworth, UK: Penguin.

Trading Economics. (2018). *China average yearly wages.* Retrieved from https://tradingeconomics.com/china/wages

Treas, J. (1995). *Older Americans in the 1990s and beyond* (Population Bulletin 50). Washington, DC: Population Reference Bureau.

Treiman, D. (1977). *Occupational prestige in comparative perspective.* New York, NY: Academic Press.

Troeltsch, E. (1931). *The social teaching of the Christian churches* (2 vols.). New York, NY: Macmillan.

Truman, J. L., & Langton, L. (2015). *Criminal victimization, 2014.* Washington, DC: Bureau of Justice Statistics.

Tucker, B. P. (1998). Deaf culture, cochlear implants, and elective disability. *Hastings Center Report, 28*(4), 6–14.

Tucker, M. J., Berg, C. J., Callaghan, W. M., & Hsia, J. (2007). The black–white disparity in pregnancy-related mortality from 5 conditions: Differences in prevalence and case-fatality rates. *American Journal of Public Health, 97,* 247–251.

Tung, I., Lathrop, Y., & Sonn, P. (2015). *The growing movement for $15.* National Employment Law Project. Retrieved from http://www.nelp.org/content/uploads/Growing-Movement-for-15-Dollars.pdf

Turner et al. (2013). *Housing Discrimination Against Racial and Ethnic Minorities.* U.S. Department of Housing and Urban Development. Retrieved from https://www.huduser.gov/portal/Publications/pdf/HUD-514_HDS2012.pdf

Turner et al. (2016, April 18). Why America's schools have a money problem." *NPR.* Retrieved from http://www.npr.org/2016/04/18/474256366/why-americas-schools-have-a-money-problem

Turk, A. (2004). Sociology of terrorism. *Annual Review of Sociology, 30*(1), 271–286.

Turnbull, C. (1983). *The human cycle.* New York, NY: Simon & Schuster.

Tyson, A., & Maniam, S. (2016, November 9). *Behind Trump's victory: Divisions by race, gender, education.* Pew Research Center. Retrieved from http://www.pewresearch.org/fact-tank/2016/11/09/behind-trumps-victory-divisions-by-race-gender-education/

UC University of California–Berkeley. (2013). *Diversity snapshot.* Retrieved from http://diversity.berkeley.edu/sites/default/files/diversity-snapshot-web-final.pdf

Uggen, C., & Blackstone, A. (2004). Sexual harassment as a gendered expression of power. *American Sociological Review, 69,* 64–92.

UNAIDS. (2016a). *Global AIDS update 2016.* Retrieved from http://www.unaids.org/sites/default/files/media_asset/global-AIDS-update-2016_en.pdf

UNAIDS. (2016b). *Haiti: HIV and AIDS estimates, 2015.* Retrieved from http://www.unaids.org/en/regionscountries/countries/haiti

UNAIDS. (2018). *Fact Sheet July 2018: 2017 global HIV statistics.* Retrieved from http://www.unaids.org/en/resources/documents/2018/UNAIDS_FactSheet

UN Department of Economic and Social Affairs. (2017). *World population prospects: The 2017 revision.* New York, NY: United Nations. Retrieved from https://esa.un.org/unpd/wpp/

UN Department of Economic and Social Affairs. (2018). *World urbanization prospects: The 2018 revision.* Retrieved from https://esa.un.org/unpd/wup/

UNESCO. (2013). *Adult and youth literacy: National, regional and global trends, 1985–2015.* Retrieved from http://www.uis.unesco.org/Education/Documents/literacy-statistics-trends-1985-2015.pdf

UNESCO. (2015). *The United Nations world water development report 2015: Water for a sustainable world.* Retrieved from http://unesdoc.unesco.org/images/0023/002318/231823E.pdf

UNESCO. (2016). *Education.* Retrieved from http://data.uis.unesco.org/Index.aspx?queryid=120

UNESCO. (2018a). *Institute for Statistics: Education, February 2018 release.* Retrieved from http://data.uis.unesco.org/Index.aspx

UNESCO. (2018b). *The United Nations world water development report 2018.* Retrieved from http://unesdoc.unesco.org/images/0026/002614/261424e.pdf

UNFCCC. (2015). Historic Paris agreement on climate change: 195 nations set path to keep temperature rise well below 2 degrees Celsius. Retrieved from http://newsroom.unfccc.int/unfccc-newsroom/finale-cop21/

UNICEF. (2012). *Measuring child poverty: New league tables of child poverty in the world's rich countries* (Innocenti Report Card 10). Florence, Italy: UNICEF Innocenti Research Centre.

UNICEF. (2013). *Pakistan.* Retrieved from http://www.unicef.org/pakistan/

UNICEF. (2015). *Undernutrition contributes to nearly half of all deaths in children under 5 and is widespread in Asia and Africa.* Retrieved from http://data.unicef.org/nutrition/malnutrition.html

UNICEF. (2016). *Female genital mutilation/cutting: A global concern.* New York, NY: UNICEF. Retrieved from https://www.unicef.org/media/files/FGMC_2016_brochure_final_UNICEF_SPREAD.pdf

Union of Concerned Scientists. (2016). *UCS satellite database.* Retrieved from http://www.ucsusa.org/nuclear-weapons/space-weapons/satellite-database#.VvMSTHAmiHk

Union of International Organizations. (2005). *Yearbook of international organizations—Guide to global civil society* (42nd ed.). Munich, Germany: K. G. Saur.

United Nations. (2011). *Violence against women.* Retrieved from http://www.un.org/en/women/endviolence/pdf/pressmaterials/unite_the_situation_en.pdf

United Nations. (2012). *World urbanization prospects: The 2011 revision.* Retrieved from http://esa.un.org/unup/pdf/WUP2011_Highlights.pdf

United Nations. (2013a). *Haiti country profile: Human development indicators.* Retrieved from http://hdrstats.undp.org/en/countries/profiles/HTI.html

United Nations. (2013b). *The Millennium Development Goals Report 2013.* Retrieved from http://www.un.org/millenniumgoals/pdf/report-2013/mdg-report-2013-english.pdf

United Nations. (2013c). *World population prospects: The 2012 revision.* Retrieved from http://esa.un.org/unpd/wpp/Documentation/pdf/WPP2012_HIGHLIGHTS.pdf

United Nations. (2015). *The world's women 2015: Trends and statistics.* Retrieved from http://unstats.un.org/unsd/gender/downloads/WorldsWomen2015_report.pdf

United Nations Department of Economic and Social Affairs, Population Division. (2009). *Trends in international migrant stock: The 2008 revision.* Retrieved from http://esa.un.org/migration/p2k0data.asp

United Nations Department of Economic and Social Affairs. (2014). *Least developed countries criteria.* Retrieved from http://www.un.org/en/development/desa/policy/cdp/ldc/ldc_criteria.shtml

United Nations Department of Economic and Social Affairs. (2015a). World population projected to reach 9.7 billion by 2050. Retrieved from http://www.un.org/en/development/desa/news/population/2015-report.html

United Nations Department of Economic and Social Affairs. (2015b). *World population prospects: The 2015 revision.* Retrieved from https://esa.un.org/unpd/wpp/Publications/Files/WPP2015_Volume-II-Demographic-Profiles.pdf

United Nations Department of Economic and Social Affairs. (2015c). *World urbanization prospects: The 2014 revision.* Retrieved from https://esa.un.org/unpd/wup/Publications/Files/WUP2014-Report.pdf

United Nations Department of Economic and Social Affairs. (2016a). *International migration report 2015: Highlights.* Retrieved from http://www.

un.org/en/development/desa/population/ migration/publications/migrationreport/docs/ MigrationReport2015_Highlights.pdf

United Nations Department of Economic and Social Affairs. (2016b). *List of least developed countries (as of February 2016)*. Retrieved from http://www. un.org/en/development/desa/policy/cdp/ldc/ ldc_list.pdf

United Nations Department of Economic and Social Affairs. (2017). *World population prospects: The 2017 revision*. New York, NY: United Nations. Retrieved from https://esa.un.org/unpd/wpp/

United Nations Department of Economic and Social Affairs. (2018). *World urbanization prospects: The 2018 revision*. Retrieved from https://esa. un.org/unpd/wup/

United Nations Development Programme. (2006). *Human Development Report 2006*. Retrieved from http://hdr.undp.org/sites/default/files/ reports/267/hdr06-complete.pdf

United Nations Development Programme. (2015). *Human Development Report 2015: Work for human development*. Retrieved from http://hdr. undp.org/sites/default/files/2015_human_ development_report_1.pdf

United Nations Development Programme. (2016). *Human Development Report: 2015*. Retrieved from http://hdr.undp.org/sites/default/files/ hdr15_standalone_overview_en.pdf

United Nations Development Programme. (2017). *Human Development Report 2016*. Retrieved from http://hdr.undp.org/sites/default/files/2016_ human_development_report.pdf

United Nations Economic and Social Commission for Western Asia. (2011). *The demographic profile of Kuwait*. Retrieved from http://www.escwa. un.org/popin/members/kuwait.pdf

United Nations Environmental Programme/ World Meteorological Organization. (2014). *Scientific assessment of ozone depletion 2014*. Retrieved from http://www.esrl.noaa.gov/csd/ assessments/ozone/2014/

United Nations Food and Agriculture Organization. (2004). *The state of food insecurity, 2004*. Retrieved from http://www.fao.org/documents/ show_cdr.asp?url_file=/docrep/007/y5650e/ y5650e00.htm

United Nations Food and Agriculture Organization. (2005). *Armed conflicts leading cause of world hunger emergencies*. Retrieved from http://www. fao.org/newsroom/en/news/2005/102562/ index.html

United Nations Food and Agriculture Organization. (2015). *The state of food insecurity in the world: Meeting the 2015 international hunger targets: Taking stock of uneven progress*. Retrieved from http://www.fao.org/3/a-i4646e.pdf

United Nations Food and Agriculture Organization. (2016). *Food security updates July 2016*. Retrieved from http://www.fao.org/3/a-c0335e.pdf

United Nations Global Initiative to Fight Human Trafficking. (2008). *Human trafficking: The facts*. Retrieved from http://www.unglobalcompact. org/docs/issues_doc/labour/Forced_labour/ human_trafficking_-_the_facts_-_final.pdf

United Nations High Commission on Refugees. (2017, April 11). UNHCR says death risk from starvation in Horn of Africa, Yemen, Nigeria growing, displacement already rising. Retrieved from http://www.unhcr.org/afr/news/ briefing/2017/4/58ec9d464/unhcr-says-death-risk-starvation-horn-africa-yemen-nigeria-growing-displacement.html

United Nations High Commission on Refugees. (2018). *Syria regional refugee response*. Retrieved from http://data.unhcr.org/syrianrefugees/ regional.php

United Nations Joint Programme on HIV/ AIDS. (2010). *Global report*. Retrieved from http://www.unaids.org/globalreport/ documents/20101123_GlobalReport_full_ en.pdf

United Nations Joint Programme on HIV/AIDS. (2013a). *Global report: UNAIDS report on the global AIDS epidemic 2013*. Retrieved from http://www. unaids.org/en/media/unaids/contentassets/ documents/epidemiology/2013/gr2013/ unaids_global_report_2013_en.pdf

United Nations Joint Programme on HIV/AIDS. (2013b). *Haiti*. Retrieved from http://www. unaids.org/en/regionscountries/countries/ haiti/

United Nations Office on Drugs and Crime. (2013). *2013 World Drug Report*. Retrieved from http:// www.unodc.org/unodc/secured/wdr/wdr2013/ World_Drug_Report_2013.pdf

United Nations Population Fund. (2005a). *Gender-based violence: A price too high. State of the world population, 2005*. Retrieved from http:// www.unfpa.org/swp/ 2005/english/ch7/ index.htm

United Nations Population Fund. (2005b). *Violence against women fact sheet*. Retrieved from http:// www.unfpa.org/swp/2005/presskit/factsheets/ facts_vaw.htm

United Nations Women. (2016). *Facts and figures: Leadership and political participation*. Retrieved from http://www.unwomen.org/en/what-we-do/leadership-and-political-participation/ facts-and-figures#notes

United Nations Women. (2017). *Women in politics: 2017*. Retrieved from http://www.unwomen. org/en/digital-library/publications/2017/4/ women-in-politics-2017-map

United States Conference of Mayors. (2008). *Hunger and homelessness survey: A status report on hunger and homelessness in America's cities*. Retrieved from http://usmayors.org/pressreleases/ documents/hungerhomelessnessreport_ 121208.pdf

United States Election Project. (2018). *National turnout rates: 1787–2016*. Retrieved from http:// www.electproject.org/home/voter-turnout/ voter-turnout-data

United States Institute of Peace. (2013). *Guide for participants in peace, stability, and relief operations*. Retrieved from http://www.usip.org/ node/5599

University of Michigan Institute for Social Research. (2008). Chore wars: Men, women and housework: Study confirms wives do most household chores. Retrieved from National Science Foundation website: http://www. nsf.gov/discoveries/disc_summ.jsp?cntn_ id=111458

University of Warwick. (2014, August 5). Girls feel they must "play dumb" to please boys, study shows. *Science Daily*. Retrieved from http://www.sciencedaily.com/ releases/2014/08/140805090947.htm

UNOC. (2017). *Executive summary: Conclusions and policy implications*. Retrieved from https://www. unodc.org/wdr2017/field/Booklet_1_ EXSUM.pdf

Upton-Davis, K. (2015). Subverting gendered norms of cohabitation: Living apart together for women over 45. *Journal of Gender Studies, 24*, 104–116.

U.S. Bureau of Justice Statistics. (2012a). *Correctional populations in the United States, 2011*. Retrieved from http://www.bjs.gov/content/pub/pdf/ cpus11.pdf,

U.S. Bureau of Justice Statistics. (2012b). *Prisoners in 2011*. Retrieved from http://www.bjs.gov/ content/pub/pdf/p11.pdf

U.S. Bureau of Justice Statistics. (2012c). *State corrections expenditures, FY 1982–2010*. Retrieved from http://www.bjs.gov/content/pub/pdf/ scefy8210.pdf

U.S. Bureau of Justice Statistics. (2013a). *Criminal victimization, 2012*. Retrieved from http://www. bjs.gov/content/pub/pdf/cv12.pdf

U.S. Bureau of Justice Statistics. (2013b). *Correctional populations in the United States, 2012*. Retrieved from http://www.bjs.gov/content/pub/pdf/ cpus12.pdf

U.S. Bureau of Justice Statistics. (2013c). *Female victims of sexual violence, 1994–2010*. Retrieved from http://www.bjs.gov/index. cfm?ty=pbdetail&iid=4594

U.S. Bureau of Justice Statistics. (2014). *Capital punishment, 2013—Statistical tables*. Retrieved from http://www.bjs.gov/content/pub/pdf/ cp13st.pdf

U.S. Bureau of Justice Statistics. (2015a). *National Prisoner Statistics (NPS) Program*. Retrieved from http://www.bjs.gov/index. cfm?ty=dcdetail&iid=269

U.S. Bureau of Justice Statistics. (2015b). *Prisoners in 2014*. Retrieved from http://www.bjs.gov/ content/pub/pdf/p14.pdf

U.S. Bureau of Justice Statistics. (2016). *Correctional populations in the United States, 2014*. Retrieved from http://www.bjs.gov/content/pub/pdf/ cpus14.pdf

U.S. Bureau of Labor Statistics. (2016). *Labor force statistics from the Current Population Survey: Labor force participation rate*. Retrieved from http://data.bls.gov/timeseries/LNS11300000

U.S. Bureau of Labor Statistics. (2017a). *American Time Use Survey*, Table 1: Time spent in primary activities and percent of the civilian population engaging in each activity, averages per day by sex, 2016 annual averages. Retrieved from https://www.bls.gov/news.release/atus. t01.htm

U.S. Bureau of Labor Statistics. (2017b, August). *Highlights of women's earnings in 2016*. Retrieved from https://www.bls.gov/opub/reports/ womens-earnings/2016/pdf/home.pdf

U.S. Bureau of Labor Statistics. (2017c, October). Projections overview and highlights, 2016–26. *Monthly Labor Review*. Retrieved from https://www.bls.gov/opub/mlr/2017/article/projections-overview-and-highlights-2016-26.htm

U.S. Bureau of Labor Statistics. (2017d, October). *Labor force characteristics by race and ethnicity, 2016*. Retrieved from https://www.bls.gov/opub/reports/race-and-ethnicity/2016/pdf/home.pdf

U.S. Bureau of Labor Statistics. (2017e, November). *Women in the labor force: A databook*. Retrieved from https://www.bls.gov/opub/reports/womens-databook/2017/home.htm

U.S. Bureau of Labor Statistics.(2017f, April). *A Profile of the Working Poor, 2015*. Retrieved from https://www.bls.gov/opub/reports/working-poor/2015/pdf/home.pdf

U.S. Bureau of Labor Statistics. (2018a). *American Time Use Survey*, Table 1: Time spent in primary activities and percent of the civilian population engaging in each activity, averages per day by sex, 2017 annual averages. Retrieved from https://www.bls.gov/news.release/atus.t01.htm

U.S. Bureau of Labor Statistics. (2018b, August). *Highlights of women's earnings in 2017.*Retrieved from https://www.bls.gov/opub/reports/womens-earnings/2017/home.htm

U.S. Bureau of Labor Statistics. (2018c). *Labor force statistics from the Current Population Survey: Employed persons by detailed occupation, sex, race, and Hispanic or Latino ethnicity*. Retrieved from https://www.bls.gov/cps/cpsaat11.htm

U.S. Bureau of Labor Statistics. (2018d). *Labor force statistics from the Current Population Survey: Employment status of the civilian noninstitutional population, 1947 to date*. Retrieved from https://www.bls.gov/cps/cpsaat01.htm

U.S. Bureau of Labor Statistics. (2018e). *Labor force statistics from the Current Population Survey, Table 5: Employment status of the civilian noninstitutional population by sex, age, and race*. Retrieved from https://www.bls.gov/cps/cpsaat05.htm

U.S. Bureau of Labor Statistics. (2018f). *Labor force statistics from the Current Population Survey, Table 7: Employment status of the civilian noninstitutional population 25 years and over by educational attainment, sex, race, and Hispanic or Latino ethnicity*. Retrieved from https://www.bls.gov/cps/cpsaat07.htm

U.S. Bureau of Labor Statistics. (2018g). *Labor force statistics from the Current Population Survey*, Table 37: Median weekly earnings of full-time wage and salary workers by selected characteristics. Retrieved from https://www.bls.gov/cps/cpsaat37.htm

U.S. Bureau of Labor Statistics. (2018h, January 19). Union members summary: 2017. Retrieved from https://www.bls.gov/news.release/union2.nr0.htm

U.S. Bureau of Labor Statistics. (2018i, February 9). Work stoppages involving 1,000 or more workers, 1947–2017. Retrieved from https://www.bls.gov/news.release/wkstp.t01.htm

U.S. Bureau of the Census. (1999). Employment status of the civilian population: 1970 to 1998. *Statistical abstract of the United States, 651*. Retrieved from https://www.census.gov/prod/99pubs/99statab/sec13.pdf

U.S. Bureau of the Census. (2000). *The changing shape of the nation's income distribution*. Retrieved from http://www.census.gov/prod/2000pubs/p60-204.pdf

U.S. Bureau of the Census. (2001). *Asset ownership of households: 1995*. Retrieved from http://www.census.gov/hhes/www/wealth/1995/wlth95-1.html

U.S. Bureau of the Census. (2003a). *Characteristics of the foreign born by world region of birth*, Table 3.1. Retrieved from http://www.census.gov/population/www/socdemo/foreign/ppl-174.html#reg

U.S. Bureau of the Census. (2003b). *Statistical abstract of the United States 2000*. Washington, DC: U.S. Government Printing Office.

U.S. Bureau of the Census. (2004). *Interim projections of the U.S. population by age, race, sex, and Hispanic origin*. Retrieved from http://www.census.gov/population/projections/files/methodology/idbsummeth.pdf

U.S. Bureau of the Census. (2005). America's families living arrangements: 2004. *Current Population Survey*. Retrieved from http://www.census.gov/population/www/socdemo/hh-fam/cps2004.html

U.S. Bureau of the Census. (2009a). Foreign-born population of the United States. *Current Population Survey*. Retrieved from http://www.census.gov/population/www/socdemo/foreign/cps2009.html

U.S. Bureau of the Census. (2009b). Voter turnout increases by 5 million in 2008 presidential election. Retrieved from https://www.census.gov/newsroom/releases/archives/voting/cb09-110.html

U.S. Bureau of the Census. (2010a). *Children by presence and type of parent(s), race, and Hispanic origin: 2010*. Retrieved from http://www.census.gov/population/www/socdemo/hh-fam/cps2010.html

U.S. Bureau of the Census. (2010b). *Household relationship and living arrangements of children under 18 years, by age and sex: 2010*. Retrieved from http://www.census.gov/population/www/socdemo/hh-fam/cps2010.html

U.S. Bureau of the Census. (2010c). *Income tables*. Retrieved from http://www.census.gov/hhes/www/income/data/historical/people/index.html

U.S. Bureau of the Census. (2010d). *Marital status of people 15 years and over, by age, sex, personal earnings, race, and Hispanic origin, 2010*. Retrieved from http://www.census.gov/population/www/socdemo/hh-fam/cps2010.html

U.S. Bureau of the Census. (2010e). *Projections: Population under age 18 and 65 and older: 2000, 2010, and 2030*, Table 5. Retrieved from http://www.census.gov/population/www/projections/projectionsagesex.html

U.S. Bureau of the Census. (2010f). Selected social characteristics in the United States. *American Community Survey: 2010*. Retrieved from http://www.culvercity.org/~/media/Files/Planning/Census2010/US%20Census%20DP-02%20Selected%20Social%20Char.%202010.ashx

U.S. Bureau of the Census. (2011a). *Statistical abstract of the United States*, Table 1156. Retrieved from http://www.census.gov/compendia/statab/2011/tables/11s1156.pdf

U.S. Bureau of the Census. (2011b). *Age and sex composition: 2010*. Retrieved from http://www.census.gov/prod/cen2010/briefs/c2010br-03.pdf

U.S. Bureau of the Census. (2011c). *Families below poverty level and below 125 percent of poverty by race and Hispanic origin: 1980 to 2008*. Retrieved from http://www.census.gov/compendia/statab/2011/tables/11s0714.pdf

U.S. Bureau of the Census. (2011d). *Historical income tables: Households*. Retrieved from http://www.census.gov/hhes/www/income/data/historical/household/index.html

U.S. Bureau of the Census. (2011e). *Median value of assets for households, by type of asset owned and selected characteristics: 2010*. Retrieved from http://www.census.gov/people/wealth/files/Wealth_Tables_2010.xls

U.S. Bureau of the Census. (2011f). *The older population: 2010*. Retrieved from http://www.census.gov/prod/cen2010/briefs/c2010br-09.pdf

U.S. Bureau of the Census. (2011g). *Overview of race and Hispanic origin: 2010*. Retrieved from http://www.census.gov/prod/cen2010/briefs/c2010br-02.pdf

U.S. Bureau of the Census. (2011h). *Net worth and asset ownership of households: 2011*, Table 1. Retrieved from http://www.census.gov/people/wealth/index.html

U.S. Bureau of the Census. (2012a). *2012 national population projections: Summary tables*, Table 2. Retrieved from http://www.census.gov/population/projections/data/national/2012/summarytables.html

U.S. Bureau of the Census. (2012b). *Age and sex of all people, family members and unrelated individuals iterated by income-to-poverty ratio and race: 2011*. Retrieved from http://www.census.gov/hhes/www/cpstables/032012/pov/POV01_100_3.xls

U.S. Bureau of the Census. (2012c). *America's families and living arrangements: 2012*, Table A1. Retrieved from http://www.census.gov/hhes/families/data/cps2012.html

U.S. Bureau of the Census. (2012d). *Educational attainment in the United States: 2012—detailed tables*, Table 3. Retrieved from http://www.census.gov/hhes/socdemo/education/data/cps/2012/tables.html

U.S. Bureau of the Census. (2012e). *Educational attainment of the population 18 years and over, by age, sex, race, and Hispanic origin: 2012*. Retrieved from http://www.census.gov/hhes/socdemo/education/data/cps/2012/tables.html

U.S. Bureau of the Census. (2012f). *The foreign born population in the United States: 2010*. Retrieved from http://www.census.gov/prod/2012pubs/acs-19.pdf

U.S. Bureau of the Census. (2012g). Growth in urban population outpaces rest of nation, Census Bureau reports. Retrieved from http://www.census.gov/newsroom/releases/archives/2010_census/cb12-50.html

U.S. Bureau of the Census. (2012h). *Historical income tables: households*, Table H-2. Retrieved from http://www.census.gov/hhes/www/income/data/historical/household/

U.S. Bureau of the Census. (2012i). *Historical income tables: households*, Table H-3. Retrieved from http://www.census.gov/hhes/www/income/data/historical/household/

U.S. Bureau of the Census. (2012j). *Historical income tables: Income inequality*. Retrieved from http://www.census.gov/hhes/www/income/data/historical/inequality/

U.S. Bureau of the Census. (2012k). *Households and families: 2010*. Retrieved from http://www.census.gov/prod/cen2010/briefs/c2010br-14.pdf

U.S. Bureau of the Census. (2012l). *Income, poverty, and health insurance coverage in the United States: 2011*. Washington, DC: U.S. Government Printing Office.

U.S. Bureau of the Census. (2012m). *Poverty status, by type of family, presence of related children, race and Hispanic origin*. Retrieved from http://www.census.gov/hhes/www/poverty/data/historical/hstpov4.xls

U.S. Bureau of the Census. (2012n). *Poverty status of people by family relationship, race, and Hispanic origin: 1959 to 2011*. Retrieved from http://www.census.gov/hhes/www/poverty/data/historical/hstpov2.xls

U.S. Bureau of the Census. (2012o). *Poverty status of the foreign-born population by sex, age, and year of entry: 2011*. Retrieved from http://www.census.gov/population/foreign/data/cps2012.html

U.S. Bureau of the Census. (2012p). *Statistical abstract of the United States*. Retrieved from http://www.census.gov/compendia/statab/2012/tables/12s0064.pdf

U.S. Bureau of the Census. (2012q). *The two or more races population: 2010* (2010 Census Briefs C2010BR-13). Retrieved from https://www.census.gov/prod/cen2010/briefs/c2010br-13.pdf

U.S. Bureau of the Census. (2013a). *America's families and living arrangements: 2012*. Retrieved from http://www.census.gov/prod/2013pubs/p20-570.pdf

U.S. Bureau of the Census. (2013b). *America's families and living arrangements: 2013: Children*, Table C3. Retrieved from http://www.census.gov/hhes/families/data/cps2013C.html

U.S. Bureau of the Census. (2013c). *America's families and living arrangements: 2013*, Table A1. Retrieved from https://www.census.gov/hhes/families/data/cps2013A.html

U.S. Bureau of the Census. (2013d). *Current population survey data on families and living arrangements*. Retrieved from http://www.census.gov/hhes/families/data/cps.html

U.S. Bureau of the Census. (2013e). *The diversifying electorate—Voting rates by race and Hispanic origin in 2012 (and other recent elections)*. Retrieved from http://www.census.gov/prod/2013pubs/p20-568.pdf

U.S. Bureau of the Census. (2013f). *Educational attainment in the United States*. Retrieved from http://www.census.gov/hhes/socdemo/education/data/cps/2012/tables.html

U.S. Bureau of the Census. (2013g). *Frequently asked questions about same-sex couple households*. Retrieved from http://www.census.gov/hhes/samesex/files/SScplfactsheet_final.pdf

U.S. Bureau of the Census. (2013h). *Historical income tables: Households*. Retrieved from http://www.census.gov/hhes/www/income/data/historical/household/

U.S. Bureau of the Census. (2013i). *Households and families: 2010*. Retrieved from http://www.census.gov/prod/cen2010/briefs/c2010br-14.pdf

U.S. Bureau of the Census. (2013j). *Household relationship and living arrangements of children under 18 years, by age and sex, 2013*. Retrieved from http://www.census.gov/hhes/families/data/cps2013C.html

U.S. Bureau of the Census. (2013k). *Income, poverty, and health insurance in the United States: 2012*. Retrieved from http://www.census.gov/prod/2013pubs/p60-245.pdf

U.S. Bureau of the Census. (2013l). *Mean earnings of workers 18 years and over, by educational attainment, race, Hispanic origin, and sex: 1975 to 2011*. Retrieved from http://www.census.gov/hhes/socdemo/education/data/cps/historical/tabA-3.xls

U.S. Bureau of the Census. (2013m). *Median value of assets for households, by type of asset owned and selected characteristics: 2011*. Retrieved from http://www.census.gov/people/wealth/files/Wealth_Tables_2011.xlsx

U.S. Bureau of the Census. (2013n). *Percent of people 25 years and over who have completed high school or college*, Table A-2. Retrieved from http://www.census.gov/hhes/socdemo/education/data/cps/historical/index.html

U.S. Bureau of the Census. (2013o). *Poverty status of people by family relationship, race, and Hispanic origin: 1959 to 2012*. Retrieved from http://census.gov/hhes/www/poverty/data/historical/hstpov2.xls

U.S. Bureau of the Census. (2013p). *Social and economic characteristics of currently unmarried women with a recent birth: 2011*. Retrieved from http://www.census.gov/prod/2013pubs/acs-21.pdf

U.S. Bureau of the Census. (2013q). *Statistical abstract of the United States*, Table 67: Family groups with children under 18 years old by race and Hispanic origin: 2000 to 2011. Washington, DC: U.S. Bureau of the Census.

U.S. Bureau of the Census. (2013r). *Urban, urbanized area, urban cluster, and rural population, 2010 and 2000: United States*. Retrieved from http://www.census.gov/geo/reference/ua/urban-rural-2010.html

U.S. Bureau of the Census. (2013s). *Years of school completed by people 25 years and over, by age and sex: Selected years 1940 to 2012*. Retrieved from http://www.census.gov/hhes/socdemo/education/data/cps/historical/tabA-1.xls

U.S. Bureau of the Census. (2013t). Characteristics of the foreign-born population by year of entry, Table 2.13. *Current Population Survey*. Retrieved from http://www.census.gov/population/foreign/data/cps2013.html

U.S. Bureau of the Census. (2015a). *Characteristics of same-sex couple households: 2014*. Retrieved from http://www.census.gov/hhes/samesex/

U.S. Bureau of the Census. (2015b). *Voting and registration: Historical CPS time series tables*. Retrieved from http://www.census.gov/hhes/www/socdemo/voting/publications/historical/A1.xls

U.S. Bureau of the Census. (2015c). *Annual social and economic (ASEC) supplement*, Table HINC-01. Retrieved from http://www.census.gov/hhes/www/cpstables/032015/hhinc/toc.htm

U.S. Bureau of the Census. (2015d). *Families and living arrangements: Households*. Retrieved from http://www.census.gov/hhes/families/data/households.html

U.S. Bureau of the Census. (2015e). *Families and living arrangements: Families*. Retrieved from http://www.census.gov/hhes/families/data/families.html

U.S. Bureau of the Census. (2015f). *Families and living arrangements: Marital status*, Table MS-2. Retrieved from http://www.census.gov/hhes/families/data/marital.html

U.S. Bureau of the Census. (2015g). *Income: Historical income tables, income inequality*, Table H-3. Retrieved from http://www.census.gov/data/tables/time-series/demo/income-poverty/historical-income-inequality.html

U.S. Bureau of the Census. (2015h). *Income and poverty in the United States: 2014*. Retrieved from http://www.census.gov/content/dam/Census/library/publications/2015/demo/p60-252.pdf

U.S. Bureau of the Census. (2015i). *Median age at first marriage: 1890 to present*. Retrieved from http://www.census.gov/hhes/families/files/graphics/MS-2.pdf

U.S. Bureau of the Census. (2015j). *Families and living arrangements: Living arrangements of Children*. Retrieved from https://www.census.gov/hhes/families/data/children.html.

U.S. Bureau of the Census. (2015k). *Historical income tables: Households, all races: 1967-2014*: Table H-2, share of aggregate income received by each fifth and top 5 percent of households. Retrieved from http://www.census.gov/data/tables/time-series/demo/income-poverty/historical-income-households.html

U.S. Bureau of the Census. (2016a). *American Community Survey 5-year estimates, 2010–2014: Population 65 years and over in the United States*. Retrieved from http://factfinder.census.gov/faces/tableservices/jsf/pages/productview.xhtml?src=CF

U.S. Bureau of the Census. (2016b). *Measuring America: 30-year-olds: Then and now*. Retrieved from http://www.census.gov/library/visualizations/2016/comm/30-year-olds.html

U.S. Bureau of the Census. (2016c). *Quick facts: California*. Retrieved from https://www.census.gov/quickfacts/table/PST045215/06,00

U.S. Bureau of the Census. (2016d, December 8). New census data show differences between urban and rural populations. Retrieved from https://www.census.gov/newsroom/press-releases/2016/cb16–210.html

U.S. Bureau of the Census. (2017a). *Educational attainment in the United States: 2017* [Table].

Retrieved from https://www.census.gov/data/tables/time-series/demo/educational-attainment/cps-historical-time-series.html

U.S. Bureau of the Census. (2017b, December 11). Educational attainment in the United States: 2017, Table 1. Educational attainment of the population 18 years and over, by age, sex, race, and Hispanic origin: 2017. Retrieved from https://www.census.gov/data/tables/2017/demo/education-attainment/cps-detailed-tables.html

U.S. Bureau of the Census. (2017c). *Historical families tables, Current Population Survey, 2017 annual social and economic supplement*. Retrieved from https://www.census.gov/data/tables/time-series/demo/families/families.html

U.S. Bureau of the Census. (2017d). *Historical marital status tables, Current Population Survey, 2017 annual social and economic supplement*. Retrieved from https://www.census.gov/data/tables/time-series/demo/families/marital.html

U.S. Bureau of the Census. (2017e). *Households tables, Current Population Survey, 2017 annual social and economic supplement*. Retrieved from https://www.census.gov/data/tables/time-series/demo/families/households.html

U.S. Bureau of the Census. (2017f). Poverty status of people, by age, race, and Hispanic origin [Table]. Retrieved from https://www.census.gov/data/tables/time-series/demo/income-poverty/historical-poverty-people.html

U.S. Bureau of the Census. (2017g). *Voting and registration in the election of November 2016*. Retrieved from https://www.census.gov/data/tables/time-series/demo/voting-and-registration/p20-580.html

U.S. Bureau of the Census. (2017h). *Historical income tables: Households, table H-5. Race and Hispanic origin of householder: Households by median and mean income*. Retrieved from https://www.census.gov/data/tables/timeseries/demo/incomepoverty/historical-incomehouseholds.html

U.S. Bureau of the Census. (2018a). *2017 national population projection tables: Projected foreign-born population by selected ages*, Table 2. Retrieved from https://www.census.gov/data/tables/2017/demo/popproj/2017-summary-tables.html

U.S. Bureau of the Census. (2018b). *America's families and living arrangements: 2017, Current Population Survey, 2017 annual social and economic supplement*. Retrieved from https://www.census.gov/data/tables/2017/demo/families/cps-2017.html

U.S. Bureau of the Census. (2018c). *Poverty status in 2016: Current Population Survey detailed tables for poverty*. Retrieved from https://www.census.gov/data/tables/time-series/demo/income-poverty/cps-pov.html

U.S. Bureau of the Census. (2018d). Poverty status in 2016: POV-04. Families by age of householder, number of children, and family structure. Retrieved from https://www.census.gov/data/tables/time-series/demo/income-poverty/cps-pov.html

U.S. Bureau of the Census. (2018e). *National Population by Characteristics: 2010–2017*.

Retrieved from https://www.census.gov/data/tables/2017/demo/popest/nation-detail.html

U.S. Department of Agriculture. (2013). How is rural America changing? Retrieved from http://www.census.gov/newsroom/cspan/rural_america/20130524_rural_america_slides.pdf

U.S. Department of Agriculture. (2017). *Rural America at a glance: 2017 edition*. Retrieved from https://www.ers.usda.gov/webdocs/publications/85740/eib-182.pdf?v=43054

U.S. Department of Education. (2006). *The condition of education*. Retrieved from http://nces.ed.gov/pubs2006/2006071.pdf

U.S. Department of Education. (2008). *1.5 million homeschooled students in the United States in 2007*. Retrieved from http://nces.ed.gov/pubs2009/2009030.pdf

U.S. Department of Education. (2011). *Digest of education statistics 2011*. Retrieved from http://nces.ed.gov/pubs2012/2012001_0.pdf

U.S. Department of Education. (2013). *Statistics about nonpublic education in the United States*. Retrieved from http://www2.ed.gov/about/offices/list/oii/nonpublic/statistics.html#homeschl

U.S. Department of Education. (2015, March 13). Secretary Duncan, Urban League President Morial to spotlight states where education funding shortchanges low-income, minority students. Retrieved from https://www.ed.gov/news/media-advisories/secretary-duncan-urban-league-president-morial-spotlight-states-where-education-funding-shortchanges-low-income-minority-students

U.S. Department of Health and Human Services. (2008). *Summary: Child maltreatment: 2008*. Retrieved from http://archive.acf.hhs.gov/programs/cb/pubs/cm08/cm08.pdf

U.S. Department of Health and Human Services. (2010). *Healthy people 2020*. Retrieved from http://www.healthypeople.gov/2020/Consortium/HP2020Framework.pdf

U.S. Department of Health and Human Services. (2012a). *Child maltreatment 2011*. Retrieved from http://www.acf.hhs.gov/sites/default/files/cb/cm11.pdf

U.S. Department of Health and Human Services. (2012b). *Profile of older Americans*. Retrieved from http://www.aoa.gov/AoARoot/Aging_Statistics/Profile/2012/10.aspx

U.S. Department of Health and Human Services. (2012c). *Results from the 2011 National Survey on Drug Use and Health: Mental health findings*. Rockville, MD: Substance Abuse and Mental Health Services Administration.

U.S. Department of Health and Human Services. (2016). *Health, United States, 2015: With Special Feature on Racial and Ethnic Health Disparities*. Hyattsville, MD.

U.S. Department of Health and Human Services. (2018). *Child maltreatment 2016*. Retrieved from https://www.acf.hhs.gov/cb/research-data-technology/statistics-research/child-maltreatment

U.S. Department of Housing and Urban Development. (2012). *The 2011 Annual Homeless Assessment Report to Congress*. Retrieved

from https://www.onecpd.info/resources/documents/2011AHAR_FinalReport.pdf

U.S. Department of Housing and Urban Development. (2017). *The 2017 Annual Homeless Assessment Report (AHAR) to Congress*. Retrieved from https://www.hudexchange.info/resources/documents/2017-AHAR-Part-1.pdf

U.S. Department of Justice. (2014). *Rape and sexual assault among college-age females, 1995–2013*. Retrieved from http://www.bjs.gov/content/pub/pdf/rsavcaf9513.pdf

U.S. Department of Labor. (2013). *Minimum wage laws in the states—January 1, 2013*. Retrieved from http://www.dol.gov/whd/minwage/america.htm

U.S. Department of Labor. (2016). *The economic status of Asian Americans and Pacific Islanders*. Retrieved from https://www.dol.gov/_sec/media/reports/AsianLaborForce/2016AsianLaborForce.pdf

U.S. Elections Project. (2016). *2016 November general election turnout rates*. Retrieved from http://www.electproject.org/2016g

U.S. Elections Project. (2018). *National general election VEP turnout rates, 1789–present*. Retrieved from http://www.electproject.org/national-1789-present

U.S. Energy Information Administration. (2016). *Independent statistics and analysis*. Retrieved from http://www.eia.gov/beta/international/

U.S. Equal Employment Opportunity Commission. (2017). *2014 job patterns for minorities and women in private industry (EEO1)—2014 EEO-1 national aggregate report by NAICS—4 code, motor vehicle manufacturing (3361)*. Retrieved from https://www1.eeoc.gov/eeoc/statistics/employment/jobpat-eeo1/2014/index.cfm

U.S. FLSA. (2011, May). *The Fair Labor Standards Act of 1938, as Amended* (WH Publication 1318), §202(a). Washington, DC: U.S. Department of Labor, Wage and Hour Division. Retrieved from https://www.dol.gov/whd/regs/statutes/FairLaborStandAct.pdf

U.S. House of Representatives. (2013a). *Black Americans in Congress: Member profiles*. Retrieved from http://history.house.gov/Exhibitions-and-Publications/BAIC/Black-Americans-in-Congress/

U.S. House of Representatives. (2013b). *House press gallery: Hispanic Americans*. Retrieved from http://housepressgallery.house.gov/member-data/demographics/hispanic-americans

U.S. Statistical Abstracts. (2012). Table 1023, Manufacturing corporations—Assets, and profits, by asset size. Retrieved from http://www2.census.gov/library/publications/2011/compendia/statab/131ed/tables/12s1023.xls

van der Veer, P. (1994). *Religious nationalism: Hindus and Muslims in India*. Berkeley, CA: University of California Press.

Van de Velde, S., Bracke, P., & Levecque, K. (2010). Gender differences in depression in 23 European countries: Cross-national variation in the gender gap in depression. *Social Science and Medicine, 71*(2), 305–313.

Vandewalker, I. (2015). *Election spending 2014: Outside spending in senate races since Citizens United*. Brennan Center for Justice. Retrieved from https://www.brennancenter.org/publication/election-spending-2014-outside-spending-senate-races-citizens-united

van Gennep, A. (1977). *The rites of passage*. London, UK: Routledge & Kegan Paul. (Original work published 1908)

Vanneman, R., & Cannon, L. W. (1987). *The American perception of class*. Philadelphia, PA: Temple University Press.

Vardi, N. (2014, August 6). The mergers and acquisition boom hits a roadblock. *Forbes*. Retrieved from http://www.forbes.com/sites/nathanvardi/2014/08/06/the-mergers-and-acquisitions-boom-hits-a-roadblock/

Vaughan, A. (2015, December 7). Global emissions to fall for first time during a period of economic growth. *Guardian*. Retrieved from http://www.theguardian.com/environment/2015/dec/07/global-emissions-to-fall-for-first-time-during-a-period-of-economic-growth

Vera. (2016). *Prison spending in 2015*. Retrieved from https://www.vera.org/publications/price-of-prisons-2015-state-spending-trends/price-of-prisons-2015-state-spending-trends/price-of-prisons-2015-state-spending-trends-prison-spending

Versey, S. (2014). Centering perspectives on black women, hair politics, and physical activity. *American Journal of Public Health, 104*, 810–815.

Vespa, J., Armstrong, D. M., and Medina, L. (2018). Demographic turning points for the United States: population projections for 2020 to 2060. *Current Population Reports, U.S. Census Bureau*. Retrieved from https://www.census.gov/content/dam/Census/library/publications/2018/demo/P25_1144.pdf

Villarosa, L. (2018, April 11). Why America's black mothers and babies are in a life-or-death crisis. *New York Times Magazine*. Retrieved from https://www.nytimes.com/2018/04/11/magazine/black-mothers-babies-death-maternal-mortality.html

Vitali, S., Glattfelder, J. B., & Battiston, S. (2011). The network of global corporate control. *PLoS ONE, 6*(10), e25995.

Vitello, P. (2011, July 2). You say God is dead? There's an app for that. *New York Times*. Retrieved from http://www.nytimes.com/2010/07/03/technology/03atheist.html?_r=0

Von Kaenel, L. (2016, March 31). Antarctica meltdown could double sea level rise. *Scientific American*. Retrieved from http://www.scientificamerican.com/article/antarctica-meltdown-could-double-sea-level-rise/

Wacquant, L. (1993). Redrawing the urban color line: The state of the ghetto in the 1980s. In C. Calhoun & G. Ritzer (Eds.), *Social problems* (pp. 448–475). New York, NY: McGraw-Hill.

Wacquant, L. (1996). The rise of advanced marginality: Notes on its nature and implications. *Acta Sociologica, 39*(2), 121–139.

Wacquant, L. (2010). *Deadly symbiosis*. Cambridge, UK: Polity Press.

Wacquant, L., & Wilson, W. J. (1993). The cost of racial and class exclusion in the inner city. In

W. J. Wilson (Ed.), *The ghetto underclass: Social science perspectives* (Updated ed., pp. 25–42). Newbury Park, CA: Sage.

Wagar, W. (1992). *A short history of the future*. Chicago, IL: University of Chicago Press.

Wagner, R. (2011). *Godwired: Religion, ritual and virtual reality*. New York, NY: Routledge.

Wagstaff, K. (2012, February 15). Men are from Google+, women are from Pinterest. *Time*. Retrieved from http://techland.time.com/2012/02/15/men-are-from-google-women-are-from-pinterest/#ixzz2hiZG4YA7

Wagstaff, K. (2013, November 9). China's massive pollution problem. *The Week*. Retrieved from http://theweek.com/article/index/252440/chinas-massive-pollution-problem

Waldron, I. (1986). Why do women live longer than men? In P. Conrad & R. Kern (Eds.), *The sociology of health and illness* (pp. 34–44). New York, NY: St. Martin's.

Walker, I., & Smith, H. J. (2002). *Relative deprivation: Specification, development, and integration*. Cambridge, UK: Cambridge University Press.

Walker, R. E., Keane, C. R., & Burke, J. G. (2010). Disparities and access to healthy food in the United States: A review of food deserts literature. *Health & Place, 16*, 876–884.

Waller, W. (1938). *The family: A dynamic interpretation*. New York, NY: Gordon.

Wallerstein, I. (1974a). *Capitalist agriculture and the origins of the European world-economy in the sixteenth century*. New York, NY: Academic Press.

Wallerstein, I. (1974b). *The modern world-system*. New York, NY: Academic Press.

Wallerstein, I. (1979). *The capitalist world economy*. Cambridge, UK: Cambridge University Press.

Wallerstein, I. (1990). *The modern world-system II*. New York, NY: Academic Press.

Wallerstein, I. (1996a). *Historical capitalism with capitalist civilization*. New York, NY: Norton.

Wallerstein, I. (Ed.). (1996b). *World inequality*. St. Paul, MN: Consortium Books.

Wallerstein, I. (2004). *World-systems analysis: An introduction*. Durham, NC: Duke University Press.

Wallerstein, J., & Kelly, J. (1980). *Surviving the breakup: How children and parents cope with divorce*. New York, NY: Basic Books.

Walmsley, R. (2009). *World prison population list* (8th ed.). London, UK: International Centre for Prison Studies, Kings College.

Wang, W. (2012). *The rise of intermarriage: Rates, characteristics vary by race and gender*. Pew Research Center. Retrieved from http://www.pewsocialtrends.org/2012/02/16/the-rise-of-intermarriage/

Wang, W. (2015, June 12). *Interracial marriage: Who is "marrying out"?* Pew Research Center. Retrieved from http://www.pewresearch.org/fact-tank/2015/06/12/interracial-marriage-who-is-marrying-out/

Wang, W., & Parker, K. (2014). *Record share of Americans have never married: As values, economics and gender patterns change*. Pew Research Center. Retrieved from http://www.pewsocialtrends.org/files/2014/09/2014-09-24_Never-Married-Americans.pdf

Wang, W., Parker, K., & Taylor, P. (2013). *Breadwinner mothers are the sole or primary provider in four-in-ten households with children; public conflicted about the growing trend*. Pew Research Center. Retrieved from http://www.pewsocialtrends.org/files/2013/05/Breadwinner_moms_final.pdf

Wang, Y. C., McPherson, K., Marsh, T., Gortmaker, S. L., & Brown, M. (2011). Health and economic burden of the projected obesity trends in the USA and the UK. *Lancet, 378*, 815.

Warner, S. (1993). Work in progress toward a new paradigm for the sociological study of religion in the United States. *American Journal of Sociology, 98*(5), 1044–1093.

Warren, B. (1980). *Imperialism: Pioneer of capitalism*. London, UK: Verso.

Waters, M. (1990). *Ethnic options: Choosing identities in America*. Berkeley, CA: University of California Press.

Waxman, L., & Hinderliter, S. (1996). *A status report on hunger and homelessness in America's cities*. Washington, DC: U.S. Conference of Mayors.

Webber, J. R. (2009, October 1). A theory of globalized capitalism. *Monthly Review*. Retrieved from https://monthlyreview.org/2009/10/01/a-theory-of-globalized-capitalism/

Weber, M. (1946). Politics as a vocation. In H. H. Gerth & C. Wright Mills (Eds.), *From Max Weber: Essays in sociology* (pp. 77–128). New York, NY: Oxford University Press. (Original work published 1921)

Weber, M. (1947). *The theory of social and economic organization*. New York, NY: Free Press.

Weber, M. (1963). *The sociology of religion*. Boston, MA: Beacon Press. (Original work published 1921)

Weber, M. (1977). *The Protestant ethic and the spirit of capitalism*. New York, NY: Macmillan. (Original work published 1904)

Weber, M. (1979). *Economy and society: An outline of interpretive sociology* (2 vols.). Berkeley, CA: University of California Press. (Original work published 1921)

Weeks, J. (1977). *Coming out: Homosexual politics in Britain, from the nineteenth century to the present*. New York, NY: Quartet.

Weitzman, L., Eifler, D., Hokada, E., & Ross, R. (1972). Sex-role socialization in picture books for preschool children. *American Journal of Sociology, 77*, 1125–1150.

Wellman, D. T. (1977). *Portraits of White Racism*. New York: Cambridge University Press.

Wellman, B., Carrington, P., & Hall, A. (1988). Networks as personal communities. In B. Wellman & S. D. Berkowitz (Eds.), *Social structures: A network approach* (pp. 130–184). New York, NY: Cambridge University Press.

Wellman, B., Salaff, J., Dimitrova, D., Garton, L., Gulia, M., & Haythornthwaite, C. (1996). Computer networks as social networks: Collaborative work, telework, and virtual community. *Annual Review of Sociology, 22*, 213–238.

Wessler, F. (2014, June 20). Poll: Fewer Americans blame poverty on the poor. *NBC News*. Retrieved from http://www.nbcnews.com/feature/in-plain-sight/poll-fewer-americans-blame-poverty-poor-n136051

West, C., & Zimmerman, D. H. (1987). Doing gender. *Gender & Society, 1*(2), 125–151.

Westbrook, L., & Schilt, K. (2014). Doing gender, determining gender: Transgender people, gender panics, and the maintenance of the sex/gender/sexuality system. *Gender & Society, 28,* 32–57.

Western, B. (1997). *Between class and market: Postwar unionization in the capitalist democracies.* Princeton, NJ: Princeton University Press.

Western, B., & Beckett, K. (1999). How unregulated is the U.S. labor market?: The penal system as a labor market institution. *American Journal of Sociology, 104*(4), 1030–1060.

Whalen, J. (2016, Dec. 15). The children of the opioid crisis. *Wall Street Journal.* Retrieved from https://www.wsj.com/articles/the-children-of-the-opioid-crisis-1481816178

Whitehurst, G. J., Joo, N., Reeves, R. V., & Rodrigue, E. (2017, November 17). *Balancing act: Schools, neighborhoods and racial imbalance.* Brookings Institution. Retrieved from https://www.brookings.edu/wp-content/uploads/2017/11/es_20171120_schoolsegregation.pdf

Widom, C. S., Czaja, S. J., & DuMont, K. A. (2015). Intergenerational transmission of child abuse and neglect: Real or detection bias? *Science, 347*(6229), 1480.

Will, J. A., Self, P. A., & Datan, N. (1976). Maternal behavior and perceived sex of infant. *American Journal of Orthopsychiatry, 46*(1), 135–139.

Williams, C. L. (1992). The glass escalator: Hidden advantages for men in the "female" professions. *Social Problems, 39,* 253–267.

Williams, S. (1993). *Chronic respiratory illness.* London, UK: Routledge.

Williams, C. L. (2013). The glass escalator, revisited: Gender inequality in neoliberal times. *Gender & Society, 27,* 609–629.

Wilson, E. O. (1975). *Sociobiology: The new synthesis.* Cambridge, MA: Harvard University Press.

Wilson, J. Q., & Kelling, G. (1982, March 1). Broken windows. *Atlantic Monthly,* pp. 29–38.

Wilson, W. J. (1978). *The declining significance of race: Blacks and changing American institutions.* Chicago, IL: University of Chicago Press.

Wilson, W. J. (1987). *The truly disadvantaged: The inner city, the underclass, and public policy.* Chicago, IL: University of Chicago Press.

Wilson, W. J. (1991). Studying inner-city social dislocations: The challenge of public agenda research. *American Sociological Review, 56,* 1–14.

Wilson, W. J. (1996). *When work disappears: The world of the new urban poor.* New York, NY: Knopf.

Wilson, W. J. (2011). Being poor, black, and American: The impact of political, economic, and cultural forces. *American Educator, 35*(1), 10–23, 46.

Wilson, W. J., Aponte, R., Kirschenman, L., & Wacquant, L. J. D. (1987, August). *The changing structure of urban poverty.* Paper presented at the annual meeting of American Sociological Association, Chicago, IL.

Winckler, O. (2013). The "Arab Spring": Socioeconomic aspects. *Middle East Policy, 20*(4). Retrieved from http://www.mepc.org/arab-spring-socioeconomic-aspects

Winkleby, M., Jatulis, D. E., Frank, E., & Fortmann, S. P. (1992). Socioeconomic status and health: How education, income, and occupation contribute to risk factors for cardiovascular disease. *American Journal of Public Health, 82*(6), 816–820.

Winner, E. (1997). Exceptionally high intelligence and schooling. *American Psychologist, 52,* 1070–1081.

Wirth, L. (1938). Urbanism as a way of life. *American Journal of Sociology, 44*(1), 1–24.

Witkowski, S., & Brown, C. (1982). Whorf and universals of color nomenclature. *Journal of Anthropological Research, 38*(4), 411–420.

Women in National Parliaments. (2013a). *Women in parliaments: World and regional averages.* Retrieved from http://www.ipu.org/wmn-e/world.htm

Women in National Parliaments. (2013b). *Women in parliaments: World classification.* Retrieved from http://www.ipu.org/wmn-e/classif.htm

Wong, M. (2016, Dec. 8). Today's children face tough prospects of being better off than their parents, Stanford researchers find. *Stanford News.* Retrieved from https://news.stanford.edu/2016/12/08/todays-children-face-tough-prospects-better-off-parents/

Wong, S. (1986). Modernization and Chinese culture in Hong Kong. *China Quarterly, 106,* 306325.

WorldatWork. (2011). *Telework 2011: A WorldatWork special report.* Retrieved from http://www.worldatwork.org/waw/adimLink?id=53034

World Bank. (1997). *World development report 1997: The state in a changing world.* New York, NY: Oxford University Press.

World Bank. (2000–2001). *World development report 2000–2001: Attacking poverty.* Retrieved from http://poverty.worldbank.org/library/topic/3389/

World Bank. (2012a). *Knowledge Economy Index (KEI) 2012 rankings.* Retrieved from http://siteresources.worldbank.org/INTUNIKAM/Resources/2012.pdf

World Bank. (2012b). *Mortality rate, infant (per 1,000 live births).* Retrieved from http://data.worldbank.org/indicator/SP.DYN.IMRT.IN

World Bank. (2012c). *An update to the World Bank's estimates of consumption poverty in the developing world.* Retrieved from http://siteresources.worldbank.org/INTPOVCALNET/Resources/Global_Poverty_Update_2012_02-29-12.pdf

World Bank. (2012d). *Women are less likely than men to participate in the labor market in most countries.* Retrieved from http://data.worldbank.org/news/women-less-likely-than-men-to-participate-in-labor-market

World Bank. (2013a). *How we classify countries.* Retrieved from http://data.worldbank.org/about/country-classifications

World Bank. (2013b). *Income share held by highest 10%.* Retrieved from http://data.worldbank.org/indicator

World Bank. (2013c). *Remarkable declines in global poverty, but major challenges remain.* Retrieved from http://www.worldbank.org/en/news/press-release/2013/04/17/remarkable-declines-in-global-poverty-but-major-challenges-remain

World Bank. (2013d). *World development indicators.* Retrieved from http://data.worldbank.org/indicator

World Bank. (2014a). *GDP growth (annual %).* Retrieved from http://data.worldbank.org/indicator/NY.GDP.MKTP.KD.ZG

World Bank. (2014b). *GNI per capita, Atlas method (current US$).* Retrieved from http://databank.worldbank.org/data/views/reports/tableview.aspx

World Bank. (2014c). *Poverty overview.* Retrieved from http://www.worldbank.org/en/topic/poverty/overview

World Bank. (2014d). *Investing in people to fight poverty in Haiti: Reflections for evidence-based policy making.* Retrieved from http://www.worldbank.org/en/topic/poverty/publication/beyond-poverty-haiti

World Bank. (2014e). *Mortality rate under-5 (per 1,000 live births).* Retrieved from http://data.worldbank.org/indicator/SH.DYN.MORT/countries/XM-XD?display=graph

World Bank. (2014f). *World development indicators.* Table 2.13: Education completion and outcomes. Retrieved from http://wdi.worldbank.org/table/2.13

World Bank. (2014g). *World development indicators.* Table 2.11: Participation in education. Retrieved from http://wdi.worldbank.org/table/2.11

World Bank. (2014h). *World development indicators.* Table 1.1: Size of the economy. Retrieved from http://wdi.worldbank.org/table/1.1

World Bank. (2015). *World bank forecasts global poverty to fall below 10% for first time; major hurdles remain in goal to end poverty by 2030.* Retrieved from http://www.worldbank.org/en/news/press-release/2015/10/04/world-bank-forecasts-global-poverty-to-fall-below-10-for-first-time-major-hurdles-remain-in-goal-to-end-poverty-by-2030

World Bank. (2016a). *Birth rate, crude (per 1,000 people).* Retrieved from http://data.worldbank.org/indicator/SP.DYN.CBRT.IN

World Bank. (2016b). *Death rate, crude (per 1,000 people).* Retrieved from http://data.worldbank.org/indicator/SP.DYN.CDRT.IN

World Bank. (2016c). *GDP growth (annual %).* Retrieved from http://data.worldbank.org/indicator/NY.GDP.MKTP.KD.ZG

World Bank. (2016d). *Mortality rate, infant (per 1,000 live births).* Retrieved from http://data.worldbank.org/indicator/SP.DYN.IMRT.IN

World Bank. (2016e). New country classifications by income level [Blog]. Retrieved from http://blogs.worldbank.org/opendata/category/tags/news

World Bank. (2016f). *Poverty: Overview.* Retrieved from http://www.worldbank.org/en/topic/poverty/overview

World Bank. (2016g). *World DataBank.* Retrieved from http://databank.worldbank.org/data/home.aspx

World Bank. (2016h). *World development indicators.* Table 1.1: Size of the economy. Retrieved from http://wdi.worldbank.org/table

World Bank. (2016i). *World development indicators.* Table 4.4: Structure of merchandise exports. Retrieved from http://wdi.worldbank.org/table/4.4

World Bank. (2016j). *World development indicators.* Table 4.6: Structure of service exports. Retrieved from http://wdi.worldbank.org/table/4.6

World Bank. (2016k). *World development indicators.* Retrieved from https://openknowledge. worldbank.org/bitstream/handle/10986/ 23969/9781464806834.pdf

World Bank. (2017a, December 19). *China economic update—December 2017: Growth resilience and reform momentum.* Retrieved from http://www. worldbank.org/en/country/china/publication/ china-economic-update-december-2017

World Bank. (2017b, October 16). The 2017 global poverty update from the World Bank [Blog]. Retrieved from http://blogs.worldbank.org/ developmenttalk/2017-global-poverty- update-world-bank

World Bank. (2018a). *China: Overview.* Retrieved from http://www.worldbank.org/en/country/china/ overview

World Bank. (2018b). *Data for upper middle income, lower middle income, high income, low income.* Retrieved from https://data.worldbank. org/?locations=XT-XN-XD-XM

World Bank. (2018c). *GDP growth.* Retrieved from https://data.worldbank.org/indicator/NY.GDP. MKTP.KD.ZG?locations=XD-US-4E

World Bank. (2018d). *Improved sanitation facilities (% of population with access).* Retrieved from https://data.worldbank.org/indicator/ SH.STA.ACSN?end=2015&locations=XD- XM&start=2010&view=chart&year_ high_desc=false

World Bank. (2018e). *World Bank country and lending groups.* Retrieved from https:// datahelpdesk.worldbank.org/knowledgebase/ articles/906519

World Bank. (2018f, May 3). *World development indicators.* Structure of output. Retrieved from http://wdi.worldbank.org/table/4.2

World Bank. (2018g). *World development indicators.* Table 2.18: Mortality. Retrieved from http://wdi.worldbank.org/table/2.18

World Bank. (2018h). *World development indicators.* Table 3.12: Urbanization. Retrieved from http://wdi.worldbank.org/table/3.12

World Bank. (2018i). *World development indicators.* Table 4.4: Structure of merchandise exports. Retrieved from http://wdi.worldbank.org/ table/4.4

World Bank. (2018j). *World development indicators.* WV.1: Size of the economy. Retrieved from http://wdi.worldbank.org/table/WV.1

World Bank. (2018k). *World development indicators.* WV.2: Global goals: Ending poverty and improving lives. Retrieved from http://wdi. worldbank.org/table/WV.2

World Bank. (2018l, March). *World development indicators. Table 2.1: Population dynamics.* Retrieved from http://wdi.worldbank.org/ table/2.1

World Commission on Environment and Development. (1987). *Our common future: The world commission on environment and development.* New York, NY: Oxford University Press.

World Economic Forum. (2017). *Why Iceland ranks first.* Retrieved from https://www.weforum. org/agenda/2017/11/why-iceland-ranks-first- gender-equality/

World Health Organization. (2000). *What is female genital mutilation?* Retrieved from http://www. who.int/mediacentre/factsheets/fs241/en/ print.html

World Health Organization. (2010). *The World Health Report: Health systems financing: The path to universal coverage.* Retrieved from http://whqlibdoc.who.int/ whr/2010/9789241564021_eng.pdf

World Health Organization. (2013). *Global and regional estimates of violence against women: Prevalence and health effects of intimate partner violence and nonpartner sexual violence.* Retrieved from http://apps.who.int/iris/ bitstream/10665/85239/1/9789241564625_ eng.pdf?ua=1

World Health Organization. (2014). *Urban population growth.* Retrieved from http://www.who.int/ gho/urban_health/situation_trends/ urban_population_growth_text/en/

World Health Organization. (2015a). *Trends in maternal mortality: 1990–2015.* Retrieved from http://apps.who.int/iris/bitstream/ 10665/194254/1/9789241565141_eng. pdf?ua=1

World Health Organization. (2015b). *World malaria report 2015.* Retrieved from http://apps.who.int/ iris/bitstream/10665/200018/1/9789241565158_ eng.pdf?ua=1

World Health Organization. (2016a, November). *Maternal mortality: Fact sheet.* Retrieved from http://www.who.int/mediacentre/factsheets/ fs348/en/

World Health Organization. (2016b). *Violence against women: Key facts.* Retrieved from http://www. who.int/mediacentre/factsheets/fs239/en/

World Health Organization. (2017, November). *Violence against women fact sheet: Intimate partner and sexual violence against women.* Retrieved from http://www.who.int/mediacentre/ factsheets/fs239/en/

World Health Organization. (2017a). *Global Health Observatory data: Prevalence of obesity among adults.* Retrieved from http://www.who.int/ gho/ncd/risk_factors/overweight_obesity/ obesity_adults/en/

World Health Organization. (2018, April). *Malaria fact sheet.* Retrieved from http://www.who.int/ mediacentre/factsheets/fs094/en/

World Steel Association. (2016). *Crude steel production: 2016–2015.* Retrieved from http:// www.worldsteel.org/statistics/crude-steel- production0.html

World Steel Association. (2017). *Steel statistical yearbook 2017.* Retrieved from https://www.worldsteel. org/en/dam/jcr:3e275c73-6f11-4e7f-a5d8- 23d9bc5c508f/Steel+Statistical+Yearbook+ 2017_updated+version090518.pdf

World Trade Organization. (2018). About WTO: The WTO. Retrieved from https://www.wto.org/ english/thewto_e/thewto_e.htm

World Vision. (2016). *What you need to know: Crisis in Syria, refugees, and the impact on children.* Retrieved from http://www.worldvision.org/ news-stories-videos/syria-war-refugee-crisis

World Vision. (2018, July 13). *Syrian refugee crisis: Facts, FAQs, and how to help.* Retrieved from https://www.worldvision.org/refugees-news- stories/syrian-refugee-crisis-facts

Worrall, A. (1990). *Offending women: Female lawbreakers and the criminal justice system.* London, UK: Routledge.

Wortham, J. (2013, February 26). Tinder: A dating app with a difference [Blog]. *New York Times.* Retrieved from http://bits.blogs.nytimes. com/2013/02/26/tinder-a-dating-app-with- a-difference/?_r=0

Wray, L., Herzog, A. R., Willis, R. J., & Wallace, R. B. (1998). The impact of education and heart attack on smoking cessation among middle-aged adults. *Journal of Health and Social Behavior*, 39(4), 271–294.

Wright, R. (2015, September 1). Japan's worst day for teen suicides. *CNN.* Retrieved from http:// www.cnn.com/2015/09/01/asia/japan-teen- suicides/index.html?eref=rss_topstories

Wrigley, E. (1968). *Population and history.* New York, NY: McGraw-Hill.

Wulfhorst, E. (2017, March 21). Global progress leaves behind millions of poor, hungry, illiterate: UN report. *Reuters.* Retrieved from https://www.reuters.com/article/ us-global-development-report/global- progress-leaves-behind-millions-of-poor- hungry-illiterate-u-n-report-idUSKBN16S2GQ

Wuthnow, R. (1988). Sociology of religion. In N. J. Smelser (Ed.), *Handbook of sociology* (pp. 473–510). Newbury Park, CA: Sage.

Yardley, J. (2011, December 28). In one slum, misery, work, politics and hope. *New York Times.* Retrieved from http://www.nytimes. com/2011/12/29/world/asia/in-indian- slum-misery-work-politics-and-hope. html?pagewanted=all&_r=0

Yeung, K. T., and J. L. Martin. (2003). "The Looking Glass Self: An Empirical Test and Elaboration," *Social Forces*, 81(3): 843–879.

Yongqiang, G. (2013, September 25). The cost of cleaning China's filthy air? About $817 billion, one official says. *Time.* Retrieved from http:// world.time.com/2013/09/25/the-cost-of- cleaning-chinas-filthy-air-about-817-billion- one-official-says/#ixzz2pMUvXP3t

Yoon, S. S., Fryar, C. D., & Carroll, M. D. (2015). *Hypertension prevalence and control among adults: United States, 2011–2014* (NCHS Data Brief No. 220). Hyattsville, MD: National Center for Health Statistics.

Yousafzai, M., & Lamb, C. (2013). *I am Malala: The girl who stood up for education and was shot by the Taliban.* New York, NY: Little, Brown.

Youthkiawaz.com. (2010). *Dowry in India: Putting the institution of marriage at stake.* Retrieved from http://www.youthkiawaaz.com/ 2010/ 08/dowry-in-india-putting-the-institution-of- marriage-at-stake/

Zald, M. N., & McCarthy J. D. (1979). *The dynamics of social movements: Resource mobilization, social control and tactics.* Cambridge, MA: Winthrop Publishers.

Zakaria, F. (2008). *The post-American world.* New York, NY: Norton.

Zee News. (2007). 1 Dowry death every 4 hrs in India. Retrieved from http://www.zeenews.com/news414869.html

Zerubavel, E. (1979). *Patterns of time in hospital life.* Chicago, IL: University of Chicago Press.

Zerubavel, E. (1982). The standardization of time: A sociohistorical perspective. *American Journal of Sociology, 88*(1), 1–23.

Zillman, C. (2016, November 9). Hillary Clinton had the biggest voter gender gap on record. *Fortune.* Retrieved from http://fortune.com/2016/11/09/hillary-clinton-election-gender-gap/

Zimbardo, P. (1969). The human choice: Individuation, reason, and order versus deindividuation, impulse, and chaos. In W. J. Arnold & D. Levine (Eds.), *Nebraska symposium on motivation* (Vol. 17, pp. 237–307). Lincoln, NE: University of Nebraska Press.

Zimbardo, P. (1972). *The psychology of imprisonment: Privation, power and pathology.* Stanford, CA: Stanford University.

Zimbardo, P., Ebbesen, E. B., & Maslach, C. (1977). *Influencing attitudes and changing behavior.* Reading, MA: Addison-Wesley.

Zimmerman, E. (2016, June 22). Campuses struggle with approaches for preventing sexual assault. *New York Times.* Retrieved from https://www.nytimes.com/2016/06/23/education/campuses-struggle-with-approaches-for-preventing-sexual-assault.html

Zimmerman, G. M., & Messner, S. F. (2010). Neighborhood context and the gender gap in adolescent violent crime. *American Sociological Review, 75,* 958–980. doi:10.1177/0003122410386688

Zong, J., Batalova, J., & Hallock, J. (2018, February 8). *Frequently requested statistics on immigrants and immigration in the United States.* Migration Policy Institute. Retrieved from https://www.migrationpolicy.org/article/frequently-requested-statistics-immigrants-and-immigration-united-states

Zosuls, K. M., Ruble, D. N., Tamis-LeMonda, C. S., Shrout, P. E., Bornstein, M. H., & Greulich, F. K. (2009). The acquisition of gender labels in infancy: Implications for gender-typed play. *Developmental Psychology, 45,* 688–701.

Zuboff, S. (1988). *In the age of the smart machine: The future of work and power.* New York, NY: Basic Books.

Credits

Index

ideology, 18
IGOs (international governmental organizations), 516
illiteracy, 230, 361
    see also education and literacy
illness, *see* health and illness
illness work, *437*, 437–38
imagination, sociological, *see* sociological imagination
IMF (International Monetary Fund), 237, 242, 499, 516, 528, 529
immigrants and immigration
    aging and, 92
    anti-immigrant sentiment against, 296, 396
    Asian, 299, 301, 302, 305
    definition of, 292
    European, 299–300
    homelessness and, 218
    Latino/Hispanic, 299, 301, 302, 303–5
    Mexican, 296, 301–2, 303–4, 314
    North African, 56
    roles in immigrant families, 326
    undocumented, 191, 302, 303
    in the United States, 299–302, *300*
Immigration and Customs Enforcement (ICE), 191, 192
Immigration and Nationality Act Amendments, 301–2
Immigration Reform and Control Act, 302
impression management, 108–10
impressions, 104
incarceration, 177–78, *178*, 180
    costs of, 177, 185–86
    of men, 172
    rates of, 158, 177, *179*
    of women, 172, *172*
incest, 57
income, 199–200, *200*
    African Americans and, 307, 309, 312, *312*
    Asian Americans and, 312, *312*
    concentration of, 526
    definition of, 199
    education and, 201–2, *202*, 370
    gender and, 260, 311–12, *312*
    gender inequality in, 260–61, *263*, 263–64, 268, 269, 278
    inequality
        global, *see* global inequality
        health and, 442
        racial and ethnic inequality and, 307, 309
        in the United States, *205*, 208–9, *209*, 219
    Latino/Hispanic Americans and, 208, *209*, 309, 312, *312*
    race and, 213, 311–12, *312*
    real, 199
    whites and, 208–9, *209*, 307, 312, *312*
India, 231
    call centers in, *514*
    caste system in, *194*, *194*, 198, 368
    climate change and, 477
    crude death rate in, 481
    eating disorders in, 432
    economic growth in, 461
    gender in, 258
    Gender Inequality Index (GII) and, 267
    as lower-middle-income country, 226
    pollution in, 460, *461*, 461–62
    population of, 461, 479, *485*

poverty in, 226
religion in, 380
self-rule and, 62
urbanization in, 476
violence against women in, 271–72
Indian Americans, 305
IndieGogo, 147
individualism, 44, 46, 331, 522
Indonesia, 234
Industrial and Commercial Bank of China, 416
industrial conflict, 413–14, *414*
industrialization, 60–63
    in China, 461–62, *463*
    definition of, 60
    education and, 357–59, *358*
    modern society and, 10, 60–63
    urbanization and, 464–65, *465*
industrialized societies, 8, 10, 60–62, *61*
Industrial Revolution, 11, 13, 60, 153, 461
industrial work, 412–15, *413*, *414*
inequality, 196
    education and, 362–70, *365*, *369*
        between-school effects and, 364
        intelligence and, 367–68
        Kozol on, 362–64
        reform in the United States, 368–70, *369*
        resegregation of American schools, 364–65, *365*
        social reproduction of inequality, 366–67, *367*
        tracking and within school effects, 365–66
    gender, *see* gender inequality
    global, *see* global inequality
    global cities and, 474
    global divisions and, 525–27, 528, *528*
    globalization and, 525–26, 527, *528*, 528–29
    in health and illness, 441–47
        gender-based, 444–45
        race-based, 309–10, 426–28, *427*, *443*, 443–44
        social class-based, 442
    income, *205*, 208–9, *209*, 219
    intelligence and, 367–65
    race and, *see* racial inequality
    social, *208*, 208–11, *209*, 218–19
    social reproduction of, 366–67, *367*
    social stratification and, *see* social stratification
    see also specific kinds of inequality
Inequality by Design (Fischer et al.), 368
infanticide, 272
infant mortality rates, 227, 232, 309, 481
infectious diseases worldwide, socioeconomic disparities in, 445–47, *446*
informal economy, 412, 476, *476*
informal networks, 142
informal relations within bureaucracies, 142–43
information flows, *514*, 514–15
information revolution, 204
information technology, 147, 148, 150, 153
informed consent, 32, 34
INGOs (international nongovernmental organizations), 516
in-groups, 131, *131*
inner cities, 466, 471–72
innovators, 164, *164*
Instagram, 19, 505
instincts, 51

institutional capitalism, 417
institutionalization, aging and, 94
institutional racism, 294–96, *295*
institutional review boards (IRBs), 32
institutions, 103
intelligence
    definition of, 367
    inequality and, 367–68
intelligence quotient (IQ), 360, 367
interaction
    face-to-face, 21, 112, *112*, 118–20
    focused and unfocused, 104–5
interactional vandalism, 113–14
interactionist theories on crime and deviance, 165–67, *166*
interest groups, 404–5, *405*
intergenerational mobility, 209
Intergovernmental Panel on Climate Change (IPCC), 489, 491
interlocking directorates, 417
International Campaign to Abolish Nuclear Weapons (ICAN), 513
International Day Against Homophobia, 456
international governmental organizations (IGOs), 516
International Labour Organization, 233, 266
International Monetary Fund (IMF), 237, 242, 499, 516, 528, 529
international nongovernmental organizations (INGOs), 516
international organizations, 64
International Telegraph Union, 516
International Union for Conservation of Nature, 489
International Union of Geological Sciences, 493
Internet, 78, 97, 153
    access, 138, *139*, 140
    communication and, 118–20
    cyberbullying and, 4, 5, 19, 454
    democratization and, 398–99
    digital divide and, 138, *139*, 197
    global culture and, 64, *65*, 66–67
    impression management and, 110
    interactional vandalism and, 113–14
    nonverbal communication and, 106–7
    protests and, 498
    social interaction and, *see* social interaction
    social movements and, *512*, 512–13
    as social network, 138, 140
    space and time and, 118
    telecommuting and, 147, 148
interpersonal aggression, 252
interracial marriage, 284, 291, 332
intersectionality, 250, 255, 258, 278–79
intersex, 258, 259
intimate partner violence (IPV), 271, 327, 343–45
intragenerational mobility, 209
Inuit of Greenland, 44
involvement, 167
IPCC (Intergovernmental Panel on Climate Change), 489, 491
iPhones, 47, 151, 241
IPV (intimate partner violence), 271, 327, 343–45
IQ (intelligence quotient), 360, 367
IQ-based explanations for racial inequality, 315–16
Iran, 379–80
Iraq, 396, 463

volunteer work, 351
voter turnout, 151–52, 402, 403, 404
voting
    politics and, 402, 404
    women's right to vote, 395, 405

W

Wade, Roe v., 406
wages, *see* income
Waller, Willard, 325
Wallerstein, Immanuel, 239, 241, 345
Wallerstein, Judith, 340
Wall Street, 249, 264
Walmart, 150, 392, 420
Wang Fuman ("frost boy"), 222, 223, 224, 245
Ward, Lester Frank, 11
war on drugs, 174
War on Poverty, 211, 214
Warren, Elizabeth, 406, 406
Washington Post, 297
Washington University, 31, 32
waste, 489, 489
Waters, Maxine, 401, 401
Watsi, 147
Way We Never Were, The (Coontz), 328
wealth
    definition of, 200
    global inequality and, 525–27, 528, 528
    power and, 278–79, 410
    racial disparity in, 201, 201, 208–9, 209
    in the United States, 200–201, 201
"wealth gap," 208
Wealth of Nations, The (Smith), 412
wearable tech, 440, 440
"weathering" of the body, 428
Weber, Max, 13–14, 14, 15t, 150
    on bureaucracy, 14, 141–42, 143, 149
    on capitalism, 235
    on class, status, and power, 195, 198–99
    on democracy, 406–7
    on organizations, 140, 141–42, 143, 145
    on religion and social change, 374–75, 375, 376, 377, 502, 508
WebMD, 439, 441
Webster v. Reproductive Health Services, 406
WeChat, 505
We Copwatch, 182
weightless economy, 515
Weinstein, Harvey, 249, 265, 278
welfare, 212, 214, 216
welfare capitalism, 417
welfare state, 396
West Africa, 230, 232
Western countries, definition of, 6
Westheimer, Ruth, 36–37
westoxification, 67, 503
West Virginia, 370
WhatsApp, 66
white Americans
    childless, 351
    education and, 202, 368
    families and, 332, 334, 335, 338, 338
    health and, 309–10

    income and, 208–9, 209, 307, 312, 312
    residential segregation and, 310, 472
    in the suburbs, 471
    voter turnout and, 402, 403
white-collar crime, 174, 174–75
white-collar work, 421, 471
Whitehurst study, 365
white privilege, 298
white supremacists, 294
WHO (World Health Organization), 231, 446, 460
Whorf, Benjamin Lee, 46
Wii, 78
Williams, Christine, 264
Williams Institute, 348
Wilson, Darren, 157
Wilson, Edward O., 50–51
Wilson, William Julius, 316, 335, 471
Windsor, Edith, 321–22, 322, 351
Windsor, United States v., 322, 322
winner take all, 400
Wirth, Louis, 465, 466, 467
Wisconsin, 415
within-school effects, 365–66
Witterick, Kathy, 81
women
    childless, 351
    crime and, 171–73, 172
    early civil rights of, 395
    eating disorders and, 430, 430–31
    health and, 426–28, 427
    housework and, 268–69, 269, 325
    marriage, *see* marriage
    political participation of, 269–71, 270, 405–6, 406
    public harassment of, 120–21, 121
    social interaction and, 120–22, 121
    trafficking of, 272
    violence against, 248, 249, 271–74, 273, 278, 279, 327
    voting rights of, 395, 405
    in the workplace, 261–66, 262, 268, 278, 331
        in corporate America, 143, 144, 144, 250
        in management, 143, 144, 144
        race and, 311–12, 312
        sexual harassment and, 249–50, 264–66, 278
    see also feminism and feminist theory; gender; gender inequality
Women's March
    in January 2017, 510–11, 511
    in Los Angeles (2018), 248, 249
Wood, Laura, 285
work and the workplace
    alienation and, 413
    automation of, 411, 411, 419–20
    contingent workforce, 422
    definition of, 410
    future of, 423
    gender inequalities in, 262–66, 263, 264, 267, 268
    globalization and, 522–23
    hostile work environment, 265
    income and, *see* income

    labor unions, 414–15
    sexual harassment and, 249–50, 264–66, 278
    as socializing agent, 78–79
    social significance of, 410, 412–15
        division of labor, 412
        industrial, 412–15, 413, 414
        paid and unpaid, 410, 412
    technology and, 411, 411, 419–21, 420
    unemployment, *see* unemployment
    women in, 261–66, 262, 268, 278, 311–12, 312, 331
working class, 206–7, 219
working poor, 212, 212
World Bank, 225, 226, 227, 421–22, 487, 499, 516, 526, 529
World Economic Forum, 242, 259
World Health Organization (WHO), 231, 446, 460
World Meteorological Organization, 489
World Revolution in Family Patterns (Goode), 329
world-systems theory, 239–41, 240, 241, 242–43
World Trade Organization (WTO), 237, 242, 502, 516, 520, 526, 529
World War II, 30, 293, 305, 328, 408
World Wide Web, 138
writing, 46, 47, 58
WTO (World Trade Organization), 237, 242, 502, 516, 520, 526, 529

X

XBox, 78
Xi Jinping, 224

Y

Yahoo, 144
Yale University, 40–42, 44, 47, 190, 192, 201
Yang, Jerry, 204
Yanukovych, Viktor, 498
Yemen, 244, 259, 382, 497, 498
Yik Yak, 19
Yoon, JeongMee, 83
YouCaring, 147
young adulthood, 86, 87
young old, 92
Yousafzai, Malala, 354, 355, 355–56, 356, 372, 387
youth and crime, 173–74
Youth Development Study, 265
Youth Risk Behavior Study (2017), 5
Youth Risk Behavior Survey (YRBS), 257
YouTube, 67, 283, 399
YRBS (Youth Risk Behavior Survey), 257
Yugoslavia, former, genocide in, 293

Z

Zerubavel, Eviatar, 118
Zhou, Min, 313
Zhou Shengxian, 462
Zimbardo, Philip, 29, 29–30, 183
Zimmerman, George, 157, 166–67
Zuckerberg, Mark, 204